KT-177-927
LADY MARGARET HALL LIBRARY
OXFORD
302209047R

ORIGIN OF SEDIMENTARY ROCKS

ORIGIN OF SEDIMENTARY ROCKS

HARVEY BLATT
University of Oklahoma

GERARD MIDDLETON
McMaster University

RAYMOND MURRAY
Rutgers University

Prentice-Hall, Inc., Englewood Cliffs, New Jersey

10 9 8 7 6 5 4 3 2

ISBN: 0–13–642702–2

Library of Congress Catalog Card Number: 71–37405
Printed in the United States of America

Prentice-Hall International, Inc., London
Prentice-Hall of Australia, Pty. Ltd., Sydney
Prentice-Hall of Canada, Ltd., Toronto
Prentice-Hall of India Private Limited, New Delhi
Prentice-Hall of Japan, Inc., Tokyo

To

Betty, Muriel, and Elaine

CONTENTS

FOUR

SEDIMENT MOVEMENT BY FLUID FLOW

FIVE

SEDIMENTARY STRUCTURES

SIX

FACIES MODELS

185

PART THREE

TERRIGENOUS CLASTIC SEDIMENTS

SEVEN

WEATHERING PROCESSES AND PRODUCTS

217

ELEVEN

MUDROCKS

374

PART FOUR

CARBONATE ROCKS AND EVAPORITES

TWELVE

ORIGIN OF LIMESTONES

409

THIRTEEN
DIAGENESIS AND CLASSIFICATION OF LIMESTONES
456

FOURTEEN
DOLOMITE
477

FIFTEEN
EVAPORITES AND NATIVE SULFUR
501

PART FIVE
OTHER SEDIMENTARY ROCKS

SIXTEEN
CHERT
531

SEVENTEEN
PHOSPHATES
543

EIGHTEEN
ZEOLITES
558

PREFACE

This book was written as a text for the first course about sedimentary rocks taught at the advanced undergraduate or beginning graduate level. In addition, we hope it will have value for those professionals in geology and related fields who seek a modern treatment of the subject. The idea of writing it began when we realized that there was no book that brought together the many profound ideas and new data about sedimentary rocks that have appeared in the last ten years. We realized that a synthesis of this material was needed by the student both as a summary of what appears to be known and as a guide to the directions in which research on sedimentary rocks is now moving.

The three authors of this book differ in background, training and research interest. We have tried to make this book the product of all of us, and not simply a collection of single-authored chapters. We feel this is important because it should be possible for the reader to see the inherent similarities and differences between different types of sedimentary rocks and processes. Rocks such as limestones and sandstones should not be regarded as completely separate entities because of their differences in mineralogy.

We were faced with two main problems of selection: overall approach and particular rock-types or phenomena. Our approach has been to emphasize mechanisms and processes of sedimentation, both physical and chemical. We assume only the standard undergraduate preparation in mineralogy and paleontology, combined with an elementary working knowledge of chemistry, physics and the calculus. Using this background knowledge, we have tried to show where possible, how an understanding of sedimentary processes can be developed and leads in turn to an understanding of the origin of sedimentary rocks.

We chose this approach, rather than the alternatives of a description of sedimentary rock-types or a discussion of sedimentary rocks in terms of environments or facies models, because we believe there now exists a sufficient understanding of processes that this, somewhat deductive, approach is possible. As far as consistent with clarity of expression, however, we have tried to keep systematization and classification to a minimum. Too often textbooks give the impression that the problems are all solved. We are so far from understanding all the ways in which sediment is produced, transported, deposited and modified after deposition that it seems premature to attempt any complete systematization, whether of processes or sediment types or of environmental models.

The choice of phenomena to be discussed reflects our viewpoint that the material in a book such as this should be confined to topics of major importance in the field. Emphasis has been placed on the processes that produce *common* sedimentary rocks. Unfortunately, not all the common sedimentary rocks have been very extensively studied, but we have tried to avoid giving an extended discussion of trivial phenomena just because there is a large literature on the subject.

Text citations to published work have been kept to a minimum. We hope this will not offend our colleagues whose ideas we may have used without specific reference to their contribution. The references at the end of each chapter include not only those cited in the text and figures but also a small group of bibliographic, classic or recent papers and books chosen to lead the interested reader quickly into the literature on the subject.

All texts are largely compilations and syntheses with few, if any new data. Certainly this book is no exception. Our object has been to collate for the reader the main results and insights of modern sedimentologic research and to present the results in a way that will both interest the reader in carrying out his own investigations and prepare him for his future work.

We hope that some of our readers will send us their comments and suggestions for improving this book.

H. B., G. V. M., R. C. M.

ACKNOWLEDGMENTS

We wish to thank all those who have assisted us with the preparation of this book. The list is a long one and includes secretaries, technicians and colleagues. Among those to whom we are most indebted are the following who read substantial parts of the manuscript: Frank Beales, Brian Burley, Kenneth Deffeyes, Carl Dutton, Murray Felsher, Hans Füchtbauer, Ronald Gulbrandsen, Anthony Hallam, Robert Harriss, Richard Hay, Hugh Hendry, Heinrich Holland, Alan Jopling, Sheldon Judson, James Kramer, Jean Lajoie, Charles Mankin, Robert Mathews, Lawrence McKague, Barry Mellon, Richard Olsson, Martha Pendleton, Michael Piburn, M. Dane Picard, Mikkel Schau, Stanley Schumm, Henry Schwarcz, Bennett Smith, Ronald Surdam, Roger Walker, Theodore Walker.

ACKNOWLEDGMENTS

PART ONE

AIMS AND METHODS IN THE STUDY OF SEDIMENTARY ROCKS

Field work forms the basis for the study of sedimentary deposits, both modern and ancient. To make best use of the limited time and resources available for field studies, the geologist must first give some attention to defining what properties he will observe and measure and how he will select his observations and specimens for further study in the laboratory. What are the fundamental properties of sediments? How do we measure them? How can we take a sample that will be truly representative of the unit being studied?

CHAPTER ONE

INTRODUCTION

1.1 INTRODUCTION

The three main classes of rocks, igneous, sedimentary, and metamorphic, were first clearly distinguished in the mid-nineteenth century. At first, the origin of sedimentary rocks was much better understood than that of the other two classes. Thus geologists turned their attention to such rocks as basalt, granite, and schists, whose origin was more controversial.

For years, the only questions asked about sedimentary rocks were How old are they? or How can they be correlated with each other? These are important questions but they distracted geologists from paying much attention to the rocks themselves and led them to concentrate on rock sequences and the fossils that could be used to compare one sequence with another. Interest in sedimentary rocks themselves was revived when geologists asked the question: How did they form? and were not satisfied with simple answers. Sandstones may be just hardened sands but why do some show cross-bedding and others graded bedding? Why are some hard and others soft? Some clean and well sorted and others muddy and poorly sorted?

To answer these questions, geologists started to look at modern sediments. As geologists moved from the land to the less accessible parts of the oceans, they began to discover things never before suspected, and apparently inexplicable. There were mysterious coral atolls and shallow platforms covered with fine calcareous mud: limestones in the making, no doubt, but why were the atolls that shape? And where did the mud come from? There were grand canyons winding down the continental slopes, drowned under thousands of feet of seawater, and there were layers of sand and sand ripples on the bottoms of the oceans, where there should have been only fine mud and ooze. What formed the canyons and how was the sand carried into the deep sea?

As knowledge of modern sediments grew, it was noted that there seemed to be no modern equivalents of some common sedimentary rocks, such as cherts, dolomites, and ironstones. Were the ancient seas different from modern ones? Or perhaps some of these rocks were not formed on the sea bottom but in the sediment after burial. Just what went on in sediments after deposition, and how were soft muds and sands changed into hard sedimentary rocks?

Only a few years ago it did not seem to be possible to set about finding answers to these questions. But now areas once inaccessible have been opened to study by modern transportation and communication techniques. Research ships are exploring every corner of the world's oceans, and sediments in the deepest waters can be sampled, photographed, or observed on television or from deep submersibles. On land, oil well drills bring up samples and core from sediments buried as much as 20,000 ft below the surface and these same drilling techniques have been adapted to recovering core from 1000 ft below the ocean bottoms.

Advances in other sciences have helped point the way to solutions of some long-standing problems. Low-temperature, dilute solutions have been studied in the laboratory to the point where even such complex solutions as seawater and groundwater can be treated theoretically with success. Engineers have studied the movement of sand in large experimental channels (flumes) and produced structures that seem very like the ones that geologists see preserved in the field. X-rays, the electron microscope, and the microprobe have opened for study the hitherto intractable field of very fine grained materials.

In this book, we examine what is known about the origin of sedimentary rocks after some 50 years of fairly intensive study. We shall use what is already known to answer some of the many questions that can be asked about sedimentary rocks. For some of these questions we think the answers are now becoming clear but other questions remain to be answered. We hope this will be a "forward-looking" book. The next few years will undoubtedly see changes in the ideas we now have and new ideas will be introduced.

1.2 THE IMPORTANCE OF FIELD WORK IN SEDIMENTOLOGY

Field work forms the basis of all sedimentological studies. Even for experimental studies, the problems to be investigated are first defined in the field. For example, a great many laboratory studies have been made of the physical chemistry of calcium and magnesium carbonates dissolved in seawater. These studies have attempted to answer such questions as Are the ocean waters saturated in calcium carbonate? What is the stable carbonate phase that should precipitate out of seawater—aragonite, calcite, or dolomite? These questions would not have aroused the interest of investigators if they had not been raised by field studies of carbonate sediments and rocks. The solution chemistry of carbonates has no particular interest for a pure laboratory science and, consequently, these questions were never thoroughly investigated in the laboratory by chemists. Only after it became apparent from field studies that the solution chemistry of carbonates was fundamental to an understanding of carbonate rocks were laboratory investigations carried out by geochemists.

Laboratory studies have themselves resulted in a renewed interest in field investigations. For example, modern concepts of carbonate classification and diagenesis stemmed mainly from microscopic studies. These concepts, however, have so improved the description of carbonate rocks in the field that the entire study of carbonate rocks, including their regional stratigraphy, has been revitalized. Another example is provided by the concept of turbidity currents, a concept derived very largely from the classic laboratory studies carried out by Kuenen in the 1940's and 1950's. The ideas provided by these laboratory studies have made possible many field investigations of supposedly "monotonous" sandstone-shale formations that had previously been neglected by most geologists because they contained no fossils or recognizable marker horizons.

1.3 DESCRIPTION OF OUTCROP SECTIONS

A geologist's ability to describe an outcrop is closely related to his understanding of the observable phenomena. It is extremely difficult to see and describe accurately what one does not understand.

To illustrate this point, consider the example of sole marks on the base of sandstone beds. Before 1950 there were almost no descriptions of these structures. Even in areas where they are abundant, their existence was rarely noted in geological reports. In the same areas, where ripple marks are generally much less abundant, their presence was always carefully recorded. Field geologists in those days understood how ripples were formed but sole marks were so little understood that it took a geologist of exceptional observational powers even to notice them at all.

All geologists make use of at least a mental "checklist" in the field, to make sure that all major aspects of the rocks have been recorded. For sedimentary rocks, the major features include dip and strike, the sequence and thickness of beds measured either bed-by-bed or by lithological groupings, the lithology described by using some standard field classification, the major sedimentary structures, including the orientation of those that can be used to reconstruct ancient current directions (paleocurrents) and the presence and kind of fossils.

As noted above, a geologist can never expect to produce a "complete" description of an outcrop because his limited understanding of the origin of sedimentary rocks limits his ability to describe them. As understanding develops, it is necessary to return again and again to the same outcrops in order to make new observations. Nevertheless, some geologists design comprehensive graphic logs that can be used in the field in order to make sure that a full description of the rocks is recorded. Such logs can be valuable aids, especially when they are designed with a particular project in mind.

Bouma (1962) has described an elaborate graphical logging technique for recording "flysch"-type sand/shale sequences in detail. Separate columns in the log are assigned to recording the rock type, bedding type and bedding plane properties, measured paleocurrent orientation, types of internal sedimentary structures, grain size, and carbonate content. All of these properties are estimated in the field, though the estimates may be revised after study of samples collected. The reason for such a detailed study was to determine different types of "flysch" facies as part of a larger study to establish a limited number of facies typical of the great majority of all stratigraphic sections. The larger project is, of course, still incomplete.

The disadvantage of such logs is that to be reasonably compact they must be designed anew for each major type of sedimentary rock studied and that even so the logs generally become so involved that they no longer serve a purpose as readily comprehensible graphic displays of the basic field data. For this reason, many workers have attempted to devise simpler versions. For example, Walker (1967) proposed a data form for recording "flysch" sequences, which is based on the idealized sequence of sedimentary structures in turbidites recognized by Bouma (1962: see also discussion in Chap 5, p. 122). The main observations recorded are the thickness of beds and of each of Bouma's three main subdivisions of an ideal turbidite bed: (a) massive division, (b) plane-laminated division, and (c) cross-laminated division, as well as the nature of the bedding planes. By greatly reducing the number of parameters to be measured in the field, Walker was able to measure a large number of beds in many stratigraphic sections.

One advantage of such techniques is that, by recording quantitative data or phenomena classified into one of a limited number of categories for each locality and stratigraphic level, the data are very readily transferred from the field log to punched cards and can then be processed by computers. The disadvantage of such recording techniques is that they may lead to neglect, in the field, of those aspects of the rocks that do not fit neatly onto the data form that has been prepared. Not all aspects of sedimentary rocks lend themselves to quantitative measurement or instant categorization. There is, however, no reason why schematic descriptions cannot be supplemented by the usual approach using a field notebook and a camera, where appropriate.

1.4 SAMPLING AND NUMERICAL ANALYSIS OF DATA

The observations made in the field are generally only a selection of the total number that might be made. Because of limitations of time and economic resources or because of limited exposure, the geologist is restricted to examining only a small part of the sedimentary unit that is being studied. Thus only a selection of measurements can be made in the field and only a small number of specimens can be carried back for laboratory measurement from among the many that could conceivably be collected. In order that the measurements should be truly representative of the whole unit under investigation, it is important that considerable attention be given to the technique of sampling.

The basic concept in sampling is that of a *random sample*, which consists of a small number of observations chosen to be generally representative of a larger number that constitutes the *population* that is being studied. A random sample is one in which every member of the population has an equal likelihood of being chosen. The mathematical science of statistics is the science of drawing valid inferences from samples to populations, and its theory is based almost entirely upon the assumption of random sampling. The theory can be very useful to the geologist because he is often faced with the problem of making inferences about rock bodies from observations made on only small parts of them, but in order to make valid use of the theory it is essential that the geologist obtain a random sample.

Taking a random sample is not so easy as it sounds. The sample must not be chosen haphazardly: thought and planning are required to select a truly random sample. Two essential steps must be followed:

1. Define the population; i.e., answer the question "What exactly is being studied?"
2. Make sure that every member of the population is equally likely to be chosen to form part of the sample.

A slight modification of random sampling, which preserves its theoretical advantages, is often useful. Through prior knowledge and observations, a geologist is generally able to subdivide a population into subgroups, each of which is more homogeneous internally than the population as a whole. In statistics these subdivisions are called sampling strata. In geology, they may be actual strata (beds) but they may also be any well-defined units, such as members, formations, or geographic areas. A random sample is taken within each subgroup and the whole sample is called a *stratified random sample*. The sample is random within each subgroup but *systematic* within the whole population.

The basic unit of sampling, suggested by Otto (1938), is the *sedimentation unit*. It was defined as that part of a deposit that was laid down under essentially uniform chemical and physical conditions. For example, a bed of sand might constitute a sedimentation unit. If it was later covered by another bed of sand, it would be important not to take a sand specimen that included parts of both beds. Such a sample could not provide a meaningful estimate of some property of the sand, such as grain size, because it would include sand from two different populations.

An immediate problem with this definition of a sedimentation unit is the difficulty in determining the limits of such a unit. If there are no clear internal breaks, how may a bed be subdivided into sedimentation units? For example, a graded bed is an easily identifiable unit in the field but this bed was not deposited under constant physical conditions. Thus it is difficult to define a sedimentation unit precisely.

An important distinction between *target populations* and *sampled populations* was introduced in statistics by Cochran, Mosteller, and Tukey and applied in geology by Krumbein (1960). Suppose the target population is a sand bed and the objective is to study the feldspar content of hand specimens taken from this bed. For a random sample, there must be an equal chance of selecting all possible hand specimens that could be taken from the bed. But part of the bed may no longer be available for sampling—it may have been eroded away—and part of the bed may be buried by other rocks and therefore inaccessible. The target population is the population that the study is aimed at but the sampled population is that part of the population that is accessible to sampling. Strictly, statistical inferences can be drawn only about the sampled population. Extension of the inferences to the target population is, of course, inevitable but it must be realized that the statistical theory no longer applies to this extrapolation.

At the beginning of any investigation of a sedimentary rock unit, some thought should be given to the problem of sampling as part of the larger *experimental design*. The purpose of this branch of statistics is to enable the investigator to distinguish the various sources of variability, to control or reduce sources such as experimental error, and to estimate quantitatively the contribution of each source to the total variability shown by the sample of measurements. Experimental design is too large a topic to be considered in this book: It is discussed in detail in the books by Griffiths (1967) and by Krumbein and Graybill (1965).

After the investigation has been properly designed and the data have been obtained, various statistical techniques may be necessary to reduce the data to more manageable proportions or to test various hypotheses. Rarely, the results of an investigation are so clear that any further calculation of statistics would simply be a waste of time. Usually, statistical techniques are valuable, if only for purely descriptive purposes, for example, graphical displays of data or calculation of means, correlation coefficients, and regression lines to data. These techniques basically all summarize in a few numbers (called *statistics*) the properties of the many numbers in the data sample. The method of calculation of some simple statistics is explained in Chap. 3.

The statistics calculated from the sample may also be used as estimates of the corresponding properties of the population (called *parameters*). The theory of statistics can be used to obtain a quantitative statement of the precision of these estimates.

Most sedimentological data consist of not one but several different types of measurements made on each of a number of specimens (or several measurements made at each of a number of localities, etc.). In other words, sedimentologic data are inherently *multivariate*. Two, or even three, variables may be studied simultaneously by graphic methods (graphs, maps, triangular diagrams, etc.) but for more than three variables multivariate numerical and statistical methods must be used for complete analysis of the data. These relatively advanced statistical techniques, such as discriminant functions and factor analysis, are being used increasingly in sedimentology because they are powerful methods of handling complex data and because the many calculations involved may be readily carried out by using digital electronic computers.

1.5 FUNDAMENTAL AND DERIVED PROPERTIES OF SEDIMENTS

Of the many properties of sediments that can be measured in the field or in the laboratory, it is useful to distinguish some as being more fundamental than others. The choice of which properties are considered fundamental and which are considered to be derived from the fundamental properties is to some extent a matter of convention. In a sedimentary rock body or specimen, however, those properties may be considered fundamental that lead most directly to an unique and complete description of the rock mass. We may also regard as fundamental those properties that relate in the most direct way to the general laws of physics and chemistry that may be used to explain the various aspects of the rock mass.

Griffiths (1961) has suggested two criteria that may be used as a guide to choosing the set of fundamental properties: (a) The set must contain all properties necessary to define a rock specimen uniquely; (b) the set must not contain more properties than are necessary for an unique definition. Because most sedimentary deposits are clearly made up of parts (grains or crystals), it seems logical to define the properties of the whole rock in terms of the properties of the parts and the way they are put together. This approach has resulted in the recognition of five fundamental properties: (a) composition, i.e., the kinds of grains and their abundance; (b) the sizes of the grains; (c) the shapes of the grains; (d) the orientations of the grains; and (e) the packing of the grains. In conventional description of rocks a distinction is drawn between structure and texture. Structure is not chosen as a fundamental property by Griffiths because structures are nothing more than the large-scale expression of variations in the five fundamental properties listed.

The fact that only five properties of sediments are regarded as fundamental does not mean that only five measurements are needed to define the properties. For example, grain shape is an elusive property, generally considered to have at least two aspects (roundness and sphericity, see Chap. 3).

For particular applications, some other (derived) property of the rock may be more important than any of the fundamental properties. It may be much easier to measure the derived property than to attempt to predict it from a knowledge of the five fundamental rock properties. Permeability, which measures the ease with which fluid moves through a sedimentary rock, is such a property. It has been found experimentally that the rate q at which fluid flows through a unit cross-section of rock material is inversely proportional to the fluid viscosity μ and is directly proportional to the pressure gradient dp/dx in the direction of flow (Darcy's law):

$$q = \frac{k}{\mu}\frac{dp}{dx}$$

The coefficient of proportionality k is called the permeability. It might, therefore, be claimed that the fundamental property of rocks, as far as the flow of fluids through them is concerned, is permeability. Permeability must, however, be an expression of the "average" size of pore connections, which itself depends on the four textural properties listed above. For some very simple sediments it has been found possible to predict the permeability accurately from a knowledge of the grain size distribution. For most sediments it is still very difficult to predict permeability satisfactorily from the five fundamental properties so that direct measurement is the simplest way to determine permeability. But in attempting to predict the occurrence of permeability in sedimentary rocks, it seems that progress will be achieved not only by seeking empirical correlations between permeability and other rock properties but also by attempting to understand how permeability is related to the fundamental properties of the rock. When this understanding has been achieved, it will be possible to relate permeability, through the fundamental properties, to the processes that lead to the formation of sedimentary rocks.

Many aspects of sedimentary rocks that can be observed in the field, such as sedimentary structures and weathering characteristics, are related to very subtle changes in the fundamental properties. It may be easier to study such changes by observation in the field than by detailed analysis in the laboratory but none of the fundamental properties of sedimentary rocks can be described adequately on the basis of field examination alone. In general, laboratory work is necessary as a supplement to field observations in order to give more information on the distribution of fundamental properties in a rock body.

1.6 GRAIN PROPERTIES

The basic building blocks of most sedimentary rocks are either clastic grains or chemically precipitated crystals. When, as in the case of rock salt, gypsum, or nodular manganese, crystals make up the framework of the rock, they may have resulted from several processes such as (a) precipitation followed by sedimentation, (b) precipitation directly on the bottom, or (c) precipitation within the sediment by displacement or replacement of earlier deposited material. Analysis of such chemically precipitated rocks proceeds in much the same way as for igneous and metamorphic rocks. The appropriate textural terms are those that describe "grains" in terms of the ideal crystal forms for a mineral of that composition. The shape, for example, may be described as euhedral or anhedral, and the occurrence of crystals of two distinct sizes may be described by using the textural terms "porphyritic" or "porphyroblastic." The sequence of mineralogical and textural changes revealed by petrographic studies becomes the basis for interpreting the depositional and diagenetic history of the rock. In addition, studies of preferred crystallographic orientation of crystals may contribute to our understanding of the depositional process or later events that modify the rock.

Clastic sedimentary rocks and other rocks formed by fragmentation are also made up of crystals but it is not the individual crystals that are the most important textural elements. Nor do the ideal crystal forms provide a basis for description of the shapes of the particles. The elementary units of clastic rocks are the clastic particles, defined in this book as particles moved from the place in which they were formed to the place of deposition. Clastic particles originate in many different ways. They may be produced by weathering from preexisting igneous, metamorphic, or sedimentary rocks; they may consist of fragments or whole parts of organic skeleta; they may be produced locally by chemical or biochemical activity; they may be molded out of mud by burrowing organisms. In whatever way they are formed, they owe their characteristics to two main influences: the nature of the processes that originally produced them and the modification they have suffered during transport. In general, the modification of a grain population during transport takes place in two fundamentally different ways: by mechanical modification of individual grains, in some cases aided by chemical corrosion, and by hydraulic sorting of grain assemblages.

As a consequence of the basic distinctions between the mechanisms forming chemical and clastic rocks, different standards are necessary for the description of clastic particles and their textures from those adopted for the description of chemically produced crystals and their textures. The ideal geometric standard of comparison for both the size and the shape of clastic particles is not the ideal crystal form but the sphere because it is generally assumed that long continued abrasion of a structurally isotropic particle will produce a sphere. Even if this were not the case, the sphere is the simplest form to consider from both a geometric and dynamic point of view. The importance of hydraulic sorting in clastic sediments results in more attention being given to the variation in size and shape of particles in a clastic rock than is commonly the case for particles in a chemical rock. Individual particles in clastic rocks are very frequently not single mineral crystals but aggregates of either the same mineral or different minerals. Even in the case of grains made of single crystals, such as many sand-size quartz grains, the orientation of the particle is not determined so much by crystallographic properties as by hydrodynamic properties of the particle. So dimensional orientation becomes more important than crystallographic orientation.

Most clastic sedimentary rocks are, however, only partly clastic. They consist of clastic particles bound together by a chemically precipitated *cement*. Moreover, two size fractions of clastic particles can be distinguished in many sedimentary rocks. In sandstones, the particles larger than about coarse silt size are called *grains* and the finer particles, of mud sizes, are called *matrix*. The exact size dividing grains from matrix varies somewhat with the particular investigator. It is commonly set at about 0.02 mm. Below this size it is difficult to give a precise description of particles using an optical microscope. The grains commonly constitute the mechanical *framework* that supports the weight of the sediment and the interstices are partly filled with matrix. The rest of the space initially consists of voids or pore spaces occupied by fluids. Later the voids may be eliminated by compaction or filled with chemically precipitated cement. One of the current problems in the petrology of sandstones is the origin of much of the fine grained micaceous or chloritic "matrix" that may actually be cement produced by the alteration of some of the grains after burial.

Chemical and mechanical processes acting after deposition (i.e., during *diagenesis* of the sediment) may greatly modify the textures of an originally clastic rock. Grains may be corroded or completely dissolved away or they may grow in size by overgrowth of a mineral in crystallographic continuity with the mineral composing the clastic grain. Grains may be fractured or plastically distorted by the forces of compaction. In extreme cases, the rock may be so recrystallized that it is only convention and the absence of mineral assemblages indicating high temperatures that keep petrologists from calling it a metamorphic rock. More frequently the properties of the original grains can be reconstructed only by considerable interpretation on the part of the sedimentary petrologist.

1.7 PROPERTIES OF GRAIN AGGREGATES

The actual properties of any sedimentary rock composed of grains depend not only on the character of the grains themselves but on the way in which the grains are related to each other in space. This arrangement is termed the *rock fabric*. Three elements are normally considered in evaluating the fabric of a particulate sedimentary rock: (a) Distribution of particles of different sizes within the rock. (b) Orientation of grains within the rock. (c) Packing of grains within the rock.

The *distribution of particle sizes* has profound effect on the bulk properties of the rock. The presence of small grains or matrix within the interstices between larger grains has the general effect of decreasing the porosity and permeability. Alternatively, fine scale interlayering of coarser and finer sediment such as seen in laminated sandstones and rocks with shale partings may not affect the porosity of either the sand or shale but has an important effect on the permeability of the rock, especially in the direction perpendicular to the lamination.

The *orientation of grains* within the rock is especially important to both the properties of the rock and to interpretations of the origin of the rock. Most particulate grains are not spherical but are ellipsoidal, platy, or bladed. When deposited by currents, these grains commonly are oriented with the long dimension parallel or perpendicular to the direction of the current. In addition, platy grains such as mica flakes may settle with the short dimension perpendicular to the depositional surface. Dimensional orientation of grains imparts a directional fabric to the pore space in the rock, thus producing a direction of relatively increased permeability or electrical conductivity. In addition, a preferred orientation of grains may determine the way in which a rock breaks, giving a plane of fissility or jointing parallel to the grain orientation. Under some circumstances grains may be deposited with their long axes or platy surfaces dipping at an angle to the stratification. This fabric is termed *imbrication*. Imbrication complicates the fabric of the rock with respect to bulk properties but is very useful in determining the direction from which the depositing current flowed because imbricated particles almost invariably dip upstream.

Packing refers to the spacing or density pattern of particles in a rock. For grains that are spheres of uniform size there are six ways in which they can be arranged in space so that each sphere is in contact with four or more adjacent spheres and no positions are vacant. With cubic packing the porosity is 47.6% and with rhombohedral packing the porosity is 26%. In rocks the grains are seldom, if ever, spheres and almost never of uniform size. In addition, vacant sites occur because of dissolution of grains or through development of larger voids by keystone arching of grains during deposition, by organic disruption, or by early diagenesis. Random heaping of grains may produce almost any porosity. In sediments composed of detrital silicate or carbonate grains, porosities between 30 and 80% have been measured. Contacts between grains are generally tangential but it is possible that the surface of one grain may come to rest adjacent to a similar surface of an adjoining grain. In this case the grains when observed in thin section will appear to have a long mutual grain contact. Compaction and pressure solution generate similar contacts that may be only tangential at grain boundaries or may be actually sutured and thus interpenetrating. These effects have the result of decreasing the porosity and altering the sonic velocity of the rock.

1.8 CLASSIFICATION AND THE INTERPRETATION OF SEDIMENTARY ROCKS

As a result of the complex nature of sedimentary rocks and of the events that produce them, sedimentary rocks may be interpreted or explained in several different ways. It was noted above that the choice of the five fundamental properties of sedimentary rocks is to some extent arbitrary. Each of the five properties is generally described by several different numerical parameters. The observed values of these fundamental properties of sedimentary rocks are in turn a result of several processes differing in scale and nature and acting over a long period of time. The choice of which genetic or historical aspects of a sedimentary rock are considered to be most fundamental is therefore even more arbitrary than the choice of fundamental descriptive properties.

The problem of how to explain sedimentary rocks is closely related to the problem of how to classify them. It has long been recognized that there are two aspects of any classification. The more primitive aspect of classification is called *descriptive* and consists of the grouping together of phenomena that have something in common. In the case of sedimentary rocks, it seems logical to group together rocks that are similar in respect to the five fundamental properties. Since this involves comparing rocks on the basis of some twenty to thirty or more numerical parameters, even a descriptive classification raises complex problems. Three lines of action are possible: (a) Certain of the fundamental properties may be regarded as more basic than the others in order to reduce the complexity of the system. Most sedimentologists regard mineralogical composition and grain size as more important than other properties. (b) An attempt may be made to devise several independent classifications, each one being based on a single aspect of one fundamental property. Most sedimentologists describe clastic sediments by a series of terms, such as silt, fine sand, and coarse sand, that refer to the mean size and carry no implications about mineralogic composition, shape, orientation, packing, or even size sorting. (c) An attempt may be made to recognize groupings of sedimentary rocks, based on measurements of many parameters. For two parameters such groupings may be recognized by plotting the observed values on a scatter diagram and searching for groups of points set off from other groups by regions with only a few points. It seems reasonable to choose the dividing boundaries for a descriptive classification in such a way that they pass through regions with a low density of observed values and around the regions in which points cluster. For more than three variables, the numerical techniques of factor and cluster analysis are now available to achieve analogous results.

The advantages of the clustering method over the first two methods of classification are (a) it reduces the number of categories in the classification by combining together many descriptive measures into a few factors suitable for discriminating between the main naturally occurring groups and (b) it focuses attention on natural groups. Natural groups presumably occur because of the operation of different processes in nature and, consequently, it may be expected that such groups have genetic significance.

At a more advanced level than the descriptive classifications are those called *genetic*, which group together sedimentary rocks that have a common origin. There are many difficulties in constructing such classifications because almost all sedimentary rocks have complex origins and it is difficult to single out one aspect of the history of the rock as more or less important than another. Classifications can be devised that generally will group together rocks derived from the same type of source, rocks with a similar history of transportation and abrasion, rocks deposited in the same environment, or rocks deposited by the same physical or chemical mechanism.

Ideally a classification should be both genetic and descriptive but the ideal can rarely be attained. In practice, there are classifications that are descriptive so that a rock may always be designated by its appropriate name or symbol, but generally some rocks with dissimilar origins are grouped together. Alternatively, there are genetic classifications but the criteria for assignment of a rock to its proper origin are generally insufficiently well understood so that mistaken assignments are common.

The appropriate place to discuss specific classifications is in those chapters that are concerned with the major groups of sedimentary rocks. Although there have been many attempts to devise general classifications for all sedimentary rocks, none have achieved widespread acceptance. In general, sedimentologists distinguish the major groups of rocks on the basis of mineralogic composition. A second major grouping of the clastic rocks is based on average grain size. Most of the debates about classification are not about major groupings but about the subdivision of a few of the groups, such as sandstones and limestones.

Now that the stage of purely qualitative description of sedimentary rocks is past, the value of most new classifications is rather small. Only rarely does a new classification appear that incorporates the results of research leading to a more fundamental understanding of the origins of sedimentary rocks.

It is our opinion that much of the debate about terminology in sedimentology is an unnecessary distraction from the major problems in the science. These problems are, in general, to devise more effective methods for the quantitative description of the fundamental properties of sedimentary rocks, to interrelate derived and fundamental properties, and to conduct studies both in the laboratory and in the field that lead to further understanding of the natural associations of sedimentary rock types.

In attempting to understand the origin of sedimentary rocks, it seems that, as a general rule, the interpretation should begin at the smaller scale and extend to the larger scale only by the synthesis of many small-scale observations and interpretations. In particular, it appears to be unwise to attempt to interpret single examples of sedimentary rock types, textural characteristics, or sedimentary structures in terms of environments, climate, or tectonics. In the case of clastic rocks, a particular texture or structure is generally to be interpreted in terms of a particular mechanism of deposition. Only very rarely does a particular mechanism operate only in one environment. Even an association of several textures and structures in a single locality may not be diagnostic of any one environment. A facies of shales interbedded with sandstones showing an association of graded bedding with flutes, grooves, and other sole markings is generally thought to be diagnostic of deposition in relatively deep water by turbidity currents (see Chap. 5). But very similar associations have been described in rocks demonstrated by other evidence to be fluvial. In most cases, environmental reconstruction or inferences about tectonics or climate must be based on a study of the relationships displayed by many different sections within the same stratigraphic unit. Even then, the past history of stratigraphy and sedimentology indicates that the probability of correct interpretation is rather low.

In the pages that follow we shall therefore emphasize mainly studies that have led to the more accurate description of sedimentary rocks and structures and to a better understanding of the mechanisms that formed them.

REFERENCES

Bouma, A. H., 1962, *Sedimentology of Some Flysch Deposits. A Graphic Approach to Facies Interpretation.* Amsterdam: Elsevier Pub. Co., 168 pp. (Describes graphic logs and gives many examples.)

———, 1969, *Methods for the Study of Sedimentary Structures.* New York: John Wiley & Sons, Inc., 458 pp. (Comprehensive description of field and laboratory methods, including coring, peel, and X-ray techniques.)

Griffiths, J. C., 1961, "Measurement of the Properties of Sediments," *Jour. Geol.*, **69**, pp. 487–498. (A fundamental paper—for a more extended discussion see the book listed below.)

———, 1967, *Scientific Method in Analysis of Sediments.* New York: McGraw-Hill Book Co., 508 pp. (Not a laboratory manual, but a careful and pessimistic scrutiny of some common techniques using statistical methods. Good discussion of sampling and experimental design.)

HARBAUGH, J. W., and D. F. MERRIAM, 1968, *Computer Applications in Stratigraphic Analysis.* New York; John Wiley & Sons, Inc., 282 pp. (Strong on subsurface methods, trend surface analysis, multivariate methods, and simulation.)

KING, J. L., 1969, *Statistical Analysis in Geography.* Englewood Cliffs, N. J.: Prentice-Hall, Inc., 288 pp. (Not written for geologists, but discusses clearly and concisely most of the multivariate techniques useful in sedimentology.)

KOTTLOWSKI, F. E., 1965, *Measuring Stratigraphic Sections.* New York: Holt, Rinehart and Winston, Inc., 253 pp. (A guide to stratigraphic field techniques.)

KRUMBEIN, W. C., 1960, "The 'Geological Population' as a Framework for Analysing Numerical Data in Geology," *Liverpool Manchester Geol. Jour.*, **2**, pp. 341–368.

KRUMBEIN, W. C., and F. A. GRAYBILL, 1965, *An Introduction to Statistical Models in Geology.* New York: McGraw-Hill Book Co., 475 pp. (Strong on experimental design and, especially, trend analysis.)

OTTO, G. H., 1938, "The Sedimentation Unit and its Use in Field Sampling," *Jour. Geol.*, **46**, pp. 569–582. (A classic paper on sampling.)

WALKER, R. G., 1967, "Turbidite Sedimentary Structures and Their Relationship to Proximal and Distal Depositional Environments," *Jour. Sedimentary Petrology*, **37** pp. 25–43. (Simplified field logging technique for turbidites.)

PART TWO

PHYSICS OF SEDIMENTARY PROCESSES

The origin, transport, and deposition of clastic particles can be considered as a result of interaction among tectonic, meteorologic, and hydraulic factors. In what way do the amount and distribution of relief control the sediment load of streams? What is the most fruitful way to analyze the size and shape characteristics of clastic particles? How are various sedimentary textures and structures produced and are any of them unique to a single depositional environment? Is there a group of sedimentary textural or structural features that is diagnostic of certain regional facies?

CHAPTER TWO

THE GEOLOGIC CYCLE

2.1 INTRODUCTION

A basic concern in sedimentology is the origin, volume, and rate of supply of sediment eroded and transported to the basin of deposition. Many problems arise in an analysis of the factors involved; these include relief, climate, rock type, and time. At one extreme, we must consider the rate of uplift of the crust, a very large-scale feature. At the other extreme, it may be necessary to consider the angle of impact of raindrops on exposed clots of soil, a very small feature. Our purpose in this chapter is to overview some of the more important factors and data dealing with uplift and erosion and to assess the nature of the equilibrium relationship between the two.

The concept fundamental to an understanding of the relationships among uplift, erosion, and sedimentation is an extremely old one, first stated clearly by James Hutton in *Theory of the Earth*. Hutton realized that erosion of the continents implies continual uplift, otherwise the land areas would soon be reduced to sea level. Yet an examination of continental rocks shows that they are not "primitive" but that most of them have themselves been formed from sediment derived from yet earlier rocks. The sequence of events revealed by study of continental rocks is long and complex and demonstrates the continual operation of what has been called the "geologic cycle." Material is continually being shifted by erosion and transport processes from parts of the crust undergoing uplift to parts undergoing subsidence. Here, sediment is accumulated and buried, in some cases to great depths in the crust. The sediment is then changed into rock by diagenesis, metamorphism, or even melting and uplifted again to supply material to some other area of deposition.

The concept of the geologic cycle implies at least some degree of equilibrium among the different parts of the cycle. The amount of material removed by erosion must equal the amount supplied to areas of deposition, though some of the material may not be deposited immediately. Some of the sediment is deposited at points in the drainage basin where accumulation of sediment rather than erosion is taking place. The proportion of sediment delivered from the source to any specified downslope location has been called the "delivery ratio" and studies in a number of drainage basins have shown that it generally is less than 100% and decreases with increasing size of drainage basin, reaching values of about 10% for drainage basins with areas of more than 100 sq mi. Of course, from the point of view of overall denudation of the continents, sediment deposited in the drainage basin is not removed from the continents and, hence is not "eroded." It is important to distinguish between true local erosion rates and denudation rates averaged over large areas, which may include smaller areas of both erosion and deposition.

There is evidence that in areas intensely cultivated for animal husbandry or growth of crops the amount of both clastic and dissolved substances carried by streams has been increased considerably in historic times, perhaps by as much as an order of magnitude (Meade, 1969). On the other hand, man has reduced clastic sediment load in some streams by building dams that act as sediment traps. Thus data on denudation rates derived from measuring sediment in streams draining regions in Europe, America, and much of Asia that are significantly affected by human activity are difficult to interpret in terms of ancient denudation rates.

Menard (1961) compared modern and ancient rates of denudation. His method consisted of using field data to estimate the volume of clastic rock derived from Late Mesozoic–Cenozoic drainage basins during known intervals of geologic time. He then compared these data with present denudation rates calculated using stream load data from the same drainage basin. His results (Table 2-1) indicate

TABLE 2-1 COMPARISON BETWEEN MODERN AND ANCIENT DENUDATION RATES IN THREE LARGE DRAINAGE BASINS.

Region	Present rate of denudation (ft/1000 yr)	Past rate of denudation (ft/1000 yr)
Appalachian	0.026	0.2
Mississippi	0.14	0.15
Himalaya	3.3	0.7

that differences of as much as an order of magnitude are to be expected between modern and ancient denudation rates obtained in this manner.

2.2 THE LOAD OF STREAMS

There are many agents of denudation of the land surface. Many are of relatively minor importance and most result ultimately in delivery of eroded material into a river system. Historical observations of coastal erosion show that the amount of material supplied directly from coasts to basins of deposition, though it may be important locally, is negligible on a world scale. Therefore most discussion of rates of denudation and the factors that affect it center on the material transported from the land surface to the oceans by streams.

The total material, other than water, that is transported by a stream is called its load. The materials may be moved in solution (*solution load*), in suspension (*suspension* or *wash load*), or by traction along the bed (*traction* or *bed load*). The rate at which sediment is moved is called the *sediment discharge* and is defined as the sediment load moved past a given cross section of the stream in unit time. The unit sediment discharge is similar but is measured per unit width of the stream. Measurement of sediment load and rate of movement are relatively easy for solution and suspended load but are very difficult for bed load. The only accurate values for bed load transport rates are those obtained by measuring the sediment trapped in reservoirs and those obtained at a few specially constructed "flumes" on rivers, where sufficient turbulence is generated to take the entire bed load into suspension. Estimates of bed load good to within an order of magnitude may also be obtained by use of theoretical "bed load formulas," by use of sediment traps, or by monitoring the movement of subaqueous dunes in rivers. In most studies of denudation rates, however, the bed load is ignored or is conventionally considered to comprise a certain fraction (commonly 10 %) of the total load.

The sediment load and discharge are controlled by four main variables. In probable decreasing order of importance the four are relief, climate, vegetation, and bedrock geology. In general, the relationships among these variables are poorly understood, as evidenced by the many dams that are nearly full of sediment many years before the life expectancy predicted by engineering analyses. It is only during the past 20 years that quantitative geomorphic studies have begun to shed some light on this dark corner of sedimentology and the importance of these studies to the national economy guarantees an increasing number of investigations in the future. Many published studies are of an interdisciplinary nature and are not readily accessible in the usual geologic journals. Fortunately, there are a number of recent review papers; these are listed in the references at the end of the chapter.

2.3 EFFECT OF RELIEF

Relief may be defined in two main ways: (a) the average elevation of an area above sea level and (b) the difference in elevation between highest and lowest points within a specified area. Obviously some regions, such as plateaus, that have high average elevation may have low local relief.

Ruxton and McDougall (1967) studied the effect of local relief, defined as the vertical distance between major ridge crests and the adjacent major valley bottoms, on an andesitic stratovolcano in northeast Papua. The advantages of this study are (a) that the effect of rock type was held constant and (b) that the rate of erosion was not obtained by sampling river load but by reconstructing the original shape of the volcano and calculating average rates of erosion from differences between the present and original topographic surfaces. The original topographic surface is known from potassium-argon dating of the lavas to have formed about 650,000 years ago. The measured rates of erosion are, therefore, values averaged over several hundred thousand years as opposed to the short-term rates measured in most streamload studies. The denudation rates calculated in this way show a linear relationship with relief and vary from 0.26 ft/1000 yr for a relief of 200 ft to 2.5 ft/1000 yr for a relief of 2500 ft. These are very high rates of denudation due to the high relief, intense annual rainfall (varying from 90 in. at sea level to more than 120 in. above 3000 ft), and the humid tropical climate.

Schumm (1963) studied sediment yield in small drainage basins in western United States, using data from both stream load and reservoir measurements. He found that sediment yield was an exponential rather than linear function of relief, but he defined relief as the ratio of maximum relief in the basin to basin length.

Unfortunately, it is not possible to compare these two studies directly because they made use of different definitions both of erosion (Schumm's study did not include dissolved load) and of relief (Schumm measured basin relief; Ruxton and McDougall measured local relief).

Ahnert (1970) studied data from twenty river basins in America and Europe. The basins vary widely in size, climatic setting, and probably rock type. The mean relief was determined as the average of the maximum difference in elevation within sample areas of 20 sq km. The data indicate a linear relation between the average local relief, so defined, and the denudation rate, with a correlation of 0.98. Denudation rates varied from 0.05 ft/1000 yr to 1.4 ft/1000 yr. Ahnert found that Ruxton and McDougall's data were compatible with his results.

One consequence of the importance of relief is that much of the sediment in a river may be derived from a small part of the total drainage basin. For example, a detailed investigation of the Amazon River Basin by Gibbs (1967) revealed that 82% of the suspended solids discharged by the Amazon River is supplied from the 12% of the basin located in the Andes Mountains.

Data on variation of types of load with relief are scarce but in general it appears that streams draining areas of high relief have the highest proportion of bed load. The example of the Amazon suggests that most suspended load is also derived from mountainous areas, although the presence of the load is most obvious in lowland streams.

It seems that the 5 to 10% of the earth's surface that is mountainous supplies 80% or more of the clastic sediment in modern basins of deposition. This does not mean, however, that the detritus, and particularly the bed load portion, makes a nonstop trip from the highlands to the depocenter (e.g., from the Black Hills of South Dakota to the Gulf of Mexico). Grains usually make many stops along the way on terraces, in floodplains, in lakes, etc., but nonetheless their presence in these areas reflects the presence within the drainage basin of upland areas. The rate at which bed load is flushed out of a basin is very slow. For example, part of the headwaters of the Rhine was diverted by glacial action during the Pleistocene but sands whose heavy mineral assemblage derives from the "lost" part of the headwaters are still being delivered to the mouth of the river. For this reason, it is doubtful that large river systems can abruptly change the mineralogy of the sands transported, unless a supply of sediment types previously very rare within the whole basin is suddenly made available. The bedrock in lowland areas supplies little clastic sediment to depositional basins. It functions mainly as a pavement on which sediment from mountainous areas is transported to the sea.

2.4 EFFECT OF CLIMATE AND VEGETATION

The main meteorologic controls on climate are latitudinal and are correlated with the major features of atmospheric circulation. Belts of relatively high precipitation are located between 0 and 5 degrees (60 to 80 in./yr at present) and 50 and 60 degrees (10 to 20 in./yr at present); belts of low precipitation are between 5 and 30 degrees (0 to 5 in./yr at present) and 70 and 90 degrees (5to10 in./yr at present). Unfortunately, these figures cannot be applied to the geologic past because of continental fragmentation and drift, because of the abnormal present climate resulting from the current glacial-interglacial alternations, and because of the severe changes in the basic latitudinal rainfall pattern caused by Late Tertiary mountain ranges. The present mean annual rainfall is 25 to 30 in./yr for the earth's land area but even this figure must have been different in the geologic past. For example, paleogeographic reconstructions for the Permo-Carboniferous period indicate almost no land between and 0 and 30 degrees north and south latitudes and a land cluster (Gondwanaland) around the geographic South Pole (Fig. 2-1). Late Paleozoic rocks in South America must reflect a mean annual rainfall different from that of Tertiary rocks simply as a function of latitude.

With this caution in mind, it nevertheless is worth investigating the effects of climate on modern streamloads, for heat and water are the factors that turn bed load into suspended load plus dissolved load. Through the group of processes called weathering, the relative amounts of clastic and nonclastic rocks are closely related to climate.

The yield of detrital sediment from a drainage basin can be correlated with the amount of stream runoff, and runoff is a function of both precipitation and temperature. To obtain a specified amount of runoff requires higher precipitation at higher temperatures because more of the water is evaporated directly back into the atmosphere instead of increasing streamflow (Fig. 2-2). Based on data

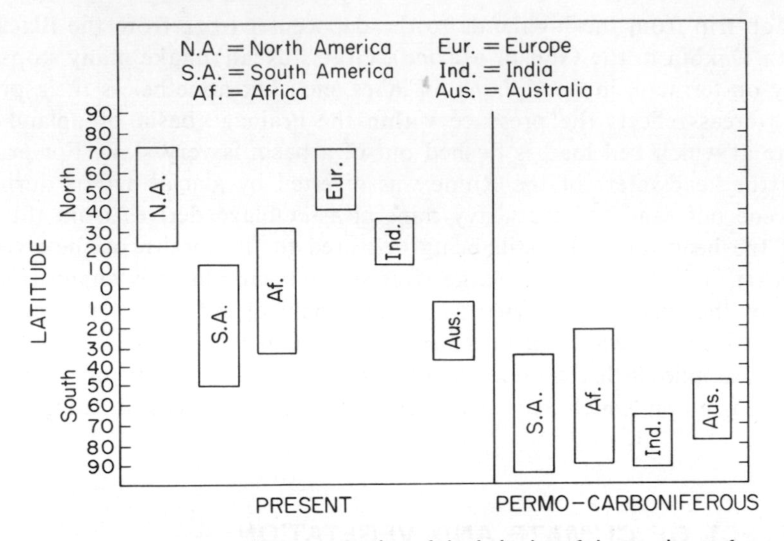

Fig. 2-1 Diagrammatic representation of the latitudes of the continents for the present and for the Permo-Carboniferous (from Girdler, 1964).

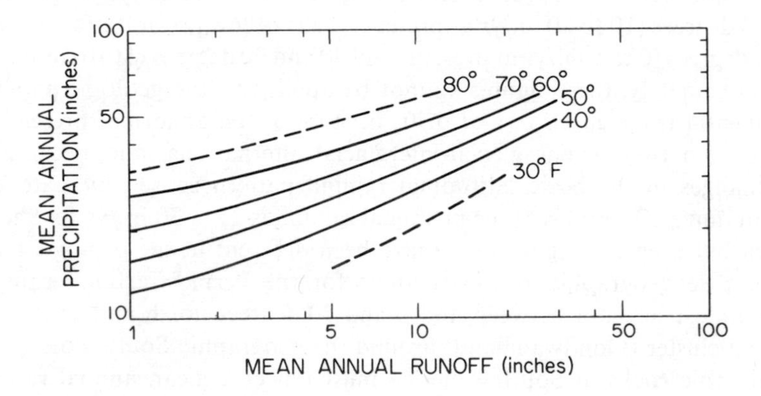

Fig. 2-2 Relationship between mean annual runoff, precipitation, and temperature (from Langbein et al., 1949).

from stream gauging stations and reservoir sediment surveys in the central United States, the relationship between mean annual precipitation and sediment yield is determined to be as shown in Fig. 2-3; the curve for 50°F bears a distinct relationship to vegetation pattern. The peak sediment yield between 10- and 14-in. annual rainfall falls at the point where desert shrubs and bunchgrass are replaced by grass as the dominant type of plant and a flattening of the curve occurs at an annual rainfall of 30 in., the ecologic point at which grassland is replaced by forest vegetation. Mean annual rainfalls of less than 10 in. produce too little runoff for long distance sediment movement. In arid regions, sediment is trapped close

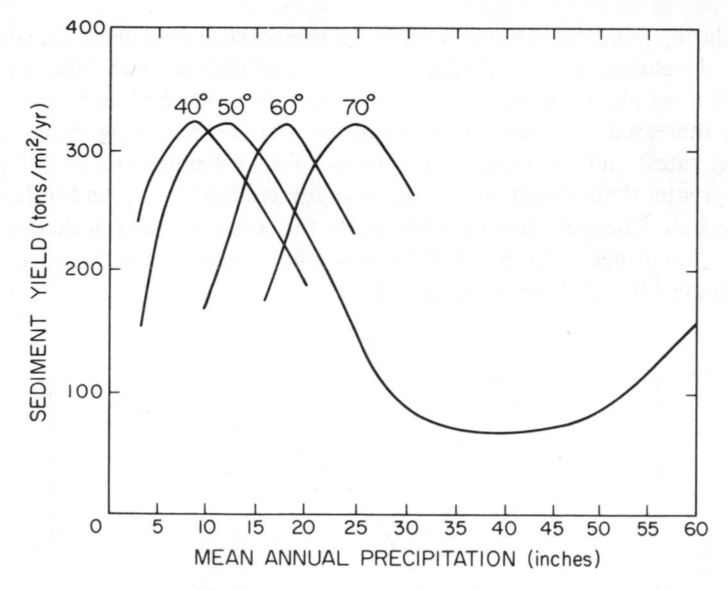

Fig. 2-3 Relationship between mean annual precipitation and sediment yield (modified after Schumm, 1965). Present mean annual temperature in the U. S. is 50°F, but probably has been higher through most of geologic time.

to the source in alluvial fans or moved slowly by wind action. Sediment yields decrease at precipitation totals between 14 and 30 in. because of the stabilizing effects of extensive vegetation cover. At mean annual precipitation values in excess of about 45 in., sediment yield in areas of high relief rises rapidly because of a marked increase in the rate of mass movement. In such cases unweathered bedrock as well as highly weathered materials may be eroded.

On a world scale, several authors have attempted to generalize about the effects of climate but have to come to somewhat different conclusions. Fournier (1960) and Strakhov (1967) have produced maps showing the world distribution of rates of denudation. The maps show maximum rates in the seasonally humid tropics, declining in the equatorial regions where the seasonal effect is lacking, and becoming low in the arid regions where there is little runoff. Attempts to produce more detailed interpretations frequently lead to conflicting conclusions, possibly because of the different climatic measures used as well as the confusing effects of other variables besides climate.

The close relationship among rainfall, vegetation, runoff, and sediment yield raises the important question of the nature of erosion patterns and sediment yields in pre-Devonian times when land vegetation was presumably lacking.

Plant roots in the soil serve two functions: (a) They increase the rate of decomposition of surficial rock and decrease the size of the sediment produced, (b) They bind this sediment and retard its erosion, decreasing the denudation rate. It is difficult to picture the nature of geomorphic interactions and rates of erosion in the absence of vegetative cover but some of the possibilities have been discussed in a stimulating paper by Schumm (1968). In the absence of vegetation, soils would be poorly developed (or absent, depending on the definition of "soil") and most precipitation would be transmitted overland as runoff. Sediment yields would be greatly increased over their present amounts (after removing man's influence on erosion rates) and the value of the ratio of gravel plus sand to silt plus clay would be greater than at present. Schumm suggested that the spread of land plants during the Late Paleozoic may be responsible for some of the cyclothemic deposits of Pennsylvanian age. Figure 2-4 illustrates his conception of the effect of vegetation changes through time on sediment yield.

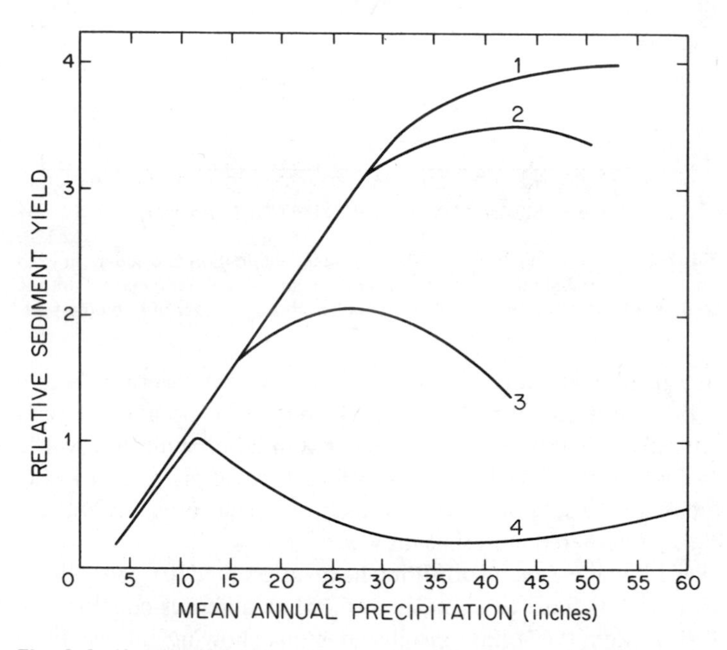

Fig. 2-4 Hypothetical series of curves illustrating the relationship between precipitation and relative sediment yield during geologic time. Curve 1 : before the appearance of land vegetation ; curve 2 : after the appearance of primitive vegetation on the land surface ; curve 3 : after the appearance of flowering plants and conifers ; curve 4 : after the appearance of grasses. The peak of modern sediment yield is shown at a relative sediment yield of 1 (from Schumm, 1968).

2.5 EFFECT OF BEDROCK

The effect on erosion rate of lithologic differences within a drainage basin is not well known. In large basins, variations in relief and climate commonly are great enough so that the effect of changes in rock type is small compared with changes in the other parameters, but in basins located entirely within plains areas lithology may be quite important. A good example of lithologic control on erosion is provided by the drainage basin upstream of the Angostura Reservoir in the southwestern corner of South Dakota. The area involved is 9000 sq. mi. in areal extent (120 by 75 mi) and includes parts of eastern Wyoming, southwestern South Dakota, and northwestern Nebraska. Rocks exposed are all sedimentary; they range in age from Devonian to Quaternary; and they include shales, sandstones, and carbonate beds. Because rainfall is sparse, averaging 14 in./yr, many small stock reservoirs have been built by the local ranchers and these reservoirs have accumulated sediment at markedly different rates, depending on the lithologic characteristics of the rocks within each small drainage area. Rates of sediment accumulation in the reservoirs range from 470 tons/sq mi/yr to 6470 tons/sq mi /yr, a range in denudation rate from 0.20 ft/1000 yr to 2.80 ft/1000 yr. As expected, claystones and shales have the higher rates; sandstones, the lower rates.

Several sedimentologically important conclusions can be drawn from these data. First, erosion rates are extremely high for shales, even in plains areas where relief is minimal; shales in plains areas may erode more rapidly than sandstones in mountainous areas. If we assume that bed load in the ephemeral streams in the area is supplied only by erosion of the sandstones and that sandstones and shales are present in subequal amounts in this area, it can be concluded that bed load totals 5 to 10 % of detrital load. Shale fragments disintegrate so rapidly that shale is transported as suspended load and, as most of the major shale units in this part of the Great Plains are nearly devoid of sand, the assumption that bed load is closely correlated to the rate of erosion of sandstone is a valid one. A second sedimentologically important conclusion based on these data is that the generation of bed load may be confined to a small part of a drainage basin not only because of the distribution of relief but also because of the distribution of lithologies in the basin. The sandy Wasatch Formation (Eocene) is located entirely in the extreme western part of the drainage basin behind the Angostura Reservoir and is exposed over 22 % of the basin. This stratigraphic unit contributes a large part of the bed load but only 1 % of the detrital load carried into the reservoir. These facts are quite significant for paleogeographic and paleogeologic interpretations based on sandstone petrology.

The effect of lithology on the dissolved load of streams results from the different solubilities of the various minerals in each sedimentary unit and from the relative areal exposures of these units. Obviously, a stream draining evaporites or carbonate rocks will have a much larger ratio of dissolved load to clastic load than one draining chert or granite. But even for silicate rocks this ratio may exceed unity (Cleaves, et al., 1970). We postpone until Chap. 10 consideration of the variety of ions in natural waters, as our interest at present is only in the total amount of dissolved load. The dissolved loads of streams vary from 2 to more than 7900 ppm with the more extreme values occurring in relatively small streams draining narrowly defined suites of rocks containing either very insoluble or very soluble minerals. For example, the rate of degradation by solution on the northeast flank of the Wind River Range in Wyoming was found to be twice that on the southwest flank despite the fact that stream runoff on the northeast flank is only two-thirds that on the southwest. The explanation is that the northeast side of the range contains subequal amounts of granite and sediments while the southwest side exposes only the relatively insoluble granite.

2.6 RATES OF UPLIFT COMPARED TO RATES OF EROSION

Uplift of the land may result from two fundamentally different causes, called *orogenesis* and *epeirogenesis* by G. K. Gilbert. Orogenesis includes those earth movements produced by folding, thrusting, and metamorphism, the type of earth movements commonly found in mountain chains. Epeirogenesis includes large-scale, gentle, upwarping or downwarping of the crust, as well as predominantly vertical movement of large crustal blocks along faults. There also are isostatic movements taking place in response to changing load on the crust, such as upward movements following the melting of large glaciers or disappearance of large lakes.

Although orogenesis is the fundamental process causing most mountain chains, existing relief in mountainous areas is by no means always produced by this type of earth movement. In many cases there is clear evidence that the uplift that produces mountains such as the Alps or the Appalachians is not directly related to the orogeny that produced the existing rock structures. The uplift, in other words, is postorogenic and is of an epeirogenic type, although the exact mechanism responsible for uplift remains a matter for debate. To avoid confusion about mechanisms, in this section we refer to "rates of uplift in mountains" rather than to "rates of orogenic uplift."

Measurements of rates of uplift are based either on comparison of accurate topographic surveys, repeated after a number of years, or on inferences from geomorphologic and stratigraphic evidence. The former are, of course, instantaneous rates in terms of geologic time and may not be typical of more extended periods or (for that matter) of other periods of geologic history.

Most current rates of uplift in mountains have been measured in California and have values ranging between 13 and 42 ft/1000 yr with an average of 22 ft /1000 yr (Schumm, 1963). Measurements in Japan range from 3 to 250 ft/1000 yr, with an average of 15 ft/1000 yr. Two measurements in the tectonically active Persian Gulf area yielded values of 10 ft/1000 yr and 33 ft/1000 yr. Based on this scattering of data, an average rate of uplift of approximately 20 ft/1000 yr seems reasonable for modern mountainous areas thought to be orogenically active. Such areas occupy only 5 to 10% of the earth's land surface, however, and there is little reason to suppose a greater percentage for past geologic times. Some geologists believe that the present is an unusually tectonically active time in geologic history and, therefore, the present areal extent of such areas should be considered a maximum.

Epeirogenic uplifts are areally more widespread but generally of much lesser magnitude than orogenic ones. Measurements along nonorogenic coasts yield values between 0.3 and 12 ft/1000 yr and average 3.2 ft/1000 yr. Therefore, it appears from very limited data that, excluding effects of glacial rebound during the past 10,000 years, the maximum rate of uplift is 20 to 25 ft/1000 yr over 5 to 10% of the crust and 3 to 4 ft/1000 yr in epeirogenically active areas. The areal extent of epeirogenically active areas is unknown because some parts of the crust may be at stillstand.

The Colorado Plateau is an example of an area that has undergone epeirogenic uplift for a long period of geologic time. It is estimated that, in the last 40 million years, the total uplift has exceeded 15,000 ft, giving a long-term average of about 0.4 ft/1000 yr.

Uplift resulting directly from isostasy, however, can be very rapid. Data for Hudson Bay indicate a maximum average value of 130 ft/1000 yr as the rate of postglacial uplift. Current rates of uplift in Scandinavia reach at least 30 ft/1000 yr. Lake Bonneville has been uplifted a maximum of 210 ft in about 16,000 years, giving a maximum average of 13 ft/1000 yr. These values are almost equal to the largest recorded in mountain regions.

Many of the rates of uplift given above are considerably in excess of the largest rates of erosion recorded earlier. Based on the relationship between relief and rate of erosion determined by Ahnert, it may be calculated that in order to produce a rate of erosion equal to a "typical" mountain rate of uplift of 30 ft/1000 yr, it would be necessary to have a local relief of the order of 150,000 ft, whereas the highest mountains reach only a fifth of this value in total elevation. It is clear, therefore, that such high rates of uplift cannot persist for long periods of geologic time and that the rates of uplift and erosion are not in equilibrium while rapid uplift is taking place. This conclusion has implications both for geomorphology and sedimentology. It suggests that an "equilibrium" theory of landforms must be severely limited in scope. Probably uplift commonly takes place in a series of rapid pulses, followed by periods of relative inactivity. It is not unreasonable, therefore, to call upon such pulses of uplift to explain repeated influxes of coarser sediment observed in the stratigraphic record. Such an explanation should, however, be used with restraint. We shall see in later chapters that there are many possible alternative explanations.

REFERENCES

AHNERT, FRANK, 1970, "Functional Relationship between Denudation, Relief, and Uplift in Large Mid-Latitude Drainage Basins," *Amer. Jour. Sci.*, **268**, pp. 243–263. (Denudation rate is directly proportional to local relief.)

AMER. SOC. CIVIL ENG. Task Committee on Preparation of Sedimentation Manual, 1970, "Sediment Sources and Sediment Yields", *Amer. Soc. Civil Eng. Proc.*, **96**, No. HY6, pp. 1283–1329. (Data-packed review of the interdisciplinary literature on erosion and sediment supply.)

BLATT, HARVEY, 1970, Determination of Mean Sediment Thickness in the Crust: A Sedimentologic Method," *Geol. Soc. Amer. Bull.*, **81**, pp. 255–262. (Sedimentologic approach to the balance between sediment generation by tectonic activity and sediment destruction by metamorphism through geologic time.)

CHORLEY, R. J. (ed.), 1969, *Water, Earth, and Man*. London; Methuen and Co., 588 pp. (A collection of well-documented essays on the role of water in geomorphology.)

CLEAVES, E. T., A. E. GODFREY, and O. P. BRICKER, 1970, "Geochemical Balance of a Small Watershed and its Geomorphic Implications," *Geol. Soc. Amer. Bull.*, **81**, pp. 3015–3032. (Denudation by solution of silicate rocks is probably as important as mechanical erosion, even in temperate climates.)

FAIRBRIDGE, R. W. (ed.), 1968, *The Encyclopedia of Geomorphology*. New York: Reinhold Book Corp., 1295 pp. (An excellent source for general information and data concerning geomorphology, geography, and sedimentation.)

FOURNIER, F., 1960, *Climat et erosion: la relation entre l'erosion du sol par l'eau et les precipitations atmospheriques*. Univ. Paris, France, 201 pp. (Discusses relation between climate and erosion and gives world maps of denudation rates.)

GIBBS, R. J., 1967, "The Geochemistry of the Amazon River System: Part I. The Factors that Control the Salinity and the Composition and Concentration of the Suspended Solids," *Geol. Soc. Amer. Bull.*, **78**, pp. 1203–1232. (A comprehensive sedimentologic and geochemical study of the world's largest drainage basin. A landmark investigation.)

GILLULY, JAMES, 1964, "Atlantic Sediments, Erosion Rates, and the Evolution of the Continental Shelf: Some Speculations," *Geol. Soc. Amer. Bull.*, **75**, pp. 483–492. (Semiquantitative evaluation of erosion rates from northeastern United States and southeastern Canada lead to the conclusion that the mean denudation rate has not changed significantly in this area since Triassic time.)

GIRDLER, R. W., 1964, "The Paleomagnetic Latitudes of Possible Ancient Glaciations," in *Problems in paleoclimatology*. A. E. M. Nairn, ed., New York: John Wiley & Sons, Inc., pp. 115–118. (Short note summarizing current ideas concerning the positions of the continents during each of the four recognized glacial periods from Precambrian to Pleistocene, based on paleomagnetic data.)

HADLEY, R. F., and S. A. SCHUMM, 1961, "Sediment Sources and Drainage Basin Characteristics in Upper Cheyenne River Basin," *U.S. Geol. Sur. Water-Supply Paper 1531-B*, 62 pp. (Detailed study of erosion rates in a relatively unpopulated semiarid region, relating sediment yield to relief, bedrock geology, vegetation, and drainage channel character.)

HARDIN, G. C., JR., 1962, "Notes on Cenozoic Sedimentation in the Gulf Coast Geosyncline, U.S.A.," in *Geology of the Gulf Coast and Central Texas*, E. H. Rainwater and R. P. Zingula, eds., *Houston Geol. Soc.*, pp. 1–15. (Discussion of sedimentation rates in space and time during the Cenozoic Era in a modern geosyncline, based on abundant subsurface well data.)

JUDSON, SHELDON, and D. F. RITTER, 1964, "Rates of Regional Denudation in the United States," *Jour. Geophys. Res.*, **69**, pp. 3395–3401. (Quantitative evaluation of erosion rates in the United States, in terms of its division into seven drainage basins with differing relief and climate.)

LANGBEIN, W. B., et al., 1949, "Annual Runoff in the United States," *U.S. Geol. Sur. Cir. 52*, 14 pp. (Detailed distribution and map of runoff in the United States.)

LEOPOLD, L. B., M. G. WOLMAN, and J. P. MILLER, 1964, *Fluvial Processes in Geomorphology*. San Francisco: W. H. Freeman and Co., 522 pp. (The first of a "new breed" of geomorphology texts, stressing quantitative aspects of fluvial processes. Included are analyses of streamflow, sediment transport, and sculpturing of the land surface.)

LUSTIG, L. K., 1965, "Sediment Yield of the Castaic Watershed, Western Los Angeles County, California—A Quantitative Geomorphic Approach," *U.S. Geol. Sur. Prof. Paper 422-F*, 23 pp. (A good example of the way quantitative geomorphology interacts with sedimentology in a small drainage basin.)

MEADE, R. H., 1969, "Errors in Using Modern Stream-Load Data to Estimate Natural Rates of Denudation, "*Geol. Soc. Amer. Bull.*, **80**. pp. 1265–1274. (Man's activities cause much more rapid erosion than natural processes and this must be taken into account in interpretations of streamloads in terms of denudation rates.)

MENARD, H. W., 1961, "Some Rates of Regional Erosion," *Jour. Geol.*, **69**, pp. 154–161. (Calculates ancient erosion rates from stratigraphic data.)

RITTER, D. F., 1967, "Rates of Denudation," *Jour. Geol. Ed.*, **XV**, pp. 154–159. (General summary of the state of knowledge as of 1967.)

RUXTON, B. P. and I. MCDOUGALL, 1967, "Denudation Rates in Northeast Papua from Potassium-Argon Dating of Lavas," *Amer. Jour. Sci.*, **265**, pp. 545–561. (Local erosion rates estimated from physiographic reconstruction of a large volcano. Rates are directly proportional to local relief.)

SCHUMM, S. A., 1963, "The Disparity between Present Rates of Denudation and Orogeny," *U.S. Geol. Sur. Prof. Paper 454-H*, 13 pp. (Summary of the quantitative relationship between uplift and erosion, based on modern data. Modern rates of uplift are uncommonly high.)

———, 1965, "Quaternary Paleohydrology," in *The Quaternary of the United States*, H. E. Wright and D. G. Frey, eds. Princeton, N. J.: Princeton Univ. Press, pp. 783–794. (An interesting attempt to use modern hydrologic and sedimentologic data to infer hydrologic conditions during glacial epochs.)

———, 1968, "Speculations Concerning Paleohydrologic Controls of Terrestrial Sedimentation," *Geol. Soc. Amer. Bull.*, **79**, pp. 1573–1588. (A thoughtful and stimulating discussion concerning possible patterns of erosion and sedimentation under different climatic conditions and in the absence of vegetation.)

SCHUMM, S. A. and R. F. HADLEY, 1961, "Progress in the Application of Landform Analysis in Studies of Semiarid Erosion," *U. S. Geol. Sur. Cir. 437*, 14 pp., (A quantitative study of the interrelationships among stream runoff, several drainage basin parameters, and sediment yield in semiarid areas of western United States.)

STRAKHOV, N. M., 1967, *Principles of Lithogenesis*, vol. 1. New York: Consultants Bureau, 245 pp. (see Chap. 1, pp. 2–45). (The first volume of the translation of three books by a leading Russian student of historical geology and the formation of sedimentary rocks.)

CHAPTER THREE

SEDIMENTARY TEXTURES

3.1 INTRODUCTION

In Chap. 2, we examined some of the broader aspects of the geologic cycle, particularly those factors operating to produce detrital or clastic materials by erosion of rocks exposed in areas of uplift. The eroded materials are transported by a variety of mechanisms to an area of deposition, where they become buried and ultimately converted into detrital sedimentary rocks. Detrital materials may also be produced within the basin of deposition itself by erosion of recently deposited, partly lithified sediment or by chemical or biochemical precipitation of dissolved materials and subsequent mechanical reworking of the precipitated or skeletal materials. Whatever their origin, detrital materials are modified by mechanical abrasion and by hydraulic sorting, and both processes depend as much on the agents of transport (wind, water, mud, ice) as on the nature of the detrital materials themselves.

In later chapters, we revert to a consideration of the mineralogic and chemical composition of the detrital particles. In this chapter we consider the composition of the particles only as one factor in the complex interaction of particle and fluid properties that determines the behavior of detrital materials during transportation and deposition. We consider first the fundamental properties of size, shape, fabric and then, in Chap. 4, those aspects of fluids necessary to an understanding of the mechanisms of transportation and deposition by wind and water.

The specific features of a sediment or sedimentary rock, such as the size, shape, orientation, and packing of its grains, and the nature of the sedimentary structures are determined mainly by the small scale *mechanisms* operating during transportation, deposition, and early compaction and deformation of sedimentary materials. Examples of such mechanisms include suspension, saltation, traction, formation of ripples and dunes, avalanching, liquefaction, and so on. Such mechanisms frequently operate as part of some larger-scale geological *process* such as flow in a meandering channel. Several processes of deposition, perhaps including a dozen small-scale mechanisms, may be at work within a single physiographic *environment*.

In the past, there have been some attempts to develop shortcuts to environmental interpretations of ancient sediments: typical has been the attempt to identify environments of deposition by simple criteria that can be measured in a hand specimen (for example, textural or compositional parameters) or observed in a core or outcrop (for example, sedimentary structures). These attempts have had some limited success. Many criteria that at first seemed successful have proved on further investigation to be misleading, particularly in cases where the basic mechanism remained poorly understood. The geologist's first task, after careful observation and description, should be to interpret his observations in terms of mechanisms. A number of such interpretations of related strata suggest larger-scale interpretations in terms of processes, and a final synthesis of all the observations and interpretations might suggest an interpretation in terms of environments and the influence of major variables such as tectonics and climate.

3.2 FREQUENCY DISTRIBUTIONS AND PARAMETERS

In this section we describe some basic descriptive statistical techniques that can be applied to a wide range of phenomena, particularly to size and shape analysis. We then discuss the measurement of size and shape and the application of statistical techniques to these measurement data before returning to the basic problem of how to interpret this type of data in physically and geologically meaningful terms.

Many geological observations consist of measurements made on a large number of specimens, for example, measurement of the maximum diameters of many grains in a hand specimen or the azimuth of dip of many cross-bedded sets in an outcrop. These observations may be of interest in their own right or they may be regarded by the geologist as a sample from a larger population that is really the object of investigation. In the latter case, the techniques of statistical inference may be used to calculate from the sample the probable properties of the population, provided the sample has been selected in an appropriate way, as discussed in

Chap. 1. In either case, the first step in an interpretation is to summarize the data by drawing diagrams or calculating a few *summary statistics* from the original measurements.

Histograms and Cumulative Curves

The main graphic devices used to display sedimentological data are *histograms*; they represent frequency by area (Fig. 3-1), *cumulative curves* that show the percentage frequency greater than a particular value (Fig. 3-3), and *scatter*

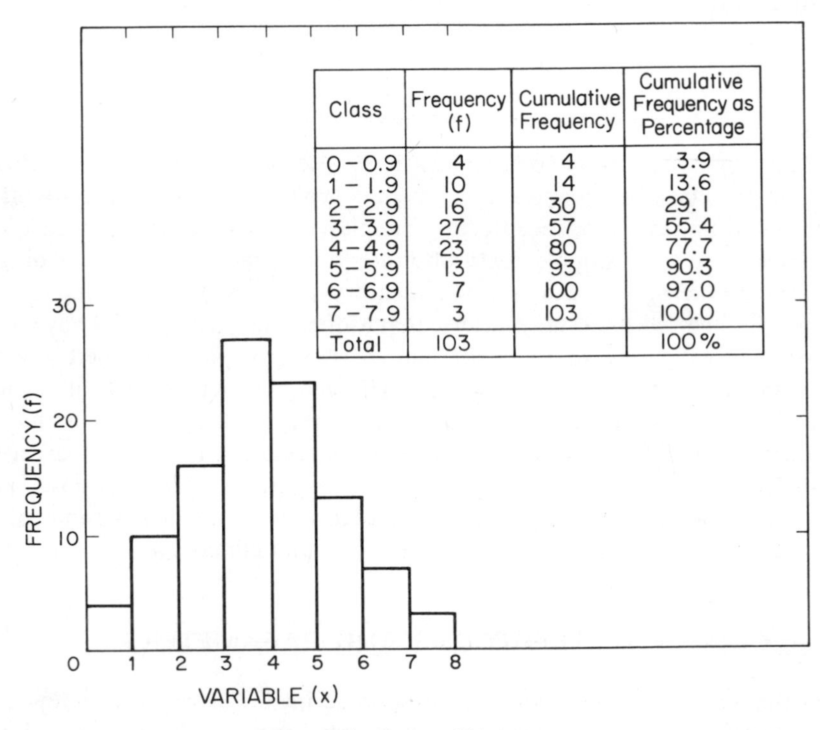

Class	Frequency (f)	Cumulative Frequency	Cumulative Frequency as Percentage
0-0.9	4	4	3.9
1-1.9	10	14	13.6
2-2.9	16	30	29.1
3-3.9	27	57	55.4
4-4.9	23	80	77.7
5-5.9	13	93	90.3
6-6.9	7	100	97.0
7-7.9	3	103	100.0
Total	103		100%

Fig. 3-1 Data and histogram.

diagrams (for example, Fig. 3-15). The value of histograms is illustrated by Fig. 3-2, which shows three different types of size grading in sandstone beds. Trends in the frequencies of particular classes, or in the position of the most frequent class (the mode), are revealed more clearly by histograms than by any other graphic devices. Cumulative curves are commonly plotted with the cumulative frequency represented on a "probability" scale, which is so designed that if the variable has a particular type of frequency distribution, known as the Normal (or Gaussian) frequency distribution, the cumulative curve will plot as a straight line. Other types of "probability" scales may be designed that give the same effect for frequency distributions other than the Normal distribution; for example, the Rosin distri-

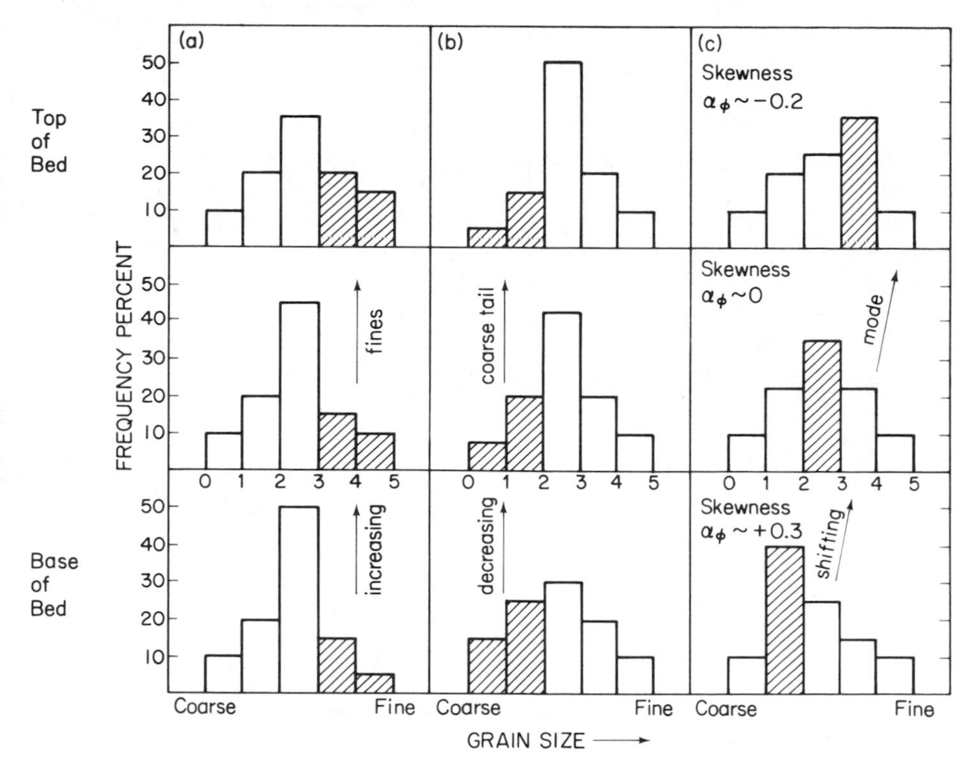

Fig. 3-2 Three types of size grading, illustrated by histograms.

bution, shown by the grain size of many crushed and weathered materials, plots as a straight line on "Rosin's law paper" (Fig. 3-4).

The advantages of cumulative curves include the following: (a) The cumulative curve is an unique representation of the data and may be drawn for both ungrouped and grouped data. Its form does not depend to any great extent on the class intervals used for grouping the data. (b) The *percentiles* of the distribution may be read directly off the cumulative curve. The nth percentile is defined as the value of the variable that divides the sample so that n % of the sample is greater (or less) and $(100 - n)$ % is less (or greater) than that value. The percentiles are useful descriptive measures of the frequency distribution and may be used as such (particularly the median, or 50 percentile, and the two quartiles, or 25 and 75 percentiles) or may be combined to give other descriptive statistics such as the graphic mean, standard deviation, skewness, and kurtosis (see Table 3-1). (c) Plotting the cumulative curve on probability paper permits easy visual comparison of the sample distribution with a Normal distribution.

The disadvantages of the cumulative curve are that the meaning of the curve is not so immediately apparent to the eye as in the histogram and in particular it is difficult to determine the mode or modal class from the curve. The interpre-

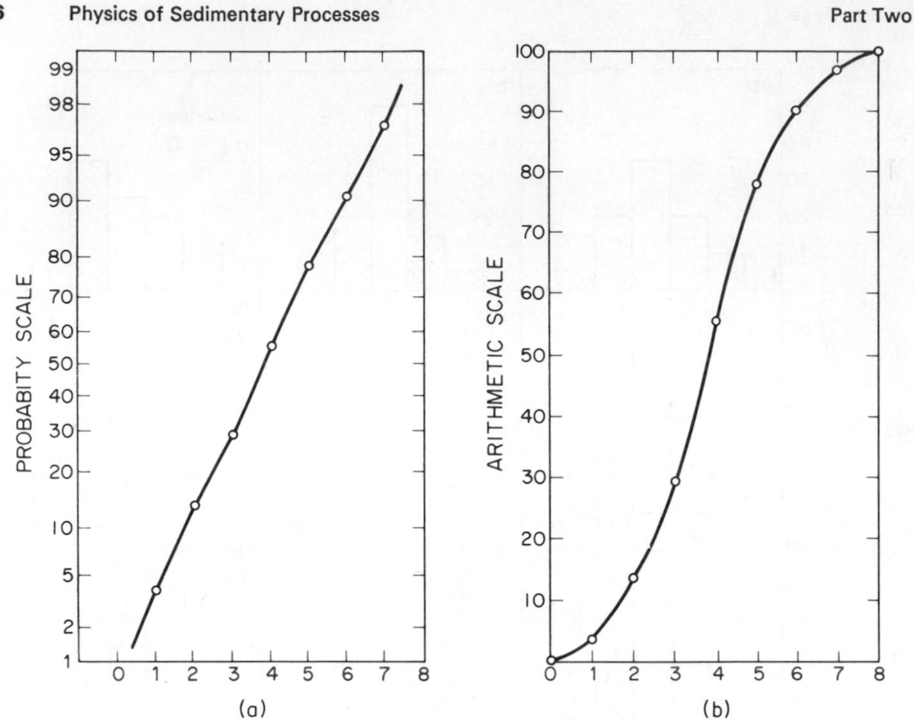

Fig. 3-3 Data of Fig. 3-1, plotted as cumulative curves on (a) probability scale, (b) arithmetic scale.

TABLE 3-1 DESCRIPTIVE MEASURES OF SEDIMENT-SIZE DISTRIBUTION ACCORDING TO SEVERAL AUTHORS

Measure	Trask†	Inman	Folk and Ward
Median	$Md = P^{\ddagger}_{50}$	$Md_\phi = \phi^{\S}_{50}$	$Md_\phi = \phi_{50}$
Mean	$M = \dfrac{P_{25} + P_{75}}{2}$	$M_\phi = \dfrac{\phi_{16} + \phi_{84}}{2}$	$M_z = \dfrac{\phi_{16} + \phi_{50} + \phi_{84}}{3}$
Dispersion (sorting)	$So = \dfrac{P_{75}}{P_{25}}$	$\sigma_\phi = \dfrac{\phi_{84} - \phi_{16}}{2}$	$\sigma_I = \dfrac{\phi_{84} - \phi_{16}}{4} + \dfrac{\phi_{95} - \phi_5}{6.6}$
Skewness	$Sk = \dfrac{P_{25}\,P_{75}}{Md^2}$	$\alpha_\phi = \dfrac{M_\phi - Md_\phi}{\sigma_\phi}$ $\alpha_{2\phi} = \dfrac{\frac{1}{2}(\phi_5 + \phi_{95}) - Md}{\sigma_\phi}$	$Sk_I = \dfrac{\phi_{16} + \phi_{84} - 2\phi_{50}}{2(\phi_{84} - \phi_{16})} + \dfrac{\phi_5 + \phi_{95} - 2\phi_{50}}{2(\phi_{95} - \phi_5)}$
Kurtosis	$K = \dfrac{P_{75} - P_{25}}{2(P_{90} - P_{10})}$	$\beta_\phi = \dfrac{\frac{1}{2}(\phi_{95} - \phi_5) - \sigma_\phi}{\sigma_\phi}$	$K_G = \dfrac{\phi_{95} - \phi_5}{2.44(\phi_{75} - \phi_{25})}$

† The formula for kurtosis was proposed by Krumbein and Pettijohn. Many workers have used the square root of *Sk*, rather than *Sk* itself, as a measure of skewness.

‡ *P* indicates a percentile measure, measured in millimeters.

§ ϕ indicates a ϕ percentile.

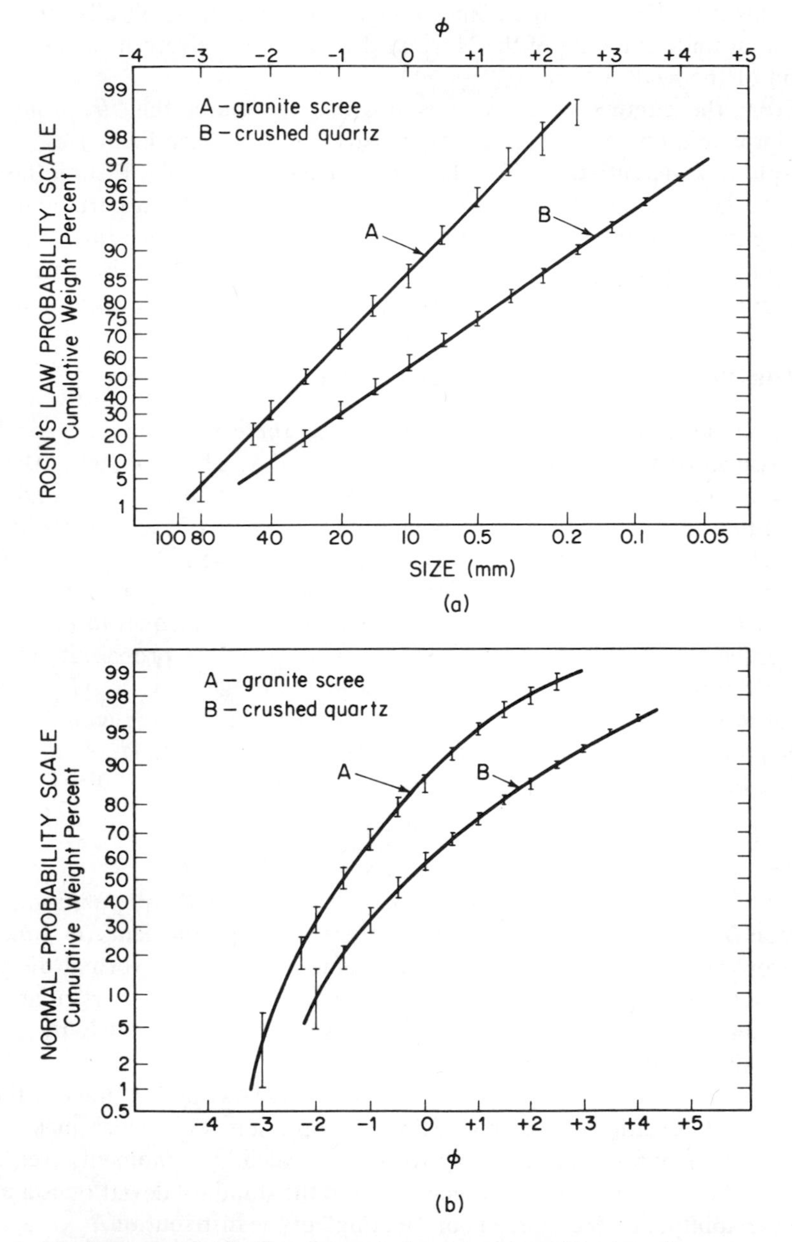

Fig. 3-4 Size distributions for crushed quartz and weathered granite scree, Plotted on (a) Rosin's law probability paper, (b) Normal probability paper. From Kittleman (1964).

tation of cumulative curves plotted on probability paper is easier than when arithmetic graph paper is used but there are many pitfalls for those inexperienced in the use of this type of graph paper. Since understanding of these pitfalls depends in its turn on an understanding of the Normal distribution, we return to the matter at the end of the next section.

From the cumulative curve it is possible to obtain the "*frequency curve*" (an approximation to the probability density function—see below) by numerical or graphical differentiation. The "frequency curve" resembles a smoothed histogram and has certain advantages over the histogram itself: In particular the frequency curve, like the cumulative curve (but unlike the histogram), is an unique representation of the observations and also permits more precise determination of modes (peaks in the frequency curve) than is possible from a histogram.

The Normal (Gaussian) Distribution

For a population of any given variable x, there is a definite probability p to be associated with any given range of values of x. The probability that x will lie within the total possible range of variation is defined as unity, and we will suppose that the probability that a single random sample of the variable lies within a restricted part of the range can be determined (it is generally greater that zero but less than unity). A curve, called the probability density function $f(x)$, can then be drawn, such that the total area under the curve is equal to unity and the area between any two ordinates x_1 and x_2 is equal to the probability of a single random sample from the population being in the range x_1 to x_2 (Fig. 3-5). The integral of the probability density function is identical with the cumulative curve for the population and is called the cumulative distribution function.

The best known probability density function is the Normal distribution:

$$f(z) = \frac{\exp(-z^2/2)}{\sqrt{2\pi}} \tag{1}$$

where $z = (x - \mu)/\sigma$, μ is the population mean, and σ is the population standard deviation. The mean and standard deviation are the two coefficients (or *parameters*) in the probability density function that define a particular Normal distribution (as opposed to the general class of Normal distributions). Any particular Normal distribution, abbreviated as $f(x) = N(\mu, \sigma^2)$, is described completely by specifying its mean and standard deviation (or variance, σ^2).

The theoretical Normal population is infinitely large and in practice the mean and standard deviation are estimated from sample statistics, which may be either the graphical measures described above or the calculated moments (see below). The mean is a measure of the average value and the standard deviation is a measure of the variability, i.e., the "spread" or "sorting" of the distribution.

In sedimentology it is found that many measured variables approximate more or less closely to a Normal or Lognormal distribution (see Table 3-2). For many of the variables, there is no good theoretical reason known why this should be the case and, in fact, it can be demonstrated for many samples that the observed distribution is *not* normal. But the close approach to normality observed in some

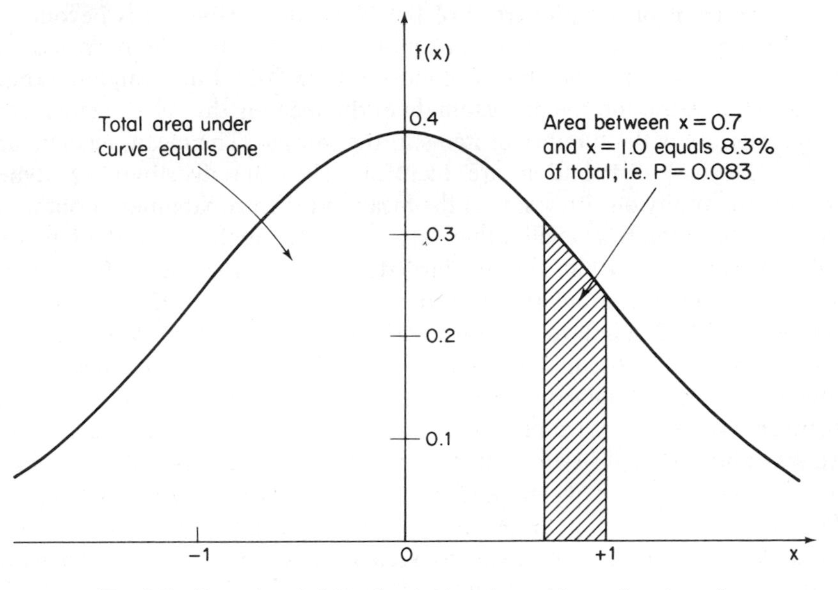

Fig. 3-5 Normal probability density function, with $\mu = 0$ and $\sigma = 1$.

cases and the theoretical properties of the Normal distribution make it a good "model" with which to compare observed distributions.

TABLE 3-2†

List of variables that tend to be normally distributed
Grain size (in ϕ units)
Grain roundness (in ρ units)
Grain sphericity
Packing density of grains in sandstone
Porosity of sandstones
Angle of dip in cross-beds
Azimuths of many paleocurrent data
Percentages of minerals in rocks (where the values lie between 25 and 75%)
Percentages of some chemical oxides in rocks
Velocity fluctuations in a turbulent flow
Errors of measurement
List of variables that tend to be lognormally distributed
Grain size (mm)
Bed thickness
Percentages of some rare components in rocks (e.g., trace elements)
River discharges

† In part after Krumbein and Graybill, 1965, p. 110, see References.

A full treatment of the properties of the Normal distribution is beyond the scope of this book but the following are most important for our purposes: (a) The distribution is symmetrical about the mean μ (Fig. 3-5). For a random sample from a Normal distribution the best estimate of the mean is the "arithmetic mean" $\bar{x} = \Sigma x/n$, where n is the number of items in the sample. The mean, median, and mode of a Normal distribution are identical. (b) The distribution extends, theoretically, to infinity on either side of the mean but the probabilities of obtaining infinitely large or infinitely small values are infinitely small. Sixty-eight percent of the distribution lies within ± 1 standard deviation of the mean, 95% of the distribution lies within 2 standard deviations of the mean, and 99.7% of the distribution lies within 3 standard deviations of the mean. In other words, in large samples drawn from a Normal population only 5% will differ from the mean value by more than 2 standard deviations. The areas of the Normal distribution lying between any two standard ordinates $+z$ and $-z$ (where z is defined as above) are given in tables printed in any statistics manual. (c) The distribution of *means* of samples (of size n) drawn at random from a Normal population is itself Normal, with the same mean as the population and with standard deviation, $\sigma_{\bar{x}} = \sigma/\sqrt{n}$.

From this brief discussion it can be seen that for the Normal distribution the 16 and 84 percentiles are the ordinates that enclose 68% of the distribution and are therefore located, respectively, 1 standard deviation below and above

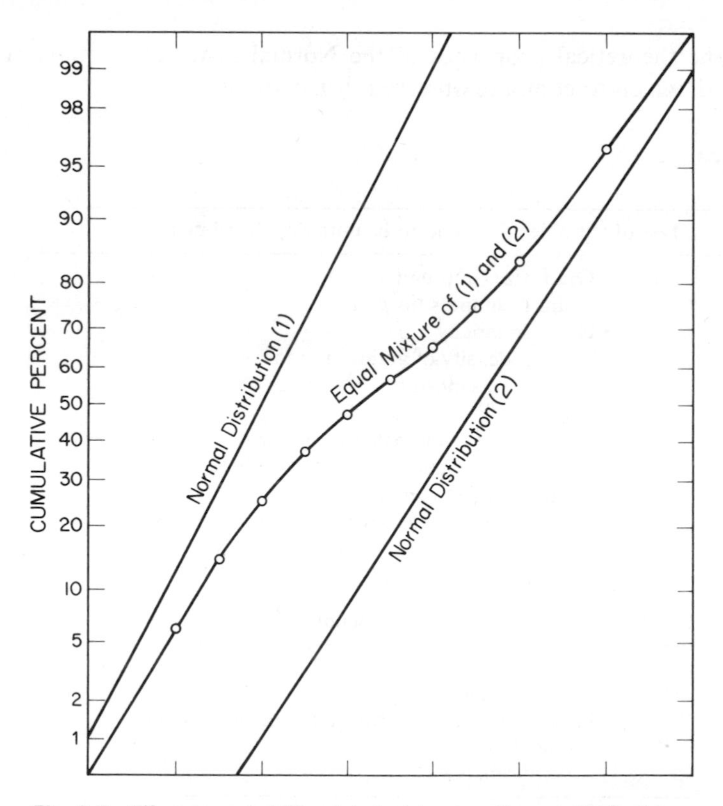

Fig. 3-6 Effect on probability plot of mixing two Normal distributions

the mean. It also happens that these percentiles define inflection points in the probability density function. Thus these two percentiles have a special significance, which explains their common use in graphic statistics (Table 3-1). It might be expected that the 2.5 and 97.5 percentiles would also be used because they are located 2 standard deviations from the mean of a Normal distribution. But, in fact, it is difficult to determine accurately the extreme percentiles (in the "tails" of the distribution) because sampling and measurement errors are greatest for the extreme parts of the distribution. So most authors compromise and use the 5 and 95 percentiles instead.

Many cumulative curves, when plotted on Normal probability paper, are not straight lines. In some cases, the "curve" appears to be made up of two or more straight line segments (see Visher, 1969). These segments are sometimes interpreted as parts of several Normal distributions mixed together. If there is little overlap, mixtures of Normal distributions do yield cumulative curves that plot as straight line segments but this is not the case if there is considerable overlap in the two distributions, as shown in Fig. 3-6. Considerable caution should be exercised in interpreting curves in terms of straight line segments (*at least* three data points are required to justify drawing a straight line!) and even more caution in interpreting segments of cumulative curves in terms of mixed Normal populations. Numerical techniques for the dissection of frequency distributions into two or more Normal distributions are available (Tanner, 1959).

Moment Measures

Use of percentiles read from the cumulative curve is not the only way to construct statistics that summarize the properties of an observed distribution. Another method, which involves much computation (now easily accomplished on an electronic computer), is the use of numerical measures.

The moments of a sample of size n have the form

$$m_p = \frac{\Sigma (x - x_0)^p}{n - 1} \tag{2}$$

where p is an integer and m_p is the pth moment of the distribution about the reference point x_0. The first moment about zero is the arithmetic mean $\bar{x} = \Sigma x/n$. Higher orders of moments are commonly computed about the mean:

Variance: $$m_2 = \frac{\Sigma (x - \bar{x})^2}{n - 1} \tag{3}$$

Third moment: $$m_3 = \frac{\Sigma (x - \bar{x})^3}{n - 1} \tag{4}$$

Fourth moment: $$m_4 = \frac{\Sigma (x - \bar{x})^4}{n - 1} \tag{5}$$

Still higher moments might be computed but they would be greatly affected by measurement and sampling errors. Conventionally, only the first four are computed and the third and fourth moments are used to compute statistics for measuring the skewness and kurtosis:

Skewness: $$x_3 = \frac{m_3}{m_2^{3/2}} \tag{6}$$

Kurtosis: $$x_4 = \frac{m_4}{m_2^2} \tag{7}$$

Both graphic and moment measures of skewness measure the asymmetry of a distribution (see Fig. 3-7). Many observed distributions are not symmetrical like

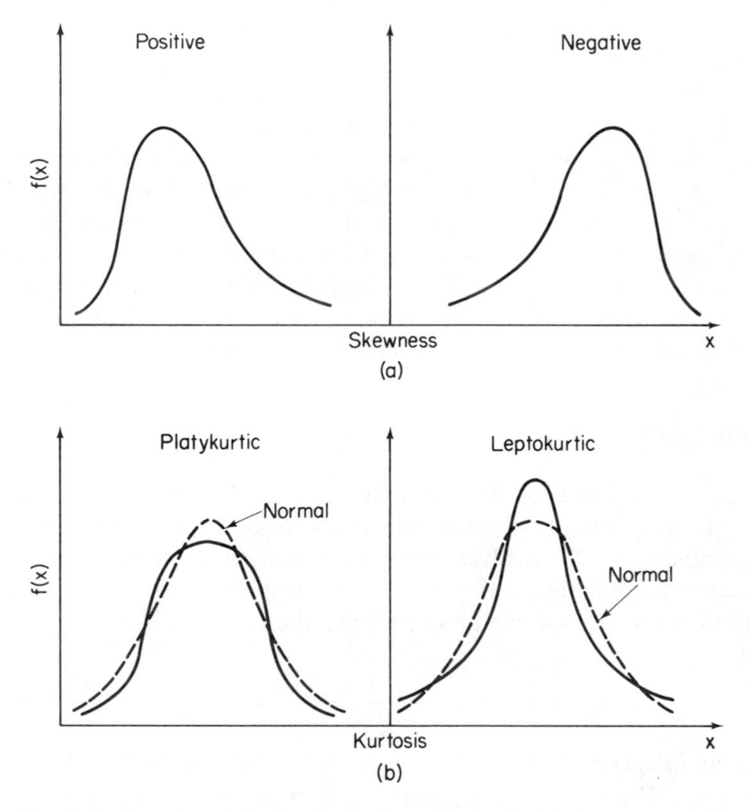

Fig. 3-7 Examples of density distributions differing in (a) skewness and (b) kurtosis.

the ideal Normal distribution and it is therefore useful to have statistics that measure the degree of asymmetry. For example, one of the types (c) of size grading illustrated in Fig. 3-2 results from a change in symmetry from positive skewness at the base to negative skewness at the top. This type of skewness trend, which is generally combined with a decrease in coarse tail (as shown in b), is very common in graded sandstone beds.

Both graphic and moment kurtosis measure the "degree of peakedness" of the distribution. The contrast between two symmetrical distributions that differ in kurtosis is illustrated schematically in Fig. 3-7(b). Kurtosis of natural distributions has been studied little except for grain size, where it has proved to have little geological importance. It appears that generally the overall shape of a distribution is adequately described by the mean, standard deviation, and skewness.

The moment-derived measures may be regarded as "better" than the graphic measures inasmuch as the moment measures are based on all the items in the distribution, not just on a few percentiles. But insofar as both are used as descriptive statistics they each have their own validity. They may be viewed simply as alternative ways of describing the distribution and they need not always give the same results. Neither moment measures nor graphic measures provide a *complete* summary of the data contained in the distribution: it is possible to have distributions that have exactly the same four moment measures and yet differ significantly in some respect.

It may well be questioned, in many applications, whether either moment or graphic measures are appropriate. For example, in size analysis, the basic data consist of the weight frequencies in each size class—these are the data yielded by the commonly used measurement techniques, such as sieve analysis. In such a case, it may be argued that the appropriate technique is not to derive further numerical measures (by either computation or graphic methods) but to make use of the basic measurement data themselves. The data may consist of the weight percent in each of a dozen or more size classes. In the past, techniques were not available for handling such a large number of variables and the purpose of computing numerical indices, such as skewness, was to express some distribution property that was considered to have more significance than the original weight percent data. But modern statistics and electronic computers permit the handling of an almost limitless number of variables. An example is the application by Moiola and Weiser (1969) of the technique of the discriminant function to discriminate between size distributions of samples of sand collected from several different environments. The basic aim of this technique is to use a set of samples whose origin is known to seek that linear combination of the variables that provides the clearest separation of the groups of different origin. Essentially the technique seeks, by numerical methods, to accomplish for p dimensions (defined by the p measured variables) what is accomplished in two dimensions by plotting a scatter diagram and drawing lines on the diagram that best separate the different groups. Further information about the technique is given in standard multivariate statistics texts (e.g., see Krumbein and Graybill, 1965).

Parameters and Statistics

In sedimentology, it has become a matter of routine to make use of simple statistical tests. Such tests may be used to establish the reliability of sample statistics as estimators of population parameters or to test for "goodness of fit" between an observed distribution and a theoretical one such as the Normal distribution.

They are described at length in the texts by Griffiths (1967) and Krumbein and Graybill (1965).

There are two important limitations regarding the application of such tests to sedimentary data.

1. Size data are commonly determined as weight percent rather than as number frequencies. Moments and graphical statistics may be used to describe such data but the theory of statistical tests is based on number frequencies and cannot be applied to weight percent frequencies. It is possible to compute roughly thc number of grains in each weight fraction. If this is done for a typical sand sample, weighing 30 g, it is found that the sample contains more than a million sand grains—i.e., the sample is so large that it may be considered to be essentially identical with the population. Statistical tests, for example, for goodness of fit are unnecessary for such large samples.
2. For application of statistical tests the sample must be drawn at random. In the case of size analysis, this means, for example, that each grain must have an equal chance of being measured and included in the sample. To sample a bed of sand properly, according to this definition, is obviously impracticable since it might be (if the sample were truly random) that one grain would come from one end of the bed and another from the other end. In practice, what is done is to collect a specimen consisting of a group of contiguous grains and to *assume* (without real justification other than visual inspection) that the bed is sufficiently homogeneous that this specimen is "typical" of the whole bed. Essentially, therefore, we substitute for a truly random sampling technique an assumption that the bed is homogeneous, i.e., that it makes no difference that the grains are all drawn from one small part of the bed. To some extent this assumption may be checked by taking more than one specimen but the practice really invalidates the use of common statistical tests, *as applied to measurements made on individual grains*. It is still possible to regard the whole specimen as a random sample of all specimens that might be collected from the bed. Thus, measurements made on the whole specimen (e.g., mean size, skewness, etc.) may indeed be regarded as random samples of the population of such measurements that might be made on all possible specimens that might be taken from the bed.

3.3 GRAIN SIZE

Basic Concept

The basic concept of size may be considered to be either (a) a linear dimension or (b) the volume of the particle. The second appears to have been the commonest ideal concept of size and has led to the definition of the *nominal diameter* (d_n)

as the diameter of the sphere having the same volume as that of the particle. The concept of linear dimension is commonly elaborated by the recognition that because of the irregular shape of most particles it is necessary to define more than one linear dimension. Commonly, three dimensions are measured: the longest, d_L; intermediate, d_I; and shortest, d_S, diameters. These three diameters must be carefully defined operationally so that they may be measured without ambiguity. One good definition (Griffiths, 1967) prescribes the following steps: (a) Establish the plane of maximum projection area. (b) The longest intercept across the particle normal to this plane is the short diameter d_S. (c) Establish, for the particle projection, the "tangent rectangle" (Fig. 3-8). (d) The short side of this rectangle

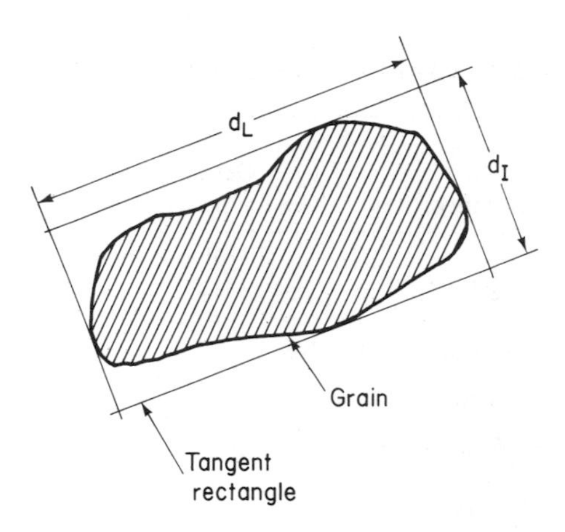

Fig. 3-8 Operational definition of intermediate and long grain diameters.

gives the intermediate diameter d_I. (e) The long side of the rectangle gives the long diameter d_L. It may be noted that the direction parallel to the long side of the tangent rectangle and the pole to the maximum projection may be used to establish the orientation of the particle in space so this operational definition is good not only for size and shape analysis but also for dimensional orientation analysis.

Grade Scales

It is commonly accepted that size should be measured on some type of geometric or logarithmic scale because, for example, a change from 1 to 2 mm is obviously a more significant change in size than a change from 101 to 102 mm.

The scale adopted almost universally by American workers was proposed by Udden (1898), who also introduced the term "grade" to refer to the sizes intermediate between two defined points on the size scale. Udden's scale was based on a center of 1 mm and a multiplier, or divisor, of 2. The names proposed by Udden to describe the size grades were later modified by Wentworth (1922) and it is this modification that is commonly used now.

The Udden-Wentworth grade scale is, however, not the only scale in common use. In Europe, many workers use a scale based on that originally proposed by

Atterberg, which has a center of 2 mm and a multiplier or divisor of 10 (for details, see Müller, 1967). Even in America, there is no uniformity of practice or nomenclature between geologists and geomorphologists or various branches of soil science and engineering. A brief summary of the different practices, and a review of the arguments in favor of the Udden-Wentworth scale has been given by Tanner (1969) on behalf of the SEPM Grain Size Study Committee. Some of the different scales are summarized in Table 3-3.

In 1934, Krumbein introduced a logarithmic transformation of the Udden-Wentworth scale, which he named the phi (ϕ) scale:

$$\phi = -\log_2 d \tag{8}$$

TABLE 3-3 VARIOUS SIZE GRADE SCALES IN COMMON USE

Udden-Wentworth	ϕ values	German Scale† (after Atterberg)	USDA and Soil Sci. Soc. Amer.	U.S. Corps Eng., Dept. Army and Bur. Reclamation‡
		(Blockwerk)		
Cobbles		—200 mm—	Cobbles	Boulders
			—80 mm—	—10 in.—
—64 mm—	−6			Cobbles
		Gravel		—3 in.—
Pebbles		(Kies)		Gravel
—4 mm—	−2		Gravel	
				—4 mesh—
Granules				Coarse sand
—2 mm—	−1	—2 mm—	—2 mm—	—10 mesh—
Very coarse sand			Very coarse sand	
—1 mm—	0		—1 mm—	
Coarse sand			Coarse sand	Medium sand
		Sand		
—0.5 mm—	1		—0.5 mm—	
				—40 mesh—
Medium sand			Medium sand	
—0.25 mm—	2		—0.25 mm—	
Fine sand			Fine sand	Fine sand
—0.125 mm—	3		—0.10 mm—	
Very fine sand			Very fine sand	—200 mesh—
—0.0625 mm—	4	—0.0625 mm—		
			—0.05 mm—	
Silt		Silt		
			Silt	Fines
—0.0039 mm—	8			
		—0.002 mm—	—0.002 mm—	
Clay		Clay	Clay	
		(Ton)		

† Subdivisions of sand sizes omitted.

‡ Mesh numbers are for U.S. Standard Sieves: 4 mesh = 4.76 mm, 10 mesh = 2.00 mm, 40 mesh = 0.42 mm, 200 mesh = 0.074 mm.

where d is the diameter in millimeters. The phi equivalents of the Udden-Wentworth scale are shown in Table 3-3. There are three main advantages to the use of the phi scale: (a) The main points on the Udden-Wentworth scale become whole numbers instead of fractions. (b) The scale is reversed so that the larger sizes, which are conventionally plotted on the left by geologists (following a practice established by Udden), become negative and the smaller sizes become positive numbers on the phi scale. (c) Use of the phi scale permits use of arithmetic rather than logarithmic graph paper and simplifies the calculation of both graphic and numerical descriptive statistics, such as mean, standard deviation, skewness, and kurtosis. The definition of the phi scale has been revised by McManus (1963) who has pointed out some confusions regarding the dimensions of the phi scale. By redefining the phi scale as

$$\phi = -\log_2 \frac{d}{d_0} \tag{9}$$

where d_0 is the standard grain diameter (i.e., 1 mm), it is made clearer that phi values are actually dimensionless numbers (also see discussion by Krumbein, 1964).

In order to allow for determination of a larger number of points on the cumulative curve than are provided by Udden grades, the grades are generally subdivided. Subdivision of each whole phi value into halves or quarters may be achieved by multiplying the size values (in millimeters) of the Udden-Wentworth grades by $\sqrt{2}$ or $\sqrt[4]{2}$, respectively.

Sieve Analysis

In most cases, no attempt is made to measure the size of grains directly but recourse is made to indirect means. For sand-size particles, the commonest technique is sieve analysis. It is not clear exactly what property of the grain is measured by sieving. It is certainly not a "pure" size but some compound of size and shape. An approximation is probably the intermediate diameter or some measure of the cross-sectional area of the particle. The matter is generally of academic interest only—the practical problem is to standardize the procedure so that whatever measure of size is being obtained, it is determined with considerable precision.

The factors affecting reproducibility are the make and precision of calibration of the screens used for the sieves, the size of the sample, the type of mechanism used to shake the sieves, and the length of time of shaking. Most American investigators use 8-in. U.S. Standard or Tyler sieves and choose those screen openings closest to those defined by the $\sqrt[4]{2}$ Udden-Wentworth scale, with a sample size of about 30 g, shaken for 10 to 15 minutes on a Ro-Tap shaker. It has been shown by Folk (1955) and Rogers (1965) that sieving following standard techniques in this way is capable of yielding results of high precision, with phi mean and standard deviation reproducible to within 0.05 phi. In a particularly thorough theoretical and experimental study of the effect of shape variation on the results obtained by sieving, Ludwick and Henderson (1968) reached somewhat less comforting conclusions about the accuracy as opposed to the precision or reproducibility of sieving. They conclude that if the screen opening is equated with particle size, the modal intermediate diameter of particles in each sieved fraction is frequently underestimated by 10 to 20%. A misnaming of the sample by one whole Udden-Wentworth grade is possible. Also some apparently polymodal distributions may result from the nature of the sieving technique.

Sedimentation Methods of Size Analysis

Because the settling velocity seems to be a more fundamental dynamic property of sediment particles than any geometrically defined measure of size, it might seem more logical to measure settling velocity than to measure size directly. Sedimentation methods are, in fact, in common use both for clay-silt and sand-size sediments. Unfortunately, sedimentation methods also suffer from considerable limitations, in regard to both the fundamental significance of the property measured and the precision of the measuring techniques.

The two techniques commonly used are the pipette method (for clay-silt sizes) and one of several settling-tube methods (for sand sizes). In order to discuss these methods and the significance of settling velocity itself, some attention must be given to the mechanics of fluid resistance or drag.

Movement of a solid body through a fluid is opposed by forces exerted by the fluid. These forces are basically of two types, viscous and inertial. Inertial resistance results because the passage of the solid body displaces a certain mass of the fluid, which must be accelerated from rest to a velocity greater than zero. These forces are, therefore, proportional to the mass density of the fluid. If only inertial forces were important, it would be possible to determine the relationship among the force of resistance and the density and velocity by means of *dimensional analysis*. A fundamental theorem of dimensional analysis states that valid equations in dynamics must be dimensionally homogeneous; i.e., the dimensions must be the same on the left- as on the right-hand side of the equation. In its most general form, we may state that there must be some function f relating the fluid force acting on the particle, F_R, to the mass density of the fluid, ρ(rho), the velocity of the fluid relative to the particle u, and the diameter of the particle d:

$$F_R = f(\rho, u, d) \tag{10}$$

In dynamics, all the the variables may be expressed in terms of their fundamental units: length $[L]$; time $[T]$; and either force $[F]$ or mass $[M]$. Use of the square bracket indicates that only the basic dimensions of the variable rather than its absolute quantity are under consideration. The basic dimensions of force in terms of mass, length, and time are readily obtained from Newton's second law:

$$\text{Force} = \text{mass} \times \text{acceleration}$$

In dimensional terms this may be written

$$[F] = [M][LT^{-2}] \tag{11}$$

The dimensions of the other variables commonly used in dynamics are listed in

Table 3-4. In the case of Eq. 10, a dimensional form of the equation is therefore

$$[MLT^{-2}] = [ML^{-3}]^a[LT^{-1}]^b[L]^c \quad (12)$$

where a, b, and c are integers. It follows from inspection of the equation that $a = 1$, $b = 2$, and $c = 2$. If this were not so, there could be no identity between the dimensions on the two sides of the equation. Thus it has been established that the function in Eq. 10 must be of the type

$$F_R = C(\rho u^2 d^2) \quad (13)$$

where C is a dimensionless constant (in other words, a pure number). Alternatively, the equation may be written in terms of a cross-sectional area of the particle A, which is proportional to d^2, as follows:

$$F_R = C_D \cdot \frac{\rho u^2}{2} \cdot A \quad (14)$$

In this form, the equation states that the force of resistance is proportional to the kinetic energy of the fluid ($\rho u^2/2$) and to the cross-sectional area of the particle. C_D is the proportionality coefficient, called the *drag coefficient*, of the particle. Equation 13 or 14 is called Newton's law of resistance, or the impact law. Dimensional analysis does not reveal the value of the drag coefficient, which must be determined from experiment.

From experiment it may also be determined that the drag coefficient is not generally a constant. The reason is that, in the derivation given above, an important variable was neglected. This variable is the fluid viscosity. Viscosity is defined by Newton's law of viscosity, which states that the rate of deformation of a fluid

TABLE 3-4 DIMENSIONS OF SOME COMMON VARIABLES IN DYNAMICS

	Symbol	Dimensions	Comment
Velocity	u, w	L/T	w is symbol for settling velocity
Acceleration	g	L/T^2	g is symbol for acceleration due to gravity
Discharge	Q	L^3/T	Volume rate of flow
Unit discharge	q	L^2/T	Volume rate of flow per unit width of channel
Mass density	ρ (rho)	M/L^3	
Specific weight	γ (gamma)	M/L^2T^2	$\gamma = \rho g$
Dynamic viscosity	μ (mu)	M/LT	
Kinematic viscosity	ν (nu)	L^2/T	$\nu = \mu/\rho$
Pressure	P	M/LT^2	Normal force per unit area of surface
Shear stress	τ (tau)	M/LT^2	Tangential force per unit area of surface
Momentum		ML/T	
Work, energy		ML^2/T^2	
Power		ML^2/T^3	

in a direction normal to a surface is proportional to the force per unit area (shear stress) applied parallel to the surface. In the case of two plates in the *xz* plane (Fig. 3-9), the relationship may be written

$$\tau = \mu \frac{du}{dy} \tag{15}$$

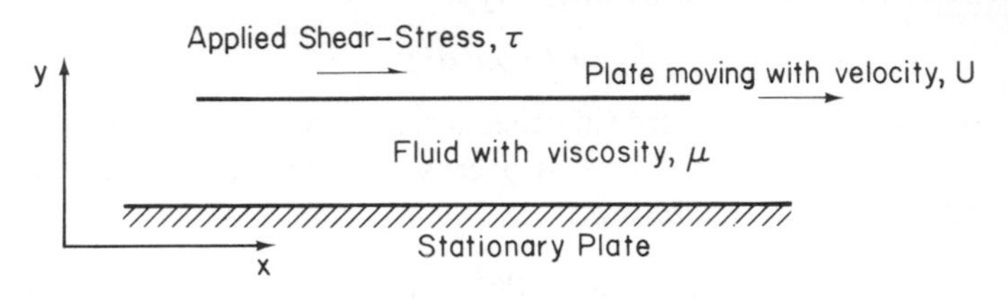

Fig. 3-9 Sketch for definition of viscosity (see text).

where τ (tau) is the shear stress, du/dy is the rate of deformation, and μ is the coefficient of proportionality, called the *dynamic viscosity* of the fluid between the two plates. The cgs unit of viscosity is the poise (grams per centimeter per second). Most common fluids, including air and water, are found to obey Newton's law of viscosity. In other words, they are Newtonian fluids. It should be noted that viscous resistance forces always act parallel to fluid surfaces.

Viscosity is not an absolutely invariable property of a particular fluid. It has its origin in intermolecular forces and consequently varies with the temperature and with presence of dissolved salts in a fluid (see discussion of the properties of water in Chap. 7). The apparent viscosity of a fluid such as water may also be increased by the presence of suspended particles, particularly clay particles.

It is found experimentally that the force of fluid resistance on a particle moving through a fluid always depends on the fluid viscosity. To phrase the matter differently, the drag coefficient is a function of the viscous forces. One way of expressing the relative importance of viscous forces is in terms of a dimensionless combination of the main properties of a moving fluid: its velocity, density, and viscosity and some length measure (such as the diameter of the particle obstructing the flow). There is only one dimensionless combination of these variables, called a Reynolds number, $R_e = ud\rho/\mu$ (further discussion of the significance of Reynolds numbers is given below). It is found experimentally that the drag coefficient is a unique function of the Reynolds number, provided the shape of the particle is held constant. A graph of this function for spheres is shown in Fig. 3-10.

To recapitulate, the force of resistance of a particle moving through a fluid depends not only on the size and velocity of the particle and the density of the fluid but also on the viscosity of the fluid. This is important for both the measurement and dynamic interpretation of grain size because the viscosity of the fluid is not a constant but varies over almost an order of magnitude under natural conditions, mainly as a consequence of temperature variation.

In the case of very fine grained spherical particles settling in water it can be

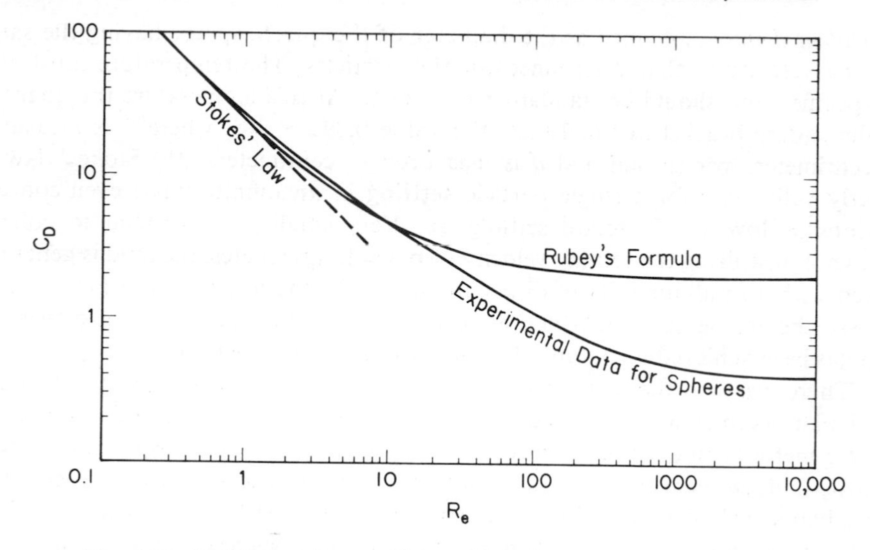

Fig. 3-10 Variation of drag coefficient (C_D) with Reynolds Number (R_e). After Graf and Acaroglu (1966).

shown theoretically and verified experimentally that at very low concentrations

$$C_D = \frac{24}{R_e} \tag{16}$$

If this equation is combined with Eq. 14 and the definition of R_e and A, we obtain

$$F_R = 3\pi\, d\mu\, u \tag{17}$$

This is known as Stokes' law of resistance. When the particle is settling in water with a constant velocity, w (called the *settling* or *fall velocity*), the force of fluid resistance must be equal and opposite to the force of gravity acting on the particle F_g. The gravity force can be calculated from the volume of the sphere $\pi d^3/6$, the density difference between the particle and the fluid $(\rho_s - \rho)$, and the acceleration due to gravity g:

$$F_g = \frac{\pi}{\sigma} d^3 (\rho_s - \rho) g \tag{18}$$

Combining Eqs. 17 and 18, with $u = w$, and solving for w, gives Stokes' law of settling

$$w = \left[\frac{(\rho_s - \rho)g}{18\mu}\right] d^2 \tag{19}$$

This law is universally used as the basis for the determination of the size of clay- to silt-size particles. These particles fall within the range of validity of Stokes' law but there are several factors that limit the application of Stokes' law in practice. (a) The particles in this size range are not spherical and their density is not known precisely. The measure of size that is obtained is, therefore, the *equivalent diameter*, defined as the diameter of the quartz sphere having the same settling velocity as that determined for the particles. The temperature must also be specified and should be standardized at 20°C. At this temperature the quantity in the square bracket in Eq. 19 has the value 0.892×10^4, where w is measured in centimeters per second and d is measured in centimeters. (b) Stokes' law is strictly valid only for a single particle settling in an infinite fluid; even concentrations as low as 1% retard settling. (c) Very small particles tend to cohere. To avoid measuring the settling velocity of particle aggregates, the mud is generally mixed with a small quantity of dispersing agent. In the finest sizes, it may be that the size being measured depends mainly on the completeness of the dispersion that has been achieved. In nature, clays are not perfectly dispersed.

There are also the usual experimental errors that limit the precision of the particular sedimentation methods used. The most common method used is the pipette method. In essence the method consists of preparing a suspension of mud of low, uniform concentration and allowing 1 liter of the suspension to stand in a graduated cylinder. Samples of the suspension are withdrawn from a level 10 or 20 cm below the top at standard intervals of time. At these times, it can be calculated using Stokes' law that all particles of a given equivalent diameter will have settled below that level. Consequently, samples should contain only finer particles. From the weight of sediment recovered in samples of known volume, the grain size distribution can be calculated.

It is clear even from this very brief description of the method that a high degree of precision cannot be achieved. The method is useful, however, for determining the approximate amounts of coarse and fine silt in a mud sample. The technique is now standardized and its limitations are well-known.

For sand-size particles, sedimentation tubes have been developed in recent years and are gaining popularity because they provide a more rapid analysis than sieves and because it is believed that settling velocity has a greater dynamic significance than size determined by sieving. Sedimentation tubes operate on the principle that the sample is introduced at one end of the tube and settles to the other end. The time taken for settling of the different sediment fractions is measured in one of three different ways (see Fig. 3-11): (a) by inspection of the sediment as it accumulates in a narrowed end portion of the settling tube (Emery tube, visual accumulation tube), (b) by recording automatically the pressure difference between the column of water and sediment and an equivalent column of pure water; this is the basis of the Woods Hole Rapid Sediment Analyzer; (Zeigler and others, 1960), (c) by recording automatically, by means of a strain gauge or electrobalance, the weight of sediment accumulating on a pan suspended at the bottom of the tube. It has been claimed that both methods (b) and (c) are capable of giving results of considerable precision.

Some limitations of settling tubes for analysis of sands are as follows: (a) The sand sizes do not follow Stokes' law. Instead they fall into a range of settling velocities that cannot be accurately expressed by an analytical expression. Rubey (1933) developed a formula that has often been used for settling velocities outside the Stokes' range but the formula is not accurate for spheres (see Graf and Acaroglu,

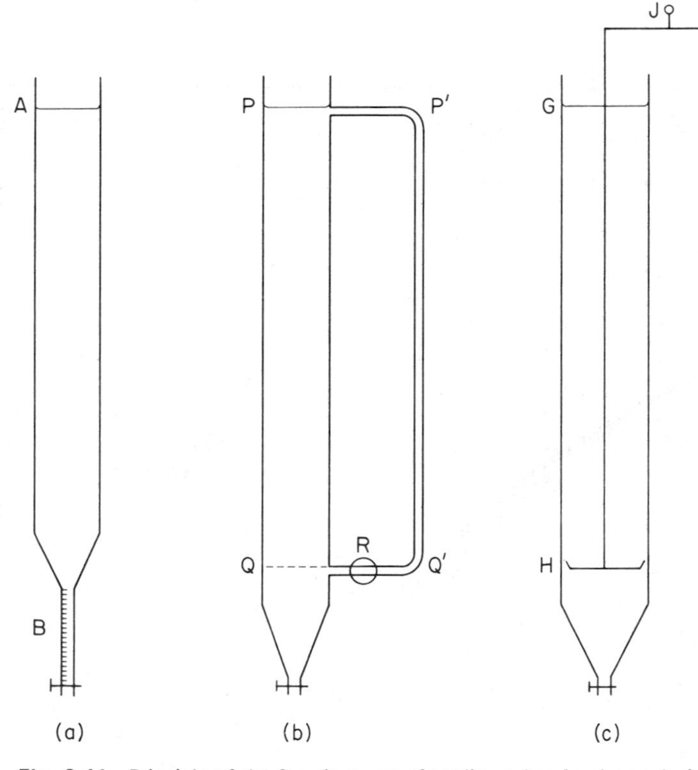

Fig. 3-11 Principle of the 3 main types of settling tubes for size analysis of sands. In type (a) the sample is introduced at A and its rate of accumulation in the narrowed portion of the tube, B, is measured. In type (b) the sample is introduced at P. So long as grains have not settled past Q, their suspended weight contributes to a pressure difference between columns PQ and P′Q′ and the pressure difference is recorded by a transducer at R. In type (c) grains introduced at G settle onto a pan, H, which is suspended from an electrobalance at J.

1966, Gibbs et al., 1971 and Fig. 3-10). Accurate empirical curves for the settling velocity of quartz spheres have been compiled by Rouse (1937) and are shown in Fig. 3-12 so this objection is not a fundamental one. Also Rubey's equation has been modified to give more accurate results by Watson (1969). Further, it may be argued that a fall velocity, even when measured at a standardized temperature instead of the temperature (generally unknown to the geologist) that prevailed during sediment deposition, is more meaningful than any other type of size measure. (b) It is difficult to record the tails of the size distribution precisely in any of the commonly used sedimentation tubes. It is believed by many geologists that the information most useful for genetic interpretation is contained in the tails of the size distribution. (c) The largest single limitation of the method is the group of effects produced at the time of introduction of the sediment at the top of the tube. Whatever method is used, it is very difficult to avoid disturbances caused by the

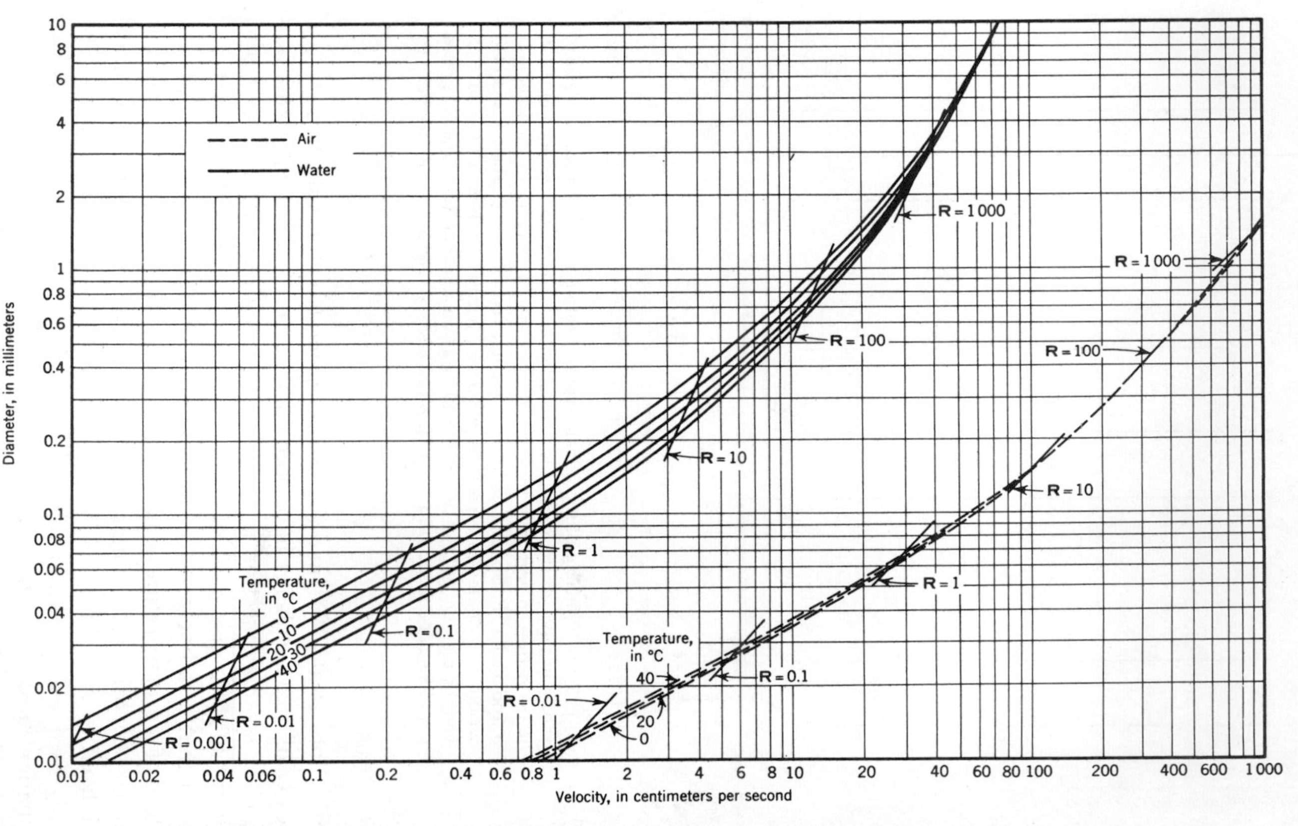

Fig. 3-12 Settling velocity of quartz spheres in water and air. After Rouse (1937).

relatively large concentration of sediment formed temporarily at the top of the tube. The results of this high sediment concentration include mass settling of grains, with the smaller grains being dragged down by the larger grains. A small vertical density plume or "turbidity current" may be formed. The high concentration might be expected to hinder settling but this effect does not appear to be important in the upper part of the tube, though it may be important in the lower part of tubes with narrow accumulation portions. A consequence of the mass-settling of grains is that the results obtained using a settling tube depend strongly on the size of sample used. Therefore, a standard, small size of sample should be used and the tube must be calibrated using samples of spheres of known density and size.

Despite these limitations several investigations have shown that settling tubes are capable of giving results of reasonably high precision and are adequate for the routine determination of mean size and sorting. For example, Schlee et al. (1965) obtained graphic statistics from fifteen samples of Cape Cod beach and dune sand analyzed by both sieves and a Woods Hole settling tube. Absolute differences averaged 0.09 ϕ units for mean size, 0.03 units for standard deviation, and 0.37 units for skewness. The skewness differences were large, with the settling tube giving values systematically larger than the sieve values. Calibration of the tube, however, could substantially reduce these differences.

Thin Section Techniques

A section cut across a population of uniform spheres randomly arranged in space will not yield a uniform size distribution. The distribution of apparent sizes is shown in Fig. 3-13. Only the maximum size observed corresponds to the true size of the parent population. If the population of spheres is not uniform but has a size distribution whose moments are known, the moments of the observed section-size distribution may be predicted. The theory has been extended to the case of perfect ellipsoids by Wicksell (1925, 1926).

In practice, however, the shape of the particles is irregular and unknown and the problem is not to predict the section-size distribution from the original size distribution but to infer the original size distribution from the observed section-size distribution. The problem is further compounded by the fact that almost all sediments composed of nonspherical particles show anisotropic dimensional fabrics (i.e., preferred orientation of the grains) so that the results of measuring the section size depend on the direction in which the section is cut relative to the grain fabric and on the degree of anisotropy of the fabric.

It is easy to show that, without gross simplifying assumptions, there is no unique solution to the problem of inferring the parent (three-dimensional) size distribution for the observed (two-dimensional) distribution. Unfortunately, this has not prevented the publication of many papers on the subject by sedimentologists. Because the problem is theoretically unsolvable, it would appear best if sedimentologists were to follow one of two procedures. (a) Adopt a standardized method for measurement of size in thin section and treat this as valid form of

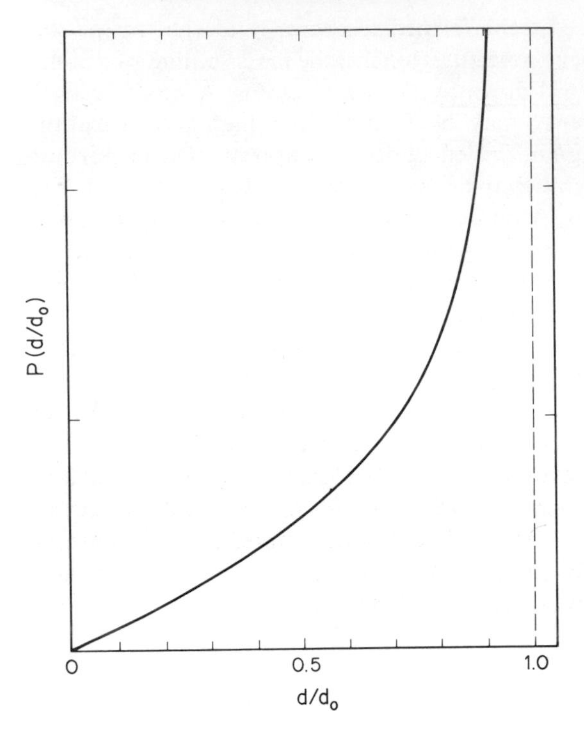

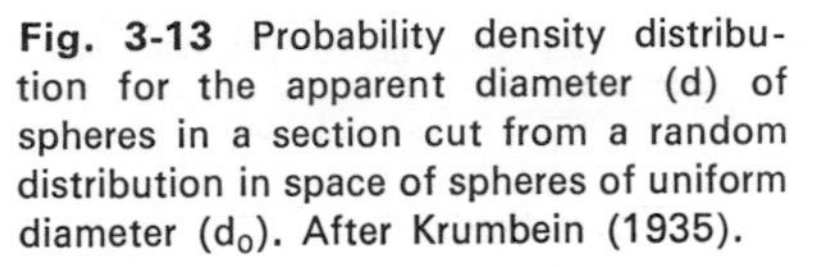
Fig. 3-13 Probability density distribution for the apparent diameter (d) of spheres in a section cut from a random distribution in space of spheres of uniform diameter (d_0). After Krumbein (1935).

measurement in its own right. Such measurements may be compared with other measurements made using the same procedure but not with size determined by sieving or sedimentation. This is not a very radical suggestion because, as pointed out above, it is already scarcely possible to make meaningful comparisons between size distributions determined by sieving and sedimentation. (b) Correct from thin section to sieve sizes using empirical correction factors developed by measuring a set of "average" sands by both methods (Friedman, 1958).

In either case, it is necessary to adopt standardized procedures for thin section analysis. These procedures must include cutting the thin section in a standard orientation and choosing grains for measurement according to a well-defined sampling plan. Because the direction of grain orientation is generally unknown, the thin section should be cut parallel to the bedding (though this is not the best orientation for mineralogical studies). The method used to select the grains to be measured is also extremely important because different methods result in different types of weightings of the size distribution.

The probability that a grain will be cut by a random section is proportional to the size of the grain (Fig. 3-14). Thus, size distributions observed in thin or polished sections are not pure number frequencies but are number frequencies already weighted by the size of the grains. Selecting grains encountered on a traverse across the slide adds another weighting by size so that the frequencies ob-

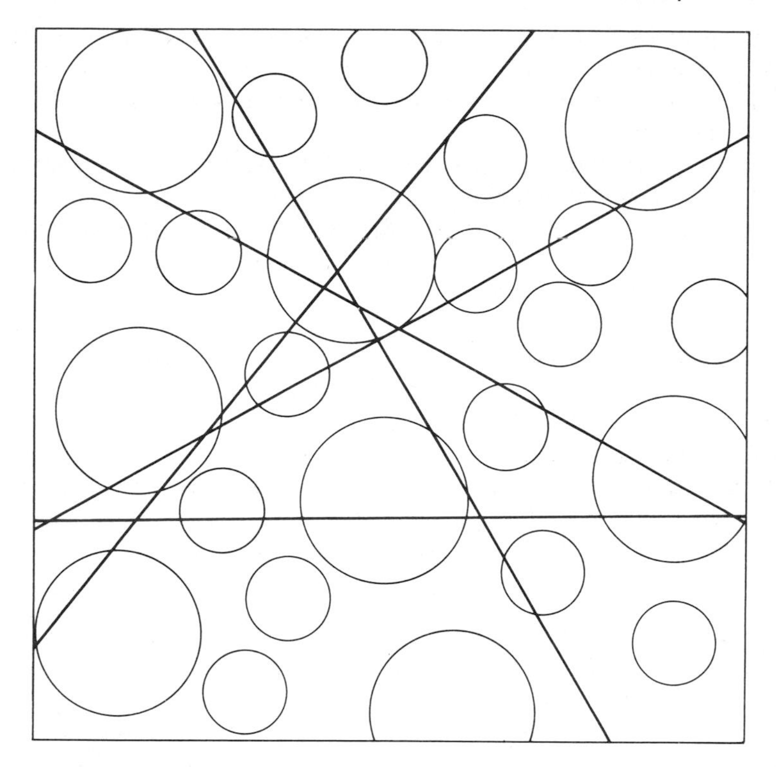

Fig. 3-14 Effect on traverse sampling of variations in particle size. In the area shown there are twice as many small as large circles. The large circles have a diameter twice that of the small circles. In five random traverses across the area, the large grains were intersected 13 times and the small grains only 10 times. In the long run, it is expected that the number of intersections will be equal for large and small grains, even though the small grains are more abundant.

tained by this method are weighted by cross-sectional area. If grains are selected at grid points (Glagolev, or "point-counting," procedure), the frequencies are weighted by the volume of the grains. This latter type of frequency weighting is similar to that obtained in sieve or sedimentation analysis where the frequencies obtained are weight frequencies. Thus, it seems to be the most appropriate method to adopt in thin section work.

Interpretation of Grain Size Analyses

There are four main reasons for grain size analysis. (a) The grain size is a basic descriptive measure of the sediment; therefore some attempt at precision in recording the size is justified. (b) Grain size distributions may be characteristic of sediments deposited in certain environments. (c) Detailed study of observed grain size distributions may yield some information about the physical mechanisms acting during deposition. In some cases, grain size may also reveal information about diagenetic processes. (d) Grain size may be related to other properties, such as permeability, and variation in these properties may be predicted from variation in grain size.

The origin of the grain size distribution of sediments must be related to the nature of the source materials, the process of weathering (or of organic decay and size reduction by organisms, in the case of bioclastic sediments), the abrasion and corrosion of the grains during sediment transport, and the sorting processes acting during transport and deposition.

Several studies have shown that the materials produced by weathering of crystalline rocks have a size distribution similar to that produced by crushing of rock materials (Fig.3-4). A mathematical model for the origin of this size distribution can be constructed based upon the assumption that the material is isotropic and that the probability of fracture of any grain in any particular place or direction is equal to the probability of fracture elsewhere (Bennett, 1936). These probability assumptions lead to the generation not of a Lognormal distribution but of a probability distribution known as Rosin's law:

$$y = 100(1 - e^{bx^n}) \tag{20}$$

where y is the weight percent passed by a sieve of mesh size x, b is the reciprocal of the average size; and n is a numerical coefficient analogous to the standard deviation of a Normal distribution.

Sedimentary materials that have been transported for even a short distance do not show a Rosin's size distribution law. Transported materials are generally believed to approximate to a Lognormal distribution, though very few samples show a very close approximation to the ideal distribution. A few authorities (e.g., Bagnold, 1968) deny that the lognormal law can be considered an ideal standard of comparison but the practice of displaying size analyses by means of cumulative plots on probability paper is now very generally adopted, and many authors go considerably further toward adopting a lognormal model, by attempting to interpret all observed size distributions in terms of mixtures of Lognormal distributions (or parts of such distributions). A fully satisfactory mathematical model for the origin of the Lognormal distribution in sediments has not yet been proposed. One of the problems concerns the relative importance of size reduction during transport by abrasion and continued weathering processes and hydraulic sorting of grains.

Studies of changes in grain size downstream in rivers suggest that mechanical abrasion is a significant factor for the larger grain sizes (pebble to very coarse sand sizes), the softer rocks or minerals, and the steeper gradient streams. By studying the change in relative proportions of resistant and soft or friable materials downstream, as well as the change in grain size, it is possible to make some estimate of the relative importance of abrasion and hydraulic sorting in any particular river.

As long ago as 1875, Sternberg observed that the weight of the largest pebbles in the River Rhine between Basel and Mannheim decreased exponentially with distance downstream (see Humbert, 1968). "Sternberg's law" may be expressed as the formula:

$$W = W_0 \exp\,[-a(x - x_0)] \tag{21}$$

where W is the weight of the largest particle at a distance x from the origin; W_0 is the weight at some reference point, x_0; and a is a constant for the particular river in question. It has been shown that Sternberg's law holds approximately for other rivers, including the Middle Rhine, the Meuse, the Mur, and the Emms. There are, however, many rivers for which it does not hold and Sternberg was certainly wrong in his belief that the decrease in size was due wholly to abrasion and that size in its turn was the main factor controlling the stream gradient (which also decreases exponentially downstream in many rivers). In most cases, the downstream decrease in grain size is to be explained to a great extent by the selective deposition of the larger particles in alluvial deposits. The slope of the river is controlled not only by the maximum grain size that can be moved (competence) but also by the rate at which the bed material can be moved (capacity). These factors are discussed in more detail in Chap. 4.

Because of the difficulty of separating the effects of abrasion from those of sorting in rivers and on beaches, there have been many attempts to study the processes of size reduction experimentally in the laboratory. Generally, changes in shape are studied in the same experiment so a discussion of these experiments is deferred until the end of Sec.3.4. Whatever the effects of abrasion (and it seems that for the sand sizes, they are generally very small), the details of size distribution of any particular sample of river sediment must certainly be explained mainly in terms of hydraulic sorting processes.

Several studies of the size distributions of clastic sediments have shown that there are general, statistical relationships observed between the different size parameters such as mean size, sorting, and skewness. The relation between mean size and sorting is particularly well established and many studies have shown that the best sorted sediments are generally those with mean size in the fine sand grade (see Griffiths, 1967). The explanation of this relationship lies in the nature of hydraulic sorting mechanisms and is discussed further in Sec.4.6. Sorting is measured by standard deviation of size but a useful verbal scale has been suggested by Folk (Table 3-5).

There have been many attempts to relate grain size distributions observed in recent sediments directly to the environment of deposition. Most of these studies have made use of moment measures or their graphic equivalents. Friedman (1961, 1967) studied samples of relatively fine grained, unimodal sands taken from many different localities around the world and representative of eolian dunes, beaches, and river channels. The most effective distinction of sands from these three environments was shown by a diagram of moment standard deviation (sorting) plotted against moment skewness (Fig.3-15). Beach sands are clearly distinguished by their combination of negative skewness and good sorting (low standard deviation) from river sands, which are generally positively skewed and less well sorted. Coarse river sands, however, may be negatively skewed. Dune

TABLE 3-5 VERBAL SCALE FOR SORTING, BASED ON MEASURED STANDARD DEVIATIONS USING FOLK AND WARD STATISTICS (after Folk 1968)†

Standard deviation	Verbal description
0 –0.35	Very well sorted
0.35–0.50	Well sorted
0.50–0.71	Moderately well sorted
0.71–1.00	Moderately sorted
1.0 –2.0	Poorly sorted
2.0 –4.0	Very poorly sorted
4.0 –	Extremely poorly sorted

† Sands with more than 5% clay are caled *immature* by Folk: Those with less than 5% clay but standard deviation over 0.5 are *submature* and those with standard deviation less than 0.5 are called *mature*. Well-sorted sands that have well-rounded grains are described as *supermature*.

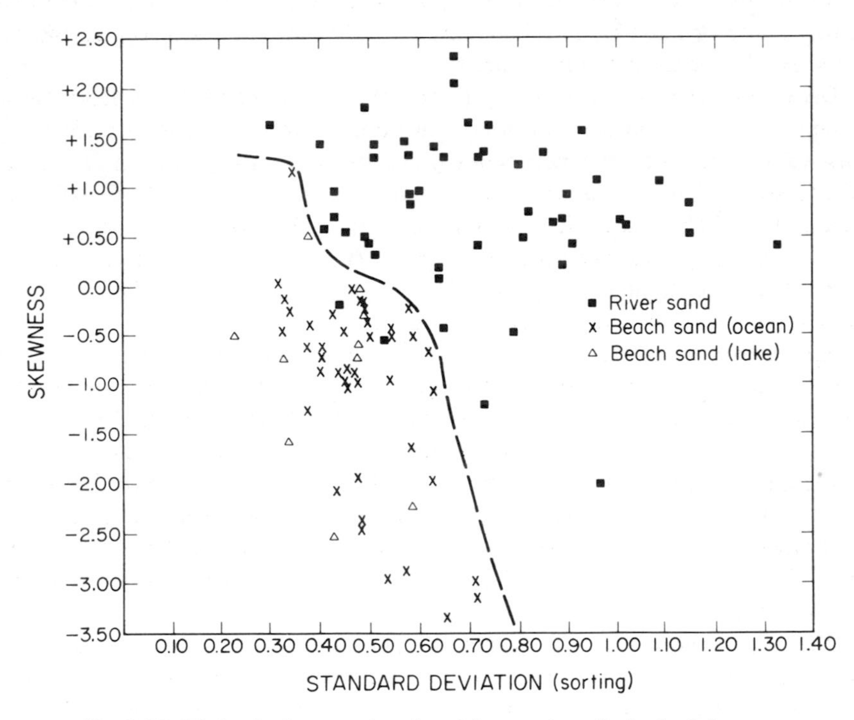

Fig. 3-15 Distinction between beach and river sands on the basis of skewness and standard deviation calculated from moments. From Friedman (1961).

sands have positive skewness and are generally somewhat finer grained than beach sands.

Use of moment measures, though it quantifies the distinctions involved, somewhat obscures the basic differences between the sands from these three environments. River sands are generally positively skewed because of the presence of a fine tail (more than 0.2% finer than 62 μ). The fines are present because the river water generally contains a fairly high concentration of suspended clay and silt. They are not present in beach and dune sands because of winnowing by both wave and wind action, which removes the fines and deposits them elsewhere. Winnowing of the fines on a beach results in negative skewness because wave action does not remove the coarse tail. Wind action, however, excludes the coarse tail even more completely than it winnows out the fines because the wind strength is generally incompetent to move the coarser grains, which are left behind as a lag deposit. Consequently dune sands have positive skewness. Note that the positive skewness of dune and river sands arises from two different mechanisms—presence of a fine tail in the case of river sands but truncation of the coarse tail in the case of dune sands.

An alternative approach was suggested by Passega (1957, 1964; see also Rizzini, 1968). Passega makes use of only three properties of the grain size distribution: (a) the one percentile (C) to measure the coarsest grains in the deposit, (b) the median (M), and (c) the percentage (L) finer than 31 μ, to measure the fine tail. Discrimination between environments is not made on the basis of a single sample but by examining the patterns produced by plotting a number of samples from the same unit on C–M or L–M diagrams (Fig.3-16).

None of the methods suggested is an infallible guide to discrimination of environments. Determination must be based on sieve analyses of several samples and the size distributions must not have been modified by diagenesis.

Most of the studies making use of size distribution to discriminate environ-

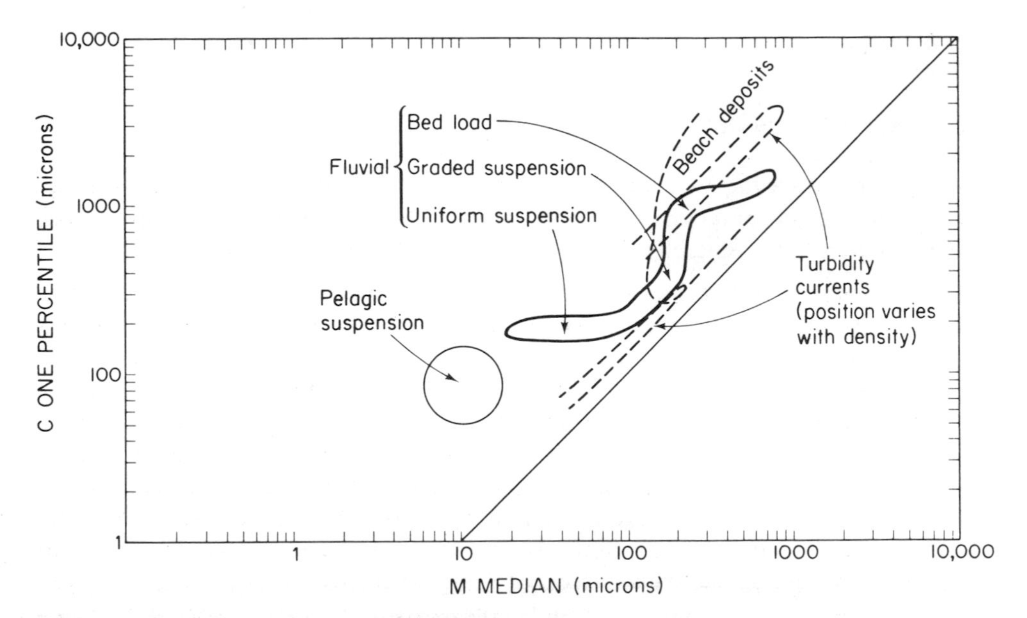

Fig. 3-16 C-M patterns of sediments deposited in different environments. After Passega (1957, 1964).

ments have been on recent sediments. The methods have been little tested on ancient sediments. In the discussion of Friedman's work it was noted that one of the most diagnostic properties of the sands is the percent of fine fraction. Unfortunately, it is just this part of the size distribution that it is most difficult to estimate accurately in lithified sands. Disaggregation tends to break soft grains, producing spurious contributions to "fines" and in thin section it is very difficult to estimate the amount of sandstone matrix (see Chap. 9). One of the advantages of Passega's scheme is that it makes use mainly of the coarsest and median sizes, both of which may be fairly accurately estimated in lithified sandstones.

Even in unlithified sands the results may be misleading because some environments may not be "typical." An example is the beach of Padre Island, Texas, where the sands are positively skewed because there are two grain size modes within the fine sand range. The two modes are related to sand derived from two sources, the Rio Grande and longshore drift. Another example occurs on the Louisiana coast where beach sands contain unusually large fractions of fines because of the local concentration of mud derived from the Atchafalaya River (Friedman, 1967).

Since the publication of Friedman's paper, there have been many similar studies, most of which generally support his results. A few typical studies are listed in the references for this chapter. Not all modern sand environments have yet been carefully studied. In particular, further information is required about the size characteristics of sands on the continental shelf and in tidal sandbar or offshore bar environments.

3.4 GRAIN SHAPE

Basic Concept

Shape is a complex property of a grain and it is difficult to describe shape precisely except for grains that approximate regular geometrical shapes. Sedimentologists have generally distinguished several aspects of shape.

1. *Surface texture* includes those aspects of shape whose scale is so small that they do not appear to affect the overall shape of the particle. Commonly distinguished surface textures include a polished surface, a greasy surface, and frosting of the grain. By using electron or scanning electron microscopes it is possible to resolve many more aspects of surface textures than is possible with the optical microscope. It appears that certain types of surface texture may be diagnostic of quartz grains modified by certain mechanisms. Criteria for several environments have been suggested by Krinsley and Donahue (1968). Some typical textures are shown in Fig. 3-17. The technique of electron microscopy has been used by only a few workers to date and although the results appear promising, more work is necessary to determine the reliability of surface texture as an environmental indicator.

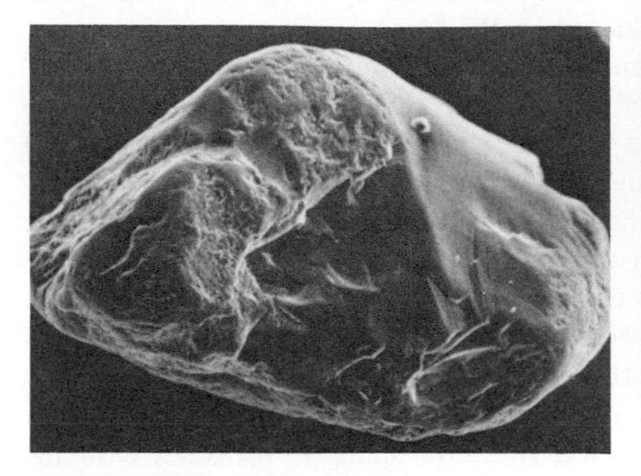

(a)

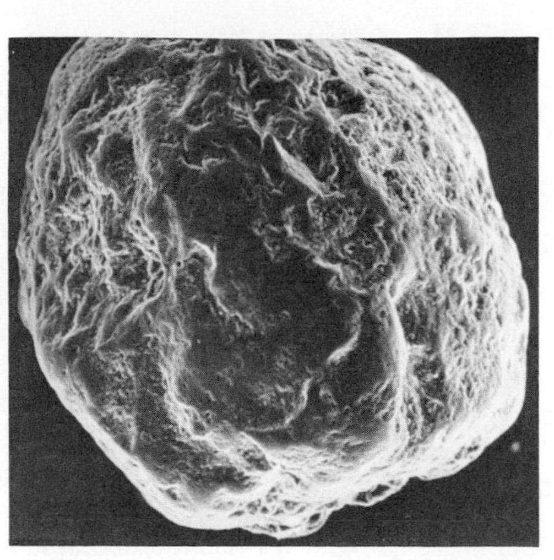

(b)

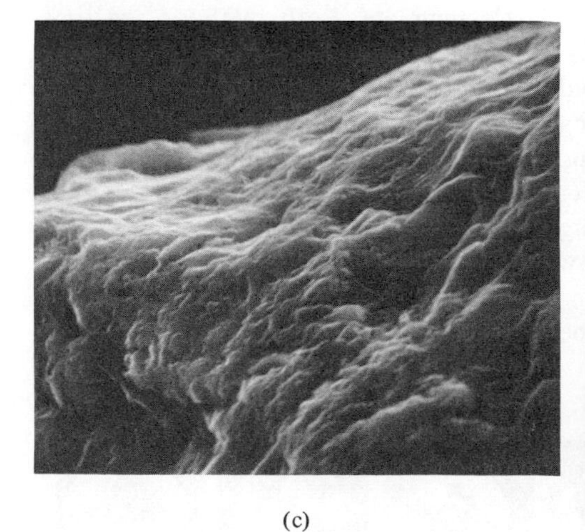

(c)

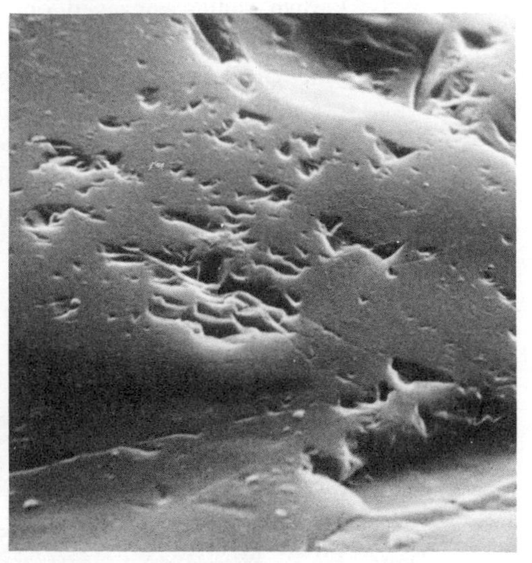

(d)

Fig. 3-17 Scanning electron micrographs of quartz grains, showing surface textures. (a) Sand grain from cross-bedded periglacial eolian sand. Right side of grain has typical glacial features. Left side of grain shows parallel plates, typical of eolian action, superimposed on glacial breakage patterns. (b) Grain from same sample as (a), but completely covered with parallel plates. (c) Part of a quartz sand grain from modern dunes in front of glacial ice, Alaska. Upturned plates are typical of eolian action. (d) Sand grain from shallow marine environment. Typical V-shaped percussion features probably indicate working on beach. (Micrographs taken by D. H. Krinsley.)

In applying the technique to ancient deposits, it must also be remembered that surface texture may be greatly modified by diagenesis (Margolis, 1968), particularly in sands older than the Mesozoic.

2. *Roundness* refers to the sharpness of the corners and edges of the grain. It therefore refers to aspects of the grain surface that are on a larger scale than those classed as surface texture but still smaller than the overall dimensions of the grain. Roundness was quantified by Wentworth (1919) and his definition was modified by Wadell (1932). Wadell's definition of roundness has been almost universally adopted as the ideal definition. It states that the roundness is the ratio of the average radius of curvature of the corners to the radius of the largest inscribed circle. Wentworth's definition was similar except that only the radius of the sharpest corner was measured. His definition has recently been revived in preference to Wadell's definition because it is less time-consuming to measure. Dobkins and Folk (1970) suggest use of the radius of the sharpest corner divided by the radius of the largest inscribed circle. For precise measurement of small particles by either method the projected image of the grain must be enlarged to a standard size and the radii measured using a set of circles of known radius engraved on a plastic template. Both methods are time-consuming and ideally what is required is a method for achieving essentially the same result by a rapid automated technique. No such technique presently exists but a promising theoretical approach that should be amenable to automation has been described by Schwarcz and Shane (1970). Until a satisfactory automated method has been developed, the most common method of estimating roundness will no doubt continue to be visual comparison of grains with standard images of grains of known roundness. One such set of images was prepared by Powers (1953) and is shown in Fig.3-18. The roundness classes are given descriptive names from

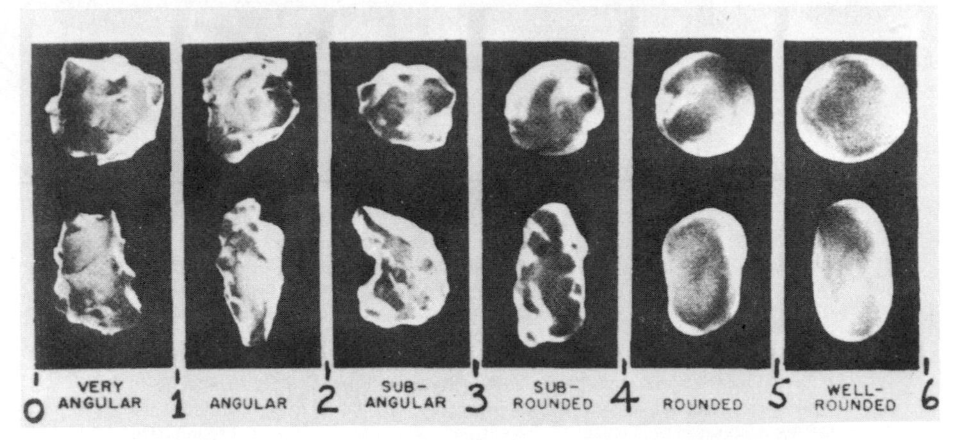

Fig. 3-18 Images of grains for the determination of roundness. From Powers (1953).

very angular to well-rounded, and the midpoints of the class intervals have values of roundness ranging from 0.15 to 0.85. Folk (1955) suggested a logarithmic transformation of roundness analogous to the ϕ scale for size. He named the scale the ρ scale. Rho values extend from 0 (perfectly angular) to 6 (perfectly rounded) and each unit corresponds to one class on Powers scale of roundness.

It has been shown that the accuracy of visual comparison methods is so low that, for the same sample of grains, different operators may estimate mean roundness values that differ by a whole roundness class. Even precision (reproducibility by the same operator) is often poor. The result is that few reliable studies of sand grain roundness have been made. Experimental studies indicate that roundness is related to the degree of abrasion or wear suffered by the particle. It should therefore be related to the type and length of transport or reworking and have considerable geological significance and it is regrettable that a more accurate, rapid method of measuring roundness has not yet been developed.

3. *Sphericity* is a concept developed by Wadell (1932). Sphericity measures the degree to which a particle approaches a spherical shape. It was defined by Wadell as the ratio between the diameter of the sphere with the same volume as the particle and the diameter of the circumscribed sphere. For large particles, the volume can be measured directly by displacement of water or mercury. Usually, however, sphericity is determined by measuring the three linear dimensions of the particle discussed above (under size). If it is assumed that the shape is a regular triaxial ellipsoid, the volume of the particle is $(\pi/6)d_L d_I d_S$. The volume of the circumscribing sphere is $(\pi/6)d_L^3$. It follows that the sphericity, ψ(psi) of the particle is

$$\psi = \sqrt[3]{\frac{d_S d_I}{d_L^2}} \tag{22}$$

Sneed and Folk (1958) have objected to the use of the Wadell sphericity on the grounds that although it correctly measures the degree to which a particle approaches a sphere, it does not correctly express the dynamic behavior of the particle in a fluid. This is because particles tend to orient themselves with their maximum projection area normal to the flow and this is true both for particles settling in fluids and for particles resting on the bottom. In either case, the fluid drag is almost the same as the force on a sphere with the same projection area as the particle whereas the gravity force is the same as on a sphere with the same volume as the particle. Defining the *maximum projection sphericity* as the ratio of the diameters of these two spheres yields the equation

$$\psi_P = \sqrt[3]{\frac{d_S^2}{d_L d_I}} \tag{23}$$

This definition proves to be very similar to that of the Corey Shape Factor

(S.F.), which was earlier used by engineers to express the main effect of shape on settling velocity.

$$\text{S.F.} = \frac{d_S}{\sqrt{d_L d_I}} \tag{24}$$

For most applications, it appears that the maximum projection sphericity is preferable to sphericity as originally defined.

Classification of Shape Types

Definition of the roundness and sphericity of a particle does not define the shape uniquely, even for the case of ideal ellipsoids. Several other indices have been proposed and used by sedimentologists to define other aspects of the shape. Only four of the most commonly used indices are described below.

Zingg (1935) proposed the use of the two shape indices d_I/d_L and d_S/d_I. The indices are used to construct a diagram, commonly called the Zingg diagram (Fig. 3-19), which is divided into four fields, defining four main shape classes: oblate (tabular or disk-shaped), equant, bladed, and prolate (rod-shaped). Lines may be constructed on the same diagram to show equal Wadell sphericity fields.

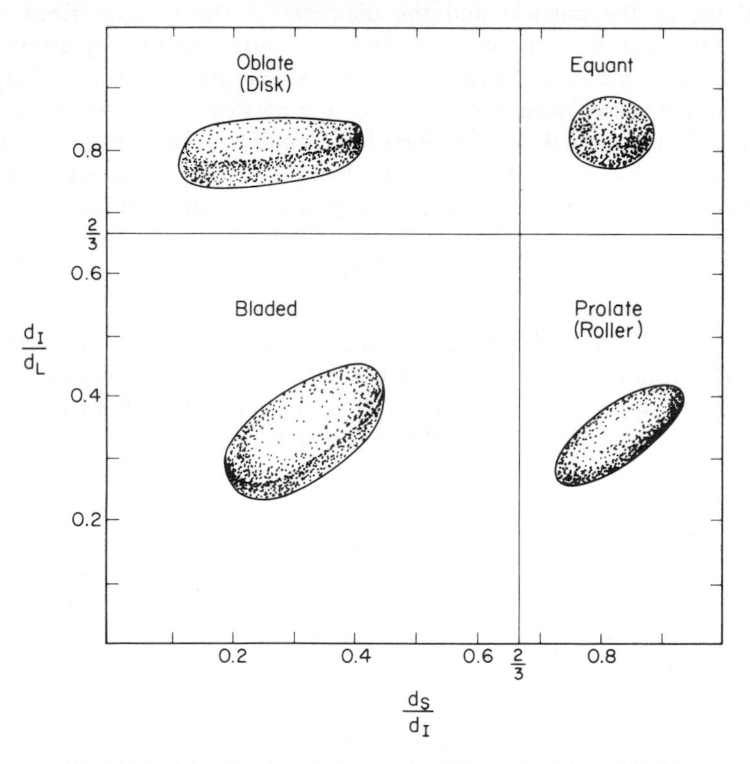

Fig. 3-19 Classification of shapes of pebbles. After Zingg (1935).

Equant particles have highest sphericity but oblate, bladed, and prolate particles may all have the same sphericity values. It might be expected, however, that these different shapes would behave differently; for example, prolate particles ("rollers") might roll much more easily than oblate or bladed particles. Many subsequent studies have confirmed that these shape distinctions are important in describing natural pebble deposits.

Sneed and Folk (1958) proposed combining the two indices d_S/d_L and $(d_L - d_I)/(d_L - d_S)$ on a triangular diagram (Fig. 3-20). The diagram is divided into a number of shape fields and lines of equal maximum projection sphericity may be drawn on it.

It may be observed that almost all these shape indices are based on combinations of the three principal diameters of the particle. In many studies, the variation in the indices with some other variable is then studied. For example, the indices may be used to characterize the shapes observed in different environments or the shape variation with distance downstream in a river. In these cases, it seems that a valid alternative method might be to make use of the original measurements rather than arbitrary computed indices. Multivariate statistical tech-

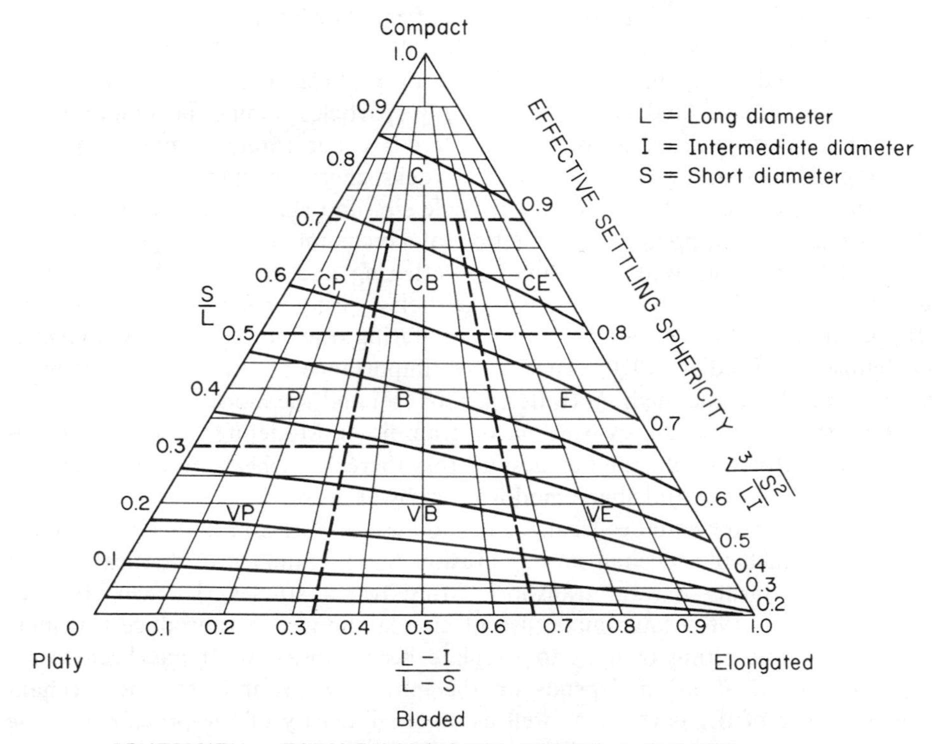

SPHERICITY — FORM DIAGRAM FOR PARTICLE SHAPES

Fig. 3-20 Classification of shapes of pebbles. After Sneed and Folk (1958).

niques such as multiple regression, factor analysis, and the discriminant function can then be used to reveal which size and shape indices are most appropriate to the data rather than an arbitrary choice of index being made before analysis of the data is begun.

Interpretation of Shape

The interpretation of shape variation in sedimentary particles is complicated by a number of factors. The shape of grains depends initially on the nature of the source rock and the weathering process. Subsequently the shapes of particles are changed by abrasion, corrosion, and breakage. Finally, the shape of the particles making up a particular sedimentary deposit is at least partially the result of hydraulic shape-sorting mechanisms. It follows also that shape interacts strongly with grain size. The composition and mechanical properties of particles, as well as the mode of transport and deposition and the mechanism of abrasion and corrosion, all vary from large to small particles. It is probable that the effect of shape varies for particles of a single size depending on the size of the associated particles. For example, a spherical pebble might roll easily over a sand bed but become lodged firmly in a bed composed of pebbles of equal or larger size.

In all studies of shape in sediments, therefore, it is essential that comparisons should be made only between the shapes of particles comparable in mineral composition, fabric, and size grade. Ideally, particles should be composed of structurally isotropic materials. Such materials occur rarely in nature but the best approximations are massive crystalline rocks, chert, or quartz.

Mechanisms causing changes in particle size and shape include (a) splitting, (b) crushing, (c) chipping, (d) formation of percussion scars, (e) grinding, and (f) sand blasting, as well as chemical weathering and solution. Movement of pebbles and sand is intermittent, with particles spending large periods of time trapped in local deposits so that weathering during transportation is by no means negligible (see Bradley, 1970). The relative importance of the mechanisms listed above depends on the medium (water, air, or ice) and the mode of transportation (sliding, rolling, saltation, suspension, or grain flow). Modeling these mechanisms in the laboratory is difficult but despite this there have been numerous experimental studies of size and shape modification by abrasion.

Early laboratory studies frequently made use of abrasion mills and tumbling barrels, in which the mechanism of abrasion was certainly not identical to that taking place during natural transport. More recent studies (Kuenen, 1955 to 1964, Humbert, 1968) have made use of circular flumes to reproduce transport by currents and tilting troughs to simulate beach abrasion. It has been found that the rate of abrasion depends on the initial size, roundness, and mechanical resistance of the particle, as well as on the intensity of the process, the size of the associated fragments, and the nature of the floor over which transportation is taking place. Most investigators have found for rolling pebbles that the mean weight or size decreases as a negative exponential function of distance, a function similar in form to Sternberg's law. Some typical results are shown in Fig.3-21. The roundness increases at first rapidly and then progressively less rapidly as the particle approaches perfect roundness. The sphericity increases at a slow, almost constant rate.

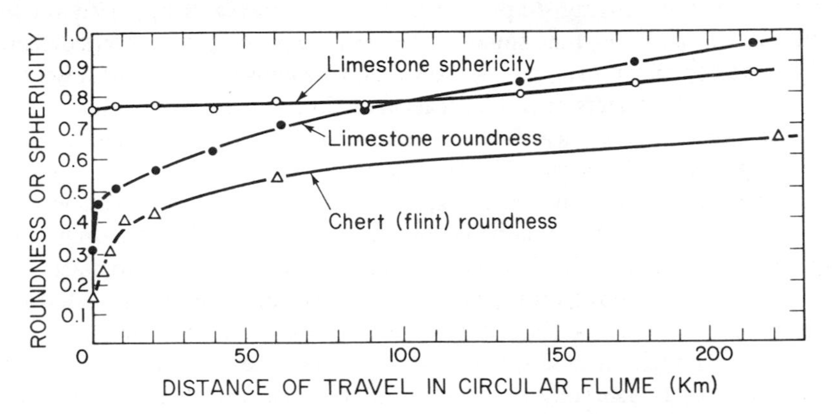

Fig. 3-21 Rounding of broken limestone and chert (flint) fragments in a circular flume. The floor of the flume consisted of rounded flint pebbles. Initial weights of pebbles were about 40g, final weights 15–30g. Velocity of movement 70 cms/sec. Data from Humbert (1968, p. 71).

In experiments performed by Kuenen (1964) to simulate wave action on a beach, it was more difficult to obtain precise figures for the abrasion loss per unit distance of transport but there was a tendency for the rate of loss to decrease with increasing roundness of the pebbles, as with rolling pebbles. Kuenen also concluded from his experiments that pebbles are flattened very little by sliding up and down the experimental "beach," so that the observed flatness of beach pebbles must be caused by shape sorting. It is doubtful, however, that these experiments were able to reproduce natural wave action very effectively.

Experiments on fluvial abrasion of smaller particles by Kuenen indicated that the mechanical abrasion of small pebbles is slight and becomes extremely slow for quartz grains less than 2 mm in diameter. For subangular quartz sand the weight loss was less than 0.1% in 1000 km of transport. Kuenen concluded that to reduce a quartz cube 0.4 mm in diameter to a sphere would require transport of several million kilometers. Weight loss values for fresh feldspar were only about twice those for quartz sand. In another series of experiments using a circular wind tunnel, Kuenen found that wind abrasion of quartz grains was 100 to 1000 times as effective as abrasion by river transport. The loss per kilometer decreased as the grains became more rounded and it also decreased with grain size, becoming negligibly small for diameters below 0.05 mm. In a sample of mixed sizes the larger grains were abraded more slowly and the smaller grains faster than the same sizes in uniform samples. The experiments failed to reproduce the typical frosted appearance of desert dune sands and it was concluded that such frosting is caused by chemical rather than mechanical action.

Although the results of experiments on particle abrasion conducted in circular flumes are of great significance, care must be taken not to apply these results to natural processes without regard to other possible factors that may play an important role in modifying shape of grains. For example, it has been shown for sand-size quartz particles that some of the roundness may be inherited from the parent igneous or metamorphic rock. Some may be caused by chemical weathering in the soil. Also, the effects of partial fracturing in the source rock and of solution during transport are unknown. It will be shown in Chap. 8 that there are reasons for believing that destruction of quartz sand grains must be possible in nature at rates larger than those indicated in experiments on mechanical abrasion.

In experimental studies sphericity does not increase with increasing abrasion as rapidly as roundness. Geologists who have studied pebbles in natural environments still do not agree on the relative importance of abrasion, solution, and shape sorting. It is clear that shape sorting is effective for pebbles and it is probably effective also, though to a lesser degree, for sand-size particles. The sphericity of beach pebbles is generally lower than that of comparable river pebbles. In a carefully controlled study of basalt pebbles in rivers and on adjacent beaches on the island of Tahiti-Nui, Dobkins and Folk (1970) have made a strong case for abrasion as the principal agent responsible for the flatness of beach pebbles.

Certain types of shapes of pebbles are generally considered to be typical of wind abrasion (pebbles faceted by sandblasting) and of glacial action ("flatiron" shapes with striated surfaces).

In contrast to the many studies of pebble shapes in recent sediments and in the laboratory, there have been few studies of pebble shapes in ancient conglomerates. Yet those that have been made show the potential of such studies. For example, it has been shown that pebble roundness can be used as an indicator of source direction in ancient alluvial fan conglomerates see (Fig. 6.2). Both McBride (1966) and Lucchi (1969) studied pebble shape in conglomerates associated with turbidites and showed that the shapes were those characteristic of river rather than beach deposits. They concluded that these deposits were "resedimented" by some downslope mass-transport mechanism from an original fluvial gravel deposit. Reworking on a beach was not an important aspect of their sedimentary history.

For quartz sand grains, evidence from both experiments and petrographic studies of natural sands (e.g., Russell and Taylor, 1937) indicates that the sphericity is not significantly modified by abrasion but is inherited from the parent rock. This suggests that sphericity of quartz grains might be used as an indicator of provenance in sandstones but unfortunately numerous observations on two-dimensional sphericity (elongation) of quartz grains observed in thin section show that the average "axial ratio" of apparent short over long diameters only ranges from 0.61 to 0.73 (Griffiths, 1967). Variations in axial ratio almost as large are observed in thin sections cut at different orientations from the same sandstone specimen, so the probability of obtaining very meaningful results from two-dimensional studies does not appear to be large. Too few studies of the three-dimensional sphericity of quartz grains in sediments have been made as yet to allow for generalizations in this case. Experiments indicate a large error for sphericity based on direct measurement of grain axes (standard deviation, 0.08). This source of error poses a problem in technique for further studies.

3.5 FABRIC

The term *fabric* is reserved for "the manner of mutual arrangement in space of the components of a rock body and of the boundaries between these components" (International Tectonics Dictionary). It therefore includes both grain orientation and packing.

Grain orientation includes both crystallographic and dimensional orientation. In clastic rocks, the dimensional orientation is most important but a dimensional orientation may imply a crystallographic orientation if the grain shape is correlated with its crystallography. For example, most quartz grains of sand size are somewhat elongated in the direction of the crystallographic *c* axis and, therefore, many sandstones show a primary (nontectonic) preferred orientation of quartz *c* axes. Dimensional grain orientation is difficult to study in three dimensions except by making use of this correlation between dimensional and crystallographic orientation. In most cases, however, the dimensional orientation may be studied in two planes: the plane of the bedding or lamination and a plane normal to the bedding plane. Once the direction of preferred orientation in the bedding plane has been established, the orientation may be studied in a plane normal to the bedding plane and parallel with the direction of preferred orientation. Orientation in this second plane is of interest because it reveals the way in which bladed or platy grains are packed against each other, to form *imbrication*. Grains are commonly imbricated so that they dip upcurrent. The formation of preferred grain orientation and imbrication is discussed in Chap. 4.

The least studied aspect of fabric is *packing*, "the spacing or density pattern of mineral grains in a rock" (AGI Glossary). The meaning of packing, and its distinction from other aspects of fabric such as orientation, is most clearly seen for the case of a sediment composed of perfect spheres uniform in size. Even in this highly idealized case, it has been shown that there are six different systematic ways of arranging the spheres so that each sphere is in contact with four or more adjacent spheres and there are no vacant positions. The arrangements vary from the "loosest" cubic packing, with a porosity of 47.6%, to the "tightest" rhombohedral packing, with a porosity of 26.0%. The six regular packings do not exhaust the number of ways that spheres may in fact be packed because in nature an infinite number of combinations of the six and of "random" packings may also be developed. Prediction of the average geometrical properties of "random heaps" of uniform spheres is a difficult problem in probability theory but some progress has been made in such studies in recent years (see Smalley, 1964a).

Real sands, no matter how well rounded and well sorted the grains, differ substantially from uniform spheres. Pryor (1968) found that the porosity of beach and dune sands (which tend to be very well sorted) varies from 30 to more than 65%, with an average of 45%. The very high porosities found in some natural sands are related mainly to the very loose packing that is developed locally under some conditions of deposition.

The effect of packing in sediments is seen in the change in bulk properties of the sediment, such as its density, porosity, and permeability. It is difficult to measure packing directly but it is important to do so in order to obtain a fundamental understanding of the bulk properties which depend not only on packing but also on grain size, shape, and orientation. Apart from the overall spacing of the grains, the most important aspects of packing have to do with the frequency with which grains touch each other and the shape of the contacts between them. Kahn (1956) devised two numerical measures for use in thin section studies.

1. The *packing density* is the ratio of the sum of the lengths of grain intercepts to the total length of the traverse across the thin section. It is a measure of the porosity of a cement-and matrix-free sand or of the "matrix-cement-free porosity" of a sandstone that has some matrix and cement.
2. The *packing proximity* is the ratio of the number of grain-to-grain contacts (encountered in a traverse across the thin section) to the total number of contacts of all kinds encountered in the same traverse (Fig.3-22). If the grains have only small areas of contact with each other, most of the contacts observed in a thin section will be contacts between a grain and matrix or cement, so the packing proximity will be small. In a rock where there has been compaction without the introduction of much cement, most of the grain contacts observed will be grain-to-grain contacts and the packing proximity will be large. An alternative technique for measuring packing has been suggested by Smalley (1964b).

The type of contact between grains may also be studied in thin section. In the ideal case of packed spheres, the only observed contacts between grains would be tangential ones. But in the case of nonspherical grains or where compaction has taken place, three other types of contacts may be observed (Taylor, 1950). The four possible types of contacts are (a) tangential; (b) long, i.e., a contact that appears as a straight line in the plane of section; (c) concavoconvex; and (d) sutured. The frequency of concavoconvex and sutured contacts relative to that of other types of contacts has been used as a measure of the intensity of compaction of sands.

The most important bulk properties of sedimentary rocks are probably porosity and permeability. There have been several investigations of the relationship between these important rock characteristics and the fundamental textural

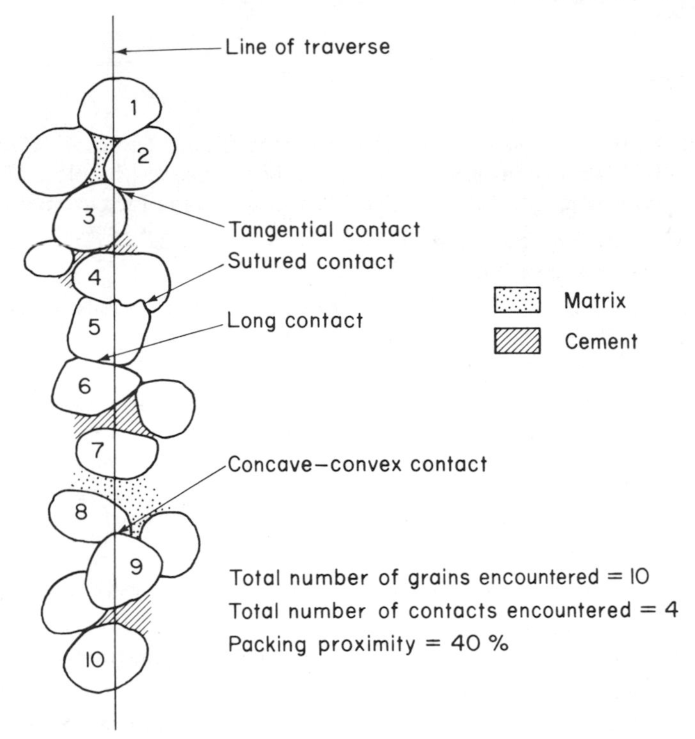

Fig. 3-22 Definition of grain contact types (after Taylor, 1950) and of packing proximity (after Kahn, 1956).

characteristics (size, shape, orientation, and packing) particularly for uncemented sands. For such sediments it has also been found that there is an empirical relationship between porosity and permeability, called the Kozeny relationship:

$$k = \frac{P^3}{KS^2} \tag{25}$$

where k is the permeability; P is the fractional porosity; S is the specific surface area, i.e., the surface area of the pore space in unit volume of the rock; and K is a constant. For sediment composed of loose grains, S is inversely proportional to the square of the grain size.

By considering the pores in the rock to be analogous to a bundle of capillary tubes of complex shape, it is possible to derive from the fundamental law for flow through a capillary (the Hagen-Poiseuille law) both Darcy's law of permeability and the Kozeny relationship. The model of a porous rock as a bundle of capillary tubes is obviously not an adequate one. For example, in such a rock, flow would be possible in only one direction. But the model has proved useful in studies of porous media (Scheidegger, 1960).

REFERENCES

BAGNOLD, R. A., 1968, "Deposition in the Process of Hydraulic Transport," *Sedimentology*, **10**, pp. 45–56. (Maintains grain sizes should be represented by plots of log proportion versus log size, and that most size distributions are not log normal.)

BENNETT, J. G., 1936, Broken coal. *Jour. Inst. Fuel*, v. 10, pp. 22–29 (Develops mathematical model for size distribution.)

BRADLEY, W. C., 1970, "Effect of Weathering on Abrasion of Granitic Gravel, Colorado River (Texas)," *Geol. Soc. Amer. Bull.*, **81**, pp. 61–80. (Experiments indicate most of size decline over 160 mi of transport is due to weathering.)

CURRAY, J. R., 1960, "Tracing Sediment Masses by Grain Size Modes," *Repts. 21st Innatl. Geol. Cong., Norden*, pt. 23, pp. 119–130. (Derives density distribution curves from cumulative curves and uses the former to trace movement of sediment in northern Gulf of Mexico.)

DOBKINS, J. E., Jr., and R. L. FOLK, 1970, "Shape Development on Tahiti-Nui," *Jour. Sed. Pet.* **40**, pp. 1167–1203. (Demonstrates that pebbles of basalt are more discoidal and have lower sphericity on beaches than in rivers supplying pebbles to the beaches.)

FOLK, R. L., 1955, "Student Operator Error in Determination of Roundness, Sphericity and Grain Size," *Jour. Sed. Pet.*, **25**, pp. 297–301. (Demonstrates low error in grain size and high error in roundness determinations.)

———, 1966, "A Review of Grain-Size Parameters," *Sedimentology*, **6**, pp. 73–93. (Cogent discussion of properties of various descriptive parameters.)

FOLK, R. L., 1968, *Petrology of Sedimentary Rocks*. Austin, Texas, Hemphill's, 170 pp. (A "syllabus" periodically revised, to be used in conjunction with a regular laboratory manual by students at the University of Texas. It has become widely used and cited because of its lively style and many practical hints on technique).

FOLK, R. L., and W. C. WARD, 1957, "Brazos River Bar: A Study in the significance of Grain Size Parameters," *Jour. Sed. Pet.*, **27**, pp. 3–27. (A classic paper on the use of grain size data.)

FRIEDMAN, G. M. 1958, "Determination of Sieve-Size Distribution from Thin Section Data for Sedimentary Studies," *Jour. Geol.*, **66**, pp. 394–416. (Gives empirically determined curves for converting thin section to sieve size data.)

———, 1961, "Distinction between Dune, Beach and River Sands from Their Textural Characteristics," *Jour. Sed. Pet.*, **31**, pp. 514–529. (Textural critera for attempting to determine environment of deposition.)

———, 1967, "Dynamic Processes and Statistical Parameters Compared for Size Frequency Distribution of Beach and River Sands," *Jour. Sed. Pet.*, **37**, pp. 327–354. (Distinction of environment based on grain size of several hundred specimens from recent sediments.)

GIBBS, R. J., M. D. MATTHEWS, and D. A. LINK, 1971, "The Relationship Between Sphere Size and Settling Velocity," *Jour. Sed. Pet.*, **41**, pp. 7–18. (Experimental study that indicates the inadequacy of the Rubey equation.)

Graf, W. H., and E. R. Acaroglu, 1966, "Settling Velocities of Natural Grains," *Bull. Innatl. Assn. Sci. Hydrology*, **11**, no. 4, pp. 27–43. (Compares experimental determinations of settling velocity and shows inadequacy of Rubey equation.)

Griffiths, J. C., 1967, *Scientific Method in Analysis of Sediments*. New York, McGraw-Hill Book Co., 508 pp. (Detailed discussion of sediment textures, with emphasis on statistics.)

Humbert, F. L., 1968, "Selection and Wear of Pebbles on Gravel Beaches," Dissertation, Univ. Groningen, Netherlands, 144 pp. (Reviews studies of pebble morphogenesis and describes new experimental and field observations.)

Inman, D. L., 1952, "Measures for Describing the Size Distribution of Sediments," *Jour. Sed. Pet.*, **22**, pp. 125–145. (A useful source of information on size distribution measures.)

Kahn, J. S., 1956, "The Analysis and Distribution of Packing in Sand-Size Sediments," *Jour. Geol.*, **64**, pp. 385–395. (A good discussion of packing.)

———, 1959, "Anisotropic Sedimentary Parameters," *N. Y. Acad. Sci. Trans.*, Ser. 2, **21**, pp. 373–386. (Describes thin section techniques for studying fabric.)

Kittleman, L. R., Jr., 1964, "Application of Rosin's Distribution to Size-Frequency Analysis of Clastic Rocks," *Jour. Sed. Pet.*, **34**, pp. 483–502. (Describes Rosin's law and applies it to size distribution of crushed materials, regoliths, pyroclastic, and glacial debris.)

Krinsley, D., and J., Donahue, 1968, "Environmental Interpretation of Sand Grain Surface Texture by Electron Microscopy," *Geol. Soc. Amer. Bull.*, **79**, pp. 743–748. (Summary of results obtained in earlier studies.)

Krumbein, W. C., 1934, "Size Frequency Distributions of Sediments," *Jour. Sed. Pet.*, **4**, pp. 65–77. (Proposes ϕ scale: See also comments by Krumbein, 1964 in *Jour. Sed. Pet.*, **34**, pp. 195–196.)

———, 1935, "Thin Section Mechanical Analysis of Indurated Sediments," *Jour. Geol.*, **43** pp. 482–496. (Derives theoretical moment corrections for sectioning spheres: See also papers by Wicksell.)

Krumbein, W. C., and F. A. Graybill, 1965, *An Introduction to Statistical Models in Geology*. New York; McGraw-Hill Book Co., 475 pp. (Chap. 3 to 7 give an extended discussion of the nature of geological populations and the problem of sampling.)

Kuenen, P. H., 1955, 1956, 1964, "Experimental Abrasion of Pebbles: 1. Wet Sand Blasting," *Leidse Geol. Meded.*, **20**, pp. 142–147. 2. "Rolling by Current," *Jour. Geol.*,**64**, pp. 336–368. 4. "Eolian Action," *Jour. Geol.*, **68**, pp. 427–449. 6. "Surf Action," *Sedimentology*, **3**, pp. 29–43. (Series of experimental studies on pebble wear.)

Landim, P. M. B. and L. A. Frakes, 1968, "Distinction between Tills and Other Diamictons Based on Textural Characteristics," *Jour. Sed. Pet.*, **38**, pp. 1213–1223. (Till is generally more poorly sorted than alluvial fan material and has smaller mean size: Use of discriminant function based on four moments efficiently separates till, outwash, and fan deposits.)

Lucchi, F. Ricci, 1969, "Composizione e morfometria di un conglomerato risedimento nel flysch miocenico romagnolo," *Gionale di Geol., Ann. Mus. Geol. Bologna*, Ser. 2a, **36**, pp. 1–47. (In Italian with English summary: Describes application of shape measures to determining origin of resedimented Miocene conglomerate.)

LUDWICK, J. C. and P. L. HENDERSON, 1968, "Particle Shape and Inference of Size from Sieving," *Sedimentology*, **11**, pp. 197–235. (Theoretical and experimental study of limitations of sieving as a method for determining grain size distributions.)

MARGOLIS, S. V., 1968, "Electron Microscopy of Chemical Solution and Mechanical Abrasion Features on Quartz Sand Grains," *Sedimentary Geol.*, **2**, pp. 243–256. (Describes etch features observed on natural sand and produced experimentally in the laboratory.)

MCBRIDE, E. F., 1966, "Sedimentary Petrology and History of the Haymond Formation (Pennsylvanian), Marathon Basin, Texas," *Texas Bur. Econ. Geol. Rept. Inv.* 57, 101. pp. (One of the few studies of pebble shape in ancient conglomerates.)

MCMANUS, D. A., 1963, "A Criticism of Certain Usage of the Phi Notation," *Jour. Sed. Pet.*, **33**, pp. 670–674. (Redefines ϕ to make clear the dimensionless nature of the ϕ transformation.)

MOIOLA, R. J. and D. WEISER, 1969, "Environmental Analysis of Ancient Sandstone Bodies by Discriminant Analysis," *Amer. Assoc. Petroleum Geol. Bull.* **53**, p. 733 (abstract). (Good results obtained by using weight percent as variable in the discriminant function.)

MÜLLER, GERMAN, 1967, *Methods in Sedimentary Petrography*. New York; Hafner Pub. Co., 283 pp. (Good review of techniques, with special reference to European practice.)

PASSEGA, R., 1957, "Texture as Characteristic of Clastic Deposition," *Amer. Assn. Petroleum Geol. Bull.*, **41**, pp. 1952–1984.

———, 1964, "Grain Size Representation by CM Patterns as a Geological Tool," *Jour. Sedimentary Petrology*, **34**, pp. 830–847. (Discusses interpretation of grain size in terms of variation in median and coarsest 1 percentile.)

POWERS, M. C., 1953, "A New Roundness Scale for Sedimentary Particles," *Jour. Sedimentary Petrology*, **23**, pp. 117–119. (Presentation of a roundness scale.)

PRYOR, W. A., 1968, "Reservoir Inhomogeneities of Recent Sandbodies," Final Rept. Amer. Petroleum Inst. Res. Project 91-B, 1 July 1966–30 June 1968, Dallas, Amer. Petroleum Inst. (Determination of *in situ* porosity and permeability of many samples.)

RIZZINI, ANTONIO, 1968, "Sedimentological Representation of Grain Sizes," *Mem. Soc. Geol. Italiana*, **7**, pp. 65–89. (In English: General review, with emphasis on Passega's methods.)

ROGERS, J. J. W., 1965, "Reproducibility and Significance of Measurements of Sedimentary-Size Distributions," *Jour. Sed. Pet.*, **35**, pp. 722–732. (Study of operator error.)

ROUSE, HUNTER, 1937, "Nomogram for the Settling Velocity of Spheres," *Div. Geol. Geo. Exhibit D, Rept, Comm. Sedimentation* **(1936–1937)** *Natl. Res. Council*, Washington, D.C., pp. 57–64. (Based on best experimental data.)

RUBEY, W. W., 1933, "Settling Velocities of Gravel, Sand and Silt," *Amer. Jour. Sci.*, **25**, p. 325–338. (Proposes a widely used, but unfortunately not too accurate equation: See Watson, and Gibbs et al.)

RUSSELL, R. D. and R. E. TAYLOR, 1937, "Roundness and Shape of Mississippi River Sands," *Jour. Geol.*, **45**, p. 225–267. (Classic study: Demonstrates downstream decreases in size and slight decrease in roundness and sphericity over 900 mi of transport; positive correlation between grain size, roundness, and sphericity of grains in samples.)

SCHEIDEGGER, A. E., 1960, *The Physics of Flow through Porous Media*. Toronto, Ont.: Univ. Toronto Press, 2nd ed., pp. 1–313. (Chapter I gives a concise review of the geometrical properties of porous materials. Chapter IV discusses permeability and Darcy's law. The Kozeny theory is discussed in Chap. VI.)

SCHLEE, JOHN, E. UCHUPI, and J. V. A. TRUMBULL, 1965, "Statistical Parameters of Cape Cod Beach and Eolian Sands," *U. S. Geol. Sur. Prof. Paper 501-D*, pp. 118–122. (None of Folk and Ward's graphic measures adequately separated beach from dune sands: Both are nearly symmetrical or have slight negative skewness as measured by sieves.)

SCHWARCZ, H. P. and K. C. SHANE, 1970, "Measurement of Particle Shape by Fourier Analysis," *Sedimentology*, **13**, pp. 213–231. (Describes new technique, capable of being automated, for measurement of both roundness and "sphericity.")

SMALLEY, I. G., 1964a, "Representation of Packing in a Clastic Sediment," *Amer. Jour. Sci.*, **262** pp. 242–248. (A good source on how to represent packing.)

———, 1964b, "A Method for Describing the Packing Texture of Clastic Sediments," *Nature*, **203**, pp. 281–284. (Reviews modern papers on random packing of spheres, describes a method for measuring packing using the radial distribution function—the number of particles as a function of distance from an arbitrary center—and gives some examples of measurements on sandstones and gravels.)

SNEED, E. D. and R. L. FOLK, 1958, "Pebbles in the Lower Colorado River, Texas, a Study in Particle Morphogenesis, *Jour. Geol.*, **66**, pp. 114–150. (Proposes new shape indices and shows that shape interacts strongly with grain size. Roundness increases downstream, both quartz and limestone pebbles reaching a limiting value of about 0.65 —this value is reached in a few miles for limestone and in 150 mi for quartz.)

TANNER, W. F., 1959, "Sample Components Obtained by the Method of Differences" *Jour. Sed. Pet.*, **29**, pp. 408–411. (Method for dissecting a cumulative curve into Normal components.)

———, 1969, "The Particle Size Scale," *Jour. Sed. Pet.*, **39**, pp. 809–812. (Compares engineering, highways, and soil science scales with Udden-Wentworth and argues for advantages of the latter.)

TAYLOR, J. M., 1950, "Pore-Space Reduction in Sandstones," *Amer. Assoc. Petroleum Geol. Bull.*, **34**, pp. 701–716. (Classifies types of contacts between grains and shows that concavo-convex and sutured contacts increase with depth of burial.)

UDDEN, J. A., 1898, "Mechanical Composition of Wind Deposits," *Augustana Library Publ.*, No. 1. (Original proposal of grade scale.)

VISHER, G. S., 1969, "Grain Size Distributions and Depositional Processes," *Jour. Sed. Pet.*, **39**, pp. 1074–1106. (Claims that subpopulations may be recognized as straight line segments on probability plots of grain size analyses: Subpopulations are interpreted in terms of depositional mechanisms.)

WADELL, HAAKON, 1932, "Volume, Shape, and Roundness of Rock-Particles," *Jour. Geol.* **40**, pp. 443–451. (Definition of roundness and sphericity.)

WATSON, R. L., 1969, "Modified Rubey's Law Accurately Predicts Sediment Settling Velocities," *Water Resources Res.*, **5**, pp. 1147–1150. (See Rubey, p. 76.)

WENTWORTH, C. K., 1919, "A Laboratory and Field Study of Cobble Abrasion," *Jour. Geol.*, **27**, pp. 507–521. (Pioneer study of roundness: Shows downstream increase in stream transport.)

——— 1922, "A Scale of Grade and Class Terms for Clastic Sediments," *Jour. Geol.*, **30**, pp. 377–392. (Proposes now accepted Udden-Wentworth scale.)

WICKSELL, S. D., 1925, 1926, "The Corpuscle Problem," *Biometrika*, **17**, pp. 84–99; **18**, pp. 151–172. (Definitive theoretical treatment of problem of size distribution that results from sectioning a population of spheres or ellipsoids.)

ZEIGLER, J. M., G. G. WHITNEY, and C. R. HAYES, 1960, "Woods Hole Rapid Sediment Analyzer," *Jour. Sed. Pet.*, **30**, pp. 490–495. (Describes settling tube for size analysis of sands: Several different modifications of this tube have been built in different laboratories.)

ZINGG, T., 1935, "Beiträge zur Schotteranalyse," *Schweiz. min. pet. Mitt.*, **15**, pp. 39–140. (Pioneer work on shape analysis.)

CHAPTER FOUR

SEDIMENT MOVEMENT BY FLUID FLOW

4.1 FUNDAMENTALS OF FLUID FLOW

Introduction

Before discussing the transport and sorting of sediment and the formation of sedimentary structures, some attention must be given to the part of the dynamic environment often neglected by the geologist—the fluid. The term *fluid* includes both liquids and gases. A fluid is a substance that is deformed by a shear force, no matter how small the force may be; i.e., it is a substance that has no strength.

The forces that act on solid or fluid bodies are vectors that may be resolved into components normal to and parallel with the surface of the body. The components normal to the surface are called *pressure* and those parallel to the surface are called *shear stress* (or simply, shear). In addition, it is convenient to distinguish certain *body forces* that act equally on every particle composing the body, e.g., gravity or inertia. Gases, including air, respond to change in pressure by expansion or contraction; i.e., they are compressible fluids and the density cannot be treated as constant. Liquids, however, are generally only slightly compressible and for a given temperature the density may be considered to be constant.

Apart from the density, the other main property of fluids controlling the way the fluid flows is the dynamic viscosity. As noted earlier, this is defined as the coefficient in the equation relating the shear stress acting on a fluid to its rate of shear (Chap. 3, Eq. 15). In many dynamic equations the ratio of dynamic viscosity to density (μ/ρ) appears and this ratio is called the *kinematic viscosity* ν (nu). These parameters have dimensions as indicated in Table 3-4.

Air and water are the two fluids of greatest geological importance. They differ substantially in their density and dynamic viscosity, with water being some 800 times as dense as air and having a much larger dynamic viscosity. Curiously, the kinematic viscosities of air and water are almost equal, having a value (in cgs units) of about 0.01 sq cm/sec. Both air and water are fluids that obey Newton's law of viscosity:

$$\tau = \mu \frac{du}{dy} \tag{1}$$

For pure water or air, the dynamic viscosity, μ, is a constant at constant temperature (see discussion in Chap. 3). Water may, however, become mixed with substantial concentration of clays, for example in mud flows. High concentrations of clay not only greatly increase the viscosity (Fig. 4-1) but also change the way in which the suspension responds to shear stress so that the coefficient of viscosity is no longer a constant. Such substances are described as *non-Newtonian fluids.*

At high sediment concentrations, muds may acquire strength so that they can no longer be sheared by very low shear stresses. At shear stresses in excess of this strength, such muds behave like viscous fluids. Substances that behave in this manner are described as *pseudo-plastics*. Some of their properties are discussed further in Chap. 5.

The behavior of Newtonian fluids is described by the equations of fluid dynamics, based upon Newton's law of viscosity and the laws of Newtonian dynamics. The basic equations are (a) the equation of *continuity*, which simply expresses the law of conservation of mass for a fluid, and (b) the three *Navier-Stokes* equations, or equations of motion, that express how Newton's dynamic laws must apply to a fluid. Together these make up a system of four partial differential equations that express in principle how a viscous fluid must behave in any and all circumstances in which Newtonian dynamics are valid.

The ideal explanation of any fluid phenomenon (including all sedimentation phenomena) is to show how the phenomenon may be deduced from the four equations, plus a statement about the shapes of the fluid boundaries (the "boundary conditions") and the forces, such as gravity, acting on the fluid and its boundaries. Nature is generally much too complicated to permit the solution of these equations, though it can be done for a few simple cases, including the case of very slow movement of a sphere through a fluid. For this case, Stokes' law, discussed on p. 51, was actually derived from the basic equations of motion by Stokes in 1851.

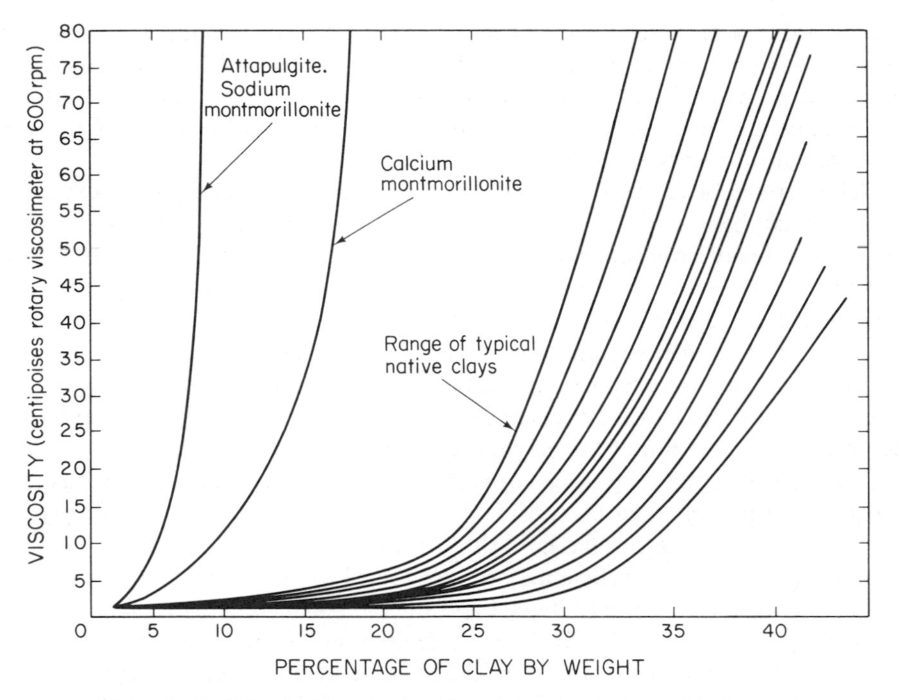

Fig. 4-1 Variation in dynamic viscosity of clay suspensions with clay concentration and mineralogy. From Grim "Applied Clay Mineralogy" McGraw-Hill Book Co. (after data in "Drilling Mud," Baroid Division National Lead Co., 1953).

Dimensionless Numbers

One very important application of the basic equations is to derive the proper criteria for making scale models of fluid flow that correctly represent phenomena at a different (usually larger) scale. In the equations if we express all the lengths as ratios of some reference length, all velocities as ratios of some standard velocity, and the other variables similarly and then rearrange the terms, we find for the equations that apply to flow of fluids with a free surface subject to gravity that two coefficients appear (see Daily and Harleman, 1966). In other words, in order to make all the variables in the equations dimensionless (scaled in terms of some reference length L and velocity U), it is necessary to introduce into the equations the following two dimensionless coefficients:

$$\mathrm{R_e} = \frac{UL\rho}{\mu} \qquad \text{(Reynolds number)} \tag{2}$$

$$\mathrm{F_r} = \frac{U}{\sqrt{gL}} \qquad \text{(Froude number)} \tag{3}$$

This means that if there are two situations (the original or *prototype* and the *model*) that have the same shaped boundaries and if the Reynolds number and Froude number of the model are equal to the Reynolds number and Froude number for the prototype, then the two situations will be exactly similar as far as all aspects of the fluids are concerned. The only differences will be in the scale of the phenomena, not in their basic nature.

The derivation of Reynolds and Froude numbers sketched above cannot be given in detail here. It is the most fundamental way however, to derive these two numbers, that play a prominent part in the discussion of most hydraulic phenomena, including many aspects of sedimentation. The significance of the numbers may be illustrated by reference to some specific applications.

The *Reynolds number* appeared previously in the discussion of particle settling given on p. 50. In that case, the representative length L may be taken as the diameter of the particle d. It was observed experimentally that the drag force acting on the particle is a function of the Reynolds number. Only at low Reynolds numbers (less than unity) is the drag force correctly given by Stokes' law. The viscous force acting on the particle may be considered to be proportional to the viscosity, the velocity gradient at the surface, and the surface area, i.e., to $\mu(U/d)d^2$. The inertial forces acting on the particle may be considered to be proportional to the mass and deceleration of the fluid "impacting" on the particle, i.e., to $Ud^2\rho \cdot U$ (a cylinder of fluid of volume Ud^2 is decelerated an amount proportional to U in unit time by the resistance of the particle). The ratio between inertial and viscous forces is, therefore, proportional to $U^2 d^2 \rho / U d\mu$, or $U d\rho/\mu$, which is the Reynolds number in this case. Thus, the Reynolds number may be interpreted as being proportional to the ratio between inertial and viscous forces.

In other cases, a different choice must be made for the representative length and velocity. For flow in pipes, for example, the Reynolds number may be defined as $R_e = UD\rho/\mu$, where D is the diameter of the pipe and U is the average velocity of flow through the pipe. It was in experiments on the flow through pipes that Osborne Reynolds first discovered another aspect of the significance of the Reynolds number. Reynolds was able to show, by making use of pipes of different diameters and different types of fluids flowing at different velocities, that there is a fundamental difference in the type of flow in the pipe at Reynolds numbers less or greater than a certain critical value. At Reynolds numbers less than about 2000, flow is *laminar* or streamline; i.e., the different layers or particles of fluid appear to slide smoothly past each other and there are no irregular eddies producing diffusion from one layer of fluid to another. At Reynolds numbers greater than about 2000, however, flow is *turbulent*, with eddies producing diffusion of fluid (and also of anything that is carried by the fluid, such as dye or sediment) from one layer to another.

The critical value of the Reynolds number for the transition from laminar to turbulent flow depends on the choice of representative length and velocity and on the geometry and some other properties of the flow system. In the case of flow past a settling spherical particle, flow is laminar up to a Reynolds number of about unity. Above this value, a wake filled with eddies is formed in the lee of the particle. The patterns of flow (for the closely similar case of flow past a cylinder) observed at progressively higher Reynolds numbers are shown in Fig.4-2. It can be seen that at first the eddies in the wake have a regular geometry but the eddies gradually become more irregular in nature until the wake is fully turbulent ($R_e > 300$). The shedding of eddies regularly from one side and then the other of a settling particle (formation of a "Karman vortex street") gives rise to the side-to-side ("falling leaf") motion observed over a certain range of Reynolds numbers.

The *Froude number* is analogous to the Reynolds number in that it too may be

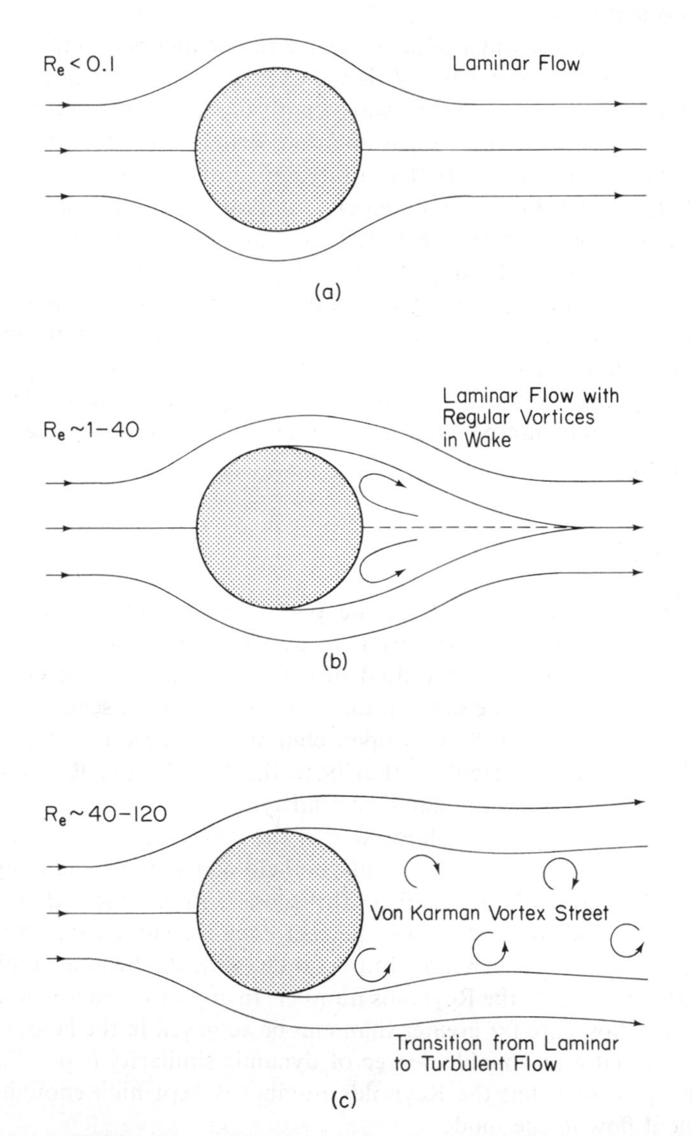

Fig. 4-2 Progressive development of a wake in the lee of a cylinder, for increasing Reynolds number of the flow.

considered to be a ratio between two types of forces: in this case, inertial and gravity forces. For a unit mass of fluid, moving with a velocity U, the inertial force is equal to the force required to decelerate the mass to rest, in a distance that may be arbitrarily chosen to be proportional to some characteristic length L. The time required is therefore proportional to L/U and consequently the rate of deceleration, or force acting on unit mass, is proportional to $U/t = U^2/L$. The gravity force acting on unit mass is equal to g, so the ratio of inertial to gravity forces is proportional to U^2/gL. The Froude number is defined by most engineers as the square root of this quantity; i.e., $F_r = U/\sqrt{gL}$.

In the special case of water of depth D, flowing in an open channel, the Froude number is therefore $U/\sqrt{gD}$, where U is the average velocity. It can be shown that the velocity of gravity waves whose wavelength is long compared with the depth of water is equal to $\sqrt{gD}$. This suggests another significant aspect of the Froude number: If the number is greater than unity, it is not possible for waves to travel upstream because the downstream velocity of flow is greater than the upstream velocity of the waves. For this reason, there are some fundamental differences in the type of flow (called *tranquil*, streaming, or subcritical) found at Froude numbers less than unity and the type of flow (called *rapid*, shooting, or supercritical) at Froude numbers greater than unity. The transition from tranquil to rapid flow (frequently observed at a point where the channel becomes steeper) may be smooth but the opposite transition (from rapid to tranquil flow) is always accompanied by a "hydraulic jump," i.e., by a sudden increase in depth accompanied by much turbulence.

Scale Models

In the phenomenon of settling, the particle is completely enclosed by the fluid. The only role played by gravity is in causing the particle to move. Gravity does not affect the motion of the fluid directly. In this case, therefore, only the Reynolds number need be the same in the original and in the scale model for complete dynamic similarity. In flow in open channels all three forces (gravity, inertial, and viscous) are important so that both the Froude and Reynolds numbers must be the same for perfect dynamic similarity.

A little consideration soon shows that if the length scale is to be changed, the Froude and Reynolds numbers can only be held constant by changing the properties of the fluid from the original to the model. In practice, this is generally not feasible so models do not achieve perfect dynamic similarity. Experience has shown that it is generally much more important to scale the Froude number correctly than to attempt to scale the Reynolds number. In any case, the Reynolds numbers of large natural flows are far greater than can be achieved in the laboratory. Luckily, it happens that a reasonable degree of dynamic similarity is possible in small-scale models provided that the Reynolds number is kept high enough to achieve fully turbulent flow in the model.

In studying models of sediment transport, however, further complications arise. Ideally, not only must the overall aspects of the flow be correctly modeled but also the interaction of the flow and the sediment particles. In Sec. 4.3 it is shown that many different variables affect the interaction of the flow with the sediment; this interaction in turn affects the geometry of the channel boundaries so that true-to-nature small-scale models are almost never possible. The realization of this fact has led experimenters to construct very large models. For example, the large tilting flume operated by the U.S. Geological Survey at Fort Collins, Colorado, is 8 ft. wide and about 200 ft. long and very large wave channels are found in several hydraulics laboratories. Even the largest models, however, cannot reproduce all the phenomena observed in nature.

4.2 FLOW IN PIPES AND CHANNELS

Flow in Pipes

Most of the flows of interest to a geologist are either channel flows such as rivers, debris flows, and flows in tidal channels or "unconfined" flows such as the large-scale flows in the oceans and atmosphere. Progress in understanding the laws governing fluid flows was achieved first by studying flow in pipes, however, so that it is difficult to understand the hydraulic formulas used for flow in channels without at least some understanding of the hydraulics of pipes. For this reason, we introduce here a very brief discussion of pipe hydraulics before proceeding to open channel flows.

Experimental studies of flow through pipes have shown that there are several factors that determine the nature and quantity of the flow. It has already been noted that the Reynolds number is the criterion for determining whether the flow in the pipe will be laminar or turbulent. Different velocity distributions within the pipe are found in the two cases: In a longitudinal section through the center of the pipe the laminar velocity distribution is parabolic; i.e., the velocity is proportional to the square of the distance from the wall y. For flows with equal average velocities, the turbulent flow has a "blunter" velocity profile, with the velocity proportional to log y close to the wall. The velocity gradient in turbulent flows is steeper close to the wall and less steep in the center of the pipe than it is for laminar flows.

In the case of turbulent flow, it has been found that the exact shape of the velocity profile depends on the nature of the wall. A distinction can be drawn between boundaries that are hydrodynamically smooth and those that are hydrodynamically rough. For smooth boundaries, studies using dye introduced into the flow close to the wall have shown that there is a thin region close to the boundary where the flow is laminar even when the flow in the rest of the pipe is turbulent. This region is called the *laminar sublayer*. For sufficiently rough boundaries the laminar sublayer is destroyed by the large irregularities that project through it and shed eddies into the flow.

The resistance of the boundary to the flow is commonly measured by a coefficient f in the empirical formula, called the *Darcy-Weisbach equation*, that relates the difference in head h at two points separated by a distance L on a pipe of diameter D to the average velocity in the pipe U.

$$h = f\frac{LU^2}{D2g} \tag{4}$$

For smooth pipes and both laminar and turbulent flows, it is found that f depends on the Reynolds number and f is smaller at larger Reynolds numbers. For turbulent flows in rough pipes, f does not vary significantly with the Reynolds number but depends on the *relative roughness* of the pipe, which may be defined as the ratio between the average height of the roughness projections k and the diameter of the pipe. Graphs, sometimes known as Moody diagrams, that show the relationship between f and Reynolds number or relative roughness have been prepared by hydraulic engineers and are reproduced in most handbooks of hydraulic engineering.

Flow in Open Channels

The concepts of fluid resistance and rough and smooth boundaries have also been applied to flow in open channels. Let us consider the basic forces operating during flow in an open channel, such as that shown in Fig. 4-3. We assume that the

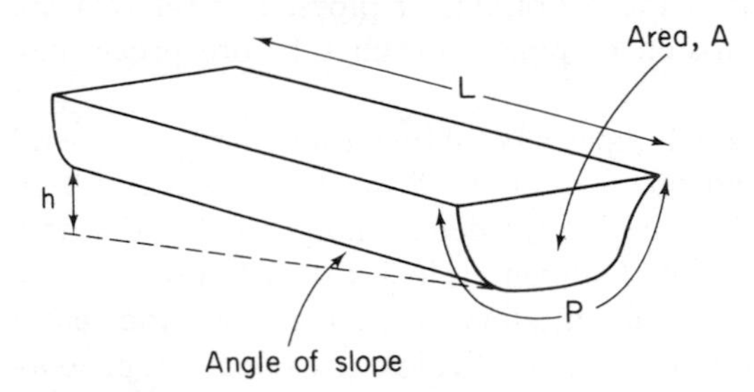

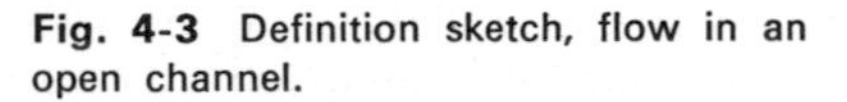
Fig. 4-3 Definition sketch, flow in an open channel.

average depth does not change down the channel (*uniform flow*), that the flow conditions are not changing with time (*steady flow*), and that the slope S is constant and small so that the sine of the angle of slope is approximately equal to the slope itself ($S = h/L$). In this case the gravity force acting on a volume of water in the channel of length L and average cross-sectional area A is ρgAL and the downslope component is $G = \rho gALS$. This force must be balanced by the resistance of the boundary, which is equal to the product of the average shear stress at the boundary τ_0, and the area of the boundary in contact with the volume of river water. This area is the product of the length L and the "wetted perimeter," P. Consequently,

$$\rho gALS = \tau_0 LP \tag{5}$$

or

$$\tau_0 = \rho g\frac{A}{P}S \tag{6}$$

An equivalent argument for pipes would show that the quantity A/P in Eq. 6 is the equivalent of half the radius of the pipe; hence, it is called the *hydraulic radius* R. For a wide channel, such as most river channels, it is almost equal to the mean depth of the channel. For a pipe of circular cross section,

$$R = \frac{A}{P} = \frac{\pi D^2/4}{\pi D} = \frac{D}{4}$$

Using this result and combining Eq. 6 with the Darcy-Weisbach equation (Eq. 4), we obtain the following results:

$$\tau_0 = \frac{f\rho U^2}{8} \tag{7}$$

and

$$U = \sqrt{\frac{8g}{f}}\sqrt{RS} \tag{8}$$

$$= C\sqrt{RS} \qquad \text{(the Chézy equation)} \tag{9}$$

The Chézy equation is a well-established empirical equation for river flow and essentially what has been done in the derivation given above is to establish a relationship between the empirical coefficient C and the better understood Darcy-Weisbach coefficient f.

$$C = \sqrt{\frac{8g}{f}} \tag{9a}$$

Many engineers prefer to use a slightly different empirical formula for river flow, the Manning equation:

$$U = \frac{1.49}{n} R^{2/3} S^{1/2} \tag{10}$$

where n is the Manning roughness coefficient. For the Manning equation to be valid with the numerical coefficient 1.49 given in Eq. 10, U and R must be measured in foot-second units. For rivers, the value of the various resistance coefficients (Chézy's C, Manning's n, or Darcy-Weisbach's f) depend in practice not only on the roughness or relative roughness of the boundary but also on the shape (especially sinuosity) of the channel. Books on hydraulic engineering give tables and pictures of typical channels to help engineers estimate the correct value of the coefficients. Typical values lie in the following ranges:

$$\frac{C}{\sqrt{g}} = 7 \text{ to } 40 \qquad n = 0.01 \text{ to } 0.15 \qquad f = 0.10 \text{ to } 0.005$$

Not only the average velocity but also the velocity distribution depends on the roughness of the boundary. For *two-dimensional flows*, in which the flow pattern remains constant in the direction of the third dimension, the velocity profile may be expressed in terms of a dimensionless velocity parameter U/U^*, where U^* is the *shear velocity* defined as $U^* = \sqrt{\tau_0/\rho}$. The shear velocity has the dimensions of velocity and may be thought of as a small velocity that exists in the flow somewhere near the boundary. In wide rivers the *maximum velocity* is found very close to the surface and the *average velocity* occurs at a depth of about $0.6D$ from the surface.

It was noted above that there is a difference in the factors affecting the resistance coefficient f, depending on whether the boundary is smooth or rough. Similarly, it is found that there are different equations relating the velocity or the velocity ratio U/U^* to the distance from the boundary y, depending on whether the boundary is smooth or rough. In both cases the velocity is proportional to log y but the nature of the function changes

For *smooth boundaries,*

$$\frac{U}{U^*} = a + b \log \frac{yU^*}{\nu} \tag{11}$$

For *rough boundaries,*

$$\frac{U}{U^*} = a' + b \log \frac{y}{k} \tag{12}$$

where k is the height of the roughness elements on the bed. In other words, for smooth boundaries the velocity is proportional to the logarithm of a Reynolds number but for rough boundaries it is proportional to the logarithm of the relative roughness. The coefficients a and a' are not the same but the coefficient b is the same in the two cases and has the value $2.3/\kappa$ where κ (kappa) is called the Von Karman constant. The value of the "constant" may be determined experimentally by plotting the velocity against the logarithm of y as in Fig. 4-4. The slope of the line

$$m = \frac{U_1 - U_2}{\log y_1 - \log y_2}$$

and the value of U^* determined from the known depth and slope may be then used to estimate κ because from Eq. 12, $\kappa = (2.3/m)U^*$. It has been found in laboratory experiments that κ has a value 0.4 for clear flows but that the value is lowered to 0.2 or less for flows that have a high concentration of sand-size sediment in suspension. The reason for the decrease in κ, which otherwise seems to be remarkably constant, is that the suspended sediment in the flow alters the structure of turbulent diffusion. Contrary to expectation, therefore, for two flows with the same depth and slope, the flow with suspended sediment will have the *higher* velocity (even after allowing for the slightly greater density).

Most natural river beds are hydraulically rough. The roughness "elements," whose height is measured by k, are made up either of the particles themselves (in the case of a plane bed) or of both the particles and the bedforms (such as ripples and dunes) molded by the flow. If there are large bedforms such as dunes, the depth of the channel contracts and then expands again as the flow moves over the crest of the dune. The flow, therefore, is not strictly uniform and the ideal logarithmic velocity profile of Eq. 12 is not developed. For this reason, it is difficult to determine whether or not the velocity profiles observed in rivers, as shown for example in Fig. 4-4, correspond to the experimentally determined law. It is gener-

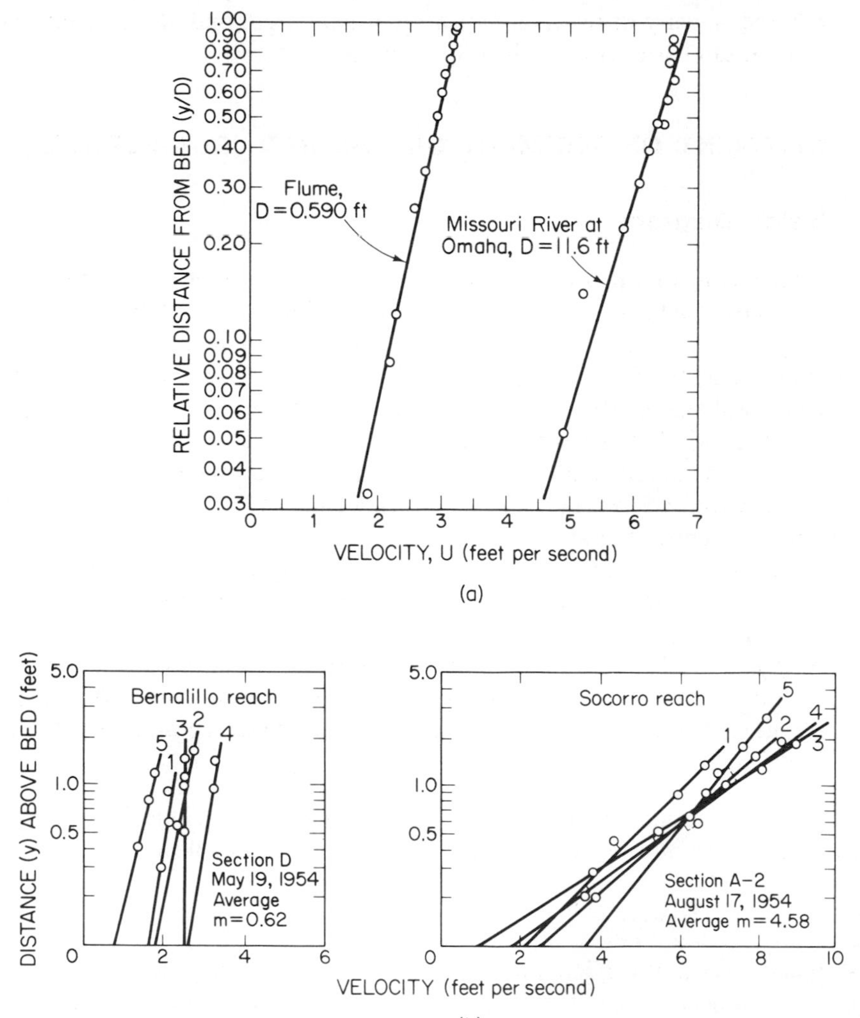

Fig. 4-4 Velocity profiles for flumes and rivers, plotted on semilogarithmic paper. (a) shows a comparison between the velocity profiles in a small flume and in a large river. (b) shows measured profiles for the Rio Grande River in New Mexico. The slope of the profile may be used to estimate Von Karman's κ (see text). In the Bernalillo reach (with dunes) κ is larger than 0.4. In the Socorro reach (with plane bed) κ is less than 0.4 and varies with suspended sediment concentration. After A.S.C.E. (1963) and Nordin and Dempster (1963).

ally assumed that they do, although it has been claimed that there are systematic differences from the theoretical law for very deep channels (see Toffaletti, 1965).

Because of the relationship between the roughness of the bed and the velocity profile, the presence of bed forms is a major factor in rivers in controlling the average depth and velocity of the river (for a given discharge) and also in controlling the amount of sand-size sediment in suspension.

4.3 MOVEMENT OF SEDIMENT ON THE BED (COMPETENCE)

Shields' Diagram

Sediment lying on the bed of a stream will not begin to move unless the stream achieves a certain intensity of flow. Alternatively, for a given flow there will be some maximum size of sediment that the flow can move. This size is said to define the *competence* of the flow. The problem of predicting the competence of flows, or of inferring the properties of a flow from the largest size of particle in a given sedimentary deposit, has attracted the attention of both geologists and engineers for many years. A reasonably satisfactory solution to the problem was achieved by the German engineer Shields (a student of Prandtl) in 1936 for the simplified case of a sediment of uniform grain size.

In this case, it may be expected that the beginning of movement of particles on the bed will be related to the size of the particle d, its submerged specific weight $g(\rho_S - \rho)$ or $(\gamma_S - \gamma)$, the shear stress acting on the bottom τ_0, and the viscosity μ, and density ρ of the fluid. It is a principle of dimensional analysis that all dynamic equations must be balanced with the same dimensions on one side of the equation as on the other. It follows that it is possible to write a general equation of the type

$$f[d, (\gamma_S - \gamma), \tau_0, \mu, \rho] = 0 \tag{13}$$

Making use of the information about dimensions given in Table 3-4, the terms may be rearranged into two dimensionless combinations as follows:

$$\frac{\tau_0}{(\gamma_S - \gamma)d} = f\frac{(dU^*)}{\nu} \tag{14}$$

This equation was first verified, and the nature of the function f determined experimentally by Shields, and all subsequent experimental work has verified Shields' result, which is shown in Fig. 4-5. Shields' diagram shows that movement begins over a range of conditions rather than at a well-defined flow criterion. In most cases of interest in nature, the fluid is water and the solid is quartz so that it is possible to draw a curve relating τ_0 directly to d (Fig. 4-6). It can be seen that for sand and gravel the critical shear stress is roughly proportional to the grain size.

As an example of the use of Shields' diagram, consider the case of sediment of diameter 7.2 mm, shown in Fig. 4-7. What shear stress τ_0 or shear velocity $U^* = \sqrt{\tau_0/\rho}$ is needed to move sediment of this size? From Fig. 4-6, the required shear stress is 70 dynes/sq cm or a shear velocity of 8.3 cm/sec. Alternatively, this result might be calculated from Fig. 4-5.

The two dimensionless parameters in Shield's diagram (Fig. 4-5) have a fundamental interpretation. The parameter on the left-hand side of Eq. 13, called Shields' β (beta), may be interpreted as proportional to a ratio of two forces: the

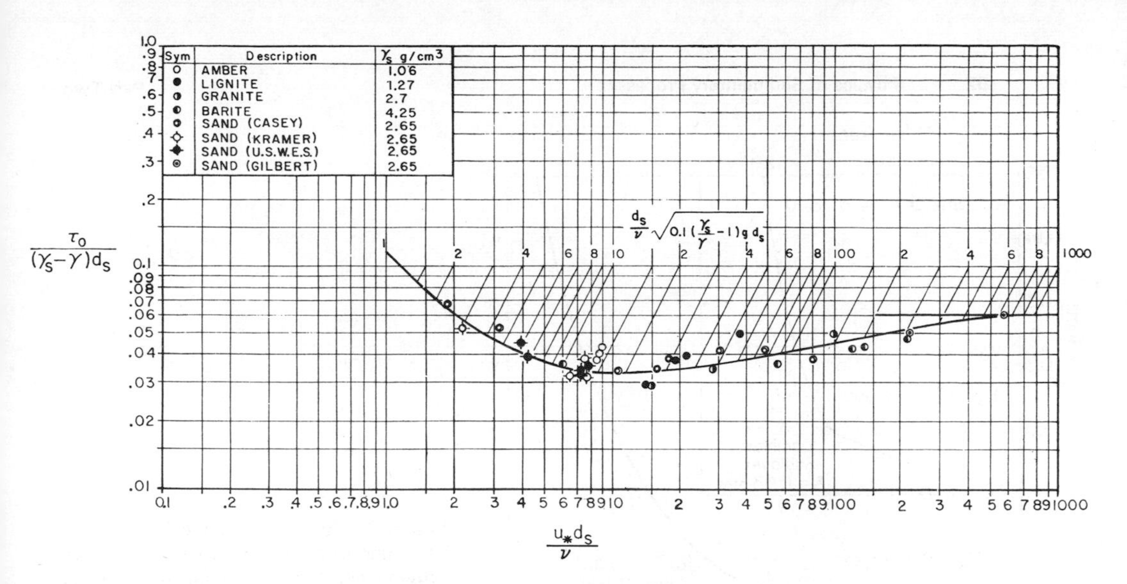

Fig. 4-5 Shields' diagram, as modified by Vanoni (1964, Caltech W. M. Keck Lab. Hydraulics Rept. KH-R-1). To calculate shear-stress required to move a given sediments, calculate $(d/\nu)\sqrt{0.1[(\gamma_s/\gamma)-1]gd}$, locate this value on the scale given in the center of the diagram, find the intersection of the Shields curve with the projection of this value along the diagonal lines on the diagram and read off the value of β on the ordinate.

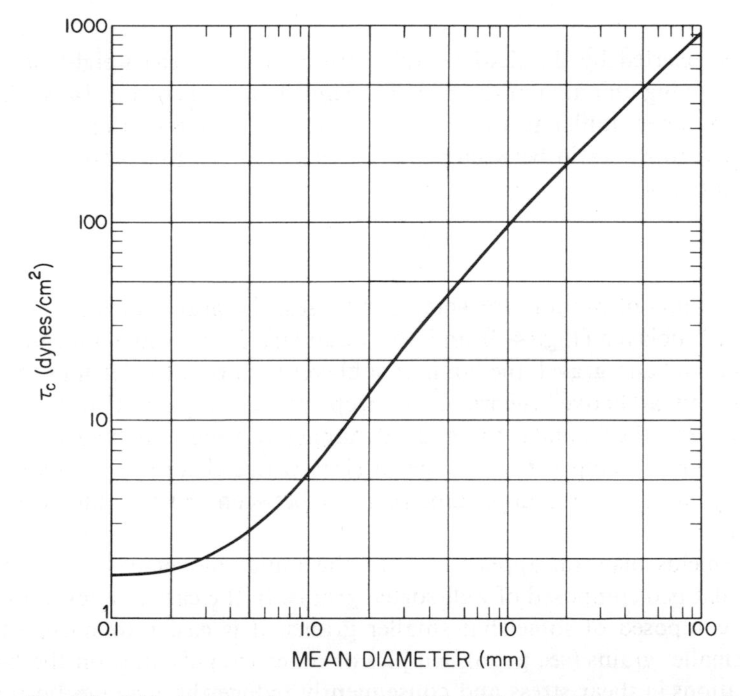

Fig. 4-6 Critical bottom shear-stress (dynes/cm^2) for initiation of movement of quartz sand on a plane bed (water temperature 16°C). Calculated from Shields' diagram.

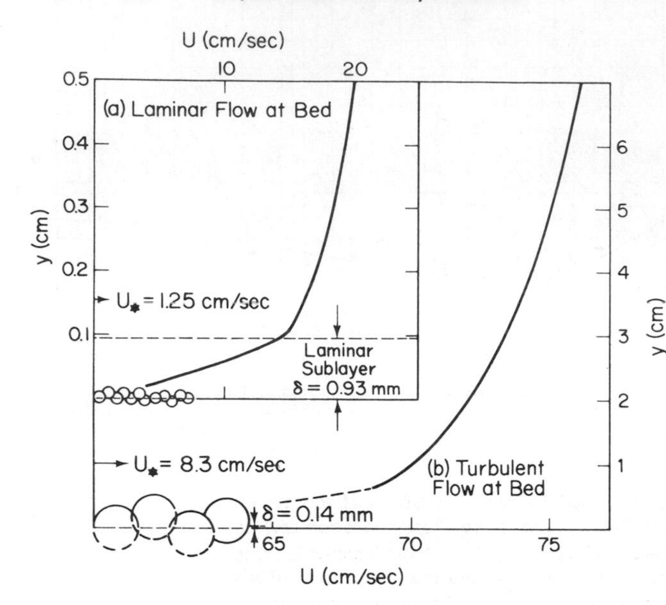

Fig. 4-7 Conditions at the bed for the beginning of sediment movement, for two extreme conditions: (a) laminar flow at the bed (smooth bed, grain diameter 0.16 mm), (b) turbulence at the bed (rough bed, grain diameter 7.2 mm). The value of Shields' β is 0.06 in both cases.

shear stress exerted by the fluid on the bottom, τ_0, and the weight of a layer of grains underlying unit area of the bed. The right-hand side of Eq. 14 is a boundary Reynolds number similar to that in Eq. 11. It can be shown that this number is proportional to the ratio between the grain size and the thickness of the laminar sublayer, δ (delta).

$$\frac{dU^*}{\nu} = \frac{12d}{\delta} \tag{15}$$

At the beginning of motion, for small sand sizes, the grain is entirely enclosed in the laminar sublayer (Fig. 4-7) and the boundary is hydrodynamically smooth. For coarse sand and gravel, the laminar sublayer is so thin that it must be considered to be only a "fictive" theoretical concept, the grain projects out into the turbulent flow and the boundary is hydrodynamically rough. It may be noted that true ripples are developed only for grain sizes where there is a laminar sublayer. Presumably there is some important relation between the two but it is not well understood.

The Shields diagram applies only to the initial movement of grains on an originally flat bed composed of well-sorted grains. In the case of a few larger grains on a bed composed of somewhat smaller grains, it is easier to move the larger than the smaller grains (see p. 96). Ripples or other irregularities on the bed cause local variations in shear stress and consequently reduce the average bottom shear stress needed to start movement of the grains. Variation in shear stress may also be caused by large turbulent eddies in the flow. Such eddies cause local shear stresses up to three or more times the average and for this reason caution is needed in extrapolating the Shields criterion of competency from flows in small laboratory channels (where the maximum size of the eddies is limited by the size of the channel) to large-scale natural flows.

It has been found experimentally that, for air, a value of β of about 0.01 is a better criterion for movement of grains than the value 0.06 indicated by Shields' diagram (see Raudkivi, 1967). The reason for this discrepancy is not yet well understood.

Hjulstrom's Diagram

In the past, most geology texts have expressed competency in terms of a critical velocity rather than a critical shear stress. The diagram relating critical velocity to grain size is often called the "Hjulstrom diagram" after its originator (Fig. 4-8).

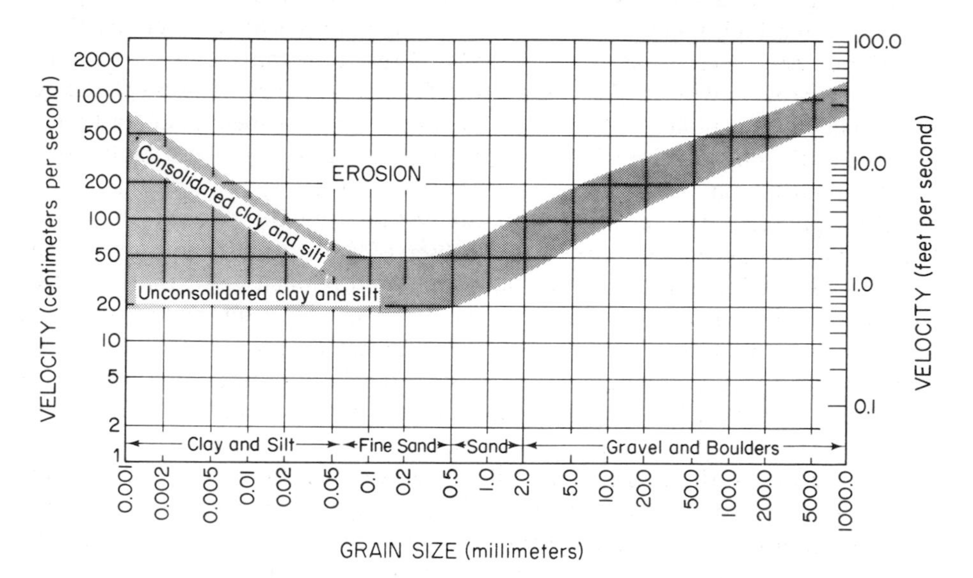

Fig. 4-8 Hjulstrom's diagram, showing critical velocity for movement of quartz grains on a plane bed at a water depth of one meter, as modified by Sundborg (1956). The shaded area indicates the scatter of experimental data. There are very few reliable data in the clay and silt region.

Strictly, the Hjulstrom diagram gives the critical velocity only for flows of 1 m depth. The Shields criterion is better established both theoretically and experimentally.

There is one area of both the Shields and Hjulstrom diagrams for which there is still no theoretical agreement or good experimental data; that is the region of grain sizes smaller than sand sizes. Hjulstrom's original diagram indicated that the finer sizes require larger velocities for movement than sand. The version revised by Sundborg (1956), shown in Fig. 4-8, indicates that a higher velocity is needed to move fine sediment only if the fine sediment is cohesive. Cohesive forces are well developed in most muds, particularly after some compaction, so that mud is generally more difficult to erode than sand. This is true for muds composed of both clay minerals and fine grained carbonate minerals and it is well illustrated by studies of the effects of hurricanes (see Chap. 12).

Mechanism of Grain Movement

The discussion given above has centered on experimentally derived criteria for the beginning of grain motion but it gives little insight into the physical mechanisms involved. Two main mechanisms have been invoked: the fluid drag on an exposed grain, which tends to roll it over the surrounding grains, and the hydrodynamic lift force, which tends to lift the grain up vertically from the bottom and project it into the fast-moving part of the flow. Studies of the movement of grains by wind, using high-speed motion photography, have shown that grains do, in fact, rise up almost vertically from the bed and that they move in a succession of leaps by a process described as *saltation*. Movement of grains by saltation has been studied by Bagnold (1941), who showed that the movement is at least partly triggered by the impact of grains striking the bottom at the end of a leap, but there is little doubt that hydrodynamic lift is the main force responsible for the initial rise of the particles. The term saltation was first used by Gilbert (see Chap. 5) to describe the similar leaping phenomenon observed when grains are moved along the bed by water but true saltation is much better developed in air than in water. The reason is that the lift force acts only while the particle is on the bottom. Once the particle has been accelerated upward, its further motion is resisted by fluid drag and it has been calculated that because of the density difference between air and water a particle of a given diameter, moved by a given bottom shear stress, will rise 800 times higher in air than in water.

To understand the origin of the lift force, we must first introduce a fundamental hydrodynamic relationship (which in turn may be deduced from the Navier-Stokes equations), the *Bernoulli equation*. The flow over a particle lying on a solid boundary may be illustrated by the simplified two-dimensional case of a cylinder lying on a plane boundary, with its axis normal to the flow, as shown in Fig. 4-9. The flow pattern is shown by the *streamlines*, which are lines everywhere parallel to the instantaneous velocity vectors in the flow. For a steady flow, the streamlines are also *path lines*, i.e., the paths followed by individual fluid particles. Where streamlines are crowded together, as over the cylinder, the velocity must increase because fluid particles do not cross streamlines.

Bernoulli's equation states that along a streamline the sum of the energy components, per unit fluid mass, due to pressure (p/ρ), head (gy, where y is the height above some datum), and velocity ($U^2/2$) must be a constant.

$$\frac{p}{\rho} + gy + \frac{U^2}{2} = \text{constant} \tag{16}$$

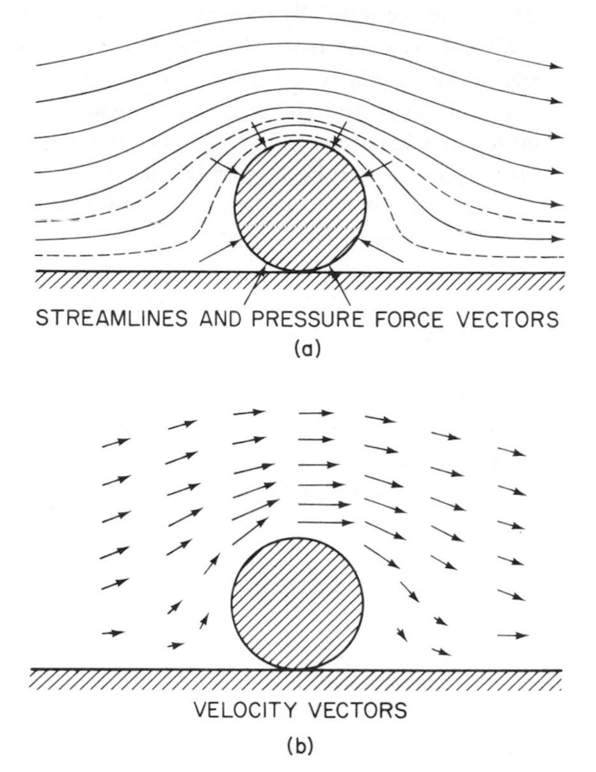

Fig. 4-9 Flow pattern computed for an ideal (inviscid) fluid moving past a cylinder lying on the bottom with its axis at right angles to the direction of flow. (a) shows streamlines and the relative magnitude of pressures acting on the surface of the cylinder. (b) shows the direction and relative magnitude of velocity vectors: note larger velocities where streamlines are closer together.

Strictly this relationship applies only to nonviscous flows, in which there is no energy loss due to viscous friction, but it holds approximately true for viscous flow also. It may therefore be concluded from Fig. 4-9 that where the streamlines are close together and the velocity is high, the pressure must be small compared to the pressure in the region away from the cylinder where the streamlines are not close together. It is assumed that the change in head may be neglected because of the small change in y. From symmetry it is clear that this pressure must act vertically upward on the cylinder; i.e., it is a lift force. It has been calculated, and demonstrated experimentally, that the lift force is probably almost as large as the drag force and is quite capable of lifting the particle up from the bed (see Raudkivi, 1967). As soon as the particle is raised from the bed, the pattern of streamlines about the particle becomes almost symmetrical (as in Fig. 4-2) and consequently there is no further lift force.

The drag force acting on a particle on the bed might be calculated from the equations for drag on a sphere developed in Sec. 3.3 if the appropriate bottom fluid velocity and the degree of grain exposure could be estimated. The drag force tends to roll the particle over the particles behind it. At the moment of initiation of movement, the drag force component acting to move the particle over its pivot must be equaled by the gravity force component tending to hold it down. In practice, an accurate theoretical analysis is not possible because it is not possible to determine theoretically the angle of pivot or the shielding effect of the surrounding grains. It is clear from the dynamic analysis, however, that the relation of the size of the grain to the neighboring grains has an important effect on both the angle of pivot and the degree of shielding. Both of these factors decrease for a large particle surrounded by smaller particles so it is probable in this case that the larger particle will be set in motion first.

Russell (1968) has suggested that the reason for the apparent scarcity of grains of granule size in river deposits is that these grains are more rapidly moved than the sand grains with which they would otherwise be associated and that they are consequently "flushed out" of the rivers and accumulated instead in the beach environment. As yet, there are few data on the size distribution of beaches to show that this is, in fact, the case.

Grains very much larger than the sand (pebbles and cobbles) are too large to be moved, even on a smooth sand bed, by currents competent to move the sand. Consequently such large grains accumulate in separate lag gravel deposits from which most of the sand has been removed rather than bypassing the sand.

4.4 RATE OF BED LOAD MOVEMENT (CAPACITY)

Once the shear stress has been raised above the critical value, sediment of a given grain size will move over the bed but the rate at which it will move is limited. The mode of sediment motion depends strongly on the grain size. Grains larger than sand move only by sliding and rolling (traction). Only very rapid flows are capable of moving such grains by saltation or suspension. Sand moves largely by saltation and intermittent suspension and silt and clay move in suspension supported by the fluid turbulence. The different modes of motion result in different mobilities of the different size fractions, which in turn result in a separation or hydraulic sorting of the sediment size (and shape) fractions. Such a sorting is especially well seen in windblown deposits because of the large density difference between the sand grains and air. The result of this large density difference is that it is almost impossible for normal winds to transport grains larger than coarse sand. Even the sand sizes are not easily taken into suspension but are transported mainly by saltation, a process which (as noted above) is much better developed in air than in water. Silt and clay sizes are easily taken into suspension by moderate winds. The silt tends to accumulate in well-sorted deposits such as loess that accumulate in areas where the wind strengths slacken.

Rivers have an almost unlimited capacity to transport clay and fine silt so the concentration of these grain sizes found in a river depends mainly on the rate of supply by erosion in the watershed rather than on the hydraulic conditions. Such material does not commonly occur in large quantities in the bed of the river and is described as *suspension* or *wash load*. In contrast, the materials in the bed of a river are commonly coarse silt to sand or coarser. Such materials are described as the *bed-material load*. Sediment moved at or near the bed (which includes most, though not all, of the bed material) is described as *bed load*. The rate of movement of bed load does depend on the hydraulic conditions. These in turn are partly controlled by the rate of supply of bed material from mass wasting. For example, a high rate of supply of coarse materials at a certain point in a river valley may lead to accumulation of material until erosion downstream has steepened the slope to the degree necessary to increase the shear stress and establish an equilibrium between the rate of supply of sediment and its rate of transport by the river.

The problem of determining an equation that relates the rate of movement of bed load to the hydraulics has not yet been solved. The best equations (e.g., the modified Einstein bed load function; see Raudkivi, 1967) give results that may be in error by over 50% of the true value. The problem is complicated by the great difficulty of obtaining accurate field measurements of bed load transport rates. Most sediment traps are unreliable because of disturbance of the bottom by bed forms. The best way to measure transport rates appears to be to measure the sediment load at a cross section of the river channel where all the sediment is locally in suspension (e.g., at a very narrow, rough section of the channel).

The sediment transport system is inherently a very complicated natural feedback system. As soon as transport begins, the flow begins to mold the bed into bed forms, which migrate and progressively change their form in response to changes in the flow. In turn, the rate of bed load movement depends on the nature of the bed forms. In rivers, the bed forms are larger than those that can be reproduced in the laboratory. The bed forms may become so large that there are serious departures in nature from a state of equilibrium and bed forms are present at low water that are at least partly inherited from high water conditions.

The number of variables that must be considered is very large. It might be thought that the list of variables given by Shields (Eq. 13) includes all the relevant ones and that the rate of bed load movement could therefore be given as a function of Shields' two dimensionless parameters, but experience has shown that this is not the case. In addition to the variables listed, the average velocity U might be added. Two of Shields' variables may be split into more than a single variable each: τ_0 splits into ρ, g, R, and S and γ splits into g and ρ. The settling velocity w might be substituted for the diameter d and some measure of grain sorting might be introduced. Many different investigators have recombined these variables into a wide variety of dimensionless combinations but none of the theoretical or empirical studies has yet proved to be entirely satisfactory.

4.5 TURBULENCE AND SUSPENDED LOAD

Nature of Turbulence

Turbulence in a flow is important both for suspension of sediment and for movement of bed load. Turbulence consists of random eddying motions superimposed on the mean flow and ranging from the smallest to the largest scales, limited only by the size of the channel. By definition, the orientation, size, and intensity of individual turbulent eddies cannot be predicted but this does not mean that it is impossible to describe turbulence at all. The appropriate type of description is obviously statistical and some progress has been made toward an adequate statistical theory of turbulence (Daily and Harleman, 1966).

Turbulence is generated at the boundaries of a flow or within the flow itself in regions of intense shear (e.g., at the borders of jets or wakes). Turbulence is generated even at boundaries that are hydrodynamically smooth. Studies of the laminar sublayer have shown that it is a mistake to think of the sublayer as having a steady, constant thickness. Instead, the sublayer is an unsteady zone that cyclically builds up and then breaks down and is swept away. Turbulence is generated at discrete "spots" on the boundary and spreads out downstream of the spot.

In the transition from flows with a laminar sublayer to flows in which turbulence extends completely to the boundary, a longitudinal pattern of flow in the boundary layer is revealed by injecting dye close to the boundary. The dye forms a ribbon-like pattern, each "ribbon" being of the order of a millimeter in width. The pattern is developed even over a perfectly smooth boundary and indicates a small-scale spiral circulation superimposed on the average flow. The cause of this "secondary" flow is apparently some basic instability in the fluid itself. It seems possible that this pattern is partly responsible for the formation of the longitudinal flow lineation and grain orientation frequently observed in fine to medium sands. The lineation is absent in coarse sands where there is no laminar sublayer.

In a turbulent flow, the instantaneous velocity component in any direction may be considered to be made up of two parts: the time-averaged component $\bar{U}$ and the instantaneous deviation U'.

$$U = \bar{U} + U' \tag{17}$$

If this substitution is made for all velocity terms in the Navier-Stokes equations and the equations are then rearranged in terms of the average velocity components, it is seen that several new terms have been introduced. These terms consist of functions of time-averaged cross products of the form $\overline{U'V'}$ (where V' is the velocity fluctuation in a direction normal to that of U'). They can be interpreted as additional shear stresses in the flow caused by turbulence and have been called the Reynolds stress terms. In laminar flow, shear stress is transmitted by the viscosity alone according to Newton's law:

$$\tau = \mu \frac{dU}{dy} \tag{1}$$

In turbulent flow, Reynolds shear stresses are also present, caused by the transfer of momentum as eddies move from one layer of the flow to another, so that Newton's law must be modified by the addition of a term η (eta) called the *coefficient of eddy viscosity*:

$$\tau = (\mu + \eta) \frac{dU}{dy} \tag{18}$$

The eddy viscosity is generally much larger than the dynamic viscosity. Unlike the dynamic viscosity, it is not a constant but varies according to the type of flow. The value of the eddy viscosity is one measure of the intensity of turbulence and it is possible to calculate its distribution, for example, in a river cross section. It is found that the largest values of the eddy viscosity are found in regions of high velocity gradient close to the boundary.

A more direct way to measure the intensity of turbulence is to measure the instantaneous velocity fluctuations. This can be done, for example, by means of a sensitive velocity probe, such as the hot wire anemometer (for air) or the hot film anemometer (for water). These instruments are capable of recording a continuous trace of the velocity component in a chosen direction at one point in the flow. The strength of the velocity component may be expressed as the root-mean-square, e.g., $\sqrt{\overline{U'^2}}$ (the equivalent of the standard deviation of the velocity component). The root-mean-square is generally expressed as a fraction of the mean velocity component and some typical results for a channel of rectangular cross section are shown in Fig. 4-10. For complete description of the turbulence, velocity fluctuation must be measured in three coordinate directions and not only root-mean-squares (standard deviations) but also covariances or correlation coefficients of the type

$$r = \overline{U'V'}/\sqrt{\overline{U'^2}}\sqrt{\overline{V'^2}} \tag{19}$$

must be calculated.

Experiments have shown that the ratio between the root-mean-square and the mean velocity component does not change much with Reynolds number, particularly in the direction normal to the average flow direction (see ASCE, 1963). As shown in Fig. 4-10, the root-mean-square of fluctuations normal to the average flow is generally 2 to 4% of the mean velocity. It was noted in Table 3-2 that tur-

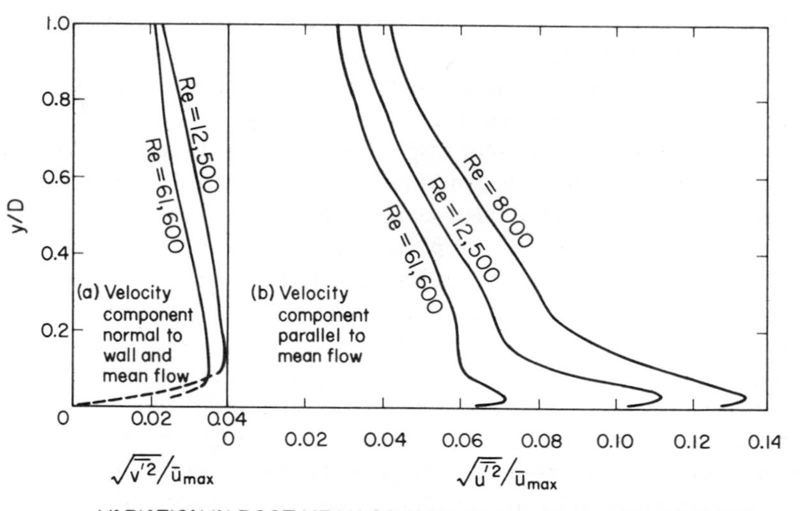

Fig. 4-10 Variation in Root-Mean-Square of the turbulent velocity fluctuation measured parallel to the flow (U') and normal to the flow and boundary (V'). Data for air flowing through a rectangular conduit, from Laufer (1951) in A.S.C.E. (1963).

bulent velocity fluctuations have been found experimentally to be normally distributed. It follows that 95% of the instantaneous velocity fluctuations normal to the mean flow must lie within 2 standard deviations of the mean; i.e., they have values less than about 8% of the magnitude of the mean flow velocity.

Suspension of Sediment

It is the velocity fluctuations normal to the mean flow direction that are responsible for holding particles in suspension and for diffusing particles across the channel. It may be predicted, therefore, that there will be a substantial quantity of sediment held in suspension only if the strongest vertical velocity fluctuations are greater than the settling velocity. Therefore, only those particles will be freely suspended that have settling velocities with magnitudes less than 8% of the mean velocity. Stated alternatively, a rough criterion for suspension, verified in several experimental studies, is that the mean velocity should be at least 12 times the settling velocity. An alternative criterion, preferred by some workers, is that the shear velocity U^* should be at least 4 times the settling velocity.

In turbulent eddies, however, not only the velocity is important but also the duration for which this velocity is maintained. An intense local velocity fluctuation for a relatively long time implies the existence of a large eddy. In fact, one way to measure the size of eddies is to calculate the correlation between two velocity fluctuation measurements made by the same probe at times separated by a small time lag. For the larger sediment sizes, only the larger eddies are capable of holding the sediment in suspension and consequently such sediment tends to be moved in suspension only intermittently. Because of the difficulty of maintaining the coarser sediment in suspension it tends to be concentrated near the bed, from which it is scoured by the impact of large eddies and to which it returns as the upward velocity of the eddy is reduced to less than the settling velocity.

The mechanism of intermittent suspension produces a region close to the bed where there is a relatively high concentration of coarse suspended sediment (as measured, for example, by a sediment sampling device). The same effect is produced by true saltation but there is an important distinction in principle between the two mechanisms. The rise of the grains in saltation is caused by hydraulic lift forces and the height to which the grains rise can be shown to be proportional to the magnitude of these forces (see Chap. 5). As the grains rise, they are caught by the mean flow and moved forward but eddying motion in the saltation layer is not sufficient to prevent the particle from dropping directly back to the bed. Consequently, the "saltation" layer has a fairly distinct upper limit defined by the maximum height to which grains can be projected by hydraulic lift forces. In contrast, both theoretical and experimental studies of the distribution of sand and silt in suspension have shown that the concentration decreases continuously and smoothly up from the bed.

The theory for the distribution of suspended sediment is derived from the older theories of turbulence, developed by Prandtl and Von Karman. In these theories the size of eddy is introduced by the concept of "mixing length" —the distance traveled by a fluid particle before it loses its identity and mixes with other fluid particles (a concept similar to that of the "mean free path" in the kinetic theory of gases). In the theory the sediment is considered to be diffused away from the bed at a rate $E_S\,(dC_S/dy)$, where E_S is a sediment diffusion coefficient proportional to the eddy viscosity and C_S is the concentration of sediment at a height y above the bed. The rate of downward diffusion, due to settling, is $C_S w$, where w is the settling velocity of a given size of sediment; at equilibrium the two diffusion rates must be equal:

$$C_S w = -E_S \frac{dC_S}{dy} \tag{20}$$

The equation may be integrated to give the concentration of sediment at any level C_y in terms of a known concentration at a reference level C_a (at $y = a$):

$$\frac{C_y}{C_a} = \left[\frac{D-y}{D-a} \cdot \frac{a}{y}\right]^z \tag{21}$$

where $z = w/\kappa U^*$. Equation 21 has been verified experimentally (for a plane bed) and also gives a reasonable approximation to suspended sediment distributions observed in natural streams (Nordin and Dempster, 1963). Figure 4-11 shows typical

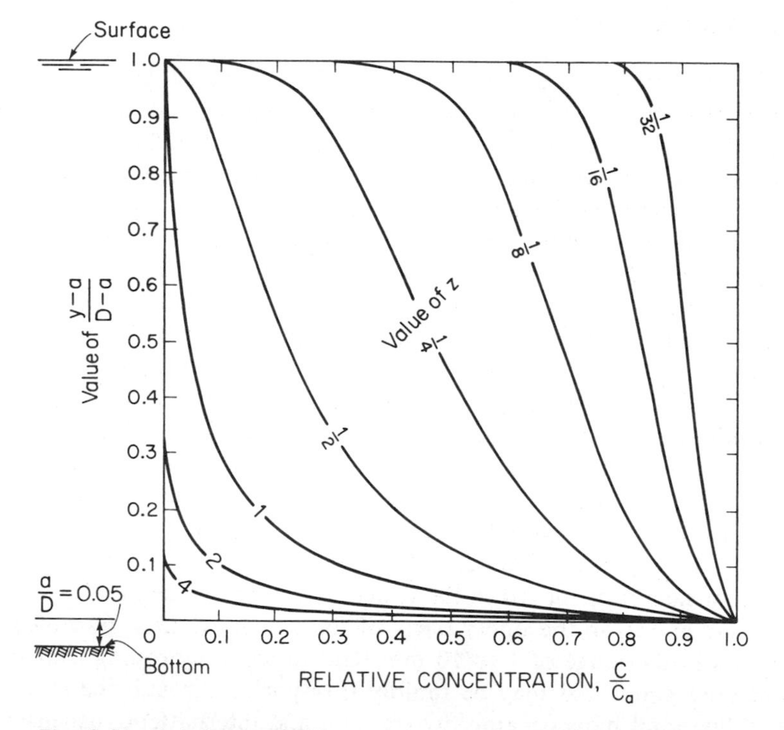

Fig. 4-11 Variation in relative concentration of suspended sediment with relative depth above the datum $y = a$, where $a = 0.05D$. Larger values of z correspond to larger values of sediment settling velocity, lower flow velocities or smaller bottom roughness. After A.S.C.E. (1963).

examples of sediment distributions calculated from Eq. 21. It can be seen that the sediment distribution becomes more uniform as z decreases. Decrease in z may result from decrease in the settling velocity so that fine sediment is more uniformly distributed than coarse sediment. Or it may result from an increase in shear velocity, which can be caused either by an increase in the mean velocity or by an increase in the roughness of the bed (refer back to Eq. 12). Suspended sediment is more uniformly distributed in high velocity flows or rough channels than in low velocity flows or smooth channels.

The value of the settling velocity commonly used in computing z is the value determined experimentally in still water. For sand-size particles, however, it has been shown experimentally that the settling velocity in turbulent water is reduced to as little as half its value in still water, even when there is no upward diffusion by turbulence. The reduction in mean settling velocity results from the inertia of the grains as they respond to the unsteady turbulent motions of the water around them (Murray, 1970).

4.6 SEDIMENT SORTING AND HYDRAULIC EQUIVALENCE

Hydraulic Sorting

Hydraulic sorting of sediment consists of the grouping together by fluid flow of particles that respond to the flow in a similar manner and, at the same time, the separation of such particles from those that respond differently to the flow. Some ways in which hydraulic sorting takes place have been discussed in Secs. 3.3, 3.4 and 4.4 and it has been pointed out that particles respond to flows of different intensities by different modes of movement. The same particle might be moved either by sliding, rolling, saltation, intermittent suspension, or continuous suspension, depending on the intensity of the flow. In addition, there are other modes of movement not yet discussed, for example, movement by particle-to-particle interaction in a highly concentrated dispersion or movement in a laminar mud flow or movement by ice or mass-wasting mechanisms. Very few of these modes of sediment movement result in the mass transportation of particles, without regard to their size, shape, and density. In most cases, only particles with a limited range of characteristics can be moved in any particular mode under the range of flow conditions normally exhibited by a given flow system.

For example, in a river system of moderate size most grains coarser than about 10 mm cannot be moved in the lower part of the system and tend to accumulate as lag gravels in the upper parts of the alluvial valleys. As noted in Sec. 4.3, particles in the range of 1 to 10 mm tend to move by rolling over the more abundant sand grains and may be rapidly transported through the river system. Coarse to fine sand moves mainly by traction and intermittent suspension, with local and temporary storage in large structures such as dunes and point bars. Silt and clay move mainly as wash load in continuous suspension and are either transported rapidly to the mouth of the river or stored on floodplains where they have been deposited by overbank flooding. Deposits from particular fluvial environments tend to be composed dominantly of sediment moved by one particular transport mechanism, with small amounts contributed by other mechanisms. Channel sands, for example, may be composed mainly of a fraction moved by intermittent suspension, with the addition of a small percentage of coarse sediment moved mainly by traction and a small percentage of fine sediments moved mainly by continuous suspension. A size analysis may reveal distinct breaks in the size distribution, showing that each fraction is distinguished by a different type of size distribution. It should not be surprising that this is the case because the different transport mechanisms characteristic of each fraction differ in the way they select grains for movement or deposition. Some authors (e.g. Visher, 1969: see reference in Chap. 3) have suggested that, for this reason and because the common range of flow intensities found in most environments is limited, most sediments are composed of a small number of grain populations that may be recognized by careful size and shape analysis.

The characteristics of the population of grains being moved by a particular mechanism have been studied little by geologists. Sampling these populations in the field and laboratory appears to be one way to test the hypotheses outlined above. At a particular depositional site, however, not all the grains being moved by a particular mechanism will necessarily be deposited. Those that are deposited will be only a part of the population of grains in motion and it will be a part selected from the total population because its characteristics are such that it cannot be transported further.

For example, suppose there is a current that is carrying suspended sediment and that is gradually losing competence and capacity because of a downstream decrease in velocity or decay in the intensity of turbulence. Ideally, grains would be deposited out of suspension in the order of decreasing settling velocity and the deposit at any one place might be expected to be composed of grains, all of which had the same settling velocity. In nature, the selection process cannot be perfect —there is always a range in the intensity and scale of turbulence with a consequent mixing in the deposit of grains with differing settling velocities. Decrease in flow intensity (as measured, for example, by a decrease in U^* and consequent increase in z in Eq. 21) decreases the capacity of the flow to suspend *all* the coarser grain sizes so that a perfect selection of grains by settling velocity is not even theoretically possible by this mechanism. Finally, grains deposited from suspension would probably undergo a period of movement on the bottom by traction. During movement by traction they would tend to be sorted into groups determined by ease of sliding or rolling, a property not necessarily related to settling velocity.

Hydraulic Equivalence

Nevertheless, the hydraulic selection process is sufficiently discriminating to make meaningful the concept of the *hydraulic equivalence* of two sediment samples; i.e., in a population of grains deposited from suspension under essentially uniform conditions, all the grains should have essentially the same range of settling velocities. The hypothesis that a given sample was deposited from suspension may therefore be tested by separating it into two or more subsamples that differ, for example, in mineral composition, with consequent differences in size, shape, and density. The settling velocity distribution of each subsample is then determined and the two distributions are compared. If the settling velocity distributions differ, this is evidence for one of the following alternatives: (a) The original conditions of settling have not been correctly reproduced; for example, the grains may have settled in water that was colder, muddier, or more turbulent than that used in the laboratory. All of these modifying factors might tend to affect one mineral fraction more than another but the difference would probably be small. A more important possibility is that the grains were originally settled in air, in which case the settling velocities of the two fractions in water should not be the same [see Fig. 4-12(a)]. (b) At least some of the grains were not deposited by settling. Grains moved by traction or saltation would not necessarily be selected according to their settling velocities. (c) The grain population has been modified by postdepositional processes, for example, dissolution of some grains or addition of grains washed in by percolating groundwater.

Application of these principles to actual sediments (e.g., Hand, 1967; White and Williams, 1967) has revealed that most sands are not simple suspension deposits. For example, in Fig. 4-12, diagram (a) shows the theoretical settling velocity distributions in water for subsamples of quartz, hornblende, garnet, and ilmenite that have an identical settling velocity distribution in air. Diagram (b) shows observed settling velocities of these minerals in a sample from a beach. The magnitude of the differences suggests strongly that processes other than suspension are operating on the beach. Diagram (c) shows observed settling velocities in water for samples from an eolian dune sand. For the larger grains, there is an approach to the theoretical [diagram(a)], except for ilmenite. It was found by Hand (1967) that despite the limited approximation to ideal hydraulic equivalence the difference in settling velocity between quartz and either garnet or hornblende was a good criterion for distinguishing between beach and dune sands.

Relation Between Size and Sorting

An important consequence of the change in common modes of transport and sorting mechanisms as the grain size changes from coarse to fine is the common correlation observed between size and sorting in clastic sediments. The correlation is best documented for the fine to coarse sand sizes, with sorting improving (decrease of standard deviation) toward the finer sand sizes. Sands from many different aqueous environments show this correlation and the fine sands are generally better sorted than very fine sands or silts. For windblown sands the best sorting appears to be found in the very fine sand grade.

These relationships were first explained by Inman (1949), who pointed out that because of the relationships exhibited by Shields' or Hjulstrom's diagrams fine sand is the most easily moved sediment. Sediments coarser than this are moved

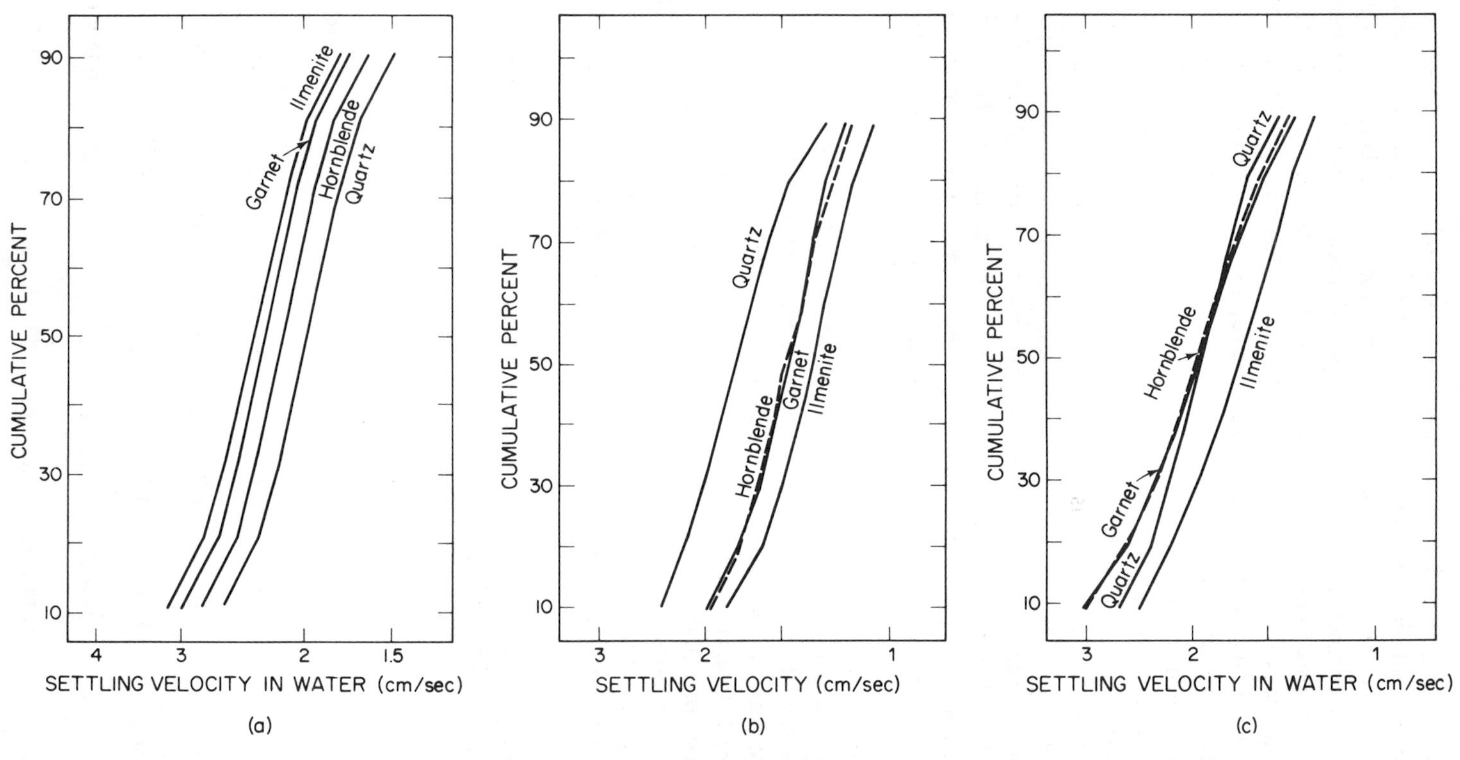

Fig. 4-12 Theoretical and measured settling velocity distributions, illustrating hydraulic equivalence. (a) shows theoretical settling velocity distributions in water of mineral grains having identical settling velocity distributions in air. (b) shows observed settling velocity distributions in water for beach samples. (c) shows observed settling velocity distributions in water for eolian dune samples. From Hand (1967).

by currents that also carry and deposit some finer materials. As competence decreases, the coarser fraction is deposited, resulting in better sorting. In many cases, the coarser sediments are skewed toward the coarse sizes, partly because of the great variability in competence exhibited because of large temporal variations in discharge in the upper parts of river systems. Thus as the coarser fraction is dropped, not only does the sorting improve but the distribution tends to become symmetrical. At the point where fine sand is deposited, a slight decrease in velocity or shear leads to deposition and a slight increase in velocity or shear leads to erosion and further transport. There is therefore a delicate balance between erosion and deposition and good sorting results. Further decrease in competence and capacity, however, leads to deposition of silt previously held in continuous suspension. This material, once deposited, is at least as difficult to erode as the very fine sand deposited with it. Either sand and silt grades are both taken back into suspension by a strong fluctuation in the current or they resist erosion and are buried in the deposit. In either case further sorting does not take place and the sorting of the sediment tends to become poorer and the sand becomes skewed to the fine sizes as more silt is added from suspension to the fine to very fine sands originally moved as part of the bed load.

Table 4-1 shows a few basic hydraulic parameters for parts of the Yellowstone-Missouri-Mississippi River system. The depth given is the depth for the mean annual discharge and the bottom shear stress has been calculated using this depth. It can be seen that upstream the mean flow is not competent to move the median size of bed material, whereas the reverse is true downstream. It is a common observation that, in small streams, bed load is moved extensively only during floods. In the lower parts of major rivers, however, bed load is moved even at low stages. Probably even at low stages a large river is competent to move all, or almost all, the material on its bed. Therefore, the main controls of sorting in bed load

TABLE 4-1 BASIC HYDRAULIC DATA FOR PARTS OF THE YELLOWSTONE-MISSOURI-MISSISSIPPI RIVER SYSTEM (after Leopold et al., 1964)

River	Slope	Mean depth (m)	Calculated bed shear stress (dynes/sq cm)	Critical diameter (Shields) (mm)	Median size, bed material (mm)
West Fork Rock Creek near Red Lodge, Mont.	0.035	0.41	1330	103	270
Rock Creek near Red Lodge, Mont.	0.021	0.79	1540	105	210
Yellowstone River at Billings, Mont.	0.0015	1.7	242	24	90
Missouri River at Omaha, Neb.	0.000012	3 (approx.)	3.9	0.7	0.29–0.15
Mississippi River at Mayersville, Miss.	0.000076	12 (approx.)	91	9.5	0.4

appear to be the competency in small streams but the capacity in large rivers. Capacity plays an important role in large rivers because the bed load is sufficiently fine grained that a substantial fraction may be held in intermittent suspension. Change in capacity leads to transfer of sediment from bed to suspended load or vice versa.

4.7 GRAIN ORIENTATION (See Johansson, 1965; Parkash and Middleton, 1970)

During settling of particles above the Stokes' range of Reynolds numbers ($R_e > 1$), the formation of a turbulent wake results in the particle adopting an orientation with its maximum projection area normal to the direction of settling. Small particles in the Stokes' range do not form a wake and do not adopt any particular orientation during settling. The fissility of shales, therefore, is a result of reorientation of particles during compaction rather than being a primary depositional phenomenon. Compaction of muds does not inevitably produce fissility because of the influence of mineralogical factors (see Chap. 11).

In viscous shear flows, particles of nonspherical shape tend to rotate continuously: The rate of rotation varies according to the orientation with respect to the mean flow so that particles have a relatively high "residence time" in orientations close to the flow direction. There are also important interactions between particles, which tend to form clusters even at extremely low concentrations. Both of these factors result in particles in a viscous shear flow (for example, a laminar sublayer) adopting an orientation parallel to the flow direction. In turbulent flow, presumably the particles in complete suspension have a random, or near random, orientation, though there are no observations to test this assumption.

The grain orientation observed in most sediments, however, must be related to mechanisms operating during the last stages of deposition rather than to processes operating during complete suspension. Various different mechanisms operate to produce different types of orientation as the particles come to rest on the bottom. Particles that can roll on the bottom may tend to assume an orientation with the long axis normal to the mean flow. Alternatively, such particles may become lodged about some point, which tends to act as a pivot so that the particle swings round until it is parallel to the flow. If particles have a streamlined shape, with a roughly elongate form and one end larger than the other, the large end will tend to point upcurrent. It has been demonstrated experimentally that such an orientation is produced in sand grains moved by traction and despite difficulty of measurement the orientation of streamlined grains has been successfully determined in ancient sandstones and shown to yield paleocurrent directions confirmed by other criteria.

Elongation of sand grains parallel to the current in the plane of the bed may also be produced by other mechanisms, such as the flow structure in the laminar sublayer discussed on p. 98, or interaction between particles in the concentrated zone close to the bottom. Whatever the mechanism, it is now clear from many experimental and observational studies that orientation parallel to the current is the most common orientation for sand grains. In some cases, orientation normal to the current has been reported, as well as bimodal orientations both parallel with and normal to the current. The grain orientation formed by plane bed, upper regime flows (see Sec. 5.2) appears to be characterized by two modes at an angle of about 15 degrees on either side of the flow direction. Orientation parallel to the flow is also produced in deposits from highly concentrated dispersions ("grain flows" and grain avalanches, see Sec. 5-10). It is, in fact, only a small proportion of sands that do *not* show a fairly well-developed orientation of grains in the bedding plane. Even massive, structureless sandstones frequently show excellent grain orientation. In some cases it has been shown that lack of grain orientation has been caused by burrowing.

The tendency for disk-shaped particles to assume a "shingled" orientation with the plane of maximum projection of the particles dipping in one direction (commonly upstream) is called *imbrication*. Imbrication is well developed, by pebbles of appropriate shape, in tractional deposits that are sufficiently well sorted to permit the particles to come into contact with each other, i.e., in deposits where the larger, platy pebbles are not separated from each other by a "matrix" of smaller particles. Pebbles on a sand bed may show very little imbrication. Imbrication with the dip upstream results because this orientation is the most stable position for closely packed particles. The flow impinging on an imbricated particle exerts a drag force that presses the particle to the bed and acts against the hydraulic lift. River gravels generally show a marked imbrication with upstream dips that are generally steep (15 to 30 degrees) but may be as low as 8 degrees.

Maps of imbrication direction on gravel bars show that there are many local deviations from the channel direction, related to local flow circulation. Orientation of pebble long axes both parallel with and normal to the current direction have been reported. On beaches, most studies report pebble long axes parallel to the shoreline. i.e., normal to the direction of swash and backwash, but the sand-size grains are aligned parallel to the backwash. Imbrication is present but varies in direction from up- to down-beach, depending on the position on the beach. The angle of imbrication is generally lower than in river deposits. It is frequently assumed by geologists that the presence of imbrication in a gravel indicates either movement by traction or wave action. This may be an unwarranted assumption, however, because it is not known that imbrication cannot be produced by other mechanisms (e.g., mass flow).

Imbrication of sands has also been studied, commonly by cutting a thin section in the direction normal to the bedding and parallel to the previously established grain orientation in the bedding plane. The grains generally show a strong preferred orientation either parallel to the bedding or with an upstream imbrication.

Grain orientation in sands may be used to help determine their environment of deposition. In a river channel deposit, for example, grain orientation is generally parallel to the current and therefore to the elongation of the sand body. In a barrier island the orientation is more variable: offshore it tends to be parallel to the shore because of rolling of grains to-and-fro by wave action or orientation by longshore currents. On the beach face, grains are oriented parallel to the backwash. Consequently they are oriented normal to the beach or at a slight angle to the normal and they have an imbrication away from the sea. In the part of the island affected by wind, orientation is very variable. In a stratigraphic section, this association of facies will tend to be found in a vertical succession and the grain orientation patterns, together with information on thc sedimentary structures and sand geometry, may be sufficiently distinctive to permit discrimination of different environments.

REFERENCES

ALLEN, J. R. L., 1970, *Physical Processes of Sedimentation.* London; George Allen & Unwin, Ltd., 248 pp. (A review of fluid mechanics and its application to the origin of sedimentary rocks.)

AMER. SOC. CIVIL ENG., Task Committee on Preparation of Sedimentation Manual, 1962, "Introduction and Properties of Sediment," *Amer. Soc. Civil Eng. Proc.*, **88**, no. HY4, pp. 77–107.

———, 1963, "Suspension of Sediment," *Amer. Soc. Civil Eng. Proc.*, **89**, no. HY5, pp. 45–76. (The preliminary versions of several sections of a "sedimentation manual" being prepared by ASCE have been published in the proceedings. These are two particularly relevant to this chapter. All are of interest to sedimentologists.)

BAGNOLD, R. A., 1941, *The Physics of Blown Sand and Desert Dunes.* London: Methuen and Co. Ltd., 265 pp. (Gives a full discussion of saltation in air.)

DAILY, J. W., and D.R.F. HARLEMAN, 1966, *Fluid Dynamics.* Reading, Mass: Addison-Wesley Pub. Co., Inc., 454 pp. (One of several good introductory texts.)

HAND, B. M., 1967, "Differentiation of Beach and Dune Sands, Using Settling Velocities of Light and Heavy Minerals," *Jour. Sed. Pet.* **37**, pp. 514–520. (Application of hydraulic equivalence concept.)

HENDERSON, F. M., 1966, *Open Channel Flow.* New York: The Macmillan Co., 522 pp. (Text on river hydraulics, with a good chapter on sediment transport.)

INMAN, D. L., 1949, "Sorting of Sediments in the Light of Fluid Mechanics," *Jour. Sed. Pet.* **19**, pp. 51–70. (Classic explanation of relation between size and sorting.)

———, 1963, "Sediments: Physical Properties and Mechanics of Sedimentation," *in* F. P., SHEPARD, *Submarine Geology*, 2nd ed., New York: Harper & Row, Pub., pp. 101–151. (Concise introduction to sediment mechanics.)

JOHANSSON, C. E., 1965, "Structural Studies of Sedimentary Deposits," *Geol. Föreningens Stockholm Förhandlingar*, **87**, pp. 3–61. (Thorough review of pebble orientation in gravels.)

LELIAVSKY, S., 1959, *An Introduction to Fluvial Hydraulics.* New York: Dover Pub. Inc., 257 pp. (Discusses classic work on sediment transport by Shields and others.)

LEOPOLD, L. B., G. M. WOLMAN, and J. P. MILLER, 1964, *Fluvial Processes in Geomorphology.* San Francisco: W. H. Freeman and Co., 522 pp. (Chapter 6 is a good summary of sediment transport in rivers, written for geologists.)

MURRAY, S. P., 1970, "Settling Velocities and Vertical Diffusion of Particles in Turbulent Water," *Jour. Geophys. Res.*, **75**, pp. 1647–1654. (Describes experiments showing a 30% decrease in settling velocity of grains, 2 mm in diameter, in turbulent water.)

NORDIN, C. F. and G. R. DEMPSTER, JR., 1963, "Vertical Distribution of Velocity and Suspended Load, Middle Rio Grande, New Mexico," *U.S. Geol. Survey Prof. Paper 462-B*, 20 pp. (One of a series of studies of sediment in the Rio Grande. See also Water Supply Paper 1498-F.)

PARKASH, B. and G. V. MIDDLETON, 1970, "Downcurrent Textural Changes in Ordovician Turbidite Greywackes," *Sedimentology*, **14**, pp. 259–293. (Grain size and orientation studied in eight beds traced for 2 mi downcurrent; contains a review of the literature on grain orientation and a bibliography.)

RAUDKIVI, A. J., 1967, *Loose Boundary Hydraulics*. New York: Pergamon Press, 331 pp. (A comprehensive review of literature on sediment transport, written for civil engineers.)

RUSSELL, R. J., 1968, "Where Most Grains of Very Coarse Sand and Fine Gravel Are Deposited," *Sedimentology*, **11**, pp. 31–38. (On beaches, according to Russell. Unfortunately most of the size analytical data come from beaches composed of volcanic fragments, which is not typical source material for sand. Extensive data from North Carolina do provide more convincing support. His theory of bypassing needs further investigation.)

SHAPIRO, ASCHER H., 1961, *Shape and Flow: the Fluid Dynamics of Drag*. New York: Doubleday & Co. Inc., Anchor Books, 186 pp. (A small masterpiece of simple yet sophisticated presentation of some important topics in fluid mechanics, written at the high-school level. There is also a series of motion picture films on the same topic, distributed by Encyclopedia Britannica.)

STERNBERG, R. W., 1968, "Friction Factors in Tidal Channels with Differing Bed Roughness," *Marine Geol.*, **6**, pp. 243–260. (One of the few studies of hydraulic roughness in natural channels other than rivers. The same author has also studied conditions for sediment movement and ripple migration in large tidal channels and confirmed the applicability of Shields' criterion to this environment and scale: see *Marine Geol.*, **5**, pp. 195–205, and **10**, pp. 113–120.)

SUNDBORG, ÅKE, 1956, "The River Klarälven, a Study in Fluvial Processes," *Geografiska Annaler*, **38**, pp. 125–316. (A river study with chapters giving a useful summary of hydraulics and sediment transport theory.)

SUTHERLAND, A. J., 1967, "Proposed Mechanism for Sediment Entrainment by Turbulent Flows," *Jour. Geophys. Res.*, **72**, pp. 6183–6194. (Describes sediment movement in experiments with small vortices striking a sand bed and applies these observations to initial movement of sand on the bed in a turbulent flow.)

TOFFALETTI, F. B., 1965, "Deep River Velocity and Sediment Profiles and the Suspended Sand Load," *Proc. Federal Inter-Agency Sedimentation Conf. (1963)*, USDA Misc. Pub. 970, pp. 207–228. (Presents data for the Mississippi and Atchafalaya Rivers: The velocity distribution is logarithmic but does not correspond with Eq. 12 in this chapter. Use of Eq. 21 leads to overestimation of suspended load.)

WHITE, J. R., and E. G. WILLIAMS, 1967, "The Nature of the Fluvial Process as Defined by Settling Velocities of Heavy and Light Minerals," *Jour. Sed. Pet.*, **37**, pp. 530–539. (Application of hydraulic equivalence concept to formation of cross-laminations.)

CHAPTER FIVE

SEDIMENTARY STRUCTURES

5.1 STRATIFICATION

Classification and Nomenclature

The terms *stratum* and *bed* are approximately synonymous and refer to a layer of sedimentary rock that is distinguishable from the layers above and below by virtue of some discontinuity in rock type, internal structure, or texture. Stratum is a general term referring to all thicknesses and types of layers. Beds are commonly defined as layers greater than 1 cm in thickness and layers less than 1 cm in thickness are called *laminae*. Beds are frequently but not neccessarily separated from each other by bedding plane joints. In the field, the average thickness of beds is frequently described by terms such as "thick bedded" and there have been several attempts to define a scale of such terms, none of which has achieved universal acceptance. One scale that appears to have some merit was proposed by Ingram (1954) and is shown in Table 5-1. Though a verbal scale may be convenient (just as it is for grain size), it is preferable for geologists to specify bedding thickness in length units.

The upper and lower boundaries of a bed may be sharp or gradational.

TABLE 5-1 SCALE OF STRATIFICATION THICKNESS (after Ingram, 1954)

Very thickly bedded	Thicker than 1 m
Thickly bedded	30–100 cm
Medium bedded	10–30 cm
Thinly bedded	3–10 cm
Very thinly bedded	1–3 cm
Thickly laminated	0.3–1 cm
Thinly laminated	Thinner than 0.3 cm

Sharp boundaries can be formed by a sudden change in depositional conditions, by erosion, or by diagenetic accentuation of an originally gradational boundary. The lower boundary of a bed is also called the *sole* and it may display sedimentary structures called sole marks, casts, or molds. Strictly, a *mark* is an original structure, formed at the sediment surface (also called the "depositional interface"). Burial of a mark forms a *mold* on the sole of the bed above and such molds are frequently seen on the soles of sandstone beds above shale beds, where the shale has been removed by erosion. Some authors have used the term *cast* as synonymous with mold, though strictly a cast is formed by infilling of a mold. For example, the imprint of a fossil is a mold. If the original fossil material is dissolved away and the imprint is filled with introduced sediment or cement, the resulting structure is a cast. The term *counterpart* has been used by some authors instead of mold (or cast). The upper boundary of a bed may also display sedimentary structures, some of which may be original bed forms, such as ripples, only slightly modified before burial.

Many beds also display internal structures, such as strata (beds or laminae) inclined at an angle to the bed boundaries (cross-beds or cross-laminae). A single group of cross-strata, bounded by bedding planes, is called a *set*; a group of similar sets not separated from each other by any major discontinuity is called a *coset* (McKee and Weir, 1953). A set with an upper surface that preserves the shape of an original bed form is called a *form set*. Cross-bedding may be classified by means of several criteria, which include (a) the scale of sets, (b) the shape and attitude of cross-strata, and (c) the shape and nature of the lower and upper bounding surfaces of sets. A distinction is frequently made among *tabular* sets, bounded by essentially plane, parallel surfaces; *wedge-shaped* sets, bounded by plane, nonparallel surfaces; and *trough* sets, whose lower bounding surface is trough- or scoop-shaped (Fig. 5-1).

A very large number of sedimentary structures have been described and named, including many that are not primary (syndepositional) but formed after deposition by diagenetic processes. A classification of primary sedimentary structures that lists some of the more important types is shown in Table 5-2. Many of these structures cannot be adequately described or illustrated in this text and the reader is referred for further information to the monographic works listed in the bibliography.

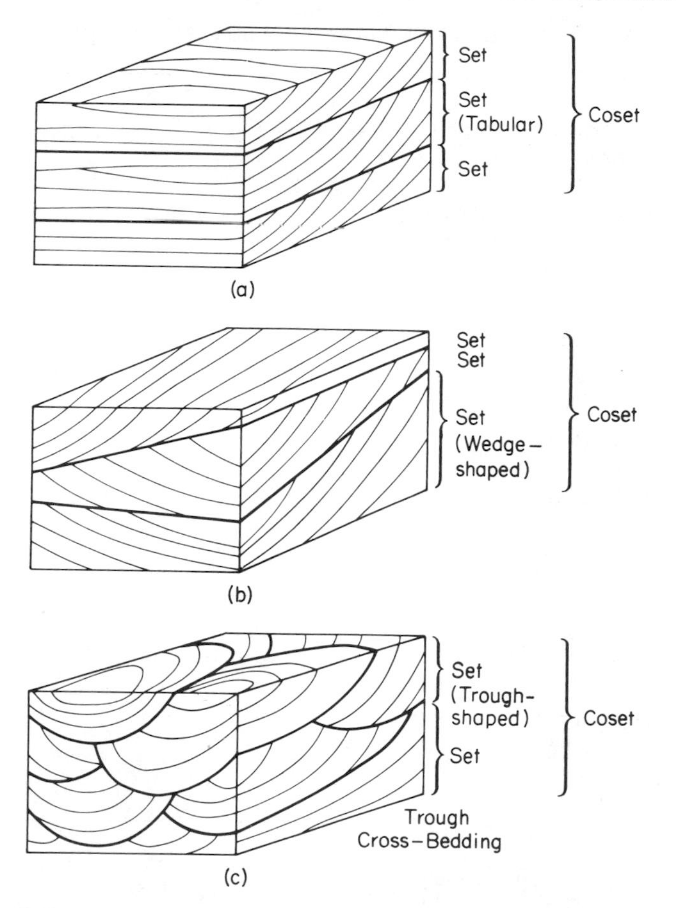

Fig. 5-1 Terminology for cross-bedding. In (a) and (b) the cross-strata are tangential to angular in shape. In (a) the sets are tabular, but in (b) they are wedge-shaped. In (c) both cross-strata and sets are trough shaped. Terminology after McKee and Weir (1953).

The classification shown is neither wholly descriptive nor wholly genetic. This is characteristic of almost all classifications that have been proposed (see Conybeare and Crook, 1968, for a review). As far as possible, geologists now use nongenetic terms to describe particular structure types, for example, "flute" or "cross-bedding" rather than the older "flow mark" or "current bedding," but genetic terms are still commonly used for higher categories of classification. For example, a distinction is commonly drawn between structures formed by the action of tools, dragged or impacted on the bottom, and those made by the action of fluid scour on the bottom. A category of "bed forms" has been added in Table 5-2 to include all those structures formed by fluid flow over a bed composed of movable sediment.

Several detailed classifications have been proposed for cross-stratification but it is less easy to categorize the various internal structures of beds grouped

TABLE 5-2 CLASSIFICATION OF PRIMARY SEDIMENTARY STRUCTURES

I. Stratification

A. *Bedding and lamination* (including graded bedding)

B. *Cross-stratification*

C. *Irregular stratification*

including soft sediment folding
convolute lamination
load structures
ball and pillow
sedimentary sills and dikes
mud cracks
bioturbation; burrows, roots, and other biogenic modifications of stratification

II. Bedding Plane Structures

A. *Tool marks*

including striations
grooves
brush, prod, bounce, and roll marks

B. *Scour marks*

including flutes
large scours ("cut and fill," channels)
rill marks, crescentic marks ("current crescents")

C. *Bed forms*

including ripples, dunes, antidunes
grain lineation, harrow marks
swash marks and other wave marks

D. *Biogenic marks*

together as "irregular stratification" in Table 5-2. These structures include those formed by deformation soon after deposition because of gravity loading, sliding down a slope, biogenic activity, generation of gas in a sediment, or loss of strength of clays (thixotropic behavior) or sands (quicksands).

Origin of Bedding

Little attention has been given by geologists to the origin of simple bedding, presumably because of two reasons: (a) The structure is so common and familiar that it tends to be accepted as a basic, inevitable property of sedimentary rocks, whereas less common structures suggest the need for an explanation, and (b) it is difficult to know how to approach the problem of an explanation for bedding. There must clearly be a very large number of factors, many of which cannot be fully reconstructed, that determine the exact thickness and extent of a particular sedimentary stratum.

Bagnold (1941) distinguished among three main mechanisms of formation of sedimentary deposits: (a) true *sedimentation*, from suspension; (b) *accretion*, i.e., deposition from a moving bed load because of a change of capacity or competency due either to a change in the flow itself or to a change in the bed roughness, which in turn produces a change in the flow, (see Chap. 4); and (c) *encroachment*, i.e., deposition in the lee of an obstacle, including deposition on the lee slope of a dune or small delta. True sedimentation may be either slow or rapid, varying from several feet per year in major deltas to a fraction of a millimeter in a thousand years in parts of the deep oceans. Accretion and encroachment are, however, relatively rapid depositional processes.

Many strata must have been deposited very rapidly. In terms of geological time they represent essentially instantaneous events. In absolute time units, they were deposited by hydrodynamic events, such as floods, that had durations ranging from a few seconds to several hours. In a few cases, the rapidity of deposition can be proved by the evidence of sedimentary structures. For example, a single cross-stratified bed (set) without major internal discontinuities must have been deposited in a period of time not exceeding a few hours. The bed may, of course, represent the final stage in a long history of reworking.

Probably most beds are removed by erosion before they can be buried and thus preserved in the geological record and, in examining processes of sedimentation and recent sediments, the *preservation potential* of the deposit should be constantly kept in mind (Allen, 1968, p. 54). The sedimentary structures or deposits most prominent in the laboratory or in modern sediments are those at the sediment surface, which is commonly examined under relatively quiet conditions. The probability of preservation of many of the features seen is small, whereas the probability of preservation of a feature formed by an unusual event of large magnitude may be much greater. For example, unusually deep scours are formed in river deposits during floods or in tidal sand deposits during storms or unusually high tides. As flood or storm or tide recedes, such scour holes tend to be filled with sediment, which is unlikely to be disturbed again under normal hydraulic conditions. Hayes (1967) has described the effect of two hurricanes on Texas coastal sediments. They included erosion of eolian dunes and deposition of sediments on a broad, flat hurricane beach, on washover fans, and in runoff channels parallel to the beach. Strong currents spilled out of channels cut into the islands as the storm surge receded and carried sandy sediment out to sea where it was deposited on the bottom as a graded bed.

The importance of the unusual event can, however, be overstressed. For example, spectacular changes in beaches may be produced by a few days of violent storms but profiling frequently reveals that many of the storm deposits are reworked by wave action during the much longer periods of calm weather.

Study of the *thickness* of strata has revealed that many deposits are characterized by statistical regularities that may not be apparent at first sight (see Pettijohn, 1957, for a review). In a given stratigraphic unit, bed thickness measurements frequently show a close approximation to a log-normal frequency distribution. There is commonly a correlation between bed thickness and mean or maximum grain size. In some fluvial sandstones, it has been possible to demonstrate that not only grain size but also bed thickness decreases progressively in the down-current direction. The same relationship is often postulated, though rarely proved to exist, in beds deposited by turbidity currents. There is still no fully satisfactory explanation for these relationships.

Origin of Lamination

The origin of lamination has received somewhat more attention than the origin of bedding and a number of different types of laminae have been described: (a) Laminae composed of alternating light and dark layers, the light layers being generally coarser than the dark layers, which are enriched in fine grained organic matter. This type of lamination is the type most typical of glacial *varves*, i.e., lamination caused by the seasonal supply of sediment to glacial lakes. Glacial varves not infrequently contain drop stones, which are larger sediment particles released by melting ice floes. (b) Laminae composed of layers of approximately equal grain size, but of differing mineral composition. Individual layers may be rich in detrital mica or heavy minerals. In some cases, there is a strong segregation of different minerals in different layers. (c) Laminae composed of layers that differ in grain size characteristics. Such laminae may show sharp or gradational boundaries and in some cases they show grain size *grading*, with the coarse grains at the base of a lamina, or *reverse grading*, with the coarse grains at the top of a lamina. (d) Laminae alternately rich and poor in clay but otherwise similar in the grain size of the coarser fraction. (e) Laminae in mudstones or shales that appear to differ in color only.

Not all of these lamination types have the same mode of origin. It is known that laminae may be formed in muds by periodic (in some cases seasonal) changes in the physical or chemical conditions of deposition. Such lamination may depend for its formation on certain characteristics of the environment. For example, in many areas burrowing organisms destroy any lamination formed by other processes (bioturbation). In the Mississippi Delta (Moore and Scruton, 1957) regularly laminated muds are confined to parts of the shallow sea floor adjacent to active distributaries, where there is moderately rapid deposition and there are few bottom-dwelling organisms. Further seaward the muds become irregularly layered and then mottled or homogeneous, commonly because of bioturbation. Laminae in these shallow marine environments are produced by seasonal fluctuations in sediment supply or periodic stirring of the bottom by wave action. Mud may, however, be deposited so rapidly by flocculation at the mouth of a delta distributary that the deposit shows no lamination. Muds deposited in freshwater lakes commonly show good lamination because little flocculation of the clay takes place and there are comparatively few burrowing organisms.

In the tidal mud flat environment, Reineck and Wunderlich (1969) studied the formation of laminae by marking the surface with different colored dyes at high and low tides and monitoring depth and current velocity. At one small tidal channel studied in this way, it was possible to show that mud layers were deposited at both high and low tide and sand layers during both the flood and ebb tides. The layers deposited at low water and during the flood tide were extremely thin and those deposited during the slack water at high tide (a mud layer) and during the ebb tide (a sand layer, showing traces of cross-lamination) were much thicker. The total thickness deposited during one tidal cycle was less than 1 cm. In other localitics, the sand layer deposited by the flood tide was thicker than that deposited by the ebb.

In contrast to lamination in muds, lamination in sands appears to be produced rapidly. Known mechanisms include swash and backwash on beaches, traction of sediment by steady flows (particularly under upper regime, plane bed conditions Sec. 5.2), and avalanching down the lee face of ripples or dunes. In addition, laminae may be produced by upstream migration of antidunes, by sediment settling out after breaking of the wave associated with an antidune, or by migration of ripples.

Some types of lamination result from accumulation of thin lag deposits, associated with erosional processes that winnow out the most easily moved sediment. An example is a type of lamination produced by migration of ripples, which may leave behind a thin lag that is enriched in heavy minerals or coarser grains and that shows traces of cross-lamination. Lamination produced by the action of swash and backwash on beaches also frequently shows strong segregation of heavy minerals, as well as reverse size grading. Individual waves may deposit laminae as much as 2 cm thick. In one example studied in detail by Clifton (1969), a lamina a few millimeters thick was about one-quarter ϕ unit coarser at the top and contained 10% heavy minerals at the base, as compared with 2% at the top. The mechanism causing such compositional and grain size sorting is still not well understood.

In the past, there has been a tendency to interpret each lamina as produced by a separate sedimentation event, for example, a tidal cycle, the swash and backwash from a single wave, or a single bed load avalanche. It is now clear, however, that laminae may also be produced by steady flow, particularly during traction on a plane bed in the upper flow regime. The mechanism is not well understood but it appears to be related to factors that cause grains of similar size and shape to lodge together in patches on the bed. In the case of grains of mixed size, the larger grains tend to roll over smaller grains (for reasons discussed in Chap. 4) but lodge against grains that are larger or of equal size. It is not easy to understand how a finer than average lamina is deposited but laminae of alternating grain size have been produced under conditions of steadily decelerating flow in laboratory experiments (Kuenen, 1966).

Origin of Cross-Stratification

The mechanism that produces most cross-stratification is encroachment by avalanching down the lee slope of dunes, ripples, bars, fans, or small deltas.

Avalanching down the lee face of dunes and ripples takes place in response to oversteepening produced by deposition of bed load at the brink of the ripple (see Allen, 1968, Chap. 16 to 19). At low rates of bed load movement, avalanching takes place at more or less regular intervals. In the intervals between avalanches, finer grained sediment accumulates by settling on the lower part of the lee slope (foreset). In this case the cross-lamination shows a distinct alternation of grain sizes. Sorting of sediments by size is also produced by the avalanching process itself: Many individual cross-laminae are coarser toward the base and show a slight reverse grading across the lamina. Such sorting processes are most marked at low rates of avalanching and become less effective at high rates. At the highest rates of bed load movement, sediment avalanches more or less continuously down the lee slope and the coarsest particles are found at the top, not the bottom, of cross-laminae. The exact mechanism that produces size sorting during avalanching is still not well understood.

Cross-stratification may also be formed by the other mechanisms of deposition. Sedimentation onto the surface of a bed form produces inclined laminae. Sedimentation alone ultimately results in burial of the bed form and formation of a plane surface but in combination with some bed load movement it may result in the formation of structures such as climbing ripple-drift cross-lamination (see Sec. 5.3). Accretion may form cross-lamination on the upflow (stoss) side of bed forms such as ripples, dunes, and antidunes. Accretion deposits tend to be well packed as compared with encroachment deposits. Cross-bedding on a very large scale, with beds inclined at angles of only a few degrees, is formed by accretion of sands on point bars during the migration of meanders (see p. 136). The cross-stratification formed on a beach face is also a type of accretionary deposit.

Absence of Lamination ("Massive" Beds)

The great majority of sands and sandstones show some internal lamination. It has been demonstrated that some sandstones that appear in the field to lack lamination (commonly described as "massive") do have lamination or cross-lamination, as revealed by etching, staining, or X-radiographs of slabs in the laboratory (Hamblin, 1962). In a few cases, a nearly complete absence of internal lamination is produced by extensive bioturbation. In other cases, sand beds genuinely lack all internal structures or show only very faint, diffuse lamination. The lower part of many graded beds (the lowest division in the Bouma sequence: see below) and certain entire thick to very thick beds, that themselves may or may not show grading but are generally associated in the field with graded beds, are almost completely lacking in lamination. The absence of lamination suggests the lack of a tractional phase during deposition of these sands because traction appears to lead invariably to the formation of some type of lamination. It has been suggested that beds lacking any lamination are formed either by very rapid deposition from suspension or by deposition from very highly concentrated sediment dispersions (see p. 169).

5.2 REGIMES OF FLOW

As soon as sediment transport begins in a channel, the bed material is molded into a number of bed forms. These forms may be classified as ripples, dunes (or megaripples), standing waves, antidunes, and chutes and pools (Fig. 5-2). Ripples and dunes are similar in shape but differ in size, with the distinction drawn at an average bed form height of 5 cm. Various combinations of bed forms are possible; for example, ripples may occur on the backs of dunes. Particular bed forms or combinations are found in the laboratory to be characteristic of certain ranges of hydraulic conditions and to succeed each other in a definite sequence as the hydrau-

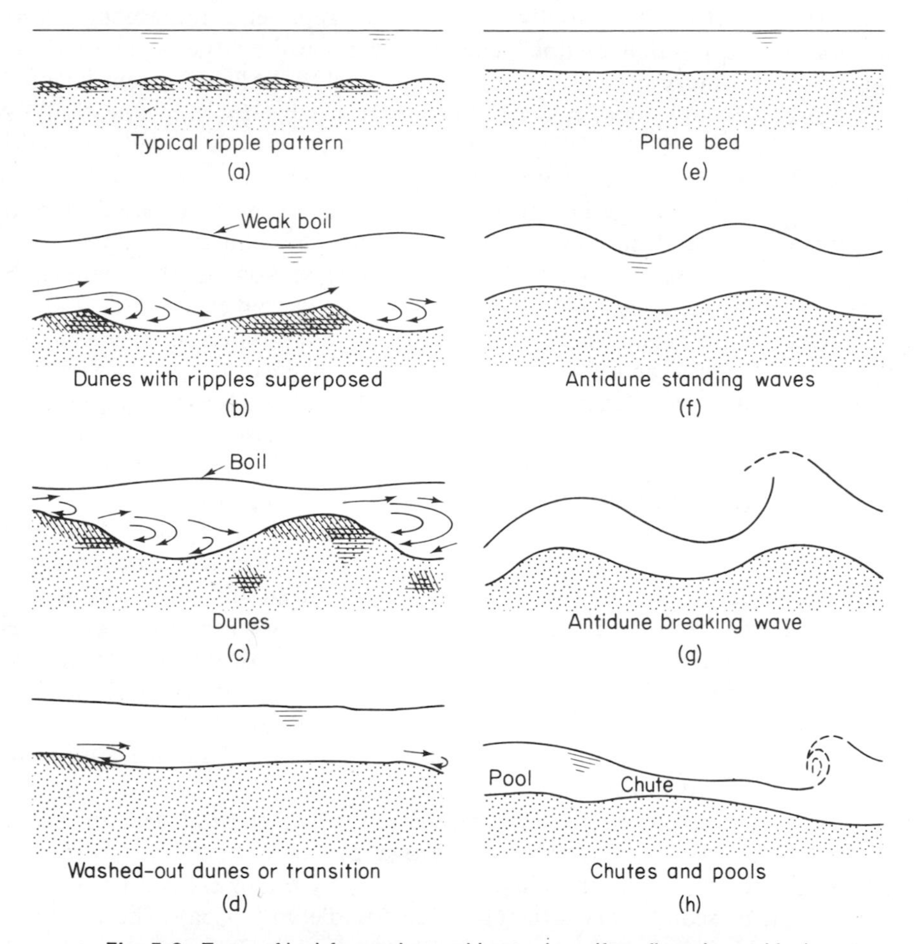

Fig. 5-2 Types of bed-forms observed in steady, uniform flows in sand bed channels (after Simons and Richardson, 1961). The sequence (a) to (h) represents the idealized sequence observed in flumes as the stream power is increased (see Fig. 5-3). Forms (a) to (d) define the "lower flow regime" and the plane bed (e) and forms (f) to (h) define the "upper flow regime".

lic conditions are changed. This observation, first made by Gilbert in 1914, was the basis for a classification by Simons and Richardson (1961) of flow in alluvial channels into a number of *flow regimes*, illustrated in Fig. 5-2.

If water is allowed to flow over a flat bed of sand in a laboratory channel ("flume") and the discharge or slope is gradually increased to the point where sediment begins to move, *ripples* begin to form on the bed. In a large enough flume, further increase in discharge leads to the formation of *dunes*. These two bed forms characterize the *lower flow regime*, in which Froude numbers are generally substantially less than unity and the waves at the water surface, negligible for ripples but quite noticeable for dunes, are out of phase with the undulations on the bed. Further increase in discharge or slope results in disappearance of the dunes and formation of a *plane bed*. Strictly speaking this represents the absence of any bed form but deposition on the plane bed does lead to the formation of a distinctive sedimentary structure: a lineation parallel to the flow (current lineation) or, in cemented sandstones, a distinctive linear parting (parting lineation). At Froude numbers above about 0.8, *antidunes* may form, with a roughly sinusoidal form. Antidunes are in phase with water waves, which have the same form and greater amplitude, and the bed forms migrate upstream by erosion from the downstream side and deposition on the upstream side of the immediately downstream bed form. *Chutes and pools* are alternating sections of the channel, the chutes being characterized by shooting flow and nearly plane bed, followed downstream by a hydraulic jump and a deeper channel section, known as a pool. Plane bed, antidunes, and chutes and pools are classified as *upper flow regime*.

The concept of flow regimes is essentially derived from laboratory studies but some of these have been on a substantial scale and may reasonably be extrapolated to rivers, though complications arise because of the great width of rivers and the presence in larger rivers of bed forms much larger than those that can be produced in the laboratory. Flow regimes are defined by the nature of the bed form but each bed form is associated with a distinctive set of both hydraulic and sedimentation phenomena. With experience, the flow regime of an alluvial stream may be easily recognized from the appearance of the surface flow, even in cases where the water is too muddy for the bed form to be observed easily. Dunes, for example, can be recognized by the slight disturbance of the water surface and the large swirling "boils" that rise to the surface. The sheet-like rapid flow of the upper regime with a plane bed and the symmetrical near-sinusoidal surface waves associated with antidunes are equally distinctive.

Migration of bed forms generally results in the formation of structures in the sand bed. Following the laboratory investigations by Simons and co-workers, the flow regime concept has been used extensively by geologists as a basis for the interpretation of sedimentary structures (see Middleton, 1965a). The concept is essentially a classification of phenomena observed in the laboratory so that the interpretation of a field occurrence of cross-bedding, for example, as "formed by flow in the upper part of the lower flow regime" means that the cross-bedding has been interpreted as formed by the migration of dunes of a type similar to those observed in large laboratory flumes.

The general relationship between different flow regimes is known from laboratory studies. The problem of establishing the hydraulic criteria for each flow regime is similar to the problem of predicting the rate of bed load movement and it is not completely solved. The graph shown in Fig. 5-3 is one of the best suggested to date. In this diagram the equivalent diameter (diameter of the quartz sphere with the same average settling velocity as the sediment) is plotted against the stream power, defined as the product of the average velocity U and shear stress on the bed τ_0. The stream power may be considered to be equal to the rate at which the

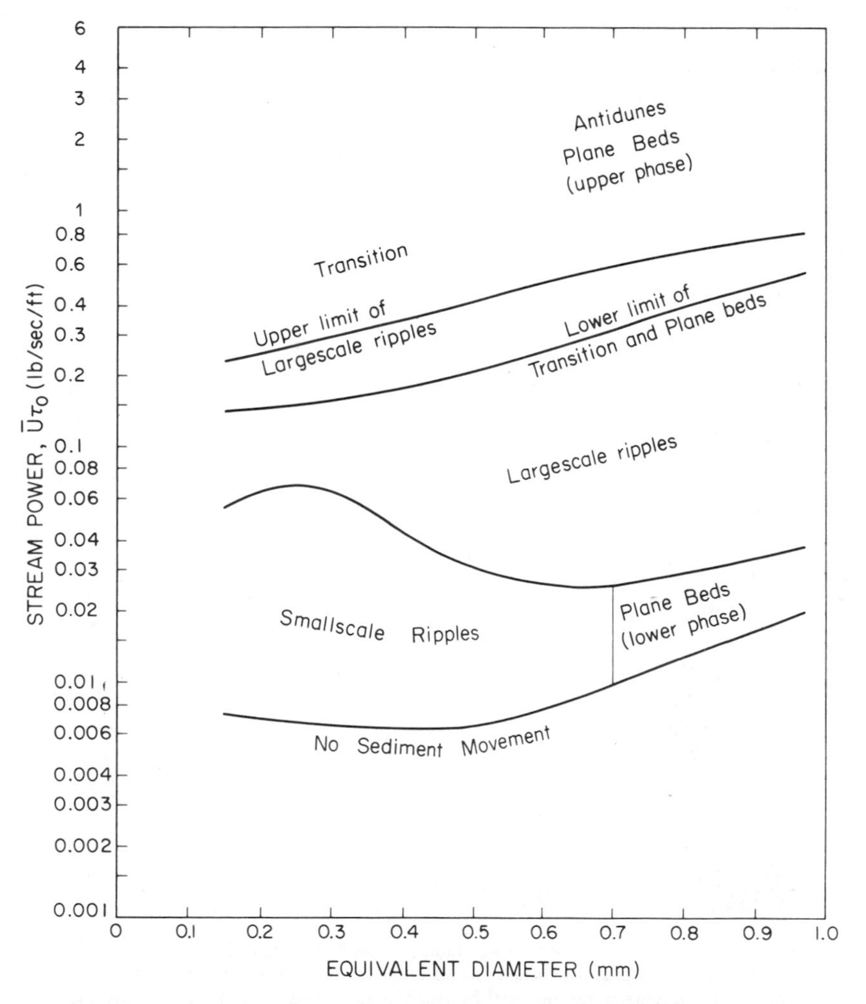

Fig. 5-3 Hydraulic criteria for bed-forms. From Allen (1968) using data of Simons and coworkers.

stream does work on a unit area of its bed. It has been found that it is possible to draw lines on this diagram that separate the conditions characteristic of the main bed forms for steady uniform flow. From this diagram it is clear, for example, that as the stream power is lowered the following sequence of bed forms may be expected: antidunes, plane bed, dunes, ripples. A sequence of sands laid down by a waning current might therefore be expected to display a sequence of structures formed by this sequence of bed forms: antidune structures (if preserved), plane lamination, large-scale cross-bedding (formed by migration of dunes), small-scale cross-bedding (formed by migration of ripples). With the exception of antidune structures, many such sequences have now been observed and have been interpreted in terms of changing flow regimes (for example, Fig. 5-4 and 5-5).

In some sequences, such as the ideal sequence of sedimentary structures observed in turbidites (and now often called the "Bouma sequence" after the geologist who first emphasized its generality: see Bouma, 1962), certain of the structures are missing or replaced by others. In the Bouma sequence (Fig. 5-5) there is a massive division below the plane lamination that is in turn followed by small-scale

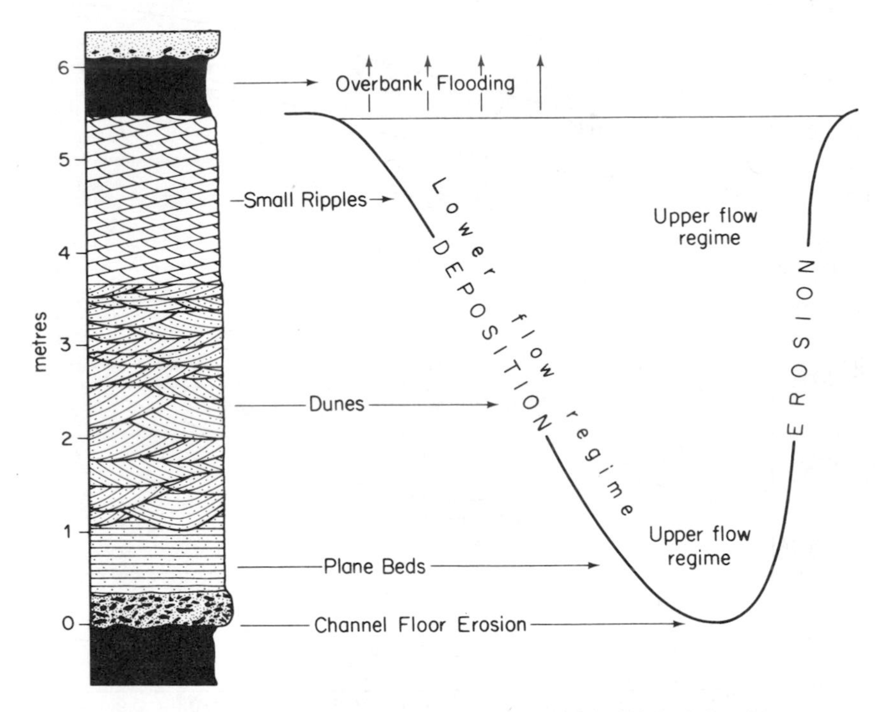

Fig. 5-4 Interpretation by Allen (1963) of a British Old Red Sandstone cyclothem in terms of flow regimes in a meandering channel model. Highest upper flow regime is represented by the erosional lower contact and basal lag conglomerate. Because of lateral migration of the channel, successively higher deposits correspond to sediments deposited successively higher on the point bar, under progressively lower flow regime conditions.

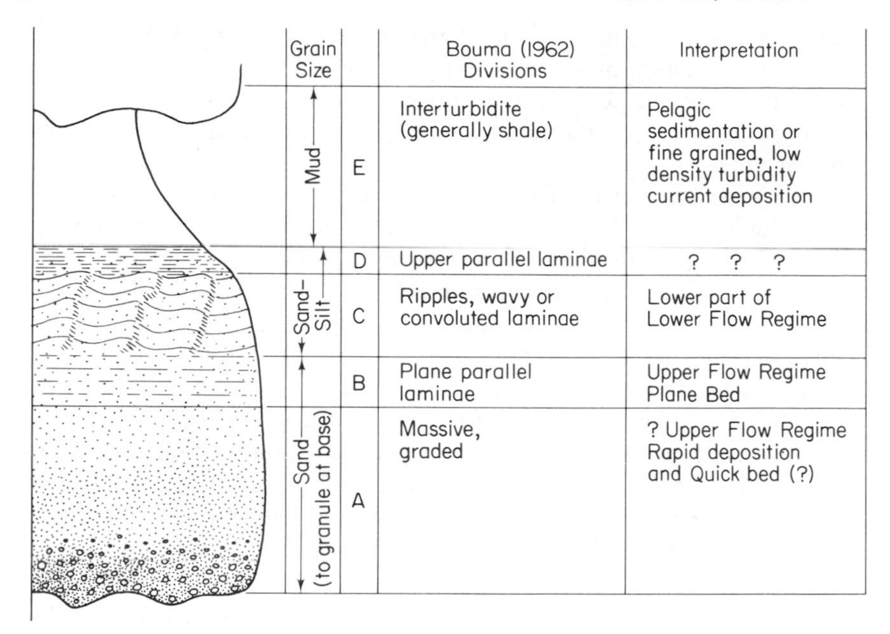

Grain Size
Bouma (1962) Divisions
Interpretation
Mud
E
Interturbidite (generally shale)
Pelagic sedimentation or fine grained, low density turbidity current deposition
D
Upper parallel laminae
? ? ?
Sand-Silt
C
Ripples, wavy or convoluted laminae
Lower part of Lower Flow Regime
B
Plane parallel laminae
Upper Flow Regime Plane Bed
Sand (to granule at base)
A
Massive, graded
? Upper Flow Regime Rapid deposition and Quick bed (?)

Fig. 5-5 Ideal sequence of structures in a turbidite bed. After Bouma (1962) with interpretation after Walker (1965), Walton (1967) and Middleton (1967).

cross-bedding. The large-scale cross-bedded unit is absent. Apart from these exceptions, the Bouma sequence fits an interpretation in terms of gradually reducing flow regime and such an interpretation was actually first given by Pearl Sheldon in 1928 by making use of Gilbert's experimental data even before the development of the turbidity current hypothesis or the formalization of the flow regime concept by Simons and Richardson. The exceptions have stimulated further research and sequences have now been described in which large-scale cross-bedding is present at the expected position in the sequence. Convincing examples of antidune structures have also been described from the base of some turbidites. Problems still remain but the ability of a concept to suggest further lines of research is a sure sign of its practical value, if not its ultimate validity.

There are certain advantages to the use of the concept of flow regime in interpretation rather than some simple hydraulic variable such as velocity, shear stress, or even stream power. The flow regime concept emphasizes the fact that the assemblages of characteristic bed forms are controlled not by a single hydraulic variable but by a complex of variables including most of those discussed in earlier sections. For example, it has been demonstrated in the laboratory that a change in flow regime, from plane bed to dune or vice versa, may be brought about by changes in the water temperature or clay content, even without a change in other variables. This occurs because changes of temperature or clay content change the viscosity of the water, which in turn controls the settling velocity of the sediment. In most cases, the geologist does not know which of the many possible hydraulic variables is responsible for a change in the sedimentary structures observed in a stratigraphic section so that it is not possible to identify the hydraulic variables except to say that they combined to give the conditions necessary for the development of a certain flow regime.

The flow regime concept has undoubted advantages for geologists but there are some disadvantages. The flow regimes defined in laboratory studies have been equilibrium states; i.e., the flow was approximately uniform and steady and there was no net deposition or erosion of sediment. In nature, these conditions are uncommon. The most important deviations from the theoretical flow regimes are probably caused by locally high rates of sediment supply with a high rate of fallout from suspension and by rapidly changing flow conditions that do not permit enough time for a full development of certain bed forms, particularly large-scale forms such as dunes. Both of these factors have been suggested as explanations of the absence of large-scale cross-bedding in the Bouma sequence (Walker, 1965; Walton, 1967).

Not all types of cross-bedding are formed by the migration of dunes. Cross-bedding may be formed by building of small "deltas" or bars into local depressions scoured by the flow or across small pools formed by a variety of mechanisms (for example, by the shifting of channels in a braided stream). Water swept over a shallow marine bank during storms or high tides builds "wash-over fans" in which the cross-beds are formed by avalanching at the advancing slip face of the fan (Ball, 1967). Cross-bedding, generally of a relatively low angle variety, may be formed by true sedimentation or accretion. And, finally, some types of structures are formed by mechanisms quite different from those of channel flow, for example, wave-formed ripples. Obviously, structures not formed by the migration of bed forms should not be interpreted in terms of flow regimes in the manner described above.

The flow regime concept also suffers from the disadvantage that it interprets structures only in terms of generalized hydraulic conditions and throws no light on the mechanisms by which sedimentary structures are formed. In Secs. 5.3 and 5.4 we discuss in more detail the different types of stratification and cross-stratification, the relation between bed forms and stratification types, and the mechanisms of sediment transport responsible for the formation of the different bed forms.

5.3 RIPPLES AND DUNES

Ripple Geometry

Studies of the thickness distributions of cross-bedded sets commonly reveal the presence of a distinctly bimodal distribution, with modal thicknesses at about 2 to 3 cm and 30 cm (Allen, 1968, p. 100). The formation of the small-scale cross-lamination (sometimes called micro-cross-lamination) has been attributed at least since the work of Sorby in the late nineteenth century to the migration of ripples. The fact that much of the medium- or large-scale cross-bedding was also formed by the migration of large ripple-like bed forms (dunes or megaripples) has become apparent only during last the 30 years. Before this time the abundance of dunes in the subaqueous environment had not been made clear and there had been few laboratory investigations made in flumes sufficiently large to reveal the fundamental distinctions between ripples and dunes.

Both ripples and dunes have similar geometry so that, despite the important distinctions between them, the same terms may be used to describe their morphology. In this book we adopt the nomenclature proposed by Allen (1968) and illustrated in Figs. 5-6 and 5-7. The gently sloping side (upstream of the summitpoint) of the ripple is called the stoss side, and the steep side (downstream of the summitpoint) is called the lee side. The lowest point on the ripple profile is called the troughpoint and the highest is called the summitpoint. The point separating the crestal part of the ripples from the steep lee slope is called the brinkpoint: In many cases the summit and brinkpoints are the same. In plan, the lines connecting the respective points are called the troughlines, crestlines (rather than summitlines), and brinklines. The ripple is divided into two parts: The part with an elevation of less than half the height is called the *trough* and the part with an elevation greater than half the height is called the *crest*. The distance from troughpoint to troughpoint, called length by many authors, is called *chord* by Allen. The *vertical form index* (often called simply the ripple index) is the chord divided by the ripple height. A variety of other descriptive indices have been introduced to describe the plan geometry of ripples and Allen has distinguished among five main schematic forms (Fig. 5-7): *straight*, *sinuous*, *catenary*, *linguoid*, and *lunate*. Further descriptive terms have been introduced to describe the geometry of ripple trains (assemblages of several ripples). Of these the most important is a distinction between in phase and out of phase relationships between adjacent ripples. True linguoid trains consist of linguoid ripples arranged out of phase (or in an "en

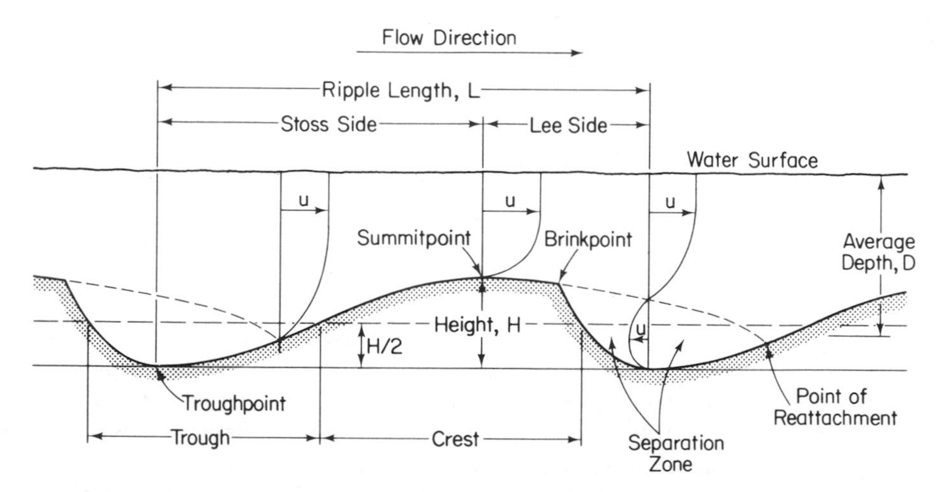

Fig. 5-6 Schematic cross-section of a ripple with nomenclature after Allen (1968).

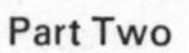

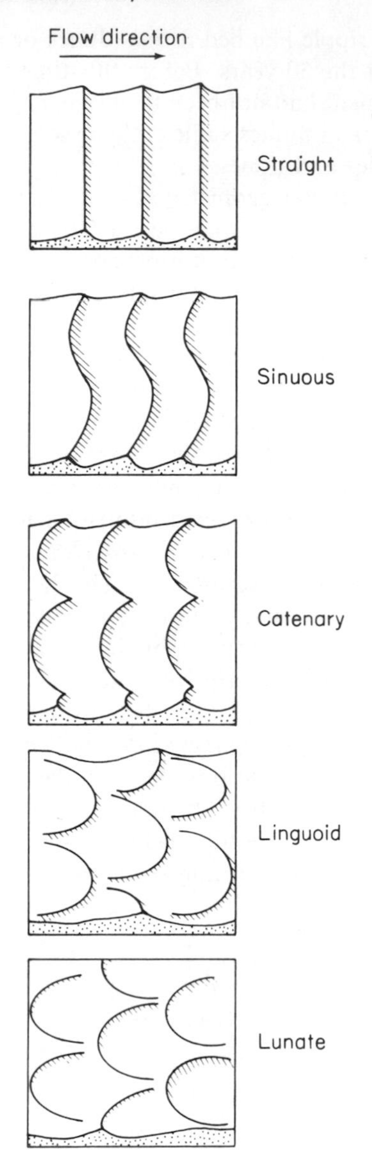

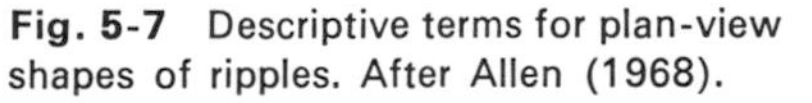

Fig. 5-7 Descriptive terms for plan-view shapes of ripples. After Allen (1968).

echelon" pattern). Linguoid ripples arranged in phase (in rows parallel to the flow) are described as *cuspate*. All of these terms are idealized. In nature, ripple forms are much more irregular than the idealized forms but the terms are nevertheless useful for description in the field. The significance of the different types of ripples is as yet not fully understood (see below).

The geometry of the ripple has an important relation to the type of cross-bedding that results from migration of the ripple. Tabular sets result mainly from migration of straight or slightly sinuous ripples. Trough sets result mainly from the migration of less regular ripple forms. Probably most small-scale trough cross-lamination was formed by the migration of linguoid ripples and most large-scale trough cross-bedding was formed by the migration of sinuous or lunate dunes (see below).

Fluid Mechanics

The mechanism of formation of cross-stratification has been illuminated by several recent studies of the fluid mechanics of ripples, dunes, and micro-deltas (see especially Jopling, 1965). In all of these bed forms, sediment is transported in suspension or as bed load up to the brink, where the flow separates form the bed to form a wake or zone of reverse circulation (Fig. 5-6 and 5-8). The ripple form is itself a result of the development of a wave-like instability between the fluid and the granular bed but the distinctive cross-lamination produced by avalanching on the lee slope of the ripple is a result of *flow separation* at the brink of the ripple. Separation of flow at the brink of the ripple is hydrodynamically similar to separation in the lee of any "negative step" and produces a large *separation eddy* (or "roller"). The flow passing over the step resembles a half jet, with a *zone of no diffusion* in the upper part of the flow retaining a velocity unaffected by turbulent friction, and a *zone of mixing* caused by turbulent mixing between the half-jet and the water below. Below the zone of mixing there is a *zone of backflow*. The circulation in the separation eddy is induced by the friction at the border of the jet above. Downstream from the point of separation, at a distance that is commonly 6 to 8 times the height of the step, the zone of diffusion has expanded to the extent that the flow becomes *reattached* to the bottom.

There are several aspects of this pattern of flow separation, expansion by turbulent diffusion, and reattachment that have important consequences for formation of sedimentary structures. Flow separation results in a physical separation of the transported sediment into two fractions: (a) the bed load that accumulates at the brink of the ripple until the lee slope exceeds the angle of repose of the sediment and avalanching takes place and (b) the suspended load that is transported on downstream. The coarser fraction in the suspended load settles down through the zone of diffusion and is deposited in the region of reverse flow in the lee of the ripple. Velocities in the zone of reverse flow may be large enough to sweep part of the suspended material up the bottom of the lee slope, where it is deposited as fine laminae between the coarser laminae formed by avalanching. In large dunes, the reverse flow may even form "*regressive ripples*" that migrate from near the point of flow reattachment toward the base of the avalanche slope (see Fig. 5-8 and Boersma, 1967). These ripples are buried under the accretionary foresets as the dune migrates downstream.

The shape of cross-strata therefore depends on the balance between bed load avalanching down the lee slope and materials being deposited on this slope by settling from suspension (Fig. 5-8). In the case where there is little suspended load

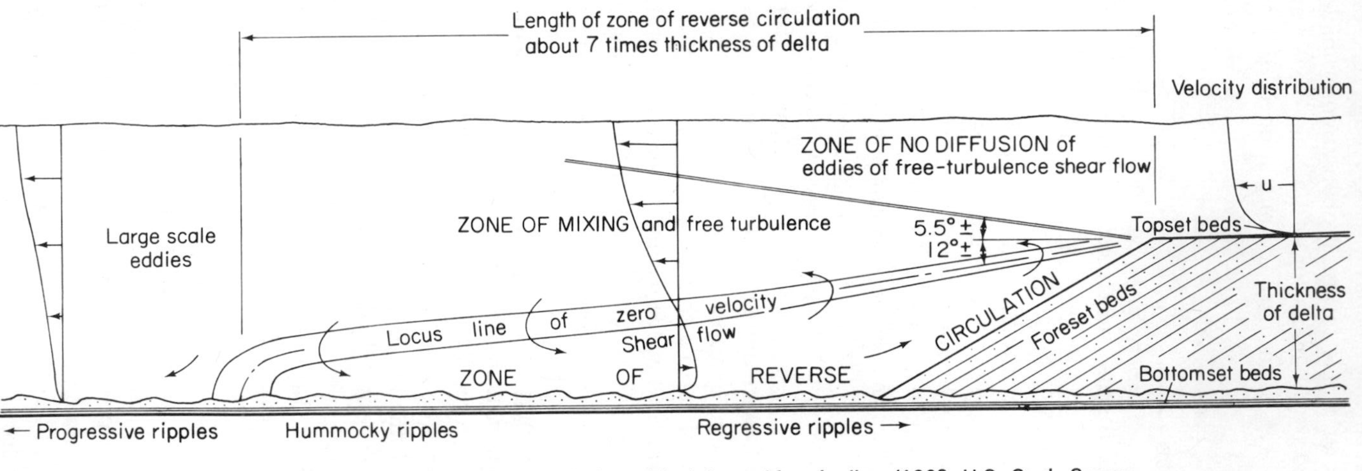

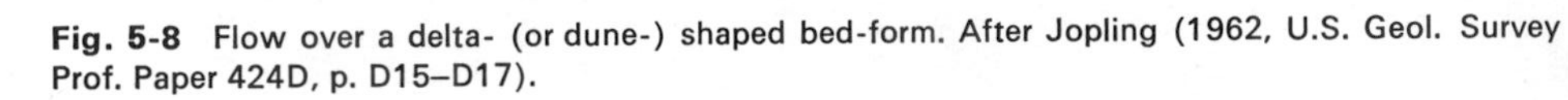

Fig. 5-8 Flow over a delta- (or dune-) shaped bed-form. After Jopling (1962, U.S. Geol. Survey Prof. Paper 424D, p. D15–D17).

(because the sediment is too coarse to be taken into suspension by the flow) the foresets will be steep, straight, and inclined at the angle of repose of the sediment throughout their length. The same effect can also result from the height of the lee face being large compared with the total flow depth. In this case, almost all the sediment settling from suspension will be deposited onto the lee slope itself, where it will be buried by avalanche deposits. In either case, the contact between the foreset and bottomset laminae will be *angular*. In contrast, if there is a large amount of sediment in suspension or if the height of the lee slope is small compared with the total depth, the lower part of the lee slope will be built up by sediment from suspension rapidly enough to keep pace with the growth of avalanche deposits and the cross-strata will be curved in shape, with a slope to the angle of repose at the top, decreasing asymptotically to the horizontal at the bottom. The contact between the foreset and the bottomset laminae will be *tangential* rather than angular.

The point where the flow reattaches to the bed can be identified only as an average position. The zone of backflow constantly fluctuates in size and shape so that the flow reattaches at a position that is constantly changing. At the point of reattachment, intense turbulent eddies, generated by shear at the boundary of the half-jet, strike the bottom and erode the sediment. In the case where the ripple crest is not straight, the flow tends to return to the bed with particular force at certain points, with the result that relatively deep scours are eroded at these points. The area of maximum scour migrates downstream ahead of the ripple crest so that the scours are generally trough- or spoon-shaped, with their long axes parallel to the flow direction. The troughs are then filled in by cross-strata as the ripple advances over them (Fig. 5-9). The formation of such scours is therefore a natural consequence of the formation of ripples with irregular crest lines and this seems the most reasonable explanation of both small- and large-scale trough cross-stratification. Large eddies, swirling about a near-vertical axis ("kolks") are also observed in rivers flowing over a dune-covered bed and it has been suggested that trough-shaped scours are frequently cut by such eddies and are not necessarily related to the large eddies with a horizontal axis formed in the lee of a dune (see Harms and Fahnestock *in* Middleton, 1965a).

The factors that control the plan-view shape of ripples or dunes are not yet fully understood, but it is clear that a very important role is played by *secondary flows*. It is generally assumed that in a uniform channel the flow is everywhere parallel to the average flow, i.e., to the channel boundaries. In many flows, however, the real flow consists not only of this "primary" flow but also of one or more superimposed "secondary" flows. Secondary flows are not really distinct from the main flow but it is often convenient to consider the total flow field to be made up of primary and secondary components. In the case of flow over a step, it has been shown that if the step is oblique to the average flow, the circulation in the zone of reverse flow will take on the pattern of a spiral (helicoidal) secondary flow. In the case of a complex ripple pattern, such as that shown by a field of linguoid ripples, there is also a complex secondary flow pattern near the bed superimposed

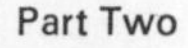

Fig. 5-9 Trough cross bedding in modern sands of the Brazos River near Richmond, Texas. Note the different appearance in sections cut parallel to the flow [which was toward the right] and normal to the flow. A thin zone with small-scale cross stratification can be seen at the top of the section. [Photo courtesy H. A. Bernard, C. F. Major, Jr., and R. J. LeBlanc.]

on the average downstream flow. Apparently, certain secondary flow patterns are more stable than a simple two-dimensional flow so that two-dimensional (straight) ripple forms tend to be replaced by certain types of three-dimensional forms. It has been demonstrated in the laboratory that the first ripples formed on a plane sand bed are always straight but that they become sinuous with time, even if the flow conditions are held constant. Allen (1968, p. 93) has suggested that depth of flow is also a controlling factor with complex forms more readily developed in shallow water. In natural environments it has been observed that as the water shallows the following sequences occur: for ripples, straight to sinuous to symmetric linguoid to asymmetric linguoid; for dunes, straight to sinuous to catenary to lunate.

Ripple Migration and Cross-Bedding

The type of sedimentary structure formed by migration of a ripple depends not only on the geometry of the ripple and the surface over which it moves but also on the relative importance of scour and deposition during ripple migration. Migration of any bed form generally implies both scour and deposition locally. If there is no *net* deposition, the amount of scour is equal to the amount of deposition and only sediment deposited in particularly deep scours has a high probability of being preserved. The same holds for any rate of deposition that is substantially smaller than the rate of migration of the bed forms. Under these conditions it is not to be expected that the bed form itself, or most of the sediment originally contained within the bed form, will be preserved in the sedimentary deposit.

Only in cases where the rate of aggradation is large compared with the rate of migration of the bed forms can the bed form itself, or a substantial amount of the sediment originally contained within the bed form, be preserved. Such high rates of aggradation must be essentially temporary or local in development because over the greater part of any sediment transport system (a river or beach, for example) there must be a near-equilibrium between sediment supply and removal. If this were not the case, a period of rapid erosion or deposition would result, tending to restore the system to near-equilibrium conditions.

These principles are illustrated by the structure called *ripple-drift cross-lamination* (Jopling and Walker, 1968). As the rate of net sediment deposition increases, ripple cross-lamination changes in character from (a) a trough cross-lamination characterized by many scoured surfaces and very incomplete preservation of the ripple form to (b) climbing ripple-drift cross-lamination characterized by tabular form and increased preservation of the bed form but with the laminae still truncated by erosion at the stoss side of the ripple to (c) climbing ripple-drift with complete preservation of cross-laminae (Fig. 5-10). The type showing complete preservation

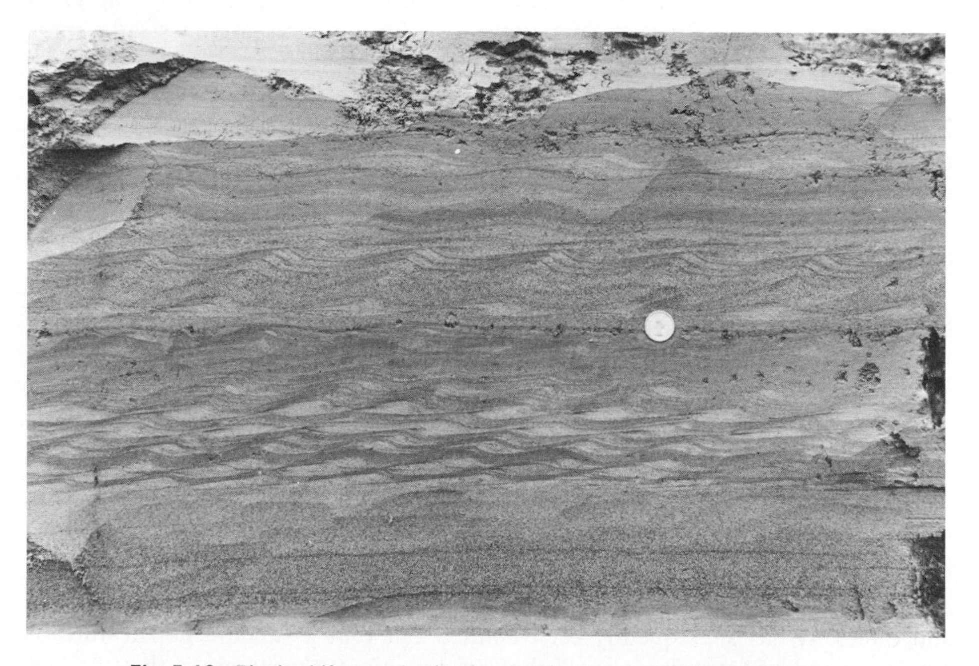

Fig. 5-10 Ripple-drift cross-lamination. In this example, from the Pleistocene near Bruce Mines, Ontario, there is preservation of the stoss-side laminae (Type B of Jopling and Walker, 1968), particularly in the sets above the coin. Plane lamination is seen at the base of the section. (Photo courtesy of Roger Walker.)

indicates very rapid supply of sediment from suspension ("fall out") combined with just enough traction to produce the ripple bed form but not enough to result in complete removal of laminae from the stoss sides of the ripples. This type of ripple-drift is found only in a few types of deposits, notably in turbidites, point-bar deposits, and fluvial flood deposits, all of which are characterized by very rapid supply of sediment from suspension.

Differences between Ripples and Dunes

Despite their many similarities there are some important differences between ripples and dunes besides size (see Simons et al., *in* Middleton, 1965a). Ripples respond only slightly if at all to changes in the depth of flow but their size is directly related to the grain size of the sediment. In water, ripples will not form in sediments with an equivalent diameter larger than about 0.7 mm. The size of dunes is not strongly related to the grain size, though dunes in coarse sand tend to be shorter and steeper than those in fine sand. In flumes, the size of dunes is controlled strongly by the total water depth. It appears that, given sufficient time, dunes tend to grow to a height limited only by the total depth of the flow. In nature, other factors may limit the size of dunes. Where dunes that show profiles displaying a gentle stoss side and steep lee side have been observed, the height generally ranges from about 20 cm to 2 m. Very thick sets of cross-bedding formed by migration of large subaqueous dunes do not appear to be common in the stratigraphic record.

Ripples migrate only in one plane with a relatively uniform ripple height: They do not climb up the backs of other ripples unless there is net addition of sediment from suspension. In contrast, dunes may migrate up the backs of other dunes so that dune height is variable and migration takes place on more than one plane. The result is that cross-bedding sets are frequently considerably thinner than the maximum size of the dune that formed them.

The term *dune* or *megaripple* is commonly applied only to those medium-scale bed forms that bear a close resemblance to those observed in large laboratory flumes. In nature, much larger asymmetrical bed forms are observed but these appear to differ from laboratory dunes and they are discussed in Sec. 5.5.

5.4 ANTIDUNES

The name antidune was coined by Gilbert (1914) for wave-like bed forms that migrate upstream, by accretion on the upstream side and erosion of the downstream side, instead of migrating downstream in the manner typical of ripples and dunes. It has subsequently been pointed out that a fundamental feature of antidunes and of "standing waves" that do not migrate upstream is that the bed form is in phase with the surface water wave, whereas in ripples and dunes there is either no appreciable surface wave or it is out of phase with the bed form (Fig.

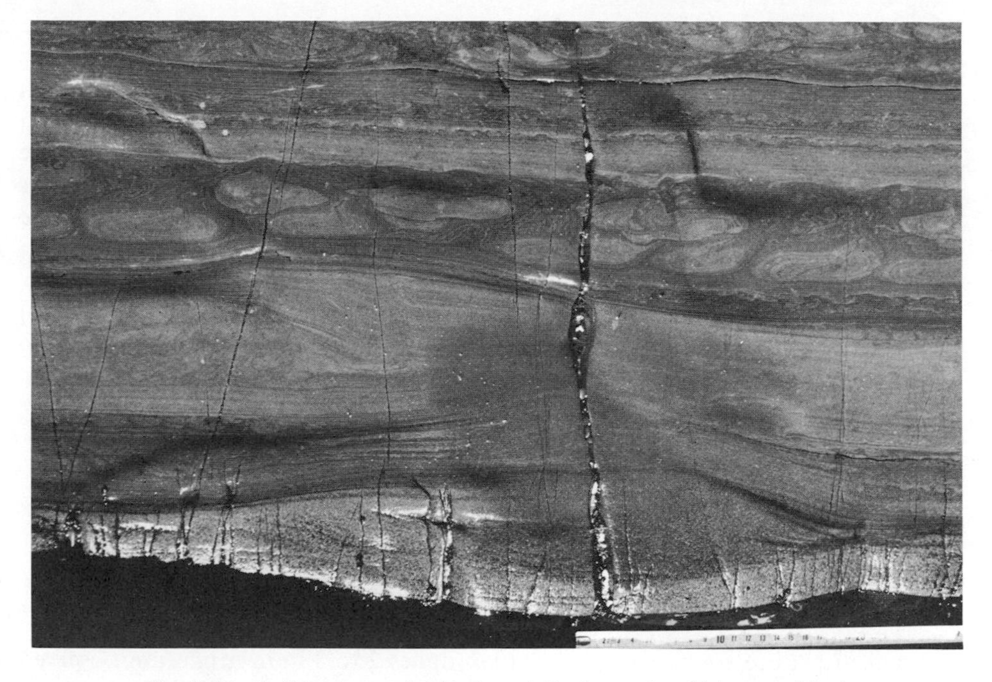

Fig. 5-11 Antidune cross-lamination at the base of a thick graded bed, Cloridorme Formation, Gaspé. Scale in cms. Flow was from right to left as proved by flutes (not shown) and ripple drift, just above the antidunes. The upper part of the bed shows interbedded divisions with plane lamination and pseudonodules formed by loading. (Photo courtesy of Keith Skipper.)

5-2). For some authors, antidunes are bed forms that are in phase with the surface wave, whether they migrate upstream, downstream, or not at all.

The bed form in antidunes is a subdued copy of the surface wave so that the depth is larger over the crest than over the trough of the antidune. Hand (1969) has shown that both water and sediment waves have a shape similar to that of a trochoidal wave, i.e., a type of relatively sharp-peaked wave form that is traced out by a point in the interior of a rolling wheel. Because antidunes migrate by accretion on the stoss side, they form faint, poorly defined laminae that are also trochoidal in shape. The characteristic shape should assist in the recognition of antidune cross-lamination in ancient deposits (Fig. 5-11). Other characteristic features are the low dip of the cross-lamination (less than 10 degrees), the association with plane lamination formed in the upper flow regime, and the inclination of the cross-lamination in a direction generally opposite to that of other paleocurrent indicators. It should be noted, however, that the water waves associated with antidunes tend to move slightly upstream of the bed form and break. Breaking of the water wave produces local scour and throws much sediment into suspension.

As the sediment settles, it may produce laminae that dip downstream as well as upstream (Middleton, 1965b).

The characteristics of antidune cross-lamination have been described from the laboratory only since 1965. Already, several examples have been described from ancient sediments and it appears that, even though well-preserved antidunes are rare in the geologic record, antidune cross-lamination is not as uncommon as assumed in the past. The wave length (L) of antidunes is directly related to the average flow velocity U by the equation

$$U^2 = \frac{gL}{2\pi} \tag{1}$$

Thus if the bed form is preserved or if the wavelength can be reconstructed from the shape of the cross-lamination, the ancient flow velocity can be estimated.

5.5 LARGE BED FORMS: SAND WAVES, RIDGES, AND BARS

Sand Waves and Ridges

Asymmetrical bed forms much larger than dunes 2 to 3 m in height are observed in rivers and elsewhere but they differ from laboratory dunes in that the lee slope angle is much less than the sediment angle of repose. Such large forms are also commonly covered by fields of "normal" dunes (in much the same way in which dunes may be covered by ripples). These observations suggest that although there seems to be no sharp break in size or shape between them, the large forms differ from dunes genetically. The large forms will therefore be distinguished as *sand waves*. Little is known about the larger bed forms and the terminology used to describe them is correspondingly confused.

Most authors use "megaripple" as a synonym for dunes but Coleman (1969), in an important study of the Brahmaputra River, has distinguished between megaripples (the dunes of Simons and co-workers) and larger forms, with heights of 5 to 25 ft and lengths of 140 to 1600 ft, which he calls "dunes." He finds that the height of these "dunes" is independent of water depth, except in very shallow water. They have irregular crest lines, with poor lateral continuity and the surface is smooth, without superimposed smaller bed forms. Large "boils" at the water surface result from flow over these "dunes" and the eddies producing the boils cut scours high on the stoss side of the next bed form downstream. Generally these scours are filled from one side by migration of a "dune" from upstream.

Coleman reserves the term "sand wave" for even larger bed forms, 25 to 50 ft high and 600 to 3000 ft long. These structures may have smaller bed forms on their backs. They are formed at flood stages of the river and are inactive at low stages. When active, they produce large separation eddies and boils over 100 ft in diameter and these scour depressions more than 10 ft deep. Linear sand waves elongate parallel to the flow, with heights of 2 to 5 ft and spaced 25 to 100 ft apart, are also present in the Brahmaputra. The rates of migration of large bed forms in rivers are very variable, depending on the discharge. In the Brahmaputra, the migration rates during flood discharges are very large, with megaripples moving at least 750 ft/day, "dunes" moving as much as 200 to 300 ft/day, and "sand waves" moving as much as 1500 ft/day.

Another type of large-scale, dune-like bed form is the *tidal sand ridge* (also called sand wave by some authors). A distinguishing feature of such ridges is that they are generally straight or curved, symmetrical or asymmetrical in cross section, and oriented parallel with the principal direction of the tidal currents (Off, 1963). In the case of asymmetrical forms, the gentle slope is inclined at an angle of less than 5 degrees and is frequently covered by straight or sinuous-crested dunes oriented almost normal to the currents and therefore to the ridge crest. The steep slope is inclined at an angle of about 5 to 10 degrees and is frequently free from large bed forms. Houbolt (1968) has suggested that these tidal ridges are formed by large secondary flows, for example, as shown in Fig. 5-12. The effect of the sec-

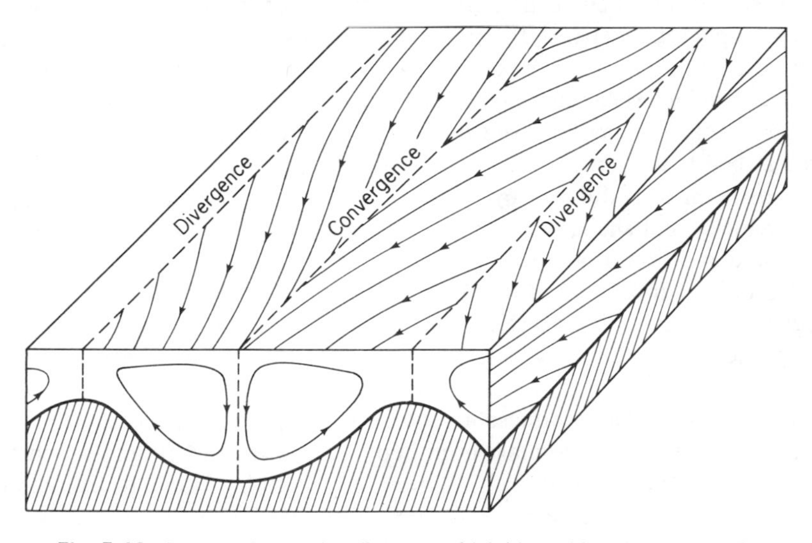

Fig. 5-12 Pattern of secondary flow over tidal ridges. After Houbolt (1968). In the case of asymmetrical ridges, it seems probable that one of the spirals is more fully developed than the other.

ondary spiral circulation is to scour sediment out of the troughs and pile it onto the ridges. In the asymmetrical forms there is some evidence that the form is not migrating in the direction of the steeper slope but that sediment is moved predominantly in one direction at a slight angle up the gentle slope and is then moved predominantly in the other direction along the steep slope. Sediment therefore circulates around the ridges but the ridge itself is almost stationary in position.

In the North Sea, the fact that some of the ridges are almost stationary has been proved by surveys extending back over 200 years.

Point Bars and Braid Bars

Many of the large-scale accumulations of sand or gravel found in rivers may also be considered to be a type of large bed form. They include point bars, formed on the inside bends of meanders, and braid bars, which separate the branches of braided channels. Meanders are characterized by a strong spiral secondary flow that results from the action of centrifugal forces (Fig. 5-13). The forces tend to deflect the moving fluid particles toward the outer bank of a meander and the deflecting force is much larger at the surface, where the velocity is high, than near the bed, where the velocity is low. Consequently the secondary flow is spiral, with the transverse component toward the outer bank near the surface and toward the inner bank near the bottom. The spiral reverses in direction with every reversal in curvature (crossover) of the meandering channel. The maximum velocity in the channel is found not at the center but near the outer bank. Consequently, there is also a strongly asymmetrical distribution of shear stress on the bottom and sides of a curved channel, with the maximum shear stress found in the bend near the outer bank. The high shear stress results in erosion of the bottom and undermining of the outer bank. Sediment derived from erosion and collapse of the outer banks is moved as bed load by the secondary circulation onto the inner bank of the next bend downstream or is diffused by turbulence as suspended load into the regions of lower velocity and shallow water and deposited close to the inner banks. The result is the deposition of a crescent-shaped bar of sand at the inner bend of the river. This point bar has a surface that slopes into the channel at an angle of a few degrees and grows by accretion in an oblique downstream direction. The sloping surface of the bar is generally covered by dunes, which are particularly active at times of flood.

Migration of the bar produces a sequence of sedimentary structures roughly equal in total thickness to the maximum depth of the channel (Fig. 5-4 and 5-13). The sequence of structures observed is commonly from lag gravel to trough cross-bedding in sets 20 cm to 1 m thick formed by migration of dunes on the lower part of the point bar, to small-scale trough cross-lamination formed by ripples on the higher parts of the point bar, where the water is too shallow and the velocities too low for the formation of dunes. Plane lamination, formed by upper regime flows, may be found below the trough cross-bedding, as shown in Figs. 5-4 and 5-13. In other cases, it may not be present in this position in the sequence, either because it was never formed or because it was reworked by the deep scouring that takes place during the migration of sinuous or lunate dunes. Frequently, plane lamination is found above the trough cross-bedding (Bernard et al., 1970) or interbedded with small-scale cross-lamination near the top of the sequence. Some of this plane lamination proves on close examination to have been formed by ripple

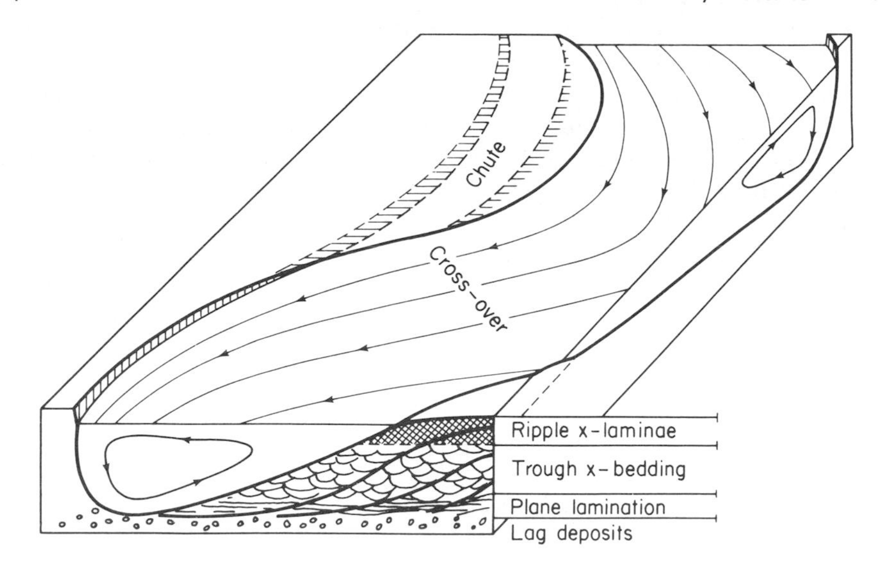

Fig. 5-13 Model of a meandering river showing secondary flow pattern and internal structure of point bar deposits.

migration but much of it was formed by plane bed upper regime flows. Such flows can be observed to form locally on the upper parts of point bars during floods, at the same time as the regime in the deeper part of the channel is the lower flow regime with dunes. Presumably the upper flow regime higher on the bar is a result of a higher slope here than in the deeper parts of the channel.

Allen (1970) has statistically analyzed the sequence of structures observed in many ancient fluvial sediments and has confirmed that there is a very low probability that large-scale cross-bedding will overlie small-scale cross-bedding but that the relation between the two types of cross-stratification and plane-laminated units is less predictable. To explain this, Allen developed a theoretical model based on the stream-power/grain size relationships observed in flumes (Fig. 5-3) and on the computed stream power for different positions within the channel and for different degrees of sinuosity. Increase in the overall power of the stream increases the relative abundance of large-scale as compared with small-scale cross-stratification and Allen suggested on the basis of his model that decrease in the sinuosity of the channel increases the relative abundance of plane lamination.

The formation and characteristics of braid bars have been reviewed by Smith (1970). High regional slopes, variable discharges, abundant supply of coarse sediment, and predominance of easily erodible bank materials are probably all factors in the development of a braided stream pattern. Longitudinal braid bars aggrade downstream and laterally from a nucleus of coarse sediment that the stream is temporarily unable to transport. The bars may become stabilized by the growth of vegetation on the highest point, which is commonly close to the upstream end.

Upper flow regimes are common but lower flow regime conditions and the formation of dunes may take place and ripples may form in local accumulations of finer grained sediment formed at low water or in relatively protected parts of the channel. Most bedding is crudely horizontal or is trough cross-stratification formed by the migration of dunes or bars across scours formed at high discharges. Braiding may also result from formation and dissection of transverse bars, which are tabular bodies of the "micro-delta" type. The internal stratification of these bars includes tabular sets of planar cross-beds as well as some trough and horizontal bedding. Smith (1970) demonstrated for the Platte River that transverse bars were finer grained than longitudinal bars and tended to replace them in the lower part of the river system.

Both the migration of point bars and the growth of braid bars can form "*fining upward sequences*." The distinction between the sequence formed in a meandering and in a braided channel may be made on the basis of the geometry of the deposit, the grain size, the assemblage of sedimentary structures, and their orientation. Less common coarsening-upward sequences are known in braided channel deposits, probably formed by infilling of abandoned channels. Such sequences appear to be rare in the deposits of meandering rivers. In some large rivers, large channel islands form and display many of the sedimentary textures and structures typical of meander point bars. Thus, the distinction between point bar and braid bar deposits is not always clear-cut. Relatively few studies have yet been made of the deposits of large sandy braided rivers, in contrast to the numerous studies of meandering streams.

5.6 STRUCTURES FORMED BY SCOUR

The basic principles controlling the extent and rate of scour in sand and other coarse, noncohesive sediments are as follows (after Laursen, 1953): (a) Scour cannot take place until the flow is competent to erode the sediment, as discussed in Sec. 4-3, (b) The rate of scour at any area on the bed is equal to the rate of transport away from that area, less the rate of transport into the area. If the two rates of transport are equal, there may be much sediment movement but no scour; i.e., there exists an "equilibrium" condition. For example, relatively little scour can take place in a river channel during flood, averaged over a long section of the channel, because little sediment is held in suspension and as much sediment is supplied from upstream as is removed downstream. Depth soundings and observations on markers implanted in the channel frequently indicate local or temporary scour of many meters but these local scours must be balanced by local storage of sediment elsewhere, in large bed forms or in wider sections of the channel or on the floodplain as a result of crevassing of the levee (Colby, 1964). (c) The rate of scour decreases as the local flow section is enlarged by erosion. Consequently there is generally a limiting extent of scour for a given initial condition and the limit is approached asymptotically with time. (d) Removal of sediment by erosion frequently leads to the accumulation of the coarsest fraction as a lag, which armors the surface and prevents further scour.

The formation of deep local scours in river and other types of channels is generally possible mainly because of the existence of a secondary flow pattern that intensifies the flow locally at the bed. Sediment eroded as a result of the flow intensification is deposited downstream as the power of the secondary flow is dispersed or the sediment is diffused both downstream and cross-stream into regions of lower speed and turbulence to form local accumulations.

An example illustrating these principles is the formation of scours ("*current crescents*") and local sand accumulations ("*sand shadows*") around local obstructions in the flow, such as pebbles, boulders, or artificial pilings (see Karcz, 1968). The fluid pressure at the surface of the obstruction has a maximum value at the *stagnation point*, where the velocity is reduced to zero. At this point, it follows from Bernoulli's theorem (Chap. 4, Eq. 16) that the pressure is $\rho U^2/2$. Because of the velocity gradient close to the bed, the fluid pressure at the boundary of the obstruction increases upward so that if the boundary is vertical, the result is a secondary flow directed toward the bed. In the case of an upright cylindrical obstruction, the result of the pressure difference is a strong downward jet, with maximum intensity at the upstream side of the cylinder, that extends with decreasing intensity downstream as a spiral secondary flow flanking the zone of flow separation behind the cylinder. Erosion by the secondary flow produces a horseshoe-shaped scour around the cylinder with sediment accumulating as a small elongate mound in the zone of flow separation behind the cylinder. Similar flow patterns are characteristic of other types of obstructions.

Scour of cohesive sediment (mud) differs in some important principles from the scour of noncohesive sediment discussed above. The main differences are that the cohesion of the mud and therefore the difficulty of erosion commonly increase downward and that the mud, once eroded, is taken into suspension and not deposited again in the near vicinity of the scour itself. Alternatively the mud may erode by flaking into pebble-size mud flakes that are deposited downstream as intraformational conglomerates.

Many marks, formed on mud surfaces but generally preserved as molds on the soles of sandstone beds, were formed by erosion of freshly deposited mud. Slight irregularities originally present on the mud surface, or formed during the process of erosion itself, resulted in flow separation and secondary currents that scoured the bed and formed the scoop-like structures called *flutes* or *flute marks* (Fig. 5-14). Allen (1969) has shown convincingly by studying the minute striations inside experimentally formed flutes and by studying the flow pattern inside plaster casts of flute molds, that flutes are produced by flow separation with formation of a trapped eddy rotating about a horizontal axis. The mechanism is therefore similar to the mechanism producing scoop-shaped scours downstream of a ripple, except that there is no bed load to fill in the flutes as they are formed. He has also argued that, for the formation of flutes, the strength of the secondary flow must be sufficient to sweep out of the flute any sediment diffused into it from the flow above.

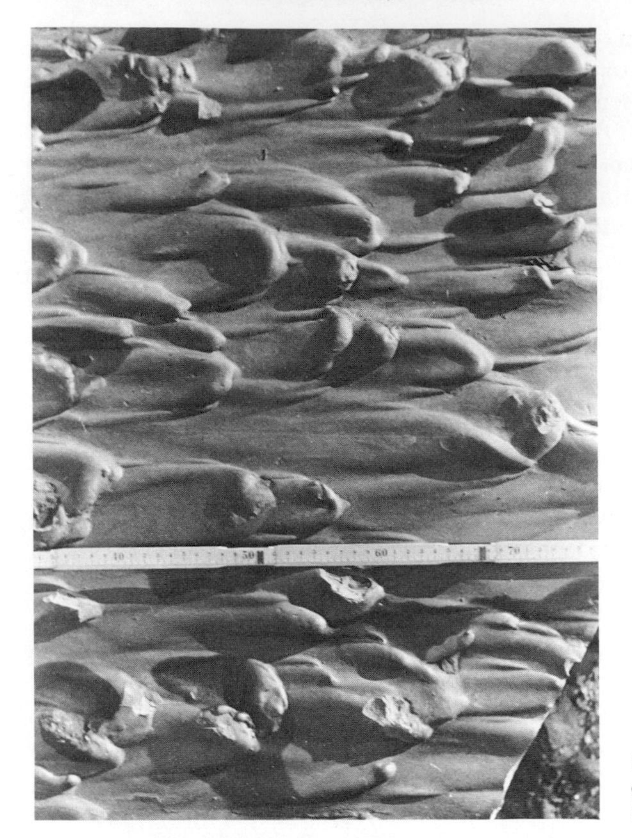

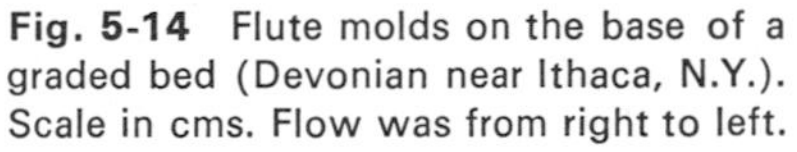

Fig. 5-14 Flute molds on the base of a graded bed (Devonian near Ithaca, N.Y.). Scale in cms. Flow was from right to left.

As the flow that eroded the flute decreases in strength, the strength of the eddy is also reduced and a point is reached where the coarse fraction of the suspended sediment (if present) begins to accumulate in the flute.

5.7 WAVES AND WAVE-FORMED STRUCTURES

Waves

In the discussion of bed forms and stratification given above, it has generally been assumed that the flow forming the bed form is unidirectional. Sediment is also moved by flows that reverse in direction periodically and that are generated by waves or tides. To most people, water waves imply surface waves of the type that can be observed breaking against the seashore or lakeshore but wave phenomena important to the sedimentologist are much more diverse. They include not only surface water waves generated by wind but also tides, tsunamis, seiches, bores, and internal waves in density-stratified water bodies. The topic is too large to be covered adequately in a text on sedimentology and we discuss only a few aspects of wave motion that are particularly relevant to the movement of sediment and formation of sedimentary structures. For further information, see the books by Bascom (1964), King (1959), Tricker (1964), and Zenkovich (1967).

Waves on the surface of the sea or lakes are generated by the wind. They may be described by various idealized wave theories, of which the oldest and best known is the theory of *oscillatory waves*, developed in the nineteenth century by Airy, Stokes, and others. In these waves (Fig. 5-15) the water particles oscillate in circular orbits about a mean position. The properties of the wave are defined by its length L, height H, period T, and the depth of water D. The definition of D, L, and H are indicated on Fig. 5-15: The period T is the time taken for two successive wave crests to pass a given point. Thus the velocity, or better *celerity c*, of the waveform (as opposed to the velocity of any particular fluid particle) is defined by the equation

$$L = cT \tag{2}$$

The celerity is given by the equation

$$c^2 = \frac{gL}{2\pi} \tanh \frac{2\pi D}{L} \tag{3}$$

In deep water, $D > L/2$, tanh $(2\pi D/L)$ is approximately equal to unity so the celerity is given by

$$c^2 = \frac{gL}{2\pi} \tag{4}$$

In a single period T, a water particle moves around one full circular orbit. At the surface, this orbit has a diameter of H and a circumference of πH (see Fig. 5-15), consequently the orbital velocity in deep water is

$$U = \frac{\pi H}{T} \tag{5}$$

For large waves these velocities may be several meters per second. It has been found that in deep water, the diameter of the orbit decreases exponentially downward, as indicated in Fig. 5-15. For depths below the surface greater than half the wavelength, the orbital diameter is less than 5% of the surface diameter so that disturbance of the water is generally negligible, in the sense that there is relatively little shear stress exerted on the bottom and, therefore, no sediment movement and little frictional deceleration of the wave motion.

Deep water waves undergo a number of changes in character and properties as they enter shallow water. It is apparent from Eq. 3 that for $L > 2D$ the celerity of the wave is progressively reduced as the water shallows. The change in celerity, which generally affects one part of a wave front before another, produces the phenomenon of wave refraction, which obeys Snell's law:

$$\frac{\sin \alpha_1}{\sin \alpha_2} = \frac{c_2}{c_1} \tag{6}$$

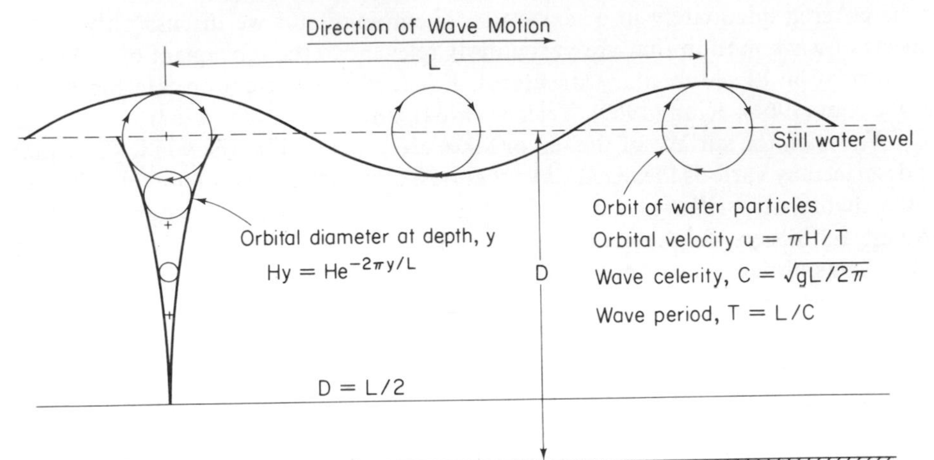

Fig. 5-15 Definition diagram for oscillatory wave in deep water.

where α is the angle between the local wave crest and the isobath (line of equal depth) and c is the celerity for that depth. In addition to the reduction in celerity, the following changes take place as the wave shallows: (a) decrease in wavelength; (b) increase in steepness (H/L); (c) simplification of the wave system as a result of the dying out of the short, irregular wave elements found in deep water; (d) change of water particle orbits from circular to elliptical, particularly close to the bottom; (e) change in waveform from symmetrical to asymmetrical, steepening of the front, oversteepening, and breaking. One characteristic of the wave remains constant: the period. Breaking of the waves takes place when the depth is approximately equal to $\frac{4}{3}H$ where H is the deep water height of the wave.

In shallow water, the classical theory that neglects viscosity is not always satisfactory. It has been found in some studies that where the depth/length ratio is small ($D/L < 1/50$) an alternative theory, *solitary wave theory*, predicts orbital velocities better (Zeigler, 1964). The orbital velocities close to the bed determine the movement of sediment in the region offshore and the development of wave-generated ripples. If the orbital paths of water particles were closed so that the particles returned to the same point as each wave passed, there would be no net drift of water produced by the wave. Even in deep water, however, the orbits do not close and there is a surface drift in the direction of wave propagation equal to a few percent of the wave celerity. In shallow water the particles are retarded by the bottom friction as they move away from shore under the wave trough (see Fig. 5-15) so that there is a much larger net drift of water toward the shore. Water moved toward the shore must be returned seaward, either by a slow drift seaward or by localized currents, such as rip currents. Longshore currents are also caused at least in part by the mass transport and theories have been devised to relate the velocity of longshore currents to the wave and shoreline characteristics based on mass transport equations.

Much of the energy of a wave is expended when it breaks. The mode of breaking varies and depends on the initial (deep water) character of the wave and the bottom conditions (Galvin, 1968). Three types of breakers have been described: (a) *spilling waves*, where the wave breaks and re-forms several times so that a considerable portion of the energy is expended before the wave reaches the beach; (b) *plunging waves*, where the wave oversteepens so that the crest plunges vertically down and the waves break with a crash. Following breaking, the oscillatory wave is transformed into a wave of translation and the water sweeps up the beach (*swash*) and then returns as a relatively thin sheet flow (*backwash*); (c) *surging waves*, which occur only when the beach is very steep so that the wave does not either spill or plunge but moves as a steep-fronted surge up the beach face. Spilling waves are characteristic of gentle offshore slopes or strong onshore winds (which blow over the wave crests). The zone inshore of the *breaker zone* where the waves first break has been called the *surf zone* and it extends inshore to the edge of the *swash zone*, defined by the effective seaward limit of backwash (see Fig. 5-19). The surf zone is best developed for gentle offshore slopes. For intermediate slopes, the zone may expand at low tide and contract at high tide as the water shallows and deepens. The most intense longshore currents are developed in the surf zone and studies using dyed sand released in the surf zone have demonstrated how rapidly sand may be moved in this zone (see Ingle, 1966). Sand is also moved along the beach by the action of the swash, which moves sand obliquely up the beach face if the waves approach at an angle.

Different shallow water wave regimes, developed in response to different deep water wave and shoreline characteristics, result in local net deposition or erosion of sand or gravel. Changes on many beaches are cyclic with sand accumulating on the beach face in response to certain long, flat waves (mainly in the summer) and being moved offshore by steep storm waves. Many beaches also undergo smaller cyclic changes related to daily or semimonthly tidal cycles (Strahler, 1966). For example, as the tide rises, the first effect felt by a point on the beach previously untouched by wave action is slight accretion due to deposition from the swash. At higher levels the main effect of the deeper part of the swash is to produce slight scour. The sediment scoured by the swash is deposited as a "step" at the base of the beach face. Then as the tide moves even higher, accretion takes place again at the point previously scoured by the swash. The whole cycle is then repeated in reverse as the tide falls.

In most water bodies, the interaction of waves, tides, and sediment is extremely complex and results in a continual state of change in response to fluctuating conditions. Not only do the characteristics of the waves change but they interact in complex ways with the wind, the tides, secular changes in sea level, and temporal fluctuations in sediment supply from rivers and other sources.

In rivers, grains move more or less directly down the river, with long "rest" periods while the grain lies temporarily buried in dunes, sand waves, or bars. In beaches, however, grains probably move up and down the beach at least 10 times as far as they are moved along the beach by longshore drift and the direction of longshore drift may be reversed from time to time depending on the wave conditions. Thus sediments on beaches are probably subject to at least 100 times as many sorting and abrasional events per unit of net transport as sediments in rivers.

Many sediments deposited along the shore display sedimentary structures not directly related to wave action but to flows, such as longshore currents, that are indirectly related to waves. Even in the case of structures directly formed by wave action, such as wave ripples, there may be a strong drift or general water movement combined with the oscillatory motion of the waves to give all transitions between wave- and unidirectional current-formed structures. The lamination and other structures formed on the beach face may be at least partly analyzed in terms of the downslope flow of the backwash, which gives rise to upper regime plane bed or even antidunes on the slopes characteristic of most beaches.

Bed Forms and Sedimentary Structures Formed by Waves

The oscillatory motion of the water produced by waves results in a corresponding motion of particles on the bed and the formation of ripples. It is possible to predict the conditions for the beginning of sediment movement on the bottom from a combination of the theory for sediment movement and the hydrodynamic theory of waves (for reviews, see King, 1959; Stanley, 1969; and Zeigler, 1964). There is a region offshore where the depth is too great and the water movement consequently too feeble to move sediment of a given size for a given set of wave conditions. Inshore of the *incipient point*, at which sediment movement begins, thoretically, there is a region where the slope of the bottom produces a zone of offshore net sediment motion. Further inshore, the greater strength of the onshore surge just equals the effect of bottom slope so that there is a *null point* where there is no net movement either onshore or offshore. Inshore of the null point, movement is onshore. This theoretical model has been elaborated in considerable detail and verified in laboratory experiments but attempts to verify the model on real beaches have given mixed results. It should be noted that the model predicts that on a given beach the null line position depends on the grain size, being closer to the shore for larger grain sizes. The result should be a gradual increase in grain size from offshore to onshore.

The mechanism of ripple formation by waves differs substantially from the mechanism described on p. 127 for current ripples. As before, the ripples form initially in response to a little understood instability of the sediment-water interface. As the crest of the wave passes, onshore flow of water over the ripple produces a lee eddy similar to that formed in the lee of current ripples. When the flow velocity slackens, the eddy rises (Fig. 5-16) and may carry with it sediment brought into suspension by the onshore surge. The offshore surge, which results from the passing of the wave trough, may carry this suspended sediment further offshore. The coarser sediment, which was moved as bed load by the onshore surge and which formed avalanche cross-laminae inclined toward the shore, is thus moved onshore at the same time that the finer sediment is separated out and moved in

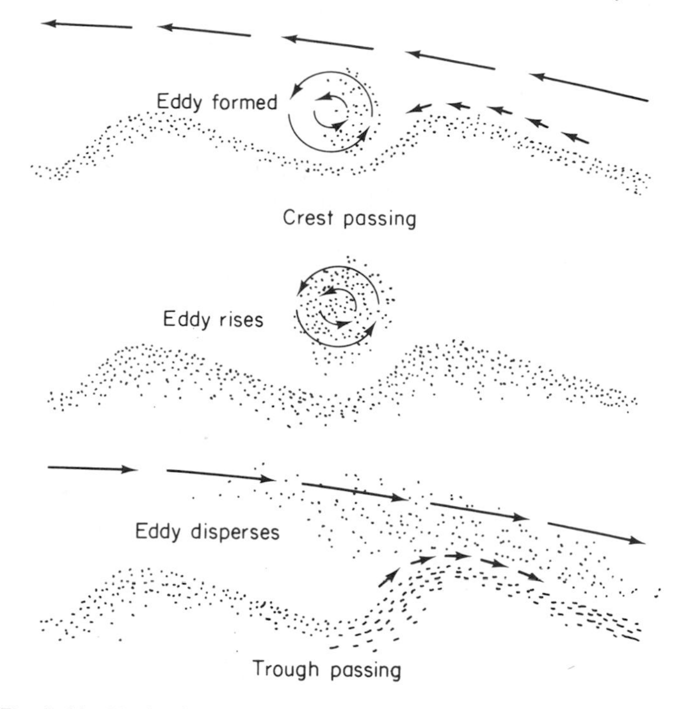

Fig. 5-16 Mechanism of sand movement over wave ripples. From Keulegan (1948, Beach Erosion Board Tech. Rept. 3).

the opposite direction. The result is that littoral sediments are generally well sorted, as well as decreasing in grain size progressively from the breaker zone offshore.

Furthermore, dip of wave-formed ripple cross-laminae is an inadequate indicator of the direction of sediment dispersal. Experiments using dyed sand have demonstrated that, even where the cross-lamination in wave ripples is uniformly inclined toward the shore, the sand is dispersed both toward and away from the shore, as well as parallel to the shore by longshore drift (Ingle, 1966).

Ripples formed by wave action are rarely entirely symmetrical and there is generally a predominant migration of the ripple onshore, with the preserved cross-laminae inclined in only one direction. In cases where there is a strong drift of sediment produced either by wave action itself or by a combination of waves and currents, such ripples may be somewhat similar to current-formed ripples (Harms, 1969). In cases where there is little net drift of sediment, the ripples are characterized by long, straight crests; symmetrical, rounded profiles; and uniform height and spacing. Ripple crests may be considerably sharper than troughs so that preserved form sets of wave ripples may be used for way-up determination. The vertical form index (ripple index) of wave-formed ripples varies from about 6 for coarse sand to 10 for fine sand, considerably less than the values of 8 (coarse sand) to 22 (fine sand) typical of current ripples, though individual current ripples are

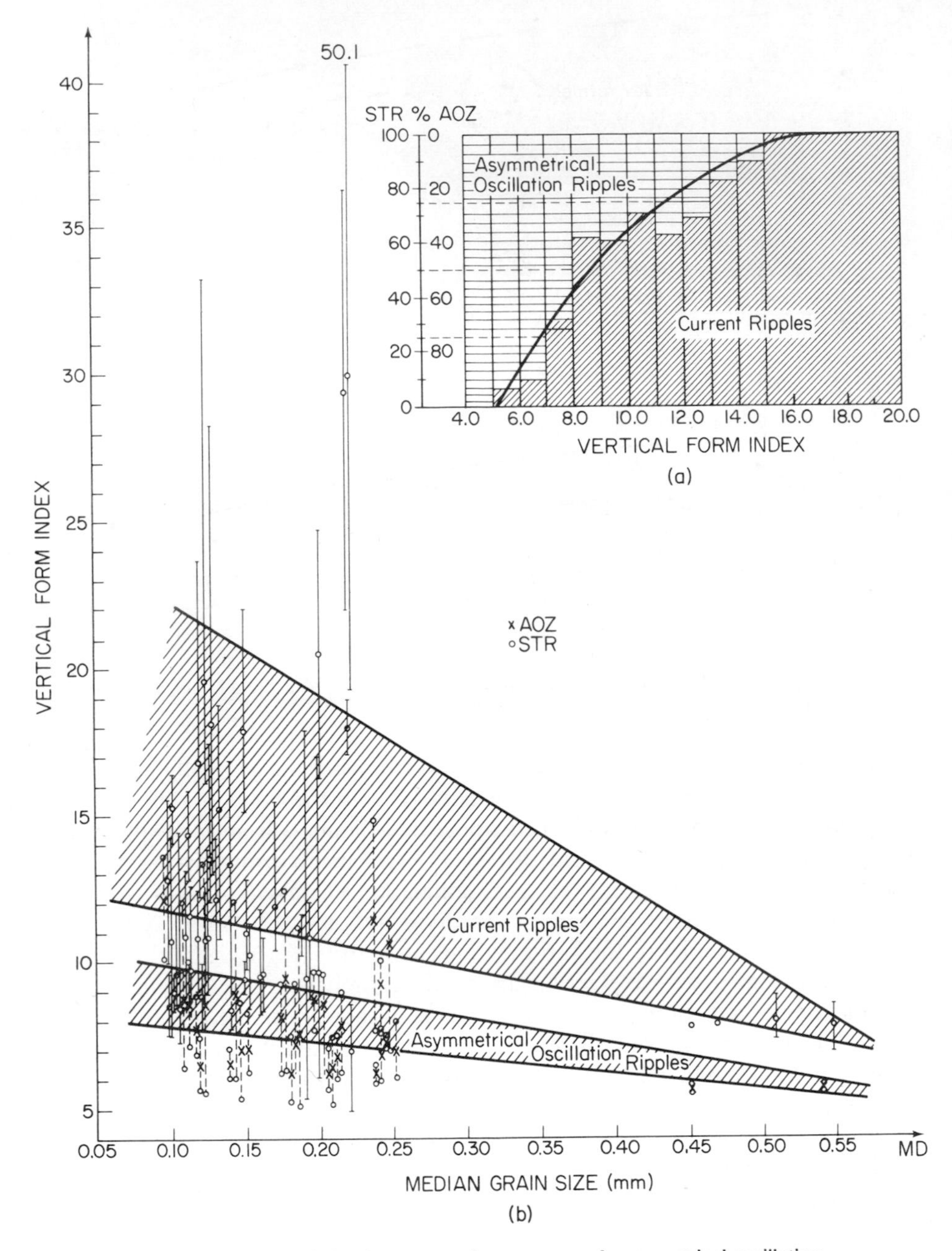

Fig. 5-17 (A) Relative frequency of occurrence of asymmetrical oscillation ripples (AOZ) and current ripples (STR) in North Sea, for different values of the vertical form index. (B) Relation between vertical form index and grain size for the two types of ripples. From Reineck and Wunderlich (1968b).

more variable and current ripples may have much smaller indices than the average (Fig. 5-17: see Reineck and Wunderlich, 1968b).

The chord of wave ripples is considerably more variable than that of current ripples. In current ripples the chord depends mainly on the grain size and to a lesser extent on hydraulic factors. For wave ripples, however, the chord is proportional to the horizontal displacement of water at the bottom. Consequently, it depends on the length and height of the waves and on the depth of water. The chord of natural wave ripples is known to vary from a few millimeters to at least 2 m.

Where the amount of wave energy is relatively low, such as the Texas coastline of the Gulf of Mexico, it appears that the main sedimentary structures formed by waves are the gently inclined beach face laminae and small-scale ripple cross-laminae at the upper part of the sandy zone (shoreface zone) seaward from the beach (Bernard, et al., 1970).

Offshore from the beach, in areas where there is a low tidal range, there may be a series of two to three submerged *breaker bars*, also called "offshore bars" (Fig. 5-18). These structures are generally parallel to the beach and their formation is related to the breaking of the waves. The depth and relief of the bars increases with distance from the beach. The largest waves break over the bar farthest offshore but smaller waves pass over this bar to break over one of the bars closer to the shore. The bars are generally asymmetric in profile, with the steeper side facing the shore. They change in position and degree of development with the wave conditions. Their internal sedimentary structures are still little known but in one

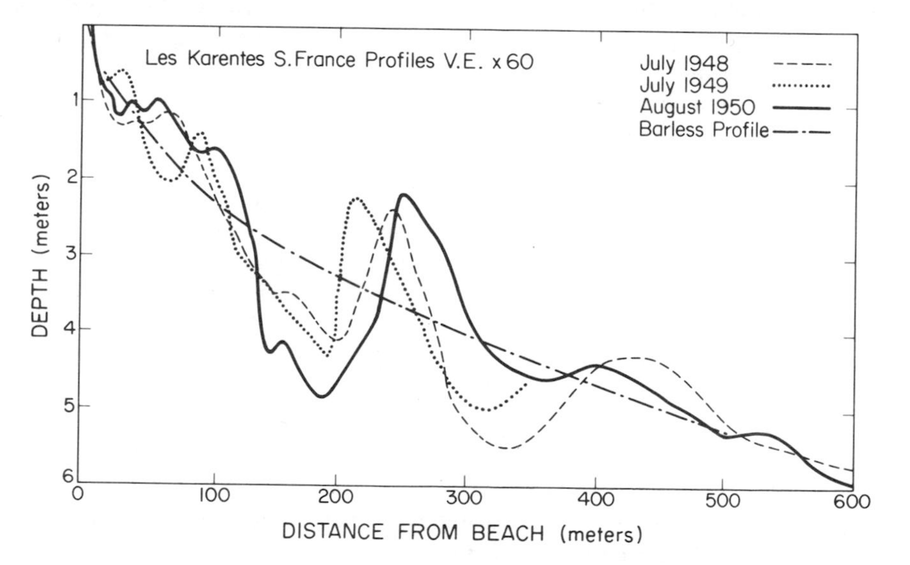

Fig. 5-18 Profiles of breaker bars off the beach at Les Karentes, south France. Note the presence of three main bars: one at a little more than one meter depth, the main bar at a little over 2 meters depth, and a third, incompletely formed bar at 4½ meters depth. From King (1959).

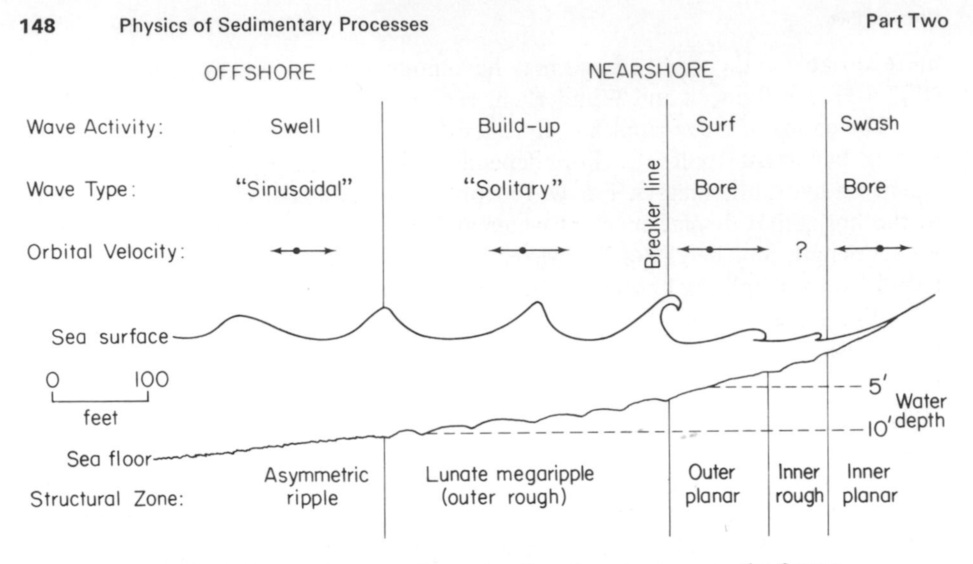

Fig. 5-19 Zones of wave action and sedimentary structures on the Oregon coast (after Clifton, Hunter and Phillips, 1970).

case in the Baltic studied by Seibold (1963) box cores revealed medium-scale cross-bedding inclined obliquely toward the shore.

The sedimentary structures formed along a coastline of high wave energy have been studied by Clifton, Hunter, and Phillips (1970). They found that offshore from Oregon beaches they could recognize a series of zones of sedimentary structures, corresponding roughly to a series of zones of changing wave conditions (Fig. 5-19). Seaward of the breaker line in water about 6 to 10 ft deep, there was a zone of lunate megaripples that migrated landward to form intermediate scale landward-dipping cross-bedding. A change in "flow regime" analogous to that seen in rivers can be recognized: from small ripples offshore to megaripples to a plane bed in the outer part of the surf zone. Between the outer surf zone and the swash zone is another zone of large-scale bottom irregularities, complex in origin, where intermediate scale seaward-dipping cross-bedding is formed. The general shoreward increase in "flow regime" is due to increases in the orbital velocity of the shoaling waves combined with the shoreward decrease in water depth.

The action of swash, backwash, and the tidal cycle in forming lamination on the beach surface is discussed on p. 143. It has been shown in several studies (e.g., Bascom, 1951) that the angle of slope of the beach face is related to the grain size and also to the degree of exposure (or energy level) of the beach. Coarser or less exposed beaches are steeper than finer or more exposed beaches. Dolan and Ferm (1966) showed that the slope was strongly correlated with the speed at which the swash rushes up the beach face.

There is also a grain size variation across the beach, with the coarsest material found not on the beach face but either above it on the berm or below it at the plunge point of the waves. Schiffman (1965) found that there were markedly bimodal size distributions in the transition zone between swash and surf zones.

5.8 TIDES AND TIDE-FORMED STRUCTURES

Tides

The tides are produced by the gravitational attraction of the sun and moon (see Clancy, 1968, and Tricker, 1964). At the earth's surface the gravitational attraction of the sun is only 0.46 as strong as that of the moon so the effect of the moon is dominant. At full and new moon, the attractive force of the sun and moon reinforce each other and the tidal range is large (*spring tides*) but when the moon is in its quarters, the effect is to decrease the tide-raising forces and the tidal range is low (*neap tides*). Most areas have semidiurnal tides, i.e., high tides approximately every $12\frac{1}{2}$ hr.

Tides produced on a perfectly spherical water-covered earth would have only a small amplitude: Tidal ranges on islands in the open oceans are in fact only about a foot. Tides in areas with a broad continental shelf, and particularly in partly enclosed gulfs open to the oceans at one end, are much higher because of resonant amplification of the tidal motion. For simplified rectangular shapes, the length and depths of gulfs (open to the ocean at one end) and seas (open to the ocean at both ends) for which resonant amplification takes place may easily be calculated (Tricker, 1964).

The Bay of Fundy, for example, has a length of roughly 160 mi, an average width of 40 mi, and an average depth of about 40 fathoms and even simple calculations indicate that it might be expected to have high tides. In fact, more detailed calculations show that the Bay of Fundy has a resonance period of 12.58 hr, which corresponds very closely with the lunar semidiurnal period of 12.42 hr. Thus the Bay of Fundy has the highest tides in the world, reaching 50 ft in range. Many other areas also show large tidal ranges (see the world map in Off, 1963). Fully enclosed water masses, such as the Great Lakes, or water masses with only a very small connection with the open ocean, such as the Mediterranean, have very low tidal ranges. Many ancient seas covered broad shelves and were connected with the ocean so that it might be expected that there would be large tides, though it has been suggested that, in very broad, shallow seas, tides would be damped out by friction.

Tidal currents may reach considerable speeds, particularly in areas where the flow is locally constricted by headlands or channelized. Detailed depth recordings have shown that many areas of the continental shelf are either swept free of sediment (as most of the English Channel or parts of the Bay of Fundy) or are covered with ridges of sand molded by the tidal currents (as much of the North Sea or Georges Banks).

Tidal Environments

A peculiarity of tidal action, as opposed to wave action, is that it tends to move fine sediment onshore. In areas protected from strong wave action, the mud accu-

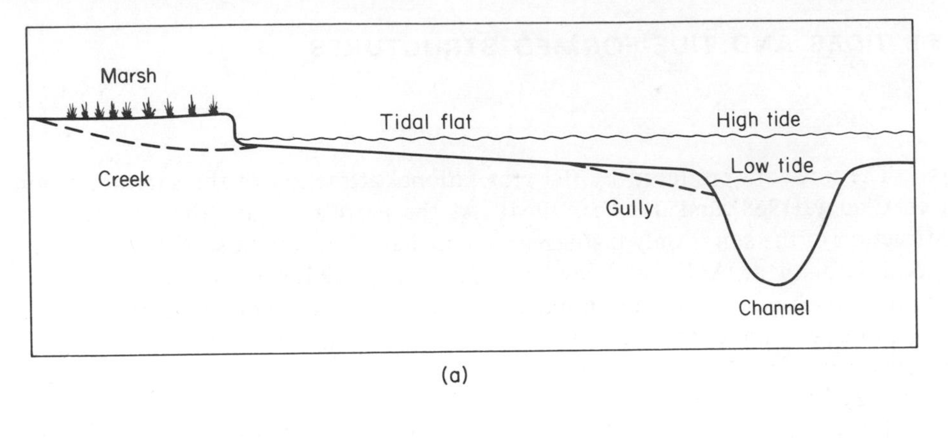

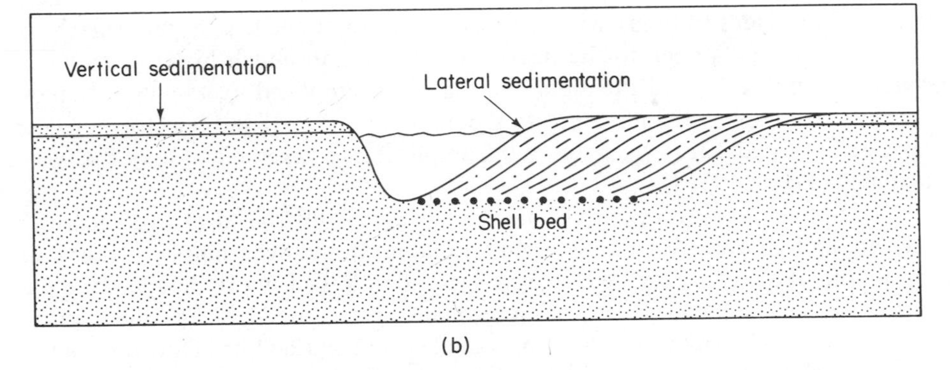

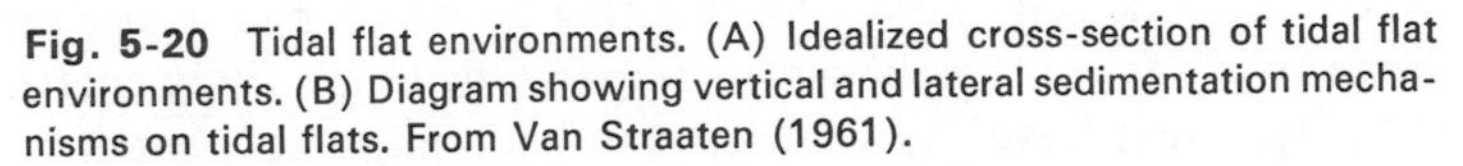
Fig. 5-20 Tidal flat environments. (A) Idealized cross-section of tidal flat environments. (B) Diagram showing vertical and lateral sedimentation mechanisms on tidal flats. From Van Straaten (1961).

mulates as intertidal flats drained by a network of tidal channels (Fig. 5-20). Several contributory factors have been suggested to explain the onshore movement of mud by tidal action but the most important appear to be the inshore decrease in tidal current velocity combined with the action of settling and scour lag (Van Straaten and Kuenen, 1957). Suspended fine sediment moved in by the flood tide does not settle immediately so that it is carried onshore even after the flood current velocity has slackened below its critical velocity for deposition ("settling lag"). At the ebb tide, scour does not take place until the velocity has risen above the critical erosion velocity, which is slightly larger than the critical velocity for deposition ("scour lag"). Thus the ebb current does not return the particle to its starting point, but to a point somewhat closer to the shore, and continual repetition of this process through many tidal cycles results in the accumulation of mud near the shore, unless it is removed by waves or local currents.

Sand is confined to the deeper parts of the tidal channels or to offshore regions where the tidal currents are sufficiently strong to move it. Inshore and on the shallow water of the mud flats, the tidal currents are too weak to transport much sand. The duration of high tide slack water is also longer than the duration of low tide slack water and the *average* depth over the flats at high tide is less than the average depth in the channels at low tide, so it is easier for fine sediment to settle out at high tide on the flats than at low tide in the channels. In many tidal flats there are regions of intermediate depth where fine sand is moved during the flood. Tidal currents and wave action form ripples that migrate over the mud deposited during high tide slack water. The resulting repeated interbedding of mud and rippled sand forms distinctive bedding types described below.

In humid regions, the highest parts of mud flats are colonized by vegetation and converted into salt marshes, which are flooded only at very high tide. Below the salt marshes a series of depth zones is commonly developed, each zone characterized by a distinctive assemblage of faunal and sediment types, down to the lowest parts of tidal channels that never emerge, even at spring tides (Evans, 1965, Van Straaten, 1961; Reineck *in* Lauff, 1967). Somewhat different zones have been described from several different mud flats in different parts of the world and the differences appear to be related to differences in the climate, tidal range, sediment supply, and degree and type of biological activity. Tidal flats in tropical regions are frequently composed of carbonate muds. These mud flats are generally similar to silicate mud flats but display distinctive associations of textures and structures due to the action of marine algae, evaporation, and diagenetic processes. They are discussed in more detail in Chapters 12 and 15 on carbonates and evaporites.

In regions where tidal or wave action is too intense to allow the formation of tidal mud flats, tidal sand bars or flats may be formed, with individual bars separated by a network of large tidal channels. Common morphological types for tidal sand bodies include ridges arranged parallel to the tidal currents, large transverse tidal sand belts, tidal deltas, and tidal point bars associated with the larger tidal channels (Coastal Research Group, 1969; Reineck, 1963).

Tidal deltas are commonly found in estuaries or associated with passages between barrier islands, where a tidal current spreads out after passing through a relatively narrow channel. Sand may be supplied by longshore drift from an adjacent island or spit. Tidal deltas are commonly dominated by either flood or ebb currents. Flood-dominated deltas are most common, if only because they build into lagoons or estuaries or onto shallow banks where they are protected from destruction by wave action. They are generally covered by dunes and ripples whose type and orientation are controlled by the depth, the nature of the tidal currents, and the topographic shielding provided by the nearby land or barrier islands.

Transverse tidal sand belts have been described by Ball (1967) from the carbonate sandbanks of the Bahamas, where they are composed of oolitic sands. The sand belts are formed from sand supplied from organic sources near the edge of the shallow water banks and moved into the shallow interior of the banks by a combination of wave and tidal action. Sand is moved across the belts during storms to accumulate as spillover lobes building into the protected water behind the sand belt. The spillovers terminate in a lobe-shaped "microdelta" with a steep slip face several feet in height. Cross-beds formed in these spillover lobes are generally inclined normal to the trend of the sand belt and toward the interior of the banks. Sand is moved by tidal currents and normal wave activity at the crest of these sand belts, but the overall geometry and many of the sedimentary structures appear to result from storm wave rather than tidal action.

Tidal ridges are common on open shelves where there is strong tidal action, such as at Georges Bank and the North Sea (Fig. 5-21). They are generally curved, several miles in length, and oriented parallel to the tidal currents. The cross section may be symmetrical or asymmetrical and the surface (the gently sloping surface only in the case of strongly asymmetrical ridges) is generally covered by dunes oriented with their crests transverse to the tidal currents. Spillover lobes may be present, oriented at a small angle to the crest of the ridge. It has been suggested that such ridges are formed by systems of secondary currents in the tidal channels (see Fig. 5-12 and discussion on p. 135).

Measurement of tidal currents over tidal sand ridges and in the large tidal channels between them has shown that the flood and ebb currents are frequently unequal in both duration and magnitude. In many cases, the surface of tidal ridges or deltas is dominated by flood currents and the channels divide into two groups, one dominated by flood and the other by ebb currents. Several factors cause inequalities of ebb and flood currents. Coriolis forces (due to the rotation of the earth) or topographic features deflect currents, causing flood currents to move up one side of an estuary or bay and down the other side. In an originally sinuous channel, inertia leads to formation of separate flood and ebb channels, with shortcuts across the channel bends being followed by the flood and the original more sinuous path being followed by the ebb. Water continues to drain off the higher tidal flats into the channels after the tide has turned so that some channels are dominated by ebb flows. Sediment transported by flood currents onto tidal deltas tends to accumulate at the landward end and these accumulations may divert ebb currents around the boundaries of the delta so that the delta surface is not affected by strong ebb currents. All of these factors may lead to dominance of flood currents over sand flats but this is not invariably the case and some flats are dominated by ebb rather than flood currents (e.g., Klein, 1970).

Bed Forms and Sedimentary Structures Formed by Tides

The bed forms produced by tides are generally similar to those formed by unidirectional flows. Waves are also important in the intertidal zone, with wave ripples locally abundant and wave action modifying both large- and medium-scale bed forms, particularly during stormy weather. Emergence of some tidal bars at low water results in deflection or even reversal of the direction of currents

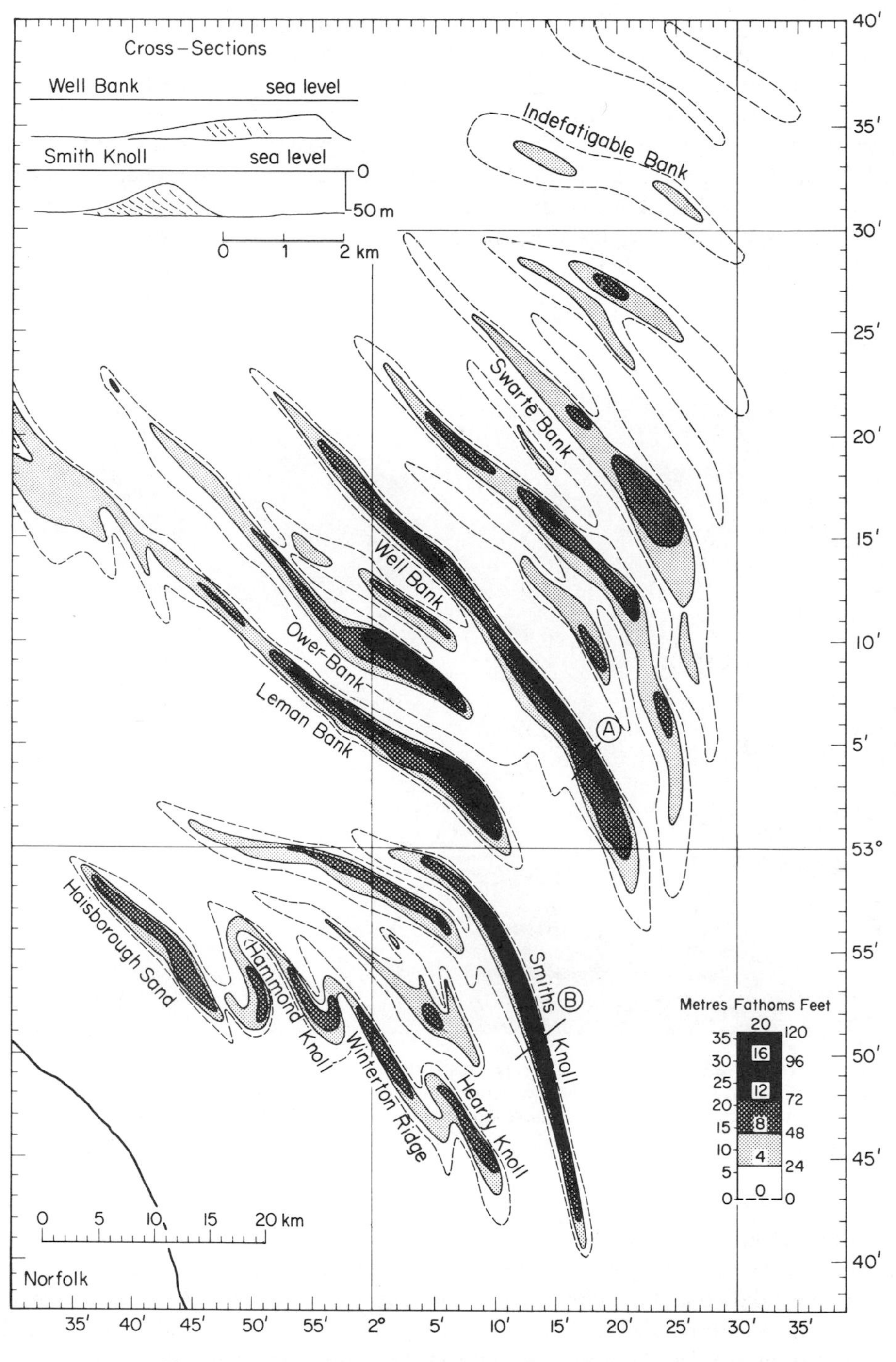

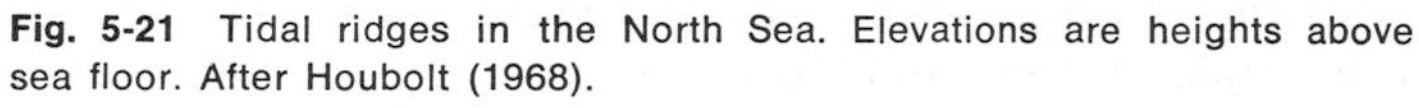

Fig. 5-21 Tidal ridges in the North Sea. Elevations are heights above sea floor. After Houbolt (1968).

during low water stages, with a consequent increase in the diversity of orientation of the bed forms and their associated sedimentary structures.

Because many parts of tidal ridges are dominated by either flood or ebb currents, preserved cross-stratification is frequently dominantly of one orientation. Even in cases where the distinctive "herringbone" pattern of cross-stratification with modes 180 degrees apart is not developed, the cross-stratified sets are rounded off by the opposed current and are of a type that should be readily recognizable in ancient rocks (Reineck, 1963; Boersma, 1969; Klein, 1970). Most dunes developed on tidal bars appear to have relatively straight or sinuous crests, probably as a result of the periodic reversal in flow direction (Fig. 5-22). Ripples, which

Fig. 5-22 Dunes with sinuous crests, formed by tidal action, Schelde Estuary, Holland. Note shallow scours downcurrent from the lee side of the dunes, falling-water level marks on the lee side of the dune, and oblique orientation of "ladder-back" ripples formed by later stage runoff. (Photo courtesy of Roger Walker.)

form and migrate much more rapidly than the larger features, are frequently of the fully developed linguoid, as well as sinuous and straight-crested varieties. In the intertidal zone, both ripples and dunes show features characteristic of emergence: These features include, on dunes, horizontal falling water level marks and small current ripples developed by late stage runoff transversely across the lee slopes, on ripples, flattops and "ladder-backs" (Fig. 5-22: see Tanner, 1960). Probably most of these structures, however, have little preservation potential and are destroyed by the next flood tide.

It appears that most of the sedimentary structures preserved in tidal sandbars are related to ripples and dunes and not to migration of the larger-scale "sand waves" or tidal ridges themselves. Cross-bedding produced by growth of micro-deltas (spillover lobes) is probably important in some environments.

On the intertidal mud flats, lamination is produced by the tidal cycle, as discussed above. Burrows and organic structures are extremely common in some climates and ecologic zones. Rippled sand migrating across a muddy substrate produces the types of bedding known as lenticular bedding (in German, Linsenschichtung), wavey bedding, and flaser bedding (in German, Flaserschichtung: see Fig. 5-23). The "flasers" are incomplete mud laminae trapped in ripple troughs during periods of slack water and all transitions are seen from predominantly sandy beds, with ripple cross-laminated sets separated by flasers, to predominantly shaly beds, with isolated ripple form sets (lenticular bedding). Such bedding types appear to be most common in tidal flat deposits, though lenticular bedding is also found in deltaic, lake, and sheet-flood deposits.

Migration of meandering tidal channels produces vertical sequences in mud-flat deposits (Fig. 5-20). At the base of such sequences, there is frequently a thin coarse lag deposit composed entirely of shell materials. Very coarse terrigenous sediments are commonly lacking, though they may be supplied by ice-rafting from beaches, as in the Bay of Fundy. Above the lag deposit, the sequence may be entirely muddy or sandy and then muddy, depending on the position of the channel relative to the facies zones developed in the intertidal zone. In the lower, sandy parts of tidal channels, migration of point bars produces a sequence of sedimentary structures and a "fining upward sequence" much like that produced by migration of point bars in meandering rivers.

5.9 WIND ACTION AND DEPOSITS

Introduction

Some aspects of wind movement of sediment have been discussed in Secs. 4.3 and 4.4. The large difference in mass density between mineral particles and air has the result that grain inertia is more important in air than in water, particularly for sand sizes. Bagnold (1941) distinguished between three principal mechanisms of sediment movement: suspension, saltation, and surface creep. Suspension and saltation by wind are discussed in Chap. 4. Saltation is much more highly developed in wind than in water and saltating grains frequently rise to a height of up to 50 cm. As the saltating grains return to the bed, they are accelerated by the wind as well as by gravity and they strike the bed with considerable force. Part of the impact energy triggers the saltation of other grains but most is dissipated in striking grains that do not saltate but are simply driven forward by the impact. The continual bombardment of the bed therefore results in a creep of grains along the surface.

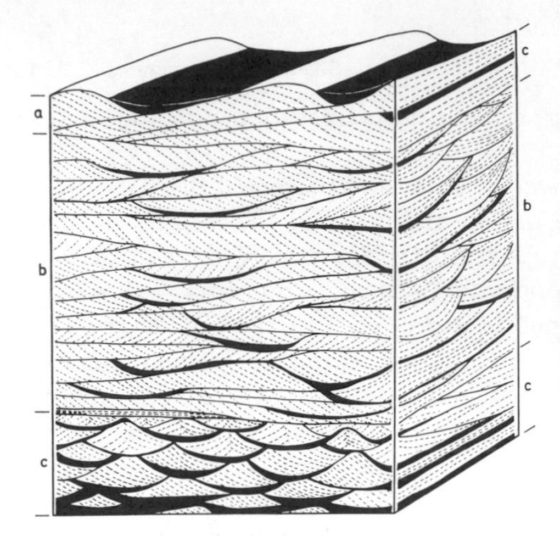

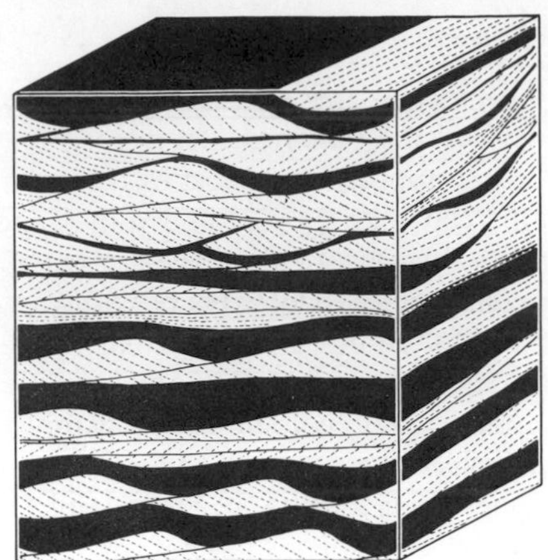

(a) Flaser bedding

a. formed from current ripples with straight crests

b. formed from current ripples with curved crests

c. formed from oscillation ripples

(b) Wavy bedding

(c) Lenticular bedding

Fig. 5-23 Flaser, wavy and lenticular bedding. After Reineck and Wunderlich (1968a). In each diagram, the upper part is drawn for current ripples and the lower part for oscillation ripples. (a) Flaser bedding: *a* formed from current ripples with straight crests, *b* formed from current ripples with curved crests, *c* formed from oscillation ripples. (b) Wavy bedding. (c) Lenticular bedding.

Observations on the velocity profile close to the sand bed have shown that movement of grains by saltation causes a marked increase in the apparent roughness of the bed. The height of rise of the saltating grains must be approximately equal to $U_0^2/2g$ where U_0 is the velocity at which the grain rises from the bed. U_0 is known to be roughly equal to the shear velocity U^*. When the velocity ratio U/U^* is plotted against the logarithm of the ratio of the height y to the thickness of the saltation layer $U^{*2}/2g$, a straight line results (Owen, 1964). This demonstrates that the velocity profile is of the type expressed by Eq. 12 in Chap.4 where the roughness height k is proportional to the thickness of the saltation layer. The significance of this observation is that during saltation, little shear stress is exerted directly by the wind on the bed. Instead, the shear stress is transferred to the saltating grains which in turn transfer it to the bed and to the air near the bed. The air within 1 or 2 cm of the bed is moved by the grains rather than vice versa. A consequence of this is that it is very difficult for the wind to move particles by surface creep that are too large to be moved by the inertial impact of saltating grains. Such large particles tend to accumulate on the bed to form a lag deposit (desert pavement).

Bed Forms and Sedimentary Structures

The predominance of saltation in the movement of sand by wind also results in important differences between the mechanisms forming wind and water ripples. Ripples in air are formed and maintained by surface creep of grains resulting from bombardment of the surface by saltating grains. Slight unevenness of the sand surface tends to protect grains to the lee side, causing accumulation, and to expose grains on the stoss side, causing erosion. The surface unevenness tends to grow into ripples and the ripples migrate downwind and gradually become larger until they reach an equilibrium size. Bagnold (1941) noted that the ripple wavelength tends to be of the same order of magnitude as the characteristic length of the saltation jump. Both tend to increase up to a critical shear velocity where the ripples flatten out and disappear. Sharp (1963) claimed that the length of the saltation jump is not the fundamental mechanism controlling ripple length or index. Wavelength increases with wind strength because increasing wind strength lowers the angle at which saltating grains strike the ground, though some of the increase in wavelength due to an increased "shadow zone" is probably lost because of a corresponding reduction in ripple height. Increasing grain size also increases ripple size because larger grains can be piled higher than smaller grains before being knocked away from the crest by the impact of saltating grains. The material composing the ripples may be somewhat coarser than the sand over which the ripples are migrating and many workers also report that the ripple crests, which are the most exposed parts, are also the coarsest (the opposite of the sorting observed in current ripples in water).

The vertical form index (ripple index) for well-sorted sands is generally 30 to 70 for fine sands but may be as low as 10 for poorly sorted sands. It appears to vary inversely with grain size and directly with wind velocity. Eolian ripples are almost invariably straight-crested. Linguoid and other three-dimensional forms are not found, presumably because the ripples are formed by saltation impact not by traction induced by flow; therefore, the lee eddy and other secondary flows that lead to the development of complex ripple forms in water are of little importance in eolian ripples.

The larger-scale bed forms produced by wind are more directly related to the pattern of wind flow than are the small-scale ripples. Eolian dunes are therefore more closely related to dunes in water but there are some important differences. Separation of flow takes place at the brink of the dune but in most cases there appears to be no single major zone of backflow formed in the lee of the dune. Presumably because of the lack of a well-developed lee eddy, scoop-shaped scours similar to those associated with dunes in water are not commonly associated with eolian dunes. The size of the dune is not restricted by flow depth in air as it commonly is in water and the direction of flow is generally more variable in air than it is in water. Dunes may be either transverse or longitudinal to the predominant wind direction and large isolated lunate dunes (*barchans*) are formed in regions of relatively low sand availability. Relatively small-scale straight, sinuous, or lunate dunes of the type common in water are rarely formed by wind. Consequently eolian cross-bedding is not generally found in medium-scale tabular or trough sets. The typical cross-bedding of both barchan and longitudinal dunes consists of large-scale, wedge-shaped, or tabular sets, with individual cross-strata extending as much as 10 m or more and inclined at high angles (24 to 34 degrees, see McKee, 1966; McKee and Tibbitts, 1964). In some coastal eolian dunes, angles of up to 42 degrees (greater than the angle of repose) are not uncommon, probably because the angle of repose is increased by dampness of the sand. The high-angle cross-strata are formed by avalanching in the lee of the dune crest and the low-angle planes bounding the sets form as erosion planes developed on the upwind side of the dunes. Accretionary cross-strata, with a relatively low angle of dip, are also common but less abundant than high-angle cross-strata. This type of cross-bedding is typical of that produced by large dunes and is not necessarily restricted to eolian dunes but there are no well-documented examples of similar structure produced by subaqueous dunes.

Transverse dunes are formed by winds that blow predominantly in one direction and the high-angle cross-strata within them commonly dip toward directions contained within an arc of 120 degrees. Longitudinal dunes, however, are formed by winds that blow from two different directions and consequently they have high-angle cross-strata with two direction modes, separated by almost 180 degrees. Coastal sand dunes may be formed by winds from one or several directions and consequently may display either unimodal or polymodal cross-bedding direction distributions.

5.10 MASS FLOWS

Types of Mass Flows: Debris Flows

In most of the modes of sediment movement considered in earlier sections, sediment movement results from fluid movement and the interaction between the moving particles is relatively unimportant. Even in the case of saltation, where impacts from saltating grains trigger other saltations and produce surface creep, the basic saltating movement appears to result from the action of hydraulic lift forces and aerial collisions between grains in the zone of saltation appear to be relatively rare and unimportant.

Downslope movement of sediment, however, does not always result from the action of forces produced by fluid flow. In many types of mass movements, fluid is present but its role is to reduce friction and motion of the fluid-solid mass is produced mainly by gravity acting on the solid materials. A spectrum of mass movements may be recognized, from those little affected by the presence of fluid, to those in which the fluid is as important as the solid. In the subaerial environments, this spectrum includes three basic types: falls, slides, and flows (Varnes, 1958). It seems probable that the same types also occur in subaqueous environments. *Rock-falls* form accumulations of scree. Presumably the probability of preservation of a subaerial scree is small but screes formed adjacent to subaqueous scarps might well be preserved in the geologic record. The authors know of no described example, however. In *slumps* and *slides*, there is movement along discrete shear planes, with relatively minor internal flow deformation. The two terms are more or less synonymous but slumping suggests rotation of the fragments and a greater degree of internal deformation. In both slides and slumps, the size of the moving blocks is very variable from extremely large to a centimeter scale. All transitions appear to exist between subaerial rock or debris slides (or slumps) and rock or debris or mud flows. In the subaqueous environment, there is also a complete spectrum of phenomena.

Slumps and slides are produced by failure of the rock mass, when its strength is exceeded along certain planes. The mechanism of failure in recently deposited sediments is discussed in Sec. 5.12 but an important factor is the reduction in strength of rapidly deposited sediment because of *excess pore pressures*. In coarse grained sediments, or in fine grained sediments accumulated very slowly, the weight of the sediment particles is supported entirely by other solid particles and is not transferred to the pore fluids. The pore pressures are therefore hydrostatic. In fine grained, rapidly deposited sediment, however, some of the weight of the particle is supported by an "excess" fluid pore pressure. Excess pore pressures may be formed even in sand-sized or coarser sediment if the sediment has an unstable packing, which is disturbed by a sudden shock. The result of a breakdown in the sediment fabric is a sudden increase in pore pressure, which results in loss of strength, causing "*liquefaction*" of the sediment. A similar type of mechanism may take place in certain clays, which lose their property of strength when sheared beyond a certain shear stress. The excess pore pressures created by liquefaction persist only until the pore fluids escape or until the unstable sediment fabric is reestablished, but while they exist they may cause the sediment to flow easily down relatively gentle slopes.

In subaerial arid or semiarid environments, sediment accumulates on slopes or in dry gullies by mass wasting processes. As long as the sediment is dry, the internal friction of the sediment particles may prevent movement downslope but wetting produced during one of the rare but intense storms typical of semiarid regions or by rapid melting of winter snow may cause mass flowage to take place. The general class of such rapid, relatively watery mass flows is called *debris flow* and it includes types composed mainly of either sand or mud called *sand flows* or *mudflows*, respectively. A variety of relatively slow, water-poor mass flows have also been described, including *solifluction* (downslope movement of water-saturated soils), rock or debris glaciers, and slow creep of sand.

Because of the very high concentration of solid particles in mass flows, the behavior of the fluid-solid mixture differs substantially from that of water or of any Newtonian fluid. Concentrations of up to 30% by volume of inert particles increase the apparent viscosity of the dispersion but the behavior remains essentially Newtonian. Higher concentrations of inert particles, or much lower concentrations of particles, such as clays, that form structured dispersions, cause departures from Newtonian behavior (Fig. 5-24). Several types of different departures are possible: (a) The dispersion may be a *non-Newtonian fluid* that has no strength but that has

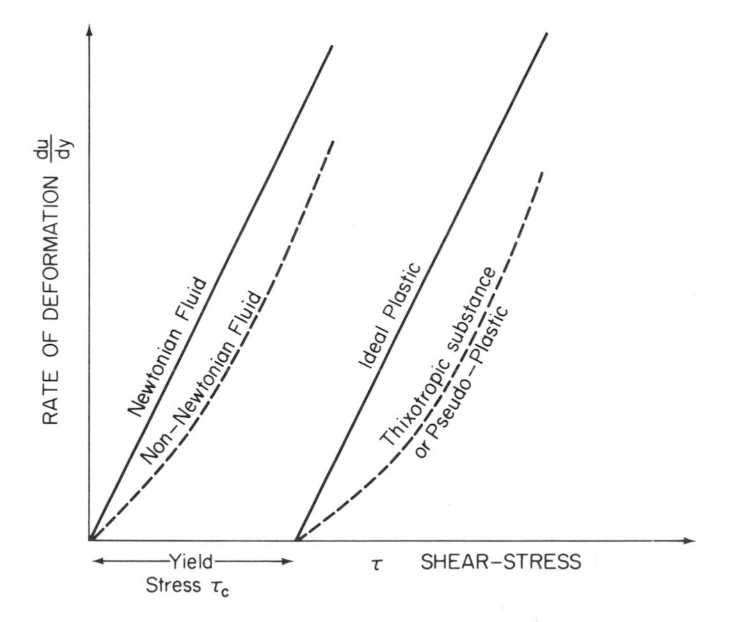

Fig. 5-24 Rate of deformation *vs* shear-stress for fluids and plastics.

a viscosity that varies with the rate of shear. Commonly the viscosity decreases at higher rates of shear as the structure of the dispersion, caused by interparticle forces, is broken down. (b) The dispersion may have strength so that no deformation takes place until a critical shear stress for yield has been exceeded. If, after the strength is exceeded, the dispersion behaves as a substance with constant viscosity, it is called an ideal plastic, or *Bingham plastic*. A more common case is that the dispersion behaves when it flows as a substance with variable viscosity, or *pseudo-plastic*. (c) A class of substances, called *thixotropic*, have strength until they are sheared, when the property of strength is lost and the substance behaves like a fluid (commonly a non-Newtonian fluid) until it is allowed to rest. After a period of rest the property of strength is reestablished. Dispersions of montmorillonitic clays (bentonite) commonly display this property of thixotropy. It is this property that makes them valuable as drilling muds. When circulation of the mud stops, the fluid "jells" and its strength prevents the settling of the cuttings suspended in the mud.

The high viscosity and pseudo-plastic or thixotropic behavior of a concentrated muddy dispersion strongly influences the nature of mass flows and their deposits. Subaerial *debris flows* have densities of 2.0 to 2.4, indicating a water content of as little as 40% by volume. Debris flows commonly advance in a series of surges, with maximum velocities of the order of 1 to 3 m/sec. The thickness of the flow is of the order of 1 m and the slope may be as low as 1 degree, though it is commonly of the order of 5 degrees. The viscosity of the flow is not known from direct measurement but, by assuming viscous flow and knowing the thickness, slope and surface velocity, Sharp and Nobles (1953) calculated a probable value of the order of 1000 poises (as compared with 0.01 poise for pure water). Thus it appears that the Reynolds number is quite low (of the order of 10 to 100) so that it is probable that the flow is indeed laminar in nature.

Debris flow deposits are commonly poorly sorted and bimodal with cobbles and boulders imbedded in a finer matrix (see Landim and Frakes, 1968: reference in Chap.3). Beds deposited by individual flows lack internal stratification. The larger fragments are relatively easily transported in debris flows because they are supported in part by high buoyancy forces due to the high density of the mud in which they are submerged. The concentration of fragments is so large that they interact with each other during flow to reduce greatly selective deposition of the coarser materials. The muddy matrix itself may retain some strength, even while shearing is taking place, so that if the submerged weight of the larger fragments does not exceed the strength of the matrix, there is no settling of the fragments through the mud. Deposition of the whole flow takes place at the point where flow is no longer possible because the maximum shear stress in the flow no longer exceeds the yield stress of the mud (Hooke, 1967).

Debris flows commonly show a general decrease in both size and amount of coarse materials downflow, showing that there is a progressive, though poorly effective, elimination of the coarser materials. Apart from this, there is little sorting of material during flow although there is commonly a concentration of the largest fragments at the front and sides of individual flows. The mechanism responsible for this concentration is not understood. Hooke (1967) has suggested that some deposits on alluvial fans result from a process called "sieve deposition." If the material at the surface is sufficiently coarse and permeable, the entire flow may infiltrate before reaching the toe of the fan, leaving a lobe of coarse debris on the surface.

Debris flows are the best studied types of subaerial mass flows but only a few flows have been directly observed. For other types of rapid mass gravity transport, for example large rockfall avalanches, the mechanism must be inferred from a study of the deposit, combined with the accounts of a few eyewitnesses (Shreve, 1968). Subaqueous rapid mass flows have never been directly observed and even the location and characteristics of modern examples are in doubt. For this reason geologists have attempted to make use of theory in interpreting suspected mass flow deposits in the geological record.

Theory of Concentrated Grain Dispersions

The best known general theory for the behavior of concentrated dispersions of inert grains under shear is that of Bagnold (1956, 1966). According to this theory, at concentrations as low as 9% by volume, the interactions between the particles themselves are far more important for determining the properties of the flow than interactions between fluid and particles (such as turbulence). Two shear regimes were distinguished by Bagnold: a viscous regime in which particles do not necessarily actually collide but interact with each other because of the viscous shear stresses set up by the near approach to each other of particles moving at different speeds, and an inertial regime in which actual grain collisions are important and fluid viscosity no longer plays a significant part.

In the viscous regime, the important parameters are the shear stress on the dispersion T, the viscosity of the fluid μ, the rate of shear du/dy, the diameter of the particles (assumed uniform in size) d, and the average spacing of the particles S. In the inertial regime, the viscosity of the fluid may be neglected but the density of the grains ρ_S must be included in the analysis. It is possible to derive equations for the two regimes from dynamic or dimensional analysis and the exact nature of these functions and of the conditions defining viscous and inertial regimes were determined by Bagnold in a series of experiments with dispersions of neutrally buoyant grains sheared between revolving drums.

Bagnold also pointed out that, corresponding to the shear stress T, there must be a normal stress, or "dispersive pressure" P, produced by grain interactions. He found that the ratio between P and T depended on concentration and is defined by an angle of internal friction α, where

$$\frac{T}{P} = \tan \alpha \tag{7}$$

From his experiments, Bagnold found that in the inertial regime $\tan \alpha = 0.32$

for moderate concentrations ($d/S < 12$) and was higher for higher concentrations. In the viscous regime, $\tan \alpha = 0.75$.

The theory briefly sketched above, and considerably elaborated by Bagnold, still needs further experimental verification. It may tentatively be used to study the conditions under which concentrated dispersions of inert particles will flow. For example, by determining the T and P from the downslope and normal-to-slope components of gravity acting on the grains and setting $T/P = \tan \alpha$ as the condition for flow downslope, it is possible to determine that the minimum slope for subaqueous flow of a grain-water mixture is 18 degrees in the inertial regime. The slope might be decreased (a) by presence of a denser interstitial fluid than water, for example, a mud matrix; (b) by the existence of an excess pore pressure acting against the gravity normal stress; (c) by addition of an additional downslope shear stress imposed from a flow above; and (d) by turbulence.

In subaerial debris flows, the slope on which flow can take place is considerably decreased by the presence of water or mud in the interstices of the coarser particles. In subaqueous mass flows, a reduction in slope angle to about 10 degrees is theoretically possible for grains with a dense mud matrix, moving in the viscous regime. The slope may be reduced to about 5 degrees if the flow is inertial. It seems probable, however, that in the subaqueous environment a more important role must be assigned to excess pore pressures than simply to the presence of a mud matrix.

Grain Flows

Mass flow of cohesionless grains was called *grain flow* by Bagnold. Submarine grain flows have been suggested by several geologists as an explanation of certain very thickly bedded, relatively well-sorted, almost structureless sandstone beds (see Stauffer, 1967, and Sanders *in* Middleton, 1965a). It is not yet clear, however, what importance factors (b) to (d), in the list given above, have in modifying grain flows in nature.

Several examples of grain flows with excess pore pressure produced by "retrogressive flow slides" of sands have been described. Such flows form by the collapse of locally steep slopes in sand. The collapsed sand liquefies and flows away from the base of the slope so that it again becomes oversteepened and collapses and so on. Flows of this type were observed on slopes of 8 to 11 degrees in Trondheim Harbor (Norway) and on slopes of 4 to 5 degrees in Helsinki Harbor (Finland). On a somewhat smaller scale, such flows have been observed to form as a result of dredging for sand in the Netherlands (see Middleton *in* Stanley, 1969).

Supposed grain flow deposits display the following distinctive characteristics: (a) outsize clasts in a sand matrix; (b) thick, ungraded, sharply bounded beds, generally massive internally; (c) presence of faint, dish-shaped laminae, 4 to 50 cm long, oriented parallel with bedding (the origin of this "dish structure" is poorly understood), (d) scarcity of sole marks; (e) lack of structures such as lamination or cross-lamination typically formed by traction (Stauffer, 1967). In some examples, these characteristics are modified by the presence of size grading and large flutes on the sole. Both grading and flutes suggest that the flow may have been turbulent. Calculations based on reasonable assumptions about the thickness, velocity, and viscosity of large grain flows also suggest that the flow may be turbulent. At present, there is little agreement among geologists about whether these sandstones and some associated conglomerates were formed exclusively by a type of grain flow mechanism or by highly concentrated turbulent turbidity currents (see below) or possibly by some other mechanism.

5.11 TURBIDITY CURRENTS

Definitions and Hydraulics

In the derivation of the Chézy formula for river flow given on p. 86, there are two implicit assumptions. One is that the air above the water has a negligible density compared with that of water. Strictly, we should write for the gravity force acting on the water not $G = \rho gALS$ but $G = (\rho - \rho_a)gALS$, where ρ_a is the density of the air. Another assumption is that all the resistance to flow comes from the solid boundary and none from the interface between the water and the air above. Both assumptions are well justified for the case of water and air but they would not be justified if the two fluids did not differ greatly in density.

Currents that are kept in motion by gravity acting on relatively small differences in density between different fluids (gas or liquid) or different parts of the same fluid mass are called *density currents*. The terms "stratified flows" and "gravity currents" are also used by some writers. The relatively small density differences that cause density currents may be caused by differences in *composition* (e.g., a density current of methane flowing above air, up the roof of a mine passage) or salinity (e.g., fresh water spreading out over the top of the more saline seawater seaward from a delta or estuary or more saline water flowing below a layer of less saline seawater) or *temperature* (e.g., cold glacial meltwater entering a warmwater lake) or *dispersed sediment*. The case of density currents caused by dispersed sediment is sufficiently important to geologists that these density currents have been distinguished by a separate name and are called *turbidity currents*. All these different types of density currents share some important hydraulic properties and these are discussed first, before considering some of the peculiarities of turbidity currents.

Consider first the case of the uniform steady flow of a density undercurrent moving down a slope S (Fig. 5-25). If the density of the fluid above is ρ and the density of the underflow is $(\rho + \Delta\rho)$ then the force of gravity acting on a volume AL is $G = \Delta\rho gALS$. The resistance is composed of two parts, the resistance at the solid boundary R_0 and the resistance at the fluid interface R_i, where $R_0 = \tau_0 PL$ (as before), and $R_i = \tau_i WL$, where W is the width at the interface of the volume AL and τ_i is the shear stress acting at the interface. To simplify, consider the two-

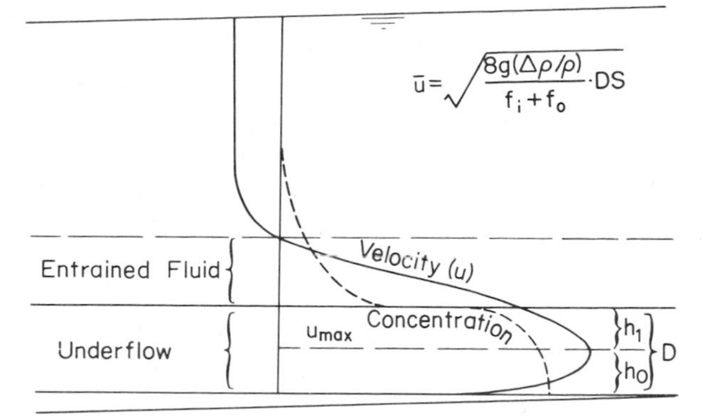

Fig. 5-25 Uniform steady flow of a density underflow, showing typical velocity and sediment (or salt) concentration profiles. Note that above the underflow there is a zone of entrained fluid, which is also a zone of mixing between the underflow and the water above. A very slow reverse flow is necessary in the upper part of the channel to replace the water entrained by the undercurrent.

dimensional case (a wide channel or sheet flow) where $P = W$. Then, equating the forces of gravity and resistance and solving for the combined shear stress at the interface and bottom,

$$\tau_0 + \tau_i = \Delta\rho g \frac{A}{W} S$$

$$= \Delta\rho g D S \tag{8}$$

where D is the thickness of the underflow. Making use of equations similar to Eq. 6 of Chap. 4 (with $\Delta\rho$ substituted for ρ), we obtain for the average velocity of the underflow

$$U = \sqrt{\frac{8g}{f_0 + f_i}}\sqrt{\Delta\rho D S} \tag{9}$$

where f_0 and f_i are resistance coefficients for the bottom and fluid interface, respectively. Equation (9) is the equivalent for density currents of the Chézy formula for rivers. It may be noted that the average velocity of a density current will be less than that of an equivalent sized river because of two factors: the reduced force of gravity $g' = \Delta\rho g$ and the presence of an additional force of resistance caused by friction at the upper interface.

One of the surprising aspects of density currents is the small amount of mixing between the density current and the fluid above. Two separate aspects of this interface may be considered: its hydrodynamic stability and the amount of actual mixing across it (if it is not stable). The criterion for stability has been investigated both theoretically and experimentally. It has been concluded from these investigations that the interface is unstable for all large fast-moving density currents (Middleton, 1966). Instability leads first to the formation of waves at the interface and then to a zone of mixing (Fig. 5-25). It has been found experimentally that the amount of mixing across the interface depends mainly on the densiometric Froude number:

$$\mathrm{F_r} = \frac{U}{\sqrt{g'D}} \tag{10}$$

Provided this number is considerably less than unity, there is actually very little mixing; therefore there is little extra resistance caused by the interface (or f_i is smaller than f_0) and the density current can flow for many miles without becoming completely mixed with the surrounding water and so losing its indentity.

There are many examples of density currents in nature that flow more or less continuously (ocean currents, density currents in some lakes) but others appear to be transient surges, set off by the sudden release of a mass of denser fluid (e.g., a large volume of relatively dense suspension created by slumping of unconsolidated sediment at the head of a submarine canyon). In these surges, the current may be considered to have a head, body, and tail. The body of the current is almost uniform and the average velocity may be predicted from Eq. 9, but the head is at least twice as thick as the body and has distinctive characteristics that differ from the rest of the current.

The general character of density current surges may be illustrated by a simple experiment in which a denser fluid contained in a lock is released suddenly, by removal of the gate of the lock, into a horizontal channel filled to the same height as the lock with lighter fluid. The movement of the fluid at successive intervals of time is illustrated by Fig. 5-26. This figure is of interest for another reason,

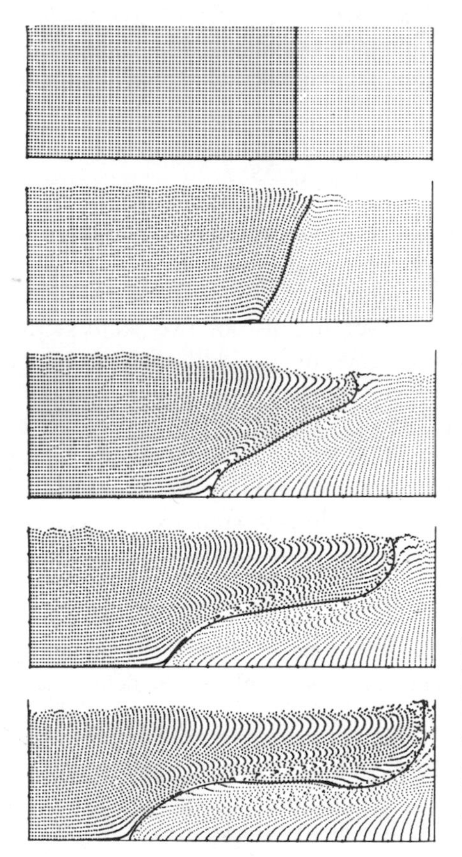

Fig. 5-26 Computer model of water particle movement in a density current surge. The top diagram shows the initial state. The lock on the right contains fluid which is 1.2 times as dense as the fluid in the channel to the left of the lock gate. After the gate is removed, the dense fluid flows to the left under the lighter fluid. After Daly and Pracht (1968, Phys. Fluids, v.11, p. 115–130).

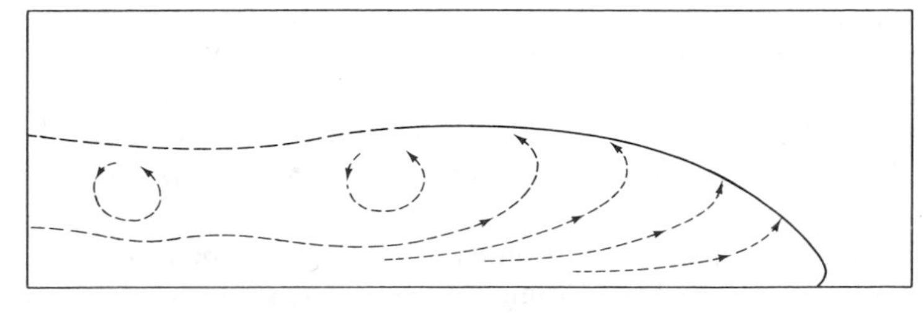

Fig. 5-27 Shape of the head of a density (or turbidity) current, showing the pattern of flow within the head. Note upward diverging flow pattern within the head and eddies at the rear of the head. After Middleton (1966).

too. It is actually not the record of a "real" experiment but of an experiment performed in a large electronic computer, by numerically solving the Navier-Stokes equations for each small particle of fluid through a succession of small time intervals (only a few of which are shown in Fig. 5-26). It has been found both theoretically and experimentally (Middleton, 1966) that the velocity of the head is given by the following equation:

$$V = 0.7\sqrt{\frac{\Delta\rho}{\rho}gD_2} \qquad (11)$$

where D_2 is the thickness of the head. The coefficient (0.7) does not change much for a small change in slope (up to about 0.05). The shape of the head and the nature of the flow inside it are shown in Fig. 5-27. Inside the head, the flow diverges and sweeps up and around the head. Behind the head, denser fluid is torn away as a series of large eddies. There are two important consequences of this pattern of flow. In a turbidity current, the head is a region of erosion, not deposition, even if deposition is taking place in the body of the flow behind the head, and denser fluid must constantly be supplied into the head from the main flow to make up for the material lost from the back of the head. Experiments have shown that the loss and supply are greatest when the slope is steep (high Froude numbers).

Mechanism of Sedimentation and Formation of Sedimentary Structures

The earliest interest shown by geologists in turbidity currents arose from studies of the undercurrent formed by the Rhone River as it entered Lake Geneva and the relation between this undercurrent and the subaqueous form of the Rhone Delta (see Daly, 1942, for a review). The Rhone River is partly derived from glacial meltwaters and it is both colder and more turbid than the lake water. The consequent density difference produces a turbidity current that flows in a subaqueous channel at least 10 kms down the delta front to a depth of over 300 m. The channel has a depth below its rim of almost 60 m and has subaqueous natural levees that rise 5 m above the surrounding part of the lake floor. Later, in the 1930's, turbidity currents were shown to be important in reservoirs. In Lake Mead, engineers were surprised to find that not all the sediment was deposited at the head of the reservoir but that much of the clay and silt was transported into the lower parts of the reservoir by underflows with velocities ranging up to about 30 cm/sec. The densities in the underflows are rarely in excess of 1.05 g/cc and the thicknesses are about 1 or 2 m.

In 1936, Daly suggested that turbidity currents might be responsible for cutting submarine canyons and producing the associated submarine channels, many of which have features somewhat similar to the subaqueous channel of the Rhone in Lake Geneva. In 1950, Kuenen reported the results of a series of experiments, started to test Daly's hypothesis. An important new feature of Kuenen's experiments was that he produced "high density" turbidity currents, i.e., turbidity currents with a density in excess of 1.1 g/cc and with sand as well as mud in suspension. The experimental "high density" turbidity current surges laid down a thin bed of poorly sorted sand that resembled a type of graded bed known in the stratigraphic record. Such graded beds were subsequently described from modern deep-sea deposits and a turbidity current origin for similar graded bedding has been the most generally accepted explanation since that time. The existence of very large, swift turbidity currents in the modern oceans was also suggested by the phenomenon of sequential cable breaks. For example, cables broken following the 1929 Grand Banks earthquake prove that following the breaking of cables by large slumps in the vicinity of the continental slope some sort of disturbance traveled south across the deep-sea floor at a speed of at least 7 m/sec. The only reasonable explanation proposed so far is that the disturbance was a large turbidity current (for a recent review, see Emery, et al., 1970). Use of Eq. 11 combined with reasonable assumptions about the density and thickness of the current suggests that the very high speeds indicated by the cable breaks are in accordance with theory.

The turbidity current hypothesis also accounts satisfactorily for many of the features commonly associated with a certain type of graded sandstone (see, for example, Dzulynski and Walton 1965; Walker, 1965; Walton, 1967). The features include (a) the sequence of internal structures (the "Bouma sequence") discussed on p. 122 in terms of decreasing flow regimes; (b) the presence of sole markings indicating scour of the bottom or formation of marks produced on the mud bottom by "tools" (clasts) carried by the flow, followed immediately by deposition; (c) evidence of very rapid deposition from suspension, with some traction to produce plane lamination or climbing ripple-drift; and (d) graded bedding. Many graded beds showing some or all of these features are interbedded with shales with a microfauna or ichnofauna (trails and burrows) indicating deposition in relatively deep water. Sedimentary structures typical of shallow water, such as mud cracks or wave-formed ripples, are generally absent. Direct observation of the formation of modern turbidity current deposits is very difficult, if not impossible, so that many details of interpretation, and even the turbidity current hypothesis itself, remain controversial (see Van der Lingen, 1969).

An unusual aspect of turbidity currents is that they are sediment transporting systems that owe their existence to the sediment rather than to the fluid. The term "autosuspension" has been suggested to describe the suspension of sediment caused by fluid turbulence, which itself is generated by the motion of the current, which in turn is impelled at least in part by the excess weight of the suspension as compared with the overlying fluid. It is not clear that perfect autosuspension, with no deposition, actually takes place but the evidence from modern cable breaks indicates that some turbidity currents can travel distances of the order of several hundred kilometers so that the conditions for autosuspension must be closely approached.

Autosuspension, like capacity, must be understood for a particular grain size or settling velocity. Also, if a turbidity current is of the transient or surge type, not all the current can be in a regime of autosuspension with respect to a given grain size. In general, we may divide a turbidity current into three main regions: the head, the body, and the tail. In the *head*, the flow pattern described above is such that one may expect that sediment will be swept off the bottom and circulated round to the back of the head, where it may either settle back into the head (presumably the fate of the coarser grains) or be lost in eddies torn from the rear of the head. In turn these eddies will be added to the zone of entrainment of the body of the flow and swept downcurrent. The coarsest sediment may be held in suspension in the head until there is a general deceleration produced by flattening of the slope or dilution of the current. There may be some erosion of the bottom, and formation of scour and tool marks, even by a current that is depositing sediment from the body. Marks produced by erosion in the head will be buried by deposition from the body.

In the *body* of the current there may be an approach to steady, uniform flow and possibly to conditions of autosuspension in some currents. Deposition from dispersion may be either extremely rapid or quite slow. The rate of deposition, the flow regime, and the grain size will determine the nature of the sedimentary structures that are formed. The lowest, massive division in the Bouma sequence may result from very rapid deposition with consequent formation of a mobile bed that deforms like a pseudo-plastic (a transient stage of grain flow with excess pore pressure) during the last stages of deposition. Less rapid deposition may permit more gradual compaction of the bed with some movement by traction and formation of structures such as plane lamination and ripple-drift. The type of ripple-drift found in graded beds is commonly of the type that indicates very rapid "fallout" of sediment from suspension. Many beds also contain a division of very fine sand showing highly contorted ripples ("convolute lamination"), which suggests rapid deposition from suspension and plastic deformation of partially liquefied sediment produced by shear stress on the bed (see below).

In the *tail* of the current, there will be a progressive reduction in thickness and density and therefore also in velocity. Sediment may be deposited by this part of the current that would be held in suspension in the body. In some currents, however, there may be a strong entrained motion of the water above so that the usual transitional top of the bed is replaced by a sharp contact between sand deposited by the body of the current and mud deposited slowly after the inertial motion of the entrained layer has been dissipated.

It may be expected that most turbidity currents will deposit beds showing some size grading and evidence of gradually reducing flow regime but the details of the deposit from a particular current at a particular place depend on many variable factors, such as the size of the current, its density, and the grain size of the source material. Consequently many different facies of deposits have been described and attributed to turbidity current deposition. It has been suggested that the deposits closest to the point of initial deposition will be thick coarse beds, generally beginning with a massive or plane-laminated division ("proximal turbidites"). Those farthest from the point of initial deposition ("distal turbidites") will be thin, fine grained beds, beginning with a ripple cross-laminated division (see review by Walker *in* Lajoie, 1970).

5.12 PENECONTEMPORANEOUS DEFORMATION

Description of Structures

Many sedimentary deposits display structures that indicate deformation of the sediment. In some cases, it can be established from the evidence of the structures themselves that the deformation must have taken place very early, either during the process of deposition itself or soon after, i.e., before the deposit was covered by more than a few feet of sediment. Such deformation, almost contemporary with sedimentation, is described as penecontemporaneous deformation.

There are four main mechanisms that may cause sediment deformation: (a) gravity forces acting on a succession of strata showing a reverse density gradient, i.e., with denser sediment layers overlying less dense layers; (b) liquefaction of sediment; (c) gravity movement of sediment deposited on a slope (slumping); (d) shear stress exerted on recently deposited sediment by a flow moving above it. In many cases it seems that two or more of these mechanisms act together to produce the observed deformation (Anketell and others, 1970).

Structures produced by sinking of heavier into lighter sediment are generally called "*load structures*." The commonest type consists of a pattern of bulbous protrusions, separated by relatively narrow ridges. The protruding material is commonly sand that has sunk down into the underlying material, which at the time of deposition must have been a soft mud. In many cases, loading modifies structures (such as flutes or grooves) that were present on the mud surface before deposition of the sand. In other cases, the internal structures of the sand indicate that it was originally present as ripples on the mud surface. Unequal loading of the mud (greater under the ripple crest) resulted in sinking of the ripple into the mud and distortion of the ripple cross-lamination. Such structures are generally observed in the field on the sandstone sole ("load structures or pockets," "load-casted flutes," "load-casted ripples").

An extreme form of this phenomenon is seen in "*pseudo-nodules*," which consist of distorted fragments of sand or silt with a lower convex surface that have become completely enclosed in mud (see Fig. 5-11). This structure has been convincingly reproduced by Kuenen (1958), who deposited a layer of sand over a thixotropic clay layer (Fig. 5-28). A shock applied to the deposit liquefied the clay and caused the sand layer to founder into the clay and form structures closely resembling natural "pseudo-nodules." "*Ball and pillow*" (also misleadingly called "flow rolls") is a similar structure, which differs from "pseudo-nodules" mainly in the minor amount of mud or shale involved. In "ball and pillow" structure a bed of fine sand is broken up at the base into "pillows" with rounded bottoms but the upper part of the bed may be relatively undisturbed and display an almost flat top. In many cases, the pillows are underlaid by only a few centimeters of shale. In such cases, thixotropic failure of the underlying clay cannot be the explanation and the structures must have been formed by partial failure of the sand itself. The presence of distorted laminae preserved within the pillows indicates, however, that complete liquefaction of the sand did not take place.

Structures produced by liquefaction of sand include sandstone dikes and sand volcanoes. *Sand volcanoes* are small mounds of sand a few centimeters in diameter formed by liquefaction of a sand bed and escape of sand before the bed was buried beneath other sediment. *Sandstone dikes* and sills were formed by liquefaction of sand after shallow burial by cohesive sediments. In some cases dikes extend up (and less commonly, down) through a thickness of several meters of sediment. Because of later compaction to shale of the enclosing mud, sandstone dikes may have irregular "ptygmatic" shapes (Dzulynski and Walton, 1965).

Some of the structures described above were attributed by early workers to downslope slumping. True "*slump structures*" are not always easy to distinguish in the field from load or liquefaction structures involving no downslope movement or from structures formed by shear stress exerted by fluid flow or from structures formed by diastrophism. Criteria for distinguishing slumps include distorted and overturned stratification, with evidence that the sediment was plastic at the time of deformation (such as marked thickening and thinning of strata), striated folds or fragments, "slump balls" consisting of rolled up fragments of strata, "pull-apart structures," and, in some cases, an eroded top to the slump structures with a thin graded bed deposited above the erosion surface. It is regrettable that it is so difficult to distinguish true slumps from other structures in ancient sediments because their presence together with the orientation of the slump folds is one of the few direct indicators of paleoslopes.

A distinctive type of sedimentary structure found mainly in very fine sands has been called "*convolute lamination*" (Fig. 5-29). Beds displaying convolute lamination have upper and lower surfaces that are generally almost planar (though the lower surface may have prominent flutes, loads, and other structures). Internally the bed displays a complex pattern of broad synclines and sharp anticlines. The folds are frequently overturned, with a predominant direction of overturn and with a rough parallelism of fold axes. The amplitude of the folds is largest

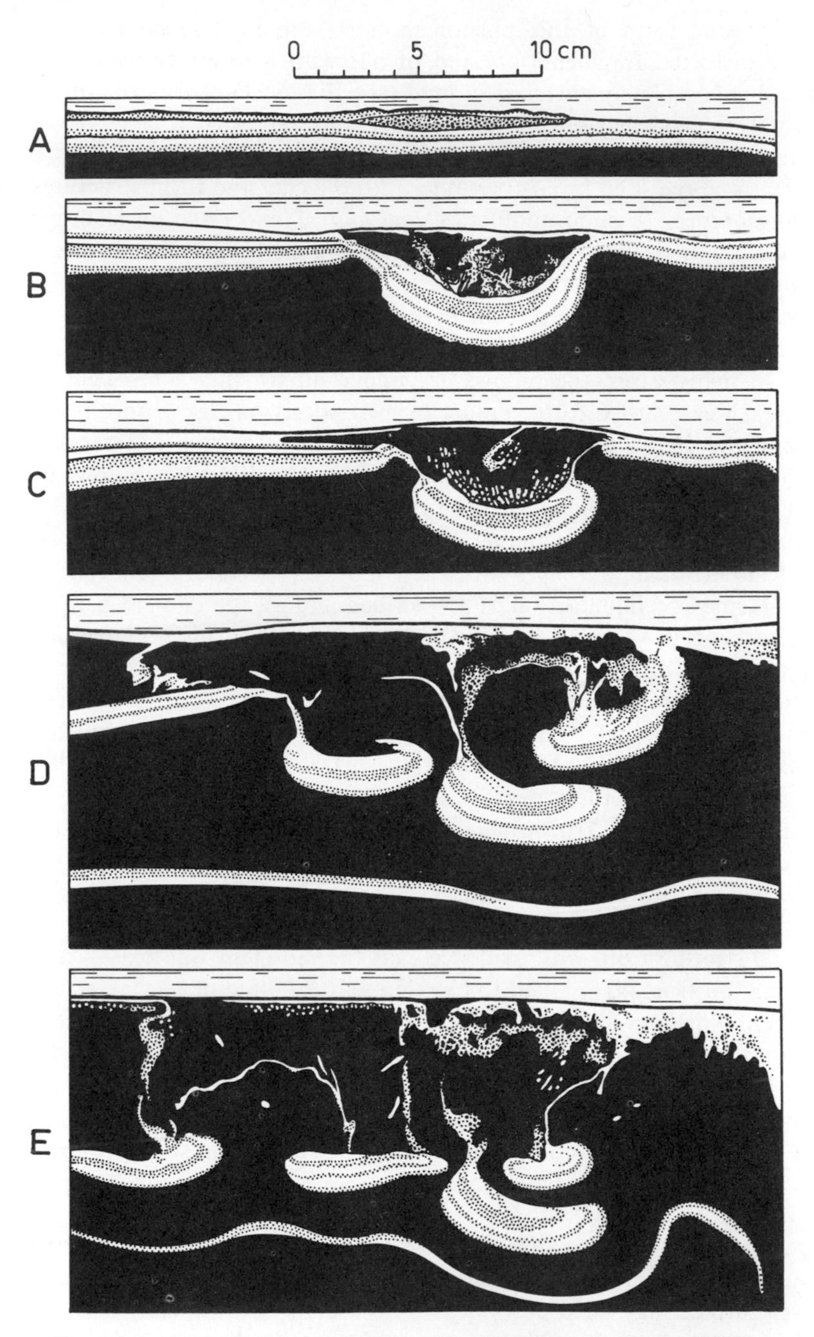

Fig. 5-28. Formation of pseudonodules in an experiment by Kuenen (1958). A layer of sand, locally thickened (A) and overlying unconsolidated clay, was vibrated. Load structures developed (B and C) and became completely enclosed in clay (D and E).

Fig. 5-29 Convolute lamination in a calcareous sandstone bed, Cloridorme Formation, Gaspé. (Photo courtesy of Roger Walker.)

in the center or upper part of the bed and is greatly reduced both toward the top and the bottom of the bed. Careful examination of some examples has revealed that the anticlines are strongly deformed ripple crests and that there are within the bed internal erosion surfaces that are themselves deformed, indicating clearly a sequence of ripple growth, deformation, and local erosion contemporaneous with deposition (see Sanders *in* Middleton, 1965a; Dzulynski and Walton, 1965). In some cases the sequence was repeated several times during deposition of a single bed. Consequently, it seems clear that beds displaying these structures (and, by inference, similar types of convolute lamination that do not display ripples and internal erosion surfaces) were formed by plastic deformation of the bed at the same time as the bed was being deposited rapidly from suspension. Evidence from associated current structures indicates for many beds that the orientation of the fold axes is normal to the paleocurrents with folds overturned predominantly in the downcurrent direction. This suggests strongly that the deformation was caused by the shear stress exerted on the bed by the current itself.

In some cross-beds, oversteepening of the upper part of the cross-laminae may be observed. The oversteepening may be extreme, with actual overturning of laminae. Presumably this phenomenon (which may be common in some sequences of cross-beds, though generally rare in most cross-beds) is also related to shear stress exerted by the current.

Mechanisms of Soft-Sediment Deformation

Freshly deposited sand has a porosity of about 45% but varies within relatively wide limits depending on the size and shape distribution and packing (see Chap. 3). The porosity of sand is commonly considerably less than that of freshly deposited clay, which has a porosity of 70 to 90%. Therefore, deposition of sand on top of freshly deposited clay will generally result in a reverse (and therefore unstable) density gradient. Formation of load structures does not necessarily take place, however, because it may be prevented by the strength or viscosity of the sediment. After deposition, *consolidation* of the sediment takes place in response to gravity forces acting on the sediment particles. These forces create pressure within the sediment mass and it is important to distinguish two parts of the total pressure: *effective pressure*, which acts on the sediment particles themselves and is available to cause compaction, and *neutral pressure*, which is pressure transmitted through the fluid. At equilibrium, the neutral pressure is equal to the hydrostatic pressure. In soil mechanics the thickness reduction produced in a bed of thickness H by any given increase in effective pressure is called the compression S. If the initial porosity is P_0 and the porosity after compression is P, then by definition

$$S = \frac{H(P_0 - P)}{1 - P} \tag{12}$$

In a sediment where the increase in effective pressure is produced rapidly (e.g., by fast sedimentation), the compression of the sediment at some time t will be less than the ultimate compression that would take place if enough time were to be allowed for the pore water to escape. The *degree of consolidation* is defined as the ratio between the compression at t and the ultimate compression. At any stage short of ultimate compression, part of the pressure of the sediment must be transmitted to the pore fluid so that the pore fluid pressure exceeds hydrostatic by an amount equal to the *excess pore pressure*. Sediments with excess pore pressures are described as *underconsolidated.*

The response of a sediment to shearing depends on the angle of internal friction (ϕ) and the cohesion of the sediment (c). Suppose the shear stress is τ and the total normal pressure is σ, then the effective pressure is $(\sigma - u)$, where u is the pore pressure; the relation between these variables is given by the following equation:

$$\tau = c + (\sigma - u) \tan \phi \tag{13}$$

Deformation of the sediment may therefore be facilitated by reducing the cohesion c, by reducing the coefficient of internal friction $\tan \phi$ (e.g., by altering the packing of the sediment), or by increasing the pore pressure u.

Some sediments (mainly sands of fine grain size) are deposited with a loose, unstable packing. This structure may alter to a more stable structure by compaction or by slow deformation of the sediment [for example "creep" down a steep slope, as described by Shepard and Dill (1966) in the heads of submarine canyons]. Alternatively, a sudden change in packing may result from application of a sudden shock produced by an earthquake or other disturbance. In this case there is a sudden generation of a very high excess pore pressure as the weight of particles previously supported by the solid framework is transmitted to the pore fluid. In a water-saturated sediment of low cohesion, the result may be to produce *liquefaction* of the sediment so that it will deform ("flow") under very small shear stress.

In finer grained sediment (mud) the low permeability of the sediment results in a low degree of consolidation for even moderate rates of deposition. In areas of rapid sedimentation, such as the Mississippi Delta, Morgenstern (1967) has calculated that the degree of consolidation probably falls well below 10%. The excess pore pressures are consequently high and the shear stress needed to cause sediment failure is small. Morgenstern calculated that in this case slumping is probable on slopes at least as low as 4 degrees. This application of soil mechanics theory helps to explain the low slopes seaward of most major deltas and is in accordance with the evidence recently revealed by refraction profiling, which shows that many delta fronts and continental slopes are characterized by an abundance of large-scale slumping phenomena.

In the previous section it was remarked that most soft-sediment deformation structures in sands are restricted to coarse silt to fine sand grades. The reason for this is now clear. These size grades are the ones where the cohesion is not large, where rapid deposition from suspension and an unstable packing are possible, and where the permeability is low enough to prevent rapid loss of pore fluids. Consequently, the fine sands are those most likely to undergo loss of strength and at least partial liquefaction.

5.13 BIOGENIC SEDIMENTARY STRUCTURES

After deposition, many sedimentary strata are disturbed by animals whose activities produce sedimentary structures. Such structures, excluding the actual skeletal remains of the organism itself, have been called biogenic structures, trace fossils, or ichnofossils (in German, *Lebensspuren*). A distinction may be made between structures produced at the sediment surface (exogenetic) and those produced below the surface (endogenetic). In practice, the two are not always easily distinguishable because organisms may burrow along the interface between sand and mud layers. After such a burrow has been formed, it may be filled by sand flowing in from the layer above and thus resemble a trail on the mud surface, later covered by a bed of sand. Structures seen on the base of a sand bed (whatever their origin) are said to be *hypichnia* or to be preserved in *hyporelief* and those on the top of the sand bed are said to be *epichnia* or preserved in *epirelief*. Other terms are illustrated in Fig. 5-30.

The morphology of trace fossils is generally more strongly controlled by the behavioral than by the anatomical characteristics of the animals. Exceptions to this statement include some tracks, particularly those of vertebrates. A consequence of the strong behavioral control is that almost identical trace fossils may be produced by animals not closely related taxonomically. Seilacher (1964) has recognized five main groups of trace fossils, based on behavioral characteristics: (a) feeding trails (*Pascichnia*), (b) crawling burrows (*Repichnia*), (c) resting tracks (*Cubichnia*), (d) dwelling burrows (*Domichnia*), (e) and feeding burrows (*Fodichnia*)

The animal's behavior might be expected to be more directly influenced by

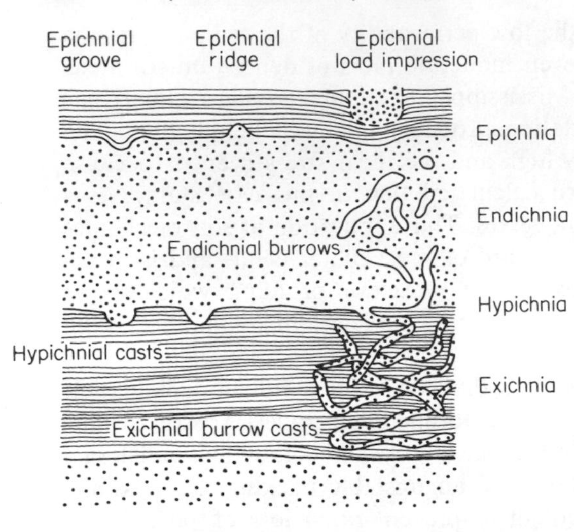

Fig. 5-30 Terminology for trace fossils, based on the position of the fossil in relation to a sandstone bed. After Martinsson (1965).

the environment than by its phylogeny and this expectation appears to be supported by recent studies, although there is evidence that not only the morphology but also the behavior of an organism may undergo long-term evolutionary changes. Thus, although some ichnofossils may be used as zone fossils, they are of special interest to the sedimentologist as paleoenvironmental indicators.

Burrowing organisms are particularly common in shallow marine environments and there have been many studies that document their effectiveness in destroying structures, such as lamination, produced by inorganic processes. All stages can be traced between well-laminated sediments, disturbed by a few burrows, to sediment that appears almost homogeneous or only slightly mottled, in which even the burrow structures themselves can hardly be distinguished. Deep vertical burrowing (30 cm) is common in the littoral zone because organisms seek within the sediment a level that is not strongly affected by the large fluctuations in temperature and salinity that are generally characteristic of the sediment surface in very shallow water. In offshore regions, such fluctuations are not so strong and the burrows are generally shallow and horizontal. In very deep waters, it appears that the most common structures are surface traces, including spiral tracks. Recognition of the various distinctive trace fossil associations (biofacies) may well prove to be one of the most reliable approaches to determining depth of deposition.

Foul bottom water (euxinic) conditions are generally identified by the presence of sulfide minerals, dark color, and absence of bottom fauna in the sediment. Sulfide minerals and dark color may, however, result from diagenetic changes below the sediment-water interface, and the skeletal parts of fossils may be removed by solution. Complete absence of trace fossils, however, is a strong indicator of euxinic conditions.

Trace fossils may also give information about the rate of sedimentation.

Sediments constantly reworked by physical processes, such as beach sands, and sediments that are deposited very rapidly contain few trace fossils. Resting tracks are produced by organisms burrowing just below the sediment surface. If such organisms are suddenly covered by rapid influx of sediment, they will burrow to the new surface and there produce a new set of resting tracks. Certain types of organisms prefer to construct dwelling burrows in well-compacted muddy sediment. Presence of a stratum with many of these burrows indicates a period of nondeposition, during which the necessary compaction of the sediment could take place. Seilacher (1962) was able to document the very rapid deposition of certain sandstones in the Spanish Cretaceous-to-Tertiary Flysch by noting that each different type of burrow was characterized by a definite maximum depth of burrow. Beds thicker than this critical depth (which varied from 7 to more than 400 cm) contained no burrows of that particular organism, indicating that the bed was deposited so rapidly that the organism was unable to burrow down to the base of the bed. Goldring (1964) defined a new species of the U-shaped burrow, *Diplocraterion.* He gave the expressive name, *yoyo,* to this species because the internal structure of the burrow indicated the way the animal moved up or down in response, respectively, to aggradation or erosion of the sediment surface (Fig. 5-31).

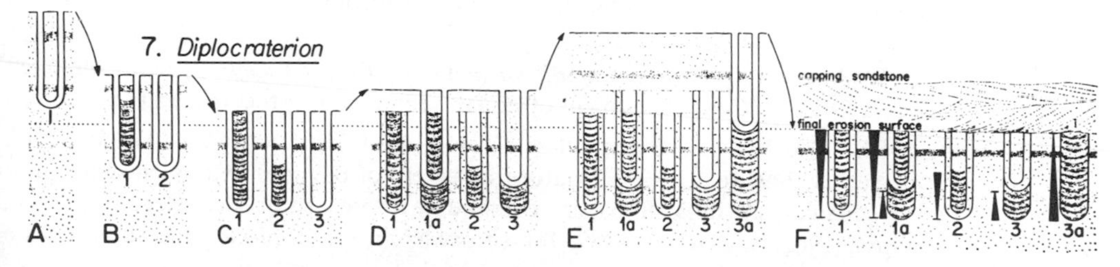

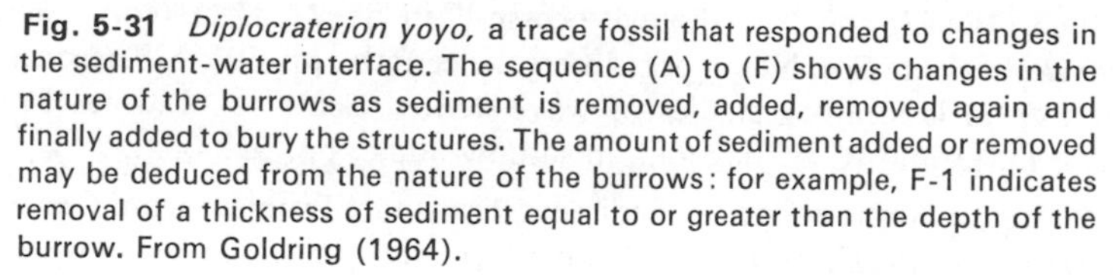
Fig. 5-31 *Diplocraterion yoyo,* a trace fossil that responded to changes in the sediment-water interface. The sequence (A) to (F) shows changes in the nature of the burrows as sediment is removed, added, removed again and finally added to bury the structures. The amount of sediment added or removed may be deduced from the nature of the burrows: for example, F-1 indicates removal of a thickness of sediment equal to or greater than the depth of the burrow. From Goldring (1964).

This particular trace fossil has been found in both Paleozoic and Mesozoic rocks and appears to be typical of the shallow subtidal environment.

Trace fossils may also be used as paleocurrent indicators. The tracks of organisms such as trilobites differ according to whether the animal was moving up, down, or crosscurrent and resting tracks may also have a preferred orientation because of a preference of some organisms to face into the current.

Finally, many burrowing organisms eat sediment (generally mud) for the organic material that it contains. Besides producing feeding burrows, such organisms mold the mud into fecal pellets. Pelleted muds produced in this way are particularly important in many areas of carbonate mud deposition (see Chap. 12).

REFERENCES

ALLEN, J. R. L., 1963, "Henry Clifton Sorby and the Sedimentary Structures of Sands and Sandstones in Relation to Flow Conditions," *Geol. en Mijnbouw*, **42**, pp. 223–228. (One of the earliest attempts to interpret sedimentary structures in terms of flow regimes. See also Allen, 1970.)

———, 1968, *Current Ripples*. Amsterdam: North-Holland Pub. Co., 433 pp. (The first six chapters give a comprehensive review of all types of small- and large-scale ripple forms. Later chapters describe experimental studies.)

———, 1969, "Some Recent Advances in the Physics of Sedimentation," *Proc. Geol. Assn. London*, **80**, pp. 1–42. (Good review of selected topics, including sediment transport, bed forms, and turbidity currents.)

———, 1970, "Studies in Fluviatile Sedimentation: A Comparison of Fining-Upward Cyclothems, with Special Reference to Coarse-Member Composition and Interpretation," *Jour. Sed. Pet.* **40**, pp. 298–323. (Statistical analysis of sequences of facies, defined by sedimentary structures, in fluvial cyclothems. Results interpreted in terms of latest experimental data. Compare with Allen, 1963.)

ANKETELL, J. M., J. GEGLA, and S. DZULYNSKI, 1970, "On the Deformational Structures in Systems with Reversed Density Gradients," *Ann. Soc. Geol. Pologne*, **15**, pp. 3–29. (An important discussion, in English, of factors controlling the geometry of different types of load structures.)

BAGNOLD, R. A., 1941, *The Physics of Blown Sand and Desert Dunes*. London: Methuen and Co., 265 pp. (Classic treatment of eolian sedimentary processes and structures.)

———, 1956, "The Flow of Cohesionless Grains in Fluids," *Roy. Soc. London, Phil. Trans.* Ser. A, **249**, pp. 235–297. (Mature statement of Bagnold's views on sedimentation mechanics. A difficult paper, demanding close study. See Bagnold, 1966, Raudkivi, 1967, and Allen, 1969, for somewhat more approachable summaries.)

———, 1963, "Beach and Nearshore Processes," in M. N. HILL (ed.), *The Sea*, **3**, *The Earth Beneath the Sea*. New York: Interscience Pub., pp. 507–528. (Concise review by a pioneer in laboratory studies of wave action. See also "Littoral Processes," by Inman, D. L. and R. A. Bagnold, in same volume, pp. 529–553.)

———, 1966, "An Approach to the Sediment Transport Problem from General Physics," *U.S. Geol. Sur. Prof. Paper* **422-I**, 37 pp. (A relatively brief clear summary of his views, which are not entirely shared by many engineers.)

BALL, M. M. 1967, "Carbonate Sand Bodies of Florida and the Bahamas," *Jour. Sed. Pet.*, **37**, pp. 556–591. (Describes tidal sand bodies.)

BASCOM, W. N., 1951, "The Relationship between Sand Size and Beach Face Slope," *Trans. Amer. Geophys. Union*, **32**, pp. 866–874. (Empirical relation established for U. S. west coast beaches.)

———, 1964, *Waves and Beaches*. New York: Doubleday & Co., Inc., Anchor Books, 267 pp. (Immensely readable, with much valuable information.)

BERNARD, H. A., C. F. MAJOR, JR., B. S. PARROTT, and R. J. LEBLANC, SR., 1970, "Recent Sediments of Southeast Texas," *Texas Bur. Econ. Geol.*, Guidebook 11, 16 pp. plus Appendices. (Describes and figures sediments of the Brazos alluvial and deltaic plains and the Galveston barrier island complex. Reprints as appendices abstracts and papers published on the topic since 1959. The description of the meandering river and barrier island sediments given in these papers has become classic.)

BOERSMA, J. R., 1967, "Remarkable Types of Mega Cross-Stratification in the Fluviatile Sequence of a Subrecent Distributary of the Rhine, Amerongen, the Netherlands," *Geol. en. Mijnbouw*, **46**, pp. 217–235. (Detailed description of several types of cross-bedding with regressive ripples.)

———, 1969, "Internal Structure of Some Tidal Megaripples on a Shoal in the Westerschelde Estuary, the Netherlands," *Geol. en Mijnbouw*, **48**, pp. 409–414. (Describes features characteristic of cross-beds formed in tidal environment.)

BOUMA, A. H., 1962, *Sedimentology of some Flysch Deposits*. Amsterdam: Elsevier Pub. Co., 168 pp. (Pages 48–54 give a description of the "Bouma sequence.")

CLANCY, E. P., 1968, *The Tides: Pulse of the Earth*. New York: Doubleday & Co., Inc., Anchor Books, 228 pp. (Good simple explanation of physics of tides.)

CLIFTON, H. E., 1969, "Beach Lamination: Nature and Origin. *Marine Geol.*, **7**, pp. 553–559. (Brief paper with interesting detailed observations.)

CLIFTON, H. E., R. E. HUNTER, and R. L. PHILLIPS, 1970, "Underwater Observations in Oregon Near-Shore Zone" *in U.S. Geol. Sur. Prof. Paper* **700-*A***, p. 100 (Preliminary report on important investigation of bed forms and structures generated by waves.)

COASTAL RESEARCH GROUP, UNIV. MASSACHUSETTS, 1969, "Coastal Environments, N. E. Massachusetts and New Hampshire," *Field Trip Guidebook*, 462 pp. (Describes tidal, beach, and dune sediments, with emphasis on sedimentary structures and textures. This work, and reports on specific areas by Daboll, Farrell, and Hartwell constitute a major contribution to sedimentation processes in a nearshore, tidal environment.)

COLBY, B. R., 1964, "Scour and Fill in Sand-Bed Streams," *U. S. Geol. Sur. Prof. Paper* **462-*D***. 32 pp. (Deep scour must be local because little sand can be held in suspension even during flood.)

COLEMAN, J. M., 1969, "Brahmaputra River: Channel Process and Sedimentation," *Sed. Geol.*, **3** (special issue), pp. 129–239. (With important observations on bed forms and sedimentary structures as well as fluvial processes in a large, braided river.)

CONYBEARE, C. E. B. and K. A. W. CROOK, 1968, "Manual of Sedimentary Structures." Canberra, Australia, *Bur. Min. Res. Geophys. Bull.* **102**, 327 pp. (Descriptions and illustrations of structures and a useful review of classifications. The sections on interpretation are too brief to be very helpful.)

DALY, R. A., 1942, *The Floor of the Ocean*. Chapel Hill, N.C.: Univ. of North Carolina Press, 177 pp. (Most of the book is now outdated but the section on the pioneer studies of turbidity currents in lakes and reservoirs is still valuable.)

DOLAN, ROBERT and JOHN FERM, 1966, "Swash processes and beach characteristics," *Prof. Geographer*, **18**, pp. 210–213. (Increasing wave height increases swash uprush speed and decreases beach slope.)

DZULYNSKI, S. and E. K. WALTON, 1965, *Sedimentary Features of Flysch and Greywackes*. Amsterdam: Elsevier Pub. Co., 274 pp. (With detailed treatment of structures, particularly sole marks.)

EVANS, G., 1965, "Intertidal Flat Sediments and Other Environments in the Wash," *Quart. Jour. Geol. Soc. London*, **121**, pp. 209–245. (Detailed classification and description of environments.)

EMERY, K. O., E. UCHUPI, J. D. PHILLIPS, C. O. BOWIN, E. T. BUNCE, and S. T. KNOTT, 1970, "Continental Rise off Eastern North America," *Amer. Assoc. Petroleum Geol. Bull.*, **54**, pp. 44–108. (Comprehensive summary of marine geology of the area, containing a bibliography and discussion of earlier work by Ewing, Heezen, and others on the Grand Banks slump and turbidity current.)

GALVIN, C. J., JR., 1968, "Breaker Type Classification on Three Laboratory Beaches," *Jour. Geoph. Res.*, **73**, pp. 3651–3659. (Describes different breaker types and gives numerical parameters for predicting their occurrence.)

GILBERT, G. K., 1914, "The Transportation of Debris by Running Water," *U. S. Geol. Sur. Prof. Paper* **86**, 263 pp. (One of the last major works by a great geologist: The experimental data are still used and the descriptions of sediment movement mechanisms are classic.)

GOLDRING, ROLAND, 1964, "Trace Fossils and the Sedimentary Surface in Shallow Water Marine Sediments," *in* Van Straaten, L. M. J. U., *Deltaic and Shallow Marine Deposits.* Amsterdam: Elsevier Pub. Co., pp. 136–143. (Discusses evidence of penecontemporaneous erosion provided by trace fossils.)

HAMBLIN, W. K., 1962, "X-Ray Radiography in the Study of Structures in Homogeneous Sediments," *Jour. Sed. Pet.*, **32**, pp. 201–210. (Also see **32**, pp. 530–533 and *Kansas State Geol. Sur. Bull.* **175**, pt. 1. Use of this tool for observing structures that are not otherwise evident.)

HAND, B. M., 1969, "Antidunes as Trochoidal Waves," *Jour. Sed. Pet.* **39**, pp. 1302–1309. (Describes the form of antidunes and reviews the hydraulics briefly.)

HARMS, J. C., 1969, "Hydraulic Significance of Some Sand Ripples," *Geol. Soc. Amer. Bull.*, **80**, pp. 363–396. (New experimental results on current, wave, and combination wave-current ripples obtained in a large flume.)

HAYES, M. O., 1967, "Hurricanes as Geological Agents: Case Studies of Hurricanes Carla, 1961, and Cindy, 1963," *Texas Bur. Econ. Geol. Rept. Invest.* **61**, 54 pp. (Describes effects of two hurricanes on Texas Gulf Coast barriers.)

HOOKE, R. LEB., 1967, "Processes on Arid-Region Alluvial Fans," *Jour. Geol.*, **75**, pp. 438–460. (Assesses the importance of debris flow, stream flow, and other processes and describes several recent alluvial fans.)

HOUBOLT, J. J. H. C., 1968, "Recent Sediments in the Southern Bight of the North Sea," *Geol. en Mijnbouw*, **47**, pp. 245–273. (Description of tidal sand ridges.)

HOYT, J. H., 1967, "Occurrence of High-Angle Stratification in Littoral and Shallow Neritic Environments, Central Georgia Coast, U. S. A.," *Sedimentology*, **8**, pp. 229–238. (High angle cross-beds are not restricted to eolian deposits.)

INGLE, J. C., JR., 1966, *The Movement of Beach Sand.* Amsterdam: Elsevier Pub. Co., 221 pp. (Describes use of dyed sand to measure movement on California beaches.)

INGRAM, R. L., 1954, "Terminology for the Thickness of Stratification and Parting Units in Sedimentary Rocks," *Geol. Soc. Amer. Bull.*, **65**, pp. 937–938. (A brief statement on terminology.)

JOPLING, A. V., 1965, "Hydraulic Factors Controlling the Shape of Laminae in Laboratory Deltas," *Jour. Sed. Pet.* **35**, pp. 777–791. (Analyzes factors responsible for formation of angular or tangential cross-laminae; gives references to several earlier studies of laboratory deltas by Jopling.)

JOPLING, A. V., and R. G. WALKER, 1968, "Morphology and Origin of Ripple-Drift Cross-Lamination, with Examples from the Pleistocene of Massachusetts," *Jour. Sed. Pet.*, **38**, pp. 971–984. (Revises an earlier classification proposed by Walker.)

KARCZ, IAAKOV, 1968, "Fluviatile Obstacle Marks from the Wadis of the Negev (Southern Israel)" *Jour. Sed. Pet.*, **38**, pp. 1000–1012. (With a theoretical analysis of current crescents; see also the paper by P. D. Richardson, *Jour. Sed. Pet.*, **38**, pp. 965–970.)

KING, C. A. M., 1959, *Beaches and Coasts.* London: Edward Arnold Ltd., 403 pp. (A very useful summary, with emphasis on geomorphology.) 2nd ed. in press.

KLEIN, G. DEV., 1970, "Depositional and Dispersal Dynamics of Intertidal Sandbars," *Jour. Sed. Pet.*, **40**, pp. 1095–1127. (Based on observations in part of the Bay of Fundy. Describes bed forms and sedimentary structures and their response to tidal currents.)

KUENEN, P. H., 1958, "Experiments in geology," *Trans. Geol Soc. Glasgow*, **23**, pp. 1–28. (Review with a description of new experiments on the origin of "pseudonodules.")

———, 1966, Experimental Turbidite Lamination in a Circular Flume. *Jour. Geol.*, **74**, pp. 523–545. (Describes lamination produced by gradual deceleration of muddy sand suspensions in a circular flume.)

LAJOIE, JEAN (ed.), 1970, "Flysch sedimentology in North America," *Geol. Assn. Canada Spec. Paper* **7**, 272 pp. (With papers describing features of flysch in different formations and discussing the origin of sedimentary structures found in flysch.)

LAUFF, G. H. (ed.) 1967, *Estuaries.* Washington, D. C.: Amer. Assn. Adv. Sci., 757 pp. (Symposium with many excellent papers on estuaries and tidal flats.)

LAURSEN, E. M., 1953, "Observations on the Nature of Scour," *State Univ. Iowa Hydraulics Conf. Proc., Studies in Eng., Bull. 34*, pp. 179–198. (Reviews fundamental principles and some experimental work.)

MARTINSSON, ANDERS, 1965, "Aspects of a Middle Cambrian Thanatotope on Oland," *Geol. Föreningens Stockholm Förhandlingar*, **87**, pp. 181–230. (Describes inorganic and biogenic structures and proposes new terminology.)

MCKEE, E. D., 1966, "Structure of Dunes at White Sands National Monument, New Mexico (and a Comparison with Structures of Dunes from Other Selected Areas)," *Sedimentology*, **7** (special issue), pp. 1–70. (Structures of dome-shaped, transverse, barchan, and parabolic dunes in an area with one predominant wind direction.)

MCKEE, E. D. and G. C. TIBBITTS, JR., 1964, "Primary Structures of a Seif Dune and Associated Deposits in Libya," *Jour. Sed. Pet.*, **34**, pp. 5–17. (Cross-strata dip steeply to two opposite directions normal to the dune crest.)

MCKEE, E. D. and G. W. WEIR, 1953, "Terminology for Stratification and Cross-Stratification in Sedimentary Rocks," *Geol. Soc. Amer. Bull.*, **64**, pp. 381–390. (Defines sets, cosets, and composite sets and classifies strata on the basis of lower surface, shape, and internal structure. The main shapes of crossbed sets are lenticular, tabular, and wedge. The main types of cross-stratification are simple, planar, and trough.)

MIDDLETON, G. V. (ed.), 1965a, "Primary Sedimentary Structures and their Hydrodynamic Interpretation," *Soc. Econ. Paleont. Mineralogists Spec. Pub.* **12**, 265 pp. (Symposium with papers on field and experimental studies.)

———, 1965b, "Antidune Cross-Bedding in a Large Flume," *Jour. Sed. Pet.* **35**, pp. 922–927. (Describes experimental formation of lamination by antidunes.)

MIDDLETON, G. V., 1966, 1967, "Experiments on Density and Turbidity Currents, I, II, and III," *Canadian Jour. Earth Sci.*, **3**, pp. 523–546, 627–637; **4**, pp. 475–505. (Reviews earlier experimental work and describes new experiments on nature of turbidity current motion and deposition of graded beds.)

MOORE, D. G. and P. C. SCRUTON, 1957, "Minor Internal Sedimentary Structures of Some Recent Unconsolidated Sediments," *Amer. Assoc. Petroleum Geol. Bull.*, **41**, pp. 2723–2751. (Structures form a continuum from regular layers to mottled or homogeneous. The main controls are rate of sediment supply and deposition, intensity of physical processes, and activity of burrowing organisms.)

MORGENSTERN, NORBERT, 1967, "Submarine Slumping and the Initiation of Turbidity Currents, in A. F. Richards, (ed.) *Marine Geotechnique*. Urbana, Ill.: Univ. Illinois Press, pp. 189–220. (Important paper on submarine slumping. Other papers in the volume also give interesting applications of soil mechanics principles to sedimentary geology.)

OFF, T., 1963, "Rhythmic Linear Sand Bodies Caused by Tidal Currents," *Amer. Assoc. Petroleum Geol. Bull.*, **47**, pp. 324–341. (Good review of tidal sand ridges.)

OWEN, P. R., 1964, "Saltation of Uniform Grains in Air," *Jour. Fluid Mech.*, **20**, pp. 225–242. (Elegant demonstration that the thickness of the saltation layer constitutes the height of the hydrodynamic roughness of the surface.)

PETTIJOHN, F. J., 1957, *Sedimentary Rocks*, 2nd ed. New York: Harper & Row Pub., Inc., 718 pp. (Gives descriptions and full bibliography of earlier work on a wide range of sedimentary structures.)

POTTER, P. E. and F. J. PETTIJOHN, 1963, *Paleocurrents and Basin Analysis*. New York: *Academic Press, Inc.*, 296 pp. (The standard monograph on directional aspects of sedimentary structures. See also the picture book *Atlas and Glossary of Primary Sedimentary Structures*, issued separately by the same authors and publishers.)

RAUDKIVI, A. J., 1967, *Loose Boundary Hydraulics*. New York: Pergamon Press, 331 pp. (Review of sedimentation mechanics written for engineers with a full review of the engineering literature on bed forms.)

REINECK, HANS-ERICH, 1963, Sedimentgefüge im Bereich der Südlichen Nordsee. *Abh. Senckenb. Naturforch. Ges.*, **505**, 138 pp. (Pioneer study of subaqueous ripples and dunes in tidal environments and the relation of dunes to cross-bedding and sediment movement: in German, with long English summary and English captions to figures. Still the most detailed study of its kind, it documents cross-bedding features resulting from migration of dunes, emergence at low tide, and reversal of tidal currents.)

REINECK, HANS-ERICH and F. WUNDERLICH, 1968a, "Classification and Origin of Flaser and Lenticular Bedding," *Sedimentology*, **11**, pp. 99–104.

———, 1968b, "Zur Unterscheidung von asymmetrischen Oszillationsrippeln und Strömungsrippeln," *Senckenbergiana Lethaea*, **49**, pp. 321–345. (Based on studies in North Sea, documents differences between wave and current ripples: in German; English summary and captions.)

———, 1969, "Die Entstehung von Schichten und Schichtbänken im Watt," *Senckenbergiana maritima*, **1**, pp. 85–106. (Detailed study of formation of laminae in tidal flat and tidal channel environments.)

SCHIFFMAN, A., 1965, "Energy Measurement in the Swash-Surf zone," *Limnol. Oceanogr.*, **10**, pp. 255–260. (Short but significant paper on observations made using a dynamometer.)

SEIBOLD, EUGENE, 1963, "Geological Investigations of Near-Shore Sand-Transport—Examples of Methods and Problems from Baltic and North Seas," *Progress in Oceanogr.*, **1**, pp. 1–70. (Summary of work by a group of German investigators—the tideless Baltic provides interesting contrasts with the tidal North Sea.)

SEILACHER, A. 1962, "Paleontological Studies on Turbidite Sedimentation and Erosion," *Jour. Geol.*, **70**, pp. 227–234. (Postdepositional sole trails of sandstone beds occur only in beds thinner than a certain thickness, proving instantaneous deposition of the beds.)

———, 1964, "Biogenic Sedimentary Structures," in J. IMBRIE and N. D. NEWELL, eds. *Approaches to Paleoecology*, New York; John Wiley & Sons, Inc., pp. 296–315. (A concise statement by an expert: Also see article in the *Treatise of Paleontology*. The symposium also has a brief review of inorganic structures by McKee.)

SHARP, R. P., 1963, "Wind Ripples," *Jour. Geol.*, **71**, pp. 617–636. (Distinguishes small sand ripples from larger granule ripples and analyzes factors controlling size and shape and rate of ripple movement.)

SHARP, R. P. and L. M. NOBLES, 1953, "Mudflow of 1941 at Wrightwood, Southern California," *Geol. Soc. Amer. Bull.*, **64**, pp. 547–560. (Analysis of a mudflow based on eyewitness accounts and study of the deposits.)

SHELDON, PEARL G., 1928, "Some Sedimentation Conditions in Middle Portage Rocks," *Amer. Jour. Sci.*, **15**, pp. 243–252. (A neglected paper, despite the attention, mostly critical, given to it by Kuenen in *Geol. en Mijnbouw*, **18**, pp. 277–283, 1956. Sheldon examined rocks, now generally regarded as turbidites, described the sequence of sedimentary structures, and interpreted the sequence in terms of Gilbert's flume experiments. It is also one of the first American studies to give paleocurrent data.)

SHEPARD, F. P. and R. F. DILL, 1966, *Submarine Canyons and Other Sea Valleys*. Chicago: Rand McNally & Co., 381 pp. and two maps. (Describes slow creep of sand and sand falls in the heads of canyons and gives information on the soil mechanical properties of the sediments.)

SHREVE, R. L., 1968, "The Blackhawk Landslide," *Geol. Soc. Amer. Spec. Paper* **108**, 47 pp. (Describes this major landslide, compares it with several others, and discusses movement and mechanisms, including an "air layer lubrication" hypothesis.)

SIMONS, D. B. and RICHARDSON, E. V. 1961, "Forms of Bed Roughness in Alluvial Channels," *Amer. Soc. Civil Eng. Proc.*, **87**, no. HY3, pp. 87–105. (The original statement of results of the USGS experimental program to study sediment transport and bedforms. See also paper by Simons, Richardson, and Nordin *in* Middleton, 1965a.)

SMITH, N. D., 1970, "The Braided Stream Depositional Environment: Comparison of the Platte River with some Silurian Clastic Rocks, North-Central Appalachians," *Geol. Soc. Amer Bull.*, **81**, pp. 2993–3014. (With bibliography and a good summary of mechanics of braiding.)

STANLEY, D. J., 1968, "Graded Bedding—Sole Marking—Graywacke Assemblage and Related Sedimentary Structures in Some Carboniferous Flood Deposits in Eastern Massachusetts," *Geol. Soc. Amer. Spec. Paper 106*, pp. 211–239. (Demonstrates formation of many structures typical of turbidites in a fluvial environment.)

STANLEY, D. J. (ed.), 1969, *The New Concepts of Continental Margin Sedimentation.* Washington, D.C.: *Amer. Geol. Inst. Short Course Notes.* (With extensive discussion of sedimentation processes in littoral, shelf, and slope environments.)

STAUFFER, P. H., 1967, "Grain-Flow Deposits and Their Implications, Santa Ynes Mountains, California," *Jour. Sed. Pet.* **37**, pp. 487–508. (Important paper on the genesis of "massive" sandstone beds. Also see paper by Sanders, *in* Middleton, 1965a.)

STRAHLER, A. N., 1966, "Tidal Cycle of Changes in an Equilibrium Beach, Sandy Hook, New Jersey," *Jour. Geol.*, **74**, pp. 247–268. (Detailed surveys reveal tidal cycle of erosion and deposition.)

TANNER, W. F., 1960, "Shallow Water Ripple Mark Varieties," *Jour. Sed. Pet.*, **30**, pp. 481–485. (Describes many kinds of ripples typical of shallow water.)

TRICKER, R. A. R., 1964, *Bores, Breakers, Waves and Wakes.* New York: American Elsevier Pub. Co., 250 pp. (Readable nontechnical discussion of physical principles.)

VAN DER LINGEN, G. J., 1969, "The Turbidite Problem," *New Zealand Jour. Geol. Geophys.*, **12**, pp. 7–50. (A review, by a sceptic, of the evidence relating to turbidity currents.)

VAN STRAATEN, L. M. J. U., 1961, "Sedimentation in Tidal Flat Areas," *Alberta Soc. Petroleum Geol. Jour.*, **9** pp. 203–226. (Good brief review with extensive bibliography.)

VAN STRAATEN, L. M. J. U., and P. H. KUENEN, 1957, "Accumulation of Fine Grained Sediments in the Dutch Wadden Sea," *Geol. en Mijnbouw*, N. S., **19**, pp. 329–354. (Explanation of inshore accumulation of mud in terms of settling and scour lag. See H. Postma *in* Lauff, 1967, for other possible mechanisms.)

VARNES, D. J., 1958, "Landslide Types and Processes," in *Landslides and Engineering Practice.* Highway Res. Board Spec. Rept. 29, U. S. Natl. Res, Council Pub. 544, pp. 20–47. (Classifies and describes slide mechanisms. These concepts were applied to submarine slides by R. H. Dott, Jr., 1963, "Dynamics of Subaqueous Gravity Depositional Processes," *Amer. Assoc. Petroleum Geol. Bull.*, **47**, pp. 104–128. See also Middleton, *in* Stanley, 1969.)

WALKER, R. G., 1965, "The Origin and Significance of the Internal Sedimentary Structures of Turbidites," *Proc. Yorkshire Geol. Soc.*, **33**, pp. 1–29. (Discusses structures in terms of flow regimes.)

WALTON, E. K., 1967, "The Sequence of Internal Structures in Turbidites," *Scottish Jour. Geol.*, **3**, pp. 306–317. (Emphasizes importance of rate of deceleration of current in formation of massive interval.)

ZEIGLER, J. M., 1964, "Some Modern Approaches to Beach Studies," *Oceanogr. Marine Biol. Ann. Rev.*, **2** pp. 77–95. (Excellent review of studies of wave action and beach sediments.)

ZENKOVICH, V. P., 1967, *Processes of Coastal Development.* New York: Interscience Pub., 738 pp. (Translated from Russian; a very comprehensive text, with emphasis on geomorphology.)

CHAPTER SIX

FACIES MODELS

6.1 GENERAL PRINCIPLES

The term *environment* is generally understood in a physiographic sense; i.e., it is a part of the earth's surface that can be distinguished from adjacent parts because of variation in the totality of all the conditions, physical, chemical, and organic, that influence the surface. The way that the environment is subdivided depends on the aspect that is of special interest to the investigator as well as on the degree of subdivision that is considered desirable. Geologists generally subscribe to the view that, even allowing for a variation in organic influences through geological time, there are a relatively limited number of basic types of sedimentary environments, which in turn give rise to a relatively limited number of associations of lithologies, faunas, and floras (facies).

There is clearly a close connection between the environment of deposition and the nature of the sediment deposited. For the geologist working with ancient sedimentary rocks, however, the primary data are the rocks and the environment must be interpreted from them. One of the geologist's first tasks is to describe the rocks and to attempt to subdivide them into a number of mappable units (stratigraphic formations). Within each formation, it is generally possible to recognize units that differ in some aspect from those around them. Such units have been called *facies*. The term has been used by geologists with a number of different meanings but perhaps the most widely accepted definition is that proposed by Moore (1949): "Sedimentary facies is defined as any areally restricted part of a designated stratigraphic unit which exhibits characters significantly different from those of other parts of the unit." According to this definition facies are restricted in extent both stratigraphically and geographically, though the same facies may be found at several levels within the same stratigraphic unit.

It is the experience of geologists, however, that facies observed in one stratigraphic unit may bear a close resemblance to those observed in units of different age or in different parts of the world. Consequently the term facies is commonly used in a more general sense for associations of sedimentary rocks that share some aspect of their appearance. For example, a group of beds may be described as belonging to a "red, conglomeratic facies" and if the geologist is confident of his interpretation, he may substitute environmental for descriptive terms, e.g., "alluvial fan facies." Such confidence has frequently proved to be misplaced so that it is to be recommended that the facies be fully described and given a purely descriptive designation before an environmental interpretation is attempted.

The combination of studies of modern sedimentary environments and of ancient stratigraphic units has led to the concept that there are certain associations of environments and environmental controls (both internal and external to the basin of deposition) that recur through geological history. These recurring combinations give rise to large-scale associations of facies that are seen many times in the stratigraphic record. Every group of stratigraphic units is, of course, unique but the similarities between certain groups may be sufficiently distinct to give meaning to the concept of a generalized *facies model*. Elaboration of the major facies models is one of the major joint tasks of sedimentology and stratigraphy. Recognition that a particular stratigraphic unit belongs to the class of units idealized in one of these facies models considerably simplifies the problem of environmental interpretation because the model has predictive value. Correct environmental interpretation of just one or two stratigraphic units, when combined with the facies model, suggests the interpretation of many other units and predicts the probable extent and overall geometry of the major units.

An important aspect of the facies distribution that must be explained by any good facies model is the vertical succession of facies. In many stratigraphic units some facies are repeated several times in the vertical sequence, suggesting the existence of *sedimentary cycles*.

A wide variety of different types of sedimentary cycles have been described by geologists and are discussed in the book by Duff, Hallam, and Walton (1967). They conclude that the terms "rhythm," "cycle," and "cyclothem" should be regarded as synonymous. Cyclic sedimentation refers to the repetition through a succession of sedimentary rock types of characteristics that are organized in a particular order. Cycles may contain two or more repeated units and the arrangement of the units may be symmetrical or asymmetrical.

Even the objective recognition and description of cycles raises considerable problems, let alone their correct interpretation. Most geologists have attempted to describe "composite" or "ideal" cycles, often without stating clearly what procedure was adopted in arriving at the cycle described. Duff and Walton (1962) distinguished the "*composite sequence*" that combines all the observed rock units in the order in which they tend to occur from the "*modal cycle*," which is that particular sequence of rock units that occurs most frequently in a given stratigraphic unit. Both the modal and composite cycles should be derived directly from observation but some authors also describe an "ideal" or "model" cycle that is an abstraction, based partly on theoretical considerations. The ideal cycle should generally be predictable from the facies model.

In ancient deposits the vertical sequence of facies is generally much easier to observe than the lateral sequence, a circumstance that is the exact opposite of the situation in modern sediments. In attempting to reconstruct environments from facies, therefore, geologists have often been guided by *Johannes Walther's Law of Succession of Facies*. This "law" is commonly stated in the English literature in the form that "facies sequences observed vertically are also found laterally" (Fig. 6-1). Walther's actual statement was more subtle: "The various deposits of the same facies area and, similarly, the sum of the rocks of different facies areas

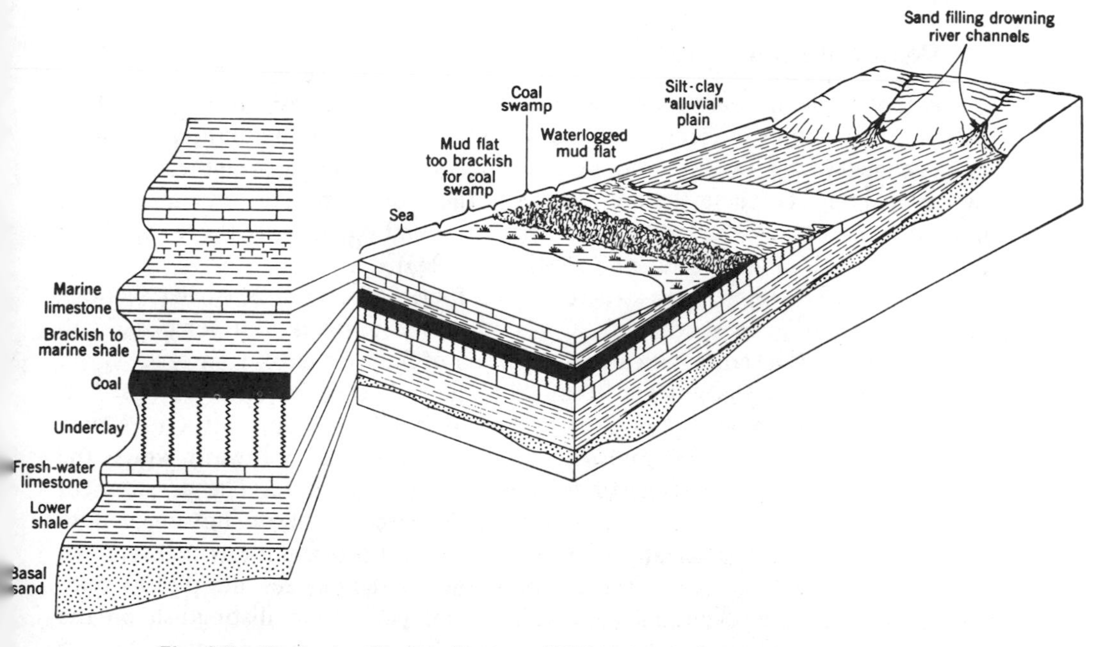

Fig. 6-1 Highly simplified application of Walther's Law to interpretation of an ideal cyclothem. After Weller (1960), from Shaw (1964).

were formed beside each other in space, but in a crustal profile we see them lying on top of each other . . . *it is a basic statement of far-reaching significance that only those facies and facies areas can be superimposed, primarily, that can be observed beside each other at the present time.*" (Walther, 1894, free translation from the German, p. 979, emphasis added).

Walther clearly understood that there must be no major breaks in the stratigraphic sequence for this principle to apply. Also, the vertical succession observed at any one section is most unlikely to display *all* the facies developed laterally. Even very important facies may be omitted from the vertical sequence. For example, many sections through a delta may omit channel sands, distributary mouth bars, or beaches, even though these are commonly regarded as most important facies within the deltaic model.

Oversimplification of the observed vertical succession tends to result in oversimplification of the lateral facies interpretation. For example, in Fig. 6-1, the vertical succession shown is the "ideal" Pennsylvanian cyclothem. The interpretation shown by the block diagram is in terms of transgression of the sea over a highly idealized coastal plain. This interpretation should be contrasted with the more complicated but more realistic explanation of cycles in terms of shifting delta distributaries discussed below and shown in Fig 6-8 and 6-9.

6.2 PALEOCURRENTS

Vectorial Data

A major role in the construction of facies models, and in the interpretation of stratigraphic units in terms of facies models, is played by the mapping of rock body geometry and of the assemblage of sedimentary features associated with each rock body. Of these features, perhaps the most important and diagnostic, besides the fossils, are the sedimentary structures and particularly those that have directional significance (Potter and Pettijohn, 1963).

Many of the sedimentary structures and fabrics discussed in earlier sections have an orientation in space that is related to the currents that formed them. In some cases the structure indicates a direction of fluid movement; in other cases, only the line of movement. The term "direction" has been commonly used in both senses so care must be taken in writing to avoid confusion. Structures such as cross-bedding, flutes, and imbrication show the direction toward which the current was flowing at the time of formation of the structure. The direction of such a structure is measured on a scale of 360 degrees. Line of movement data include features such as lineations, grooves, and grain orientations measured in the bedding plane. They are measured on a scale of 180 degrees and plot as symmetrical current rose diagrams because it is not possible to distinguish on the basis of these structures which of the two alternatives was the direction toward which the current was flowing. Directional features recorded from tilted or folded strata must be recorded as though they were in the original horizontal position.

Commonly this is done by applying a simple correction for tilt in the field but more extensive corrections must be made in areas of plunging folds.

Current directions deduced from the study of sedimentary structures or fabrics are described as *paleocurrents*. Measurements that indicate a direction are commonly called *vectors*, though some authors have objected that a vector must have magnitude as well as direction and that consequently measurements such as cross-bedding dip azimuth should be called "directionals." The magnitude of any particular directional measurement may, however, be arbitrarily assigned a value one, making all such measurements unit vectors. This procedure is necessary, whatever the merits of the term, in order to compute an average direction because the usual method of taking an arithmetic mean cannot be used for directional data. As a simple example, the mean of two measurements of 350 and 10 degrees is 180 degrees, just the opposite of the "true" average. It can also be shown that standard deviations or variances calculated from directional data change their magnitude depending on the choice of origin. To avoid these difficulties, it is standard practice to compute the vector mean and strength instead of the arithmetic mean and variance of the azimuths (Curray, 1956). Besides using the vector mean, the mode may be determined from graphical plots and used as a measure of average direction. This has the advantage that it is not strongly affected by highly skewed or polymodal distributions. Bimodal distributions are common in directional data because there were frequently currents that flowed in two dominant directions. Use of the vector mean for such data may give misleading results.

Interpretation of Vectorial Data

On the largest scale, mapping of directional sedimentary structures in many rock units has revealed that despite much local variability the average paleocurrent directions remain almost constant over very large areas and persist with only slight changes through long periods of geological time. This observation has given rise to the concept that the basic regional control of paleocurrents was the *paleoslope*. For fluvial deposits, it is clear that the average of the current directions will tend to be down the regional slope of the alluvial plain but doubts arise about the validity of the concept of the paleoslope when this interpretation is extended to marine sediments (Klein, 1967). Tidal currents in some areas are oriented normal to the coast but in many other areas they have a dominant orientation parallel to the coast and bottom slope. Large-scale current systems ("geostrophic" or "contour" currents) with speeds high enough to move fine sand flow parallel to the slope along some continental margins. Mapping sole markings has shown that in many turbidity current deposits the predominant direction of flow during the deposition of the sediments appears to have been parallel to the long axis of the sedimentary basin. The slopes along the basin axis were presumably small compared with those down the basin sides. Thus, although the structures indicate *a* bottom slope, it is not to be interpreted as the major slope in the basin of deposition.

In fluvial systems, the opposite of the average paleocurrent direction is commonly interpreted as showing the direction of the source area for the detrital sediments. Obviously, paleocurrent measurements do indicate the direction in which sedimentary materials were being moved at the time of deposition but this direction may be only indirectly related to the position of the ultimate sources of the detrital materials. Studies of many basins containing turbidity current deposits, particularly those of the Polish flysch (Dzulynski and Walton, 1965) strongly suggest that despite the predominant orientation of measured paleocurrents longitudinal to the basin, much of the sediment was supplied transversely from one or both sides of the basin. To a lesser degree, the same seems to be true of many fluvial systems.

When interpreting paleocurrents, it must always be remembered that only those currents are recorded that resulted in the deposition and ultimate preservation of sediment in the stratigraphic record. This statement applies at all scales, from that of sedimentary basins to that of ripples or grain orientation. Because of the nature of tectonic controls, it is more probable that the central parts of sedimentary basins will be preserved than the margins, particularly those parts of the margins adjacent to rising source areas. Even if the basin margin is preserved, it is likely to be an area where sedimentary units are absent because of erosion or nondeposition rather than an area whose history is preserved in a continuous sedimentary record.

It must also be remembered that not all sediment movement is recorded by directional structures. Large areas of tidal sand bars, where crossbedding is being formed, may be dominated by one part of the tidal cycle, commonly the flood currents. The channels between the bars are dominated by ebb currents but leave little evidence in the stratigraphic record. On beaches, the dominant longshore direction of movement of sand does not appear to be indicated by most of the sedimentary structures that are formed.

In mapping paleocurrents, the data yielded by different types of structures should be recorded and plotted separately. Not only are there frequently discrepancies between the average paleocurrents indicated by different structures but certain types of structures may be found to yield more consistent results than others. Comparison of the data from different structures may contribute substantially to the environmental interpretation of the formation.

6.3 PALEOHYDRAULICS

Reconstruction of the paleocurrents in a sedimentary basin may be regarded as simply the first stage in the reconstruction of *all* the hydraulic factors acting in the basin. One possible approach to the interpretation of paleohydraulics has been discussed in Sec. 5.2: the interpretation of structures in terms of flow regimes. Flow regime assemblages in turn carry implications about larger-scale processes or environments. Although all the data for a full hydraulic interpretation of sedimentary structures are probably never available to the geologist, there seems to be no reason why attempts at interpretation should be restricted to the level of paleocurrents and flow regimes. Some other possibilities have been indicated in the discussion of sedimentary textures and structures given above and are briefly illustrated below.

For fluvial systems the main variables are the discharge, width, depth, sinuosity, and slope of the channel. The relations among these variables observed for modern rivers have been summarized by Leopold et al. (1964) and Schumm (1969). Study of the geometry of an ancient river channel sand may reveal the maximum channel depth (roughly, the thickness of the channel sand if only one fining-upward sequence is represented in the preserved channel deposit), the channel width, and the radius of curvature and sinuosity of the channel meanders. Alternatively, the sinuosity may be estimated from knowledge of the silt-clay fraction in the bed and banks, using empirical relationships determined by Schumm. These basic data permit estimation of many hydraulic parameters including discharge and channel slope.

Cotter (1971) has described the fluvial Ferron sandstone in the Upper Cretaceous Mancos shale of east-central Utah. The sandstone forms lenticular channel-shaped bodies with thicknesses of about 30 ft. Paleocurrents flowing to the north are indicated by cross-bedding and by a grain size decrease from coarse to fine over a distance of some 50 mi. The fluvial channel environment is proved by the fining-upward sequences, with sharp base, lower unit of trough cross-bedding, and upper unit of ripple lamination. Formation of the sandstone bodies by lateral migration of point bars is further proved by inclined sandstone cross-bedded sets dipping at angles of 5 to 10 degrees normal to the predominant trough cross-bedding direction. From the geometry of the sandstone bodies, Cotter estimates the channel depth at 25 ft. The channel width is estimated from the size of the inclined sets formed by point bar migration to be about 300 ft. The percentage silt and clay in the channel bed and banks can be determined from sandstone samples and from these data it is possible to calculate the sinuosity and the discharge. The values obtained are a sinuosity of about 2, corresponding to a highly sinuous channel, and a mean annual discharge of 6000 to 7000 cu ft/sec, with a mean annual flood of 22,000 cu ft/sec. These relationships further suggest that the slope was less than 0.0007. Use of the Manning equation, with a value of roughness typical of dune beds, suggests a more precise slope value of 0.0003, or about 1 ft/mi. From the known values of discharge, channel depth, and width, it is easy to estimate mean velocities at bankfull discharges to have been about 2 ft/sec.

In this example, therefore, it has been possible to achieve a very complete hydraulic reconstruction of an Upper Cretaceous river. The reconstruction is based on geometry of the sandstone bodies, type of sedimentary structures, and grain size characteristics, together with empirical relationships observed on modern rivers. Several of the most important hydraulic variables can be estimated by more than one method, thus providing a check on the internal consistency of the reconstruction.

The wavelength of preserved antidunes permits a direct estimation of current velocity (Chap. 5). It is known that, for antidune formation, the Froude number must be approximately unity and consequently the depth may be estimated. The roughness or hydraulic resistance may be estimated from experimental data and observations on modern streams with antidunes; from this estimate and the estimates of velocity and depth, the slope may be determined. In this way, Hand et al. (1969) have estimated that currents responsible for forming antidune cross-bedding in a Triassic sandstone had a velocity of 1 m/sec, were 1 to 2 cm deep, and flowed down a slope of at least 2.7 degrees.

Few such attempts to analyze observed sedimentary sequences in terms of hydraulics have yet been made. It seems clear that despite the uncertainty introduced into the calculations by the assumptions that must be made, analysis of this type is a very valuable check on qualitative interpretation of sedimentary processes and environments and will ultimately lead to much greater refinement in interpretation than is now possible.

6.4 ENVIRONMENTS

A list of all the factors that influence sedimentary environments would be a long one. Attempts to reduce the complexity of the list by concentrating attention on a few major factors are generally only partly successful. To some extent the relative importance of the factors on the list depends on the particular aspect of the sediments that is under consideration. For example, a geologist interested in the relative importance of stream and debris-flow deposits on alluvial fans is likely to emphasize a different set of environmental controls from those stressed by a geologist interested in the development of redbeds in the same environment.

One of the most frequently cited controls is the "energy" of the environment, by which is usually meant some ill-defined index of the kinetic energy or "turbulence." In a few cases, the distinction between environments on the basis of "energy" can be formulated in more precise terms. For example, the amount of wave energy expended on different parts of a shoreline can be calculated from a knowledge of the deep water wave conditions and the offshore topography. But, generally, the use of the terms "high energy" or "low energy" is so vague as to be almost meaningless.

Among physical controls, the list of major factors must include the water depth, velocity, and temperature. The climate is important in controlling the temperature and also as a control of aridity or humidity (ratio of evaporation to precipitation) and the magnitude and frequency of floods and storms. In turn, such factors control organic and chemical factors, such as the salinity and the abundance and types of marine algae and organisms. The physical shape or setting of the environment may control the degree of mixing of different water masses, and consequently the salinity or oxygen content. These and many other factors combine to form a complex interacting system, with many feedback loops. It is precisely because of this interaction that there appears to be a relatively limited number of major depositional environments, each composed of a variety of closely related subenvironments. In the paragraphs that follow, a few of these major controls are isolated and criteria for recognition in ancient environments are briefly discussed.

The *depth of water* is important in seas and lakes for several reasons (Hallam, 1967). Waves disturb water down to a depth of only approximately half their wavelength. Light penetrates to a depth that depends on the light intensity and the water turbidity but in general does not exceed about 200 m. The growth of algae, and consequently also of organisms such as hermatypic corals, whose life cycle is intimately connected with algae, is restricted to water at least as shallow as this. Shallow water exposes a larger surface to the atmosphere, relative to its total volume, than deep water and consequently fluctuations in salinity due to evaporation or rainfall are more important in shallow than in deep water; temperature fluctuations are also more extreme. Criteria for water depth in marine deposits therefore include (a) signs of current and particularly wave activity, for example, cross-bedding, cut and fill structures, and symmetrical ripple marks. This evidence for shallow water must be used with caution since there can also be strong currents in deep water. (b) Signs of periodic exposure to the atmosphere, for example, mud cracks, or certain types of ripple marks modified by erosion during emergence. It has been demonstrated that mud may crack under water in the laboratory but, in nature, mud cracks appear to be a relatively reliable indicator of emergence. (c) Signs of intense biological activity and particularly traces of plant roots, algae, or structures formed by algae. Carbonate buildups appear to form only in shallow water. (d) The types of trails and burrows change with increasing water depth, from a dominance of vertical burrowers in shallow water to shallow, horizontal burrowers in deeper water to surface sediment grazers in the deepest abyssal environments. (e) Certain authigenic minerals, particularly glauconite and apatite, appear to be most typical of, though not restricted to, depths of 30 to 1000 m.

Depth of water at the time of relatively deep water sedimentation may be estimated for Tertiary rocks using paleoecologic data. The benthonic foraminifera have proved most useful. In rocks deposited earlier in geological history, paleontologic criteria are generally less reliable, though trace fossils, among others, appear to be valuable indicators. Generally, however, deep water can be inferred only from the overall paleogeographic reconstruction, the relationships with adjacent facies, and an absence of features typical of shallow water.

The *velocity*, kinetic energy, or turbulence of the water controls the movement of sediment. Fine sediment cannot settle in an environment in which the level of turbulence is always high, as in the beach environment, unless the rate of supply is extremely large. Constant movement of sediment results in hydraulic size and shape sorting, abrasion of the larger grains, and formation of sedimentary structures such as lamination or cross-stratification. A more detailed discussion of criteria for reconstructing the hydraulics of the environment is in Sec. 6.3. Substrates formed from constantly moving sand are inhospitable to sessile organisms. Sand, particularly if composed of carbonate particles that contain some organic matter, may support an organic burrowing or bottom feeding community if it is moved by currents only infrequently. Such sand rapidly becomes stabilized by the growth of grasses and filamentous algae.

The *temperature* controls the solubility of calcium carbonate and the metabolic rate of plants and animals: Cold waters rarely support large communities of carbonate-secreting or encrusting organisms. High rates of evaporation are also associated only with warm climates and consequently both carbonates and evaporites are commonly restricted to warm waters. Temperature may be estimated by use of oxygen isotope data in carbonates or by use of paleontologic criteria. Very cold waters are, of course, indicated by evidence of ice action, such as till or varved clays with ice rafted stones.

The *salinity* of an environment is an important control of biologic activity, with organisms scarce in both hypersaline and brackish environments (and particularly in environments where the salinity often changes). The salinity contrast between seawater and river water also plays an important physical role in reducing mixing and causing the fresh water in deltas and estuaries to spread out over the denser seawater. The chemical contrast causes flocculation of clay particles. Rapidly deposited flocculated muds show little lamination, in contrast to the well-developed lamination commonly shown by muds deposited in freshwater lakes. The distinction between marine and freshwater muds may be made on the basis of faunal content and by making use of trace elements, particularly boron, which tend to be enriched in marine muds. Shimp et al. (1969) studied modern muds from a variety of marine and freshwater environments and found that boron content was correlated with the content of clay finer than 2 μ. Several other trace elements besides boron were tested but it was found that a combination of boron and clay content gave the best discrimination between the two groups. The use of trace elements to distinguish between fresh and marine environments is valid only on a statistical basis and is by no means infallible but plausible results have been obtained with ancient shales as well as modern muds.

The pH (hydrogen ion concentration, or relative acidity) of the bottom water and the Eh (oxidation potential) are frequently cited as two important chemical aspects of the environment that greatly affect the type of sediment and bottom fauna. pH and Eh are, however, not primary environmental controls but depend in turn on physical controls of the circulation and mixing of water masses and on factors, such as salinity, that affect the availability of organic matter. In many cases, both pH and Eh change radically as the sediment accumulates. For example, alkaline, oxidizing conditions in the sea, just above the bottom, may rapidly become acid, reducing conditions a few centimeters or less below the sediment-water interface because of bacterial decay of the organic matter in the sediment.

The physical *setting* of the environment refers to the larger scale physiography, determined by the overall tectonic and geologic history of the area. The setting is important because it acts as a control of such factors as the slope of the land surface, the amount of wave or tidal action, and the supply of sediment and freshwater from land masses adjoining the area of deposition.

An outline classification of depositional environments is presented in Table 6-1. This classification is skeletal at best. Any experienced geologist will be able to subdivide many of the environments listed and such subdivision is necessary for an adequate interpretation of a stratigraphic unit. The purpose in presenting such a highly schematic classification is to indicate the major environments at a glance. The table can serve as a "check list" in considering possible interpretations.

Of the environments listed, those in categories I and II (terrigenous or mixed environments) are best known. Some of the categories are very large. For example, there is a very wide range of lake environments but they are all listed together in the table because it is not clear at present what subdivision of this group of environments would be most meaningful for geologists. Under category II, deltas are not listed separately because they are made up of the other major environments listed.

The shallow marine environments are possibly the most difficult to classify.

TABLE 6-1 CLASSIFICATION OF SEDIMENTARY ENVIRONMENTS

I. **Terrigenous**
   - A. Alluvial fan
   - B. Floodplain (meandering, braided)
   - C. Lacustrine (humid, arid)
   - D. Eolian desert
   - E. Swamp
   - F. Glacial

II. **Mixed or Shore-related**
   - A. River channel or distributary (and levee)
   - B. Estuary
   - C. Bay, lagoon
   - D. Marsh
   - E. Intertidal and supratidal flat, bar, and channel
   - F. Barrier island and beach
   - G. Glacial-marine

III **Shallow Marine** (Neritic, depth up to 600 ft)
   - A. Shelf banks (tidal, nontidal)
   - B. Shelf basin (unrestricted, restricted humid, restricted arid)
   - C. Graded shelf
   - D. Carbonate shelf and reef (attached or unattached to major land mass)
   - E. Evaporite basin (standing body of water)

IV. **Deep Marine** (Bathyal, 600–6000 ft; abyssal, deeper than 6000 ft)
   - A. Slope and canyon
   - B. Submarine fan
   - C. Deep ocean basin (pelagic, terrigenous)
   - D. Deep enclosed marine basins (inland seas: humid or arid)

This group of modern environments has been strongly affected by the post-Pleistocene rise in sea level and it is not certain how relevant the distinctions made on modern shelves are for the interpretation of the ancient. The modern shelves are generally divided into banks and basins, and shelves dominated by terrigenous clastics can be distinguished from those dominated by carbonate deposition. The carbonate shelves are frequently associated with organic reefs and supratidal evaporite deposits (see Chap. 12 and 15). A true "graded shelf," where there is a progressive deepening of water and decrease in grain size toward the edge of the shelf, is not known today but it has been suggested that such environments existed in the past (see Chap. 20).

A full description of all that is known about modern environments and a discussion of the problems that arise in applying this knowledge to ancient sediments goes beyond the scope of this book. The reader seeking more information will find some basic references at the end of this chapter.

6.5 CLASSIFICATION OF FACIES MODELS

It is premature at this stage of the development of sedimentology and stratigraphy to attempt a systematic classification of facies models. A list of some important facies models is presented below as a basis for further discussion.

1. Alluvial fan
2. Fluvial
3. Deltaic
4. Barrier island
5. Offshore shoal
6. Turbidite—deep basin
7. Carbonate platform
8. Lacustrine
9. Eolian desert
10. Glacial

The list given is simpler than the list of environments in Table 6-1. This apparent simplicity in part reflects incompleteness and ignorance but in part it stems from the fact that the environments listed in Table 6-1 are not independent variables. They are found together, associated in both space and time (as expressed, for example, by Walther's law discussed on p. 187). Their association is the basis of the establishment of facies models.

No attempt is made in the following pages to give a complete discussion of facies models. In some cases (e.g., carbonates and evaporites) the models cannot be fully discussed until these rock types are more fully described in Chapters 12 and 15. In other cases (e.g., glacial, eolian desert, and lacustrine) the facies are relatively uncommon in the stratigraphic record, considered as a whole, and no further discussion is given here. These facies are, however, sufficiently distinctive and important locally so that they have been extensively discussed in other books, to which the student is referred (see Selley, 1971, and Rigby and Hamblin, 1972).

6.6 ALLUVIAL FAN MODEL

This model is typified by the model developed for the Triassic fault basins of the northeastern United States and Maritime Provinces of Canada. Coalescing fans spread into the sedimentary basin from an upfaulted mountainous source area. In the distal parts of the basin (farthest removed from the source of terrigenous sediment) lacustrine sediments may be deposited, including sediments with many authigenic minerals or evaporites. The fans include the deposits of debris flows, sheetwash, and braided streams. The sediments are frequently red, probably because of a combination of a source area that supplies iron-rich minerals and a semiarid climate that produces an oxidizing vadose groundwater zone (see Chap. 10). Eolian sand deposits are common in some basins.

The deposits are characterized by great thickness (exceeding 15,000 ft in the Connecticut and New Jersey Triassic), rapid lateral facies changes, and preponderance of conglomerates and sandstones, with shales mostly silty and not so abundant as in most other facies models.

Most of the sediments can be shown, by studies of the mineralogy and petrology of the clasts and by paleocurrent studies (including mapping of pebble imbrication and pebble roundness, see Fig. 6-2) to have a local provenance. Consequently the sediments are generally both mineralogically and texturally immature.

6.7 FLUVIAL MODEL (OR ALLUVIAL PLAIN MODEL)

Two major subdivisions of this model may be recognized, corresponding to meandering and braided streams, respectively. The most common variety of this model is that of a meandering river or rivers flowing over a broad coastal alluvial plain, which is bordered on one side or at one end by the sea. The fluvial model may therefore be connected with an alluvial fan model or with a deltaic or barrier beach model. Because of the preferred association of alluvial fans with a semiarid climate and alluvial plains and deltas with a humid climate, the latter association is more common than the former.

The modern lower Mississippi River alluvial plain and its Tertiary and Cretaceous antecedents whose deposits are preserved in the "Mississippi embayment" formed the basis of one of the earliest attempts to formulate a facies model in the modern sense of the phrase (Pryor, 1961). In the same region, the Mississippian (Chesterian) sediments show much in common with the Cretaceous to modern sediments.

Few modern alluvial plains have been studied as intensively as that of the

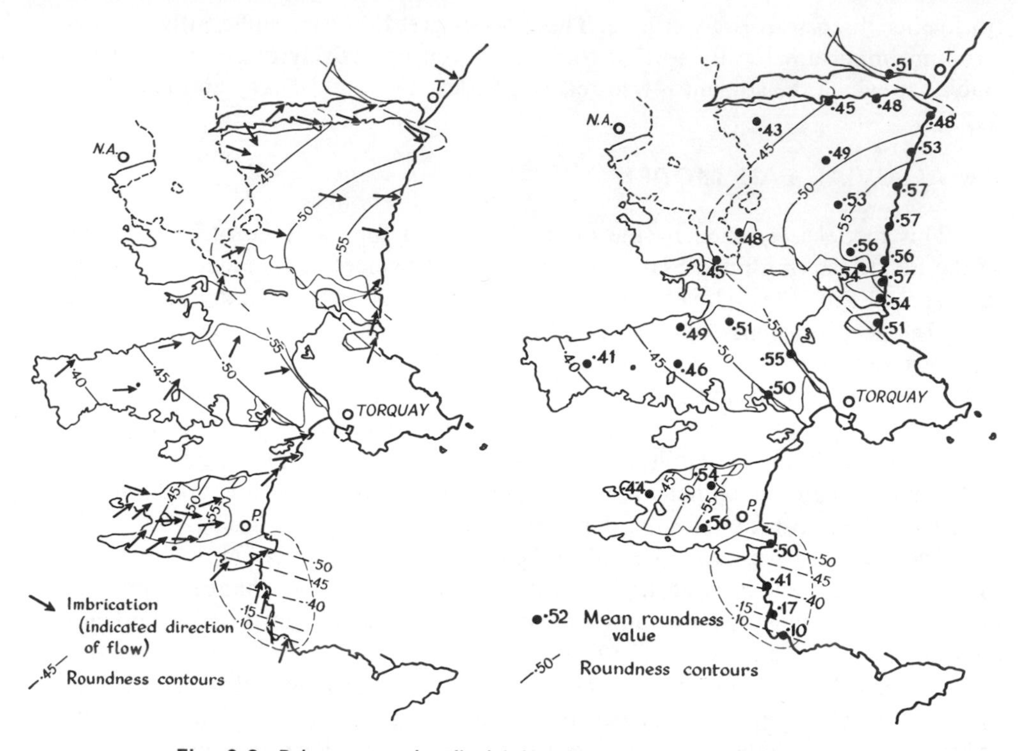

Fig. 6-2 Paleocurrents in alluvial fan deposits of the Permo-Triassic of South Devonshire, England, determined by mapping (a) pebble imbrication and (b) pebble roundness. After Laming (1966).

Mississippi but there have been many local studies of fluvial sedimentation so that the basic processes are well understood, at least for *meandering channels.* The main processes are discussed in Chap. 5 and are reviewed in papers by Allen (1964, 1965) and Beerbower (1965). The main sedimentary environments are the river channel and point bars, the levees, and the floodplain, which includes swamp, lake, and small channel environments (Fig. 6-3). Channels may be abandoned locally by cutoff of meander loops resulting from erosion of the necks of very sinuous meanders or from the formation of "chutes" by floodwaters taking shortcuts across less sinuous meanders. Abandoned channel loops form "oxbow lakes" in which fine sediment accumulates (clay plugs). During floods, levees may be crevassed, with formation of "splays" of sediment across parts of the floodplain (Fig. 6-3). Alternatively, crevassing may be so severe that an entire stretch of channel is abandoned as the river seeks a lower path across the floodplain (*avulsion*). All of these processes may be regarded as forms of either vertical or lateral accretion and all of them lead to the shifting of depositional environments and, therefore, of facies across the alluvial plain.

The characteristic deposit laid down by the migrating channel of a meandering

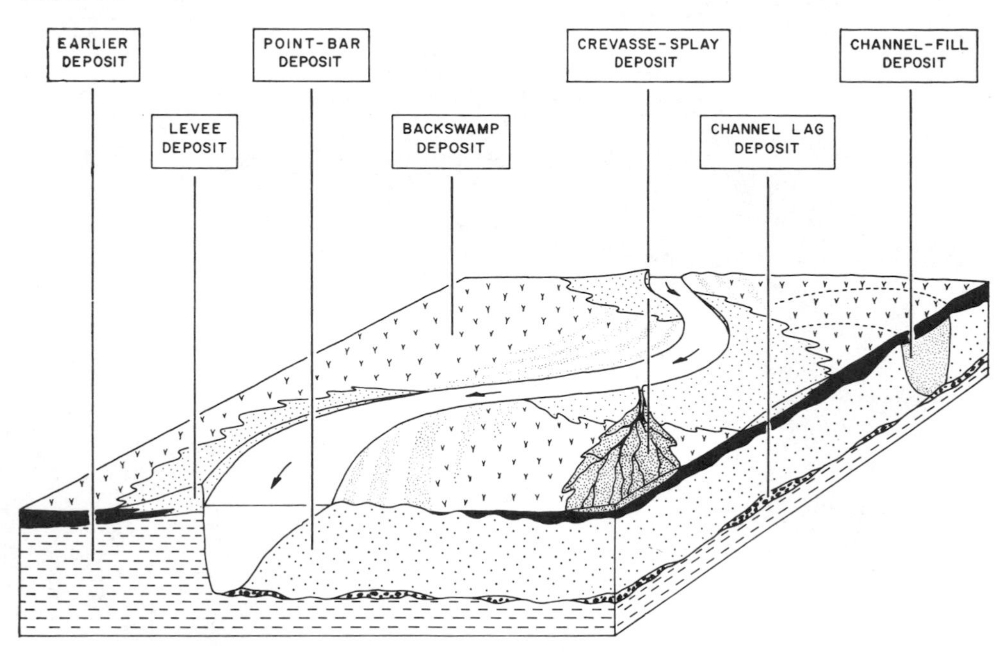

Fig. 6-3 Facies model of the floodplain of a meandering river. From Allen (1964).

river is the fining-upward sequence discussed in Chap. 5. In coarse grained point bars, the normal fining-upward sequence may be truncated or succeeded by one or two thick tabular sets of cross-beds deposited in *chute bars* (McGowen and Garner, 1970). These are lobes of sediment built out on the downstream part of the point bar by floodwaters passing through a chute that cuts across the convex bank of the meander. On the natural levees the deposits consist of silts and fine sands and commonly show ripple marks and ripple-drift cross-lamination. On low-lying areas of the floodplain, fine muds that are rich in plant remains accumulate. They contain silty or sandy laminae deposited during floods but the lamination is generally poorly preserved due to bioturbation. The muds may be inter-bedded with peat and may pass up into coarser grained splay or levee deposits.

Most *braided rivers* that have been carefully studied are depositing gravel or coarse sand (Fig. 6-4). Such rivers are very common on glacial outwash plains. Much larger braided rivers, with fine sand as the dominant sediment, are also found and are typified by the Yellow River in China and the Ganges-Brahmaputra River in East Pakistan (Coleman, 1969).

In the Brahmaputra, there are moderate floods every 4 years and catastrophic floods every 30 to 50 years. During floods, very large areas are inundated and vast quantities of sediment are moved. In only 200 years, the Brahmaputra has

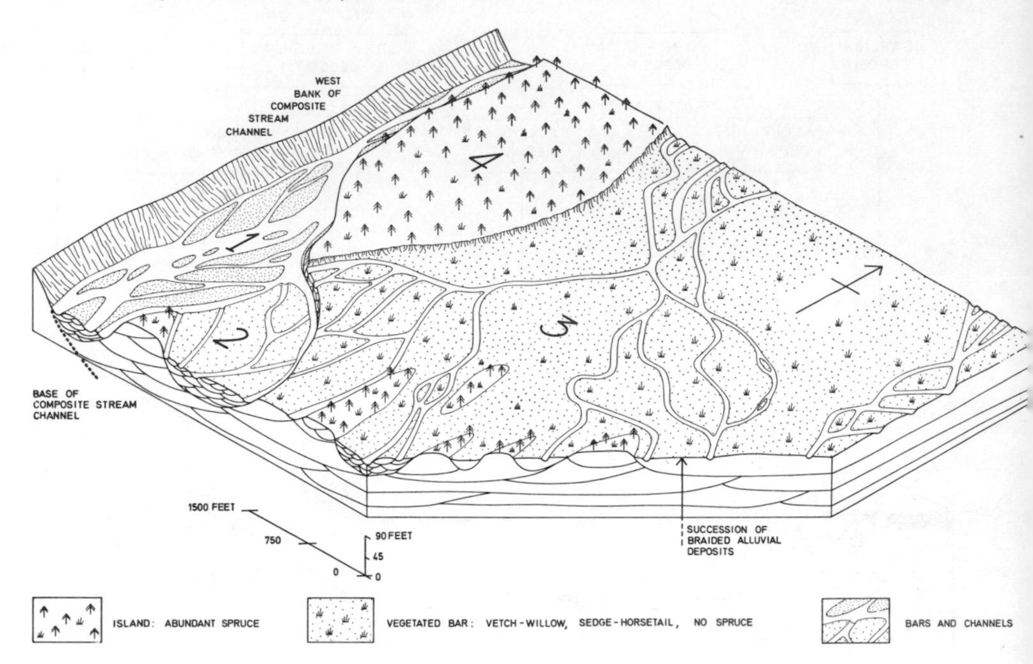

Fig. 6-4 Facies model of the floodplain of a braided river. The floodplain is dissected into four levels, numbered in order of increasing elevation. The lowest levels are most active, the highest least active and most stabilized by vegetation. From Williams and Rust (1969).

cut a valley 130 mi long and 8 mi wide, in which a blanket of sand some 60 ft thick has been laid down by migrating braid bars. "The channels have a wide and shallow bed choked with sand bars. Water flows in a number of branching and reuniting channels with one or two serving as major channels. The mid-channel islands and sand banks shift rapidly with the ever changing flow regimes.... The position of the main current in a braided stream channel is extremely unstable and causes the river course to shift in position. There is also a distinct lack of tight river bends and only gently curved thalwegs exist within the wide and shallow channels" (Coleman, 1969, p. 166).

The shifting bed forms and sandbars, on several different scales, result in the formation of both large- and small-scale cross-stratification. As a result of abundant sediment supply, "megaripple drift" with gently climbing sets up to 2 ft thick is common. The levees are composed almost entirely of overlapping splay deposits formed during floods and showing a vertical sequence with fairly massive, poorly sorted silts and sands overlaid sharply by well-developed cross-beds. The cross-beds are truncated by a thin layer of plane lamination that is succeeded by ripple cross-laminated sands and silts.

Paleocurrents derived from cross-beds in braided river deposits show relatively low variances as compared with those from meandering rivers.

6.8 DELTAIC MODEL

A large literature exists on the characteristics of both modern and ancient deltaic deposits. Deltas are a major site of deposition of terrigenous sediments at the present time and it is to be expected that they were similarly important in the past. Deltas are complex systems and several subdivisions of the deltaic model are needed to represent it adequately. A spectrum of delta types may be recognized depending on the relative importance of rate of sediment supply and its redistribution by waves and tides. The elements of the spectrum, beginning with a dominance of sediment input, are birdsfoot, lobate, cuspate, arcuate, and estuarine deltas. Many other factors besides sediment supply and waves and tides must also be taken into account. They include the relative supply of bed and wash load, the depth of water into which the delta is building, the rate of subsidence, and the climate.

An important aspect of deltaic sedimentation is the interplay of constructive and destructive phases of development. In the *constructive phases,* the delta builds out into deeper water by extension of its major distributary channels. There are generally several such channels, flanked by natural levees composed of silt or fine sand deposited rapidly from suspension during periods of flood. Sand moved by traction is normally confined to the channels and to the shallow water close to the end of the distributary, where it is deposited as one or more distributary mouth bars. The river water is generally lighter than seawater, even allowing for the river's content of suspended sediment. As the river water enters the sea, it spreads out over the surface to form a *plane jet* (Fig. 6-5). The jet spreads out and mixes with seawater at the lower surface so that velocities are checked, bed load is deposited to form the distributary mouth shoal, and the coarser suspended load settles to form broad submarine natural levees. The fine sediment is carried out beyond the bar crest and deposited on the sloping delta front or carried farther seaward to the prodelta depositional environments (Fig. 6-6). Much sediment is swept away by marine currents and carried tens or even hundreds of miles along the coast. The different clay minerals have different settling velocities and respond differently to flocculation by seawater so that there may be a mineralogic as well as size segregation in the prodelta deposits (see Chap. 11).

During periods of flood, there may be crevassing of the levees and rapid filling of the interdistributary areas with crevasse-splay deposits of sand and silt (Fig. 6-7). The interdistributary areas are normally occupied by shallow marine or brackish bays. The bays may be cut off from the sea and transformed into lagoons by the development of beaches linking the mouths of the distributaries. As the bays or lagoons fill with sediment, they become colonized by marshes (or mangrove swamps in a suitable climatic setting) and the marshes are in turn replaced by drier swamp vegetation, beginning on the flanks of the natural levees and spreading toward the center of the interdistributary areas. Thick peat deposits may be formed, interfingering with natural levee deposits (Fig. 6-6).

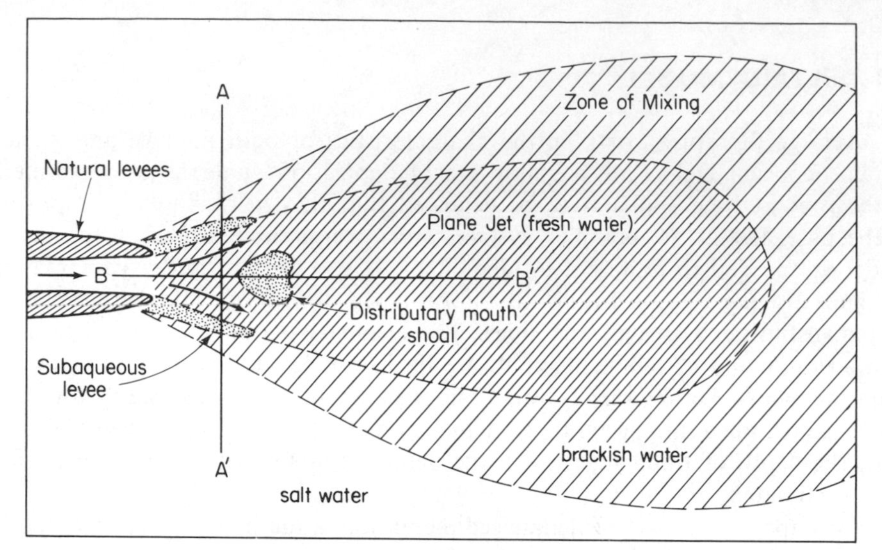

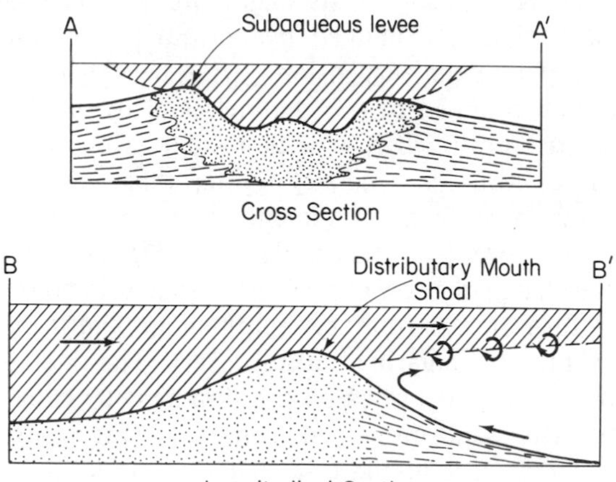

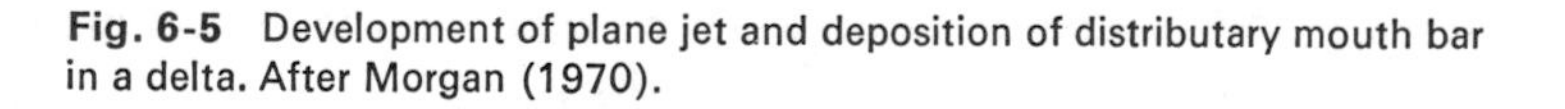

Fig. 6-5 Development of plane jet and deposition of distributary mouth bar in a delta. After Morgan (1970).

Similarly, areas of the floodplain between major levees are occupied by extensive low-lying swamps that become inundated by several feet of water during floods.

The rapid rate of supply of sediment and the lack of strong marine agents for dispersing sediment, such as tides and waves, result in some deltas being dominated by constructive activity. Examples are the modern Mississippi Delta, which is building into relatively deep water and therefore has a "birdsfoot" form, and the mid-Holocene LaFourche Delta of the Mississippi River (Fig. 6-8). The LaFourche Delta was built into shallower water than the modern Mississippi

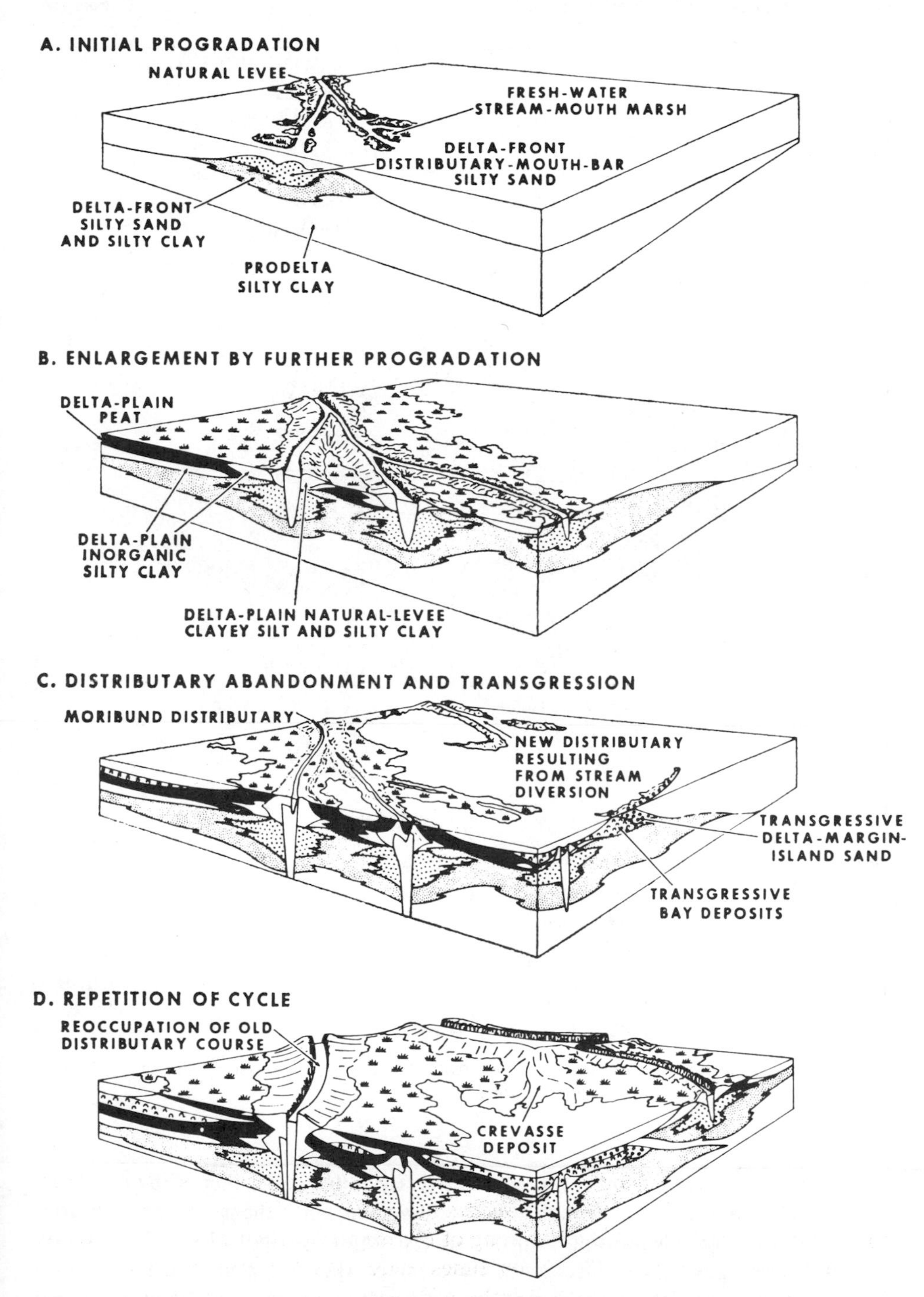

Fig. 6-6 Development of delta sequences in the Mississippi delta. From Frazier (1967).

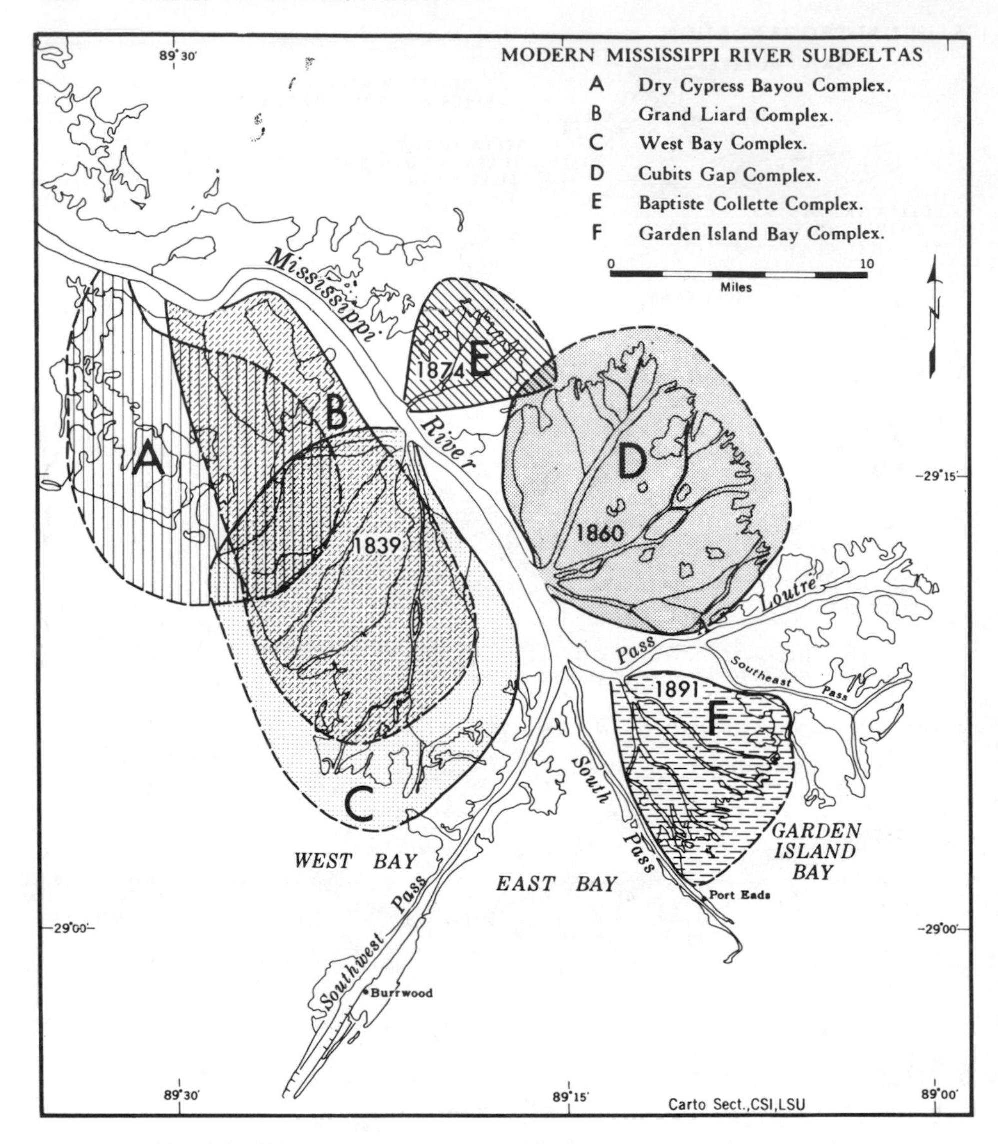

Fig. 6-7 Major crevasse-splays or subdeltas of the modern Mississippi delta. From Coleman and Gagliano, (1964).

Delta and had a more compact, lobate form. Even deltas dominated by constructional agents, however, go through a *destructive phase* when the major distributaries shift position. In the Mississippi, shifting of the major distributaries and formation of new deltas has taken place seven times since the Holocene rise in sea level (Fig. 6-8). As one delta is built up, the adjacent areas continue to subside and because they are receiving comparatively little sediment, they become low topographically. Eventually the path of the river shifts by avulsion and a new delta is

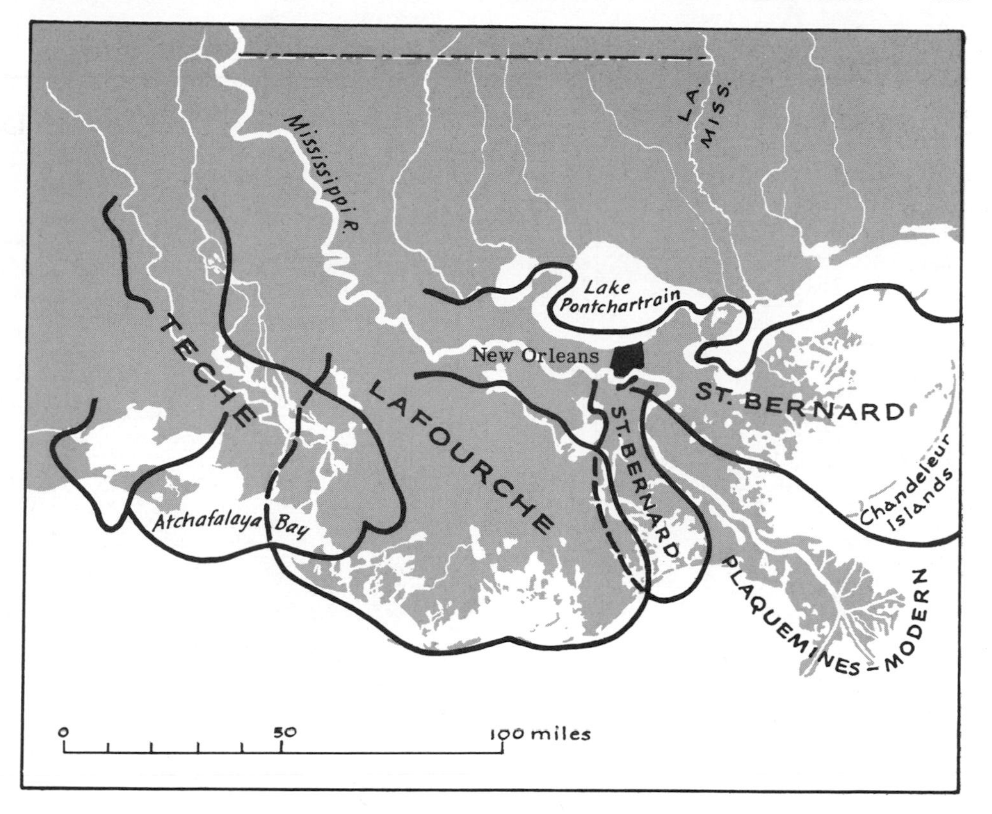

Fig. 6-8 Holocene deltas of the Mississippi River. After Frazier, 1967. Four major delta complexes are recognized on the basis of mapping, drill hole data and radiocarbon dating. Each complex consists of several lobes of slightly different age and there is some overlap in age from one complex to another. The four main complexes are: Teche (6000–4000 B. P.), St. Bernard (4700–600 B. P.), Lafourche (3500 B. P.–present), Plaquemines-Modern (1000 B.P.–present).

formed. Subsidence and both subaerial and marine forces of erosion then begin to modify the exposed parts of the old delta (Fig. 6-6). East of New Orleans, the barrier Chandeleur Islands (Fig. 6-8) have been formed by waves reworking the upper part of the old St. Bernard Delta of the Mississippi.

Delta shifting combined with subsidence and periods of marine incursion and erosion form cyclic repetitions of sedimentary facies, as illustrated in Fig. 6-9.

Not all deltas, however, are characterized by frequent shifting of the major distributaries. Many deltas enter seas characterized by strong tides or waves and in this case the marine forces may be able to redistribute the sediment as rapidly as it can be supplied by the river. Examples of such "high destructive" deltas are shown in Fig. 6-10. Another example of a wave-dominated delta is the "type" delta of the Nile River. In these deltas, waves and tidal currents tend to equalize

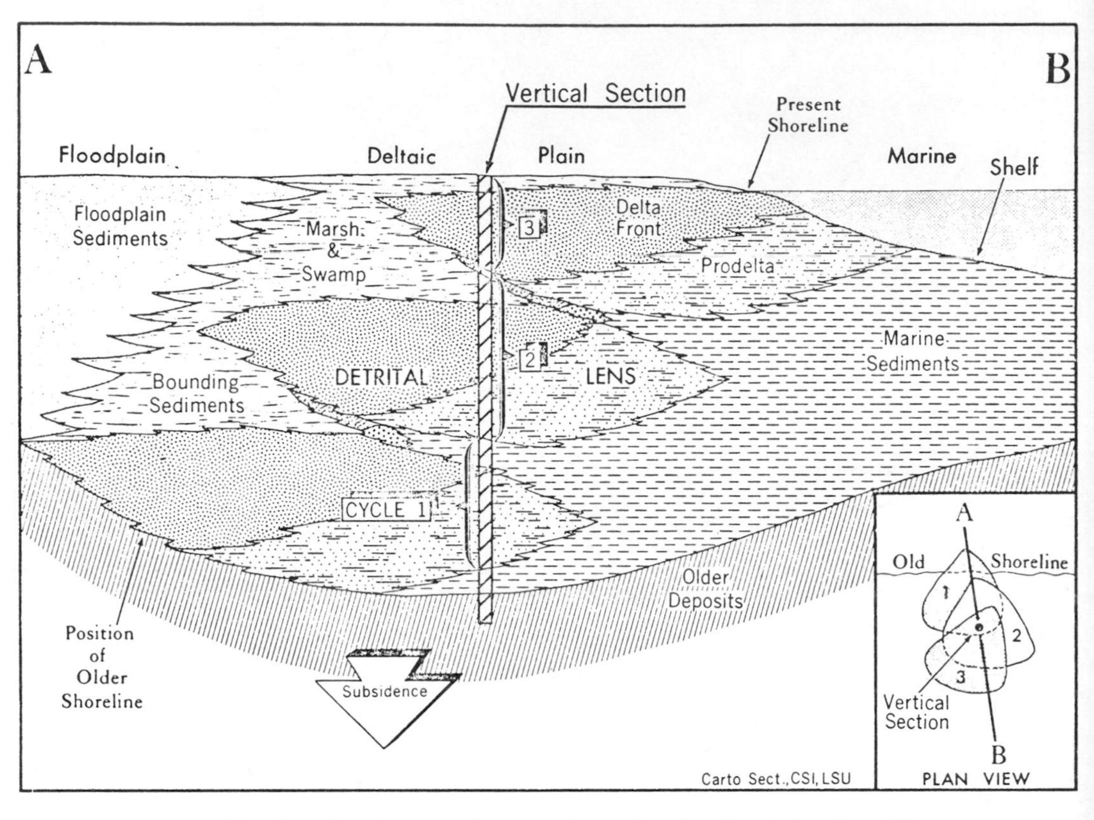

Fig. 6-9 Development of overlapping cycles of sedimentation by shifting deltas in a hypothetical delta-complex. From Coleman and Gagliano (1964).

sediment distribution and consequently the major distributaries are more stable in position than they are in high constructive deltas. But although they are less frequent, major shifts in channel position do take place in high destructive as well as in high constructive deltas.

6.9 BARRIER MODEL

The main environmental elements of this model are, from the shoreline outward, fluvial plain, marsh, tidal flat and lagoon or bay, barrier beach and dune, shoreface, nearshore, and offshore marine (Fig. 6-11). Shifting of the environments takes place in response to changes in sea level and supply of sediment, producing transgression or regression, and corresponding onlap or offlap of sedimentary units. During regression, nearly all the environments may be represented in a thick vertical sequence of sediments. In contrast, during transgression, sequences are generally thin and consist mainly of sediments deposited in the shoreface, offshore, and nearshore marine environments (see Chap. 20). Sediments deposited

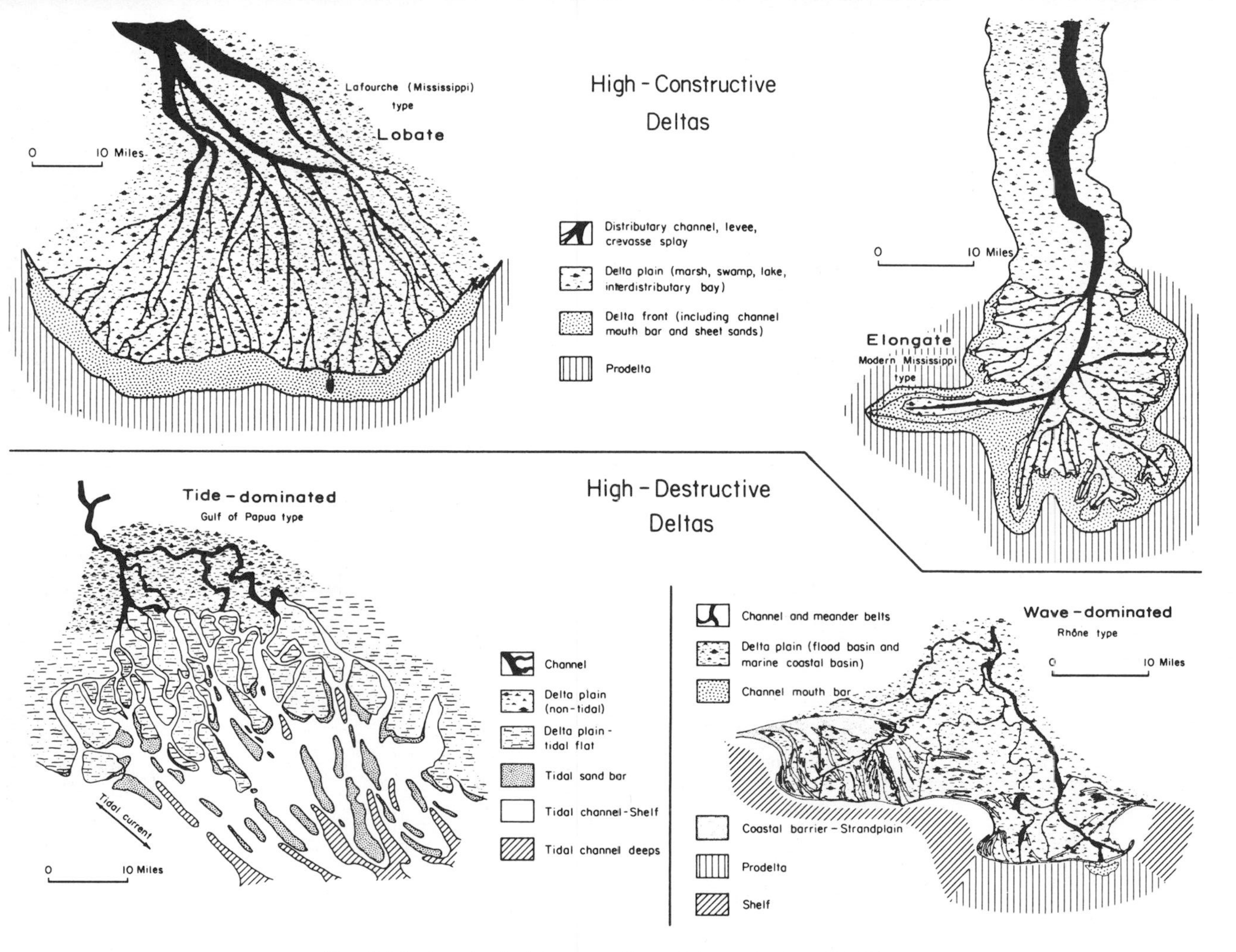
High - Constructive
Deltas
Lafourche (Mississippi)
type
Lobate
0
10 Miles
Distributary channel, levee, crevasse splay
Delta plain (marsh, swamp, lake, interdistributary bay)
Delta front (including channel mouth bar and sheet sands)
Prodelta
0
10 Miles
Elongate
Modern Mississippi
type
High - Destructive
Deltas
Tide - dominated
Gulf of Papua type
Tidal current
0
10 Miles
Channel
Delta plain (non-tidal)
Delta plain - tidal flat
Tidal sand bar
Tidal channel-Shelf
Tidal channel deeps
Channel and meander belts
Delta plain (flood basin and marine coastal basin)
Channel mouth bar
Coastal barrier - Strandplain
Prodelta
Shelf
Wave - dominated
Rhône type
0
10 Miles

Fig. 6-10 Four basic delta types. From Fisher and others (1969).

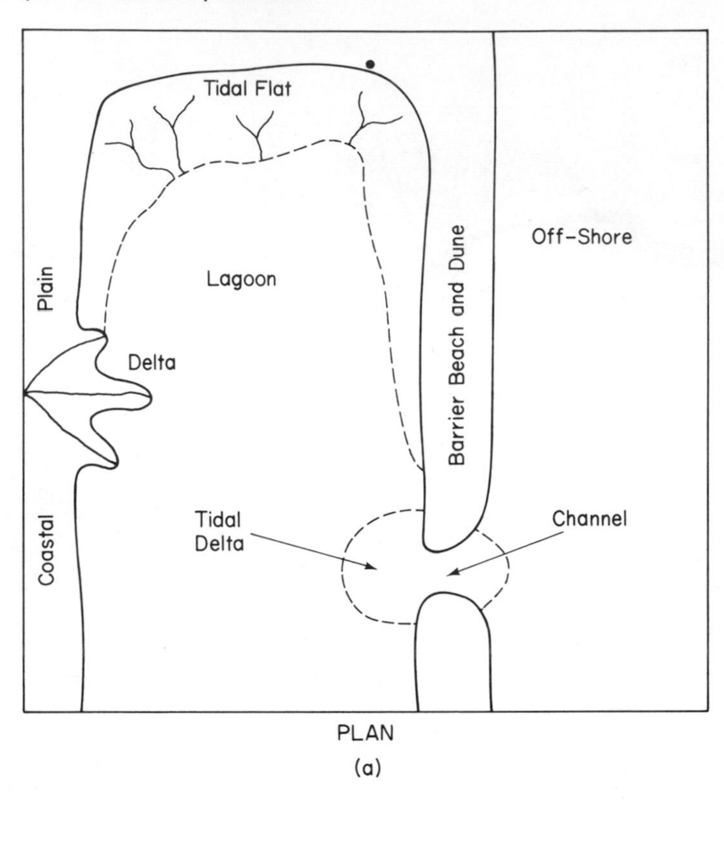

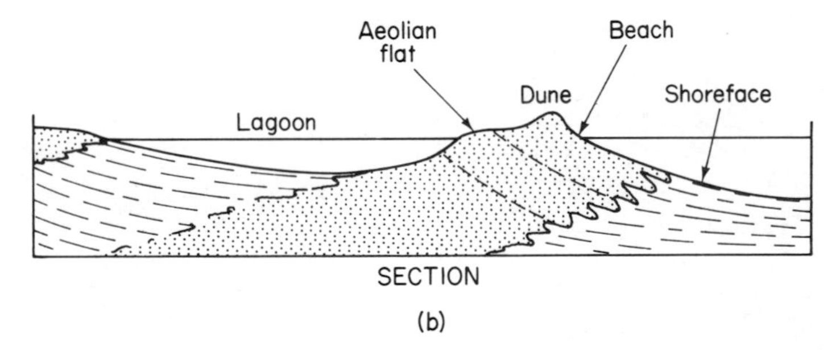

Fig. 6-11 Components of the barrier model, based on Galveston Island, Texas.

in the other environments tend to be reworked during transgression with the sediment being moved out to the deeper water. The result is that many preserved barrier sequences form asymmetric sequences with thin transgressive units alternating with thicker regressive units.

The modern sediments near Galveston Island, Texas, have served as the basis for many barrier models (see Bernard et al., 1962, and Fig. 20-1). In this barrier, the sand is fine to very fine so that there is a low beach slope. The island is capped by low, hummocky dunes, and the entire barrier is inundated during severe storms. Generally, however, wave action is not intense and in the lagoon and the marine zones below the effects of continual wave action the sediments are much burrowed by organisms. There is a size gradation from predominantly mud offshore to thin beds of silt and fine sand in the lower shoreface zone (with rare graded beds). In water shallower than 30 feet deep, the sediment coarsens, becoming better sorted with less interbedded mud. Near the top of this zone ripples may be formed and preserved if not reworked by burrowing organisms. The beach zone is characterized by low angle lamination with swash and rill marks; there may be scour and fill structures at the base of the lamination zone, recording wave plunge or longshore currents. The dunes form relatively large-scale, high-angle cross-lamination: the lamination may be partly destroyed by weathering and the action of plant roots. Behind the barrier, there is a lagoon with bioturbated muds, bordered by tidal flats and marshes, and there may be transitions to deltaic or fluvial sediments.

Considerable variation on this basic Texas coast type of model is possible. There are commonly tidal deltas associated with passages between the islands. Depending on the tidal range, tidal flats may be more or less extensively developed behind the barrier islands. Sediment of the tidal flats and lagoons depends strongly on the climate and on local sediment sources, as well as on tidal strength (G. Evans, 1970). In arid climates, the tidal flats may be a locus of evaporite deposits. In humid tropical climates the growth of mangrove swamps may be very important. Strong onshore winds may supply much blown sand into the lagoon. Coarser sand or more vigorous wave action may increase the importance of cross-lamination formed by waves relative to that of structures formed by organisms.

There are few well-documented ancient examples of the barrier model, although there are many references to it. The Cretaceous transgressive-regressive cycles of the western interior of the United States appear to fit this model (see papers listed in the references).

Among the supposed barriers in the Cretaceous is the Eagle Sandstone of Montana, described by Shelton (1965). The Eagle Sandstone is well exposed in an escarpment at Billings, Montana, and its NNW trend is known from subsurface studies. The maximum thickness is 100 ft, the width 20 to 30 mi and the known length is 40 mi. The sandstone grades into shale along its lower and lateral boundaries, coarsens upward and is characterized by very large-scale low-angle cross-stratification that dips west, normal to the trend of the sand body, and can be interpreted as accretion surfaces, corresponding to the sloping face of the barrier. It is inferred that the sandstone formed as a bar formed in some 75 ft of water, separated by more than 50 mi from land to the west. The paleogeography thus differs considerably from modern barriers. The Eagle Sandstone grew by accretion toward the land mass, rather than away from it, and there is no evidence for eolian deposition and therefore no compelling evidence that the bar was emergent rather than being simply an offshore shoal.

6.10 OFFSHORE SHOAL

This model is not well-known. It is not infrequently invoked for ancient sand bodies, where there is evidence of deposition tens of miles from the shore or where there is a lack of evidence of a restricted lagoon or of emergence of the bar above sea level. Modern analogues may include the tidal sand bodies in the North Sea, Georges Bank, and elsewhere, discussed in Chap. 5.

Ancient examples possibly include the Cretaceous Cardium Sandstone in Alberta, discussed by Michaelis and Dixon (1969) and the Cretaceous Viking sandstones of Saskatchewan described by W.E. Evans (1970).

6.11 TURBIDITE—DEEP BASIN

This model is based partly on studies of modern sediments, especially the California continental borderland, and partly on well investigated ancient sedimentary basins, such as the Polish flysch basins (Dzulynski and Walton, 1965). The basic model is of an elongate basin, with sediment supplied from submarine canyons down one or both sides or at one end. At the base of the canyons there are large overlapping submarine fans, with gently inclined surfaces and with submarine channels forming a continuation of the canyon over the surface of the fan. The channels contain sand or gravel, derived from shallow water and moved down the canyon by a variety of mechanisms, including creep and grain flow in the steeper part of the canyon, and turbidity currents in the lower part of the canyon. The submarine channels have levees composed of interbedded fine sand, silt, and mud, presumably deposited from the upper parts of turbidity currents that spilled over the sides of the channels. The lower parts of the fans and basin floors consist of interbedded fine grained muds deposited from suspension and graded sand layers deposited from turbidity currents. The turbidites may thus be divided into four main facies: *conglomeratic* or *coarse sandy deposits*, which may be grain flow deposits rather than true turbidites (some varieties have been called "fluxoturbidites") and which are generally confined to large channels; *proximal turbidites*, with a thick massive interval, poorly developed tractional structures, generally poor grading and little interbedded mud; *turbidites* of the "common" type, showing graded bedding, well-developed sole markings, and Bouma sequences, interbedded with muds; and *distal turbidites*, consisting of thin, fine grained beds without massive intervals and frequently without plane lamination but showing much ripple cross-lamination (see Walker, *in* Lajoie, 1970).

Interbedded with the turbidites, or forming sequences with few turbidites, there may be deposits formed by quiet deposition in relatively deep water, i.e., in water too deep to be disturbed by wave action, even during storms. These deposits include fine grained terrigenous muds or shales, bedded cherts, and fine grained limestones formed by the accumulations of pelagic microorganisms.

In the turbidites, sole markings frequently indicate flow predominantly parallel to the basin axis, though in some cases with transverse components. The common interpretation is that turbidity currents flowed down the side of the basin and then turned to flow parallel to the axis of the basin. Some authors have suggested that by whatever mechanism sand is moved down into the basin, it is then moved parallel to the basin axis not by turbidity currents but by deep-sea currents flowing parallel to the slope or basin axis.

6.12 CONCLUSION

It should be clear from these brief notes on a few of the major facies models that these models are both highly idealized and strongly influenced by a few well-described examples of modern sedimentary environments. In some cases, a sufficiently large number of modern examples has been investigated so that it is possible both to generalize what is essential to the facies model and to distinguish the range of varieties that are characteristic of the general model as a response to a particular combination of controls. This is probably the case for meandering river floodplains and for deltas but comparable knowledge has scarcely been achieved for any of the other major models. Very few ancient stratigraphic units have been studied in sufficient detail to permit testing of the application of modern sediment models to the geological record. For this reason, any attempt, such as that sketched in the preceding pages, to systematize facies models should be regarded by the reader with considerable skepticism. At the present time, there are no shortcuts to environmental and paleogeographic interpretation. Skill in the recognition and interpretation of sedimentary facies is achieved only by long experience in the field with ancient and modern examples and by careful study of the world literature of sedimentology and stratigraphy.

REFERENCES

General References on Facies Models

ALLEN, J. R. L., 1967, "Notes on Some Fundamentals of Paleocurrent Analysis with Reference to Preservation Potential and Sources of Variance," *Sedimentology*, **9**, pp. 75–88. (Introduces and discusses the concept of "preservation potential.")

COTTER, EDWARD, 1971, "Paleoflow Characteristics of a Late Cretaceous River in Utah From Analysis of Sedimentary Structures in the Ferron Sandstone" *Jour. Sed. Pet.* **41**, pp. 129–138. (Example of paleohydraulic reconstruction.)

CURRAY, J. R., 1956, "Analysis of Two Dimensional Orientation Data." *Jour. Geol.*, **64**, pp. 117–131. (Describes vector methods and gives graph for converting vector consistency to variance.)

DUFF, P. M. D., A. HALLAM. and E. K. WALTON, 1967, *Cyclic Sedimentation.* Amsterdam: Elsevier Pub. Co., 280 pp. (Review with many examples.)

DUFF, P. M. D, and E. K. WALTON, 1962, "Statistical Basis for Cyclothems: a Quantitative Study of the Sedimentary Succession in the East Pennine Coalfield," *Sedimentology*, **1**, pp. 235–255. (Introduces important numerical methods and concept for definition of cycles.)

GRABAU, A. W., 1906, "Types of Sedimentary Overlap," *Geol. Soc. Amer. Bull.*, **17**, pp. 567–636. (Classic development of models of transgressive and regressive sequences.)

HALLAM, A. (ED.), 1967, "Depth Indicators in Marine Sedimentary Environments," *Marine Geol.*, **5** (special issues), pp. 329–567. (Symposium on criteria for determining depth of deposition of sedimentary rocks.)

HAND, B. M., J. M. WESSEL, and M. O. HAYES, 1969, "Antidunes in the Mount Toby Conglomerate (Triassic), Massachusetts," *Jour. Sed. Pet.* **39**, pp. 1310–1316. (Example of paleohydraulic reconstruction.)

JOPLING, A. V., 1966, "Some Principles and Techniques Used in Reconstructing the Hydraulic Parameters of a Paleoflow Regime," *Jour. Sed. Pet.* **36**, pp. 5–49. (With a detailed paleohydraulic analysis of a Pleistocene micro-delta.)

KLEIN, G. DEV., 1967, "Paleocurrent Analysis in Relation to Modern Marine Sediment Dispersal Patterns," *Am. Assoc. Petroleum Geol. Bull.*, **51**, pp. 366–382. (Documents lack of control of marine paleocurrents by bottom slope.)

MOORE, R. C., 1949, "Meaning of Facies," *Geol. Soc. Amer. Mem.* **39**, pp. 1–34. (Introduction to a symposium on facies with a classic definition.)

POTTER, P. E., 1967, "Sand Bodies and Sedimentary Environments: A Review," *Amer. Assoc. Petroleum Geol. Bull.*, **51**, pp. 337–365. (Informative review of six major facies models.)

POTTER P. E., and F. J. PETTIJOHN, 1963, *Paleocurrents and Basin Analysis*. New York: Academic Press Inc., 296 pp. (Comprehensive review of paleocurrent studies and methods and a brief discussion of some facies models.)

RIGBY, J. K. and HAMBLIN, W. K. (eds.), 1972, "Recognition of Ancient Sedimentary Environments," *Soc. Econ. Paleont. Mineralogists Spec. Pub.* **16**, 340 pp. (Symposium on criteria for determining environment of deposition of sedimentary rocks.)

SELLEY, R. C., 1971, *Ancient Sedimentary Environments*. Ithaca, N. Y.: Cornell Univ. Press, 240 pp. (An introductory treatment with a good bibliography.)

SHAW, A. B., 1964. *Time in Stratigraphy*. New York: McGraw-Hill Book Co., 365 pp. (Early chapters contrast sedimentation in modern shelf and ancient inland "epeiric" seas and develop models for the latter.)

SHIMP, N. F., J. WITTERS, P. E. POTTER, and J. A. SCHLEICHER, 1969, "Distinguishing Marine and Freshwater Muds," *Jour. Geol.*, **77**, pp. 566–580. (Content of boron gives good results.)

VISHER, G. S., 1965, "Use of the Vertical Profile in Environmental Reconstruction," *Amer. Assoc. Petroleum Geol. Bull.*, **49**, pp. 41–61. (A modern statement of Walther's law and its application to six facies models.)

WALTHER, J., 1893–1894, *Einleitung in die Geologie als historische Wissenschaft*. Jena: Fischer Verlag, 3 vols. (The statement of Walther's law is in Vol. 3, Chap. 27.)

Rivers and Alluvial Fans

ALLEN, J. R. L., 1964, "Studies in Fluviatile Sedimentation: Six Cyclothems from the Lower Old Red Sandstone, Anglo-Welsh Basin," *Sedimentology*, **3**, pp. 163–198 (Detailed interpretation of cyclothems in terms of fluvial model.)

———, 1965, "A Review of the Origin and Characteristics of Recent Alluvial Sediments," *Sedimentology*, **5** (special paper), pp. 89–191. (A source of information on recent alluvial sediment data.)

BEERBOWER, J. R., 1965, "Cyclothems and Cyclic Depositional Mechanisms in Alluvial Plain Sediments," *Kansas State Geol. Surv. Bull.* 169, **1**, pp. 31–42.

COLEMAN, J. M., 1969, "Brahmaputra River: Channel Processes and Sedimentation," *Sedimentary Geol.*, **3** (special issue), pp. 129–239. (Detailed review of a large braided river.)

LAMING, D. J. C., 1966, "Imbrication, Paleocurrents and Other Sedimentary Features in the Lower New Red Sandstone, Devonshire, England," *Jour. Sed. Pet.*, **36**, pp. 940–959. (Alluvial fan and aeolian sediments.)

LEOPOLD, L. B., M. G. WOLMAN, and J. P. MILLER, 1964, *Fluvial Processes in Geomorphology*. San Francisco: W. H. Freeman and Co., 522 pp. (Review of hydraulics as well as geomorphology.)

MCGOWEN, J. H. and L. E. GARNER, 1970, "Physiographic Features and Stratification Types of Coarse-Grained Point Bars: Modern and Ancient Examples," *Sedimentology*, **14**, pp. 77–111.

PRYOR, W. A., 1961, "Sand Trends and Paleoslope in Illinois Basin and Mississippi Embayment," in *Geometry of Sandstone Bodies*, Tulsa, Okla: Amer. Assn. Petroleum Geol., pp. 119–133. (Classic paleocurrent study and formulation of facies model concept.)

SCHUMM, S. A., 1969, "River Metamorphosis," *Proc. Amer. Soc. Civil Eng., Jour. Hydraulic Div.*, **95**, pp. 255–273. (Statement of empirical relations between hydraulic and geomorphic variables in alluvial rivers.)

WILLIAMS, P. F. and B. R. RUST, 1969, "The Sedimentology of a Braided River," *Jour. Sed. Pet.* **39**, pp. 649–679. (A study of a river in the Yukon, with description of structures and paleocurrent analysis.)

Deltas

COLEMAN, J. M., and S. M. GAGLIANO, 1964, "Cyclic Sedimentation in the Mississippi River Deltaic Plain," *Gulf Coast Assn. Geol. Soc. Trans.*, **14**, pp. 67–80. (An important discussion of formation of cyclic vertical sequences by crevassing and shifting of distributaries.)

FISHER, W. L., L. F. BROWN, JR., A. J. SCOTT, and J. H. MCGOWEN, 1969, "Delta Systems in the Exploration for Oil and Gas," Austin, Texas: Bur. Econ. Geol., 78 pp. plus 168 figures and 24 pp. bibliography. (Well-illustrated summary of modern and ancient deltas.)

FRAZIER, D. E., 1967, "Recent Deltaic Deposits of the Mississippi River: Their Development and Chronology," *Gulf Coast Assn. Geol. Soc. Trans.*, **17**, pp. 287–315. (A good source of information on the Mississipi delta.)

MORGAN, J. P., 1970a, "Deltas—A Resumé." *Jour. Geol. Ed.*, **18**, pp. 107–117. (Up-to-date brief summary with emphasis on Mississippi.)

MORGAN, J. P. (ED.), 1970b, "Deltaic Sedimentation: Modern and Ancient," *Soc. Econ. Paleont. Mineral. Spec. Pub.* **15**, 312 pp. (Symposium with summaries by leading authorities of many modern and ancient deltas.)

Barriers and Offshore Shoals

BERNARD, H. A., R. J. LEBLANC, AND C. F. MAJOR, 1962, "Recent and Pleistocene Geology of Southeast Texas," Field Excursion no. 3 in *Geology of the Gulf Coast and Central Texas and Guidebook of Excursions. Houston Geol. Soc.*, pp. 175–224. (With description of Galveston Island. See also reference listed in Chap. 5.)

EVANS, GRAHAM, 1970, "Coastal and Nearshore Sedimentation: a Comparison of Clastic and Carbonate Deposition," Geol. Assoc. London Proc., **81**, pp. 493–508. (Comparison of barrier island and tidal flat sedimentation in Holland and Persian Gulf.)

EVANS, W. E., 1970, "Imbricate Linear Sandstone Bodies of Viking Formation in Dodsland-Hoosier Area of Southwestern Saskatchewan, Canada," *Amer. Assoc. Petroleum Geol.*, **54**, pp. 469–486. (Sand bodies interpreted as tidal offshore shoals.)

HARMS, J. C., D. B. MACKENZIE, and D. G., MCCUBBIN, 1965, "Depositional Environment of the Fox Hills Sandstones near Rock Springs, Wyoming," *Wyoming Geol. Assn. Guidebook 19th Field Conf.*, pp. 113–130. (Criticizes interpretation of Weimer, 1961.)

MICHAELIS, E. R., and G. DIXON, 1969, "Interpretation of Depositional Processes from Sedimentary Structures in the Cardium Sand," *Bull. Canadian Petroleum Geol.*, **17**, pp. 410–443. (Good example of reasoned inference from structures to process. Sands tentatively interpreted as offshore shoals.)

SHELTON, J. W., 1965, "Trend and Genesis of Lowermost Sandstone Unit of Eagle Sandstone at Billings, Montana," *Am. Assoc. Petroleum Geol. Bull.*, **49**, pp. 1385–1397. (Well-exposed example of possible Cretaceous barrier.)

WEIDIE, A. E., 1968, "Bar and Barrier Island Sands," *Gulf Coast Assn. Geol. Soc. Trans.*, **18**, pp. 405–415. (Summary of differences between barriers and offshore bars, with examples.)

WEIMER, R. J., 1961, "Spatial Dimensions of Upper Cretaceous Sandstones, Rocky Mountain Area," in *Geometry of Sandstone Bodies*, Tulsa, Okla.: Amer. Assn. Petroleum Geol., pp. 82–97. (An often-cited example of an ancient barrier—but see critical comments by Harms, MacKenzie, and McCubbin, 1965.)

Turbidite—Deep Basin

DZULYNSKI, STANISLAW, and E. K. WALTON, 1965, *Sedimentary Features of Flysch and Greywackes*. Amsterdam: Elsevier Pub. Co., 274 pp. (The last chapter discusses turbidite models.)

LAJOIE, J. (ED.), 1970, "Flysch Sedimentology in North America," *Geol. Assn. Canada*, Spec. Paper 7, 272 pp. (An excellent summary of knowledge on this subject.)

PART THREE

TERRIGENOUS CLASTIC SEDIMENTS

Sandstones and mudrocks form 80 to 90% of the sedimentary rocks exposed on the continents. To what extent are the formation and relative abundances of these clastic rocks controlled by organic or biochemical processes? Were Precambrian weathering processes different from those of today? The concept of an equilibrium mineral assemblage has been successfully applied to igneous and metamorphic rocks. Can the concept be applied to detrital sedimentary rocks and, if so, what modifications to the concept are necessary? In what ways can diagenetic processes alter the depositional mineral assemblage? Why are many Paleozoic and Mesozoic sandstones presently friable? Have they ever been completely lithified? Why is the average shale composed mostly of illite rather than the other groups of clay minerals?

CHAPTER SEVEN

WEATHERING PROCESSES AND PRODUCTS

7.1 INTRODUCTION

Up to this point, we have considered only the physical aspects of sedimentation. Even to produce the basic materials on which the physical processes of sedimentation operate, however, it is necessary to have a disintegration of source rocks. The main agents of rock disintegration are chemical and biological.

The net result of the weathering process is to transform solid rocks that are largely composed of silicate minerals formed at relatively high temperatures and pressures (as compared with conditions existing at the earth's surface) into three main fractions of disintegrated rock materials: (a) materials in solution; (b) "resistates," i.e., materials present in the original rock that remain chemically more or less unaltered in the weathered product; and (c) new minerals produced by weathering.

The new minerals produced during weathering are primarily those resulting from the reaction of silicates, sulfides, or oxides with water, which is more abundant in the weathering environment than in the environments in which igneous or metamorphic rocks are formed. The typical weathering products are therefore hydrated minerals, including the clay minerals and hydrated oxides, such as limonite and gibbsite. The process whereby common silicate minerals break down to form such hydrated minerals is called *hydrolysis.*

Weathering also involves reactions with two other materials found more abundantly in the hydrosphere and atmosphere than elsewhere: namely, oxygen and carbon dioxide. Hydrolysis, therefore, is commonly accompanied by oxidation (mainly of ferrous to ferric iron) and carbonation. All three of these reactions proceed most rapidly immediately below the interface between rock and atmosphere, where they are generally intimately connected with plant root growth and organic decay. The result is the development of a "mantle of weathering" between the unweathered rock and the atmosphere. Within this mantle, it is generally possible to recognize a series of layers, or horizons, each of which is characterized by certain biological, physical, and chemical processes and properties. This mantle is generally called a soil, though the term is often restricted to mantles in which organic agencies play an important part. The term soil is also used by civil engineers to describe any loose, unconsolidated material.

Chemical weathering may be defined as the approach to equilibrium of a system involving rocks, air, and water, which occurs at or near the surface of the earth (Keller, 1957). Nevertheless, it is difficult to define the equilibrium mineral assemblage to which weathering is proceeding. Generally, the geologist thinks of such substances as quartz, kaolinite, iron oxide, and perhaps a few others as the final mineral assemblage; but certainly this is not true under tropical conditions. In the humid tropics, kaolinite and quartz are demonstrably unstable. It is clear that physicochemical conditions at the earth's surface vary sufficiently both in time and space so that there may exist more than one equilibrium assemblage of substances. For example, quartz and kaolinite are as stable in the present climate of Louisiana as hydrated aluminum oxide (bauxite) was during Tertiary times in the same area. It is only when particular climatic conditions are cited that the nature of an equilibrium assemblage may be specified. During metamorphism, almandine garnet may form and be stable only a few miles from the place where epidote is stable; so also gibbsite may be stable a few miles from a stable kaolinite locality.

In practice, the concept of an equilibrium assemblage during weathering is not as critical as in metamorphism because reactions at the earth's surface are so slow that sedimentologists are nearly always dealing with mineral assemblages that are not in equilibrium with the set of chemical conditions under which they are forming. Unstable minerals are present at low temperatures not only because chemical reactions are slow at low temperatures but also because metastable substances frequently are precipitated from aqueous solutions at the earth's surface. Examples of these include aragonite and opal. Thus, from the viewpoint of a chemist, the study of chemical weathering is a nightmare in which nearly all reactions are incomplete, chemical components come and go at unpredictable times, temperature varies irregularly, and thermodynamically unstable compounds are frequently precipitated.

7.2 WATER

Because of the importance of water in almost all weathering, including "physical" weathering, it is important to review some of the properties of this extraordinary substance. The properties of water that are particularly unusual are (a) the high dipole moment of the water molecule, important for the properties of water as a solvent, and (b) the ability of water molecules to form hydrogen bonds with adjacent water molecules, important as a cause of unusual physical properties of water that affect not only weathering but also the transportation of sediment particles.

In the liquid state, a molecule of H_2O has the shape of an isosceles triangle. The spacing between O and H is 0.96 A and between H and H it is 1.51 A. Oxygen is more electronegative than hydrogen and consequently the electrons are drawn toward the oxygen atom, creating a dipole. The hydrogen end of the molecule has a slight positive charge and the oxygen end a slight negative charge. The H-O-H bond has an angle of 104.4 degrees and thus departs substantially from the 90 degree angle that might be anticipated because of the mutually perpendicular orientation of *p*-orbitals, the orbitals used in forming oxygen-cation bonds. The cause of the enlarged bond angle is, in part, the difference in electronegativities between hydrogen and oxygen. The residual positive charge on each hydrogen atom causes the atoms to repel each other, resulting in an increase in bond angle.

Oxygen is among the most electronegative of elements and therefore the dipole moment of water is high. The dipole moment of a molecule is expressed in electrostatic units and is given by the equation $\mu = Zd$, where Z is the magnitude of the charge and d is the distance between the two charges. As the electronic charge on a single electron is close to 10^{-10} electrostatic units (esu) and molecular dimensions are close to 10^{-8} cm, dipole moments are usually given in Debye units, where 1 Debye is equal to 10^{-18} esu cm. For water, the dipole moment is equal to 1.85 Debye units (as compared, for example, with only 0.95 Debyes for its sulfur analogue, H_2S). The strongly dipolar nature of the water molecule is the reason for its outstanding properties as a solvent. The force of attraction between the dipole and the ions on crystal surfaces is responsible for the destruction of the crystal—the process called solution or hydrolysis. We return to this topic later in the chapter.

The ability of the water molecule to form hydrogen bonds with the oxygen ends of adjacent water molecules is responsible for most of the other sedimentologically important properties of liquid H_2O, for example, its high viscosity, high surface tension, and high melting and boiling points. The hydrogen bonds are also believed to cause the unusual relationship between temperature and the density of water, which is greater at 4°C than at the freezing point (Kavanau, 1964).

The high dynamic or absolute viscosity of water is important in sediment transport because of its effect on the settling velocity of particles and on the generation of turbulent eddies, the swirling water masses that lift particles off the stream or sea bottom (see Chap. 4). The viscosity of water varies with temperature, decreasing from 1.8 cp (centipoise) at 0°C to 0.3 cp at 100°C. This decrease in viscosity with increasing temperature is important not only during grain transport but it also has an effect on the ease of movement of water through rock pores at depth.

The high surface tension of water is of sedimentologic importance for the same two reasons as is its viscosity—grain transport and underground water movement. Platy-shaped materials such as micas, clay flakes, and leafy plant debris may be transported for great distances held at the air-water contact because of the high surface tension of water. Its strength of 73 dynes/cm at 20°C may be compared with 42 for bromoform and 24 for acetone. Surface tension, like viscosity, almost invariably decreases with increasing temperature. The surface tension of water decreases from 75.6 dynes/cm at 0°C to 58.9 dynes/cm at 100°C and water at depth in the crust wets the pore walls more effectively than it does near the surface. Therefore water can move through smaller pores at depth than near the surface.

The hydrogen bonding among water molecules also causes the unusually high melting point of H_2O and the large temperature range through which H_2O is a liquid.

The relationship between water density and temperature is the cause of the most important type of mechanical weathering, frost wedging. Although it is generally accepted that it is hydrogen bonding that causes water to be densest at 4°C rather than at the freezing point, the nature of the change in the structure of water at 4°C is still uncertain despite the fact that water was one of the first substances studied during the development of chemistry as a science.

Hydrogen Ion Concentration (pH)

Water is slightly dissociated at all temperatures and at room temperature this dissociation results in 10^{-7} moles/liter of hydrogen or hydronium ions and an equal amount of hydroxyl ions. Hydrogen ions serve an important function during weathering processes, replacing the metal cations removed from silicate minerals by the water dipole forces. As with all dissociation reactions, the amount of hydrogen ions released by dissociation of water varies with temperature. For most dissociation reactions, the extent of dissociation increases with temperature. Hence, it is to be expected that the concentration of hydrogen ions will be greater at, say, 100°C than at 20°C. It is important to recall that *neutral pH is defined as the condition where equal numbers of* H^+ *and* OH^- *ions are present, not as the condition where pH is 7.0.* Figure 7-1 shows how neutral pH varies with temperature and it is clear that at 120°C, for example, neutral pH is 6.0 (10^{-6} moles/liter each of H^+ and OH^- ions). Therefore, a pH reading of 7 at a depth of a few thousand feet indicates a hydrogen-poor or basic solution and this should be borne in mind in discussions of the role of pH in diagenesis.

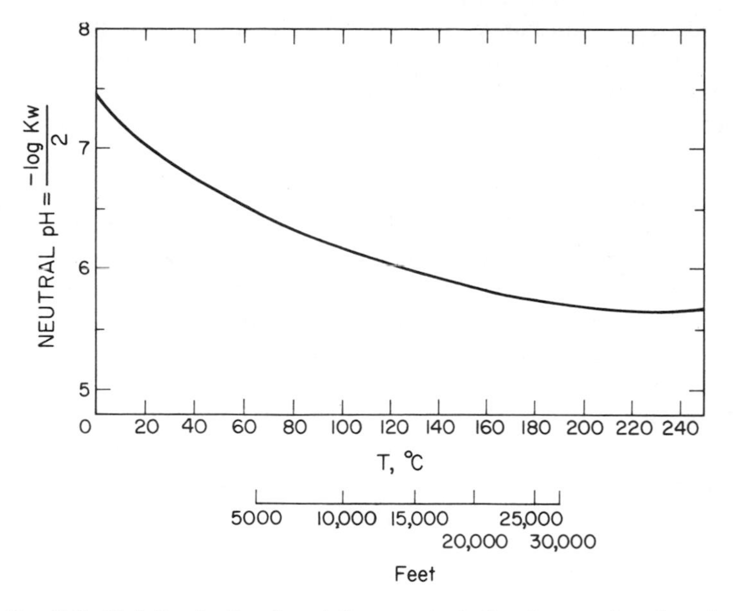

Fig. 7-1 Variation in the dissociation constant of water as a function of temperature and depth in a geosyncline.

Oxidation-Reduction Potential (Eh)

The tendency of some reactions to occur in aqueous solution can be adequately described in terms of pH. For many reactions in the sedimentary environment, however, the value of an additional parameter, Eh, must be specified. The Eh of a solution is a measure of the oxidizing or reducing capability of the solution, a factor of critical importance in reactions involving elements such as iron, manganese, and sulfur. In the sedimentary environment the control on Eh is the supply of gaseous oxygen in relation to the amount of organic matter to be decomposed (oxidized).

The oxidizing or reducing potential of an environment is determined quantitatively using electrochemical techniques. If we connect a piece of copper and a piece of zinc by a wire and immerse the two metals in a solution of cupric sulfate, the zinc dissolves and copper precipitates from the solution onto the copper fragment as electrons are transferred between the electrodes. Each metal fragment is an electrode or half-cell and the processes occurring at the electrodes may be symbolized by the equations

$$Zn \longrightarrow Zn^{2+} + 2\,e^- \qquad \text{(oxidation at the anode)}$$

$$Cu^{2+} + 2\,e^- \longrightarrow Cu \qquad \text{(reduction at the cathode)}$$

The zinc displaces copper from the solution. In a similar manner, copper would displace silver from a silver sulfate solution. In order to make such relative displacing tendencies quantitative, a standard half-cell must be chosen so that the displacing tendency of any other half-cell may be compared to it. There is no way to measure potentials for half reactions independently so that Eh numbers have meaning only in relative terms. The reaction

$$2\,H^+ + 2\,e^- \longrightarrow H_2$$

is arbitrarily given a potential of 0 volts ($E^\circ = 0.0$) at 25°C and 1 atm total pressure in equilibrium with a solution containing hydrogen ions at unit activity (1 molar or $pH = 0$). The potential (E°) of the standard zinc electrode in the Zn-Cu-$CuSO_4$ cell relative to the hydrogen reaction is -0.76 volts. A half-cell reaction written as $Zn \longrightarrow Zn^{2+} + 2\,e^-$ is a standard shorthand notation for the whole-cell reaction $Zn + 2\,H^+\ (a = 1) \longrightarrow Zn^{2+} + H_2\ (a = 1)$. According to the standard sign convention, positive values of Eh indicate oxidizing tendencies; negative values, reducing tendencies. Figure 7-2 illustrates the Eh-pH conditions in the more common sedimentary environments.

Eh-pH Diagrams

In the natural environment neither Eh nor pH is an independent variable. Both are determined by reactions occurring in the solution such as the oxidation of iron, the leaching of andesine, or the decomposition of organic matter. Nevertheless, it often is convenient and meaningful to consider the presence of various chemical species in terms of pH and Eh (Fig. 7-2). In terms of pH and Eh there are three types of reactions, all of which we have occasion to consider later in this book.

1. Reactions involving only a shift in hydrogen ions, i.e., independent of oxidation-reduction potential.

$$H_2CO_3 \rightleftharpoons H^+ + HCO_3^- \qquad \text{(Chaps. 12 and 13)}$$

$$H_3PO_4 \rightleftharpoons H^+ + H_2PO_4^- \qquad \text{(Chap. 17)}$$

$$Fe^{2+} + 2\,H_2O \rightleftharpoons Fe(OH)_2 + 2\,H^+ \qquad \text{(Chap. 19)}$$

2. Reactions involving only electron transfer, i.e., independent of pH.

$$S^{2-} \rightleftharpoons S + 2\,e^- \qquad \text{(Chap. 15)}$$

$$Fe^{2+} \rightleftharpoons Fe^{3+} + e^- \qquad \text{(Chap. 19)}$$

3. Reactions involving both pH and Eh.

$$2\,Fe^{2+} + 4\,HCO_3^- + H_2O + \tfrac{1}{2}O_2 \rightleftharpoons 2\,Fe(OH)_3 + 4\,CO_2 \qquad \text{(Chap. 10)}$$

$$Fe^{3+} + 3\,H_2O \rightleftharpoons Fe(OH)_3 + 3\,H^+ + e^- \qquad \text{(Chap. 19)}$$

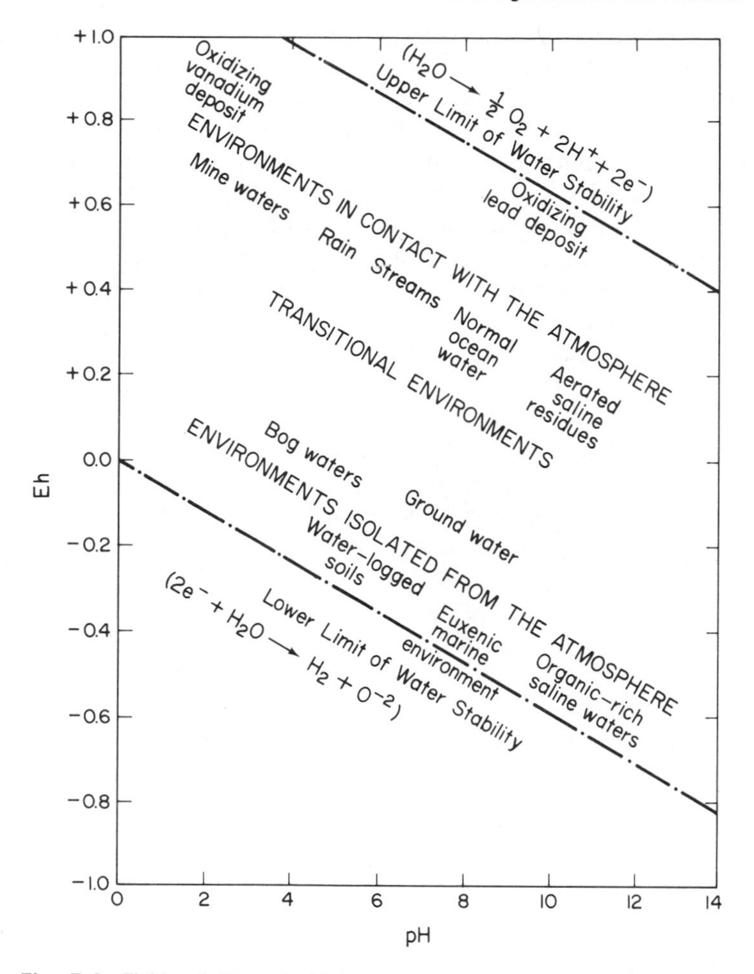

Fig. 7-2 Fields of Eh and pH for some common natural environments. Based on data in Bass Becking and others (1964).

7.3 PHYSICAL WEATHERING

Frost Weathering

In most environments on the earth's surface, mechanical weathering is of almost negligible importance when compared with chemical weathering. Probably the only exceptions to this generalization are subarctic environments and temperate latitudes at high elevations, where frost wedging may occur.

The physical basis of frost wedging is the 9.2% volume increase in H_2O during the phase change from water to ice. During the relatively warm daylight hours, precipitation falls as rain and penetrates into tortuous rock fractures and, as temperature decreases at night, this water freezes. Ice continues to expand between 0 and −20°C and it is possible experimentally to generate a force of more than 30,000 lb/sq. in. (2 kb) by lowering temperature through this range, providing the ice is maintained in equilibrium with water. Under natural conditions, however, such an equilibrium cannot be maintained and maximum pressures exerted by the volume increase during freezing do not exceed 2000 lb/sq. in. (0.1 kb). Such pressures are sufficient to crack homogeneous massive rocks such as granite, which has an average tensile strength of 600 lb/sq in.; or marble, 850 lb/sq in. The strength of most layered or foliated rocks would be less than these values. It is important that the water-filled crevice be tortuous, for water freezes from the air-water contact downward and if freezing occurs in a straight-walled V-shaped crack open to the surface, the ice cap will rupture long before sufficient stress can be generated to widen the crack and split the rock. The strength of a 1- to 2-in. crust of ice is only 20 to 40 lb/sq in.

Frost wedging is most effective in areas of recurrent freezing and thawing rather than in areas characterized by perpetually freezing temperatures. Figure 7-3(a) illustrates freeze-thaw frequencies for continental United States and highs are evident in the mountainous western areas and in the central Appalachians. Frequencies decrease in Canada [Fig. 7-3(b)] because of increase in the number of days continually below freezing. Frequencies also decrease in southern United States because of increase in number of days continually above freezing. The data on the maps represent a maximum estimate of effective frost action because the phenomenon depends also on the presence of water and in parts of the west the rocks and their upper cracks are often dry.

Diurnal Temperature Change

Many years ago it was suggested that diurnal temperature changes are the cause of rock fracture, particularly in desert areas in which daily temperature ranges may be quite wide. The basis for this hypothesis is the fact that rocks are poor conductors of heat and, therefore, surfaces exposed to the sun's rays will expand to a far greater extent than the rock a few inches or feet below the surface. This would cause the outer mass to become detached from the main body of rock, giving rise to phenomena known as spheroidal weathering and exfoliation. A similar argument has been made for disintegration of rock fragments into their constituent minerals. Minerals are of different colors and therefore absorb heat at different rates, expand differentially, and the rock crumbles. Carrying this reasoning to the extreme, it could be argued that nearly all minerals are anisotropic; therefore they have coefficients of thermal expansion that differ in different directions within the crystal and individual mineral grains could be reduced to silt or clay size by diurnal temperature changes.

Unfortunately, as reasonable as these hypotheses may sound, they are not supported either by field observations or experimental studies. Although fires promote flaking and splitting of rock, such heating is local and far above the range of the diurnal temperature fluctuations anywhere on earth. Also, there is clear evidence that even in conditions of extreme aridity and diurnal temperature

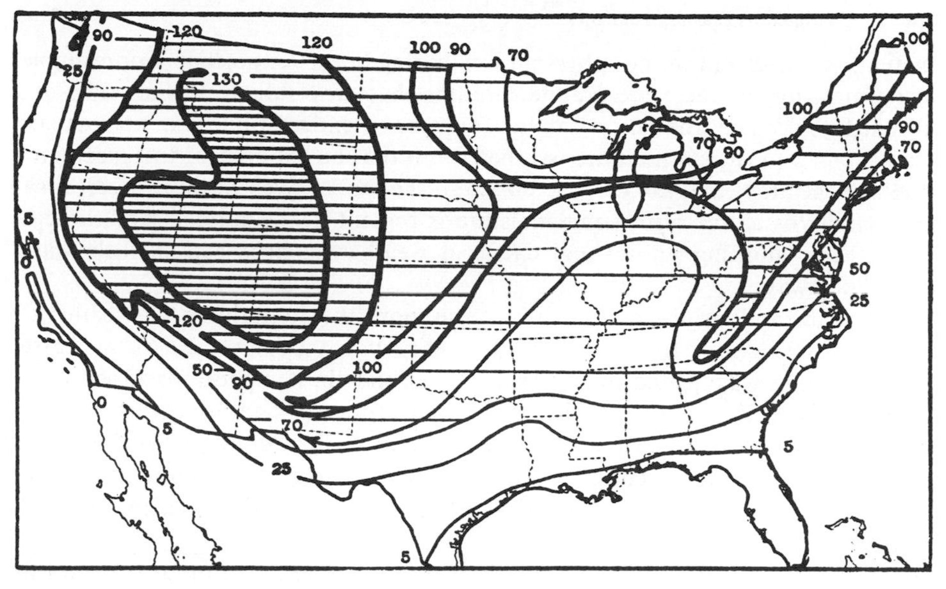

(a)

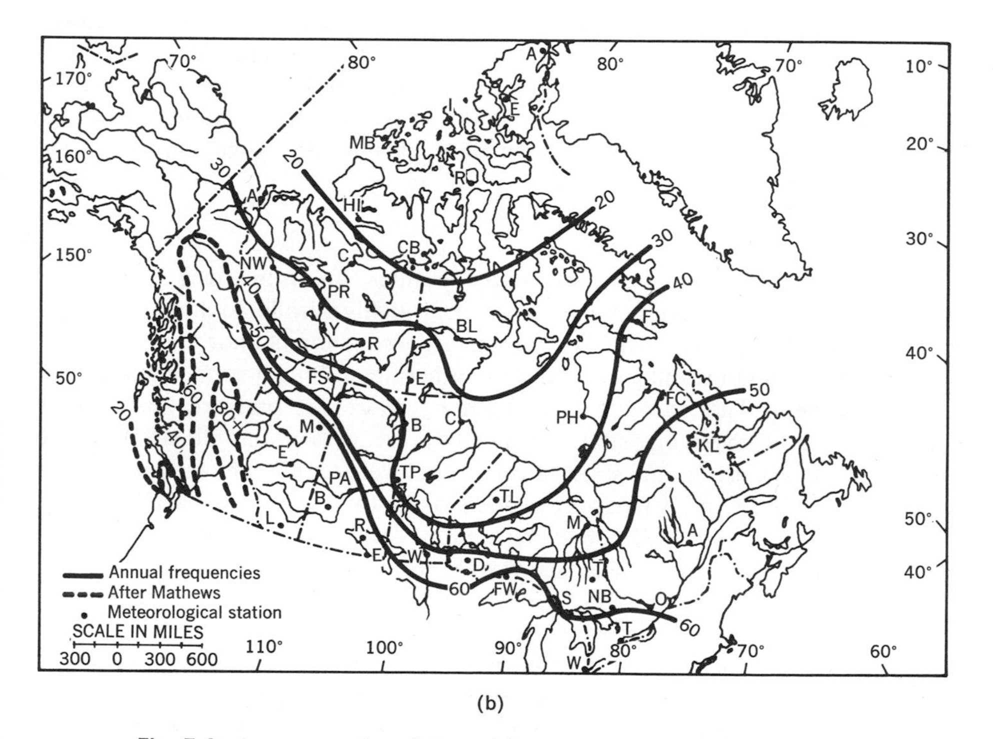

(b)

Fig. 7-3 Average number of days of freeze-thaw in conterminous United States and Canada. ((a) from Visher, 1945, and (b) from Fraser, 1959.)

fluctuation it is chemical, not physical, weathering that is important. For example, near Cairo, Egypt, there are granite columns that have fallen on their sides and are partly buried by the desert sands. Contrary to expectation, it is not the surfaces exposed to the sun that are weathered but the shaded areas. Most weathered of all are the areas that are or were buried in the sands. Even in these desert areas the humidity is above zero; some moisture is present in the surface sand, at least during early morning hours; and chemical weathering proceeds. A little water goes a long way.

Experimental evidence also fails to support the hypothesis of weathering caused by insolation. In one informative experiment (Griggs, 1936), a coarse-grained granite was heated in an electric oven and the temperature varied every 15 min between 58 and 256°F. This unearthly torture was continued for a number of cycles equivalent to 244 years of daily heating and cooling; yet, at the end of the experiment, the granite showed no signs of disintegration. This, despite the fact that a temperature variation of 200°F is far greater than occurs naturally on the earth.

It seems possible that the question of the effectiveness of diurnal temperature changes in causing physical weathering will finally be decided in the decade of the 1970's. Geologic studies on the moon are the ideal setting in which to resolve the question because the moon contains no water to promote chemical weathering and no ice to promote frost wedging, and the moon has a diurnal temperature range of about 500°F, from −250 to +250°F. In addition, as the moon has no atmosphere, temperature variation is quite rapid and daily highs and lows are separated in time by only a few hours.

Sheeting

Many types of rocks, but particularly granites, are sometimes seen in the field to be jointed into a series of massive sheets, roughly parallel to the topographic surface and cutting across normal sedimentary, igneous, or metamorphic structures. These fractures do not result from surface weathering, although surface weathering accentuates them. The fractures are caused by release of pressure that accompanies erosion of overlying strata. Thus they are present in all stratigraphic sequences but because granites are normally unlayered, they are most visible in these rocks. In flat-lying foliated or bedded rocks the jointing is parallel to the rock layering and is not visible but presumably its presence facilitates the penetration of water and resultant chemical weathering.

Biologic Factors

Although the weathering activities of living organisms are largely chemical, organisms also cause mechanical weathering. Possibly the effectiveness of biologic processes is second only to frost wedging in importance as an agent of mechanical weathering. For example, lichen hyphae (roots) are gelatinous and when they expand and contract, they are known to disintegrate shale by a plucking process. In some cases, this disintegration precedes noticeable decomposition of the rock.

Lichen hyphae also have been observed to penetrate diorite and tear away lamellar-shaped particles from the surfaces of minerals. It can at least be concluded that chemical weathering is accompanied by mechanical weathering in such cases and perhaps the two processes are inseparably interrelated.

The burrowing activity of animals, particularly worms, is effective in promoting chemical weathering by overturning the soil. While burrowing, worms ingest mineral grains up to 1 mm in size. No doubt these grains are chemically attacked during the passage through the animal's alimentary tract but the chewing and grinding actions also reduce the size of the grains. Worms are transporting agents as well as weathering agents. Some land areas have as many as 50,000 earthworms to the acre and it has been estimated that these annually transport 18 tons (one-fifth of an inch) of soil to the surface. Worms have a geologic record dating from the Precambrian and presumably nonmarine worms colonized the land surface penecontemporaneously with plants. Hence, the activities of worms in causing disintegration and decomposition of mineral grains probably has been quite significant through geologic time.

Organic Processes

Recognition of the importance in rock weathering of bacteria dates from 1890, of algae from 1891, and of lichens from 1904; and it probably is true that most weathering of rocks and minerals results from organic activity, particularly by plants of low taxonomic level. Soil scientists have calculated that the top 7 in. of temperate zone soil has a live weight of soil biota composed of algae, fungi, earthworms, bacteria, and other protistids of various types totaling 0.3 to 0.5 lb/cu ft of soil. Microorganisms average on the order of 1 billion/g of soil and experiments using albite and muscovite have shown that these minerals are decomposed twice as rapidly in soil when bacteria are present as in sterile clay. In addition, in most areas, both filimentous and coarse roots of grasses and higher plants are abundant. All of these organisms acidize the soil by excreting carbon dioxide during respiration and some also excrete organic acids as part of their metabolism. After death, all decompose to produce a great variety of organic acids, most of which are unidentified ("humic acids," "fulvic acids", etc.).

Although the mechanism by which plants cause minerals to decompose is essentially the same regardless of phylogeny, it has been found experimentally that primitive plants are particularly effective and that changes in effectiveness also are discernible among taxonomic levels within phyla. For example, in one study, the capacity to extract potassium from feldspar was determined for twenty-two seed plants from different taxonomic levels and it was determined that lower seed plants were considerably more effective in causing this element to be released into the soil. Very primitive plants such as lichens (symbiotic association of an alga and a fungus) are particularly adept at initiating mineral decomposition. Schatz (1963) conducted some informative experiments in which eight genera of lichens were grown in experimental pots in soil composed of granite fragments, muscovite, or glauconite. He found that these plants crystallize on their hyphae many types of organic acids which, being in direct contact with mineral surfaces, replace metallic cations with hydrogen ions generated by solution and dissociation of the crystals of acid.

It seems reasonable to suppose that the ability of lichens to grow on bare rock is related to their occurrence as one of the earliest colonizers of land in the Silurian Period. At the onset of colonization there was, of course, little or no organic matter present to supply the lichens with readily available nutrient ions and so the ability to obtain nutrients from bare rock was an important survival factor. Lower taxonomic groups retain that ability today.

Plant root growth involves the liberation of large quantities of organic acids and carbon dioxide and the absorption of inorganic nutrients. One of the few studies that have been made of soil gases showed that the content of carbon dioxide increased downward in the soil zone, reaching amounts as high as 30% of soil gases. The amount of carbon dioxide that is generated by plant respiration and decomposition, and which apparently is retained *below* the ground surface, is normally 10 to 100 times the amount in equilibrium with the air *at* the ground surface. Retention of such a high concentration of carbon dioxide within the soil results both from adsorption of the gas molecules on clay surfaces and from a slow rate of upward diffusion of carbon dioxide.

The roots of higher plants are negatively charged bodies that, therefore, are always surrounded with cations. In the general case, these cations are hydrogen ions and their presence serves to surround the root with the acid environment needed to hydrolyze silicate mineral grains such as feldspars and ferromagnesian minerals or to strip loosely held nutrient ions from clay or humus particles. The bonding energies between clay or humus and interlayer or adsorbed metallic cations are normally of lesser magnitude than the bonding energy of ions in a feldspar crystal. As a result, growing plant roots can have a marked effect on the chemical composition of clay minerals. Illite, the most abundant clay mineral, contains easily available potassium ions and in shales it commonly is found that illites are "degraded," i.e., do not have their full complement of potassium ions. How much of this has resulted from paleobotanical processes is uncertain but there is little question that such processes are at least partially responsible. Among organic molecules, carboxyl groups, phenols, and enols (varieties of ketones) are of particular importance as suppliers of ions to plants.

7.4 CHEMICAL WEATHERING

The best studies of weathering phenomena are those that analyze both mineralogic changes and the associated chemical changes. Most useful are studies that relate these changes to weathering intensity or depth in the weathering profile. Numerous studies of this type have been made of chemical weathering of both mafic and silicic rocks in a variety of climatic zones. In nearly all cases, essentially similar trends are observed, although the completeness of chemical and mineralogic alteration varies as a function of time and weathering intensity. With few exceptions, the sequence in which minerals weather is the same as the order of crystallization of minerals from a magma, i.e., Bowen's Reaction Series. The first to crystallize at depth are the first to decompose at the surface.

Weathering of Granitic Rocks

Massive silicic plutonic rocks and gneisses are the most abundant crystalline rocks exposed on the continental surfaces and, therefore, their weathering pattern is of particular interest. Several excellent studies of this pattern have been made since the classic study of the Morton Gneiss in Minnesota by Goldich in 1938. We shall examine the similar results obtained by Wahlstrom (1948) in a study of a weathering profile of Mississippian age developed on a granodiorite near Boulder, Colorado.

The profile is 80 ft thick. It is underlaid by the granodiorite and overlaid by the arkosic Fountain Formation. Wahlstrom collected and analyzed one sample of the granodiorite and seven samples of the paleosol at one location but at different positions in the weathering profile, with results shown in Table 7-1 and Fig. 7-4. The main trends in mineralogic and chemical changes during soil profile development are clearly evident. Hornblende disappears first and this is accompanied by the alteration of biotite to vermiculite through loss of ferrous iron during oxidation. In the upper part of the profile, oligoclase disappears

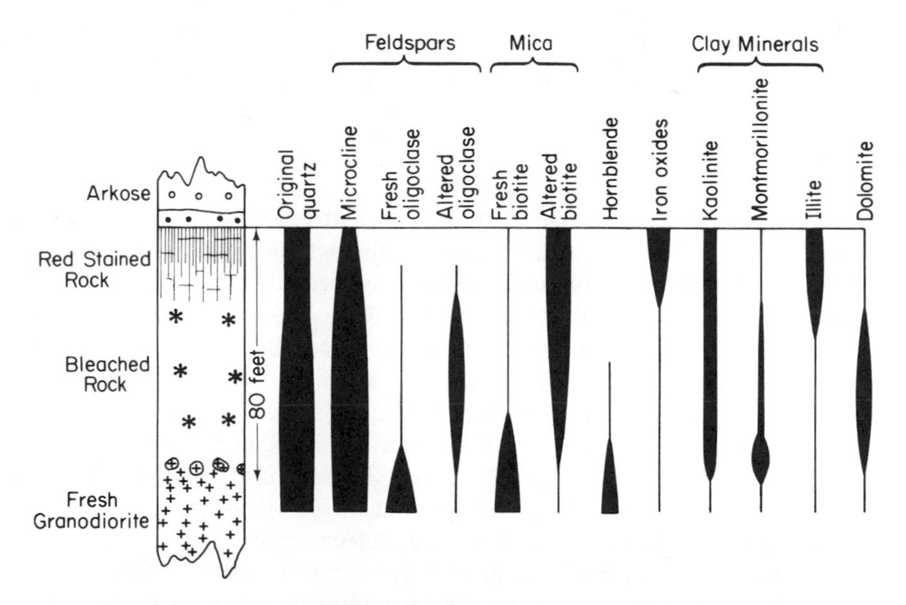

Fig. 7-4 Diagram showing persistence with depth of important minerals in pre-Fountain mantle (after Wahlstrom, 1948).

and microcline decreases in abundance by about 50%. Simultaneously with these changes, new substances appear in the paleosol, clay minerals, iron oxides, and dolomite. The dolomite was probably formed by diagenesis because dolomite is not known from modern soils on granitic rocks. The minor accessory minerals of the granodiorite, chiefly apatite, sphene, magnetite, and allanite, are not visible in any of the thin sections of the altered rocks and presumably were altered beyond recognition.

TABLE 7-1 CHEMICAL ANALYSES OF THE PARENT GRANODIORITE AND THE SOIL PROFILE DEVELOPED ON IT DURING THE LATE PALEOZOIC (from Wahlstrom, 1948, p. 1178)

Feet above base	Granodiorite	3	31	57	68	74	77	80
SiO_2	67.92	68.15	65.46	61.49	58.54	58.66	57.69	55.46
TiO_2	0.55	0.55	0.44	0.78	1.12	1.21	1.34	1.53
Al_2O_3	14.70	14.26	13.22	14.35	19.08	18.02	18.44	21.37
Fe_2O_3	0.91	2.24	2.79	4.65	6.46	6.55	6.85	8.65
FeO	2.61	1.03	0.74	0.78	0.44	0.82	1.04	0.37
CaO	2.94	1.17	2.04	1.75	0.83	0.58	0.76	0.38
MgO	0.98	1.19	1.31	2.18	2.01	1.69	2.71	0.56
Na_2O	3.31	1.95	0.53	0.42	0.61	0.57	0.54	0.49
K_2O	4.38	4.76	5.59	6.90	6.71	7.10	5.94	6.48
H_2O^+	0.80	3.54	5.64	5.81	3.25	3.32	4.03	3.58
H_2O^-	0.36	0.48	0.48	0.43	1.14	1.33	0.93	0.97
P_2O_5	0.18	0.12	0.13	0.24	0.38	0.29	0.32	0.20
CO_2	None	0.10	2.26	0.90	None	None	None	None
MnO	0.03	0.04	0.05	0.05	0.02	0.03	0.03	0.03
	99.67	99.58	100.68	100.73	100.59	100.17	100.62	100.07
Uncombined SiO_2	24	31	28	25	21	20	22	20

In thin sections of weathered granites, it is observed that mineral decomposition is initiated along surfaces exposed to the action of percolating meteoric or soil waters. All crystals have unfilled electron orbitals "dangling" from their exteriors and these orbitals will interact with hydrogen ions in the water and with dipolar water molecules. Other prime targets for attack include twin composition surfaces and cleavage planes in minerals, for although these are not fractures, the bonding forces across them are weaker than those across other planes in the crystal. Laboratory experiments reveal that the initial surficial alteration of exposed mineral surfaces is rapid but the rate of reaction slows somewhat thereafter because of the protective armor formed by the thin coat of altered substances.

The mineralogic changes caused by the Paleozoic weathering of the granodiorite studied by Wahlstrom are reflected by chemical analyses of the paleosol. Relative to their initial abundances in the granodiorite, the most altered part of the soil has lost nearly all the sodium and calcium (destruction of oligoclase) and ferrous iron (alteration of biotite and magnetite). Also lost are at least 40% of the original magnesium (destruction of hornblende) and 15% of the silicon (destruction of oligoclase and hornblende). In contrast, aluminum and potassium are relatively increased by at least 50% because illite and the other clay minerals hold them in the soil. The amount of titanium in the soil is almost tripled, for although the original sphene has been largely destroyed, titanium is normally immobilized as a polymorph of solid TiO_2, commonly anatase, which is essentially insoluble at pH values higher than 2. The quantity of ferric iron increases by an order of magnitude because of the oxidizing nature of the paleosol environment and the insolubility of ferric oxide at pH values higher than 3.

Weathering of Basaltic Rocks

Basalts and andesites are extensively exposed on the continental blocks and at many places within the ocean basins as well. From the viewpoint of weathering phenomena, basalts differ from granitic rocks in several important respects. The minerals in basalt crystallize at temperatures about 300°C higher than those in granite and in a relatively anhydrous environment. Therefore, it is to be expected that the calcic and mafic minerals in basalt will decompose more easily. In addition, the crystals in basalt are smaller than those in granite and basalts may contain a glassy groundmass that also increases the rate at which basalt decomposes. These differences between granite and basalt, coupled with the fact that the average basalt contains about 12% total iron compared to only 4% in granite, lead to the fairly rapid development in humid tropical climates of thick iron-rich crusts on basalt. These crusts are an important source of iron in many parts of the world. Hence, the weathering of basalt in tropical climates has received a great deal of study by economic geologists as well as sedimentologists.

One of the best localities in which to study the decomposition of basalt during weathering is in the Hawaiian Islands. The dates of eruption of many flows are accurately known and therefore the rate of weathering can be determined. Also, the variation in annual rainfall is exceptionally great, ranging from 10 in. or less to about 450 in. in the wettest belts. The topography also varies, from level to steep mountain slopes. The parent material is the only factor in the weathering process that is not variable, most flows being olivine basalts containing about 45% silica.

Basalts are normally altered in place directly to clay minerals, aluminum oxides, and titanium-rich iron oxides. As we noted earlier, the oxides of titanium, aluminum, and iron are the most stable residues of chemical weathering. An important difference between temperate weathering, such as occurred during the formation of the Paleozoic regolith in Colorado and humid tropical weathering in Hawaii, is the mobility of silica. In tropical areas quartz is largely dissolved and is removed from the weathering profile, possibly to be reprecipitated elsewhere.

The sequence of types of clays produced, by varying either the intensity of weathering or length of time during which it occurs, is illustrated in Fig. 7-5.

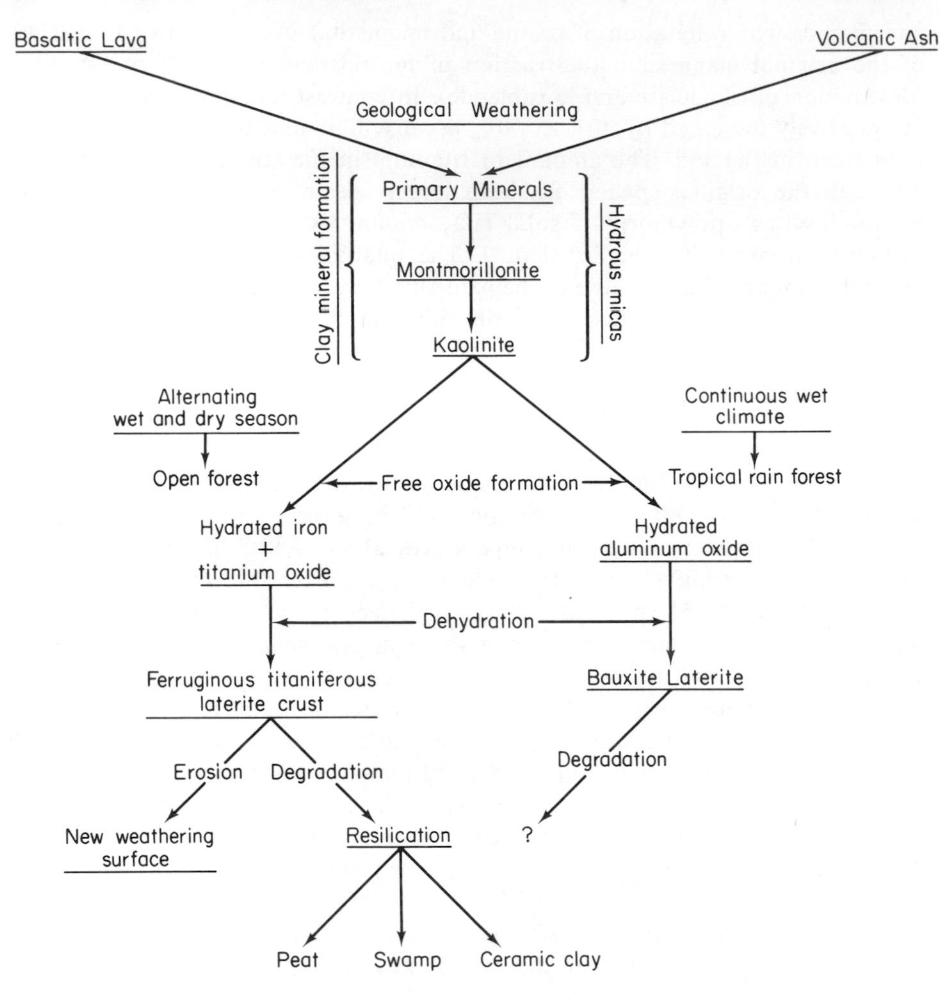

Fig. 7-5 Outline of weathering in the tropical soils of the Hawaiian Islands, showing primary, intermediate, and end products of pedological weathering. (From Sherman, 1952.)

The primary minerals of the basalt alter directly to montmorillonite, partly because there is insufficient potassium to permit the crystallization of illite and partly because of the relatively high Al/Si ratio in the parent rock. As a result of further leaching, the montmorillonite is destroyed and is normally replaced in the soil either by kaolinite or gibbsite. From this point in the weathering process, the substances produced vary, depending on the seasonality of rainfall. In a climate having alternating wet and dry seasons, iron oxides are stabilized. During the wet season, ferrous iron is leached from mineral grains to be oxidized during the dry season and precipitated as extremely insoluble hydrated ferric oxides. These substances cannot be remobilized during the succeeding wet season. The ultimate end product of weathering in a seasonally wet tropical climate is a ferruginous-titaniferous laterite horizon. In Hawaii these crusts contain 50 to 60% Fe_2O_3 and 15 to 25% TiO_2.

In areas of Hawaii in which rainfall is not seasonal and a tropical rain forest exists, the ferrous ions released from mafic minerals remain in the soluble ferrous state and are not precipitated from the soil pore waters. Consequently, an aluminous residue accumulates in the form of minerals such as boehmite, diaspore, and gibbsite. For a given parent rock such as basalt, the ferric iron/aluminum ratio is determined by soil aeration and drainage conditions; the higher the amount of aeration and the better the drainage, the higher the ratio. The tholeiitic basalts on Hawaii typically contain about 11% each of alumina and iron. We return to problems of laterite development in our consideration of soils on p. 256.

Stability of Minerals in the Zone of Weathering

The reason why some silicates break down more rapidly than others is to be found in the relative bond strengths of the various cations with oxygen. The approximate energies of formation (not free energies, as entropy values are unknown) of these bonds have been calculated from a reference state of gaseous ions (Table 7-2). The order of bond strengths nearly duplicates the order of increasing amount of covalent bond character arrived at from electronegativity considera-

TABLE 7-2 BOND STRENGTHS BETWEEN OXYGEN AND COMMON CATIONS, CALCULATED FROM A REFERENCE STATE OF GASEOUS IONS (from Keller, 1957, p. 18)

Ion	Energy of formation (kcal/mole)
K^+	299
Na^+	322
H^+ (in OH^-)	515
Ca^{2+}	839
Mg^{2+}	912
Fe^{2+}	919
Al^{3+} (4-fold coordination)	1793
Al^{4+} (6-fold coordination)	1878
Ti^{4+}	2882
Si^{4+} (in tectosilicates)	3110
(in phyllosilicates)	3123
(in double-chain inosilicates)	3127
(in single-chain inosilicates)	3131
(in sorosilicates)	3137
(in nesosilicates)	3142

tions (see Ionic Hydration, p. 236). The silicon-oxygen bond is by far the most covalent (i.e., directional) and is also the strongest in silicates; aluminum forms the only other bond with oxygen that is dominantly covalent in character and the aluminum-oxygen bond is nearly twice as strong as the strongest dominantly ionic bond. In Table 7-2, it also is significant that differences in Si–O bond strength among the various silicate structures are very small relative to the bond strengths between oxygen and other cations. Therefore, the silicates with high Si/O ratios are more stable regardless of which metallic ions balance the structure electrically. *Within* each of the six silicate groups, however, the species of metallic ion present determines the position of a silicate mineral in the weathering stability series. Using bond strengths in the table, and idealized chemical compositions of the minerals (i.e., neglecting effects of isomorphic substitution), the total bond strengths of the minerals in Bowen's reaction series can be calculated with results shown below. The energies of formation of the micas fail to coincide with their position in the reaction series, perhaps due to the presence of hydroxyl groups, whose energy effects are unknown. Also, it is clear that factors other than the bond strengths in Table 7-2 must be significant in mineral stability, for the table predicts no difference among polymorphs in resistance to hydrolysis, for example, among sanidine, orthoclase, and microcline. Nevertheless, despite some present deficiencies, the bonding energy approach is a sound one on which to initiate predictions of resistance to chemical weathering. To date, there is no established order of stability for metamorphic minerals and some minerals formed at high temperature facies (e.g., garnet) seem to be more resistant to hydrolysis than minerals of low temperature facies (e.g., epidote).

forsterite (29,789) anorthite (31,935)

augite (30,728)

hornblende (31,883)

biotite (30,475)

albite (34,335)

orthoclase (34,266)

muscovite (32,494)

quartz (37,320)

Sum of Bond Strengths Between Oxygen and Cations
of Minerals in Bowen's Reaction Series

The order of increasing strength of the oxygen-metallic cation bonds is exactly the reverse of the order of resistance of weathering of the minerals in which the cations are important. For example, excluding the micas and quartz, K-feldspars (potassium feldspars) are the most resistant phases in Bowen's Reaction Series; yet potassium-oxygen bonds are very weak. At the other extreme, forsterite weathers most rapidly of the igneous minerals; yet the magnesium-oxygen bond is relatively strong. The reason for this seeming paradox lies in the strength of the silicon-oxygen bond, which prevents the orthoclase structure from disintegrating despite the relatively rapid removal of potassium ions. Another way of stating this is that hydronium feldspars are common in weathering profiles. Forsterite, on the other hand, has relatively few Si–O bonds (Si/O $= \frac{1}{4}$) and without the stabilizing magnesium atoms the structure is very unstable and collapses rapidly.

Comparison between cation-oxygen bond strengths shown in Table 7-2 and the field data from the weathering profile obtained by Wahlstrom (Table 7-1) reveals general agreement between prediction and observation. Sodium, calcium, and magnesium are separated from the silicate framework easily, surrounded by water molecules and carried away in groundwater. Concurrently, the soil is relatively enriched in ferric iron, aluminum, and titanium. The major differences between observation and prediction are the unexpectedly small loss of potassium and large loss of silica. The small loss of potassium reflects the presence of illitic clay in the weathering profile. Potassic clay is more stable under nearly all weathering conditions than potassic feldspar is, pointing out another deficiency in the simplified bond strength approach being applied. A potassium-oxygen bond in illite does not have the same strength as a potassium-oxygen bond in orthoclase because of differences in the ionic environment surrounding the bond.

The fact that the loss of silica is greater than expected reflects mostly the ultimate breakdown of feldspar into clay minerals plus silica. The silica network depolymerizes in the undersaturated soil solution and some of it is carried away in the groundwater. That is, even orthoclase with its abundant silicon-oxygen bonds decomposes eventually and the silicon-oxygen groups are removed from the weathering profile.

The Hydrolysis Reaction

Given the relative rates of weathering of the major silicate minerals, the next problem is the nature of the hydrolysis reaction and of the decomposition products. It is possible to view weathering simply as a hydrolysis reaction resulting in liberation of metallic cations; for example,

$$Mg_2SiO_4 + 4\,H_2O \longrightarrow 2\,Mg^{2+} + 4\,OH^- + H_4SiO_4$$

But in the sedimentary environment carbon dioxide is ubiquitous and, therefore, it is more realistic to write

$$Mg_2SiO_4 + 4\,H_2O + 4\,CO_2 \longrightarrow 2\,Mg^{2+} + 4\,HCO_3^- + H_4SiO_4$$

Most silicates contain several metallic cations that have different bonding energies and, consequently, the kinetics of the hydrolysis and carbonation reactions are complex. Another complication arises when the reacting silicate contains iron, for iron immediately oxidizes when released from the mineral and forms the highly insoluble ferric oxide rather than staying in solution and being carried away by groundwaters. The association between free ions and water is considered in more detail below.

At any stage in the hydrolysis reaction, the silicate grains can be seen in thin section to be coated with an outer shell in which the crystal of the mineral is disorganized, the intermediate step between a fresh crystal and no crystal at all. This layer is much more soluble than the more unweathered parts of the crystal because the normal bonds have been disrupted by replacement of metallic cations by hydronium ions and the presence of this layer signals the beginning of the end for the crystal. In thin section, it is apparent that hydrolysis proceeds along structurally weak surfaces in the rock, such as joints, crystal boundaries, twin composition planes, and cleavages. On either side of such surfaces, alteration to clay is evident. The presence of clay requires the presence of aluminum in the system, for all the major groups of clay minerals contain this element as an integral part of their structures. As nearly all silicates are alumino-silicates, the source of the aluminum poses no problem and the hydrolysis-carbonation reaction for K-feldspar may be written as

$$\underset{\text{(feldspar)}}{5\,KAlSi_3O_8} + 4\,H^+ + 4\,HCO_3^- + 16\,H_2O \longrightarrow \underset{\text{(illite)}}{KAl_5Si_7O_{20}(OH)_4} + 8\,H_4SiO_4 + 4K^+ + 4\,HCO_3^-$$

If weathering is severe enough, this is followed by the formation of kaolinite from the illite.

$$\underset{\text{(illite)}}{2\,KAl_5Si_7O_{20}(OH)_4} + 2\,H^+ + 2\,HCO_3^- + 13\,H_2O \longrightarrow \underset{\text{(kaolinite)}}{5\,Al_2Si_2O_5(OH)_4} + 4\,H_4SiO_4 + 2\,K^+ + 2HCO_3^-$$

Laboratory experiments conducted to duplicate the natural weathering of orthoclase to clay minerals (Correns, 1963) have revealed the thin rind that forms on the outer surfaces of the feldspar to consist essentially of silica and alumina, always in the ratio of more than 5 : 1, SiO_2 to Al_2O_3. In unweathered potassium feldspar the ratio is 6 : 1. Decomposition of the potassic core takes place through the rind contemporaneously with its destruction so that the thickness of the rind remains constant at about 300 A.

Ionic Hydration

Cations

Upon release from the silicate crystal structure, ions are immediately surrounded by water molecules because the electrostatic potential gradient is many millions of volts per centimeter near the surface of an ion. Therefore, ions can be expected to interact strongly with their solvent. An exact quantitative treatment of this interaction cannot be made at present but the concepts involved may be illustrated using simplifying assumptions. Ideally (in a vacuum) the mutual electrostatic energy of a water molecule and an ion composed of a single atom will have its greatest value when the two particles are in contact and when the axis of the water dipole lies along the line joining them. When the length of the dipole is small compared with the distance between the centers of the particles, the electric potential energy for a monovalent ion is given by

$$-\frac{e\mu}{r^2}$$

where r is the distance between centers, e is the charge on an electron in esu, and μ is the dipole moment of water. The potential energy is double this for a divalent ion. Consider the force of attraction between a potassium ion ($r = 1.33$ A) and a water dipole ($r = 1.40$ A).

$$-\frac{e\mu}{r^2} = -\frac{4.80 \times 10^{-10}(1.85 \times 10^{-18})}{[(1.33 + 1.40) \times 10^{-8}]^2} = -1.19 \times 10^{-12} \text{ ergs}$$

The question of interest in sedimentology is whether the water molecule will be attracted to another water molecule more strongly than it will be attracted to the potassium ion. In other words, will potassium hydrate or be forced from solution, i.e., precipitate? The maximum force of attraction between two water dipoles is slightly different from that between a dipole and an ion and is given by

$$-\frac{2\mu^2}{r^3} = -\frac{2(1.85 \times 10^{-18})^2}{(2.80 \times 10^{-8})^3} = -0.31 \times 10^{-12} \text{ ergs}$$

Therefore, a water molecule has almost 4 times the attraction for a potassium ion than for another water molecule. Simple calculation using the ion-dipole formula reveals that a monovalent cation would need to have a radius of 3.95A for the potential energy to equal that of the water-water combination. No cation has a radius greater than 1.67 A (cesium) and therefore, assuming excess of water (i.e., adequate precipitation), all monovalent cations in water are hydrated. For a divalent cation the critical radius would be even larger than for a monovalent cation; calculation yields a value 5.17A.

As both the dipole moment of water and the charge on an electron are constants, it is apparent from the formula that the strength of the ion-water attractive force for ions of like charge varies as a function of the square of the ionic radius. Larger ions are less hydrated. The phenomenological picture is that the electronic charge on the ion must be distributed over a larger surface area for large ions, decreasing the force of attraction between the ion and the polar ends of water dipoles. Thus, the relative degree of hydration of ions can be expressed by Z/r; i.e., the hydration is directly proportional to charge and inversely proportional to ionic radius. The parameter Z/r is called *ionic potential*, a concept introduced

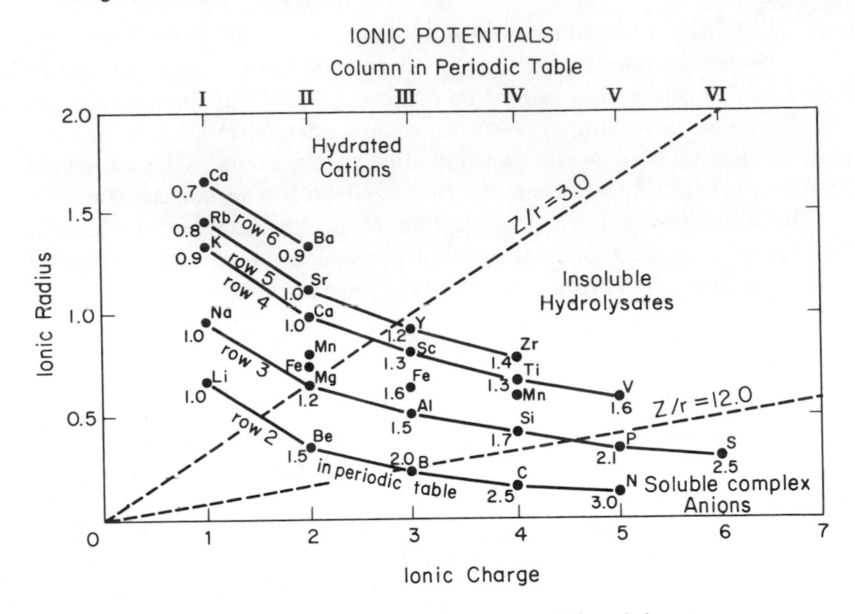

Fig. 7-6 Ionic potentials and electronegativities of elements.

by Cartledge, a chemist, in 1928, and carried into geology by V.M. Goldschmidt in 1934. Figure 7-6 is a plot of ionic charge versus ionic radius for cations of sedimentologic importance. Also on the graph are five lines connecting elements in the same rows of the periodic table and next to each element is its electronegativity value. Ionic potential is a useful concept in sedimentology because it explains much about sedimentary differentiation during weathering. On p. 237 it was calculated that a water dipole is attracted to a potassium ion 4 times as strongly as it is to another water molecule. On the graph (Fig. 7-6), it is apparent that the potassium ion has a low ionic potential and a low electronegativity; i.e., its ability to attract electrons is low. Therefore, it is to be expected that its ability to attract the negative ends of water dipoles also would be low and the hydrated radius of the ion should not be a great deal larger than its "naked" radius. This inference is supported by laboratory data. Figure 7-7 illustrates a typical set of results according to which the hydrated radius of the potassium ion is about 1.75 A, or 1.3 times its naked radius. In contrast, the hydrated radius of lithium, which has both a higher ionic potential and electronegativity, is approximately 2.5 A, or 3.7 times its naked radius.

The second-column ions of sedimentologic importance, magnesium, calcium, strontium, and barium, also are hydrated in amounts proportional to their ionic radii. Their hydrated radii increase sequentially from barium through magnesium. As with monovalent ions, those with the largest naked radii have the smallest hydrated radii. And, for ions of comparable size, divalent cations are more highly

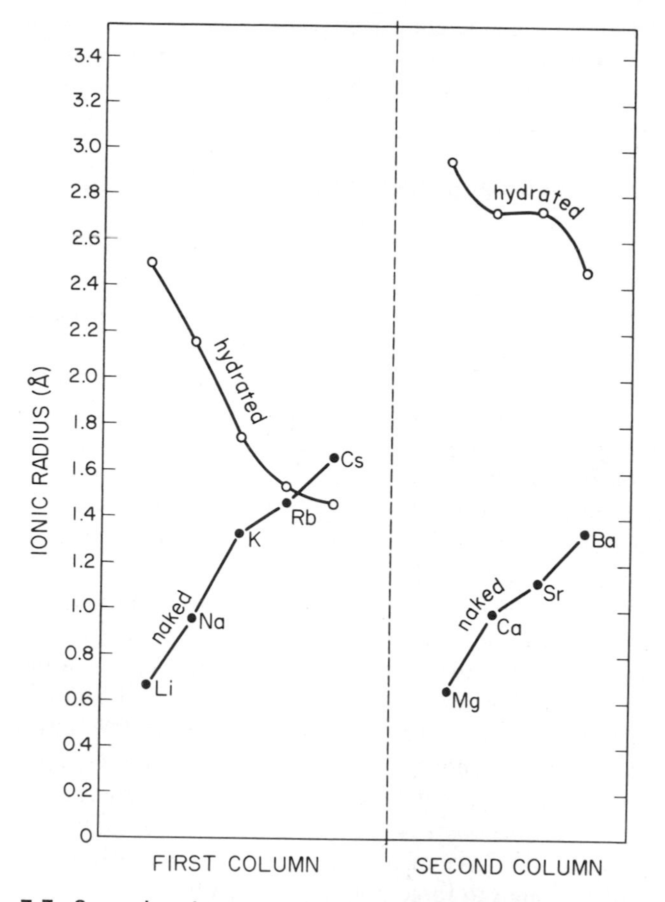

Fig. 7-7 Comparison between naked ionic radii and hydrated ionic radii. Data from C. B. Monk, 1961, *Electrolytic Dissociation*, N. Y., Academic Press. The trend toward less hydration with increasing ionic size is evident and real, but the absolute values of hydrated radii are, in part, an artifact of the measurement technique. Note that some hydrated radii are smaller than the naked radii for the same ion!

hydrated than univalent ones. For example, lithium has a naked radius of 0.68 A; magnesium, 0.66 A; yet the hydrated radius of lithium is 2.5 A and that of magnesium is 2.96 A. In ion exchange reactions, the participating ions are hydrated, not naked, so the size of hydrated ions has significance in weathering reactions.

Anions

Data are available for the hydration of numerous monatomic anions and it has been observed experimentally that hydration shells of anions are smaller than those of size- and charge-equivalent cations, a fact of considerable significance in ion exchange reactions. There is, as yet, no generally accepted explanation for this difference in hydration behavior of cations and anions and the search for an explanation constitutes one of the main concerns in the field of ion-water interactions.

The division of the graph of ionic potentials (Fig. 7-6) into three areas (separated by the two dashed lines) was suggested by Cartledge to designate areas of different behavior of the ions with respect to water. The dashed lines are drawn at ionic potentials of 3.0 and 12.0 and approximately parallel the electronegativity contours (not drawn on the graph) of 1.2 and 1.9. The electronegativity or ionic potential of a cation determines the intensity of its reaction with the oxygen end of a water dipole. For ions of like charge, the larger ones have both lower electronegativities and lower ionic potentials. The interaction of these large cations (e.g., K^+) with water dipoles is simply one of bonding in which the cation becomes surrounded with coordinated water molecules and remains in aqueous solution. When smaller cations (e.g., Al^{+3}) interact with water dipoles, a different phenomenon occurs. The decreased radius of the cation and its consequent higher electronegativity (between 1.2 and 1.9) and ionic potential (between 3.0 and 12.0) cause the electrons in the outer shell of the oxygen atom in the dipole to be pulled nearer to the cation (Fig. 7-8). This polarization or distortion of the electron cloud weakens the bond between the oxygen and hydrogen in the water molecule to the extent that the binding force between the oxygen and one of the hydrogen ions is overcome and an insoluble hydroxide [e.g., $Al(OH)_3$] is formed. The hydrogen ion "ejected" into the aqueous solution increases the acidity of the solution. The extreme case of this type of interaction is found with the smallest cations (e.g., S^{+6}), whose electronegativities are greater than 1.9 and whose ionic potentials are greater than 12.0. The force of repulsion between the cation and the hydrogen ions becomes great enough to force both hydrogens from the oxygen ion, forming a soluble oxy-anion (e.g., SO_4^{2-}) and further increasing the acidity of the solution.

It is apparent from the discussion above that the concepts of ionic potential and electronegativity are closely related. Characteristically, chemists discuss the polarizing power of an ion in terms of its electronegativity, while geologists generally discuss polarization only in terms of ionic potential. This difference in terminology poses no problems in interpretation and the reader should feel equally comfortable with either usage.

Weathering produces not only hydrated ions and insoluble hydroxides but also clay minerals. The clay minerals are not inert products "in equilibrium" with the "weathering environment." The "soil environment" is actually a complex of many different environments and, within these environments, reactions of the type called ion exchange continually modify the types and properties of the clays that are forming. In order to understand clays and ionic exchange further, it is necessary to consider the structure and reactivity of clays and the electrochemical processes occurring in a clay-water solution.

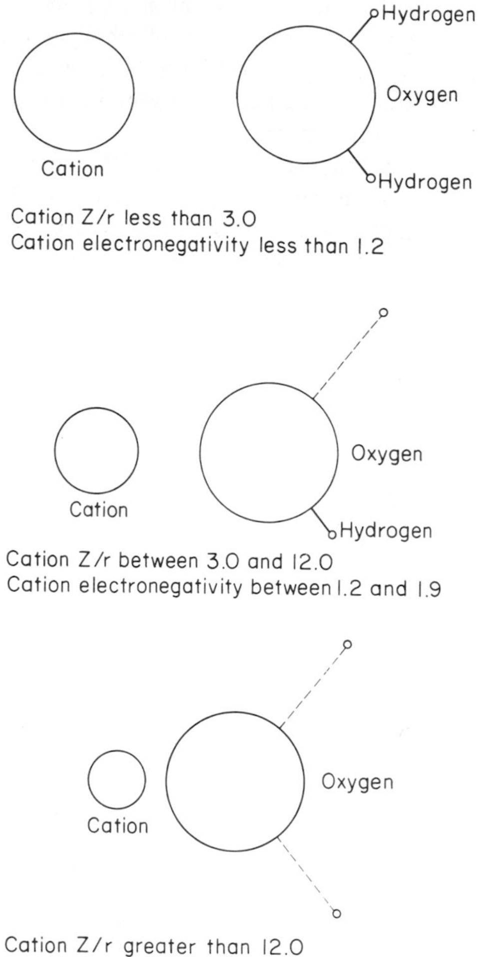

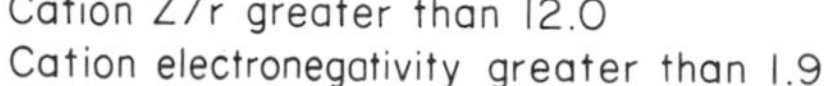

Fig. 7-8 Diagrammatic sketch of the relationship between cation size, cation—oxygen separation, and force of repulsion between cation and hydrogen ions of a dipolar water molecule.

7.5 STRUCTURE OF CLAY MINERALS

Discussion of the crystal structure of minerals is beyond the scope of this book but an understanding of the fundamentals of clay mineral structure is of sufficient importance in sedimentologic investigations to deserve review. In part, the importance of clays results from the fact that they form about 40% of all minerals in sediments; and in part their importance is due to their small size (generally less than 2 microns) and consequent chemical reactivity. The basic alumino-silicate structure of clay minerals is, compared to that of most other minerals, very stable at the earth's surface (hence its abundance). However, alterations in the chemical composition of phyllosilicates at low temperatures may occur both within the structural layers and at their surfaces. For the most part, these changes are restricted to variations in the types and amounts of metallic cations in the phyllosilicate structure.

The structure of a single clay (or mica) flake consists of two types of layers. One type is composed largely or entirely of silicon-oxygen tetrahedra with three of the oxygen atoms of each tetrahedron shared with adjacent tetrahedra (Fig. 7-9); hence, the layers are considered to be composed of Si_2O_5 units. Within these layers, substitution is restricted to replacement of up to half the silicon atoms by aluminum atoms.

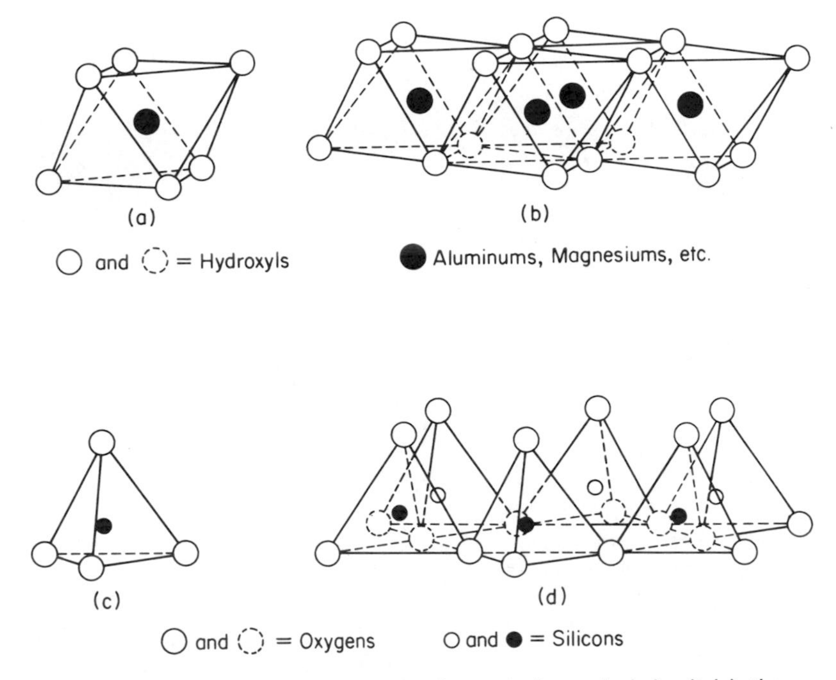

Fig. 7-9 Diagrammatic sketches showing a single octahedral unit (a), the sheet structure of the octrahedral units (b), a single silica tetrahedron (c), and the sheet structure of silica tetrahedra arranged in a hexagonal network (d). (From Grim, 1968.)

The second type of layer in phyllosilicates is formed of the cations magnesium, iron, or aluminum in octahedral coordination with oxygen and hydroxyl ions. These octahedral units are arranged in a planar network because the anions are shared among adjacent octahedra (Fig. 7-9). The mineral gibbsite consists entirely of Al–OH octahedra and, therefore, a layer of this composition in clays is referred to as a "gibbsite layer." The hydroxide brucite consists only of Mg–OH octahedra; hence an octahedral layer in phyllosilicates that is populated by magnesium atoms is called a "brucite layer." Kandites (kaolinite group of clay minerals) consist of

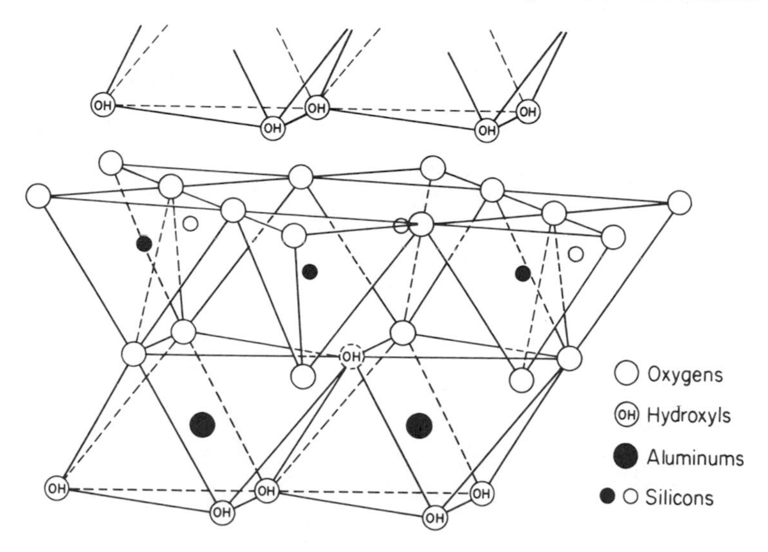

Fig. 7-10 Diagrammatic sketch of the structure of the kaolinite layer. (Modified after Grim, 1968.)

one tetrahedral layer and one octahedral (gibbsite) layer (Fig. 7-10). All other clays (and micas) consist of three layers, one octahedral layer sandwiched between two tetrahedral layers (Fig. 7-11). Thus, phyllosilicates are referred to as having 1:1 or 2:1 structures.

Because aluminum is trivalent, there must be two aluminum ions present for each six hydroxyl ions and, for this reason, the gibbsite layer is termed dioctahedral. In the brucite layer, there must be three cations for six hydroxyl ions because magnesium is divalent. Hence, this layer is trioctahedral. In classification schemes for the phyllosilicate minerals the primary groupings are in terms of the number of layers (two or three) and the valence of the cation in the octahedral layer (di- or tri-).

Bonding within tetrahedral layers is largely covalent, as required by the difference in electronegativity between silicon and oxygen or between aluminum and oxygen. Within the octahedral layer, bonding may be either predominantly covalent or ionic, depending on the relative amounts of aluminum, magnesium, and iron present. Between layers the major bonding forces are the covalent-ionic bonds between the unshared oxygen of each tetrahedron in the tetrahedral layer and the divalent and trivalent cations in the octahedral layer. There also exist hydrogen bonds between hydroxyl ions in the octahedral layer and the unshared oxygen in the tetrahedral layer but these bonds are much less important than the covalent-ionic bonds between the two layers.

Thus, the bonding, both within each layer and between layers, is largely covalent and therefore gentle disaggregation of a floccule of clay particles does not destroy the mineral structure. Disaggregation and cleavage do occur, however,

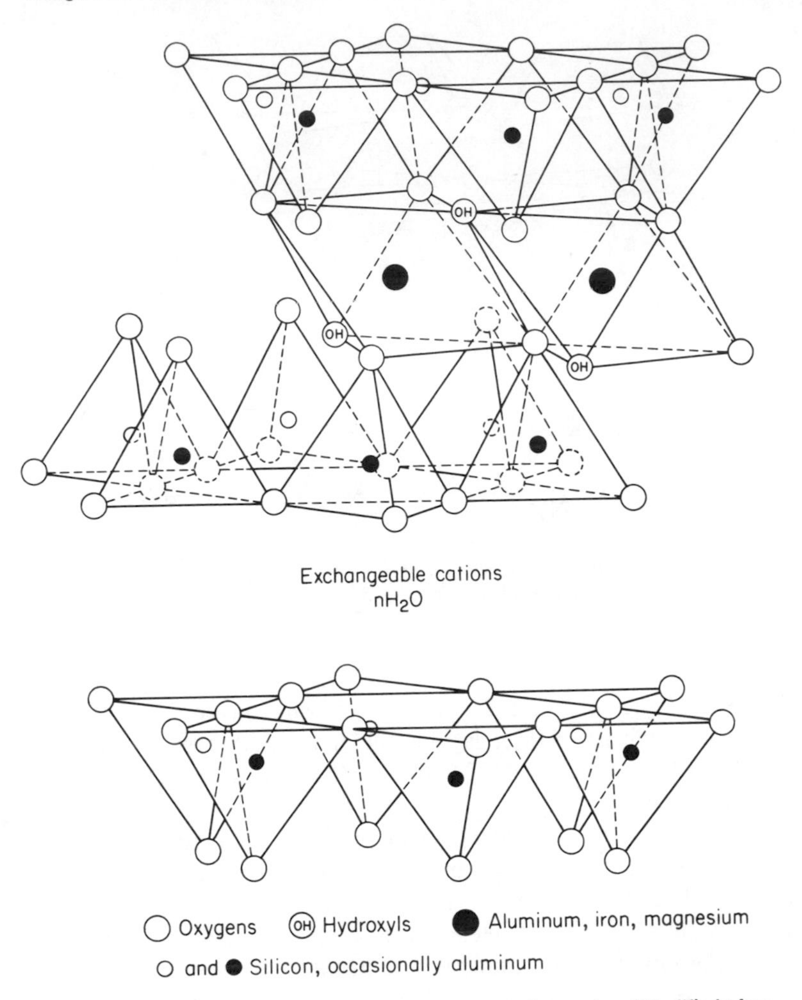

Fig. 7-11 Diagrammatic sketch of the structure of smectite. (Modified after Grim, 1968.)

between flakes. Because of charge deficiencies within the tetrahedral or octahedral layers caused by substitutions of Al^{3+} for Si^{4+} or Mg^{2+} for Al^{3+}, it is necessary for some clays to have interlayer cations. Illite includes potassium in this position; smectites (montmorillonite group of clay minerals) most commonly include sodium or calcium in this position. In addition, smectites can accommodate water molecules associated with the interlayer cations.

The bonding force between interlayer cations and the site of the charge deficiency (tetrahedral layer in illites; octahedral layer in smectites) is largely ionic, nondirectional, and therefore weak relative to other structural forces. If water is present in the interlayer position, the bonding force is a mixture of hydrogen bonding and van der Waals bonding. Van der Waals bonds are present among all atoms but are mentioned only when of quantitative importance. Because of the relative weakness of the bonds between structural units, it is not difficult for percolating waters to remove interlayer ions during weathering and diagenesis (leaching), and clays with less than stoichiometric amounts of these ions are termed "degraded." Most natural illites and montmorillonites are degraded, as are many chlorites, biotites, and muscovites in sediments.

Sericite is white mica of silt and clay size that commonly has less potassium and more iron and magnesium than muscovite. It is the micaceous mineral seen as the alteration product within decomposing K-feldspar grains in sediments. Illite is a species of clay mineral containing somewhat more iron and magnesium and less potassium than sericite and is the most abundant clay mineral in shales. Neither sericite nor illite is chemically well defined and it is more by custom than chemical knowledge that the phyllosilicate in shales is generally called illite.

A large proportion of clays in sediments are mixed-layer clays. That is, they consist of alternating layers of illite and chlorite or of illite and montmorillonite; clays of mixed layer construction may be as stable as the individual minerals themselves. Many clay beds are essentially pure deposits of the different clay minerals, however, and some of these have been analyzed chemically. Figure 7-12 illustrates the mean amounts of the major metallic cations (excluding aluminum) in the volumetrically important phyllosilicate minerals. The analyses of chlorites, biotites, and muscovites are of crystals in igneous and metamorphic rocks. Note

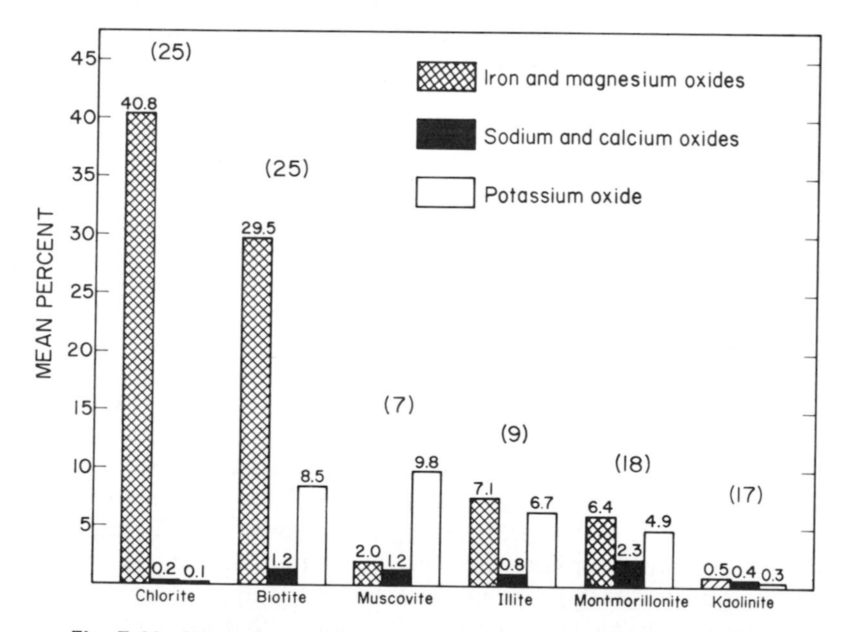

Fig. 7-12 Mean amounts of metallic cations (excepting aluminum) in the most abundant micas and clay minerals. Based on 101 analyses (in parentheses) from various sources.

in the figure the contrasting amounts of FeO-MgO and K_2O in muscovite and illite, as described previously. Montmorillonite contains three times as much sodium and calcium as illite but the FeO + Fe_2O_3 + MgO/K_2O ratio is about the same in the two minerals, based on these analyses of natural materials. The large amount of potassium in the montmorillonites analyzed probably reflects their origin as degraded illites, retaining some potassium. Kaolinite consists only of a tetrahedral layer and a gibbsite layer; hence, kaolinites are nearly devoid of metallic cations other than aluminum.

7.6 ION EXCHANGE

All solid substances have unsatisfied bonds at their surfaces and the quantity of these bonds is particularly large for amorphous materials and for crystalline materials with high surface/volume ratios. Because of this, clay minerals and organic matter have the capacity to adsorb on their external surfaces, edges, and inner surfaces (expandable clays only), large amounts of ions that may be supplied by percolating waters. These ions may later be desorbed or swapped for other ions, a process called ion exchange. There exists considerable disagreement among clay mineralogists and colloid chemists concerning the method by which ion exchange occurs, but all agree that the most important factors are the structure of the solid exchanger and ionic hydration. Present evidence points to a continuous gradation in the degree of order between perfectly crystalline clays and X-ray amorphous materials in soils. These amorphous substances in the soil probably total at least 20% of the $<2\mu$ fraction of soil. Many clay minerals in shales are known to be somewhat disordered but the amount of amorphous clay-like solids in shales is uncertain.

Relative Exchanging Power of Ions

Inorganic Ion Exchange

Ionic exchange reactions are reversible and this fact suggests the possibility of applying the law of mass action to explain quantitatively ion exchange equilibria. Despite numerous attempts, however, there is at present no satisfactory equation to cover all types of ion exchange over a range of physical conditions. The basic reason for this lack is the large number of variables involved. In addition, many natural exchangers are heterogeneous.

The exchanges of the alkali metals and alkaline earth metal cations have been extensively studied. In the great majority of cases the same general order of adsorption is found that is predicted by the concepts of ionic hydration and bond strength. Results most in accord with theory are obtained when the solution is dilute, the activities of all ions in the solution are approximately equal, and only one species of exchanger is present. It is found that divalent ions are held more strongly than univalent ones because although divalent ions are more hydrated, they form bonds with the exchanger that are more covalent than those of univalent ions. For ions of equal valence, the bond strength between ion and exchanger increases with increasing naked radius (decreasing hydrated radius). The grouping of ions on the basis of replacing power is known as the lyotropic series. In order of decreasing replacing power or increasing ease of replacement, the series is for divalent elements: Ba > Sr > Ca > Mg; univalent elements: Cs > Rb > K > Na > Li.

Studies of the exchanging power of divalent cations other than the alkaline earth metals have been made by several workers. No definite order has been found. The relationships are complicated by the fact that many heavy metal cations form complex ions in solution. In general, differences in exchange capacity between one of these ions and another are not great in comparison to differences among the divalent alkaline earth metals.

From the viewpoint of the mechanism of ion exchange, the behavior of the hydronium ion is particularly interesting. In ordinary water, hydronium and hydroxyl ions move faster than most other ions. The great mobility of hydronium and hydroxyl ions is generally believed to result from the phenomenon of fast proton transfer between the water molecules and the ions. The protons can move rapidly among neighboring H_3O molecules and, therefore, are very efficient displacers of other ions from solid ion exchangers. The hydrated proton at the clay surface causes hydration of the clay and thus hydrogen clays are necessarily the most hydrated clays.

In basic solutions, cations such as sodium, potassium, or calcium are abundant, while hydronium ions are scarce. Their scarcity in the solution lowers their exchanging power because the activity of ions produces a marked effect on exchange capability, i.e., marked departures from the lyotropic series.

The transition of a solvent-exchanger system from a basic to an acidic condition is well shown in soils. In the early stages of weathering and clay formation the soil solution is markedly basic due to the abundance of silicate minerals high in calcium, magnesium, potassium, and sodium. Calcium is adsorbed more strongly than the other metallic cations commonly present in any quantity. As organic matter accumulates, however, acids are generated by its decomposition. Hydronium ions enter the soil solution and are adsorbed onto the clay colloids, replacing the calcium ions, which, in humid regions, are then lost in drainage. The colloid becomes weakly acidic. In regions of lower rainfall, organic matter is scarce; hydronium ions are not generated; bases are not leached from the soil and caliche forms (see discussion below).

If the pH of the soil solution rises to above 10 or decreases to below 4, clay exchangers are destroyed. They become soluble in the aqueous solution because aluminum is very soluble in strong bases or strong acids. Under these conditions aluminum moves from its fixed, unexposed lattice sites to exchange sites on the exterior of the colloid. Silica also is very soluble in strongly basic solutions.

Anion exchange equilibria have not been extensively studied, although the occurrence of adsorption and exchange of anions has been recognized for a long time in clays, apatite, and other minerals. However, the amount of anion exchange in clays is thought to be small relative to cation exchange. For example, in one group of experiments it was determined that the ratio of cation to anion exchange capacity is about 0.5 for kaolinite, 2.3 for illite, and 6.7 for montmorillonite. Exchange occurs largely with exposed hydroxyl groups on the edges of clay flakes; almost none occurs at basal plane surfaces. Thus, in clay minerals in which most cation exchange is due to broken bonds, such as in kaolinite, cation and anion exchange capacities are substantially equal. In expandable minerals such as montmorillonite, cation exchange is due mostly to substitutions in the mineral structure and to surface adsorption and, therefore, anion exchange capacity can be only a small fraction of cation exchange capacity.

Organic Ion Exchange

No "lyotropic series" has been established for organic molecules and little is known concerning displacement reactions between different organic molecules adsorbed on clays. In general, it is found that organic ions and neutral molecules are strongly adsorbed on clay surfaces and edges. For example, if an amine salt is added to a clay-water suspension, the organic cation replaces the inorganic ones present on the clay surfaces and the amino groups become bonded, displacing previously adsorbed water molecules. If the clay surface is not large enough to accommodate long-chain aliphatic compounds lying flat, the chains may tilt. In expandable clays, several layers of organic cations may be superimposed. Hydrocarbon chains with a polar grouping on one end will orient normal to the clay surface with the polar end toward the clay.

In general, the number of organic molecules that can be adsorbed varies with such factors as the nature of the organic ion or molecule (protein, amino acid, ketone, etc.), aliphatic chain length, types of functional groups on the chain and their location in the molecule, and the species of clay. An important consequence of adsorption of organic ions is the fact that it causes large particles to break into smaller ones through swelling and exfoliation. This is particularly true of clays in which the organic substance and water penetrate interlayer positions and may cause an expansion that exceeds the interflake attractive forces holding the layers together. Smectites seem to have the highest adsorptive capacities for organic substances, analogous to their behavior with respect to inorganic substances.

Charged organic species are held onto clays by Coulomb forces, as would be expected from the intensity of the force field immediately surrounding charged particles. Neutral organic molecules are held by hydrogen bonds between hydrogen atoms in the molecule and oxygen atoms on the clay surfaces. With some organic species, however, van der Waals bonds are most effective and when this occurs it is found that larger molecules are more strongly adsorbed than smaller ones. Also, when van der Waals forces are quantitatively important, the number of organic molecules adsorbed often exceeds the total theoretical exchange capacity of the clay species concerned; i.e., the thickness of the adsorbed layer is greater than the thickness of a single adsorbed molecule. Both of these observations are in accord with the character of the van der Waals bond.

The studies of several investigators indicate that the resistance of some organic materials to biological (i.e., bacterial) decomposition may be increased when they are adsorbed by clay minerals. Probably hydrolysis of these materials is reduced because the molecules, when adsorbed by the clay, are oriented so that the chemically active groups on them are inaccessible to the bacterial enzyme that ordinarily destroys them. It also is possible that the enzyme itself is partly adsorbed and thereby rendered partly inactive.

7.7 RELATION BETWEEN INTENSITY OF WEATHERING AND CLAY MINERALOGY

The weathering products of the silicic granites and gneisses that form the bulk of crystalline rocks may be potassic clay (illite), trioctahedral smectite (divalent ions from ferromagnesian minerals), kaolinite, or gibbsite, depending on the intensity of weathering and the length of time during which weathering occurs. In moist, temperate climates, both illite and smectite remain in the soil. Indeed, smectite forms directly from the igneous minerals in potassium-poor mafic rocks, without the necessity of an illite precursor. But as duration or intensity of weathering increases, sodium and calcium are stripped from their interlayer positions in smectites; potassium is stripped from its interlayer position in illite; and only kaolinite remains in the soil. Sodium and calcium in smectites are bonded dominantly to the charge-deficient octahedral layer and these bonds are longer and weaker than the bonds between potassium and the charge deficient tetrahedral layer in illites. Also, the total charge deficiency to be balanced by interlayer cations is less in smectites than in illites. Therefore, the leaching of sodium and calcium from clays is more rapid than that of potassium and it is to be expected that illites resist interlayer collapse longer than smectites. Electrodialysis experiments with smectites have shown that the order of destruction of octahedral layers, in which differences in cation-anion bond length are not involved, is magnesium-rich layers first, then iron-rich layers, and finally aluminum-rich layers. This is the order predicted by the bonding strength data in Table 7-2.

Illites and smectites may be either partially or completely degraded. That is, either some or all of their interlayer potassium, sodium, and calcium ions may be leached by migrating waters and their positions occupied by a water layer. Regrading can occur, however, should weathering conditions become less severe because of climatic change. Should the severity of weathering increase, the octahedral layers of illite and montmorillonite are attacked and destroyed by the hydronium ions produced by organic decomposition. We may hypothesize that the acidity at the surface of the clay flakes is more intense than in the surrounding soil solution and the chemically aggressive hydrogen protons proceed to attack

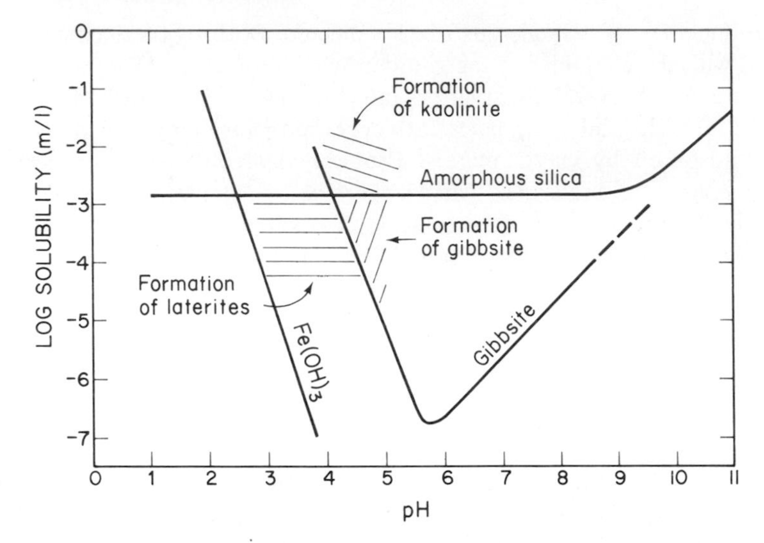

Fig. 7-13 Solubilities of amorphous silica, gibbsite, and ferric hydroxide as a function of pH. Also shown are the probable stability fields of kaolinite, gibbsite, and laterite.

the edges of octahedral clay layers. Magnesium and iron are solvated first, followed by aluminum, which is extremely soluble at pH values below 4. The "free" tetrahedral layers (Si_2O_5 sheets) remaining after the destruction of the octahedral layers are unstable and unite with colloidal Al–OH units released from the octahedral layers to form kaolinite. Figure 7-13 illustrates the solubilities of alumina and silica as functions of pH, indicating the pH range in which kaolinite may form by the series of events described. This sequence of events has not been verified experimentally, however, and therefore must be considered only as a tentative hypothesis for the formation of kaolinite from the 2 : 1 clay minerals.

Under more severe weathering conditions, such as are present in moist tropical climates, desilication of kaolinite occurs and gibbsite, $Al(OH)_3$, is produced (Fig. 7-13). It is not clear how silica is made soluble in such soils, as silica solubility is essentially unaffected by pH at values below 8.5 to 9.0. Current hypotheses center on the activity of organic matter in these soils but a satisfactory mechanism has yet to be proposed. Certainly the many occurrences of petrified wood in the geologic record suggest the existence of a chemical relationship between silica and organic matter. On the other hand, many temperate plants are known to be silica accumulators (Kutuzova, 1968); yet such plants do not generally silicify after death. Alternatively, it has been suggested that fluorine can be concentrated in tropical soils. Silica is known to be very soluble in the presence of this element and the only published analysis of lateritic soil for fluorine revealed that the more lateritized or desilicated horizons contained larger amounts of this element. Complete desilication results in the formation of either aluminous soils composed largely of gibbsite, boehmite, and diaspore or ferruginous soils composed largely of goethite and hematite (Fig. 7-13).

Based on the field evidence concerning the close relationship between leaching intensity and the stability of the various clay mineral species, we can anticipate that the types of ions in a groundwater will reflect the stability fields of the clays. For example, we would expect illite to be stable in an aqueous environment in which the K^+/H^+ ratio is high, kaolinite to be stable when the ratio is low. Gibbsite should be stable only in environments in which silica concentrations are very low. Hess (1966) extrapolated published hydrothermal data from the system K_2O-Na_2O-Al_2O_3-SiO_2-H_2O to a temperature of 25°C and, considering his results in the light of field data, constructed a diagram illustrating the apparent stability fields of the common clay mineral groups as a function of silica concentration and the concentration ratios Na^+/H^+ and K^+/H^+ (Fig. 7-14).

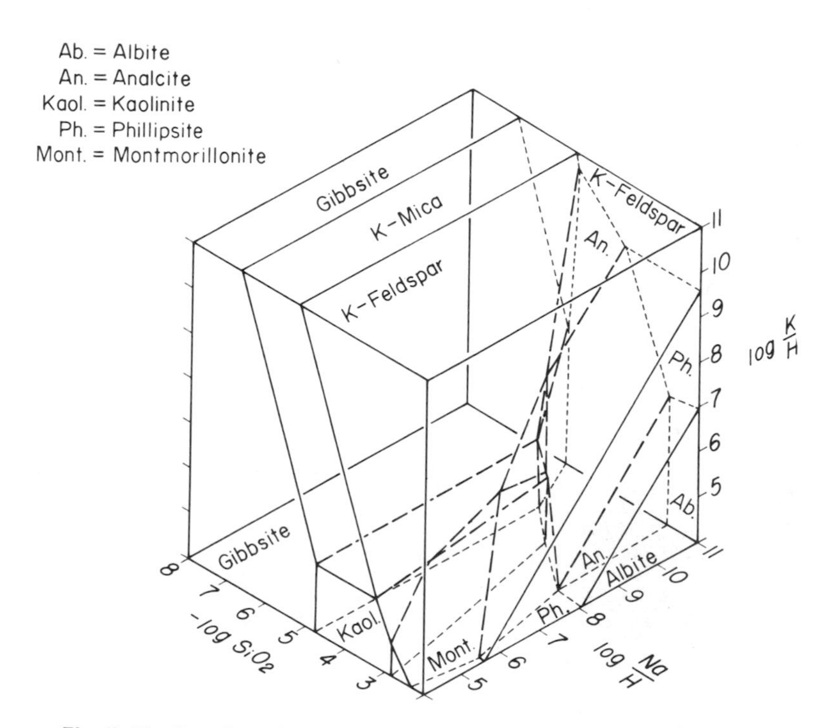

Fig. 7-14 Complete phase diagram for the system K_2O-Na_2O-Al_2O_3-SiO_2-H_2O at 25°C and 1 atmosphere. (From Hess, 1966.)

The Missouri Diaspore Deposits

The Cheltenham Formation (Early Pennsylvanian) in Missouri is an excellent example of the effects of changes in weathering intensity or the severity of leaching on the distribution of clay mineral types (Keller and others, 1954). The Cheltenham is a claystone or shale unit and at most localities is the basal detrital deposit of the Pennsylvanian Period. It rests on the upturned edges of sedimentary rocks ranging in age from Ordovician to Mississippian. Most of these beds are chert-bearing platform carbonates and in some localities a Mississippian weathering residue of chert conglomerate or sandstone separates the Cheltenham from the carbonate substrate. Cheltenham beds are thinnest in southeast Missouri and thicken toward the west and north into the Lower Pennsylvanian epicontinental sea.

TABLE 7-3 CHEMICAL ANALYSES OF CHELTENHAM CLAYS IN MISSOURI ILLUSTRATING CHANGES IN CLAY MINERALOGY FROM ESSENTIALLY PURE DIASPORE (SAMPLE NO. 1) TO MORE KAOLINITIC AND ILLITIC LITHOLOGIES. (Data from McQueen, 1943, p. 229.)

	1	2	3	4
SiO_2	2.84	10.80	43.40	56.10
Al_2O_3	77.00	67.09	37.21	24.47
Fe_2O_3	1.65	1.73	1.65	3.64
TiO_2	3.16	3.16	1.68	1.58
CaO	0.12	0.32	0.46	0.61
MgO	0.09	0.17	0.36	1.11
Na_2O	0.05	0.89	0.56	0.17
K_2O	0.57	1.13	1.94	2.89
H_2O	14.58	13.57	12.09	8.39
Others	0.09	0.68	0.15	0.31
Total	100.15	99.54	99.50	99.27

When viewed regionally, the Cheltenham Formation grades from high-alumina clay (diaspore) in its southern part into kaolinitic clays and shales and finally into illitic shales in its northern and western exposures. Chemical analyses of different clay facies of the formation are given in Table 7-3. Maximum present thickness of the diaspore deposit is 65 ft but its upper surface is erosional so that the original thickness is unknown. An additional complicating factor is that the thickest diaspore clays occur in sink holes in the underlying pre-Pennsylvanian carbonate rocks (Fig. 7-15).

The diaspore deposits are composed largely of well-crystallized diaspore and some kaolinite. In some areas boehmite occurs in place of diaspore. As indicated by the chemical analyses, iron and titanium oxides are present in the deposits in small amounts, probably as goethite, lepidocrocite, and anatase. The diaspore occurs as both disseminated microcrystalline grains and as the main constituent in pisolitic structures whose textures indicate an authigenic origin. The texture of kaolinite is commonly that of a random orientation of crystals, unlike that of

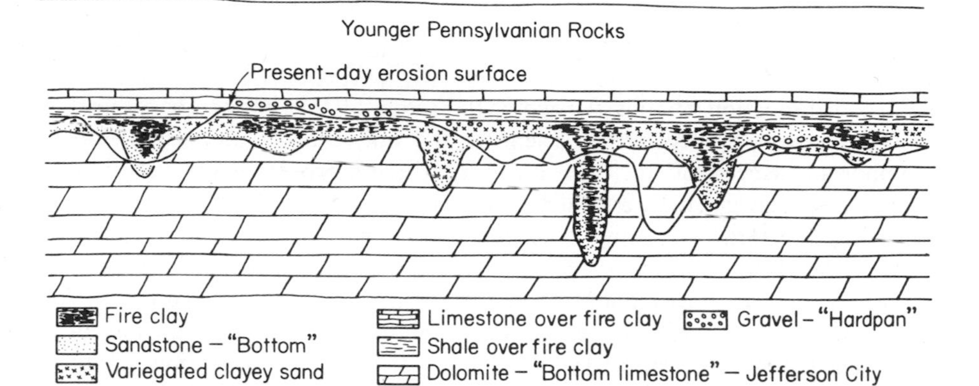

Fig. 7-15 A diagrammatic cross section through a part of the diaspore-kaolinite district of Missouri. The complete section shows the stratigraphy as it occurred shortly after deposition of the Fort Scott limestone (the limestone shown over the fire clay).

The heavy line which cuts irregularly across the dolomite and the overlying rocks (Pennsylvanian) represents the surface of the land today. The presence of a few erosion remnants of gravel, shown in two places, indicates prior erosion followed by deposition of a sheet of gravel, and subsequently renewed erosion.

Below the present erosion surface are found Pennsylvanian sediments barren of fire clay, shallow deposits of fire clay, and a few deep deposits filled with fire clay and/or sandy clay. The deep deposits are preserved as buried erosion remnants. They may be surrounded by Pennsylvanian rocks or by Ordovician dolomite, depending upon the vagaries of erosion as occurred in Cheltenham and Holocene times (Keller et al., 1954).

the laminated and fissile marine shales that overly the Cheltenham. Thus, the textures of both the aluminous and the alumino-silicate minerals suggest *in situ* origin.

Further supporting this interpretation is the absence of gibbsite in the diaspore deposits. Experimental and field data suggest that gibbsite is the stable polymorph under oxidizing conditions but that diaspore and boehmite are stable under reducing conditions such as would be common in a karst area where drainage was poor and oxygen stagnation frequent. The association of diaspore with coal, pigmenting carbon, and occasionally pyrite support the hypothesis of reducing conditions during formation of the diaspore clays. The amount of kaolinite is small in southeastern exposures of the Cheltenham but increases toward the northwest, i.e., toward the sea, presumably because drainage from the land surface was more integrated (either above or below ground level) closer to the sea. Better drainage results in less stagnation, decreasing concentration of organic matter, and less acid conditions, all of which favor formation of kaolinite rather than gibbsite or other aluminous polymorph. Further to the northwest, within the marine environment, clay rich in metallic cations occurs, specifically illite.

7.8 SOILS

A full discussion of soils is beyond the scope of this book. The subject has been exhaustively treated by soil scientists and only a brief summary of some of the important aspects of soils relevant to weathering processes and not discussed previously can be given here. Soil is generally distinguished from the mantle of weathering, the term soil being restricted to that part of the mantle that shows the presence of soil "horizons" and the direct influence of organic matter. Soil scientists recognize three main horizons: the *A horizon*, or eluvial horizon, from which material has been removed either in solution or by washing away of the finer grained weathered mineral and organic particles; the *B horizon*, or illuvial horizon, in which accumulation takes place either mechanically or by chemical precipitation; and the *C horizon*, which includes the unweathered or, in the case of deep weathering, the partly weathered material from which the soil is developed. *Eluviation* is the general term used for the transfer of material from one soil horizon to another: The term *solum* refers to the A and B horizons together. The existence of eluviation means that analyses of samples taken from different levels in a weathering profile do not generally yield a simple transition from least weathered to most weathered materials. Although all parts of the solum may have lost some constituents, others may have been added so that some levels in the soil are richer in certain elements or types of particles than others (including the parent rock).

Eluviation of clay minerals is known from field observations to be a common phenomenon in soils. Hallsworth (1963) studied this phenomenon experimentally by percolating distilled water through a plastic tube containing a layer of sand overlaid by a sand-clay mixture. Experiments were run using fine, medium, and coarse grained sand; varying amounts of either kaolinite or montmorillonite; and pH values between 2 and 13. For a fixed interval of time, more clay was moved through coarser and better sorted sand, as would be expected because these sands are more permeable. Within the pH range characteristic of soils, clay migration was greater at high pH values (up to 9 to 10) but showed no variation with type of clay. In all cases, the movement of clay was restricted as the proportion present increased and ceased when the quantity exceeded 40% for kaolinite clay or 20% for montmorillonite clay. At these clay concentrations, however, significant migration occurred down cracks that develop when the soil is dried out.

Other common examples of the migration of constituents through the soil include B horizons enriched in organic matter and iron oxides in podzols (the typical soils of humid temperate regions), soils with carbonate "hardpans" or concretionary layers, and soils with "ferricrete" or concretionary iron oxide layers (typical of lateritic soils). An important economic application of eluviation is the secondary enrichment of sulfide ores by the formation of a zone of "supergene enrichment," above which there is a leached, oxidized zone or "gossan." Such zones are frequently very thick and related to water table levels existing at some stage in the geological history of the ore body. Gossans are not generally considered

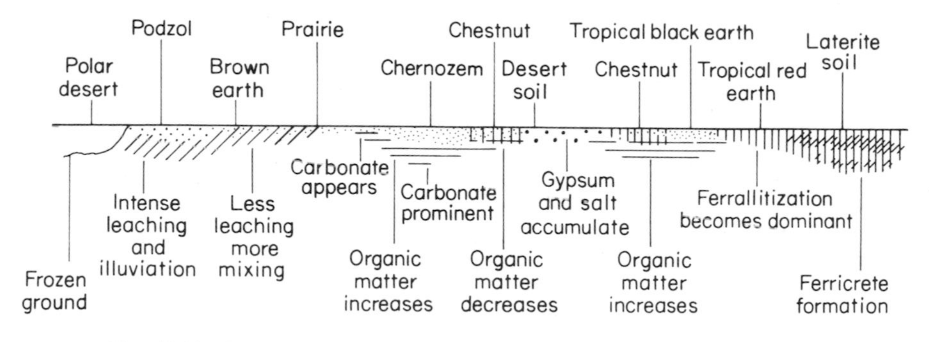

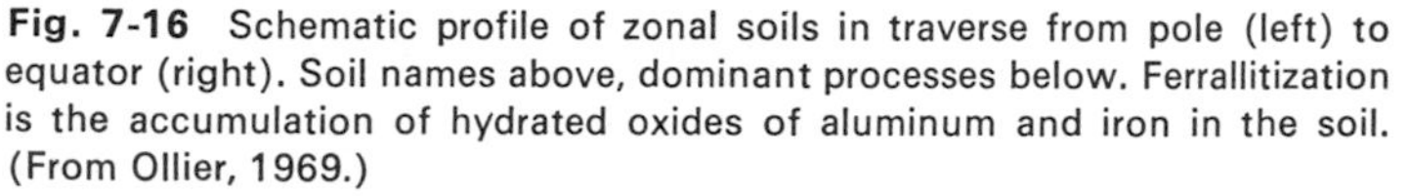
Fig. 7-16 Schematic profile of zonal soils in traverse from pole (left) to equator (right). Soil names above, dominant processes below. Ferrallitization is the accumulation of hydrated oxides of aluminum and iron in the soil. (From Ollier, 1969.)

to form part of the soil profile but they result from processes similar to those acting on a more limited scale in most soils.

Certain soils are called "zonal" because it is believed that they represent broadly defined types that depend largely on climate rather than on local conditions (including source rock types). The arrangement of zonal soils is shown in a highly schematic form in Fig. 7-16. The zonal concept dominated soil classification for many years but has been abandoned by many recent workers in favor of more objectively descriptive, less genetic classifications. The most popular of these, at least in North America, is known as the USDA "Seventh Approximation" and is intended to be a descriptive classification, usable in the field throughout the world, and capable of detailed subdivision as well as broad classification. Details of this and other systems are given in the book by Bunting (1967).

The effect of climate on the development of soils is seen not only in their overall characteristics but in the mineralogy of the clays developed in the major soil groups. Kaolinite is typical of acid, tropical soils and red/yellow podzolic soils, all of which are characterized by extensive leaching. Montmorillonite is found in soils formed under neutral conditions, including chestnut and prairie soils, in highly alkaline soils in arid regions, in poorly drained clayey soils (gleys), and in black tropical soils. Illite is found in many temperate soils, including podzols, where there has been only limited leaching. Chlorite is found in soils of arid regions. These broad generalizations are of course subject to modification from other major soil forming factors, such as parent rock type, geomorphic factors, and time.

An excellent example of a detailed study of modern soils illustrating the effect of a single climatic factor (mean annual rainfall), where not only parent rock type but also other climatic factors, such as mean annual temperature and seasonality of rainfall, were held constant, was described by Barshad (1966). His results, based on a quantitative mineralogical study of the less than 2μ fraction of more than 400 samples of the surface 6 in. of soils collected from traverses

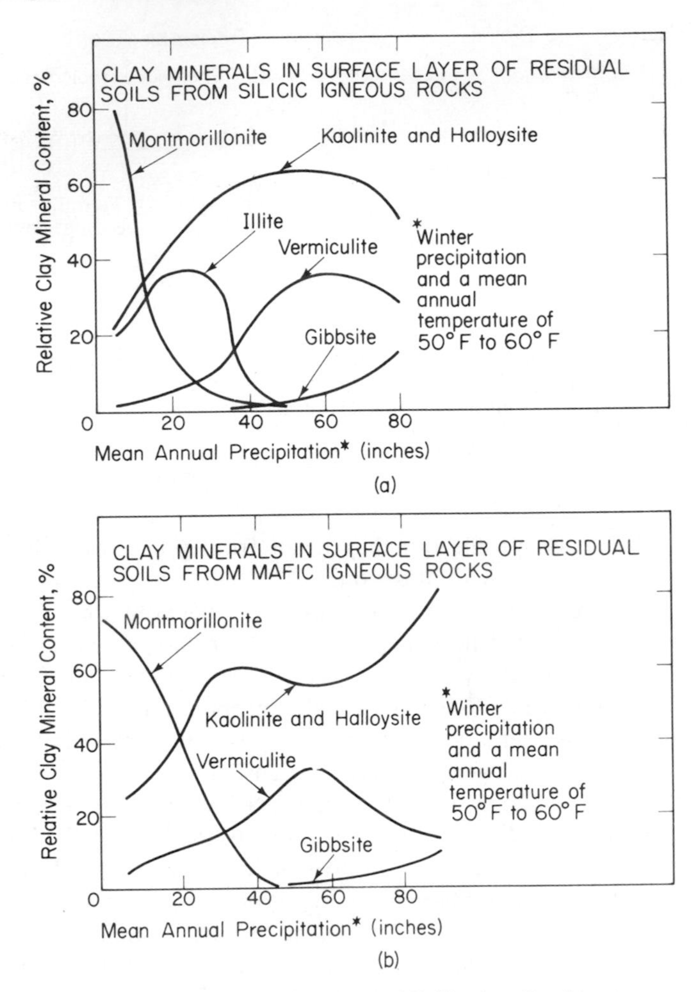

Fig. 7-17 Variation in clay mineralogy of California soils with mean annual precipitation. (a) soils on silicic igneous rocks; (b) soils on mafic igneous rocks. All samples were from locations with mean annual temperature between 50°F and 60°F, and with seasonal (winter) precipitation. (From Barshad, 1966.)

across California, are shown in Fig. 7-17 (a) and (b). The influence of parent rock as well as climate can be seen, for example, in the absence of illites in soils developed on mafic igneous rocks. But clearly, the major influence on clay mineralogy of these soils is the amount of precipitation, with montmorillonite characteristic of arid and kaolinite and vermiculite characteristic of humid conditions.

Laterite

The major group of tropical soils called *laterites* is often cited as showing the influence of climate in its most extreme form. The type laterites were named for soils in India that display "a soft earthy mass, full of cavities, (containing) a very large quantity of iron in the form of red and yellow ochres and hardening in the air..." (quotation from Buchanan, 1807, *in* Bunting, 1967). The term has been widely used for soils with horizons rich in either hydrated iron oxides (ferruginous laterite) or aluminous oxides (bauxitic laterite). Commonly these soils are pisolitic. Lateritic soils are common in India, central Africa, South America, and Australia but are uncommon in Europe and North America, largely as a function of Late Tertiary to Holocene climatic patterns. Wet tropical climates are ideal for the formation of laterites. Except that one bauxitic laterite of Precambrian age has been reported, laterites do not occur in ancient rocks until the spread of profuse terrestrial vegetation in the Devonian Period. From this fact we infer that vegetation is essential for laterite development, an inference supported by field studies discussed on p. 231. The relationship between temperature and the growth and decay of organic matter is illustrated in Fig. 7-18. Organic matter

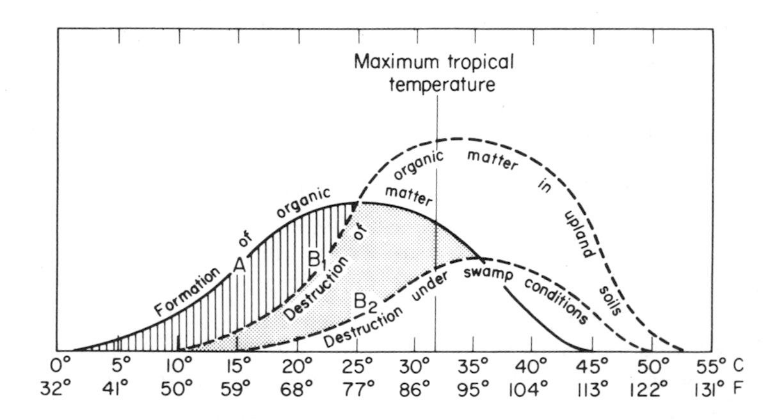

Fig. 7-18 Relationships among temperature, rate of organic growth, and its destruction by microflora. Vertical lines show where, under conditions of full aeration, the rate of formation of humus exceeds its destruction. At temperatures in excess of 25°C humus is destroyed faster than it is formed and none can exist. Also, above this temperature water moving in well aerated soil above the water table carries little organic matter in solution. (After Gordon, et al., 1958.)

serves, in some unknown way, to mobilize silica from quartz crystals and also functions as a reducing agent for iron so that it remains in the ferrous state until transported downward to the B horizon of the soil to be precipitated. Organic matter also serves as a complexing agent for iron, which may be illuviated as an organic complex. The iron is released when subsequent microbiological processes destroy the organic matter. Bacteria are known whose metabolic processes depend on energy transfers during iron oxidation or reduction and these organisms may also play a role in laterite genesis. There are many unsolved problems concerning the formation of this soil type.

Caliche

In arid or semiarid climates, soils tend to accumulate calcium carbonate, generally as calcite. These carbonate-enriched soils are termed *caliche*. Accumulation of calcium carbonate in soils is generally taking place today in the United States west of a line representing 25 in. of rainfall and a mean annual temperature of less than 40°F. If rainfall is excessive, leaching of base cations in the soil is complete and only very insoluble iron oxides can accumulate. If the climate is too arid, the amount of leaching is inadequate to mobilize calcium carbonate and only minor surficial accumulations of calcite occur.

The formation of caliche results from the combined effects of evaporation and changes in the partial pressure of carbon dioxide in the soil zone. Large amounts of CO_2 are released by plant roots during respiration and during decay of organic matter and this may increase the amount of CO_2 in soil air to at least 15 times that of the normal atmosphere. The concentration of CO_2 decreases both above and below the root zone. This high CO_2 pressure in part of the A horizon dissolves calcium carbonate and shifts the CO_3-HCO_3 equilibrium toward an increasing proportion of bicarbonate (see Chap. 13). In areas of adequate rainfall (but less than about 25 in.) these bicarbonate ions are carried down to the B horizon. In the B zone, biological activity is reduced, the partial pressure of CO_2 decreases, the CO_3/HCO_3 ratio increases, and calcium carbonate is precipitated. Illuviated clay size particles commonly accumulate with the calcium carbonate.

In more arid regions the quantity of organic matter is less and the effect of evaporation of vadose water becomes more important. Evaporation causes upward movement of soil water by capillary action. This results in a decrease in the activity of water as well as a loss of carbon dioxide gas, causing precipitation of calcium carbonate. Thus, in areas of low rainfall, the caliche zone is close to the surface. As annual rainfall increases, it moves to greater depth. When annual rainfall exceeds 40 in. (in temperate regions), caliche disappears completely from the soil profile (Fig. 7-19).

Early stages of caliche formation may produce either a powdery granular calcite or development of indurated isolated nodules of fine grained carbonate crystals. With further development, the whole soil may become indurated with the development of fibrous crystals and pisolitic structure. Characteristically, the pisoliths show truncation on their upper surfaces and accretionary growth on their lower surfaces, the result of precipitation from downward percolating vadose waters. Individual pisoliths and blocks of lithified caliche (calcrete) are often subsequently rotated or transported short distances from their place of formation, however, giving rise to a great variety of complex secondary rock textures. These textures are well displayed by the caliche capping the Ogallala Formation (Pliocene) in the high plains of west Texas and New Mexico (Swineford *et al.*, 1958).

Inhomogeneous masses of microcrystalline calcite are particularly susceptible to recrystallization by groundwaters and, therefore, it is difficult to recognize

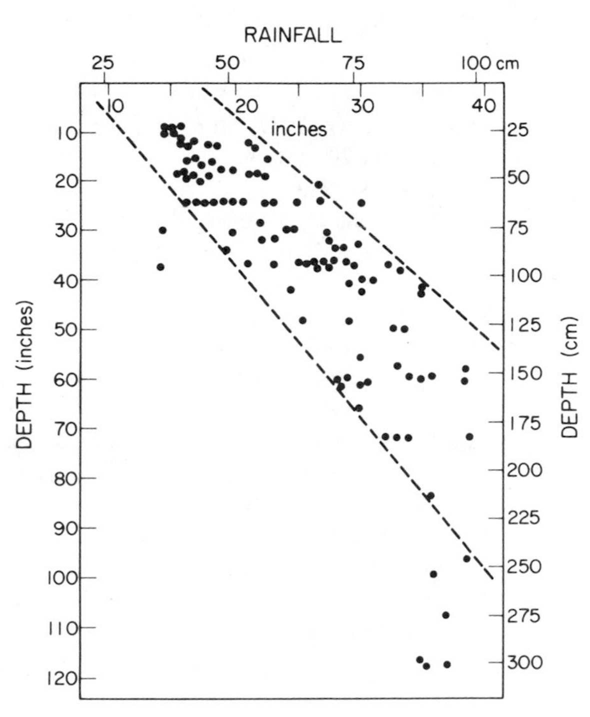

Fig. 7-19 The relation between rainfall and depth to the top of the carbonate accumulation in soils derived from loess. (After Jenny and Leonard, 1934.)

caliche in ancient rocks. Rotation of pisoliths may destroy the criterion for "facing." Coarsening of calcite crystals, partial dolomitization, and coating with ferric oxides may obscure original microtextures. In addition, ovoid structures similar in appearance to caliche pisoliths may be formed by encrustations of marine algae (see Chap. 12). In semiarid areas there often exist small ephemeral lakes that contain freshwater algae, which also can form biscuit-shaped structures resembling caliche nodules. Correct recognition of ancient caliche is an art and, as with most art, the experts often disagree among themselves concerning the criteria to be used in an evaluation.

REFERENCES

BAAS BECKING, L. G. M., I. R. KAPLAN, and D. MOORE, 1964, "Limits of the Natural Environment in Terms of pH and Oxidation-Reduction Potentials," *Jour. Geol.* **68**, p. 243–284. (Definitive study of this topic with abundant data.)

BARSHAD, ISAAC, 1966, "The Effect of a Variation in Precipitation on the Nature of Clay Mineral Formation in Soils from Acid and Basic Igneous Rocks," *Proc. Internatl. Clay Conf.*, **1**, p. 167–173. Jerusalem, Israel Prog. Sci. Translations, (Convincing study of the effect of rainfall on soil clay mineralogy.)

BEAR, F. E., 1964, *Chemistry of the Soil*, 2nd ed. New York: Reinhold Pub. Co., 515 pp. (A group of twelve articles by different specialists in soil chemistry, containing much sedimentologically useful data concerning clays, ion exchange, organic matter, etc.)

BUNTING, B. T., 1967, *The Geography of Soils*, 2nd ed. London: Hutchinson Univ. Library, 213 pp. (Concise, up-to-date review of soils and soil-forming processes.)

CARROLL, DOROTHY, 1959, "Ion Exchange in Clays and Other Minerals," *Geol. Soc. Amer. Bull.*, **70**, pp. 749–780. (Quantitative review of ion exchange phenomena in natural materials.)

———, 1962, "The Clay Minerals" and, "The Application of Sedimentary Petrology to the Study of Soils and Related Superficial Deposits," in H. B. Milner, ed., *Sedimentary Petrology*, 4th ed., **2**, pp. 288–371 and 457–518. (Valuable review of aspects of clays and soils relevant to sedimentary petrology.)

CORRENS, C. W., 1963, "Experiments on the Decomposition of Silicates and Discussion of Chemical Weathering", *Clays and Clay Minerals, Proc. 10th Natl. Conf.*, pp. 443–459. (Laboratory studies of the decay of silicate minerals and rocks.)

DELWICHE, C. C., 1967, "Energy Relationships in Soil Biochemistry," in *Soil Biochemistry*, A. D. McLaren and G. H. Peterson, ed. New York: Marcel Dekker, Inc., pp. 173–193. (A survey of metabolism and energy relationships among bacteria of both general and geological importance, e.g., sulfur and nitrogen bacteria.)

EHLMANN, A. J. 1968, "Clay Mineralogy of Weathered Products of River Sediments, Puerto Rico," *Jour. Sed. Petrology*, **38**, pp. 885–894. (Modern study of weathering of the same rock on rainy and dry sides of a mountain range in a subtropical climate.)

EVANS, W. D., 1964, "The Organic Solubilization of Minerals in Sediments," in *Adv. in Organic Geochemistry*, U. Colombo and G. D. Hobson, ed, New York: Macmillan, p. 263–270. (Experiments demonstrating that quartz can be dissolved and quartz sand lithified by organic compounds such as sodium adenosine triphosphate.)

FRASER, J. K., 1959, "Freeze-Thaw Frequencies and Mechanical Weathering in Canada," *Arctic*, **12**, pp. 40–53. (Mechanical weathering seen in Canada today dates, in part, from Pleistocene climatic conditions.)

GOLDICH, S. S., 1938, "A Study in Rock Weathering," *Jour. Geol.*, **46**, pp. 17–58. (An early, but still important, study of chemical weathering of gneiss, diabase, and amphibolite, through examination of residual soils. A "mineral stability series" was proposed that is still generally valid.)

GORDON, MACKENZIE, JR., J. I. TRACEY, JR., and M. W. ELLIS, 1958, "Geology of the Arkansas Bauxite Region," *U. S. Geol. Sur. Prof. Paper* **299**, 268 pp. (Comprehensive study of the origin of the most extensive and well-developed bauxite deposit in North America.)

GRIGGS, D. T., 1936, "The Factor of Fatigue in Rock Exfoliation," *Jour. Geol.*, **44**, 783–796. (Insolation is inadequate to effect rock disintegration.)

GRIM, R. E., 1968, *Clay Mineralogy*, 2nd ed. New York: McGraw-Hill Book Co., 596 pp. (The standard reference text for clay mineralogy. Includes sections on clay structure, occurrence, and interaction between clay minerals and their environments.)

HESS, P. C., 1966, "Phase Equilibria of Some Minerals in the K_2O-Na_2O-Al_2O_3-SiO_2-H_2O System at 25°C and 1 Atmosphere," *Amer. Jour. Sci.*, **264**, pp. 289–309. (Derives the approximate stability fields of the common clay minerals and authigenic minerals using field and laboratory data.)

Huang, W. H. and W. D. Keller, 1970, "Dissolution of Rock-Forming Silicate Minerals in Organic Acids: Simulated First-Stage Weathering of Fresh Mineral Surfaces," *Amer. Miner.*, **55**, pp. 2076-2094. (Experimental study suggesting the great importance of organic acids in the soil as a cause of mineral decomposition.)

———, 1971, "Dissolution of Clay Minerals in Dilute Organic Acids at Room Temperature," *Amer. Miner.*, **56**, pp. 1082–1095. (Stresses the importance of naturally occurring organic acids in the alteration of clay minerals.)

Jenny, Hans, and C. D. Leonard, 1934, "Functional Relationships between Soil Properties and Rainfall," *Soil Sci.*, **38**, pp. 363–381. (An early paper investigating these relationships.)

Kavanau, J. L., 1964, *Water and Solute-Water Interactions*. San Francisco: Holden-Day, Inc., 101 pp. (An advanced but nonmathematical discussion of the structure of water and its relation to cations and anions dissolved in it.)

Keller, W. D., 1957, *The Principles of Chemical Weathering*. Columbia, Mo., Lucas Bros. Pub., 111 pp. (A good elementary introduction to the topic.)

Keller, W. D., J. F. Westcott, and A. O. Bledsoe, 1954, "The Origin of Missouri Fire Clays," *Clays and Clay Minerals*, Proc. 2nd Conf., pp. 7–46. (Discussion of the origin of these interesting deposits.)

Kittrick, J. A., 1970, "Precipitation of Kaolinite at 25°C and 1 Atm," *Clays and Clay Minerals*, **18**, pp. 261–267. (The first synthesis of kaolinite from montmorillonite at room temperature.)

Krauskopf, K. B., 1967, *Introduction to Geochemistry*. New York: McGraw-Hill Book Co., 721 pp. (Chapter 4 is an elementary introduction to chemical weathering. Chapter 7 discusses clay mineral structure and soil formation.)

Kutuzova, R. S., 1968, "Silica Transformation during the Mineralization of Plant Residues," *Soviet Soil Sci.*, no. 7, pp. 970–978. (Interesting experiments on the rate and method of postmortem release and mobilization of silica from silica-secreting spruce, aspen, and birch leaves and needles.)

Loughnan, F. C., 1969, *Chemical Weathering of the Silicate Minerals*. New York: American Elsevier Pub. Co., New York, 154 pp. (Basic text on the origin of soil minerals and major element changes during weathering.)

Lovering, T. S., 1959, "Significance of Accumulator Plants in Rock Weathering," *Geol. Soc. Amer. Bull.*, **70**, pp. 781–800. (Summary of knowledge in this area and a plea, still unanswered, for further research into the interrelationship between botany and rock weathering.)

Maignien, R., 1966, "Review of Research on Laterites," *UNESCO*, Paris, 148 pp. (Comprehensive, with more than 300 references.)

McQueen, H. S., 1943, "Geology of the Fire Clay Districts of East Central Missouri," *Missouri Geol. Sur. & Water Resources*, ser. 2, **28**, 250 pp. (Detailed description of the deposits from an economic viewpoint.)

Monk, C. B., 1961, *Electrolytic Dissociation*, New York: Academic Press, 320 pp. (One of many good books on this topic.)

Nagtegaal, P. J. C., 1969, "Microtextures in Recent and Fossil Caliche," *Leidse Geol. Mededelingen*, **42**, pp. 131–142. (The occurrence and character of caliche of Upper Paleozoic and Lower Mesozoic age on the Mediterranean coast of Spain.)

OLLIER, C. C., 1969, *Weathering*. Edinburgh: Oliver & Boyd, Ltd., 304 pp. (Largely descriptive, with emphasis on geomorphologic aspects.)

PARFENOVA, E. I. and E. A. YARILOVA, 1965, "Mineralogical Investigations in Soil Science," *Israel Prog. Sci. Trans.*, Jerusalem, 178 pp. (Exposition of the approach to soil mineralogical studies in the Soviet Union by two active researchers. Contains an interesting section on biological weathering by lower plants.)

REEVES, C. C., JR., 1970, Origin, Classification, and Geological History of Caliche on the Southern High Plains, Texas and Eastern New Mexico," *Jour. Geol.*, **70**, pp. 352–362. (A recent paper concerning these extensive deposits, containing a review of theories of the origin of caliche and many references.)

RICH, C. I. and G. W. KUNZE, ed., 1964, *Soil Clay Mineralogy*. Chapel Hill, North Carolina: Univ. North Carolina Press, 330 pp. [A symposium containing valuable reviews of techniques and of the origin of clay minerals (by Keller) and the textures of soils (by Brewer).]

SCHATZ, ALBERT, 1963, "Soil Microorganisms and Soil Chelation. The Pedogenic Action of Lichens and Lichen Acids," *Jour. Agricultural & Food Chemistry*, 11, pp. 112–118. (Interesting experiments in which lichens are grown on different mineral substrates to determine their effectiveness in causing decomposition of minerals.)

SHERMAN, G. D., 1952, "The Genesis and Morphology of the Alumina-Rich Laterite Clays," *in Problems of Clay and Laterite Genesis*. New York: Amer. Inst. Mining Metal. Eng., p. 154–161. (General discussion of the weathering of basalt in the Hawaiian Islands.)

SWINEFORD, ADA, A. B. LEONARD, and J. C. FRYE, 1958, "Petrology of the Pliocene Pisolitic Limestone in the Great Plains," *Kansas Geol. Sur. Bull.* **130**, pt. 2, pp. 97–116. (Classic study demonstrating conclusively that the rock formed by soil-forming processes rather than in a subaqueous environment.)

VISHER, S. S., 1945, "Climatic Maps of Geological Interest," *Geol. Soc. Amer. Bull.*, **56**, pp. 713–736. (Useful paper containing many maps of the United States showing patterns of rainfall, temperature, etc.)

WAHLSTROM, E. E., 1948, "Pre-Fountain and Recent Weathering on Flagstaff Mountain near Boulder, Colorado," *Geol. Soc. Amer. Bull.*, **59**, pp. 1173–1190. (One of the early and better studies of the chemical weathering of granitic rock.)

WEBLEY, D. M., E. K. HENDERSON, and I. F. TAYLOR, 1963, The Microbiology of Rocks and Weathered Stones," *Jour. Soil Sci.*, **14**, pp. 102–112. (An interesting survey of the kinds of bacteria, fungi, and lichens found on and in different types of rocks and the ability of these organisms to decompose silicate minerals.)

WIMAN, STEN, 1963, "A Preliminary Study of Experimental Frost Weathering," *Geografiska Annaler*, 60, pp. 113–121. (An unusual and interesting laboratory study of the effect on mechanical disintegration of repeated freeze-thaw cycles. Rocks are immersed in compartments in a freezer and the freezer is continually cycled between -10 and $+10°C$.)

CHAPTER EIGHT

MINERAL COMPOSITION OF CLASTIC SILICATE ROCKS

8.1 INTRODUCTION

In Chap. 2 we discussed the nature of the relationship between orogenesis and the amount and type of sediment produced. We then examined the ways in which sedimentary textures and structures may be used to infer the upcurrent direction in a fluvial system, i.e., the direction in which the mountainous source area lay. In the present chapter, we examine the ways in which the mineral composition of clastic sediments may be used to estimate the relative amounts of igneous, metamorphic, and sedimentary rocks exposed within the drainage basin and to specify the types of crystalline rocks present.

All inorganic materials in the sedimentary environment are derived, either directly or indirectly, from crystalline rocks, which form approximately 95% of the earth's crust. Therefore, it is mandatory that sedimentologists be well versed in igneous and metamorphic petrology and geochemistry. Some minerals, such as staurolite or chromite, are quite diagnostic of specific origins but most minerals may originate in more than one type of crystalline rock. For example, in a sandstone, how might hornblende from a granite be distinguished from hornblende derived from an amphibolite or mafic schist? Or, how is it possible to determine whether a garnet grain in a sandstone originated in a schist, an eclogite, a skarn, or a rhyolite? In this book we cannot explore the field of igneous and metamorphic petrology but a brief overview of some points of particular importance in sedimentology is appropriate.

8.2 IGNEOUS ROCKS

Mineral Composition

Wedepohl (1969) has estimated the relative abundances of intrusive rocks in the upper part of the earth's crust to be as listed in Table 8-1.

TABLE 8-1 A TENTATIVE STANDARD SECTION OF INTRUSIVE IGNEOUS ROCKS IN THE UPPER CONTINENTAL EARTH'S CRUST (from Wedepohl, 1969)

Plutonic rocks	%
Granite and quartz monzonite	44
Granodiorite	34
Quartz diorite	8
Diorite	1
Gabbro	13
Others	1
	100

The mineral compositions of the various types of plutonic rocks have also been estimated (Table 8-2) and these data allow preliminary inferences to be made concerning the mineral composition of sands derived from plutonic igneous sources. In areas with minimal chemical weathering, such as modern or ancient deserts or areas of glaciation, it should be possible to obtain sands composed of up to two-thirds feldspar and in these sands plagioclase should be more abundant than the total of K-feldspar plus microperthite. Alkali-rich feldspars (K-feldspars, albite, and oligoclase) should be more abundant than the more calcic feldspars. (Other studies have determined that orthoclase is twice as abundant as microcline among K-feldspar crystals in intrusive igneous rocks). For large drainage basins it probably can be safely assumed that the relative proportions of plutonic igneous rocks approximate Wedepohl's average but there is no apparent way to establish a petrologically meaningful definition of "large." Clearly, small basins may contain restricted suites of crystalline rocks and may yield assemblages of feldspars that are either entirely alkali or entirely more calcic in composition.

The volume percentage of quartz in unweathered first-cycle sediments from massive plutonic rocks should average about 20% in large drainage basins but may range from 0 to 25% in smaller basins. Blatt (1967a) has shown that 85% of these quartz crystals will have undulatory extinction.

TABLE 8-2 MINERAL PROPORTIONS IN THE ABUNDANT TYPES OF PLUTONIC IGNEOUS ROCKS (from Wedepohl, 1969)

	In volume % of					
	Granite	Granodiorite	Quartz diorite	Diorite	Gabbro	Upper crust
Plagioclase	30	46	53	63	56	41
Quartz	27	21	22	2	—	21
Potassium feldspar (including microperthite)	35	15	6	3	—	21
Amphibole	1	13	12	12	1	6
Biotite	5	3	5	5	—	4
Orthopyroxene	—	—	—	3	16	2
Clinopyroxene	—	—	—	8	16	2
Olivine	—	—	—	—	5	0.6
Magnetite, ilmenite	2	2	2	3	4	2
Apatite	0.5	0.5	0.5	0.8	0.6	0.5

Amphiboles should be more abundant than pyroxenes in these large basins but in more areally restricted regions the composition of the nonopaque heavy mineral fraction may be either all amphibole or nearly all pyroxene. Among opaque heavy minerals, magnetite should be twice as abundant as ilmenite.

Types of Feldspar Twinning

In plutonic igneous rocks, a large but quantitatively uncertain percentage of orthoclase feldspar crystals are twinned. The only abundant twin type in these monoclinic crystals is the Carlsbad twin. Among plagioclase crystals in these rocks, several twin types are common and all plagioclase crystals are twinned. These facts contrast sharply with the relationships for metamorphic rocks, in which most plagioclase crystals are untwinned. Also, the relative frequency of occurrence of twin types among twinned crystals differs markedly between massive plutonic rocks and their metamorphic equivalents (Table 8-3). These data are important in determinations of the source areas of ancient sandstones.

Compositional zoning of plagioclase crystals is very common in igneous rocks, particularly in those of extrusive or hypabyssal origin. In contrast, zoning of feldspar is rare in crystals from metamorphic rocks.

Crystal Size

Feniak (1944) made length-width measurements of minerals in thin sections of 200 massive plutonic rocks with results as shown in Table 8-4. His data reveal that the modal grain size of feldspar crystals in these rocks is 1 to 2mm, or very coarse sand in the Wentworth size scale; quartz crystals are coarse sand size; and the abundant accessory minerals are common throughout the sand-size range.

TABLE 8-3 RELATIVE FREQUENCY OF OCCURRENCE OF TWIN TYPES IN PLAGIOCLASE FELDSPAR OF IGNEOUS AND METAMORPHIC ROCKS (adapted from Slemmons, 1962, p. 11)

Twin type	Comments	Approximate percentage of twins by mode of origin		Ratio $\frac{\text{Igneous}}{\text{Metamorphic}}$ origin
		Igneous	Metamorphic	
Albite	Usually with polysynthetic repetition	45	62	0.7: 1.0
Carlsbad	Usually with two or three subindividuals	25	4	6.2: 1.0
Albite-Carlsbad	Usually with two or three systems	12	3	4.0: 1.0
Pericline	Either simple or polysynthetic	10	25	0.4: 1.0
Acline	Usually polysynthetic	5	5	1.0: 1.0

It is important to note that the average size (long dimension) of zircon crystals in these rocks is precisely at the sand-silt size boundary (0.063 mm) and, therefore, counts of heavy minerals in sandstones must include silt-size grains to be representative. This is true even neglecting the reduction in size of all heavy minerals by sedimentary processes.

8.3 METAMORPHIC ROCKS

Mineral Composition

The mineral compositions of metamorphic rocks are closely controlled by the temperature and pressure at which they form, as well as by the bulk chemical composition of the preexisting rocks and metasomatic injections. Because of the wide range of pressure-temperature conditions under which metamorphic rocks may form, a much greater number of different mineral species are present than in igneous rocks. Therefore, most mineral species in sandstones (but not necessarily most mineral grains) are derived ultimately from metamorphic rocks. Further, because of the close control of mineralogy in these rocks by P-T conditions, it should be possible to make fairly detailed reconstructions of paleogeology in an ancient drainage basin from the minerals in sandstones. This problem is discussed in more detail later in this chapter.

The composition of plagioclase feldspars in metamorphic rocks correlates closely with metamorphic facies. The plagioclase in greenschist rocks has compositions in the range An_{0-7}; in epidote-amphibolite facies rocks, An_{15-30}. Compositions in the range An_{8-14} are absent in low and medium grade schists and this

TABLE 8-4 AVERAGE LENGTHS AND WIDTHS OF MINERALS IN PLUTONIC IGNEOUS ROCKS. "SILICIC" MEANS GRANITE AND NONFELDSPATHOIDAL SYENITE; "MEDIUM" MEANS FELDSPATHOIDAL SYENITE, MONZONITE, AND DIORITE; "BASIC" MEANS GABBRO AND ULTRAMAFIC ROCKS (AREAS ARE IN SQ MM; FIGURES IN PARENTHESES FOLLOWING AREAS INDICATE NUMBER OF ROCKS FROM WHICH AVERAGE IS DERIVED.) (after Feniak, 1944, p. 418)

Mineral	Silicic	Medium	Basic
Plagioclase	1.3 × 0.75 = 0.975 (87)	1.10 × 0.60 = 0.66 (48)	1.3 × 0.60 = 0.78 (43)
Microcline	1.15 × 0.75 = 0.87 (49)	2.05 × 1.15 = 2.36 (4)	(0)
Orthoclase	1.25 × 0.80 = 1.0 (76)	1.40 × 0.75 = 1.05 (21)	1.6 × 0.90 = 1.44 (5)
Microperthite	1.80 × 1.05 = 1.89 (7)	1.95 × 0.75 = 1.46 (2)	(0)
Quartz	0.85 × 0.55 = 0.468 (88)	0.49 × 0.34 = 0.166 (26)	0.39 × 0.26 = 0.101 (4)
Hornblende	0.80 × 0.48 = 0.384 (38)	0.85 × 0.50 = 0.425 (27)	0.95 × 0.55 = 0.525 (9)
Pyroxene	0.77 × 0.44 = 0.339 (3)	0.824 × 0.421 = 0.347 (27)	1.114 × 0.696 = 0.697 (35)
Biotite	0.70 × 0.31 = 0.217 (62)	0.70 × 0.375 = 0.262 (36)	0.70 × 0.38 = 0.266 (16)
Olivine	(0)	0.50 × 0.37 = 0.185 (2)	0.9 × 0.6 = 0.54 (26)
Nepheline	(0)	1.1 × 0.85 = 0.935 (10)	0.80 × 0.45 = 0.36 (2)
Allanite	0.20 × 0.108 = 0.0216 (6)	0.43 × 0.29 = 0.125 (3)	0.335 × 0.195 = 0.065 (1)
Sphene	0.23 × 0.114 = 0.0262 (20)	0.31 × 0.154 = 0.048 (20)	0.20 × 0.132 = 0.0264 (2)
Fluorite	0.35 × 0.22 = 0.077 (5)	0.17 × 0.085 = 0.0145 (1)	(0)
Apatite	0.12 × 0.05 = 0.006 (58)	0.17 × 0.07 = 0.012 (51)	0.23 × 0.10 = 0.023 (27)
Zircon	0.063 × 0.042 = 0.0026 (43)	0.066 × 0.046 = 0.003 (7)	0.057 × 0.035 = 0.002 (4)
Magnetite	0.185 × 0.123 = 0.023 (66)	0.185 × 0.14 = 0.026 (42)	0.37 × 0.28 = 0.103 (47)
Muscovite	0.445 × 0.23 = 0.103 (11)	0.26 × 0.103 = 0.027 (2)	(0)
Pyrite	0.137 × 0.088 = 0.012 (3)	0.23 × 0.17 = 0.039 (6)	0.20 × 0.132 = 0.026 (8)

is quite significant for studies of sedimentary petrogenesis. Plagioclase in this compositional range is abundant in granites. Rocks of the almandine amphibolite facies contain plagioclases in the range An_{30-50}. Granulite facies rocks are characterized by perthite and antiperthite but the small proportion of plagioclase in these rocks has compositions in the range An_{30-40}. Rocks of this facies are the only ones in the regional metamorphic sequence that contain perthites, which otherwise are essentially restricted to magmatic rocks.

Among the K-feldspars, none are formed during low grade metamorphism and those found in such rocks are relicts of the rock existing before metamorphism. In high grade metamorphic rocks, orthoclase is thought to predominate over microcline, except in migmatitic rocks, in which the reverse is true. Probably the magmatic part of migmatites contains most of the microcline.

The percentage of quartz in metamorphic rocks is thought to be approximately equal to the amount in igneous rocks, although there are no reliable data on this point. As in plutonic igneous rocks, the vast majority of the quartz is plastically deformed, i.e., has undulatory extinction. The visibility of undulatory extinction in thin section depends in part on quartz crystal size, however, and because of this fact the quartz in slates, phyllites, and fine grained schists may appear largely nonundulatory. Obviously, the crystal size of quartz in metamorphic rocks ranges widely, from predominantly silt (in some slates and phyllites) to largely gravel (in some coarse grained schists and gneisses).

8.4 SANDSTONES AND CONGLOMERATES

Introduction

As noted earlier, the mineralogy of sediments, unlike that of crystalline rocks, does not represent an equilibrium assemblage. For example, in metamorphic rocks, epidote and kyanite do not normally coexist. Nor does sanidine normally coexist with microcline in an igneous rock. However, these associations are common in sandstones. Any mineral or combination of minerals may occur in clastic rocks and, therefore, in contrast to crystalline rocks, determination of the composition of a few plagioclase grains does not permit conclusions to be drawn concerning the composition of other plagioclase grains in the sediment. The general principle to be followed during petrographic analyses of detrital rocks is that each grain is an independent variable and, as interpretations can be no better than the data on which they are based, the success of provenance determinations in clastic rocks can be no better than the characterizations of the mineral grains in them. Thus, it clearly is inadequate to characterize polygenetic mineral species simply as tourmaline or garnet. Such features as color, presence of zoning, inclusions, or content of trace elements must be specified, as these features commonly give valuable clues to the type of crystalline rock in which the grain originated.

Unfortunately, most sedimentary petrologists consider that exactness of mineralogic definition is not worth the time required to obtain it. Thus, the level of characterization of minerals is low in sedimentary studies compared to modern requirements in igneous and metamorphic petrology. In sedimentary studies "epidote" usually includes zoisite and clinozoisite; "rutile" includes cassiterite; "hornblende" may include actinolite; and sometimes "amphibole" and "pyroxene" serve as mineral categories. Rarely are compositions of plagioclase or garnet grains determined. These inexact identifications do not indicate lack of ability on the part of the researcher but are simply a reflection of the general feeling that more accurate mineral diagnoses in sediments are not worth the time they require. And, certainly, there is no doubt that continual use of the universal stage, index oils, or the electron microprobe is extremely time-consuming.

In summary, existing methodology in sedimentary petrography has developed as a compromise between rigor in mineral identification and the amount of work to be done, keeping in mind that the exact chemical composition or structure of a clastic mineral grain tells the worker nothing about any other grain in the sediment. It is important to keep in mind, however, that more exact diagnostic procedures are available than are commonly used, and these should be utilized whenever time and circumstances permit.

Recycling of Grains

In addition to contributions from the entire spectrum of crystalline rocks, clastic sediments usually contain large proportions of grains whose proximate or immediate source is older sediments. The distinction between ultimate source and proximate source is fundamental to paleogeologic interpretation; yet few techniques are now known that enable the researcher to make this distinction for most detrital grains. Although a worn secondary growth on a quartz grain is diagnostic of derivation of that grain from an older sediment, there may be no obvious way to determine this for the other grains in the rock. Certainly, if one grain in the sandstone is of recycled origin, some of the others must also be. But which ones? What percentage of the quartz grains? And what about the other minerals in the rock?

At present, we are even without an accepted definition of the term "recycled." In the ideal case, one pictures the formation of structural relief and an associated topographic relief. Detritus weathered from elevated areas is delivered by mass wasting and stream flow directly to a basin of deposition where it is buried, lithified, and perhaps someday resurfaced to begin a second sedimentary cycle. Alternatively, the lithified sediment may be metamorphosed or magmatized and be designated first-cycle debris when uplifted and eroded. In this model the start and end of a sedimentary cycle are clearly defined in both space and time. Even the most cursory glance at fluvial processes reveals that grain transport is intermittent, however, with alluvial fans, terraces, floodplains, and lakes all serving as temporary repositories for clastic detritus. How can a meaningful line be drawn between intermittent transport and the end of a "cycle"? Often grains "rest" between the source and depocenter for millions of years before continuing downslope and, as terminal basins of deposition often are geosynclinal in origin, the downslope direction may be reversed before the source area is leveled. For example, the Gulf Coast geosyncline may be uplifted before the northern Rockies, central Appalachians, and Black Hills are leveled.

Ideally, an adequate definition of the end of a sedimentary cycle might specify that the grain reach the terminal depocenter without a reversal of the paleoslope direction, but how do we specify the terminal depocenter for continental rocks? How deeply must a grain be buried before we can consider that it has been incorporated into the continental mass and has not simply taken a brief (?) rest before continuing on to the sea? At present, the answers to these questions are uncertain. In practice, the amount of polycyclic detritus in a sandstone is "guesstimated" on the basis of the percentage of quartz it contains. Sands composed almost entirely of this stable mineral and containing only relatively stable heavy minerals such as zircon and tourmaline are assumed to be composed almost entirely of recycled sediment. However, first-cycle sands in which all rock fragments and feldspars, and nearly all heavy minerals, have been destroyed may be produced by lateritic weathering processes in tropical and subtropical climates, which at present are confined to the area between approximately 25 degrees north and south latitudes. Such first-cycle pure quartz sands have been possible at least since the advent of abundant terrestrial vegetation in the Devonian and many Devonian laterites are known. Because climates were warmer and less strongly latitudinally zoned in the nonglacial geologic past, the formation of residual pure quartz sands was similarly less restricted. Bauxite pebbles of probable Early Tertiary age have been found in the Pleistocene of Massachusetts at a latitude of 42 degrees suggesting that, on a worldwide basis, climate and soil development may be very significant in determining the mineralogic character of sand grains reaching intermediate or terminal depocenters.

Mineral Composition

Quartz

The average sandstone contains approximately 65% quartz; the average shale, 30%; the average carbonate rock, 5%; the average modern pelagic sediment, 5%. From these and other sedimentologic data it can be calculated that the sediments in the earth's crust contain an average of 21% quartz, i.e., approximately the same amount as the average crystalline rock. The mean size of the detrital quartz in the earth's existing sediments averages approximately 2ϕ in sandstones, 3ϕ in carbonate rocks, 5ϕ in shales, and 6ϕ in the average ocean basin sediment (Blatt, 1970). The weighted grand mean size of detrital quartz is 4ϕ.

Nearly all quartz-bearing crystalline rocks are either massive plutonic granitoid rocks, quartzo-feldspathic gneisses, or schists. The total quartz contributed to sediments by rhyolites, veins, and other "primary" sources is trivial in comparison. Therefore, it is important for sedimentary petrologists to examine thin sections of granites, gneisses, and schists and to try to identify distinctive features of the quartz grains in these rocks for use in provenance studies of detrital rocks

(Sorby, 1877, 1880). Also, it is useful to examine the untransported disintegration products of such rocks (grus). In this way the original grain sizes and internal structures of clastic quartz grains may be determined and related to the type of crystalline rock from which they originate. Using these data, it may then be possible to recognize features characteristic of specific types of crystalline rocks in quartz grains in sandstones of mixed parentage. Unfortunately, the only quartzose grusses examined so far for factors other than grain shape have been from desert areas in southwestern United States (Blatt, 1967a). The ratio of polycrystalline to monocrystalline quartz and of undulatory to nonundulatory quartz in these grusses may not be representative of those in grusses formed in more humid climates.

Quartz from Massive Plutonic Rocks. Massive granitoid rocks disintegrate to release subequal amounts of polycrystalline and monocrystalline quartz grains. The polycrystalline grains are, of course, larger in mean size than those formed of only a single quartz crystal† and average 0ϕ (1 mm); monocrystalline grains average 1ϕ (0.5 mm). Therefore, sand-size polycrystalline quartz grains derived from granites are formed of only a very few quartz crystals, nearly always 2 to 5. The size distributions of both types of grains are nearly lognormal and have standard deviations of about 1ϕ.

An average of 80 to 90% of the crystals have been plastically deformed and show undulatory extinction in thin section (Fig. 8-1). The degree of undulatory

Fig. 8-1 Medium sand size strongly undulatory monocrystalline quartz fragment from a naturally disintegrated granite.

extinction (amount of angular separation between *c* axes in different parts of the crystal) averages 3 to 5 degrees in medium sand-size crystals and ranges up to 20 degrees. On a flat microscope stage, however, this angle often appears larger (10 to 30 degrees) because of the angular relationship between the plane of the thin section and the *c* axes in the crystal. In addition to this evidence of postcrystalliza-

†In this discussion, a grain is a clastic fragment. It may be composed of one or more crystals. The two terms are not used interchangeably.

tion strain, granitic polycrystalline quartz grains commonly have sutured intercrystalline boundaries. It is not known whether the formation of intercrystalline suturing requires more or less stress or higher or lower temperatures than the formation of undulatory extinction.

Within each granitic polycrystalline quartz grain the sizes of crystals are very similar and their shapes are subequant to moderately elongate. There is no obvious crystallographic orientation of crystals within each grain.

Quartz from Gneisses. Upon disintegration, the average gneiss yields 20 to 25% monocrystalline quartz and 75 to 80% polycrystalline quartz as its quartz population. The mean size of the polycrystalline grains is, as in granitic grus, about 0ϕ (1 mm); but the size of monocrystalline grains is much smaller, averaging 2.2ϕ (0.2 mm). Therefore, sand-size polycrystalline quartz grains derived from gneisses are commonly formed of more than five crystals. The standard deviations of the size distributions of both polycrystalline and monocrystalline quartz grains are similar, 1.1 to 1.2ϕ.

As in granites, nearly all crystals show undulatory extinction, and intercrystalline suturing in polycrystalline grains is common. Within gneissic polycrystalline quartz grains, however, the crystals frequently are morphologically elongated (or flattened) and have parallel crystallographic orientations. Also, in gneissic polycrystalline grains the quartz crystals frequently have a bimodal size distribution (Fig. 8-2). This feature reflects an arrested stage in the recrystallization of the quartz aggregate (in the parent rock), the smaller crystals being the newly developing ones. This feature has been produced experimentally.

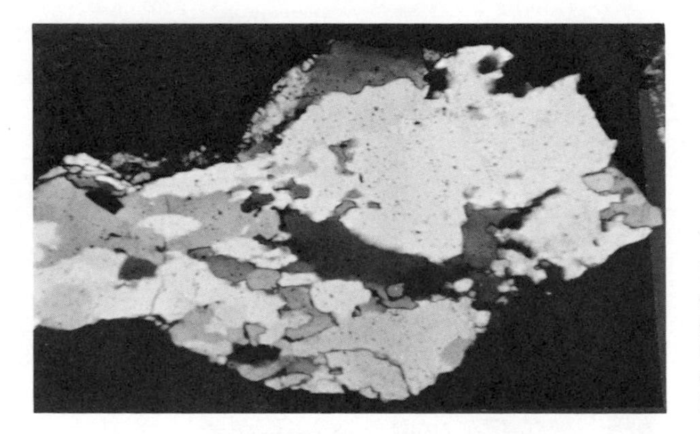

Fig. 8-2 Very coarse sand size polycrystalline grain from a naturally disintegrated gneiss. Note the strongly sutured intercrystalline boundaries, elongation of quartz crystals, and bimodal size distribution of these crystals.

Quartz from Schists. Quartz released from coarse and medium grained schists has some characteristics of granitic quartz and some of gneissic quartz. These schists yield 40% monocrystalline quartz and 60% polycrystalline quartz as their quartz population, similar to the proportions released from granites. The polycrystalline grains average 0.85ϕ (0.55 mm) in size; monocrystalline grains, about 2ϕ. Because of this size relationship, the average number of quartz crystals in polycrystalline quartz from medium and coarse grained schists is intermediate between that of granitic quartz and gneissic quartz. Schistose polycrystalline quartz is like gneissic quartz in that they both frequently contain elongate and crystallographically oriented crystals. Also, both often contain bimodal size distributions of crystals within polycrystalline grains.

In coarse and medium grained schists, most quartz crystals have undulatory extinction, and *c* axis orientations within a single strained crystal differ by an average of 3 to 6 degrees.

Fine grained schists, phyllites, and slates contain quartz populations that usually resemble in character those of the shales from which the metamorphic rocks were formed. Most of their quartz grains are silt or fine sand size and, as a result, are largely monocrystalline. Undulatory extinction is less commonly present or visible in sizes smaller than medium sand and, therefore, these rocks release a quartz assemblage relatively rich in monocrystalline nonundulatory grains. Although phyllosilicates recrystallize and increase in crystal size during the slate and phyllite grades of metamorphism, pervasive size increase of quartz occurs only during a higher grade of regional metamorphism, generally the greenschist facies.

Quartz from Volcanic and Hypabyssal Rocks. Quartz in volcanic rocks, some hypabyssal rocks, and perhaps in some massive plutonic rocks crystallizes as high temperature or beta (β) quartz. But on cooling below 573°C, the crystal structure inverts immediately to that of the low temperature form. Thus, all quartz in sediments is low temperature quartz. Quartz in granites that may have formed in the beta stability field does not develop beta crystal outlines because the grains abut against their neighbors during crystallization and growth. Quartz in extrusive and hypabyssal rocks usually crystallizes in a more fluid environment, however, permitting the beta quartz crystal form to express itself and, although the internal structure of the crystal adjusts to the alpha (α) state on cooling, the beta outward form remains (Fig. 8-3). Unfortunately, quartz crystals in volcanic rocks commonly

Fig. 8-3 Coarse sand size quartz crystal showing well developed β-quartz habit. This grain was released during weathering of a rhyolite porphyry.

are fractured as a result of the rapid change of 0.3% in the *c/a* axial ratio during inversion. Also, quartz in volcanic rocks is often partially resorbed, which causes rounded indentations in the grains, weakening them structurally. Fractured quartz grains displaying remnant beta outlines and resorption features are common in tuffs and in sandstones containing abundant silicic aphanitic volcanic rock fragments but in most sandstones they cannot be detected. Repeated fracturing during transport has destroyed the remnant beta outlines. Also, these grains are numerically swamped under by quartz from more common crystalline rocks.

Quartz from volcanic rocks usually is monocrystalline, has nonundulatory extinction, and is water-clear, lacking inclusions of any kind. Some grains contain inclusions of glass.

Detrital tridymite and cristobalite have not been reported from sediments but fibrous authigenic cristobalite (lussatite) occurs as cement in some volcanic sandstones.

Quartz from Hydrothermal Rocks. Quartz from this source may be monocrystalline or coarsely polycrystalline and commonly can be identified by an abnormally large content of water-filled vacuoles and, hence, a milky color. In thin section, these vacuoles are quite distinctive, particularly in reflected light. Unfortunately, much quartz in veins is not milky. Also, clastic grains that are milky when coarse become clear when reduced in size, presumably because the number of water bubbles in the smaller pieces is insufficient to produce the light dispersion necessary for a white color to be seen. Some hydrothermal quartz contains inclusions of vermicular chlorite (Fig. 8-4). These inclusions are thought to be distinctive of this provenance, but they are rare. Intercrystalline boundaries within polycrystalline quartz grains from hydrothermal sources commonly show cockscomb structure in thin section but this criterion must be used cautiously in identifying hydrothermal quartz, as many quartz grains from metamorphic rocks show similar internal structure.

Quartz in Sediments. *Internal Structure.* As noted earlier, the mean size of quartz grains differs significantly among the abundant types of sedimentary

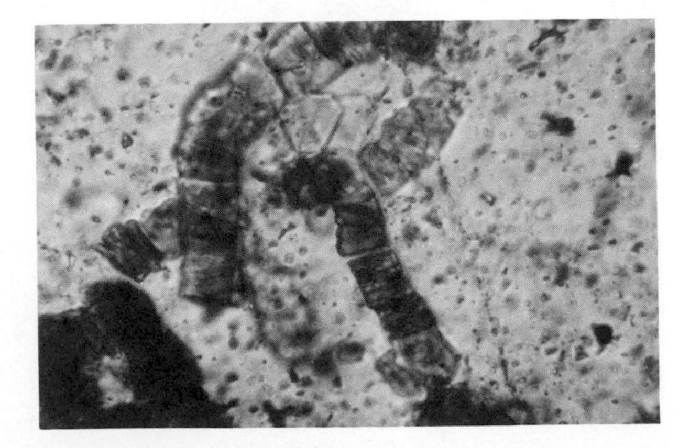

Fig. 8-4 Vermicular chlorite in a detrital quartz grain. Width of "worms" is approximately 10μ.

rocks, being coarsest in sandstones (2ϕ) and finest in ocean basin sediments (6ϕ). In the light of the characteristics of quartz in grusses, it is to be expected that sandstones will contain the highest percentages of polycrystalline quartz grains; ocean basin sediments, the least. Shales and carbonate rocks should contain little polycrystalline quartz. These expectations are confirmed by thin section study of sedimentary rocks (excluding ocean basin sediments, for which no thin section data are available). However, the quartz population of sandstones contains far less polycrystalline quartz than would be anticipated from the grus data. Less than 10% of the quartz in most sandstones is polycrystalline and most pure quartz sands contain none at all (less than 1%). The higher the percentage of quartz in a sandstone, the smaller the proportion of polycrystalline quartz present. These facts suggest that polycrystalline grains are disaggregated by weathering, transport, and diagenesis, an inference that seems reasonable considering the relative structural weakness of intercrystalline boundaries.

Monocrystalline quartz grains with undulatory extinction are thermodynamically less stable than those with nonundulatory extinction because of the difference in dislocation frequency. Therefore, we would expect sands containing higher percentages of quartz to have higher ratios of nonundulatory to undulatory quartz. Grus from crystalline rocks averages 10 to 20% nonundulatory quartz in its total quartz. Examination of sandstones containing more than 90% quartz reveals that these rocks average 43% nonundulatory quartz (Fig. 8-5), confirming

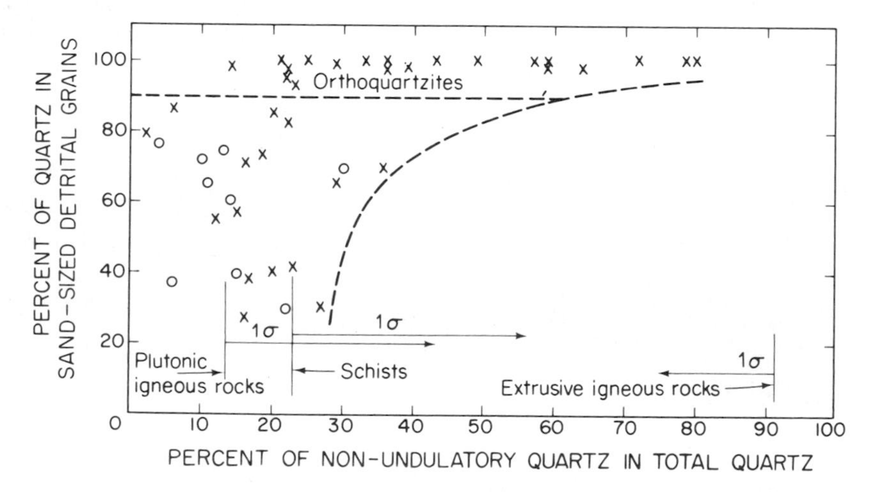

Fig. 8-5 The relationship between the total quartz percentage and the percent of that total quartz which is of the non-undulatory type in the sand-sized detrital fraction in 44 sandstones. Circles represent samples with more than 5 percent "matrix." The dashed curved line includes all sample points. The lines in the lower part of the diagram indicate the mean ratio of non-undulatory to total quartz in igneous and metamorphic rocks, together with the standard deviation of the ratio for each group shown. Note that sandstones with low total quartz percentages (that is, mineralogically immature rocks) have low percentages of non-undulatory quartz, as do crystalline rocks. Mineralogically mature sandstones (that is, almost pure quartz) contain abnormally high and widely varying percentages of non-undulatory quartz. (From Blatt and Christie, 1963.)

the prediction based on thermodynamics. In the figure, many orthoquartzites contain percentages of undulatory quartz not significantly different from the percentages in undisintegrated or disintegrated crystalline rocks. Probably many of the undulatory grains in these sediments acquired their strain as a result of folding or faulting in an earlier sedimentary cycle and are relatively recent additions to the clastic aggregate represented by the orthoquartzites studied.

Inclusions. Randomly distributed inclusions in quartz crystals were first recognized by the Greeks 2000 years ago; were first examined using a polarizing microscope by Sorby in 1858; and subsequently were reexamined by Mackie in 1896, Tyler in 1936, and by Keller and Littlefield in 1950. From the viewpoint of usefulness in provenance determinations, the results of these studies have been inconclusive. In part, this results from the fact that most mineral inclusions in quartz are on the order of a few microns in size and, hence, are nearly impossible to identify with certainty using a light microscope. Another difficulty is that most of the common mineral inclusions may originate in both magmatic and metamorphic rocks, for example, mica, rutile, and tourmaline. A third problem is that inclusions in most quartz grains in sediments are restricted to water-filled vacuoles and probably no more than one grain in a hundred in the average sandstone contains visible mineral inclusions. Katz et al., 1970, have shown that the presence of such inclusions is a sedimentologically significant structural weakness in a quartz grain.

The only generalizations concerning randomly distributed inclusions in quartz that have received widespread acceptance are (a) extremely bubbly quartz (milky color) is derived almost exclusively from hydrothermal veins; (b) quartz in schists contains few bubbles and water bubbles are rare in volcanic quartz; and (c) acicular inclusions of rutile (Fig. 8-6) are common in granitic quartz.

Color. Color in minerals may be caused by solid mineral inclusions, trace elements, or crystal defects, and quartz occurs in many colors for all of these reasons. In addition to the white color caused by the presence of abundant water

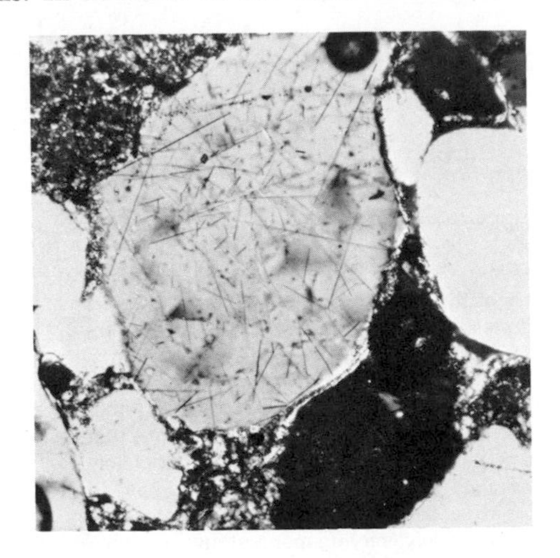

Fig. 8-6 Coarse sand size detrital quartz grain containing abundant acicular rutile crystals.

bubbles (on the order of 1 billion/ cu cm), quartz may be amethyst or yellow (due to ferric iron), smoky (due to radiation-generated crystal defects), blue or rose (due to titania or rutile inclusions), and other less common shades. Schnitzer (1957) studied Bunter and Keuper sandstones and found they contain a variety of different colors of quartz. Using a nonpolarizing microscope, he was able to use the colors of 1-to 2-mm grains for stratigraphic subdivision and provenance determinations. Each color was traced to the particular pluton in the Bohemian massif from which it came and it was also possible to trace the overlapping patterns of the alluvial fans that issued from the Triassic valleys cut into the massif.

Quartz grains have been irradiated in the laboratory and it was found that the degree of smokiness produced could be correlated with the rock type from which the quartz came (Hayase, 1961). Smokiness grade is highest in volcanic quartz, intermediate in granitic quartz, and lowest in either schist quartz or hydrothermal quartz.

Grain Size. The quartz fragments in grus from granites, gneisses, schists, and the phyllite-slate group have different grain size distributions and different proportions of monocrystalline and polycrystalline quartz in each size grade. Knowledge of the quantitative relationships among these variables permits some generalizations concerning the origin of quartz in clastic rocks.

1. Monocrystalline quartz grains of granular gravel, very coarse sand, and coarse sand size are more likely to have originated in massive plutonic rocks than in foliated metamorphic rocks. The probabilities are approximately 4.5: 1 for granular gravel size grains, 3.9: 1 for very coarse sand, and 1.6: 1 for coarse sand (Blatt, 1967a).
2. Assuming derivation directly from crystalline rocks (i.e., first-cycle origin), fine and very fine grained monocrystalline quartz grains are more likely to have originated in foliated metamorphic rocks than in massive plutonic rocks. The probabilities are 2.5: 1 for fine grained quartz and greater than 5: 1 for very fine grained quartz.
3. A large amount of monocrystalline quartz in the very fine sand and coarse silt sizes is released from fine grained schists, phyllites, and slates. But much quartz in these sizes is also formed by chipping of coarser quartz grains of any origin and, therefore, microchemical techniques must be used to determine the ultimate provenance of grains of this size.
4. Among the polycrystalline quartz grains, the distinction between those derived from massive plutonic rocks and foliated metamorphic rocks may be made by examining internal characteristics of the particle, specifically the presence of morphologic or crystallographic orientation of crystals and the unimodal or bimodal nature of the size distribution of these crystals.

Grain Shape. The form and sphericity of crystals and grains can only be determined quantitatively for loose particles. In thin sections it can only be determined whether most particles are subequant or not; quantitative distinction between rod-shaped and platy grains cannot be made. Thus, measurements of length/

width ratio of quartz particles in thin sections are of dubious value and, considering the tediousness of the procedure, it usually is not attempted. In most crystalline rocks there is no apparent qualitative difference in the shape of quartz crystals among different rock samples, although some foliated metamorphic rocks do contain crystals that are noticeably elongate in two dimensions.

In sedimentary rocks, many meaningless measurements of "two-dimensional sphericity" (i.e., width/length ratio) have been made. Based on measurements of more than 20,000 quartz sand grains, it has been determined that average width/length ratios for different sandstone formations lie in the range 0.61 to 0.71 (see also Chap. 3). However, individual grains may have values as low as 0.3 or 0.2. Measurements have also been made of the relative lengths of three mutually perpendicular axes in loose sand-size quartz grains. Most results have been expressed in terms of either Wadell or Sneed and Folk sphericity values and these nearly always range between 0.70 and 0.90. No correlation with sand grain size has been demonstrated. Silt size quartz grains in shales often have low width/length ratios and this may result from an origin in fine grained foliated metamorphic rocks. However, it may also reflect the fact that silt size quartz grains are produced by chipping of larger grains during transport. In general, quartz grain shape is a poor or ineffective criterion of provenance in sandstones.

Geochemistry of Clastic Quartz. Stable isotope studies of detrital quartz grains are potentially extremely useful because the O^{18}/O^{16} ratio of a mineral is directly proportional to its temperature of formation (see Chap. 10). Therefore, it should be possible to distinguish between magmatic and metamorphic origins for detrital grains and it may even be possible to determine the metamorphic facies from which the grains came. Quartz from igneous rocks is enriched in O^{18} by 0.94 to 1.03% compared to an accepted standard; metamorphic quartz by 1.02 to 1.67% (Savin and Epstein, 1970). The amount of enrichment in O^{18} in authigenic quartz varies from about 1.25 to 2.68% in response to the oxygen isotopic composition of the water in which it forms and the temperature at which it forms. At present, standard mass spectrometric techniques allow determination of oxygen isotopic ratio in samples as small as the volume of a single quartz grain 3 mm in size.

The trace elements in individual grains may be determined using an emission spectrograph but, so far, this has not been attempted. It is known that the types and contents of trace elements in quartz crystals vary among plutons and other crystalline rocks. These differences may be useful in provenance determination. Dennen (1967) was able to trace an assemblage of quartz grains in a sandstone to its parent granite using trace element techniques on bulk samples of the sandstone.

Feldspar

Abundance. The mineralogy and geochemistry of feldspars have received an enormous amount of attention from students of crystalline rocks but have been almost entirely ignored by sedimentary petrologists. In part, this difference in emphasis results from the much greater abundance of feldspars in crystalline rocks than in sediments. But the difference is also partly due to the lack of use by sedimentary petrologists of some of the more recently developed mineralogic techniques, particularly the electron microprobe. This instrument permits, for the first time, microchemical analysis of not only individual grains, but of small parts of grains. Thus, the microprobe is an instrument of the type sedimentologists have needed for decades, a tool to supply microchemical analyses on a grain-by-grain basis. As such, it merits extensive use by students of sedimentary mineralogy and petrology. The interest of geologists in the feldspars has provided a wealth of basic data that can and should be applied in the analysis of feldspathic sandstones.

The average sediment in the earth's crust contains approximately 5% feldspar, one fifteenth of the amount in the average crystalline rock. Data now available indicate the mean amount of feldspar in sandstones to be 10 to 15%; in shales, 4%; in carbonate rocks, less than 1%; and in modern ocean basin sediment, less than 1% except in the vicinity of intrabasinal volcanoes. The composition of detrital feldspars in sediments is almost unknown. Majority opinion in the United States considers K-feldspars more abundant than plagioclases and orthoclase slightly more abundant than microcline, on an average. But no reliable data exist concerning these suggested abundances and these statements are, therefore, little better than speculations. Only three analyses of clastic rocks have been published that show the relative frequencies of plagioclase varieties (Fig. 8-7). It is, perhaps, significant that all three of these frequency distributions are markedly non-Gaussian and two are markedly trimodal. This suggests that such analyses may be extremely informative from the viewpoint of source area determinations and should be made for all feldspathic sediments.

Origin of Arkoses. Feldspars are present in the same crystalline rocks as quartz grains but commonly in much larger quantities. Despite this initial preponderance, however, feldspars form only a small part of the average sediment because they hydrolyze rather rapidly. To form a highly feldspathic sediment ("arkose") requires high relief, rapid erosion and transportation, and burial before decomposition can occur in lowland soils. For feldspar crystals to be preserved after burial, diagenetic alteration must be minimized. Thus, the amount of feldspar in arkosic deposits (e.g., Fountain Formation of Colorado, Triassic sediments of Connecticut, and Eocene sediments of California) is determined largely by three factors: (a) the proportion of feldspathic crystalline rock in the upland part of the drainage basin, (b) the climate within and at the foot of the mountains, and (c) the amount of postdepositional destruction of feldspar grains. The combination of these factors, mixed in uncertain proportions, has resulted in 40 to 60% feldspar in the California Eocene sandstones, 40% in the Connecticut Triassic, and only 30% in the sand fraction of the Colorado Pennsylvanian. Although each of these sediments is believed to contain unusually high proportions of first-cycle detritus, their drainage basins must have contained older sedimentary rocks as well and these sediments contained less feldspar than granite and gneiss. Therefore, it is not possible to evaluate quantitatively the effectiveness of climate or diagenesis

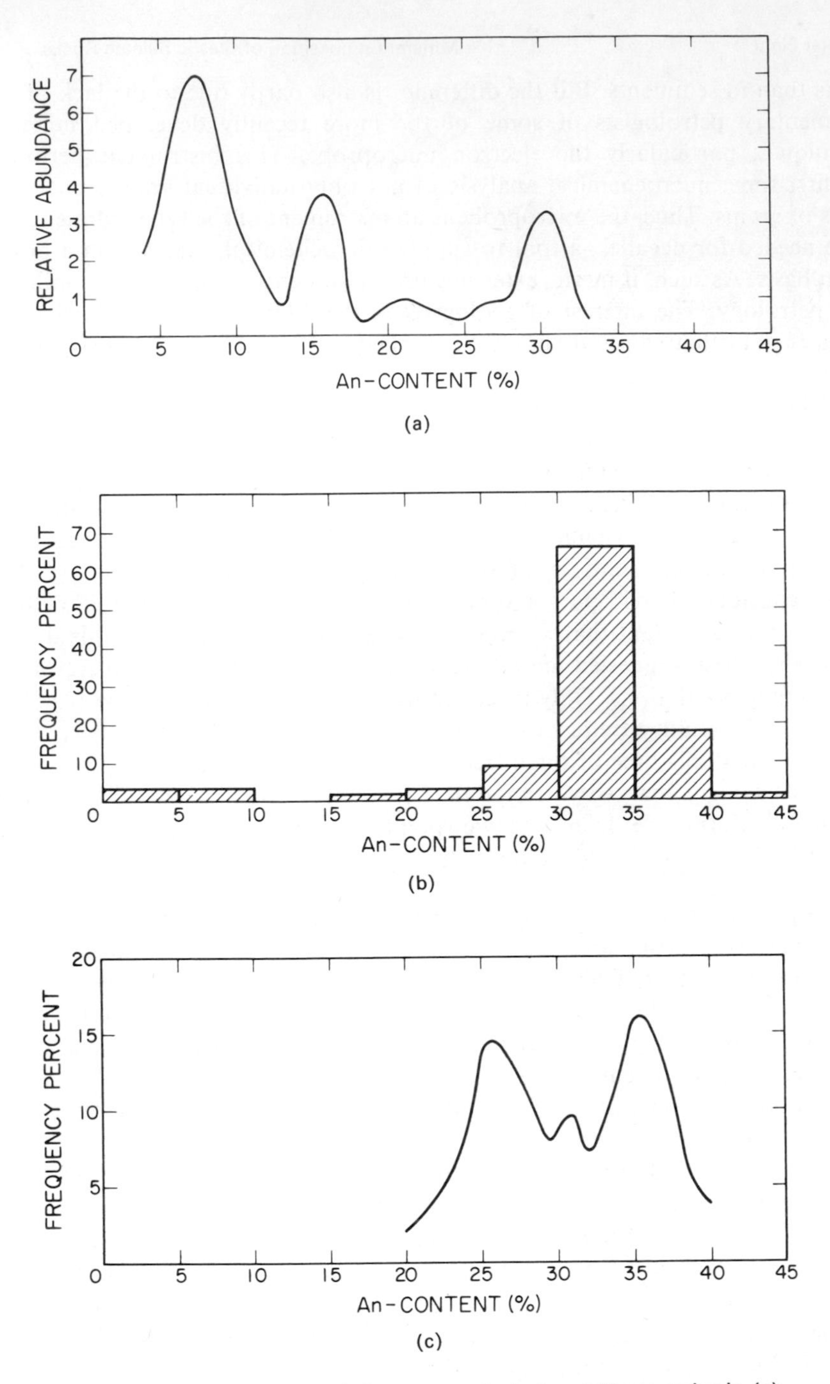

Fig. 8-7 Compositional variation among plagioclase feldspar grains in (a) Jotnian Sandstone (Proterozoic) of Finland (from Simonen and Kouvo, 1955); (b) Permian sandstone, Japan (from Mizutani, 1959); (c) Great Bay de l'Eau Cgl. (Devonian), Newfoundland (from Ermanovics, 1967).

in reducing the percentages of feldspar in the three arkoses to the values observed. Perhaps the 30 or 40% values reflect only dilution by quartz-rich sediments.

The usual way by which sedimentary petrologists attempt to resolve this problem is through study of the weathering characteristics of the feldspar grains. Feldspar may be altered (a) in the source rock by deuteric or hydrothermal activity, (b) in the soils of the upland or depositional site, (c) during diagenesis by groundwaters, or (d) on the modern outcrop. Characteristically, the first and third of these alternatives is considered of minor importance. The fourth is generally ignored if the outcrop surface seems as lithified as the rock a few inches below the surface. Thus, both the altered state of clastic feldspars and the fact that arkoses contain only one-half to two-thirds the amount of feldspar contained in crystalline rocks are usually attributed to weathering contemporaneous with the origin of the sandstone. During erosion of the crystalline source rock, turbulent stream waters cut through the weathered regolith into fresh crystalline bedrock, even in tropical climates where soils are thick, and so fresh feldspar may be present in the arkose. Most arkoses contain feldspars in all stages of decomposition.

Degree of Crystal Alteration. Normally, in feldspathic sediments, it is found that the most altered feldspars are the most calcic ones and the freshest feldspars are potassic. Microcline commonly seems particularly fresh. These observations are to be expected as they are in accord with predictions based on bonding strengths and the silicon/oxygen ratios of the minerals concerned. In some sandstones, however, reversals of this order are found. How should this be interpreted? Does the association of fresh labradorite with weathered orthoclase mean that markedly different climates existed in different parts of the paleo-drainage basin? For example, sediment in the Mississippi River comes from both the humid central Appalachians and the semiarid areas of Wyoming. Does the association mean that the orthoclase has been recycled several times and has been in several soil profiles but the plagioclase is first-cycle detritus? Or are there other ways to explain the observed reversal of the normal order of weathering? This question has been studied by T. W. Todd (1968). In Eocene sandstones exposed in northern California he found that orthoclase is consistently more weathered than andesine; yet geological evidence indicates a common geological history for the feldspar grains. Using paleobotanical and paleogeographic evidence, Todd was able to establish that the maximum elevation in the area during Eocene time was about 2000 ft, the climate was subtropical, and rainfall was abundant and nonseasonal. Under such conditions, the water table was always high. Potassium leached from orthoclase below the water table was used by the abundant vegetation and hence leaching of orthoclase was not retarded. However, no plant uses much sodium. Therefore, sodium concentrations in the water below the water table remained high, retarding the leaching intensity and alteration of the plagioclase grains. The crux of Todd's thesis is that the chemistry of waters in the vadose or weathering zone may be quite different from that of waters in the zone of saturation and this is particularly important in areas where the water table is always near the ground surface. In the vadose zone the expected order of mineral weathering may occur but below the water table the Na^+/K^+ ratio may remain high and a reversed order of weathering may develop. Todd's hypothesis should be tested by combined field and laboratory studies of modern soils and sediments.

Diagenetic Alteration. Decomposition of feldspars by percolating groundwaters during diagenesis can definitely be demonstrated for a few sandstones by the existence of clay pseudomorphs but the quantitative importance of this process is not known. As seen later in this chaper, the accessory minerals of clastic rocks are frequently etched and perhaps completely decomposed during diagenesis. Therefore, there are grounds to suspect that feldspars may be "weathered" diagenetically and perhaps completely destroyed.

Survival During Transport. Feldspar is somewhat softer than quartz, has several good cleavages, commonly is twinned, and nearly always is partially hydrolyzed. For these reasons its mechanical durability should not be great. However, the numerous modern studies of feldspar abundance as a function of distance of transport in streams uniformly fail to reveal large decreases in percentage of the mineral over transport distances of 1000 mi or more. The reason for this is not certain and several suggestions have arisen to explain the apparent mechanical durability of feldspar. Probably the most accepted explanation lies in the fact that all studies but one so far made of this phenomenon have been made in streams of low gradient, in which sharp grain-to grain impacts are minimal. Lending support to this idea is a recent study of feldspar transport in a stream of high gradient. In this study (Pittman, 1969), conducted in the Merced River of mountainous southern California, it was found that feldspar grains were significantly reduced in size by fracture along twin composition planes in a distance of 20 to 25 mi. Carlsbad twins, which are composed of only two individuals, tend to disappear, while the polysynthetically twinned plagioclase grains remain but suffer size decrease. Pittman's observations are in accord with the documented rarity of Carlsbad twins in sandstones. More studies are needed of sediment in mountain streams to answer questions such as (a) What are the relative effects on resistance to fracture of cleavages, twinning, perthite lamellae, and weakening of feldspar by hydrolysis? (b) What are the quantitative relationships among stream gradient, hydraulics, and feldspar durability? (c) Under what conditions is dilution by quartz-rich detritus more effective in reducing the feldspar content of a stream than grain fracture during transport?

Feldspar may be rapidly destroyed in turbulent environments, such as mountain streams, sand dunes, and beaches. In a comprehensive field and petrographic study of the Fountain Formation and overlying Lyons Formation, Hubert (1960) found marked differences in feldspar content in the Lyons as a function of environment of deposition, which was inferred from sedimentary structures. In fluvial facies of the formation, feldspar contents averaged 28%; in the beach-dune facies, only 8%. Such variations occurred frequently within the Lyons, indicating the speed and effectiveness of beach and dune environments as disintegrators of feldspar grains.

Recycling of Feldspar. Despite folklore to the contrary, feldspar grains are able to survive more than one sedimentary cycle, at least in many continental environments. For example, the Tertiary arkoses of southern Mexico described by Krynine in 1935 are today being eroded, carried downslope into the savanna belt, and deposited as second-cycle arkoses. Even in these highly unfavorable climatic conditions, feldspars survive a second-cycle with little loss in percentage, a fact noted by Krynine.

In another study, conducted in Illinois, many river sands were examined whose drainage basins contain only older sedimentary rocks. In these modern sands, the content of feldspar and rock fragments (which probably are less durable than feldspar grains) averages 20%. Therefore, it is clear that feldspar-rich sands are not necessarily derived directly from crystalline rocks. Additional evidence is needed to support first-cycle origin.

Rock Fragments

Rock fragments are the most important clastic particles in a sediment because they supply definitive information concerning the nature of the source rocks of the sediment. Therefore, fragments of preexisting rocks in sediments should be described in the same detail used in characterizing specimens of schist, rhyolite, or limestone. For example, the description quartz-muscovite-almandine-staurolite schist is far more useful than either "metamorphic rock fragment" or "schist." The enlarged description indicates that the source area was regionally metamorphosed to the amphibolite facies and the premetamorphic rocks probably were argillaceous. Similarly, the description of a limestone fragment in a sandstone may convey important information. For example, the presence of unfossiliferous black, pyritic, microcrystalline limestone pieces suggests a different upstream paleogeology than do pieces of well-sorted and coarsely crystalline crinoidal limestone.

Abundance. The abundance of rock fragments in sediments is extremely variable, both as a function of detrital grain size and among clastic rocks of similar grain size. Conglomerates may be composed entirely of rock fragments; shales usually contain none. The average sandstone is thought to contain 10 to 15% rock fragments among its clastic grains but many sands contain none and some sands are composed of 95 to 100% rock fragments.

The reason for the dependence of rock fragment percentage on grain size is apparent. Many rocks are so coarse that sand-size fragments of them are very unlikely to contain more than a single crystal. For example, sand-size pieces of granite, gneiss, or coarse grained sandstone are unlikely to occur. On the other hand, fragments of rocks such as basalt, slate, or chert may occur as easily in pebble conglomerates as in fine grained sandstones and this fact, coupled with the scarcity of granite and gneiss throughout the sand sizes, may lead to distorted estimates of the proportions of the different rock types in the paleo-drainage basin. It is meaningless to interpret quantitatively upstream paleogeology from rock fragments in downstream sediment unless grain sizes of the various types of fragments are specified.

Igneous Rock Fragments. Although massive plutonic rocks are much more abundant in the earth's crust than flow rocks and tuffs are, fragments of the extrusives are more abundant in sediments (Fig. 8-8). There are two reasons for

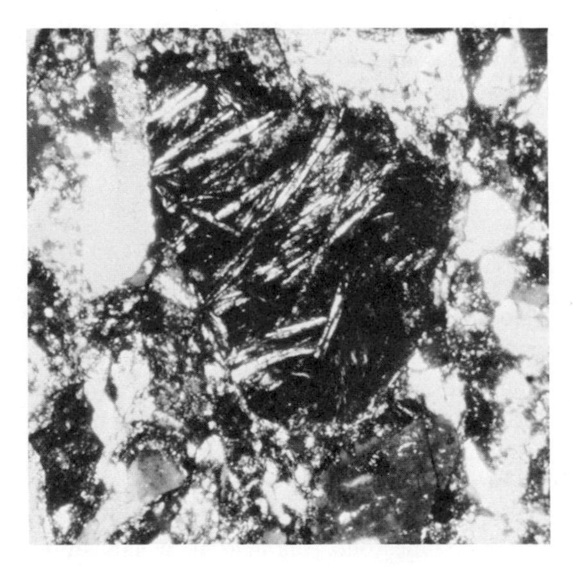

Fig. 8-8 Vitrophyric texture in a very coarse sand size volcanic rock fragment.

this reversal. The first is the control by size that we noted above. The number of sand-size clastic grains in sediments is many times greater than that of gravel-size particles. The second factor is the relative susceptibility to alteration of intercrystalline boundaries in the coarser grained igneous rocks. Weathering quickly weakens the bonds between feldspar grains in granite, causing the disintegration of the rock fragment. It is significant that, despite the abundance of granitoid rocks containing 60 to 70% feldspar, rock fragments composed entirely of feldspar crystals are only rarely reported to occur in conglomerates or coarse sandstones, even those rich in single crystals of feldspar. Granitoid fragments in clastic rocks are most commonly composed of one or two quartz crystals, several feldspar crystals in contact with the quartz crystals, and shreds of ferromagnesian minerals (Fig. 8-9). Because these rock fragments usually contain only a few feldspar crystals, correct identification of the rock type cannot be based on the plagioclase/K-feldspar ratio and must be obtained from the composition of plagioclase crystals in the fragment.

Correct identification of felsitic volcanic fragments is difficult for several reasons. Not only are individual crystals commonly too small for accurate micro-

Fig. 8-9 Very coarse sand size granite rock fragment containing quartz (white), orthoclase (gray), and plagioclase containing albite and pericline twinning.

scopic identification, but much rhyolite is optically indistinguishable from chert. Both often are composed of untwinned and unaltered subequant crystals of gray birefringence. Staining techniques normally useful for feldspar identification may be ineffective on very small crystals and, in such cases, recourse must be made to microchemical analysis.

Metamorphic Rock Fragments. Quartzo-feldspathic metamorphic fragments with granoblastic texture are often indistinguishable from plutonic igneous fragments in thin section. Commonly the only distinguishing feature of these grains is the appearance of their contained quartz crystals. In most cases the texture of quartz crystal aggregates from igneous rocks can be differentiated from textures characteristic of metamorphic rocks, as described on pp. 271-273. Occasionally, quartzo-feldspathic metamorphic fragments will contain a characteristic mineral, such as garnet or sillimanite, either as an inclusion in a quartz crystal or as a separate crystal.

The transition from shale to slate involves no visible recrystallization of phyllosilicates and slaty cleavage usually is not visible in sand-size fragments of slate. Because of this, fragments of slate and even low grade phyllite are often indistinguishable from pieces of hard shale, both in hand specimen and in thin section. This is a particularly bothersome problem, as fine grained foliated rock fragments are very common in sandstones. Also, a number of classifications of sandstones attempt to categorize fragments based on sedimentary, metamorphic, or igneous origins.

Sedimentary Rock Fragments. Fragments of older sandstone are not common in most gravels or sands because of the structural weakness of the common cements, calcite, dolomite, hematite and clay. These binding agents are generally dissolved or weakened on the outcrop to the extent that the fragments disintegrate very rapidly to separate grains. Nearly all sandstone fragments in sediments are silica-cemented and represent a very biased sample of the sandstones available in the paleo-drainage basin.

Limestone fragments are uncommon in clastic rocks (excluding clastic limestones) because of their softness. They tend either to be present in great abundance (very near their source) or else to be completely absent.

Chert grains are very common in clastic rocks and are derived either from nodules in carbonate rocks or from beds of "deep water" chert deposited in geosynclinal settings. Unless the chert contains carbonate inclusions or "ghosts" of oolites or shallow water fossils, there is no way to distinguish between these two possible origins without resort to trace element analyses. Bedded cherts usually contain several percent iron; nodular cherts generally contain less than 1%.

It is difficult to estimate the abundance of mudstone and shale fragments in sandstones. Certainly shale is far more abundant than other types of sediments in the average drainage basin but shale fragments are so mechanically unstable that they seem to be almost untransportable. Possibly most of the apparent shale and mudstone fragments in water-laid sandstones are actually fragments of flood-plain or sea bottom mud moved only a few feet from the location where they formed.

Heavy Minerals

The accessory or "heavy" minerals are extremely valuable constituents of sediments but are difficult to study. Because they only rarely are present in amounts greater than 1% of the clastic rock, they must be concentrated in the laboratory for study and even then it is uncommon to recover conveniently sufficient amounts of grains for a thin section to be feasible. Therefore, heavy mineral grains are nearly always mounted whole in an epoxy resin for examination. This technique allows reasonably rapid point counting but also increases the difficulty of accurate identification. Birefringence of anisotropic minerals varies with grain size; mineral cleavage is usually less easily recognized in whole grains than in thin section; and nonremoveable coatings on grain surfaces or extensive internal alteration may obscure characteristic properties of the mineral. Because of these difficulties a great deal of experience is necessary to achieve facility, speed, and accuracy in heavy mineral counts. Even "experts" make errors in identification, although they never are identifiable in published work. Published papers are generally well written and the impressiveness of the printed page is so great that students usually are unaware of the uncertainties inherent in the process of data collection.

All sizes of heavy minerals should be examined, including silt. Detrital heavies can be identified petrographically in sizes as small as 10 μ. In many published studies only a single arbitrarily chosen size fraction has been examined and, as minerals with higher specific gravities may be concentrated in finer size grades, a single size fraction may not be representative of the entire heavy mineral crop (Rittenhouse, 1943). The few studies made of both sand and silt size heavies in a single sediment suggest that the modal grain size of heavy minerals frequently lies in the silt range.

The best procedure for separating heavies from a loose aggregate is to use a suitable heavy liquid and a centrifuge. A centrifuge is needed because the finer heavies tend to be suspended by the surface tension of the heavy liquid and settle very slowly because of its viscosity if not made to settle by the force of centrifugation. Heavy liquids in common use (and their specific gravities) are bromoform (2.88), tetrabromoethane (2.95), methylene iodide (3.30), and Clerici solution (4.25).

Heavy Mineral Provinces. Nearly all the different mineral species found in clastic rocks exist in the heavy mineral suite and the number of species and their relative amounts may vary on several scales of observation. For example, the Gulf of Mexico Basin is receiving sediment derived in part from the cratonic interior of the United States, in part derived from Mesozoic mountain chains of Mexico, and in part derived from carbonate reefs near the Yucatan Peninsula (Fig. 8-10). Mineralogic variations on such a broad scale must be differentiated from those resulting from grain size variation or random fluctuation in mineral composition from one side of a stream channel to the other. To this end, a hierarchal model has been developed as a device for categorizing the causes of heavy mineral variation in sediments.

A *sedimentary petrologic province* is composed of a group of sediments that constitute a natural unity by age, origin, and distribution. This concept is useful in recognizing trends in the regional distribution and stratigraphic sequence of sediments. Petrologic provinces are gradational in space and time. To characterize these provinces, only the abundant minerals need to be specified and, by convention, this means minerals present in amounts of at least 5%.

A *mineral association* is an association of minerals by which a sediment is characterized. The relation between province and association is analogous to that between genus and species in biology and there may be several mineral associations within a petrologic province (Fig. 8-11). The Gulf of Mexico petrologic province contains many distinct associations because of the large variety of geologically distinct drainage basins that supply sediment to the gulf. Continental deposits sometimes contain a greater number of different mineral associations than do marine sediments because of the lesser degree of sediment homogenization that characterizes fluvial systems, i.e., different tributaries may have distinct mineral assemblages.

Granular variations are local concentrations of mineral species that result from differences in grain density, size, or shape (Fig. 8-12). Such variations may be superimposed on mineral associations but should not be considered as separate associations. To do so weakens the usefulness of the concept of mineral association.

Chance variations are those variations among heavy mineral suites resulting from random fluctuations. For example, eddies in streams may sweep a slightly different mineral assemblage to one location than to another spot a few feet dis-

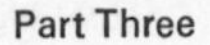

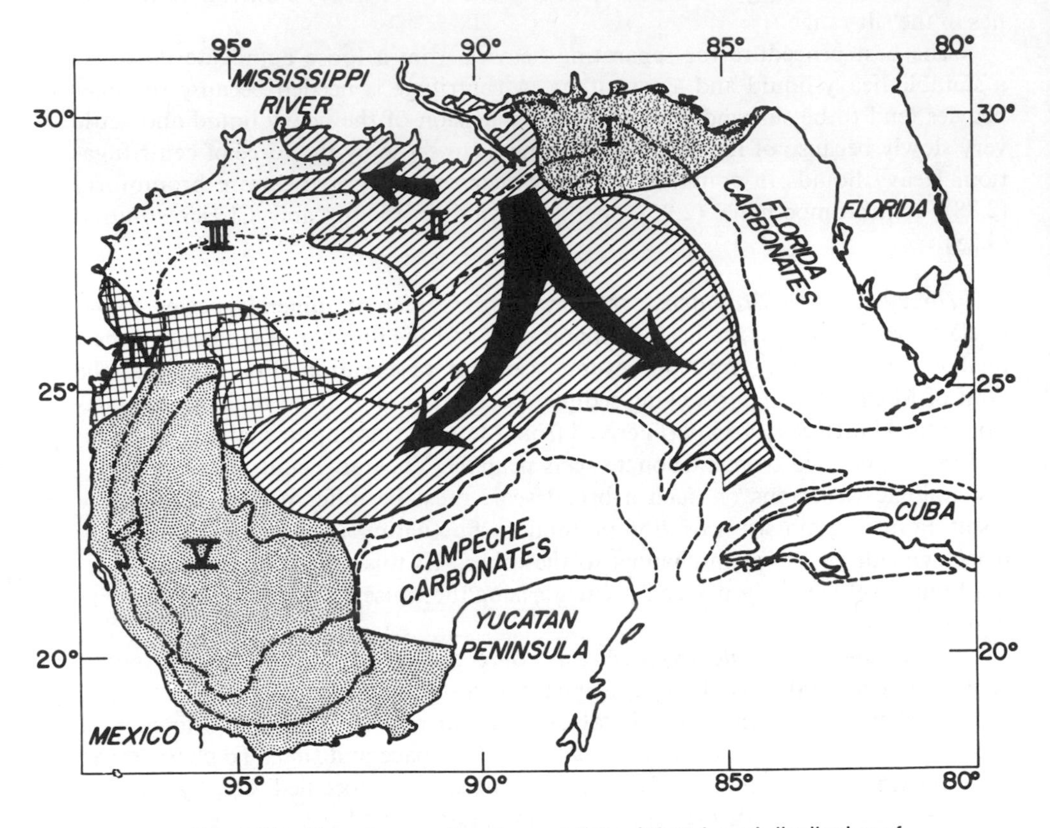

Fig. 8-10 Major heavy mineral associations, inferred areal distribution of Mississippi sediment, and principal sediment dispersal directions in the Gulf of Mexico. (I) Eastern Gulf Province; (II) Mississippi Province; (III) Central Texas Province; (IV) Rio Grande Province; (V) Mexican Province. (From Davies and Moore, 1970.)

tant. In some cases such variations may be quite large and can only be correctly interpreted by intensive sampling in a small area. Chance variations may also occur because of counting insufficient numbers of grains in petrographic analyses. For example, a count of 100 heavy mineral grains on one slide of a sand may encounter two kyanite grains; in a different split from the same sample, a count of 100 grains may encounter no kyanite fragments. In one informative study, two splits of a sample were analyzed for nonopaque heavy minerals with results as shown in Fig. 8-13. It is apparent from the figure that it is meaningless to report heavy mineral percentages to tenths of a percent, even for the abundant minerals, irrespective of the number of grains counted. In most studies a maximum of 300 nonopaque heavies are tabulated but, regardless of the number tabulated, it is important to scan the slide afterward to find "traces" of minerals not encountered

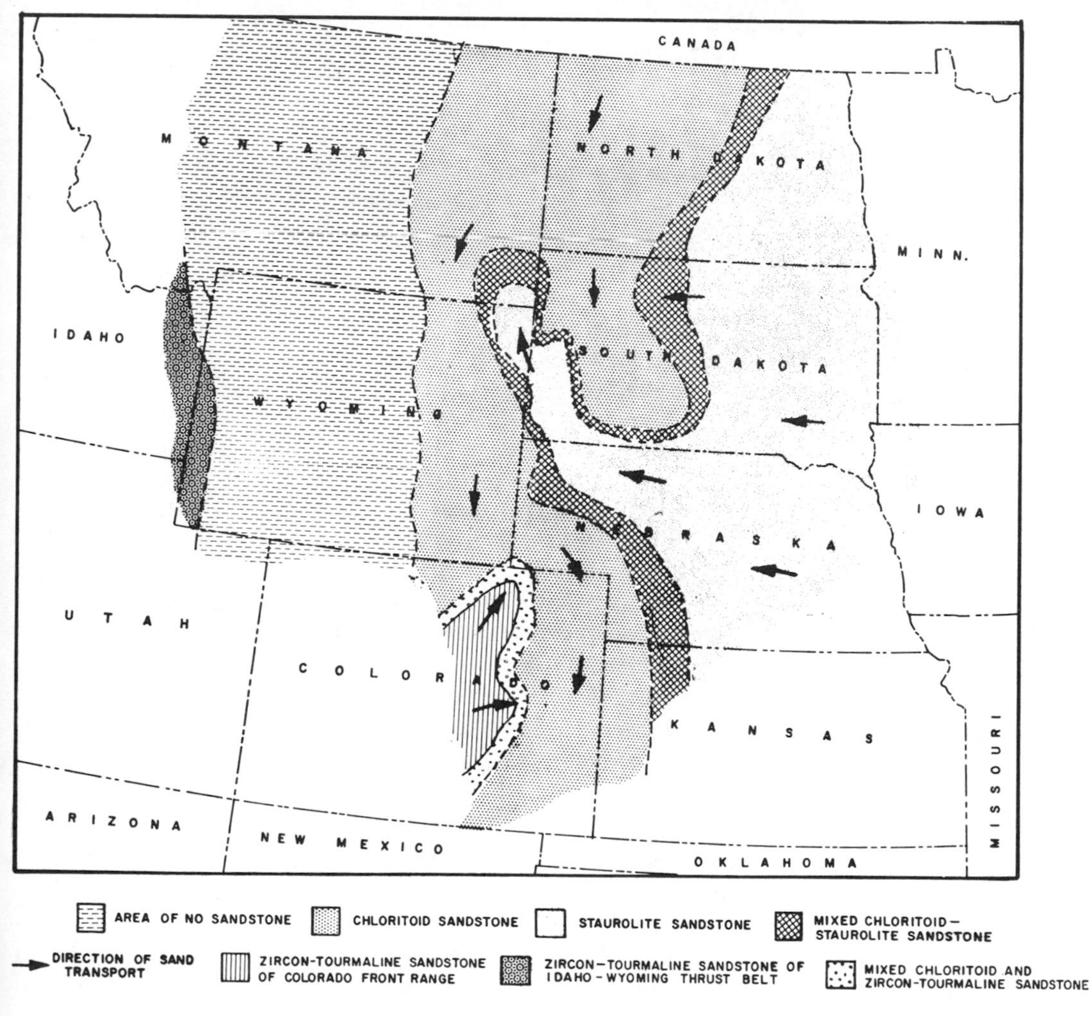

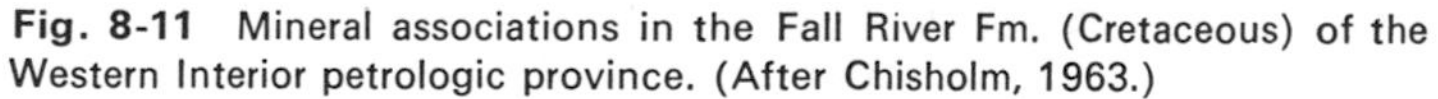

Fig. 8-11 Mineral associations in the Fall River Fm. (Cretaceous) of the Western Interior petrologic province. (After Chisholm, 1963.)

in the count. A single grain of an exotic mineral may be very informative from the standpoint of sedimentary petrogenesis.

Opaque Minerals. Opaque heavy minerals are generally considered a nuisance by sedimentary petrologists and are almost always discarded essentially unanalyzed, despite the fact that they commonly form the bulk of the heavy crop and are useful for provenance determinations. In part, the neglect of opaques results from the special and time-consuming preparatory techniques required for the examination of loose opaque grains. In part the neglect results from the

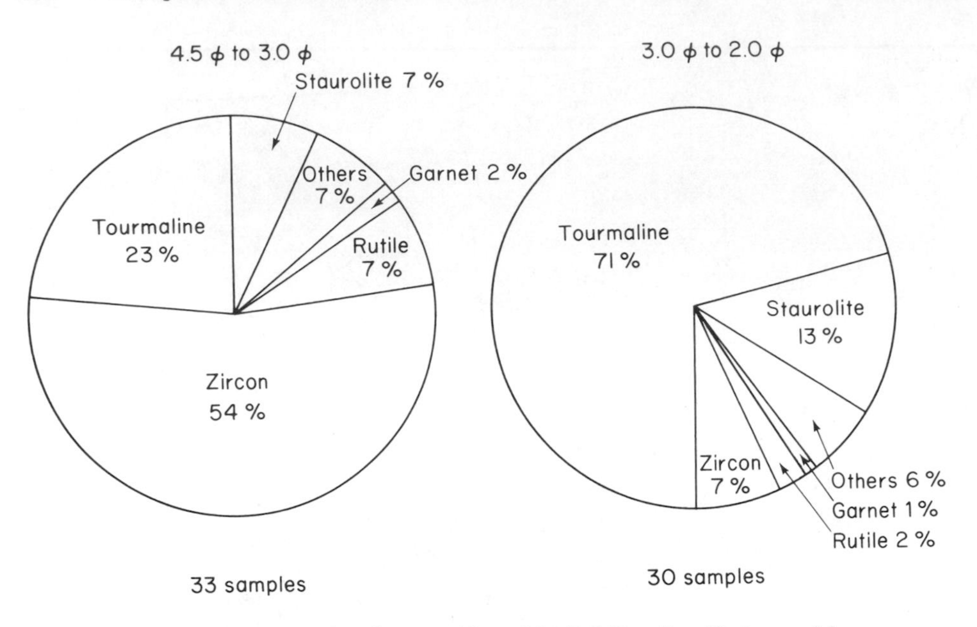

Fig. 8-12 Heavy mineral composition of the Fall River Fm. (Cretaceous) in the eastern Black Hills, South Dakota, illustrating granular variation. (After Chisholm, 1963.)

fact that few sedimentary petrologists have had training in ore microscopy. The study of opaque heavy minerals in sandstones represents an unexplored frontier in clastic petrology.

In most studies, a hand magnet is used to remove "magnetite" and the remaining opaques are called "ilmenite," despite the fact that this approach is known to be erroneous. An uncertain, but probably large, proportion of grains called magnetite or ilmenite on grounds of magnetic susceptibility are actually maghemite (magnetic Fe_2O_3) or microintergrowths of several phases, including rutile, titaniferous magnetite, ferrian ilmenite, etc. The potential usefulness of such grains for provenance determinations is illustrated by the work of Balsley and Buddington (1958). In an area of igneous and metamorphic rocks they identified five mineral phases in the iron-titanium-oxygen system and six types of intergrowths. Considering the occurrence of these phases and intergrowths in relation to rock type, they were able to conclude that magmatic rocks in this area contain mixtures of titaniferous magnetite and ferrian ilmenite. Metamorphosed sediments, migmatites, and gneisses of granitization origin generally contain mixtures of magnetite and titaniferous hematite. This study suggests that the potential of opaque minerals in studies of sedimentary petrogenesis is large.

Magnetite and ilmenite frequently coexist in igneous and metamorphic rocks, particularly in intermediate and mafic rocks. To the extent that such coexistence occurs, the percentage of titanium in magnetite may be used as a criterion of the igneous or metamorphic parentage of the grain and also as an indicator of the metamorphic temperature at which the grain was formed (Fig. 8-14). It is important

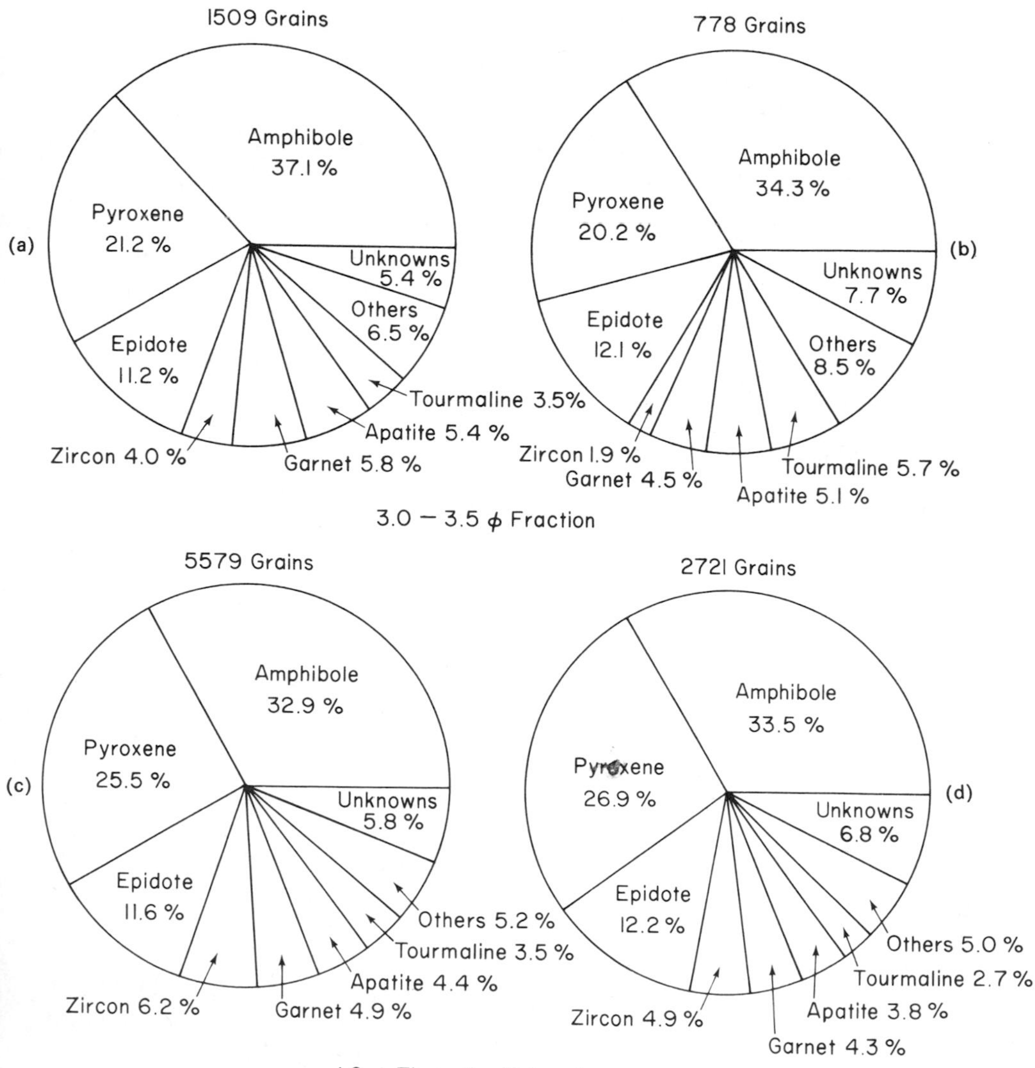

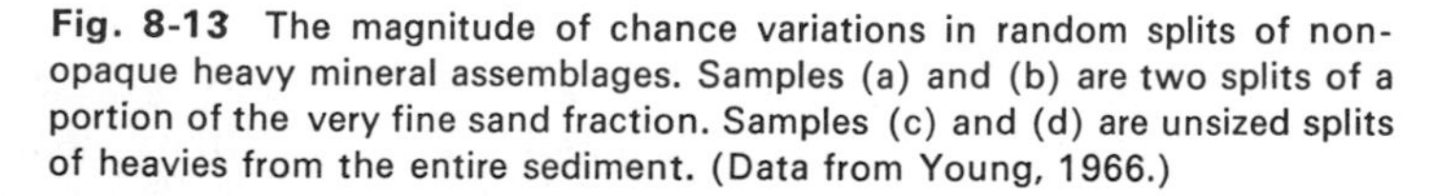
Fig. 8-13 The magnitude of chance variations in random splits of non-opaque heavy mineral assemblages. Samples (a) and (b) are two splits of a portion of the very fine sand fraction. Samples (c) and (d) are unsized splits of heavies from the entire sediment. (Data from Young, 1966.)

that the magnetite have coexisted with ilmenite in the parent rock, for otherwise the amount of titanium in the grain is, in part, also a function of oxygen fugacity (effective partial pressure) during crystallization or recrystallization. Assuming coexistence of the two minerals, the percent of titanium in magnetite of metamorphic rocks is usually 1 to 2% and is nearly always less than 4%. In magmatic rocks the range is generally from 4 to 30% but some grains may contain less than 4%

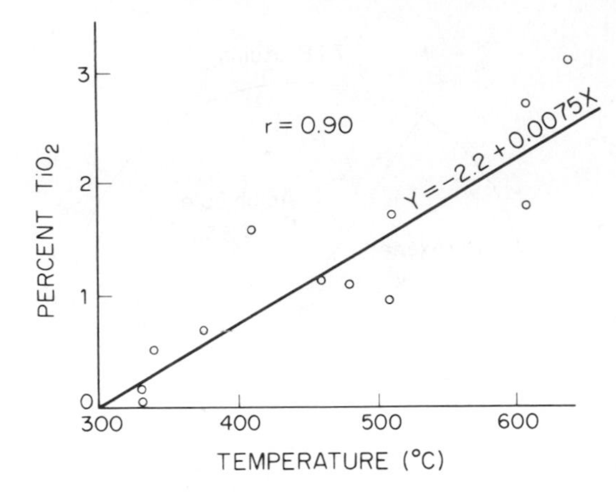

Fig. 8-14 The titania content of magnetites in metamorphic rocks from the Adirondack Mtns. (New York) and from Great Britain. In all samples magnetite coexisted with ilmenite. (Data from Abdullah, 1965.)

titanium. Magnetites in mafic rocks usually contain 15 to 30%; magnetites in silicic rocks, less than 15%. But there is much overlap in the 10 to 20% range.

Practically nothing has been done relating the occurrence of the common opaque mineral species to particular varieties or compositions of crystalline rocks. A few minerals are known to be restricted to magmatic rocks but there are no quantitative data concerning igneous versus metamorphic frequency of occurrence for most minerals. Chromite is restricted to peridotites, dunites, and other ultramafic rocks, either plutonic or hypabyssal. Columbite occurs in silicic massive plutonic rocks and in associated pegmatites. Neither of these two minerals occurs to a significant extent in metamorphic rocks.

Nonopaque Minerals. The number of different nonopaque heavy minerals that form at least 5% of the suite in a sandstone normally is less than ten. An additional five or ten are often present in lesser amounts. Although any of the thousands of known minerals of high specific gravity can be present in clastic rocks, a list of the two dozen most common ones would include nearly all heavy mineral grains reported in published works.

Among the nonopaque heavies very few are restricted to igneous rocks. Nearly all heavies reported from intrusives are also reported frequently from foliated metamorphic rocks. Exceptions are topaz and cassiterite, which characterize pegmatites and the later stages of crystallization of silicic plutons, and oxyhornblende and augite, which occur in mafic igneous rocks.

Metamorphic rocks contain many diagnostic minerals such as kyanite, staurolite, and sillimanite, which do not occur in magmatic rocks. Many distinctively metamorphic minerals are characteristic of particular types of source terranes and this information should form part of any paleogeologic interpretation. For example, wollastonite characterizes contact-metamorphosed impure limestones; sillimanite, high grade regional metamorphism of argillaceous sediments.

Unfortunately, however, most of the accessory minerals of crystalline rocks can originate in either magmatic or metamorphic conditions, for example, zircon, tourmaline, epidote, and hornblende. Probably some minerals are significantly more common in, say, magmatic rocks but relative abundances are quantitatively unknown for any mineral found in both igneous and metamorphic rocks. Because of this and because of the limited number of different heavy mineral species present in the nonopaque heavy crop of most sediments, it becomes necessary to extract provenance information from intraspecific variations such as color, isomorphic substitutions, and trace element content. Accurate interpretation of such data in terms of parent rock type requires familiarity with the geochemistry of crystalline rocks and, for this reason, sandstone petrologists should read "hard-rock" journals as well as those dealing exclusively with sedimentary rocks.

Zircon. Zircon is generally agreed to be the most ubiquitous and abundant nonopaque heavy mineral in sediments, a circumstance resulting from its occurrence in most crystalline rocks, the fact that it is harder than quartz, and its lack of good cleavage. Zircon is slightly more abundant in silicic rocks than in mafic ones but the difference in not large enough to have diagnostic value in detrital rocks.

The only feature of zircon known to be useful for determination of ultimate parent rock is shape. Study of magmatic rocks reveals that the majority of zircons are euhedral. Zircons in paraschists and paragneisses tend to be round, retaining the roundness of sedimentary zircons well up into the higher grades of metamorphism. Only in the metamorphic facies immediately below the melting temperatures of rocks do the zircons recrystallize and lose their detrital shapes. Apparently zircon is extremely resistant to recrystallization.

Although an adequate method has not yet been found for distinguishing between zircons of igneous and metamorphic origins, there are several possible ways to distinguish among zircon populations from different clastic rocks. One possibility is grain color. The color of zircons is variable, ranging from colorless through pink, yellow, brown, green, blue, purple, or color zoned, depending on the types and amounts of impurity elements. The colors are best seen in whole grains, most zircons in thin section appearing colorless. In addition to stable elements, many zircons contain uranium and thorium as impurity elements and, over long periods of time, alpha-particle radiation damage to the crystal structure may isotropize the structure, producing the distinctive zircon variety called malacon or metamict zircon.

In sediments the vast majority of zircons is colorless and free of inclusions, a circumstance probably due to selective destruction in the sedimentary environment of radiation-damaged and inclusion-weakened structures. This occurrence is entirely analogous to the selective destruction of strained quartz, polycrystalline quartz, and inclusion-rich quartz, in relation to nonundulatory quartz with few inclusions. The instability of brown or black (i.e., radiation-damaged) zircon was demonstrated by Poldervaart (1955) in a study of the Mahalapye granite and immediately overlying arkose in the Bechuanaland Protectorate (now Botswana). The granite contains about 75% of brown to black zircons in its zircon population; yet the arkose contains about 25%, indicating rapid disintegration or solution of such grains. In general, sandstones and siltstones contain few malacons but shales are said to contain higher porportions of such grains. If true (no quantitative data exist), the explanation would be the rapid size reduction of malacons.

Commonly, colorless zircon grains can be made to fluoresce in several colors by exposure to ultraviolet light. A commercial "mineralight" may be used. This technique may be useful for source area determinations or for distinguishing among zircon populations from different clastic units.

Tourmaline. Tourmaline probably is second only to zircon in abundance among nonopaque heavies and many conflicting statements have been made concerning the correlations among source rock types and the color of tourmaline grains. None of the published statements is supported by quantitative data and apparently no one has attempted to obtain such data. Many names are used for tourmaline varieties depending on chemical composition and color. Iron-rich crystals are called schorlite (ferrous iron); manganous-rich crystals are tsilaisite; magnesium-rich, dravite; calcium-rich, uvite; lithium-rich, elbaite. However, there is overlap with names based on color. For example, those crystals green because of ferrous iron are called verdelite; pink due to manganese are rubellite. The cause of blue color in indicolite is uncertain. Zoned tourmaline is known and commonly results from variations in the iron content within schorlite crystals. Some iron-poor varieties of tourmaline are colorless ("achroite").

The only correlations between color and source rock on which all writers on the subject agree are that dravites are always brown to yellow-brown and occur almost exclusively in metamorphic rocks; also, that other colors are most common in granites and pegmatites.

Only one study has been published concerning the relationship between tourmaline provenance and trace element composition in a group of areally extensive crystalline rocks. It was found that the frequency distribution of percentages of each trace element studied can be analyzed in terms of a lognormal distribution (Fig. 8-15) and that hydrothermal tourmalines are significantly lower in iron and manganese and higher in magnesium, calcium, strontium, and tin than the general population of granites examined. Chromium, nickel, scandium, and vanadium percentages in hydrothermal tourmalines were higher than in the granites but the separation between plutonic and hydrothermal origin was not so clear-cut for these elements.

Rutile. Third in abundance among the nonopaque heavy minerals may be rutile but to date nothing of use to sedimentary petrology has emerged from the few hydrothermal studies made of the mineral. However, the amounts of niobium, tantalum, or iron, all of which are common in rutile, may be useful as geothermometers or be diagnostic of provenance. As we saw earlier, the entire iron-titanium-oxygen system seems very promising as a source of provenance information in clastic petrology.

Garnet. Garnet probably ranks among the top half dozen nonopaque heavies in terms of abundance and almost certainly ranks high in potential usefulness in sedimentary petrologic investigations. The reason for this is not only the fact that

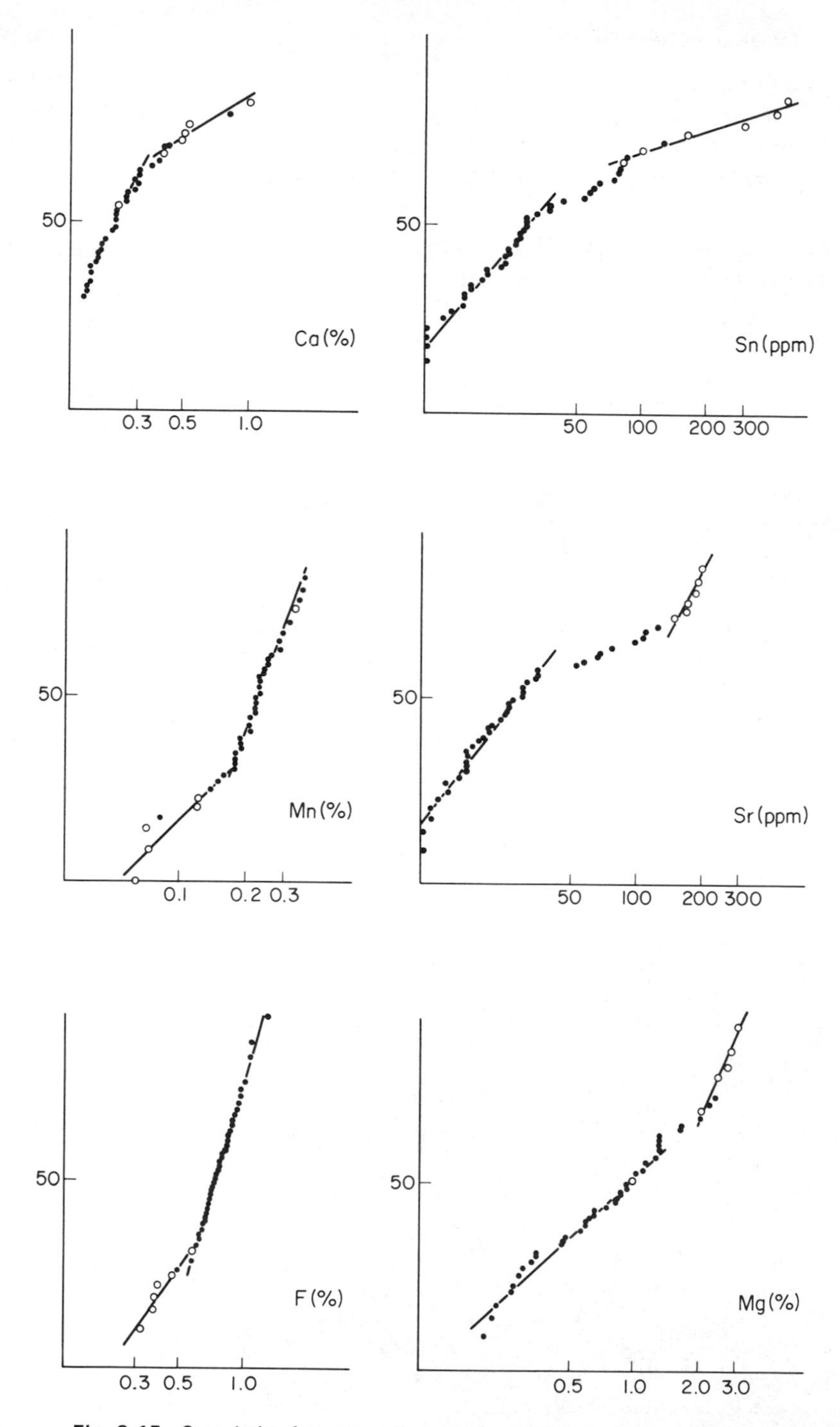

Fig. 8-15 Cumulative frequency diagrams for calcium, manganese, fluorine, tin, strontium, and magnesium in tourmalines from granites in south-west England (○ is hydrothermal tourmaline). (From Power, 1968.)

garnet occurs in a great variety of colors but also because the six chemical end-members of the garnet series vary widely in major element composition. And the major element composition of the mineral is often closely related to a specific igneous or metamorphic rock type (Table 8-5 and Fig. 8-16).

TABLE 8-5 AVERAGE PROPORTIONS OF THE FIVE MAJOR GARNET MOLECULES IN DIFFERENT ROCK TYPES (Wright, 1938)

Rock type	Almandine	Andradite	Grossular	Pyrope	Spessartine
Pegmatites	41.8	—	—	—	47.1
Granites	56.8	—	—	—	36.0
Contact-altered siliceous rocks	56.4	—	—	—	30.7
Biotite schists	73.0	—	6.0	13.8	—
Amphibole schists	53.6	—	20.7	20.3	—
Eclogites	39.1	—	18.5	37.4	—
Kimberlites and peridotites	13.4	—	9.0	72.3	—
Various basic rocks	34.4	15.6	28.7	20.7	—
Calcareous contact rocks	—	40.8	51.5	—	—

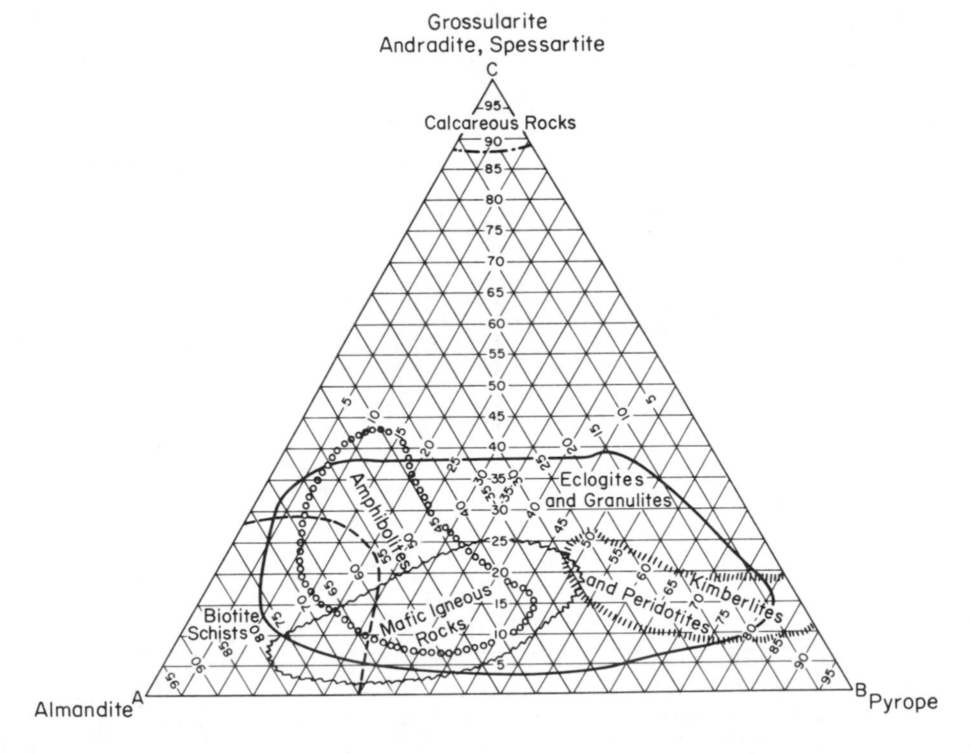

Fig. 8-16 Petrologic affinities of garnets. (From Barth, 1962.)

In addition, it has been found that it frequently is possible to determine from the ratios of the major cations in a garnet grain the metamorphic facies in which the grain originated. Specifically, investigations of many pelitic schists reveal that the ratio FeO + MgO/CaO + MnO in garnets increases markedly with metamorphic facies (Fig. 8-17). The explanation for this relationship is the decreasing molar volume of garnets (and all minerals) at higher pressures. Iron and magnesium are considerably smaller in size than calcium and manganese are and therefore are accommodated easier in the garnet structure at higher pressures. As most garnets in crystalline rocks probably are in pelitic schists, the data of Fig. 8-17 should be quite useful in provenance studies of clastic rocks.

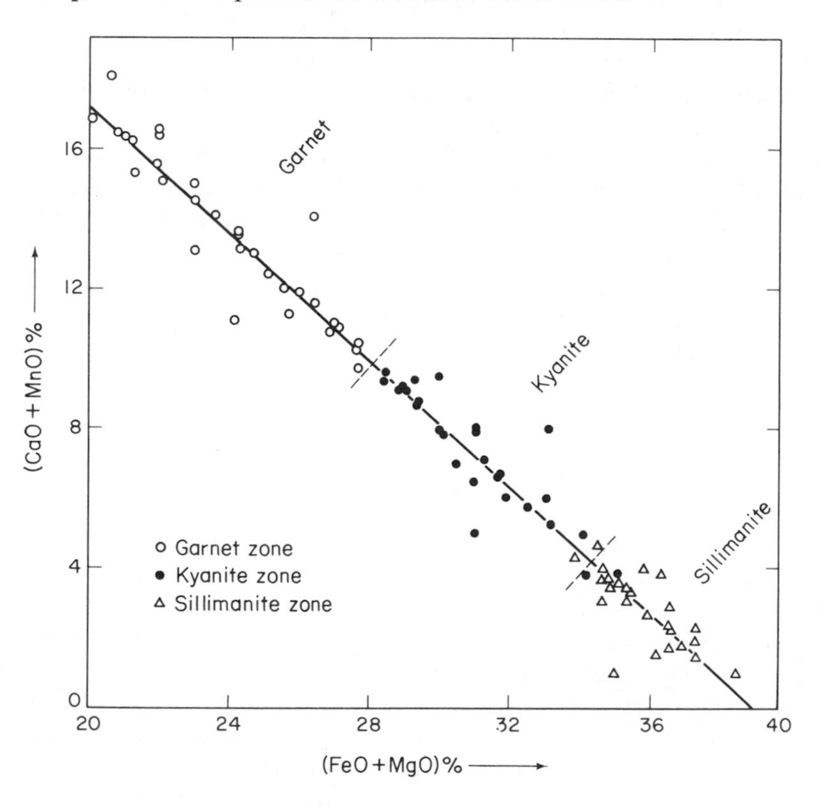

Fig. 8-17 Relationship between FeO + MgO content and CaO + MnO content in garnets as a function of metamorphic grade in pelitic schists. (From Nandi, 1967.)

Intrastratal Solution

Among the dozen or so nonopaque heavy minerals that occur fairly commonly in clastic rocks, there has been noted a correlation between frequency of occurrence and geologic age. The earliest documentation of this relationship was by Thoulet in France in 1913; in 1941, F. J. Pettijohn surveyed the published literature and confirmed Thoulet's observation. Pettijohn found that very unstable minerals (i.e., those with weak interatomic bonds) such as olivine, augite, and hypersthene seem to be restricted in occurrence to Cenozoic clastics. Also, he noted that the younger sediments generally contain a greater number of different mineral species than older rocks. The heavy mineral suites of many Paleozoic sediments consist almost entirely of only the most stable species such as zircon, tourmaline, rutile, and garnet. Pettijohn suggested that the older beds are impoverished in unstable heavies because of intrastratal (diagenetic) solution by underground waters and, as would be anticipated, the older rocks show the effects of this process to a higher degree than younger rocks.

This interpretation was challenged almost immediately by P. D. Krynine, who attributed the mineral zonation with age to tectonism and provenance. Krynine's interpretation was that, with progressive uplift, deeper levels of the crust become exposed and, as rocks formed under higher P-T conditions contain more "exotic" minerals, younger sediments contain higher proportions of unstable heavies. In some local geographic areas this explanation may be valid but as an explanation of the worldwide mineral zonation found by Pettijohn, it is inadequate. There is no evidence that exposure of deeper crustal levels correlates with passing time over continental-size areas. Orogenic uplifts are irregularly distributed in both time and space and therefore could not lead to similar mineral distributions paralleling time planes in unrelated stratigraphic sections. Also, the zonation is frequently present in Tertiary sections whose sources (mainly sedimentary) have not changed during that period (Fig. 8-18).

A third explanation for Pettijohn's data was suggested by van Andel (1959), who feels that Pettijohn's survey of the published literature represents a necessarily biased sample. According to van Andel, prior to 1941 most studies of Paleozoic sandstones were made in North America where sediments of this age are largely cratonic and polycyclic in origin. Just as their light fractions are largely quartz, so also only the most stable heavy minerals remain. Further, van Andel considers that most early studies of Tertiary sandstones were European studies of Alpine clastics, in areas where cratonic sands are absent and sediments are derived in large part directly from crystalline basement rocks. If this is true, we are unable to conclude the importance of intrastratal solution from the results of Pettijohn's survey. Although van Andel grants that destruction of heavies during diagenesis may be important locally, he denies its general occurrence.

Neither Krynine nor van Andel presented quantitative data to support their views; however, as both are experienced petrographers who have examined sediments from many parts of the world, their opinions should not be dismissed without consideration.

One method of resolving this problem may be found by examination of shales, which are relatively impervious. If intrastratal solution of heavies is significant, then there should be noticeable differences in the heavy minerals of contemporaneous sandstone-shale pairs. Those in sands should be more altered and the sands should contain a smaller number of mineral species than the shales. In a recent study (Blatt and Sutherland, 1969), nonopaque heavies were examined from contemporaneous sand-shale pairs from Tertiary sediments on the Texas Gulf Coastal

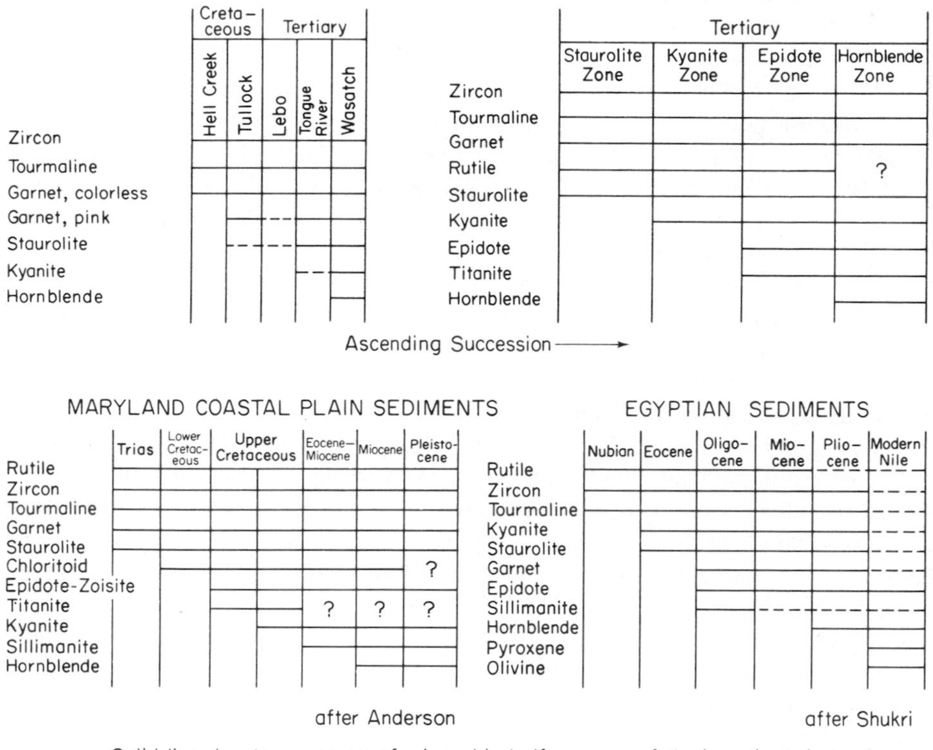

Fig. 8-18 Heavy mineral zonations in four different areas illustrating very similar rankings of order of occurrence or persistence of nonopaque heavy minerals. (From Pettijohn, 1957.)

Plain and the importance of intrastratal solution was documented. The sandstones contain much higher percentages of altered grains and fewer different mineral species than the shales. Presumably these differences between sands and shales increase in older beds but so far there has been no study of Paleozoic or Precambrian sand-shale pairs designed to investigate this question. In general, it probably is better to use the heavies in shales in stratigraphic analyses rather than those in sandstones.

Mineralogic Maturity

According to F. J. Pettijohn, "The mineralogic maturity of a clastic sediment is the extent to which it approaches the ultimate end product to which it is driven by the formative processes that operate upon it." As such, the stage of maturity of any clastic sediment is the joint product of deuteric alteration in the parent rock, weathering processes in the soil, abrasion processes during transport to and in the depocenter, and diagenetic alteration after burial. Only rarely is it possible to evaluate quantitatively the relative importance of each of these processes for a sandstone sample.

Petrographic Criteria

From the point of view of detrital mineralogy, the "ultimate end product" of destructive sedimentary processes cannot be specified without knowledge of the existing climatic conditions. Typical end product assemblages include (a) the complete absence of detrital minerals, as in laterites in humid tropical climates; (b) light fraction composed of nonundulatory monocrystalline quartz grains plus gibbsite or kaolinite with only zircon, tourmaline, and rutile in the nonopaque heavy fraction (pedalfer soils in humid temperate or subtropical climates); (c) quartzose sands containing variable amounts and types of clays, fresh and altered feldspars, rock fragments, and a large variety of heavy minerals (in less humid temperate climates); and (d) sands with low percentages of quartz, rich in fresh feldspars and rock fragments, containing many highly unstable heavy minerals (arid and arctic climates). Unfortunately, nearly all natural sandstones probably were derived from a variety of sources and represent a diverse assortment of climatic conditions. As most detrital grains are polycyclic in origin, it is tacitly assumed that all grains have spent significant lengths of time in all the abundant climatic zones. Thus, the archetypal detrital "ultimate end product" is taken to be the quartz-kaolinite assemblage in the light mineral fraction and the zircon-tourmaline-rutile assemblage in the heavy mineral fraction.

The degree of approach of sandstones to such a mineral assemblage may be measured in many ways but irrespective of the technique used, it is mandatory that similar grain sizes be compared. For example, it is meaningless to conclude that a coarse sand in an arkosic formation is mineralogically less mature than a very fine sand in the formation on the basis that the coarser sand contains a higher proportion of granite fragments. The detrital mineral composition of the light fraction and commonly of the heavy fraction of sandstones is strongly size-dependent. To overlook this fact may result in erroneous conclusions concerning provenance, climate, and the significance of recycling. Figure 8-19 illustrates the probable relationship between grain size and detrital fragment composition, based on the limited data currently available. The diagram is an attempt to answer the following question: If all the detrital silicate rocks in the crust were piled on a table, disaggregated, sieved, and examined for mineral composition, what would be found in each size fraction?

One method of evaluating the approach of a mineral assemblage to maturity is to use ratios of stable to less stable detrital constituents. For example, the ratios of quartz to feldspar, of quartz plus chert to feldspar plus rock fragments, or of monocrystalline to polycrystalline quartz are all valid estimators of mineralogic maturity. It is important to keep in mind, however, that the use of such ratios assumes that all these phases or rock fragments have been present among the ultimate sources of the sediment examined. If the crystalline source area contained little or no feldspar, the high quartz/feldspar ratio tells nothing concerning min-

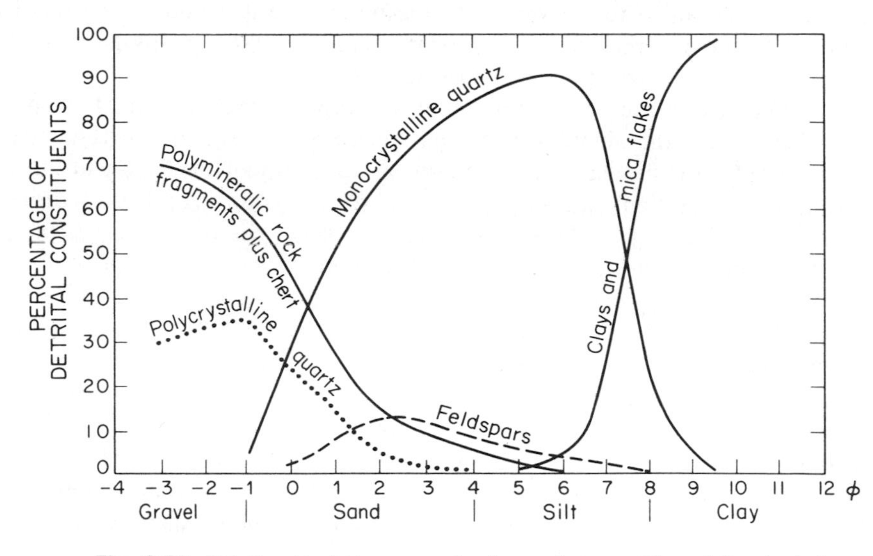

Fig. 8-19 Relationship between grain size and composition of the detrital fraction in clastic silicate rocks.

eralogic maturity. Fortunately, nearly all the abundant crystalline rocks contain feldspar. (Slate, phyllite, and metaquartzite are exceptions). Among the heavy minerals a commonly used estimator of maturity is the "ZTR index," the percentage of the nonopaque suite that is composed of zircon, tourmaline, and rutile. These three minerals are common in almost all types of crystalline rocks.

So far, our discussion of the stability of detrital grains has been general, with no distinction between mechanical and chemical stability. Such a distinction is important, however. An abundance of mechanically unstable grains in a sand permits inferences concerning distance of transport. An abundance of chemically unstable fragments may lead to inferences concerning climate or diagenesis, which are unrelated to distance of transport. Particularly needed are studies of relative rates of disaggregation of rock fragments in mountain streams whose drainage areas have been mapped geologically. Because of the neglect of this early phase of transport, absolute or relative rates of disaggregation of the different types of polymineralic rock fragments are uncertain. Suppose we examine a coarse sandstone whose light fraction is composed of 25% volcanic rock fragments, 25% granite fragments, 25% schist grains, and 25% shale or mudstone grains. How should this be interpreted? Does the equal abundance of the four fragment types imply equal exposures in the drainage basin? Or perhaps the outcrop area of volcanic rocks is small and its abundance in the sediment reflects relatively great mechanical durability. Could the mudstone grains be pieces of overbank mud accumulated by the depositing stream rather than the record of an ancient deposit? For that matter, can shale be transported at all, or are such fragments so easily disaggregated that they are essentially untransportable? At present there are few data with which to answer such questions. The question of the relative durabilities of sand-size rock fragments exemplifies one of the many gaps in our understanding of the effects of transport on sedimentary mineralogy.

The abundance in a clastic sediment of any type of rock fragment depends on three factors: (a) the abundance of the source rock upstream in the paleo-drainage basin; (b) the mechanical and chemical durability of the fragment during weathering, transportation, and diagenesis; and (c) the grain size of the sediment being examined. The first of these factors is, of course, the object of petrologic studies of sediments. The third factor is easily determined. Therefore, if factor (b) can be evaluated, paleogeological interpretations can be greatly improved. Unfortunately, few field or laboratory studies have attempted to determine the relative durabilities of the various types of common rock fragments found in sediments. In addition, distinction is rarely made between chemical durability (i.e., resistance to weathering) and mechanical durability (strength of intercrystalline bonds, mineral hardnesses, brittleness, etc.). Some rock fragments, such as chert, are very hard but brittle; others, such as basalt fragments, are mechanically resistant but very unstable chemically. Still others, such as shale fragments, are stable chemically but unstable mechanically. It is clear that any combination of mechanical-chemical resistance to disintegration or decomposition may exist in natural sedimentary particles. Thus, some types of fragments may be abundant in a source area but almost lacking in derived sediment because of weathering factors alone, almost irrespective of the distance of transport downstream. Others, such as quartzites, are essentially unalterable and their abundance in a sediment is unaffected by most climatic variations. These factors have not received the attention they deserve. There is great need for many field and laboratory investigations of the interrelationships among these factors.

The interrelated effects of weathering and transport on feldspars are poorly known. In some sandstones strongly kaolinized feldspar grains are pinched and "necked" by adjoining quartz grains and nearly split in two. The orientations of the kaolinite crystals reveal that the necking has occurred by flowage rather than by brittle fracture. Certainly the effective strength of many clastic feldspar fragments must be less than that of fresh feldspar and this may be a significant factor in the destruction of the former during transport. Other unstudied factors that contribute toward destruction of feldspars include structurally and chemically unstable surfaces such as twin composition planes, phase boundaries between perthite lamellae, and cleavage planes. Perhaps the presence of Carlsbad or albite twinning in a feldspar grain is a more important structural weakness in a detrital grain than are the two good cleavages of the mineral. Perhaps a fresh, twinned feldspar grain is less durable in transport than a sericitized untwinned grain. The answers to problems such as these would greatly increase the reliability of interpretations based on the petrology of sandstones.

In a petrographic analysis of an ancient sandstone we observe only the end product of the interactions among source rock abundance, chemical and mechanical durability, distance and type of transport, and grain size. Our success in interpreting upstream paleogeography and paleogeology from an end product assemblage is as much an art as a science. Conclusions concerning the origin and history of specific grains or of groups of grains can never be proved rigorously and the quality of an interpretation can only be judged on the bases of its consistency with pertinent experimental and field studies and its success in predicting the results of succeeding geological investigations. So far in our discussion of mineralogic maturity we have concentrated on the continuing need for laboratory and field studies of modern sediments to establish a sound base for interpretations of the mineralogy of sandstones. Let us now turn to techniques that can be used to display information from ancient rocks.

The most commonly used technique for this purpose is graphic, particularly triangles showing the approach of a grain assemblage to specific end members. One useful group of triangles is illustrated by Fig. 8-20. Existing data indicate

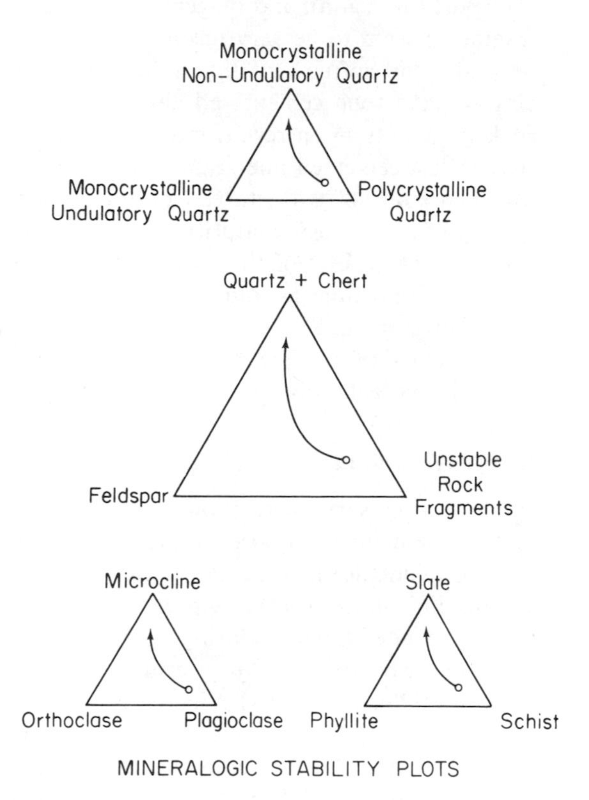

Fig. 8-20 Relative effects of sedimentary processes on sand-sized clastic grains. The circle within each triangle represents an arbitrary initial composition and the track shows the probable change in composition resulting from sedimentary processes. (From Blatt, 1967b.)

that rock fragments are destroyed more rapidly by sedimentary processes than feldspar grains and, therefore, a group of grains with the Q + C/F/URF ratio indicated in the figure would change in composition along the track indicated by the curved line in the central triangle. Each end-member of this triangle can be usefully subdivided into second-order triangles whose poles are also defined in terms of relative stabilities of the minerals or rock fragments concerned. The reasons for the relative stabilities of the three quartz types in the upper triangle were discussed on pp. 274-276 and the reason for the stability of K-feldspar in relation to plagioclases on pp. 233-235. Among the potassium feldspars, microcline is more resistant to decomposition than orthoclase because its crystal structure is more highly ordered and consequently more stable at low temperatures.

The triangle for rock fragments is speculative, as there are no data on which to establish relative stabilities. We might suspect, however, that schist would be less stable than either phyllite or slate because of its relatively coarse grain size and heterogeneous mineral composition. The physical appearance of slate suggests it may be more durable mechanically than phyllite. There is no reasonable basis on which to speculate concerning relative stabilities of other types of polymineralic igneous and metamorphic rock fragments, such as felsite, granite, or amphibolite, because the stability of each depends in a complex way on climatic (weathering), transport (abrasion), and diagenetic factors. Nearly all types of sedimentary rock fragments seem to be very unstable in the sedimentary environment. The fissility of shale; the softness of limestone; and the cleavage of calcite, dolomite, and clay in sandstone cements all cause relatively rapid destruction of sedimentary rock fragments. As a result, most older sedimentary fragments in sandstones are either silica-cemented fine-grained quartz sandstones or chert fragments.

The kinds of two-, three-, or four-component graphs that can be useful in a study of mineralogic maturity is limited only by the ingenuity of the investigator and the peculiarities of the rocks being studied. Pictorial representations of end-member proportions are difficult to devise when the number of components exceeds four, however, and it becomes necessary to express relative amounts entirely in mathematical terms. The relationships among end-members are precisely expressed by mathematical tools such as multiple correlation coefficients and factor analyses.

Chemical Criteria

The necessary correlation between the mineral composition of sandstones and their chemical composition suggests chemical analyses as an alternative method to determine the mineralogic maturity of these sediments. There are, however, serious difficulties with this approach, the most pervasive of which is the authigenic growth of new minerals during diagenesis. High percentages of calcium, magnesium, and iron resulting from cementation by diagenetic carbonate and hematite are not related to detrital mineralogy as are high percentages of these elements resulting from an abundance of detrital andesine, augite, and magnetite. Bulk chemical analyses of clastic rocks must be accompanied by petrographic analyses to be most meaningful, particularly in the case of "well-washed" or "clean" sandstones, in which authigenic cements are common.

The variable grain size of clastic particles presents a second difficulty to the interpretation of chemical analyses in terms of mineralogic maturity. As illustrated by Fig. 8-19, in a series if interbedded conglomerates and sandstones, the sandstones would appear to be more mineralogically mature simply as a function of grain size. The same problem is present with chemical analyses because a mixture of sand-size quartz and weathered orthoclase may be nearly identical chemically to an aggregate of quartz and illite, again illustrating the necessity for chemical analyses to be accompanied by petrographic descriptions. The presence of clay in a sandstone is particularly sensitive to minor variations in local environment of deposition, much more so than is the presence of medium as contrasted to fine grained sand. Therefore, clay minerals are an interface between mineralogic and textural maturity. Also, the maturity of the sand fraction cannot be assumed identical to that of the clay fraction as they may be derived from different sources. For example, the gravel and sand fractions of Oligocene clastics in parts of the Northern Great Plains are rich in granite rock fragments and clastic feldspars derived from the Black Hills. Yet interbedded muds are bentonitic, rich in glass shards and montmorillonite derived from volcanic sources located elsewhere.

A related problem with chemical analyses in relation to mineralogic maturity is the presence of rock fragments. Felsite fragments, granite fragments, and an aggregate of detrital quartz and feldspar grains may be chemically identical; yet they certainly represent different stages of mineralogic maturity. In addition, the presence of one of these three groups of grains, as opposed to the others, permits inferences concerning mechanical or chemical maturity that cannot be obtained from chemical analyses alone.

As a result of these problems, many of the published chemical analyses of sandstones are difficult or impossible to interpret in terms of mineralogic maturity. The best use of these analyses is in large-scale tectonic syntheses (e.g., eugeosynclinal versus miogeosynclinal orogens) and in research conducted for economic purposes. The relationship between tectonics and sediment chemistry is explored more fully in Chaps. 9 and 20 concerned with the classification of sandstones and with sedimentary associations, respectively.

REFERENCES

ABDULLAH, M. I., 1965, "The Iron-Titanium Oxide Phases in Metamorphism," in *Controls of Metamorphism*, W. S. Pitcher and G. W. Flinn, ed., Oliver & Boyd, Ltd., Edinburgh: pp. 274–280. (Very brief summary of phase and minor element relationships in the iron-titanium-oxygen system.)

BALSLEY, J. R. and A. F. BUDDINGTON, 1958, "Iron-Titanium Oxide Minerals, Rocks, and Aero-Magnetic Anomalies of the Adirondack Area, New York," *Econ. Geol.*, **53**, pp. 777–805. (Field and laboratory study relating the magnetic oxide assemblage in Adirondack crystalline rocks to lithology.)

BARTH, T. F. W., 1962, *Theoretical Petrology*, 2nd ed., N. Y., John Wiley and Sons, Inc. 416 pp. (Standard text).

———, 1969, *Feldspars*. N. Y.: John Wiley & Sons, Inc., 261 pp. (Summary of available data concerning the most abundant of all rock-forming minerals.)

BLATT, HARVEY and J. M. CHRISTIE, 1963, "Undulatory Extinction in Quartz of Igneous and Metamorphic Rocks and Its Significance in Provenance Studies of Sedimentary Rocks," *Jour. Sed. Petrology*, **33**, pp. 559–579. (An extensive study of source rocks and sandstones, demonstrating selective destruction of polycrystalline and undulatory quartz grains by sedimentary processes.)

BLATT, HARVEY, 1967a, "Original Characteristics of Clastic Quartz Grains," *Jour. Sed. Petrology*, **37**, pp. 401–424. (Examination of grus from granites, schists, and gneisses reveals correlations among internal structure of quartz grains, quartz grain size, and source rock type.)

———, 1967b, "Provenance Determinations and Recycling of Sediments," *Jour. Sed. Petrology*, **37**, pp. 1031–1044. (Philosophical paper stressing the importance of reworking of mineral grains in the sedimentary environment.)

BLATT, HARVEY, and BERRY SUTHERLAND, 1969, "Intrastratal Solution and Nonopaque Heavy Minerals in Shales," *Jour. Sed. Petrology*, **39**, pp. 591–600. (The most recent documentation of destruction of less stable heavy minerals during diagenesis.)

CAMERON, K. L. and H. BLATT, 1971, "Durabilities of Sand Size Schist and 'Volcanic' Rock Fragments During Fluvial Transport, Elk Creek, Black Hills, South Dakota," *Jour. Sed. Petrology* **41**, pp 565–576. (Silicic volcanic rock fragments are very durable during stream transport. Micaceous schist fragments disintegrate rapidly.)

CHISHOLM, W. A., 1963, "The Petrology of Upper Jurassic and Lower Cretaceous Strata of the Western Interior," *in Wyo. Geol. Assn.—Billings Geol. Soc. Guidebook to Northern Powder River Basin*, pp. 71–86. (An excellent paper illustrating regional mapping of heavy minerals within a sedimentary petrologic province.)

DAVIES, D. K. and W. R. MOORE, 1970, "Dispersal of Mississippi Sediment in the Gulf of Mexico," *Jour. Sed. Petrology*, **40**, pp. 339–353. (Dispersal pattern of nonopaque heavy minerals in Mississippi River sand can be used to define mineral associations and distribution patterns in the Gulf of Mexico sedimentary petrologic province.)

DENNEN, W. H., 1967, "Trace Elements in Quartz as Indicators of Provenance," *Geol. Soc. Amer. Bull.*, **78**, pp. 125–130. (A brief original study demonstrating the potential usefulness of trace elements in quartz for provenance determinations.)

ERMANOVICS, I. F., 1967, "Statistical Application of Plagioclase Extinction in Provenance Studies," *Jour. Sed. Petrology*, **37**, pp. 683–687. (Modification of the Michel-Levy method for use in determining the composition of plagioclase grains in a conglomerate.)

FENIAK, M. W., 1944, "Grain Sizes and Shapes of Various Minerals in Igneous Rocks," *Amer. Miner.*, **29**, pp. 415–421. (Measurement of these features. The only compilation of this type for crystalline rocks.)

GASTIL, R. G., MARK DELISLE, and JR. MORGAN, 1967, "Some Effects of Progressive Metamorphism on Zircons, " *Geol. Soc. Amer. Bull.*,**78**, pp. 879–906. (Changes in the character of zircon in sediments with progressive metamorphism in the aureole of a batholith.)

HAYASE, ICHIKAZU, 1961, "Gamma Irradiation Effect on Quartz. (I) A Mineralogical and Geological Application," *Kyoto Univ., Inst. Chem. Res.*, **39**, pp. 133–137. (Pioneering study of an interesting technique for provenance determinations of detrital quartz grains.)

HILL, T. P., M. A. WERNER, and M. J. HORTON, 1967, "Chemical Composition of Sedimentary Rocks in Colorado, Kansas, Montana, Nebraska, North Dakota, South Dakota, and Wyoming," *U. S. Geol. Sur. Paper* **561**, 241 pp. (Compilation of chemical analyses of sedimentary rocks from central United States.)

HUBERT, J. F., 1960, "Petrology of the Fountain and Lyons Formations, Front Range, Colorado," *Colo. School Mines Quart.*, **55**, no. 1. 242 pp. (A very detailed field and petrographic study of a classic arkose.)

KATZ, M. YA., M. M. KATZ, and A. A. RASSKAZOV, 1970, "Mineral Studies in the Gravitation—Gradient Field, 2. Changes of Quartz Sand Density Due to Natural and Experimental 'Maturation'," *Sedimentology*, **15**, pp. 161–177. (Detrital quartz grains containing inclusions are less durable than inclusion—free quartz.)

KLEIN, G. DEV., 1966, "Dispersal and Petrology of Sandstones of Stanley-Jackfork Boundary, Ouachita Fold Belt, Arkansas and Oklahoma," *Am. Assoc. Petroleum. Geol. Bull.*, **50**, pp. 308–326. (A fine study of the interrelationship between petrologic variations and dispersal patterns in a flysch trough as obtained from sedimentary structures.)

MACKENZIE, F. T. and J. D. RYAN, 1962, "Cloverly-Lakota and Fall River Paleo-Currents in the Wyoming Rockies," in *Wyoming Geol. Assoc. Guidebook*, 17th Ann. Field Conf., pp. 44–61. (Relationship between directional structures and petrographic variation in a shelf area.)

MIZUTANI, SHINJIRO, 1959, "Clastic Plagioclase in Permian Graywacke from the Mugi Area, Gifu Prefecture, Central Japan," *Jour. Earth Sci.*, Nagoya Univ., **7**, pp. 108–136. (A detailed study of the variation in composition and type of twinning in plagioclase grains for purposes of source area determination.)

NANDI, KAMAL, 1967, "Garnets as Indicators of Progressive Regional Metamorphism," *Miner. Mag.*, **36**, pp. 89–93. (Summary of data from eighty-four samples of pelitic schists with regard to garnet compositions and grade of metamorphism.)

PETERSON, M. N. A. and E. D. GOLDBERG, 1962, "Feldspar Distributions in South Pacific Pelagic Sediments," *Jour. Geophys. Res.*, **67**, pp. 3477–3492. (Most feldspar grains within the Pacific basin are relatively calcic and are derived from within the basin.)

PETTIJOHN, F. J., 1957, *Sedimentary Rocks*, 2nd ed.,N. Y.: Harper & Bros., 718 pp. (The standard textbook in sedimentology.)

———, 1963, "Data of Geochemistry, Chap. S, Chemical Composition of Sandstones—Excluding Carbonate and Volcanic Sands," *U.S. Geol. Sur. Prof. Paper 440-S*, 21 pp. (The most recent summary of the range in chemical composition of sandstones.)

PITTMAN, E. D., 1969, "Destruction of Plagioclase Twins by Stream Transport," *Jour. Sed. Petrology*, **39**, pp. 1432–1437. (A study of stream sediment demonstrating the importance of twin composition planes in feldspars as loci of structural weakness during transport.)

POLDERVAART, ARIE, 1955, "Zircons in Rocks 1. Sedimentary Rocks," *Amer, Jour. Sci.*, **253**, pp. 433–461. (Comprehensive survey of the subject.)

POTTER, P. E. and W. A. PRYOR, 1961, "Dispersal Centers of Paleozoic and Later Clastics of the Upper Mississippi Valley and Adjacent Areas," *Geol. Soc. Amer. Bull.*, **72**, pp. 1195–1250. (Probably the outstanding example of the integration of sedimentology, petrology, and statistics, in the analysis of a large area.)

POWER, G. M., 1968, "Chemical Variation in Tourmalines from South-West England," *Miner. Mag.*, **36**, pp. 1078–1089. (A pioneering study of trace element composition in tourmalines as a function of rock type in granitic plutons and their hydrothermal offshoots.)

RITTENHOUSE, GORDON, 1943, "The Transportation and Deposition of Heavy Minerals," *Geol. Soc. Amer. Bull.*, **54**, pp. 1725–1780. (The classic study demonstrating the importance and effects of granular variations in heavy mineral assemblages.)

Savin, S. M. and S. Epstein, 1970, "The Oxygen Isotopic Compositions of Coarse Grained Sedimentary Rocks and Minerals," *Geochim. Cosmochim. Acta*, **34**, pp. 323–329. (The oxygen isotopic composition of quartz and feldspar in sediments can be used to infer the temperature at which they were formed.)

Schnitzer, W. A., 1957, "Die Quarzkornfarbe als Hilfsmittel für die stratigraphische und Paläogeographische Enforschung sandige Sedimente," *Erlanger Geol. Abh.*, **23**, 13 pp. (Colors of quartz grains in Bunter and Keuper sandstones of Bavaria can be used for provenance determinations.)

Simonen, Ahti and Olavi Kouvo, 1955, "Sandstones in Finland," *Bull. de la Comm. Geologique de Finlande*, no. 168, pp. 57–88. (A survey of the origin, distribution, and petrology of the few sandstones overlying the crystalline basement in Finland.)

Slemmons, D. B., 1962, "Determination of Volcanic and Plutonic Plagioclases Using A Three- or Four-Axis Universal Stage," *Geol. Soc. Amer. Spec. Paper* **69**, 64 pp. (Standard explanation and reference for universal stage work.)

Sorby, H. C., 1877, "The Application of the Microscope to Geology," *Monthly Microscopical Jour.*, **17**, pp. 113–136. (A report of the first detailed studies of sand and clay in grain mounts and thin section using the petrographic microscope.)

———, 1880, "On the Structure and Origin of Non-Calcareous Stratified Rocks," *Geol. Soc. London Proc.*, **36**, pp. 46–92. (A most interesting early synthesis of the origin, diagenesis, and metamorphism of sand and mud.)

Stanley, D. J., 1965, "Heavy Minerals and Provenance of Sands in Flysch of Central and Southern French Alps," *Am. Assoc. Petroleum Geol. Bull.*, **49**, pp. 22–40. (Modern provenance study of some Alpine sediments stressing the use of heavy minerals.)

Todd, T. W., 1968, "Paleoclimatology and the Relative Stability of Feldspar Minerals under Atmospheric Conditions," *Jour. Sed. Petrology*, **38**, pp. 832–844. (An excellent study of the effects of climate and the position of the groundwater table on the relative rates of weathering of orthoclase and plagioclase feldspars.)

van Andel. T. H., 1959, "Reflections on the Interpretation of Heavy Mineral Analyses," *Jour. Sed. Petrology*, **29**, pp. 153–163. (The argument against the world-wide importance of intrastratal solution of heavy minerals by its chief proponent.)

van der Plas, L., 1966, *The Identification of Detrital Feldspars*. N. Y.: Elsevier Pub. Co., 305 pp. (Survey of the chemistry, petrology, and identification techniques useful for studying feldspars in sedimentary rocks.)

Wedepohl, K. H., 1969, "Composition and Abundance of Common Igneous Rocks," in *Handbook of Geochemistry*, K. H. Wedepohl, ed. **1**, Berlin: Springer-Verlag, pp. 227–249. (Compilation of basic data.)

Wright, W. I., 1938, "The Composition and Occurrence of Garnets," *Amer. Miner.*, **23**, pp. 436–449. (Chemical analyses of garnets in relation to rock type.)

Young, E. J., 1966, "A Critique of Methods for Comparing Heavy Mineral Suites," *Jour. Sed. Petrology*, **36**, pp. 57–65. (Interesting study of comparative reproducibilities of number percentages and mineral ratios.)

CHAPTER NINE

CLASSIFICATION OF SANDSTONES

9.1 INTRODUCTION

The classification of sandstones has been a subject of controversy for many years and an examination of recent issues of professional journals reveals that the different viewpoints have still not been reconciled. In this chapter we discuss some of the reasons for the lack of agreement, illustrate some of the more widely used modern schemes, and review some data on chemical composition of sandstones that have not generally been discussed but that may nevertheless be relevant to the problem.

The general principles of classification of sedimentary rocks were reviewed in Chap. 1. The characteristics of the classification system must not only lead to mutually exclusive categories, they must be precisely defined and objectively determinable. For example, it is permissible to use the amount of quartz in a rock as a basis for classification but it is not permissible to use the source of the quartz, which is speculative. A scheme based on presence or absence of well-defined sedimentary structures is permissible but one based on environment of deposition is not. We may *hope* that our classification will lead to separation of sandstones deposited in different environments but the environment cannot itself be used as a basis of a descriptive classification because it can only be inferred rather than observed. The distinction between what can be observed and what cannot is not always so clear-cut. For example, most petrographers believe they can observe the amount of "matrix" in a sandstone. They generally assume that the "matrix" is original mud—a highly debatable point—but even if a purely descriptive, nongenetic criterion is used to identify "matrix," it is often found that experienced petrographers cannot agree among themselves, for example, whether there is 5, 10, or even 15% in a given sandstone. If this is the case, it seems that the definition of matrix must be made more objective (for example, there must be a clear set of rules for distinguishing between "matrix" and "fine grained rock fragments") *or* a new technique must be used to determine matrix *or* the use of the concept "matrix" must be abandoned as a criterion for use in a practical, working classification. It is an unfortunate fact that many of the supposedly objective criteria used in sandstone classification suffer from deficiencies of this type.

Classifications, therefore, should be based on objectively determinable characteristics. But which of the almost infinitely large number of sandstone characteristics should be chosen? It seems best to select those features that we think give the most insight into rock genesis. This is fair enough but leads directly to further problems. Sandstones are not the result of a single cause but of a delicate interplay of many causative factors. Which aspect of the genesis should we choose to emphasize? Source rock? Physical environment of deposition? Chemical environment? It is no wonder, therefore, that there are many classifications. Perhaps we should not hope for a time when there will be general agreement since this might imply the acceptance of some set of dogmas about the nature and origin of sandstones. For such a diverse group of rocks, surely no single set of dogmas could ever be true.

9.2 CLASSIFICATIONS

Most attempts to devise a comprehensive and systematic classification of sandstones have been based on the mineral composition of the sand-size grains in the "light" mineral fraction. In nearly all cases, procedure has been to select three mineral species or groups of rock fragments as poles of an equilateral triangle, which has then been internally subdivided on an arbitrary basis. Each area blocked out in this manner was then named in a manner soothing to the psyche of the author, resulting in a plethora of terms whose meanings are uncertain without direct reference to the author's triangle. For example, there exist subfeldspathic lithic arenites, ultraarkoses, protoquartzites, and lithic sublabile arenites, among others. Despite occasional voices crying in the wilderness for a halt to this type of creative thinking, there is no sign at present that we are approaching the end of the outpouring. Klein (1963) and Okada (1971) have reviewed the 50 sandstone classifications published since the classic triangle devised by Krynine in 1948. Figure 9-1 displays four of the more widely used triangles.

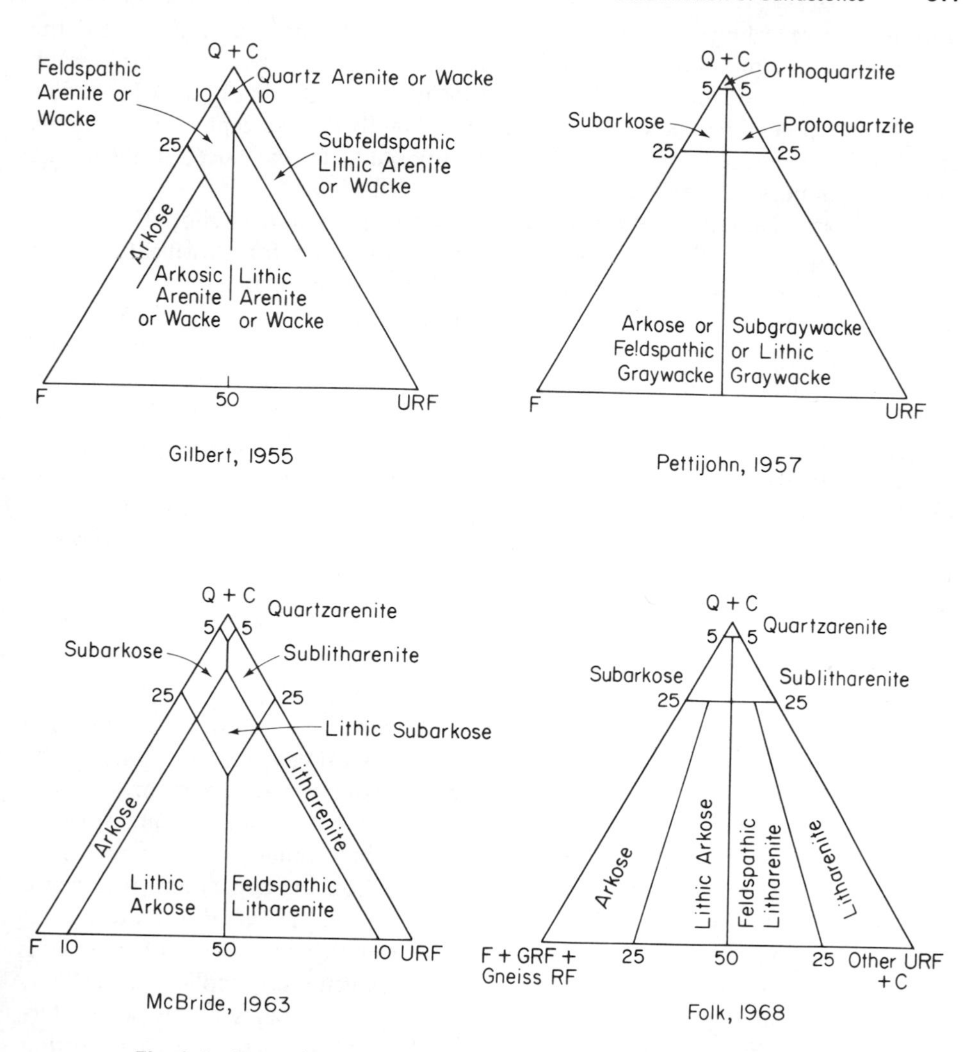

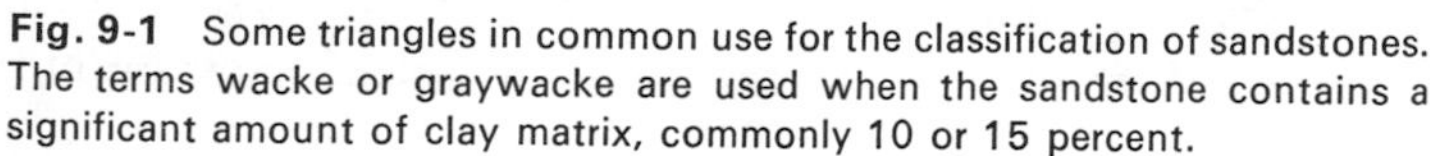
Fig. 9-1 Some triangles in common use for the classification of sandstones. The terms wacke or graywacke are used when the sandstone contains a significant amount of clay matrix, commonly 10 or 15 percent.

One bright spot may be emerging from the chaos, however. There seems to be a growing tendency on the part of classifiers to use quartz plus chert at one pole, feldspars at a second, and "unstable" or "labile" rock fragments at the third pole. These three poles are desirable from a practical standpoint because distinction among these grain types is made with little difficulty in field studies of most sandstones. Subsequent thin section studies will thus only increase the precision of the rock description rather than cause a major change in the clan name. From a theoretical or conceptual standpoint, choice of these three end-members has the advantage that it places emphasis on the very important genetic factor of sedimentary recycling. As we noted on p. 303 about sandstone mineralogy, the order of fragment stability is quartz plus chert > feldspars > unstable rock fragments. Classifications that attempt to allocate different types of rock fragments to different poles are, therefore, not in accord with the concept of mineralogic stability in addition to being difficult or impossible to use successfully in the field or in hand specimen work.

There is, of course, no obvious reason why petrographers should favor the use of triangles and therefore only three main variables for classification. The popularity of the triangle is probably attributable partly to its value for graphic display and partly to motives that only a psychoanalyst could plumb. Any such limitation in the number of main variables must inevitably result in the general choice of "poles" being inconvenient for some investigations. For example, it makes little sense to a petrographer dealing with sandstones composed essentially of quartz and chert (and little else) to tell him that he must group the two together. For the purpose of his study, he may wish to separate them and even to draw up some new triangle for graphic display. We are considering, however, only the general problem of classification of *all* sandstones and, for the general problem, it is necessary to try to reduce the number of variables.

9.3 IMPORTANCE OF DETRITAL CLAY

The presence of even a small amount of detrital clay or mud in a sandstone is of great significance for interpretation of depositional environment. This results from the fact that clay matrix is easily and rapidly winnowed from a clastic aggregate by a small amount of turbulence. Small but significant quantities of clay can be trapped with sand during deposition in a muddy environment such as a river. Larger amounts may possibly be trapped during deposition from dense turbidity currents and still larger amounts may be present in mud flows or glacial tills. Preservation of large amounts of detrital clay implies an absence of reworking of the sediment in a turbulent environment; i.e., it generally implies quiet water, such as found in floodplains, lagoons, or in the deeper parts of the sea. Alternatively, many muddy sandstones may be formed postdepositionally during diagenesis or by mixing of mud into an originally clean sand by burrowing or root growth.

It is generally accepted that the quantity of detrital clay in a sandstone is an important observation to make. It is, unfortunately, not an easy one to make in many compacted sandstones that contain micaceous rock fragments or in sandstones that have suffered a significant degree of diagenetic alteration. The controversy among classifiers concerning matrix centers around the question of whether clay content should be used as a qualifier (e.g., immature arkose), and therefore excluded from the basic triangle, or whether it should be used to determine the basic name of the sandstone (e.g., graywacke), and therefore included at one pole of the triangle.

Many workers follow R. L. Folk's system of textural maturity and refer to sandstones containing detrital clay as "immature." This system has the advantage, or disadvantage (depending on one's point of view), that the term "immature" carries the additional information that the sand grains probably are poorly sorted and angular (see Chap. 3). Other workers describe all sandstones with clay as "wackes" or "graywackes." It seems to us that the disadvantages of basing the core of the rock name on the amount of clay are great, in view of all the different factors that may be responsible for the presence of "matrix," and particularly the difficulty of achieving an accurate estimate of its abundance. In addition, the significance of a given fraction of clay is related to the mean grain size of the sandstone and is not the same for fine as for coarse sands.

In summary, it seems to us that an acceptable compromise, if not necessarily the best of all possible choices, is achieved by adopting as the end-members of a mineralogically oriented sandstone classification (a) quartz plus chert, (b) feldspars, and (c) all rock fragments other than chert and polycrystalline quartz grains, i.e., all unstable or "labile" rock fragments. With minor reservations, we consider that the concept of textural maturity is a useful one and suggest that "immature" be used instead of the terms "wacke" or "graywacke" to describe a sandstone containing clay matrix. These preferences are identical to the usage suggested, for reasons different from ours, by E. F. McBride (1963, see Fig. 9-1).

9.4 TERMINOLOGY

The terms graywacke, arkose, orthoquartzite, and lithic are in common use among sandstone petrologists but, unfortunately, not all users agree on the definitions of these terms. As in most controversies, each author has some justification for the usage he prefers based on either assumed or actual historical precedent. Thus, in order to understand the various points of view, we need to review briefly the history of each of the three terms.

Graywacke

The oldest of the four terms is graywacke, which dates from German usage in 1789 to describe Upper Paleozoic rocks in the Harz Mountains. Thirty years later, Jameson introduced the word to the English-speaking world, following a visit to Werner at Freiburg (Dott, 1964, p. 626); differences of opinion concerning the correct definition of graywacke multiplied so rapidly that by 1854 Murchison suggested that it be abandoned. "When will my valued friends, the mineralogists and geologists of Germany, abandon a word which has led to such endless confusion?" (p. 359). Murchison's plea was not heeded and new definitions of the term have continued to multiply in the succeeding century to the point where almost any sand-size sediment can be correctly called a graywacke by someone's definition.

A few contrasting or intertwining themes run through the dozens of definitions now in published works.

1. The original definition was established nearly 40 years before the invention of the polarizing microscope by Nicol. Therefore, it is clear that the term was defined and meant to be applied on the basis of field observations, not on the results of detailed thin section petrography. As a field term, graywacke connotes the gray-green sandy members of thick, rhythmically interbedded sand-shale sequences characteristic of orogenically active zones in the crust.
2. Despite the fact that the original graywackes were not described by point counts in thin section, it is the mineralogy of these sandstones that gives them their distinctive appearance. Therefore, refined definitions for laboratory usage are acceptable provided they correspond to the petrography of the Harz Mountain rocks.
3. The gray and "dirty" appearance of the Harz graywackes is caused by their high content of phyllosilicate minerals as both matrix and foliated metamorphic rock fragments. Therefore, laboratory definitions based on either characteristic are acceptable.

Most definitions of graywacke also imply that normal ones possess characteristics in addition to those on which the definition is based. Included among these accessory characters are angularity of grain, incipient metamorphism, poor grain sorting, or association with sedimentary structures such as flute casts and graded bedding.

It is significant that both those who wish to restrict the term to field usage and those who find laboratory definitions permissible rely on characteristics of the Harz rocks as their justification and for this reason it is useful to examine these rocks to establish their petrography. Huckenholz (1963) has collated the results of modal analyses of eighty-eight Harz graywackes published between 1952 and 1960, with results shown in Table 9-1. The average chemical composition is shown in Table 9-2.

The grain size distribution of three Harz rocks has been determined and averages 1% gravel, 59% sand, 37% silt, and 3% clay, a silty sandstone according to the Wentworth scale. However, muddy matrix, which is normally defined as petrographically irresolvable fine silt and clay, totals approximately 15%.

It is evident from the mineralogic and size analyses of the "type" graywackes that they are both mineralogically and texturally immature sandstones derived from many different types of source rocks and no one fragment type predominates over the rest. The only distinctive petrologic features of these rocks are their heterogeneous mineral composition and their high content of muddy matrix compared to most sandstones. The use of matrix as the defining criterion is unsatisfactory for reasons noted on p. 312.

In this book we use the term graywacke for those areally extensive and similar-looking field occurrences of synorogenic "dirty" sandstones, the rocks for which it was originally intended. The mineral composition may be specified using words such as feldspathic graywacke. However, use of the word "arkose" implies the

TABLE 9-1 MODAL COMPOSITION OF HARZ MOUNTAIN GRAYWACKES (modified from Huckenholz, 1963, p. 915)

	Range (%)		Average (%)	
Quartz	28–53		37	
Feldspars	22–48		32	
Rock fragments	10–55		31	
metamorphic		6–21		16
igneous		4–16		10
sedimentary		2–7		5

TABLE 9-2 MAJOR ELEMENT COMPOSITION OF SOME SANDSTONES

<table>
<tr><th></th><th>Q</th><th>LA</th><th>GW</th><th>AR</th><th>HGW</th><th>CHS(G)</th><th>CHS(R)</th><th>CRS</th></tr>
<tr><td>SiO_2</td><td>95.4</td><td>66.1</td><td>66.7</td><td>77.1</td><td>69.7</td><td>71.5</td><td>69.9</td><td>66.5</td></tr>
<tr><td>Al_2O_3</td><td>1.1</td><td>8.1</td><td>13.5</td><td>8.7</td><td>14.3</td><td>13.4</td><td>15.1</td><td>13.9</td></tr>
<tr><td>Fe_2O_3</td><td>0.4</td><td>3.8</td><td>1.6</td><td>1.5</td><td>1.9</td><td>1.3</td><td>4.9</td><td rowspan="2">4.7</td></tr>
<tr><td>FeO</td><td>0.2</td><td>1.4</td><td>3.5</td><td>0.7</td><td>2.4</td><td>3.6</td><td>1.5</td></tr>
<tr><td>MgO</td><td>0.1</td><td>2.4</td><td>2.1</td><td>0.5</td><td>1.8</td><td>1.0</td><td>0.8</td><td>2.0</td></tr>
<tr><td>CaO</td><td>1.6</td><td>6.2</td><td>2.5</td><td>2.7</td><td>1.3</td><td>1.0</td><td>0.6</td><td>3.4</td></tr>
<tr><td>Na_2O</td><td>0.1</td><td>0.9</td><td>2.9</td><td>1.5</td><td>3.1</td><td>2.8</td><td>2.1</td><td>2.9</td></tr>
<tr><td>K_2O</td><td>0.2</td><td>1.3</td><td>2.0</td><td>2.8</td><td>1.4</td><td>1.6</td><td>2.9</td><td>2.1</td></tr>
<tr><td>CO_2</td><td>1.1</td><td>5.0</td><td>1.2</td><td>3.0</td><td>0.9</td><td>0.6</td><td>0.2</td><td>nd</td></tr>
</table>

Q—Average orthoquartzite (26 analyses), Pettijohn, 1963, p. 15.
LA—Average lithic arenite (20 analyses), Pettijohn, 1963, p. 15.
GW—Average graywacke (61 analyses), Pettijohn, 1963, p. 15.
AR—Average arkose (32 analyses), Pettijohn, 1963, p. 15.
HGW—Average Harz graywacke (17 analyses), Pettijohn, 1963, p. 7.
CHS(G)—Average of six analyses made on two beds of greenish-grey Charny sandstone.
CHS(R)—Average of four analyses of red Charny sandstone beds.
CRS—Average Columbia River sand (68 analyses), Whetten et al., 1969, p. 1162.

NOTE: The analyses from the geosynclinal Charny sandstones have not been previously published. They are included here because these rocks share some characteristics of both arkoses and graywackes. It appears that the provenance was similar to that of a typical arkose. The differences in chemical composition between the Charny and typical arkoses may be explained partly on the basis of its chloritic matrix and partly as resulting from diagenetic metasomatism by seawater and the absence of leaching by vadose waters.

rock is not a field graywacke. As a field term the word is very useful to describe a sensibly homogeneous and stratigraphically important group of rocks. It has no place as a term for use in the laboratory, where sandstones are described precisely using thin section petrography. Petrographic classifications of sandstones should have as their basis the concept of relative fragment stability, as described on pp. 310–312. Graywackes may be highly feldspathic ("arkoses"), rich in rock fragments ("lithic" or "labile"), or (less commonly) rich in quartz. Graywacke is thus not a parallel term to arkose, lithic sandstone, or orthoquartzite but exists independently of these terms.

Arkose

The term arkose was invented by Brogniart in 1823 and he elaborated on the characteristics of this clan in subsequent works in 1826 and 1827 (Oriel, 1949). The clear essence of Brogniart's definition is that an arkose is a coarse clastic rock containing abundant feldspar (amount not specified) and he also noted that although some arkoses contain clay matrix, others do not. Despite this observation, however, many later workers have considered that arkoses should contain little or no clay and that feldspar-rich rocks containing clay belong to another clan. The basis for this contention is the supposed prior definition of the graywacke clan to include all clay-rich sandstones.

At present, most sandstone petrologists set a lower limit of 25% feldspar in the nonclay detrital fraction for a rock to belong to the arkose clan. Sandstones with lesser amounts are generally called subarkoses. "Feldspathic" and "subfeldspathic" are commonly used in place of arkosic and subarkosic.

Orthoquartzite

This word was coined by Krynine (1948, p. 149) as a clan name for detrital sediments composed largely of quartz and chert. Although the cementing material in such rocks is nearly always silica or carbonate, Krynine did not specify that a particular type of cement was required in the definition of the clan. Subsequent workers, however, frequently have considered the term orthoquartzite as a parallel to metaquartzite and added the qualification that a quartz sandstone must fracture through the detrital grains to be a true orthoquartzite. Only quartz sandstones cemented by coarsely crystalline quartz (or possibly chert) fit this restriction. It seems to the present authors rather pointless to throw out a useful term because of the vagaries of temporal and stratigraphic cementation phenomena and we recommend the use of orthoquartzite as originally defined by Krynine. The essence of the clan name is the percentage of detrital silica, not authigenic cement precipitated at some later date.

Most current workers in sandstone petrology require that the composition of the sand fraction of the sediment be at least 90% quartz plus chert to belong to the orthoquartzite clan, although Krynine suggested that values as low as approximately 75% would qualify.

Lithic

The adjective "lithic" was borrowed from igneous petrography, where it is used to denote tuffs rich in fragments of volcanic rocks, as contrasted to tuffs rich in unattached crystals or glassy particles. As applied to sandstones by Gilbert (1955, p. 292), it characterizes sandstones (irrespective of clay content) that contain less than 90 to 95% quartz plus chert and more rock fragments than feldspar.

The term "lithic sandstone" is parallel with and equal in rank to the terms arkose and orthoquartzite.

9.5 CHEMICAL COMPOSITION

Igneous petrographers attach a great deal of significance to the chemical compositions of igneous rocks and have used them as the basis of many classifications but little attention has been given to chemical composition as a basis for classifying sandstones. There are two main reasons: the high cost of chemical analyses and sedimentary petrographers' belief that mineralogical and textural distinctions visible in thin section are more fundamental and important than chemical composition. Modern techniques, however, have greatly reduced the cost of chemical analysis to the point where it actually costs more (in terms of time paid for) for a skilled petrographer to make a reliable modal analysis than it does for a skilled analyst to make a reliable chemical analysis. The result is an increasing tendency for sedimentary petrologists to make use of chemical analyses. For example, in 1963, a survey of the world literature was able to discover fewer than 200 reliable chemical analyses of sandstones but a recent paper was based on more than 100 analyses of major and trace elements in a single sandstone formation!

The sedimentary petrographer's traditional skepticism of the value of bulk chemical analyses is based on several factors, largely concerned with the effects of diagenesis. These effects include the introduction of chemical cements that can increase the amounts of silicon (siliceous cements), calcium (calcite or dolomite cements), and iron (hematite and siderite cements). The possible introduction during diagenesis of calcium is particularly bothersome because the Na/Ca ratio of a sandstone has important provenance and tectonic implications. Fortunately, the interpretive difficulties caused by the possible introduction of calcite during diagenesis can be resolved by the examination of thin sections in conjunction with the chemical analyses. The question of postdepositional introduction of iron into the rock as hematite is usually not answerable by thin section analyses, however, and this fact must be considered in the interpretation of chemical analyses of hematitic sandstones.

Diagenetic changes in bulk chemistry of a sandstone can also occur without introduction of chemical cement. Potassium is often added during regrading of illites stripped of potassium during weathering and sodium is added to many graywackes during albitization of more calcic plagioclases. As we shall see, however, the difficulties in interpretation caused by diagenetic changes in sandstones are not insurmountable. And in view of the fact that such analyses commonly yield insights into petrogenesis not otherwise obtainable, we strongly support their increasing use.

One approach to the subject of classification, much used for carbonate rocks but not normally used for sandstones, is factor analysis. This numerical technique identifies groups of variables and combines them into a few "factors," thereby objectively identifying, on the basis of the data themselves, those "end-members" most important for classification. This technique has been applied in an attempt to relate chemical composition of sandstones to the tectonic setting in which they form and the mathematical analyses led to the choice of chemical end-member "factors" shown in Fig. 9-2. Total iron (expressed as Fe_2O_3) plus magnesia are

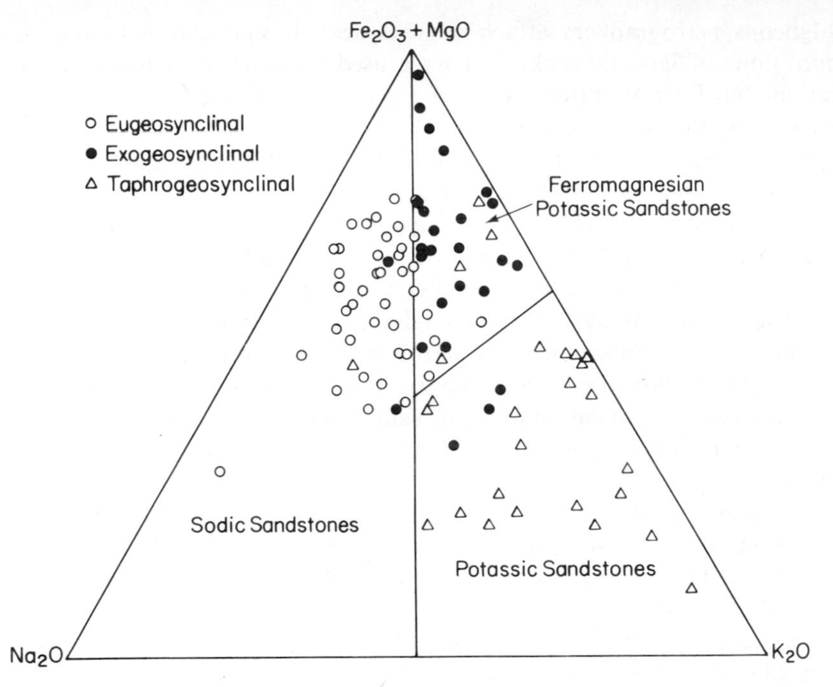

Fig. 9-2 Chemical composition of sandstones in relation to tectonic setting.

found in detrital iron oxides or ferromagnesian minerals such as biotite and hornblende. They also are found in clays such as smectite, illite, and chlorite. Some of the iron-bearing minerals such as pyrite and some of the chlorite were formed in the sandstones during diagenesis but generally the iron is simply redistributed locally so that the total content remains more or less constant after burial. The same is true of magnesium in sandstones such as graywackes, which have low permeabilities. It appears, however, that magnesium may be lost from rocks such as arkoses. There is evidence that will be reviewed in Chap. 10 that many red arkoses have been altered by groundwaters. The iron is retained in the rock as red oxides but magnesium is probably lost in solution, as shown by the very low content of MgO in most arkoses (see Table 9-2).

Potassium and sodium in sediments are present in alkaline feldspars and in muscovite and illite. These minerals are all more resistant to hydrolysis than ferromagnesian minerals or calcic plagioclases.

Sandstones with less than 5% Al_2O_3 have been omitted from Fig. 9-2 because these rocks generally contain little besides quartz and a cement. The content in such sandstones of the four elements used to construct the diagram is generally low and is likely to be strongly affected by diagenetic changes.

It can be seen that the diagram makes a fairly effective separation, with some overlap, among eugeosynclinal sandstones (mostly graywackes); taphrogeosynclinal sandstones, i.e., sandstones deposited in deep fault basins on the craton (mostly arkoses), and exogeosynclinal sandstones, i.e., sandstones forming part of the "clastic wedges" spread out over the edges of cratons from sources in peripheral folded mountains (mostly lithic sandstones). The arkoses are distinguished from the lithic sandstones by their generally lower ratio of iron and magnesium to potash (see Table 9-2). This is due partly to diagenetic factors, as noted above, but it is also related to source: Arkoses are generally derived from plutonic rocks of roughly granitic composition and these rocks are themselves relatively rich in potash and low in content of ferromagnesian components. In the light of known transitions in mineralogy and tectonic setting, it is not surprising that there are also transitions in chemical composition between arkoses and lithic sandstones.

The contrasts in chemical composition between arkoses and graywackes are surprisingly clearly defined. Arkoses have less iron and magnesium, reflecting source control. Most of the difference in iron content is due to a deficiency in ferrous iron in arkoses. The content of ferric iron is about the same in both arkoses and graywackes (Fig. 9-3 and Table 9-2). Most graywackes have labile rock fragments rich in both iron and magnesium as well as some detrital ferromagnesian minerals and/or a recrystallized chloritic matrix.

The graywackes, or eugeosynclinal sandstones, differ from almost all other sandstones in having a soda/potassia ratio greater than unity (Fig. 9-2). The excess of soda in the graywackes is difficult to explain. It is not due to the presence of sodic phyllosilicates because analyses of associated shales reveal that they generally have a soda/potassia ratio considerably less than unity. The soda must be present in graywackes mainly in the form of coarse silt and sand-size soda feldspars and soda-rich rock fragments. This conclusion is supported by petrographic observations. For example, the Tanner graywacke of the Harz Mountains has about 30% feldspar of which 85 to 90% is untwinned albite. Graywackes generally contain substantial amounts of potassium and it is generally assumed that part of this is present in potash feldspars such as orthoclase and microcline and part is present in clay minerals. This is confirmed by petrography in some cases but a surprisingly large number of graywackes have been shown by recent studies to contain practically no potassic or calcic feldspars.

The absence of calcic plagioclases is relatively easily explained. Many graywackes contain calcic authigenic silicates such as prehnite and pumpellyite, minerals characteristic of the zeolite facies of metamorphism (see Chap. 18). Many feldspars in graywackes are also partly corroded by calcite. It seems reasonable to suppose that calcic plagioclases originally present have been altered during diagenesis and very low grade metamorphism to sodic plagioclase and calcite or zeolites.

The enrichment of graywackes in soda and the absence of potash feldspars in some graywackes might be explained in terms either of provenance or of soda

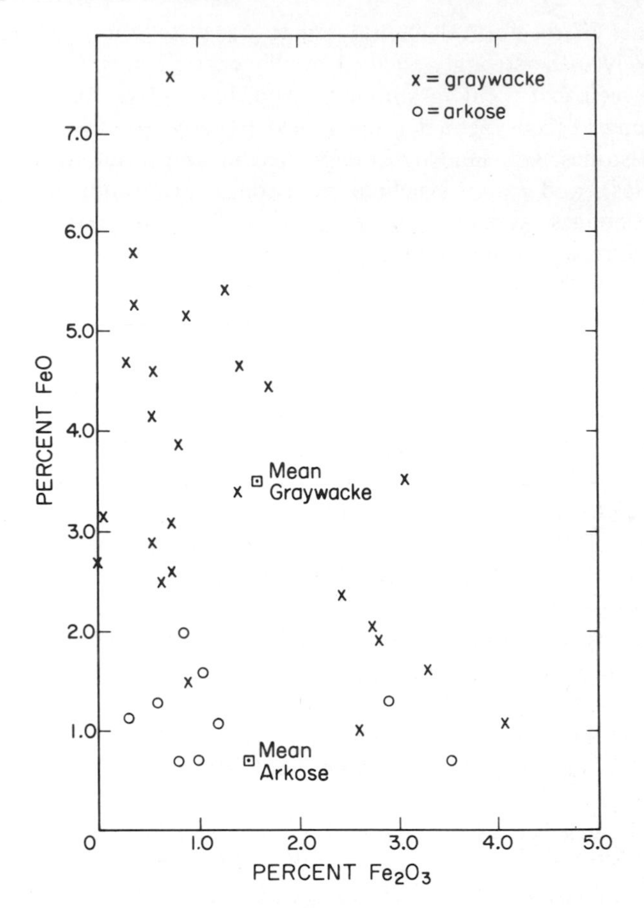

Fig. 9-3 Relative abundances of FeO and Fe_2O_3 in graywackes and arkoses. × = graywacke, ○ = arkose. (Analyses from various publications.)

metasomatism by reaction of detrital minerals with seawater during diagenesis. It is rarely possible to decide definitely between these two alternatives but source areas devoid of potash feldspar and yielding large amounts of sand-size untwinned plagioclase are difficult to envisage. The sandstones of the Charny Formation (Cambrian of Quebec) contain 20% untwinned albite, which has a fine lamellar structure with the same crystallographic orientation as perthite lamellae. In this case, therefore, it seems very probable that the albite has replaced original potash feldspar. However, pseudomorphs of perthite lamellae are rarely observed in soda-rich graywackes. Metasomatic processes are accelerated by the deep burial and consequent high temperatures characteristic of eugeosynclinal sediments. Compaction of thick associated shales provides a ready supply of water rich in sodium so that it is not unreasonable to suppose that there may be a net addition of soda to graywackes during diagenesis in such an environment. The soda-rich zeolite analcime is reported from several graywackes and may be more common than has been generally supposed.

Many graywackes, however, show signs of a soda-rich volcanic provenance. An example of a modern sediment with a volcanic provenance and a composition almost identical with that of the average graywacke is the sand in the Columbia River (Table 9-2). It remains for further research to determine the relative merits of provenance and diagenesis in determining the mineralogical and chemical characteristics of most graywackes.

REFERENCES

DOTT, R. H., JR., 1964, "Wacke, Graywacke and Matrix—What Approach to Immature Sandstone Classification?" *Jour. Sed. Petrology*, **34**, pp. 625–632. (Thoughtful commentary on the antiquity and evolution of the term graywacke.)

FOLK, R. L., 1954, "The Distinction between Grain Size and Mineral Composition in Sedimentary Rock Nomenclature," *Jour. Geol.*, **62**, pp. 344–359. (Sandstone classification descended from the an earlier one by Krynine. Folk has now abandoned this system of classification but it is in wide use.)

———, 1968, *Petrology of Sedimentary Rocks*. Austin, Tex.: Hemphill's Book Store, 170 pp. (The latest edition of the undiluted thoughts of one of the country's leading sedimentary petrologists.)

GILBERT, C. M., 1955, "Sedimentary Rocks," in *Petrography*, by Howel Williams, F. J. TURNER, and C. M. GILBERT. San Francisco, Calif: W. H. Freeman and Co., 406 pp., pp. 251–384. (Sandstone classification in common use among geologists along the Pacific Coast of the United States.)

HUCKENHOLZ, H. G., 1963, "Mineral Composition and Texture in Graywackes from the Harz Mountains (Germany) and in Arkoses from the Auvergne (France)," *Jour. Sed. Petrology*, 33, pp. 914–918. (Summary of analyses of the type graywackes and arkoses.)

KLEIN, G. DEV., 1963, "Analysis and Review of Sandstone Classifications in the North American Geological Literature, 1940–1960," *Geol. Soc. Amer. Bull.*, **74**, pp. 555–576. (One man's analysis of classifications published during a time of prolific outpouring, 1940–1960.)

KRYNINE, P. D., 1948, "The Megascopic Study and Field Classification of Sedimentary Rocks," *Jour. Geol.* **56**, pp. 130–165. (The earliest comprehensive scheme for the classification of sandstones according to their mineralogy.)

MCBRIDE, E. F., 1963, "A Classification of Common Sandstones," *Jour. Sed. Petrology*, **33**, pp. 664–669. (A fairly recent classification that emphasizes mineralogic maturity.)

MIDDLETON, G. V., 1972, "Albite of Secondary Origin in Charny Sandstones, Quebec," *Jour. Sed. Petrology*, **42**, pp. 341–349. (Demonstrates replacement of perthite by albite and discusses the chemical composition of these green and red feldspathic graywackes.)

MURCHISON, R. I., 1854, *Siluria*. London: John Murray,. Ltd. 523 pp. (Classic work on stratigraphy and paleontology of Lower Paleozoic rocks in Russia, Germany, and Britain.)

NOCKOLDS, S. R., 1954, "Average Chemical Compositions of Some Igneous Rocks," *Geol. Soc. Amer. Bull.*, **65**, pp. 1007–1032. (Summary and analysis of data.)

OKADA, HAKUYU, 1971, "Classification of Sandstone: Analysis and Proposal," *Jour. Geol.*, **79**, pp. 509–525. (The most recent summary of the literature on this topic.)

ORIEL, S. S., 1949, "Definitions of Arkose," *Amer. Jour. Sci.*, **247**, pp. 824–829. (A paper tracing the history and usage of the word.)

PETTIJOHN, F. J. 1957, *Sedimentary Rocks*. 2nd ed., N. Y.: Harper & Bros., 718 pp. (Chap. 7 contains what is probably the most widely used sandstone classification.)

———, 1963, "Data of Geochemistry, Chap. S., Chemical Composition of Sandstones—Excluding Carbonate and Volcanic Sands," *U. S. Geol. Sur. Prof. Paper 440-S*, 21 pp. (The most recent summary of the range in chemical composition of sandstones.)

ROGERS, J. J. W., 1966, "Geochemical Significance of the Source Rocks of Some Graywackes from Western Oregon and Washington," *Texas Jour. Sci.*, **18**, pp. 5–20. (A recent description of some volcanic graywackes and their meaning in terms of crustal chemistry and evolution.)

RONOV, A. B., Y. P. GIRIN, G. A. KAZAKOV, and M. N. ILYUKHIN, 1965, "Comparative Geochemistry of Geosynclinal and Platform Sedimentary Rocks," *Geochem. Internatl.*, **2**, pp. 692–708. (Presents abundant data from Mesozoic and Cenozoic rocks of Caucasian geosyncline and adjacent platform and shows trends in chemical and mineralogical composition from platform to geosynclinal facies.)

WELSH, WILLIAM, 1967, "The Value of Point-Count Modal Analysis of Graywackes," *Scottish Jour. Geol.*, **3** pp. 318–328. (Standard deviation of a single operator measuring matrix was 3%, i.e., precision was 23 ± 6%. Different operators, however, differed by larger amounts, with mean differences of 11 and 8% for two sandstone formations. Generally smaller differences were shown for most other measured constituents.)

WHETTEN, J. T. and J. W. HAWKINS, JR., 1970, "Diagenetic Origin of Graywacke Matrix Minerals," *Sedimentology*, **15**, pp. 347–361. (Experiments at 250°C and 1 kb using volcanic rock fragments and artificial sea water produced authigenic matrix in sand and silt originally lacking matrix.)

WHETTEN, J. T., J. C. KELLEY, and L. G. HANSON, 1969, "Characteristics of Columbia River Sediment and Sediment Transport," *Jour. Sed. Petrology*, **39**, pp. 1149–1166. (Includes much data on major and trace elements in this example of a modern sediment having the composition of ancient graywackes.)

CHAPTER TEN

NATURAL WATERS AND DIAGENESIS OF SANDSTONES

10.1 INTRODUCTION

The changes that take place in sedimentary rocks after deposition are called diagenetic and affect not only the mineral particles in the sediment but its connate water as well. Most diagenetic changes increase the density of the rock and tend to eliminate open space that initially existed between the individual grains. This is accomplished most commonly by precipitation of mineral cements within the open spaces or dissolution at grain boundaries (but see also Sec. 3-5). Some diagenetic processes have the net effect of increasing the amount of open space, however, by simple dissolution of grains or earlier formed cements. Most important is the fact that essentially all of these processes involve dissolution and/or precipitation of minerals from aqueous solutions that exist within the open space of the rock. Thus, availability of water within a rock and the ability to move water through a sediment or rock are critical to diagenesis. An understanding of the character of surface and subsurface waters, the chemistry of these fluids, the solid minerals that form from them, and the nature of pore space are fundamental to our understanding of how sediments become rocks.

10.2 COMPOSITION OF NATURAL WATERS

Nearly all reactions of importance in the sedimentary environment occur in an aqueous setting, either at the ground surface or in the subsurface to depths of up to perhaps 50,000 to 60,000 ft. In Chap. 7 we considered the relationship between water and weathering: mineral decomposition. We now consider some aspects of the reverse process, precipitation of cements in pore spaces of clastic rocks, particularly the coarser clastics—conglomerates and sandstones.

Stable Isotope Studies

The study of stable isotope ratios in natural waters and in minerals often yields useful and sometimes unique information concerning the mode of origin (inorganic or organic, primary or replacement) and temperature at the time of origin. The kinds of questions that may be answered by stable isotope studies include the following: (a) Do subsurface water molecules represent "connate" water or meteoric water of more recent origin? (b) Has the calcite cement in exposed Pleistocene beach rock or dune rock precipitated from seawater or meteoric water? (c) What was the temperature of the water in which Cretaceous ammonites grew shell material? (d) Was more globogerina ooze formed in the equatorial Atlantic during glacial or interglacial time, or is there no difference? Investigation of stable isotopes in natural waters and minerals is perhaps the clearest example of the power of chemistry in geology, in that this approach may give geologic information not obtainable by any other method now known.

Fractionation

Isotopic separations in natural systems occur because of the differences in chemical properties between nuclides of an element, the thermodynamics of which were initially explained by Urey in 1947. The vapor pressure is greater for the lighter nuclides of an element and hence is greater for that variety of a compound composed of the lighter nuclides. For example, at 25°C the vapor pressure of $H_2^1O^{16}$ is 1.008 times that of $H_2^1O^{18}$. Hence evaporation preferentially removes H_2O^{16} from seawater into the atmosphere and, as a result, rainwater or river water contains a higher proportion of O^{16} than seawater does. From this fact we can anticipate a correlation between water salinity and the O^{18}/O^{16} ratio. The correlation is far from perfect, however, as fresh waters are not equally enriched in O^{16}. Addition of fresh water from melted Arctic snow will dilute seawater in O^{18} 3.5 times as much as an equal volume of water from the Mississippi River. The general rule is that polar waters have lower O^{18}/O^{16} values than equatorial waters because atmospheric H_2O moves northward from equatorial areas by a multiple distillation process in which each successive rain or snow depletes the H_2O in molecules of H_2O^{18}. The same is true for hydrogen isotopes; polar waters are enriched in H_2^1O with respect to H_2^2O. In reactions where a phase change is involved, the less condensed phase usually is enriched in the lighter isotope. For example, in the one-component H_2O system at the triple point, the vapor will be most enriched in H^1 and O^{16} and the ice enriched in H^2 and O^{18}.

Both the rate and amount of equilibrium fractionation of isotopes increase with the relative difference in mass. Deuterium has double the mass of protium and, therefore, hydrogen undergoes the greatest amount of fractionation by natural processes. Although mass difference between two isotopes is a very useful guide to the amount of fractionation to be expected, other factors are involved as determinants. Often the nature of these factors is unknown. For example, the mass difference between C^{12} and C^{13} is close to that between Si^{28} and Si^{30}; yet carbon fractionations average 1 to 2% and sometimes exceed 5%; but those of silicon are about 0.3%. With present equipment, fractionations of 0.01% are easily measurable.

There are two ways to express differences in isotopic ratios among substances. The ratio can be expressed independently of an arbitrary standard or can be referred to such a standard. Carbon, oxygen, and sulfur are, by convention, referred to a standard. No standard has been established for magnesium, silicon, or calcium and the relatively few studies conducted of isotopes of these elements have expressed the results as Mg^{24}/Mg^{26}, Si^{30}/Si^{28}, Ca^{48}/Ca^{40}, or Ca^{40}/Ca^{44}. For sulfur the standard is the S^{34}/S^{32} ratio in troilite (FeS) of the Canyon Diablo Meteorite. For carbon and oxygen the standards are the C^{13}/C^{12} and O^{18}/O^{16} ratios in the calcite of a Cretaceous belemnite from the Peedee Formation in South Carolina; this standard is commonly shortened to PDB in publications. Modern seawater of 3.48% salinity, comparable to the world average seawater in which the belemnite is presumed to have lived, also is used as a standard occasionally. Isotopic ratios in this Standard Mean Ocean Water, commonly written as SMOW in publications, are different from those in calcite from the standard belemnite and a known correction factor must be applied to data based on SMOW so that it may be compared to PDB data ($\delta O^{18}_{SMOW} = 1.03\,\delta O^{18}_{PDB} + 29.5$). When compared to a standard, isotopic ratios are expressed as *del* values, written as δ, and are computed by the following formula.

$$\delta = \frac{O^{18}/O^{16}\text{ sample} - O^{18}/O^{16}\text{ standard}}{O^{18}/O^{16}\text{ standard}}(1000)$$

or alternatively,

$$\delta = \left(\frac{O^{18}/O^{16}\text{ sample}}{O^{18}/O^{16}\text{ standard}} - 1\right)1000$$

Therefore, positive δ values mean enrichment in the heavier isotope; negative values, enrichment in the lighter isotope. As 1000 is used as a multiplication factor, a δ value of $+10$ means 1% enrichment. The multiplication factor is simply a convenience to avoid dealing with small decimals. Most δ values for natural fractionations range between nearly zero and $\pm$50/mil (5%). For a sequence of reactions such as in the carbonate system, in which carbon dioxide gas is ultimately converted into carbonate ions, the total fractionation is the sum of the fractionations that occur in all steps of the conversion.

An alternative way of expressing isotopic fractionation processes is through the equilibrium constant for the fractionation reaction. For example, for the reaction in which carbon dioxide gas is bubbled through water,

$$\frac{1}{2}\,CO^{18}_{2\,(gas)} + H_2O^{16}{}_{(liq)} \rightleftharpoons \frac{1}{2}\,CO^{16}_{2\,(gas)} + H_2O^{18}{}_{(liq)} \qquad K_{25°C} = 1.041$$

This is quite a large fractionation (4.1%) because two phases are involved and, as noted earlier, the more condensed phase becomes enriched in the heavier isotope by exchanging oxygen atoms with the lighter phase. Equilibrium constants vary with temperature and data are available to illustrate this for several isotope fractionations of geochemical importance. For example, note the following reaction:

$$C^{13}O_{2(gas)} + HC^{12}O^{-}_{3(aq)} \rightleftharpoons C^{12}O_{2(gas)} + HC^{13}O^{-}_{3\,(aq)} \qquad \begin{aligned} K_{0°C} &= 1.0093 \\ K_{10°C} &= 1.0084 \\ K_{20°C} &= 1.0076 \\ K_{30°C} &= 1.0069 \end{aligned}$$

We see that the amount of fractionation decreases and approaches unity as temperature increases, indicating less fractionation at higher temperatures. A complete explanation of this phenomenon can only be given in terms of quantum theory. In general terms, however, the reason for this phenomenon is that at higher temperatures the percent difference in vibrational frequencies of atoms is reduced, although the absolute difference is constant; i.e., all atoms vibrate faster at high temperatures. Although fractionation is less effective at higher temperatures, theoretically some fractionation always occurs. In practice, fractionations of isotopes can be measured for minerals formed in magmatic as well as metamorphic environments and these data give important information concerning mineral transformations in these rocks.

Meteoric Waters

Perhaps the most convenient point at which to begin our study is with water in its purest form during the hydrologic cycle, i. e., as snow, hail, or rain. The atmosphere is essentially a five-component, three-phase system. The components are N, O_2, A, CO_2, and H_2O; the first four exist only as gases. The H_2O molecule is dipolar and attracts other molecules so that natural precipitation contains dissolved gases in amounts proportional to their concentrations in the atmosphere and their solubilities. This is illustrated in Table 10-1. The carbon dioxide percentage in rainwater is more than 6 times that in the dry atmosphere because of the relatively high solubility of CO_2 in water.

TABLE 10-1 DISSOLVED GASES IN THE ATMOSPHERE AND RAINWATER

Gas	Volume percent dry atmosphere	Volume percent rainwater	Percentage change
N	78.08	63.69	−18.4
O_2	20.95	34.17	+63.1
A	0.93	1.93	+107.5
CO_2	0.033	0.21	+536.4

Rainwater also commonly contains small amounts of other chemically active substances, including sulfuric acid, nitric acid, and hydrochloric acid. Near oceanic coastlines, sodium and chlorine ions may be abundant in precipitation and in the vicinity of industrial areas a diverse assortment of inorganic or organic substances may occur.

River Waters

Inorganic Substances

As soon as precipitation impacts on the earth's surface, it reacts with all available materials and becomes more saline but considerable variation in salinity of river water is possible both temporally and spatially. Temporal variation results from seasonality of rainfall because river water is a mixture of "new" rainwater and subsurface waters entering the streams from the groundwater table. During rainy seasons most river water is obtained from surface flow and salinity is relatively low. Underground waters are relatively more saline because they flow at a much slower rate than surface waters and, therefore, have had more extended contact with soluble mineral particles. In general, the higher the stream discharge, the lower the total salinity of the streams, and this relationship is seen most clearly in arid regions.

As an illustration of the control of river salinity by the types of rocks exposed in different parts of a single drainage basin, we can consider data from the Powder River Basin in Wyoming and Montana (Table 10-2). It is clear that differential solubilities of minerals are an important control not only on total salinity but on the proportions of the various dissolved materials as well. Coarsely crystalline silicate rocks are relatively insoluble and streams obtain only small amounts of ions during their passage over such rocks. Limestone waters are, of course, rich in calcium and bicarbonate ions and may contain abundant magnesium and silica as well if dolomite and chert are present in the limestone. Waters flowing on gypsiferous substrates are rich in sulfate ion relative to other waters and have high total salinities because of the great solubility of gypsum. Shales and mudstones rich in adsorbed ions can yield large amounts of dissolved materials to stream waters, as illustrated by the Powder River data. For example, the high proportion

TABLE 10-2 CHEMICAL QUALITY OF WATER IN STREAMS FLOWING OVER SPECIFIC ROCK TYPES IN TRIBUTARIES OF THE POWDER RIVER (from Hembree et al., 1952)

Dissolved substance	Granite substrate	Limestone substrate	Gypsum substrate	Shale substrate
$HCO_3^- + CO_3^{-2}$	15.9	86.3	101.8	163.6
Ca^{2+}	10.0	32.2	121.4	139.4
Mg^{2+}	0.9	17.1	40.0	76.8
H_4SiO_4	11.9	10.1	10.2	6.1
SO_4^{-2}	3.2	5.7	383.9	1212.0
$Na^+ + K^+$	0.5	1.9	55.3	410.1
$Cl^- + NO_3^-$	0.6	4.7	14.5	12.1
Total ppm	43.0	158.0	727.1	2020.1

of Na + K (probably mostly potassium) likely reflects illitic clay. The high sulfate content may reflect either sulfate-rich organic matter in the shales or gypsiferous layers.

Organic Substances

The land surface of the earth has been partially covered with a mantle of vegetation since Late Silurian time and when the plants die and decay, the organic substances from them go partly into solution. Hence, streams contain both dissolved and particulate organic matter. The amount of total organic matter in streams is most commonly 10 to 30 ppm and, according to Clarke (1924), organic matter may form more than 50% of the dissolved solids in river water, as illustrated by Table 10-3. In the tropical streams at the end of the table, the high percentages result from contributions from jungle swamps through which the rivers flow and in which the rates of organic productivity and decay are high. Arctic rivers commonly contain abundant particulate organic matter, a result of the low annual temperatures, which retard decomposition. There are very few data concerning the exact nature of the organic substances in streams and most descriptions of dissolved materials are confined to characterizations such as "humic acids," "potassium humates," or other general groupings. Particulate material usually is considered only in terms of molecular weight and relative proportions of hydrogen, nitrogen, carbon, sulfur, and ash. The grossness of such descriptions is unfortunate, for organic species are important substrates for the adsorption of minor elements such as iron, manganese, nickel, and copper. In addition, the organic matter in stream waters contributes significantly to the total

TABLE 10-3 DISSOLVED ORGANIC MATTER IN STREAMS AS PERCENTAGES OF TOTAL SOLIDS (from Clarke, 1924, p. 110)

River	Percentage	River	Percentage
Danube	3.25	Amazon	15.03
James	4.14	Mohawk	15.34
Maumee	4.55	Delaware	16.00
Nile	10.36	Lough Neagh, Ire.	16.40
Hudson	11.42	Xingu	20.63
Rhine	11.93	Tapajos	24.16
Cumberland	12.08	Plata	49.59
Thames	12.10	Negro	53.89
Genesee	12.80	Uruguay	59.90

amount in nearshore marine waters, which, in turn, is an important source of hydrocarbon accumulations in sediments.

Average River Water

It is evident that the larger the drainage basin of a river, the closer the chemical composition of its rocks will approach the mean chemical composition of the surface rocks of the earth. Hence, although in any large river system the proportions of dissolved salts are different in the various headwater tributaries, these local irregularities tend to cancel each other as one proceeds downstream and there is a tendency for the composition of waters in the downstream parts of rivers to resemble one another. Table 10-4 lists the most abundant dissolved species in modern world average river water according to the summary data presented by Livingstone (1963). Also tabulated is the ultimate source of each constituent and the most common proximate sources.

Soil Waters

Most water passes through the soil on its way to either nearby streams, the sea, or deeper layers of sedimentary material. Unfortunately, there are almost no quantitative data concerning the composition of soil moisture and we can only speculate concerning the magnitude of the changes in water composition that occur in this environment.

Probably the most chemically active component in soil moisture is carbon dioxide. Plants synthesize organic compounds from water and carbon dioxide from the atmosphere and release carbon dioxide during respiration and decomposition. In soil waters of humid temperate regions the percentage of CO_2 usually ranges between 0.5 and 5.0%, increasing downward in the soil zone to a reported maximum of 30%, more than 100 times the percentage dissolved in rainwater.

TABLE 10-4 COMPOSITION OF MEAN RIVER WATER[†]

Ionic species	Abundance		Proximate source	Ultimate source
	ppm	Molality		
HCO_3^-, CO_3^{2-}	58.8	0.00096	Carbonate rocks, soil gases	Volcanic emanations
Ca^{2+}	15.0	0.00037	Carbonate rocks	Plagioclase feldspars
H_4SiO_4	13.1	0.00014	Silicate minerals	Silicate minerals
SO_4^{2-}	11.2	0.00012	Rainwater, gypsum, organic matter	Volcanic emanations
Cl^-	7.8	0.00022	Rainwater, pore waters of clastic rocks	Volcanic emanations
Na^+	6.3	0.00027	Rainwater, pore waters of clastic rocks	Alkali feldspars
Mg^{2+}	4.1	0.00017	Dolomite, ferromagnesian minerals	Ferromagnesian minerals
K^+	2.3	0.00006	Illites	Alkali feldspars
NO_3^-	1.0	0.00002	Bacterial activities	Volcanic emanations
Fe^{2+}, Fe^{3+}	0.67	0.00001	Hematite, ferromagnesian minerals	Ferromagnesian minerals
$Al(OH)_4^-$	0.24	2.5×10^{-6}	Silicate minerals	Silicate minerals
F^-	0.09	5×10^{-6}	Micas, apatite, tourmaline	Volcanic emanations
Sr^{2+}	0.09	1×10^{-6}	Limestones	Plagioclase feldspars
H_3BO_3	0.1–0.01	$\approx 5 \times 10^{-7}$	Clay minerals	Volcanic emanations
Br^-	0.02	2.5×10^{-7}	Pore waters of clastic rocks	Volcanic emanations
Total	120.8	0.00235		

† Data from Livingstone, 1963, pp. 41–44.

It is apparent, therefore, that most carbon dioxide in soil water is due to plant and bacterial activities rather than directly to atmospheric sources. This implies that most of the bicarbonate ion in rivers comes from biogenic sources, at least in areas not underlaid by extensive limestone deposits. It has been calculated that the amount of bicarbonate ion produced by the carbon dioxide in soil waters is more than 10 times the amount that can be produced in meteoric water in equilibrium with the carbon dioxide pressure of the atmosphere.

Marine Waters

Inorganic Substances

Most sediments preserved in the geologic record are marine and, therefore, the initial pore water of most sedimentary rocks is seawater (Table 10-5). Several important differences are immediately apparent between the compositions of average river water and seawater (Table 10-6).

TABLE 10-5 MOLALITY OF SEAWATER†

	Amount	
Dissolved species‡	ppm	Molality
Cl^-	18,980	0.535
Na^+	10,556	0.459
SO_4^{2-}	2,649	0.028
Mg^{2+}	1,272	0.052
Ca^{2+}	400	0.010
K^+	380	0.010
HCO_3^-, CO_3^{2-}	140	2.3×10^{-3}
Br^-	65	8.1×10^{-4}
H_3BO_3	26	4.2×10^{-4}
Sr^{2+}	8	9.1×10^{-5}
F^-	1.3	6.8×10^{-5}
H_4SiO_4	1	1.0×10^{-5}
NO_3^-	0.5	8.1×10^{-6}
Fe^{2+}, Fe^{3+}	0.01	2×10^{-7}
$Al(OH)_4^-$	0.01	1×10^{-7}
Total	34.479	1.106

† From Mason, 1966, pp 194–199.
‡ All other species total less than 1ppm. Ionic strength=0.673

1. In river water the ratio mCa^{2+}/mMg^{2+} is 2.2. In seawater it is 0.2, the sharp reversal in relative abundance resulting from removal of calcium carbonate from seawater by calcareous-shelled organisms. Most of these shells do contain some magnesium in solid solution but the amount is small in comparison with the amount of calcium the organisms remove from the

TABLE 10-6 COMPARISON BETWEEN THE PROPORTIONS OF DISSOLVED SPECIES IN RIVER WATER AND SEAWATER

	Molality in		
Dissolved species	River water	Seawater	Major cause of change
HCO_3^-, CO_3^{2-}	9.6×10^{-4}	2.3×10^{-3}	Kept at equilibrium activity in contact with atmosphere
Ca^{2+}	3.7×10^{-4}	1.0×10^{-2}	Kept at equilibrium activity by calcareous shell formation
H_4SiO_4	1.4×10^{-4}	1×10^{-5}	Used by siliceous marine organisms for hard parts
SO_4^{2-}	1.2×10^{-4}	2.8×10^{-2}	No celestite precipitated in sea water; only minor usage in organic tissues
Cl^-	2.2×10^{-4}	5.4×10^{-1}	Only minor usage by marine organisms or mineral authigenesis
Na^+	2.7×10^{-4}	4.6×10^{-1}	Only minor usage by marine organisms or mineral authigenesis
Mg^{2+}	1.7×10^{-4}	5.2×10^{-2}	No dolomite or chlorite formation in sea water; only minor usage as trace element in calcite
K^+	6×10^{-5}	1.0×10^{-2}	Only limited regrading or authigenesis of illite on sea floor
NO_3^-	2×10^{-5}	8.1×10^{-6}	Used in tissues of marine organisms
Fe^{+2}, Fe^{+3}	1×10^{-5}	2×10^{-7}	Authigenesis of glauconite and manganese nodules
F^-	5×10^{-6}	6.8×10^{-5}	No significant precipitation of apatite in the sea
Sr^{+2}	1×10^{-6}	9.1×10^{-5}	Only minor usage as trace element in aragonite structure
H_3BO_3	5×10^{-7}	4.2×10^{-4}	Submarine volcanic activity
Br^-	2.5×10^{-7}	8.1×10^{-4}	Only minor usage by marine organisms or mineral authigenesis
$Al(OH)_4^-$	2.5×10^{-6}	1×10^{-7}	Adsorption and authigenesis
Total	100.0	100.0	

water. No organism at present secretes a significant amount of dolomite or magnesite skeletal parts and there is no evidence suggesting that any organism has done so in the geologic past. However, calcium-rich dolomite has been reported in the teeth of some modern sea urchins near Bermuda.

2. Silica, which exists in natural waters almost entirely as undissociated orthosilicic acid (H_4SiO_4), is only one-tenth as abundant in sea water as in river water. Chapter 16, which considers the origin of chert, discusses in detail the explanation for this difference. The decrease results almost entirely from removal of silica from the sea by protistids, particularly diatoms and radiolaria. Although diatoms also live in fresh waters, their numbers are small in comparison with those in the marine environment.

The reason for the other differences in relative amounts of the various dissolved species are given in the table. We have occasion to refer to some of them later in this book.

Ionic Activities

In dilute solutions such as most river waters it is a simple matter to determine whether the water is saturated with respect to a specific compound or mineral. It is only necessary to multiply the concentrations in the water of each ion needed in the mineral and compare the resulting value with the solubility product. If the value is less than the solubility product, the solution is undersaturated; if greater, the solution is supersaturated. In concentrated solutions such as seawater and subsurface brines, however, the movement of ions is restricted to varying degrees by electrical fields due to the presence of other ions, leading to the concept of ionic "activity" or effective concentration. Although these interionic attractions exist at all concentrations in solutions of ions, they can be ignored in dilute solutions. Activity is related to concentration by a proportionality factor called the activity coefficient.

$$\text{Activity } (a) = \text{activity coefficient } (\gamma) \times \text{concentration } (c)$$

At infinite dilution, the activity coefficient is unity but may be smaller or larger than this, depending on the ions involved and on the concentration of the solution (Fig. 10-1).

Ion-Pair Formation

The interactions among ions in solution conform to Coulomb's law.

$$F = \frac{e_1 e_2}{\epsilon r^2}$$

where e_1 = charge on cation
e_2 = charge on anion
ϵ = dielectric constant of the solvent
r = distance between ions
F = force between ions

Therefore, as the concentration of the solution increases and the distance between ions (r) decreases, the force of attraction between ions increases very rapidly, resulting in the formation of ion pairs. The existence of ion pairs is simply another way of recognizing the incomplete dissociation of compounds; ion-pair formation and dissociation constants are opposite sides of the same coin. We may define an ion pair as a species formed by the association of two or more simpler species, each of which can exist independently.

The phenomenon of ion-pair formation is unimportant for a pure solution of a strong electrolyte such as NaCl or KCl. Dissociation is essentially complete and, therefore, changes in activity coefficients for sodium and potassium in such solutions result only from the effect of the "ionic atmosphere" or surrounding electric field. In weak electrolytes or in complex solutions such as seawater, however,

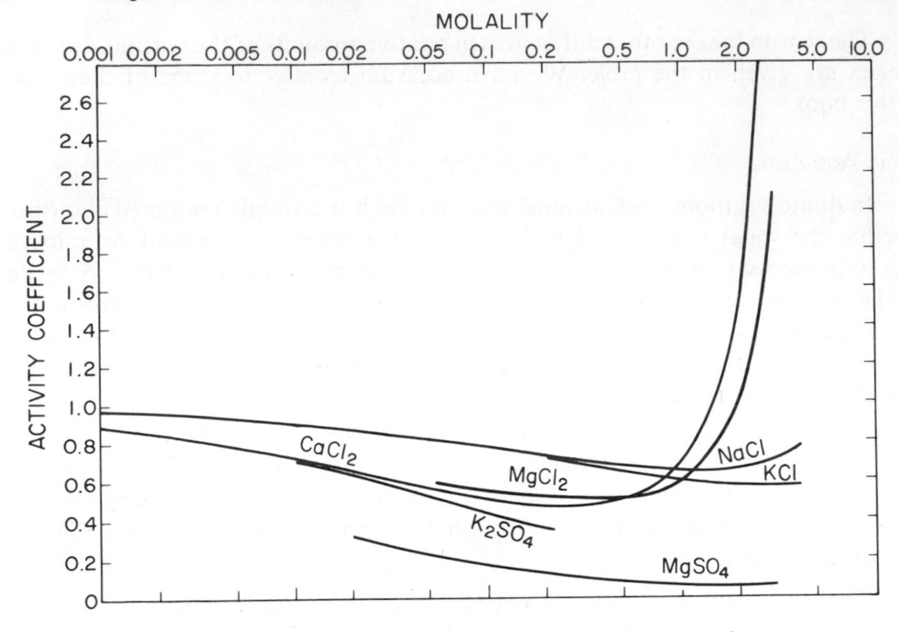

Fig. 10-1 Mean molal activity coefficients of some electrolytes important in natural waters.

the formation of ion pairs occurs and must be quantitatively evaluated to determine the proportion of "free" ions among the total in solution. For example, a sulfate ion tied up in the ion pair $NaSO_4^-$ is not available to form gypsum; carbonate ion paired as $NaCO_3^-$ is not a contributor to the ion product $[Ca^{2+}][CO_3^{2-}]$ that must be increased in the solution to cause saturation of the water with respect to aragonite or calcite. Quantitative data now available (Table 10-7) reveal that ion-pair formation of the carbonate ion in surface seawater at 25°C makes 91% of it unavailable for the formation of calcium carbonate.

The existence of ion-pair formation is evident from examination of dissociation constants (Table 10-8), which reveal that most of the carbonate occurs as HCO_3^-, $MgCO_3^0$, and $CaCO_3^0$. Note that $MgCO_3^0$ and $CaCO_3^0$ are uncharged molecular species fully in solution, not small grains of magnesite and calcite floating in the water.

As shown in Table 10-8, the most abundant ion pair is formed by H^+ and CO_3^{2-}; i.e., the bicarbonate ion is the most abundant ion pair in the sea. Note also that the chloride ion does not form ion pairs to a measurable degree; i.e., chlorides are very soluble and the chloride ion exists in the sea as 100% free ion.

In summary, to determine the amount of an ion present and available to react in a concentrated solution, we need to consider not only its molality but also its activity and the proportion of the molality that is ion-paired with another species. In seawater the activities of all ions are less than their concentrations and the percent of free ion is, excepting chlorine, always less than 100%. Anions

TABLE 10-7 DISTRIBUTION IN SEAWATER OF IMPORTANT DISSOLVED SPECIES (19% chlorinity, 25°C, pH 8.15)

Ion	Total molality (i.e., concentration)	% Free ion	% $MeSO_4$–pair	% $MeHCO_3$–pair	% $MeCO_3$–pair	Total†
Na^+	0.459	98.5	1.5	—	—	100.0
K^+	0.010	99	1	—	—	100
Mg^{2+}	0.052	90	11	1	0.3	102.3
Ca^{2+}	0.010	84	14	2	0.3	100.3

Ion	Total molality (i.e., concentration)	% Free ion	% Ca–anion pair	% Mg–anion pair	% Na–anion pair	% K–anion pair	Total†
SO_4^{2-}	0.033	39	4	19	38		100
HCO_3^-	0.0022	69	4	19	8	—	100
CO_3	0.00027	9	7	67	17	—	100
Cl^-	0.535	100	—	—	—	—	100

† Data are from several sources and, therefore, do not always total to 100%.

TABLE 10-8 DISSOCIATION CONSTANTS OF THE ION PAIRS MOST ABUNDANT IN SEAWATER, EXPRESSED AS NEGATIVE LOGARITHMS (after Garrels and Christ, 1965, with minor modification)

	OH^-	HCO_3^-	CO_3^{2-}	SO_4^{2-}	Cl^-
H^+	14.0	6.35	10.33	2.0	n.m.a.
K^+	n.m.a.†	n.m.a.	n.m.a.	0.85	n.m.a.
Na^+	−0.7	−0.25	1.27	0.98	n.m.a.
Ca^{2+}	1.30	1.26	3.2	2.31	n.m.a.
Mg^{2+}	2.58	1.00	3.4	2.40	n.m.a.

† n.m.a. is no measurable association

are much more highly ion-paired than cations, the carbonate ion being the most paired of all.

Organic Substances

The distinction between particulate (living or dead organisms) and dissolved organic matter in seawater is somewhat arbitrary from an analytical viewpoint and, therefore, the average amount of dissolved organic matter in seawater is uncertain. Published estimates of the amount of dissolved organic carbon range from 0.2 to 3.0 ppm but, as there is no estimate of an average molecular weight for these carbonaceous substances, the meaning of these figures in terms of weight percent or molality of organic matter is not known. The sources of the organic compounds are almost entirely marine organisms which, while alive and also on death, release carbohydrates, proteins, peptides, etc., into the water. Organic productivity in the sea is so much larger than in fluvial systems that fluvial contributions of organic matter are trivial except in nearshore environments.

Composition of Ancient Seawater

Geologists study present events and then assume (as do all other scientists) the constancy of physical laws through time ("uniformitarianism") to interpret ancient events. But although we may accept that physical and chemical laws are unchanging, we cannot blindly assume that the quantities and ratios of substances on which these laws have acted has not changed. That is, we cannot assume, without supporting evidence, that the chemical composition of seawater has been constant through time. Yet such an assumption is critical to our understanding of diagenesis because most formation waters were initially samples of ancient seawater and if we can not assume its composition, we are indeed "at sea" when we try to interpret subsurface water chemistry and cementation patterns.

Fortunately, however, available data do support the view that the composition of the ocean has been very similar to that of the present for at least the past few billion years (Rubey, 1951). The most cogent points favoring this assumption are, first, the similarity between the types and relative amounts of recent and ancient sediments over the past 2 to 3 billion years; second, the overwhelming predominance of calcareous organisms (algae) among organic remains in Precambrian sediments, as is true in modern rocks, which implies constancy in the partial pressure of CO_2 in the atmosphere through time. And, as carbon dioxide is the basic buffer for pH in seawater, we may infer a hydrogen ion concentration for ancient seas similar to the present value of $10^{-8.15}$ moles/liter. These and other arguments favoring the assumption of constant composition of seawater through time are detailed in Rubey's classic paper. An evaluation of more recent data bearing on the history of seawater is given by Garrels and Mackenzie (1971, pp. 285–299) who reach a conclusion similar to Rubey's.

It is also generally assumed that the ratios among stable isotopes of each element in the world ocean have not changed significantly during geologic time but this assumption is less easily proved than is the assumption of constancy of chemical composition. Circumstantial evidence is provided by the fact that paleotemperatures of Mesozoic seas, determined using O^{18}/O^{16} ratios, give results that are compatible with seawater temperatures to be expected in the absence of extensive glaciers on the land surface. Therefore, the difference in O^{18}/O^{16} ratio between ancient and modern seas probably was small, if indeed any difference existed.

Subsurface Waters

Subsurface water is nearly always a mixture of juvenile waters, meteoric waters, ocean waters, waters produced from diagenetic or metamorphic reactions, and magmatic waters. Rarely, if ever, is it possible to recognize the relative importance of each of these sources in a subsurface water sample. In addition to the interpretive difficulties generated by the diversity of sources, all formation waters are more or less continuously modified compositionally by filtration through clay membranes (salt sieving), by ion exchange reactions, by precipitation of minerals, and by solution of surrounding rock. In summary, the history of almost any sample of subsurface water is extremely complex. In most investigations we are unable to determine the sources and earlier history of each chemical constituent and must be content with simply describing its present abundance and percentage of total dissolved solids, a circumstance that greatly hinders our attempt to understand diagenetic processes in general and the areal distribution of mineral cements in particular.

Range of Ionic Ratios

The initial formation water of most sedimentary rocks was seawater and if we assume that the composition of seawater has not changed appreciably through time, we can fruitfully compare the composition of modern formation waters with this standard. It is clear that if we evaporate normal seawater, the total salinity will increase but the ratios of dissolved constituents will not change, assuming no minerals are precipitated. With this fact in mind, let us examine the relative amounts of the abundant and petrologically important ions in concentrated subsurface brines, waters whose chlorinities are greater than 19,000 ppm and have, therefore, apparently suffered least from contamination by inflow of meteoric

TABLE 10-9 COMPARISON BETWEEN MOLALITIES OF DISSOLVED SPECIES IN MODERN SEAWATER AND IN "UNDILUTED" AND NONEVAPORITIC FORMATION WATERS (data from Chave, 1960; White, 1965; and Graf et al., 1966)

Characteristic	Modern seawater	Range in nonevaporitic formation waters	Mean direction of change
Chlorinity	0.535m	Up to 7.08m	——
mNa/mCl	0.86	0.15–0.95	Decrease
mMg/mCl	0.097	0.003–0.117	Decrease
mSO_4/mCl	0.052	7×10^{-6}–0.013	Decrease
mK/mCl	0.019	0.001–0.064	Decrease
mCa/mCl	0.019	0.008–0.34	Increase
$mHCO_3/mCl$	0.004	2×10^{-5}–0.024	?
mSr/mCl	1.7×10^{-4}	0.001–0.013	Increase
mH_4SiO_4/mCl	1.9×10^{-5}	3×10^{-5}–5×10^{-4}	Increase
mCa/mMg	0.192	0.56–26.92	Increase

waters (Table 10-9). Chlorine, being an anion, is retained beneath clay membranes (see below) and does not enter in significant quantities into authigenic minerals, short of evaporite precipitation. Hence it is a good standard against which to evaluate changes in abundance of other ions.

The differences between normal seawater and ancient formation waters are rather large and clearly significant in terms of processes known to occur during diagenesis. Calcium and strontium in formation waters exceed the amount in seawater because of dissolution of evaporites, carbonate minerals, and salt sieving. Magnesium is lowered in subsurface waters by dolomitization of calcite or by regrading and authigenesis of chlorite. Silica is slightly higher in formation waters because of the net effect of dissolution of diatom tests, the tendency of silica in solution to equilibrate with quartz during diagenesis, increase of silica by diagenetic "weathering" of detrital minerals, and loss of silica from solution by precipitation of chert and quartz overgrowths during diagenesis.

Sulfate is low in formation waters as a result of bacterial reduction to sulfide. Most published values of bicarbonate ion in formation waters are unreliable because of the sensitivity of this species to the pH of the water. During sampling, the water equilibrates with surface carbon dioxide pressure, causing changes in the quantity of bicarbonate present. Potassium is generally lower in formation waters than in seawater (the high value of 0.064 in the table is much larger than any others measured) because of regrading and authigenesis of illite in the subsurface but salt sieving may also be an important factor. Sodium decrease results from salt sieving.

There is no apparent correlation between ionic ratios and the age of the rock from which the subsurface water was obtained, although total salinities do tend to be higher in older or more deeply buried sediments. The source of the high dissolved salt content of deep subsurface brines is extremely complex. Dissolution of minerals, especially the carbonate and evaporite minerals, has locally made important contributions. This is well illustrated in exaggerated fashion by subsurface brines saturated with sodium chloride within sandstones associated with salt domes in south Louisiana. It is further well illustrated by increases of 2 to 3 times in total dissolved solids in subsurface waters between the carbonate Central Basin Platform of west Texas and the immediately adjacent evaporitic rocks of the Delaware Basin. This change is reflected in the rise to the near surface of the calculated gypsum/anhydrite boundary (Fig. 15.8).

In addition to the enrichment in total dissolved salts and changes in ionic species as a result of diagenetic changes in mineralogy, there also occurs a less selective increase in salinity of formation waters because of salt sieving. The concept of salt sieving, as originally suggested in 1947 by L. U. DeSitter, envisions lutites as semipermeable membranes that permit a selective escape of water upward during sediment compaction but retain dissolved salts in water remaining behind the membrane. DeSitter's hypothesis suggests that the energy for this process is provided by gravitational forces acting on the solid phases of the overlying rocks. During rapid sedimentation and compaction, pressures on interstitial waters are higher than hydrostatic pressure at comparable depth.

Bredehoeft et al. (1963), have suggested that highly saline formation waters can also result from concentration of dissolved matter contained in dilute meteoric waters migrating down dip from outcrops of an aquifer. Near the center of a basin, the water moves upward through confining strata and salt sieving occurs with the driving force provided by the hydrostatic pressure of meteoric waters in the updip parts of the aquifer. This same hydrostatic pressure may operate alone or may complement compaction as a driving force for salt sieving of the original seawater of the formation.

Studies of hydrogen and oxygen isotopic ratios in formation waters provide strong support for the concept of salt sieving. As noted on p. 324, evaporative processes cause increases in the O^{18}/O^{16} and H^2/H^1 ratios of the residual water. Compaction and filtration by charged net membranes, however, should not noticeably affect the isotopic ratios of the waters because only a single phase is involved. Therefore, by comparing the stable isotope distribution of modern seawater and highly saline (and therefore essentially undiluted) formation waters, it is possible to determine whether syngenetic evaporation is responsible for the high salinity of the waters. Results of these studies reveal the oxygen isotope ratios of brines more concentrated than seawater to be very similar to the ratio in normal seawater (δO^{18} in seawater is zero, by definition), as shown in Fig. 10-2. Nearly all values for these brines are within 3/mil of mean seawater, a rather insignificant difference considering the large range in brine concentrations involved. Therefore, these brines were not concentrated by syngenetic evaporation. Studies of variations in the H^2/H^1 ratio in brines from many areas indicate that the δO^{18} variations also do not result from exchange with reservoir rocks. Therefore, they must be explained in terms of salt sieving.

Waters less saline than seawater have been diluted by meteoric or fresh water inflow at some time after burial of the enclosing sediment, as indicated by the negative δO^{18} values. The scatter of the plotted δO^{18} points at each salinity value

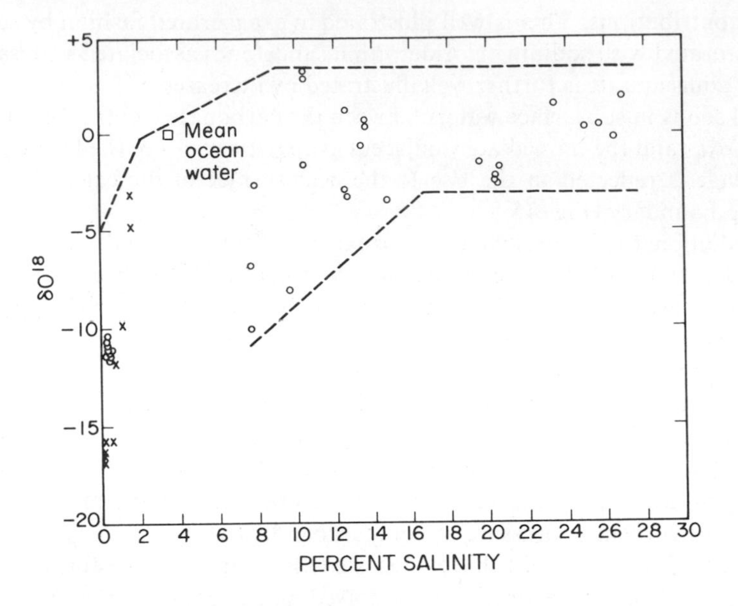

Fig. 10-2 Correlation between salinity and δO^{18} of marine (○) and non-marine (×) formation waters. (Data from Degens, et al., 1964.)

may result from several causes, among which are (a) meteoric and fresh waters do not all have the same δO^{18} values and will, therefore, cause different amounts of O^{16} enrichment of formation waters. The increasingly wide scatter of δO^{18} values at salinities less than 15% may mean that these waters were diluted by low salinity, low O^{18} waters but have since reelevated their salinities by salt sieving. Changes in O^{18}/O^{16} ratio caused by dilution, however, have a greater permanency than salinity decreases caused by dilution. In this connection, it is noteworthy that hydrochemical studies of water within a single aquifer have shown that oxygen isotopes are not fractionated during lateral movement of at least 700 mi at shallow subsurface depths. (b) Reaction occurs between formation waters and mineral grains in the enclosing rocks, altering the δO^{18} value of both. Not only do detrital minerals interact with subsurface waters, but, as we discuss in subsequent pages, early-formed cements may be dissolved and replaced with new cements of either the same or different mineral composition.

The mechanism of salt sieving depends on the fact that smectite and illite structures have internal charge deficiencies that are balanced by adsorption of exchangeable cations. During early stages of compaction when the porosity of lutites is large, channels between clay flakes are relatively large so that water and its dissolved salts pass through the sediment in proportions approximately equal to their abundance on the underside of the clay membrane. With increasing compaction however, the interflake spaces decrease in diameter and the fixed negative charges on adjacent flakes become so close together that anions in solution on the underside of the membrane are repelled and must be retained there. The adsorbed cations are still free to move between adjacent exchange sites and through the membrane but any movement of cations through the membrane must be compensated by an electrically equivalent inflow from the reverse direction. In general, the cation most easily diffusable to the underside of the membrane is hydrogen, which results in the formation of a pH difference between the topside and underside of the membrane. White (1965, p. 352) cites three waters believed on other criteria to be membrane filtered and their pH values range from 7.1 to 7.6. Two waters believed to be membrane concentrated have pH values of 6.2 and 6.8, in agreement with predictions based on semipermeable membrane theory. There are very few reliable pH measurements of subsurface waters, however, and the five values cited by White cannot be taken as proof of the correctness or adequacy of the explanation proposed. But if the suggested mechanism is correct, it makes available a significant way for pH changes to occur in subsurface waters, with its attendant effects on the solubilities of mineral cements, particularly carbonates.

The concept that metallic cations may pass through clay membranes in the subsurface may be expanded to include the concept of differential mobilities of cations (Fig. 10-3). But which ionic radius is the determinant in such considera-

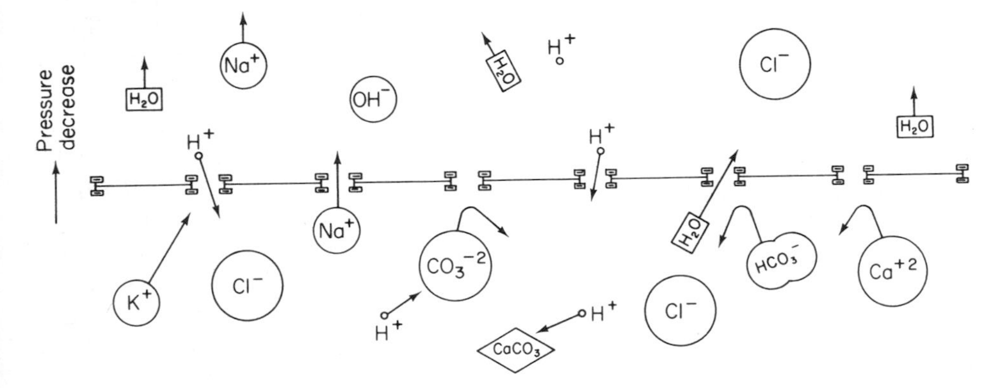

Fig. 10-3 Schematic diagram illustrating differential movement of some dissolved species of petrologic importance in subsurface waters. (Modified from White, 1965.)

tions? Is it hydrated radii, which follow the order Li > Na > K > Rb; or is it naked radii, which have the reverse size ordering (see Chap. 7). In a recent study, Billings et al. (1969) found formation waters in the western Canada sedimentary basin to retain alkali metals in the water on the undersides of clay membranes in the order Rb > K > Na > Li. The controlling mechanism for migration through the membrane, therefore, cannot be pore size because the free ions "floating" through the pores are hydrated and the retention order resulting would be the reverse of that found. Billings et al. therefore suggested that the alkali ions move through the membrane by site hopping on the clay surfaces and as the ease of replaceability during site hopping is Li > Na > K > Rb, the concentration pattern in the remaining brine should be Rb > Na > K > Li, precisely the pattern revealed by their data. As divalent cations are bonded to clay surfaces more strongly than univalent ones, we would expect enrichment of calcium in formation waters despite extensive precipitation of calcite cement during diagenesis. The enrichment of calcium will affect carbonate equilibria by increasing the ion activity product $[Ca^{2+}][CO_3^{2-}]$ in the water toward attainment of the thermodynamic solubility product required to effect precipitation of calcite cement. On the other hand, diffusion of hydrogen ions toward the underside of the membrane will cause an increase in the HCO_3^-/CO_3^{2-} ratio. Clearly, the processes affecting the availability of calcium ions, carbonate ions, and the precipitation of calcite cement from subsurface waters are complex and there is no doubt that our present understanding of them is rather primitive. We return to the geochemistry of calcium carbonate in Chaps. 12 and 13.

The chemical character of subsurface brines may vary quite markedly over very small distances laterally or stratigraphically, as can be inferred from field observations of cementation patterns in sandstones. It is common to find differences in the amount and kind of cement present in closely adjacent strata or in the same formation at nearby locations, demonstrating that chemical activity has varied greatly in beds where geological histories are otherwise practically identical. In addition to the processes of salt sieving and ion-pair formation, such differences in cementation patterns and chemistry of formation waters may result from the presence of evaporites at some levels in the stratigraphic section, from the presence of unconformities, or perhaps from nonrandom geometry and stratigraphic distribution of sandy and shaly units.

Another important but neglected mechanism for modification of subsurface waters is through mixing of highly evolved formation waters from different stratigraphic units. Such mixing may result in either undersaturation or supersaturation of the water with respect to cementing materials, with accompanying potential for leaching or precipitation. Russian geologists have reported precipitation of calcium and magnesium carbonates because of encroachment of seawater into subsurface oil field brines. They also reported an interesting example of chemical variation in formation waters with depth in the Donetz Basin as follows: calcium-bicarbonate water near the surface, superseded successively at depth by sodium-bicarbonate water, sodium-sulfate water, sodium-bicarbonate water, sodium-bicarbonate-chloride water, and possibly saline sodium-chloride water. It takes little imagination to conceive of the variety of precipitates that might be formed by mixing of these waters as a consequence of faulting, jointing, or dissolution of earlier-formed cements.

Organic Matter

The detection of organic constituents in subsurface waters is obviously a useful clue to the occurrence of petroleum in the area and this has resulted in a great deal of study of brines associated with commercial deposits of oil and gas. The recent upsurge of interest in organic pollution of groundwaters promises to result in equal attention to the more common nonpetroliferous waters but so far the organic constituents of such waters have been largely neglected.

In order to know what concentrations of dissolved hydrocarbons to expect in subsurface brines, it is necessary to know the solubility in water of each pure

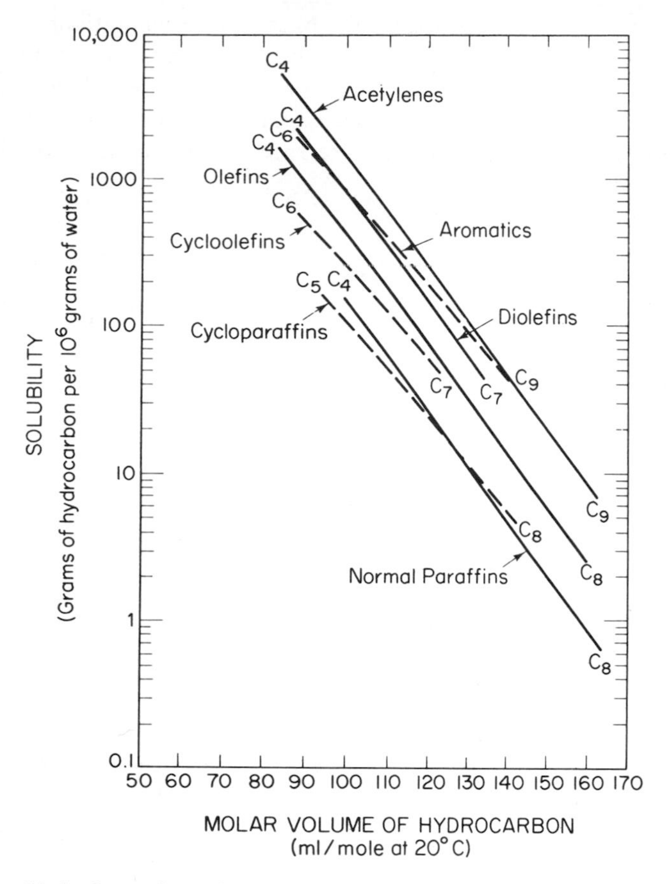

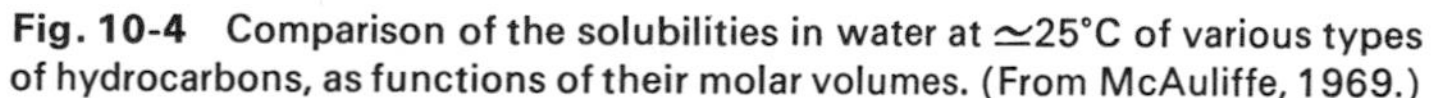
Fig. 10-4 Comparison of the solubilities in water at ≃25°C of various types of hydrocarbons, as functions of their molar volumes. (From McAuliffe, 1969.)

hydrocarbon. A subsurface brine in equilibrium with a crude oil should have dissolved in it individual hydrocarbons in proportion to their concentrations in the crude oil. Figure 10-4 illustrates the magnitudes of these solubilities as a function of various homologous series and as normal paraffins are the most abundant series in petroleum, the most abundant hydrocarbon in brines should be methane. This expectation is borne out by analyses. For example, Buckley et al. (1958) examined brines from approximately 300 wells along the Gulf Coast and from rocks ranging in age from Jurassic to Miocene. In these brines methane was dominant, as in a sample from the highly petroliferous Woodbine Formation in which the alkanes were 96.64% methane, 2.45% ethane, 0.70% propane, 0.17% butane, and 0.04% pentane by volume. The aggregate quantity of hydrocarbon gas dissolved in the 300 brines examined is enormous, averaging about 0.03% by weight or 0.0015 mol % based on methane, and exceeding the proved gas reserves at that time. In some formations the water was nearly saturated at the existing hydrostatic pressure; in others the water was substantially undersaturated. It is thought that anaerobic bacterial fermentations are the major source of the methane but the origin of the heavier alkane gases is uncertain.

Degens et al. (1964) studied the amino acids in petroleum brine waters; they found a positive correlation between amino acid content and brine salinity and no significant difference in the spectrum of amino acids between ancient waters and modern seawater.

Hydrochemical Facies

One useful way of visualizing areal or stratigraphic changes in water chemistry within a basin is by constructing contour maps, from which we can see hydrochemical facies. These facies reflect the effects of chemical processes occurring

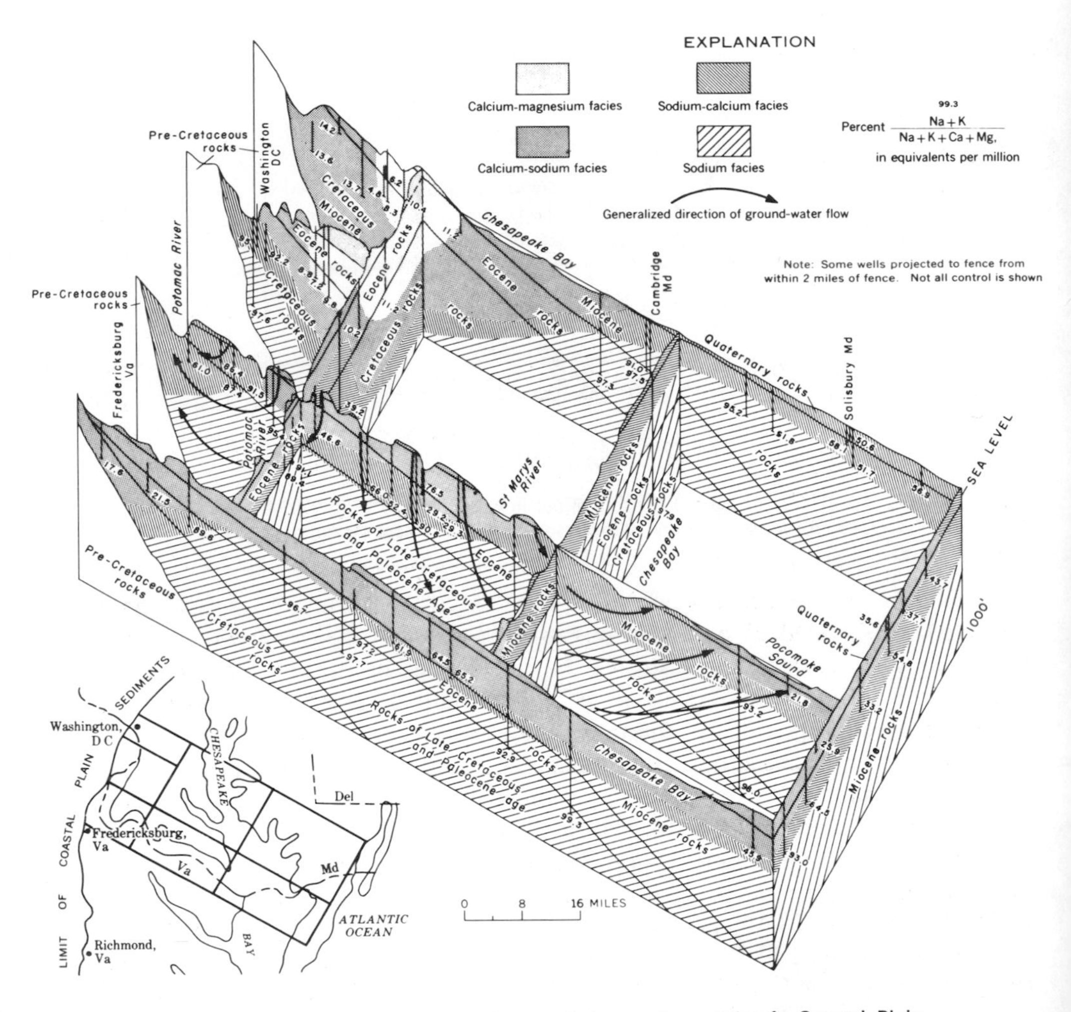

Fig. 10-5 Fence diagram of part of the northern Atlantic Coastal Plain showing hydrochemical facies. (From Back, 1961.)

between the minerals within the lithologic framework and the formation water. These processes change in magnitude and direction with time and hydrochemical facies maps are simply pictures of water quality during an instant in time. The facies boundaries are continually modified by the chemical composition of influent water in the recharge area, chemical reactions among ions in solution which were introduced from recharge areas, effects of salt sieving in underlying rocks, changes in porosity and permeability patterns in the formations, and variations in temperature with depth in the basin or with structural position in a dipping lithologic unit.

Fence diagrams are particularly informative as summaries of a great deal of data, as illustrated by Back (1961) in his study of the hydrology of the northern Atlantic Coastal Plain (Fig. 10-5). A very useful feature of such diagrams is that the relationship between lithology and formation water chemistry is clearly evident. The calcium-magnesium facies is essentially coextensive with the recharge area underlying the plateau of southern Maryland. As water enters through Miocene beds containing calcareous clays, the calcium-magnesium facies is developed. The water continues its downward and lateral movement and exchangeable sodium adsorbed on clay particles gradually converts the water through intermediate facies to the sodium facies. Further mapping of the sodium facies substantiates the conclusion, based on other evidence, that the outcrop of Cretaceous sediments functions as a discharge area along the Potomac River rather than as an area of recharge as it commonly does in the Atlantic Coastal Plain.

The summary chemical characteristics of a hydrochemical facies are commonly given in terms of the one or two most abundant cations or anions; e.g., calcium chloride waters, sodium sulfate waters, etc. The relative proportions of the ions present may be shown more specifically on triangular plots as is done in sedimentary rock classification. The most generally useful end-members seem to be calcium-magnesium-alkalis and chloride-sulfate-bicarbonate but other types of plots or mathematical representations may be more informative for particular waters.

10.3 POROSITY, PORE SIZE DISTRIBUTION, AND PERMEABILITY

Porosity

The porosity of a sandstone is defined as the ratio of the volume of void space to the bulk volume of the rock. Pore space is an element of the fabric in any sedimentary rock; it has a history and can be and should be studied as part of the rock. It is an unfortunate circumstance that few geologists outside of industry report porosity values of sandstones despite the fact that porosity changes are a prime control on lithification processes.

It is generally true that porosity values in sandstones decrease with depth of burial. Several factors are responsible for this result. (1) The solubility of quartz increases with temperature and pressure, resulting in cementation and pressure solution phenomena. (2) Most sandstones contain significant amounts of easily deformable materials such as clay and foliated lithic fragments. With increasing depth and pressure these grains are squeezed into existing voids. (3) The salinity of pore waters tends to increase with depth, enhancing the possibility of clay mineral authigenesis and recrystallization, which eliminate pore spaces.

In a well documented study involving data from more than 17,000 cores of Miocene and younger rocks from boreholes in south Louisiana, Atwater and Miller (1965) showed that porosity decreases linearly with increasing depth in the interval 2,000–20,000 feet. The defining regression line is: depth = 32,000–800 porosity (%). Assuming the average composition of these sands to be similar to modern Mississippi River sands, we may hypothesize a mean mineralogy of about 70% quartz and chert, 20% feldspar, and 10% foliated lithic fragments. The amount of detrital clay in the Tertiary sands is, however, unknown. The sands are normally unconsolidated or friable and the chief cause of porosity reduction appears to be increasing deformation with depth of less resistant lithic fragments and clay matrix. There is no doubt that early precipitation of pore-filling chemical cements, which occurs in some sedimentary basins, would greatly modify the porosity–depth relationship found by Atwater and Miller.

Pore Size Distribution

Porosity represents a bulk or general property of a rock. However, the actual size of pore space or more specifically the distribution of sizes of pores within a rock is critical to the distribution of fluid within a rock when two fluids exist. In general, the fluid with the greater wetting capability with respect to the mineral surfaces will exist within the smaller pores. The distribution of pore size also will determine the ability of fluid to flow through the rock. Distribution of pore size is a property of a rock in the same sense that distribution of grain size is a significant property of a detrital rock.

There is no simple method for determining the size of pores within a rock. Pore space for the most part represents a series of micro-cavities interconnected by smaller entryways. Toward the edges of the micro-cavities the actual size of the pore normally decreases gradually. Pores may be observed and measured in thin sections; however, the thickness of the thin section places a limit on the size of the pore that can be seen. In a poorly consolidated sandstone the size of the pores and thus the distribution of sizes may be estimated if the size of the grains and their packing is known. The laboratory method commonly used for determination of pore size distribution is mercury injection. Mercury is a nonwetting fluid with respect to most mineral surfaces and, therefore, pressure is required to force mercury into the pore space of rocks. The pressure is necessary to overcome the capillary pressure of the pores and the size of a pore that will be entered by mercury is a direct function of the pressure. For example, a pore with a size equivalent to a capillary tube having a diameter of 0.002 mm will be entered by mercury at approximately 100 psi. Thus a plot of pressure exerted against volume of mercury injected at that pressure is a good approximation of the volume of pore space existing in each size range. It must be remembered that in measuring pores we are not measuring the size of discrete objects such as grains but the amount of pore space that exists in a continuous network of pore space in which the dimensions of the space are continually changing in all directions. If a large pore such as a vug is entered by fluid through a network of smaller pores, then the large vug will be included within the volume of pore space represented by the smaller size. Only if the vugs are interconnected by pore space of similar size will their true size be approximated by the method.

Permeability

Permeability is a measure of the ease of flow of a fluid through a porous rock. It is commonly expressed by the symbol k in the Darcy's law equation (Sec. 1-5). Permeability is one of the more important bulk properties of a porous rock in that it strongly influences the economic value of a rock in terms of oil, gas, or water production. In addition, because most of the diagenetic changes that take place within a sedimentary rock result from reactions between moving water solutions and the rock, the ability of the rock to transmit these fluids in a geologically significant length of time strongly determines the rate and intensity of diagenesis.

Permeability values range widely but some generalizations may be made.

TABLE 10-10 TYPICAL PERMEABILITIES OF SEDIMENTS AND SEDIMENTARY ROCKS

Tightly cemented crinoidal limestone	10^{-6} md.
Sucrose dolomite	0.1 md.–150 md.
Cemented quartz sandstone or carbonate grainstone	10 md.–300 md.
Poorly cemented quartz sandstone or carbonate grainstone	300 md.–5000 md.
Uncemented carbonate mud or silicate mud	0.01 md.–10 md.
Unconsolidated quartz sandstone or carbonate grainstone	In excess of 1000 md. depending on size and sorting of grains

A permeability of 1 darcy or 1000 millidarcys represents 1 cc of fluid with a viscosity of 1 cp. flowing through a cross-sectional area of 1 sq cm of rock in 1 sec under a pressure gradient of 1 atm/cm of length in the direction of flow.

It is obvious that in considering the diagenesis of a sediment or rock the pressure exerted on the fluid is important in determining whether large volumes of fluid will pass through the material. As diagenetic changes that reduce the permeability take place, the amount of further diagenesis that may take place under a given set of conditions is drastically reduced. When we observe a sandstone that is completely cemented with quartz or calcite and that has little or no measurable porosity or permeability, we wonder how water could have been transmitted through such a rock. What we must remember is that the original sandy sediment had a permeability of several darcys and a porosity of 25 to 40%. Thus, it had the potential to transmit fluid, given a reasonable pressure differential.

Porosity in sandstone is originally determined by the shape, packing, and sorting of the individual grains. Any deviation from spherical shape normally produces a less efficient use of space and an increase in porosity because grains of irregular shape can be packed to use space less efficiently. When the individual grains differ widely in size (poor sorting), there exists the possibility of fitting smaller particles between larger ones, thus occupying space that would otherwise be pore space. This has the general effect of decreasing porosity. Poor sorting also has the effect of increasing the relative number of small pores and thus decreases the permeability of the rock.

In general, almost all diagenetic processes in detrital silicate rocks cause a decrease in porosity and permeability. Thus the longer and more intense the diagenetic history, the lower the porosity. Two exceptions may be cited to this generalization. If the rock originally contained carbonate particles such as shell fragments, they may be removed by dissolution, thus increasing porosity. Unless these dissolved particles are very numerous, their removal will probably have little effect on permeability. The development of fractures within the rock may have a significant effect on permeability with little effect on porosity. The addition of mineral cements and compaction have the effect of decreasing porosity, pore size, and permeability.

The development of open fractures in a rock has the effect of producing large-scale tabular conduits for fluid movement. Where the total contribution of fractures in highly fractured rock to porosity has been evaluated, the contribution is commonly less than 0.1%. Thus the storage capacity of fractures for fluid is small; however, there is no question that they may have a significant effect on the permeability of the rock. Indeed, they play an important role in the movement of diagenetic fluid and in the ease with which oil, gas, or water may be removed from a reservoir.

10.4 ORIGIN OF CHEMICAL CEMENTS

Diagenetic Temperatures

In the previous section we considered the general chemical character of formation waters. We now consider these waters in relation to the precipitation from them of the common primary cementing agents of sandstones: silica, calcite, and hematite; and, in this regard, we must concern ourselves with parameters in addition to those noted previously. These additional parameters include temperature, hydrostatic and nonhydrostatic stresses, stability fields of minerals, and time. Perhaps the most important of these, and one about which there is little reliable information relevant to sedimentary processes, is temperature. Bottom-hole temperature readings probably give fairly reliable results for a Cenozoic basin to depths of perhaps 15,000 ft. For greater depths in geosynclinal piles, published data are scarce and consist largely of temperatures logged immediately after cessation of drilling and active mud circulation and hence are of dubious value.

Maxwell (1964) collated the most reliable equilibrium temperatures measured in deep wells in sedimentary basins in the United States (Fig. 10-6) and his data

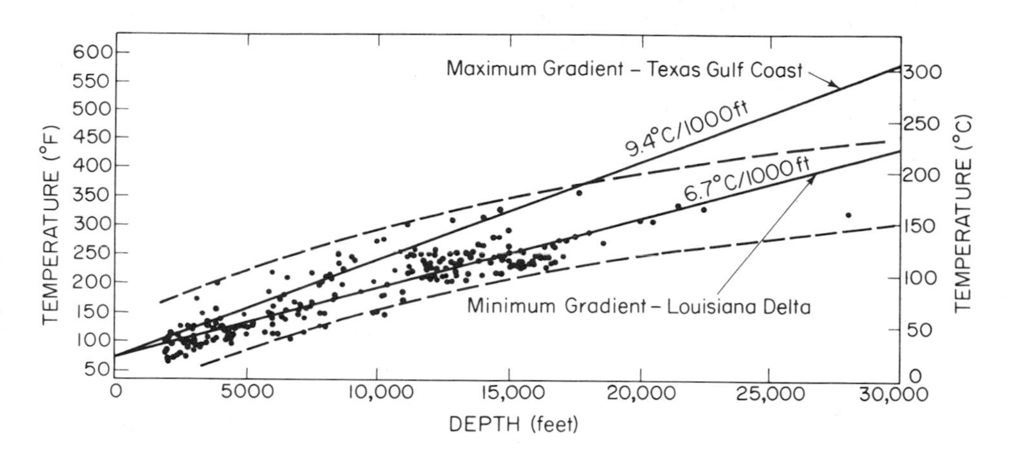

Fig. 10-6 Temperature-depth plot. Data from wells in Gulf Coast of Texas and Louisiana with a few values from Cenozoic basins in California and Rocky Mountains. See text for further explanation. (From Maxwell, 1964.)

indicate that temperatures exceeding 200°C can occur when burial depth exceeds about 20,000 ft. But are the processes occurring at 200°C to be classed as cementation in the "sedimentary environment" or as "zeolite facies metamorphism" (Chap. 18)? Perhaps a more pertinent question is "What difference does it make?" The boundary between cementation or diagenesis in general and metamorphism is gradational, as is true for most natural phenomena. There is no single criterion on which differentiation can be made that is usable for all rocks. For example, calcium carbonate in sediments may recrystallize at near-surface temperatures. Illitic clay cement may be chemically and physically upgraded to sericite and muscovite at perhaps 100°C. Diagenetic or detrital quartz recrystallizes above 150°C and recrystallization of quartz in sandstones precedes that of orthoclase. The best procedure in diagenetic studies is to be aware of the effects of physicochemical changes on the mineralogy of sediments so that these changes can be interpreted in terms of processes. It is pointless to be concerned with whether certain "diagenetic" effects are more properly considered as "metamorphism."

Jam L. et al. (1969) have made an interesting study of subsurface well temperatures in south Louisiana and reported values within the temperature-depth envelope reported by Maxwell. In addition, they constructed a contour map for the area showing isotherms at a depth of 10,000 ft (Fig. 10-7), illustrating a hot

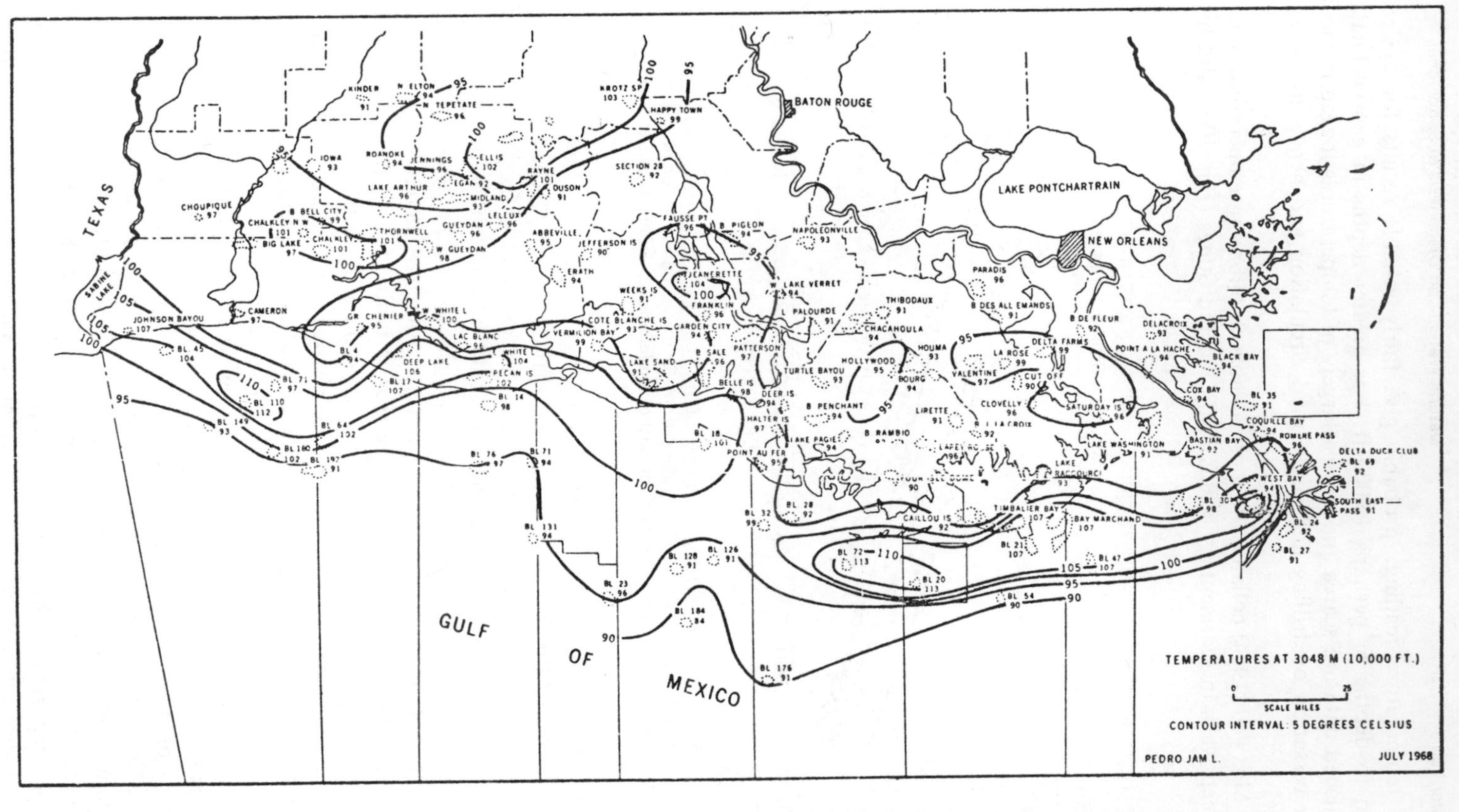

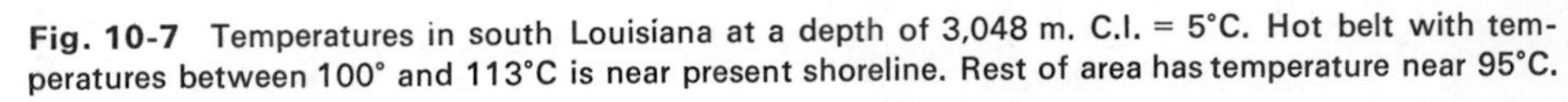

Fig. 10-7 Temperatures in south Louisiana at a depth of 3,048 m. C.I. = 5°C. Hot belt with temperatures between 100° and 113°C is near present shoreline. Rest of area has temperature near 95°C.

belt slightly south of the present coastline with temperatures 10 to 15°C hotter than the surrounding area. Extrapolating their geothermobaric curve to a depth of 50,000 ft, the thickness of the sedimentary pile under the hot belt, they concluded that the base of the section was undergoing albite-epidote hornfels facies metamorphism. Clearly, the controls on temperature at depth in a sedimentary section are complex and significant local variations may occur along iso-depth surfaces, particularly in geosynclinal sections.

Temperature and Rank of Coal

Coal is particularly sensitive to changes in temperature and is, therefore, very useful for the determination of diagenetic grade. The common type of coal (bituminous coal) is a black friable substance composed of carbon, oxygen, and hydrogen, with minor amounts of nitrogen, sulfur, water, and incombustible ash. Use of the microscope reveals that coals are composed of a number of types of material such as wood, spores, or cuticles derived from plant materials. The term *phyteral* has been used to designate the different plant forms, as distinct from the organic materials of which the phyterals and other, structureless parts of the coal are composed. These organic substances, whose composition and physical properties are fairly distinct though not fixed, are called *macerals.* They are the equivalent for coal of the minerals composing normal rocks but they are not true minerals because they lack a definite crystalline structure. As with other rocks, certain combinations of macerals and textures are recognizable, even without thin section examination, and have been named coal types or lithotypes. The best known of these are the four types of coal composing common bituminous coals, as described by Marie Stopes in 1919 (Table 10-11).

TABLE 10-11 COAL TYPES AS DEFINED BY TEXTURE (after Stopes, 1919)

Vitrain	"a coherent and uniform whole, brilliant, glossy, indeed vitreous in its texture;" occurs in thin horizontal bands, friable, breaks with a clean, conchoidal fracture.
Clarain	"has a definite and smooth surface when broken at right angles to the bedding plane and these faces have pronounced gloss or shine: the surface lustre is seen to be inherently banded."
Durain	"hard, with a close, firm texture, which appears rather granular to the naked eye; a broken face is never truly smooth but always has a fine lumpy or mat surface."
Fusain	"occurs chiefly as patches or wedges; it consists of powdery, readily detachable, somewhat fibrous strands." This is the constituent in coal that soils the fingers and has been described as "mineral charcoal."

Some of the differences between coals are due to differences in the nature of the original plant materials and environments of accumulation. These differences are, however, relatively minor in importance as compared with differences due to rank, i.e., to the relative position in the coal series. The main trends with increasing rank include (a) increase in content of carbon; (b) decrease in content of hydrogen; (c) decrease in the proportion of volatiles evolved by distillation;

(d) increase in the caloric value; (e) increase in depth of color, luster, and reflectivity of the coal constituents.

In a recent review, M. and R. Teichmüller (*in* Murchison and Westoll, 1968) have demonstrated, with many detailed examples, that changes in rank are related mainly to variations in the maximum temperature to which the coal has been exposed. The effect of pressure appears secondary and relatively unimportant. For example, in the Upper Rhine rift valley, where the geothermal gradient is high (8°C/100 m), bituminous coals of Tertiary age occur at a depth of only 2300 m. North of the Alps in Bavaria, where the geothermal gradient is low, subbituminous coals are found as deep as 4000 m. Locally anomalous occurrences of high rank coals have been shown by geophysical surveys to be related to underlying igneous intrusions. Other cases of formation of high rank coals by thermal metamorphism are known. For example, in Cretaceous coal near Madrid, New Mexico, the normal bituminous material has been increased in rank to anthracite near Tertiary dikes and sills. The role of temperature in causing increasing rank with depth is strikingly demonstrated by studying the rank variation in stratigraphic sections characterized by alternations of predominantly sandy and shaly formations. The sands conduct heat better than the shales so that the thermal gradient is less steep in sandy than in shaly formations. The gradient of rank variation is also found to be less pronounced in sandy than in shaly formations.

Temperature increase may be produced by diastrophism but studies in Germany have not supported the idea that increase in rank can be produced by tectonic pressures. For example, the more highly folded parts of coal basins do not invariably contain coals of higher rank. In some basins, where a correlation between folding and rank has been observed, the rank variation can be explained more plausibly as a result of greater temperature than as a result of tectonic pressures. In other basins, it can be demonstrated that the rank was established before folding took place.

Temperature appears not to be the only factor of importance. Rank variations exhibited by coals and coaly inclusions in the Carboniferous of the Lower Elbe Basin, Germany, and the Upper Miocene of the Gulf Coast are distinctly different, even though the temperature range and depth of burial are similar in the two cases. It is suggested by M. and R. Teichmüller that the critical difference between these two cases is the time available for coalification. In the Gulf Coast this amounts to less than 20 million years, as compared with nearly 270 million years in Germany.

A significant aspect of these studies of rank variation is the powerful tool that is provided for investigating diagenetic changes in other sedimentary rocks. A reliable index of rank may be obtained by studying the properties, such as optical reflectance, of a single coal maceral (for example, vitrinite). There is no other index of diagenetic "grade" known to be as sensitive as that of coal rank. A few studies have already shown that coal rank is correlated with the porosity of sandstones, with the bulk properties of shales (for example, density and sonic velocity), with the type and crystallinity of clay minerals, and with alterations in the detrital minerals (Kisch, 1969). Such studies are by no means restricted to coal-bearing strata because vitrinites suitable for rank determination occur also as inclusions in shales. The potential of these studies for future investigations of diagenesis appears great.

Lithification

One of the clearest effects produced by diagenesis is the change from a loose sand to a sandstone that has strength (lithification). The strength is due in part to the interlocking of crystals that results from the growth of mineral cements into pore spaces but it also results from attractive forces between crystal surfaces. Most of what is known concerning these forces has been obtained by metallurgists and it is difficult to apply their data to silicate rocks. Chemical bonding or electrical interactions within and between silicate mineral structures are largely covalent and directional in character rather than nondirectional as is true of metallic bonds. Also, the bonds of interest in the cementation of clastic rocks are commonly between minerals of entirely different crystal structure, such as quartz and calcite, feldspar and hematite, or dolomite and a volcanic rock fragment. Further complication is introduced by the asymmetry of crystal structures, which may cause differences of several orders of magnitude in surface energy between, for example, a quartz grain surface parallel to {0001} as contrasted to a surface parallel to {0011}.

The least surface energy for a solid-solid mineral boundary is obtained between clean surfaces of crystals of the same species, particularly when the two crystal structures are in parallel orientation. This is the reason quartz-cemented quartz sandstones are as apt to fracture through the clastic grains as around them. This also is the reason evaporites, limestones, and "crystalline" rocks are coherent. In clastic rocks there probably is extensive cleaning of the surfaces of detrital grains by circulating pore waters before precipitation of mineral cements. The lack of a visible boundary between host grain and secondary growth in many quartz-cemented quartz sandstones demonstrates the effectiveness of scrubbing. Electron microscopy reveals crystallographically controlled etch pits on all detrital grains and most of these markings may be produced during diagenesis.

Two general mechanisms exist for moving large quantities of dissolved substances through rocks. (a) Transfer by bodily movement of the solution through the rock pores. The physical movement of fluids through rocks is determined by the existence of pressure gradients. These may be developed because of differences in elevation of the free water surface, differences in density of the fluids, pressure on the fluid within rocks developed by compaction of the rock or introduction of new fluid within the rock by diagenetic changes such as replacement of gypsum by anhydrite. (b) Transfer as a result of diffusion gradients between adjacent parts of the water mass. Charged particles of ionic size can move through openings too small to permit movement of water (surface tension). The rate of diffusion of ions in natural waters is only 1 cm/day, in contrast to about 10 cm/day for groundwater movement. Therefore, diffusion of ions is much less effective a mechanism for effecting cementation than is subsurface water movement. It is worth noting, however, that the mean direction of ionic diffusion may differ from the direction of movement of the water mass itself. Also, although the movement of water through rock pores decreases markedly as the amount of clay matrix increases, diffusion is uninhibited. Even shales conduct electricity, indicating that ionic diffusion is not inhibited by permeability decrease.

Differences in the amount and kind of cement present in closely adjacent strata or in the same formation at nearby places demonstrate that chemical activity has varied greatly in beds where geological histories otherwise are identical. Permeability differences due to locally increased clay content, differential compaction, sedimentary structures, or initial spotty cementation may exist almost from the moment of deposition. And certainly they may be generated after deposition by the complex assemblage of processes that characterizes diagenesis.

It also is true that there is no single sequence of cementation phenomena that is applicable to all clastic rocks. As temperature, chemical composition of groundwaters, pH, and other parameters change, so also must one cement succeed a previous one. From petrographic studies it is known that any cement can be replaced by any other cement. A temporal sequence of replacements in one sandstone may be reversed in another. Our view in thin section is usually confined to the last cement, although in some rocks fragmentary evidence exists concerning the nature of previous cements. Most commonly this evidence consists of partially digested inclusions of one mineral within another, e.g., calcite shreds in quartz cement in many sandstones. Calcite is much more soluble than quartz and, therefore, is relatively easily removed to reestablish some or all of the original porosity of the clastic sediment. It seems likely that decementation is an important process in the subsurface but evidence supporting this idea is, by the nature of things, difficult to come by. Certainly decementation occurs on surface exposures of clastic rocks so there seems to be no reason to question its production by circulating subsurface waters.

Silica Cements

Occurrence

Diagenetic silica appears in sandstones in many morphologic and crystalline forms. Among the crystalline forms, cristobalite occurs as fibrous rim cement around detrital grains in volcanic sandstones and as submicroscopic crystallites in opaline cement or clastic opal grains. The fibrous morphology of cristobalite cement in sandstones makes it difficult to distinguish from chalcedony in thin section and recourse must commonly be made to X-ray diffraction analysis. Cristobalite cement (often called lussatite) has only been identified in sandstones containing obvious volcanic detrital components such as quartz with beta outlines, felsite fragments, or sanidine grains. Tridymite cement has not been reported from detrital rocks although tridymite probably occurs as clastic fragments in some volcanic terranes.

Quartz produced during diagenesis may occur as microcrystalline subhedral blocks (chert), as a coarsely crystalline mosaic of subequant blocks, as elongate fibers (chalcedony), as euhedral secondary growths on detrital quartz grains, as micro- or macro-quartz replacement of carbonate minerals (either fossils or matrix), and less commonly as replacement of glauconite or other materials. The reason why one morphology is present rather than another is not known, although vague allusions to "degree of supersaturation," number or areal distribution of nuclei, presence of foreign ions, etc., are numerous in published works. From a thermodynamic viewpoint, only macrocrystalline morphology is stable. Therefore, kinetic factors must be responsible for both the initial occurrence and also the preservation of chert and chalcedony.

Euhedral overgrowths on detrital quartz nuclei are a common feature of many friable quartz sandstones and have been produced in the laboratory at room temperature (Mackenzie and Gees, 1971). As these growths nucleate onto an existing crystal structure, it is thermodynamically required that they continue the existing structure. Consequently, secondary growths are in crystallographic continuity with the host grain even when the host grain is plastically deformed. If crystallographic continuity is not present, it may be inferred that the silica nucleated on an adjacent grain (perhaps a grain not in the plane of the thin section) rather than directly onto the observed detrital grains.

In thin section the boundary between host grain and overgrowth commonly is marked by impurities that form an incomplete coating on the detrital grain. These impurities are most often iron oxide, clay, water bubbles, or petrographically irresolvable "dirt." Many quartz grains have been "scrubbed clean" by groundwaters prior to overgrowth precipitation, however, and no host-overgrowth boundary is visible using standard microscopic techniques. Many orthoquartzites with "metamorphic" texture have been shown by use of luminescence petrography to be silica-cemented quartz sands (Sippel, 1968).

Chemistry

Amorphous silica will dissolve either in distilled or saline waters at room temperature in amounts which range between 100–140 ppm, the variability resulting from the fact that "amorphous silica" is neither entirely silica nor truly amorphous. Natural solid silica (opal) contains small amounts of impurity atoms and also contains cristobalite and tridymite crystallites in small but significant amounts. Usually, 120 ppm is cited as the equilibrium solubility of amorphous silica and this is sufficiently accurate for most geologic purposes. The solubility of amorphous silica increases with temperature and, as illustrated in Fig.10-8, the increase is large, even at shallow burial depths. Burial to only 300 ft increases solubility by 10 % and solubility is doubled when burial depth is 3,000 ft. Thus, it is clear that opaline cement in sandstones may be safely interpreted to mean cementation at shallow depths. In deeply buried sandstones, silica cement will nearly always be quartz, either microcrystalline or macrocrystalline. There are two reasons for this. (a) The solubility of quartz is only 5 % that of amorphous silica at room temperature and the solubility of quartz rises much less rapidly with temperature than does opal solubility. (b) The ease of precipitating an ordered phase (i.e., quartz as contrasted to opal) increases as temperature increases. Thus, at higher temperatures the silica-rich solution establishes equilibrium with quartz rather than amorphous silica. This is in contrast to the relationship at surface temperatures where the difficulty of precipitating crystalline quartz usually establishes

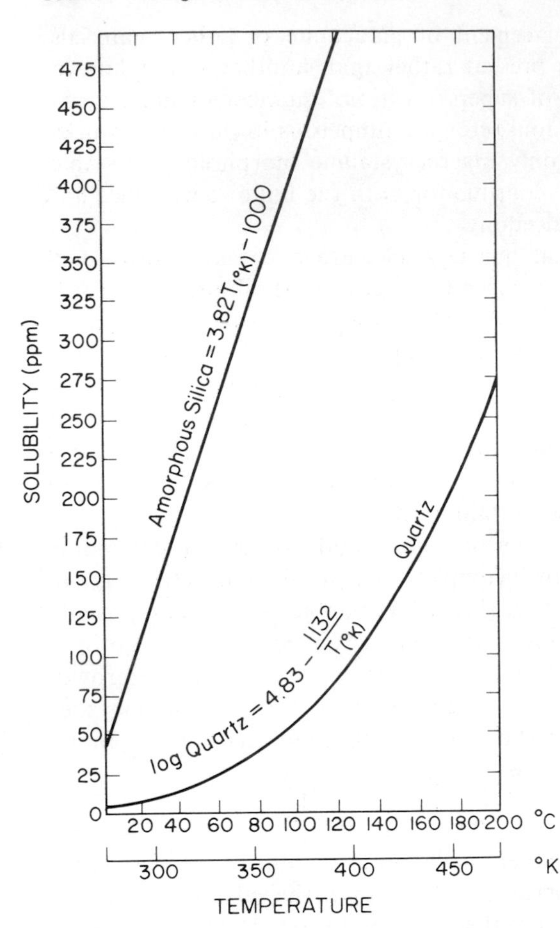

Fig. 10-8 Solubilities of amorphous silica and quartz as a function of temperature. (From data of Krauskopf, 1959, and Siever, 1962.)

opal solubility as the control on precipitation of silica from the underground waters. Therefore, the higher the diagenetic temperature, the easier it is to form quartz overgrowths as opposed to opal cement.

Silica in solution is present under normal surface conditions almost entirely in the form of orthosilicic acid, H_4SiO_4. As pH increases within the range found in sedimentary environments, however, orthosilicic acid ionizes to $H_3SiO_4^{-1}$ and $H_2SiO_4^{-2}$. The formation of orthosilicic acid and its dissociation are represented by the following equations.

$$SiO_2 + 2H_2O \rightleftharpoons H_4SiO_{4(aq)}$$

$$K = \frac{[H_4SiO_4]}{[SiO_2][H_2O]^2} = [H_4SiO_4] = 120 \text{ ppm} = 10^{-2.88} \text{ m/l}$$

$$H_4SiO_{4(aq)} \rightleftharpoons H^+ + H_3SiO_4^-$$

$$K_1 = \frac{[H^+][H_3SiO_4^-]}{[H_4SiO_4]}$$

$$H_3SiO_4^- \rightleftharpoons H^+ + H_2SiO_4^{-2}$$

$$K_2 = \frac{[H^+][H_2SiO_4^{-2}]}{[H_3SiO_4^-]}$$

The two dissociation constants have been evaluated by several investigators with results ranging from $10^{-9.46}$ to $10^{-9.93}$ for K_1, and ranging from $10^{-11.45}$ to $10^{-12.56}$ for K_2. The variation in estimate of equilibrium values results largely from three factors: (a) inaccuracy in the measurement methods used; (b) uncertainties concerning the attainment of equilibrium during experiments; (c) the equilibria are strongly temperature dependent (Fig.10-9). A change in experi-

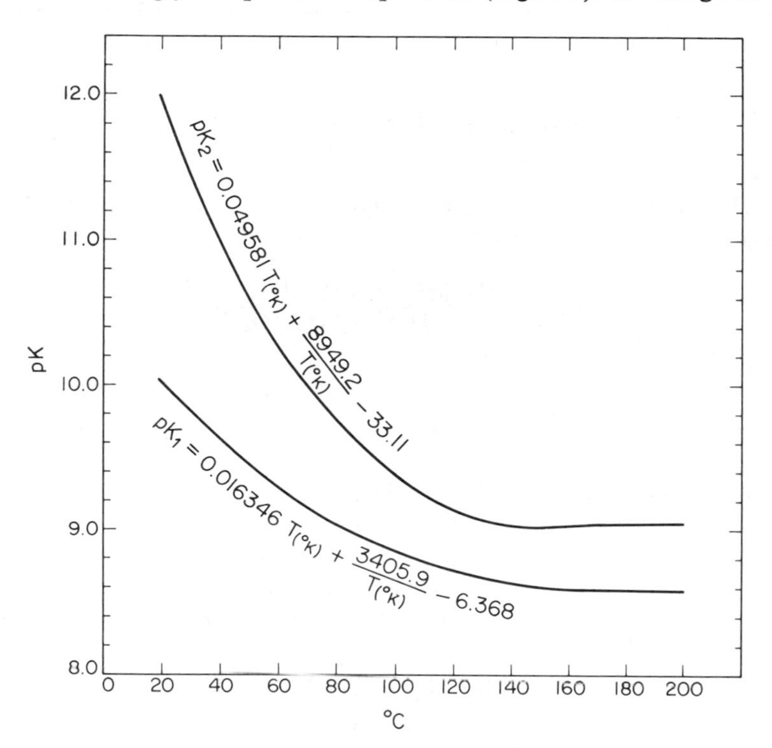

Fig. 10-9 Temperature dependence of the first and second dissociation constants of orthosilicic acid at 25°C. (Based on data in Ryzhenko, 1967.)

mental temperature of only five degrees, from 25°C to 30°C, is sufficient to cause a change in K_1 from $10^{-9.93}$ to $10^{-9.82}$ and in K_2 from $10^{-11.70}$ to $10^{-11.45}$. Table 10-12 shows the results of three sets of determinations.

Equilibrium constants for the dissociation reactions can be used to determine not only the total amount of silica in solution at any pH, but also the proportion of the silica in solution represented by each molecular or ionic species, assuming no polymerization.

$$K\,K_1 = \frac{[H_4SiO_4][H^+][H_3SiO_4^-]}{[H_2O]^2[SiO_2][H_4SiO_4]}$$

TABLE 10-12. EQUILIBRIUM CONSTANTS FOR THE DISSOCIATION OF ORTHOSILICIC ACID

Reference	25°C		30°C	
	K_1	K_2	K_1	K_2
Roller and Ervin, 1940	—	—	$10^{-9.8}$	$10^{-12.16}$
Lagerström, 1959	$10^{-9.46}$	$10^{-12.56}$	—	—
Ryzhenko, 1967	$10^{-9.93}$	$10^{-11.70}$	$10^{-9.82}$	$10^{-11.45}$

Therefore,

$$[H_3SiO_4^-] = \frac{K K_1 [H_2O]^2[SiO_2]}{[H^+]}$$

As the values are known for K, K_1, $[H_2O]$, and $[SiO_2]$, we can determine the amount of $[H_3SiO_4^-]$ at any hydrogen ion concentration (pH). Further,

$$K_1K_2 = \frac{[H^+][H_3SiO_4^-][H^+][H_2SiO_4^{-2}]}{[H_4SiO_4][H_3SiO_4^-]}$$

Therefore,

$$[H_2SiO_4^{-2}] = \frac{K_1K_2[H_4SiO_4]}{[H^+]^2}$$

As before, K_2, K_3, and $[H_4SiO_4]$ are known; hence the amount of $[H_2SiO_4^{-2}]$ can be determined for any pH. No hydrogen ions are involved in the formation of H_4SiO_4, so it is present to the extent of 120 ppm regardless of pH, assuming silica to be present in excess. Fig.10-10 shows the results of calculations using the dissociation constants at 25°C in Table 10-12. Despite the rather wide variation in values used for each constant, several facts are apparent. (a) Total silica in solution rises rapidly as pH increases above 9.0–9.5. (b) Below a pH of about 9.5 undissociated orthosilicic acid is the major species present; above 9.5 it is $H_3SiO_4^{-1}$. (c) The divalent species $H_2SiO_4^{-2}$ is of minor quantitative importance in nearly all natural environments.

Sources of Diagenetic Silica

The amount of silica dissolved in average fluvial, marine, or subsurface waters is usually much below the solubility of amorphous silica (120 ppm). The silica contents of subsurface waters are often greater than those of fluvial or marine waters, however, and values exceeding 6 ppm, the solubility of quartz, are common. Secondary growths of quartz in amounts sufficient to cause cementation of sandstones also are common, particularly in older quartz-rich sandstones. Where does the additional silica come from? Several sources are known to exist but their relative importances are uncertain.

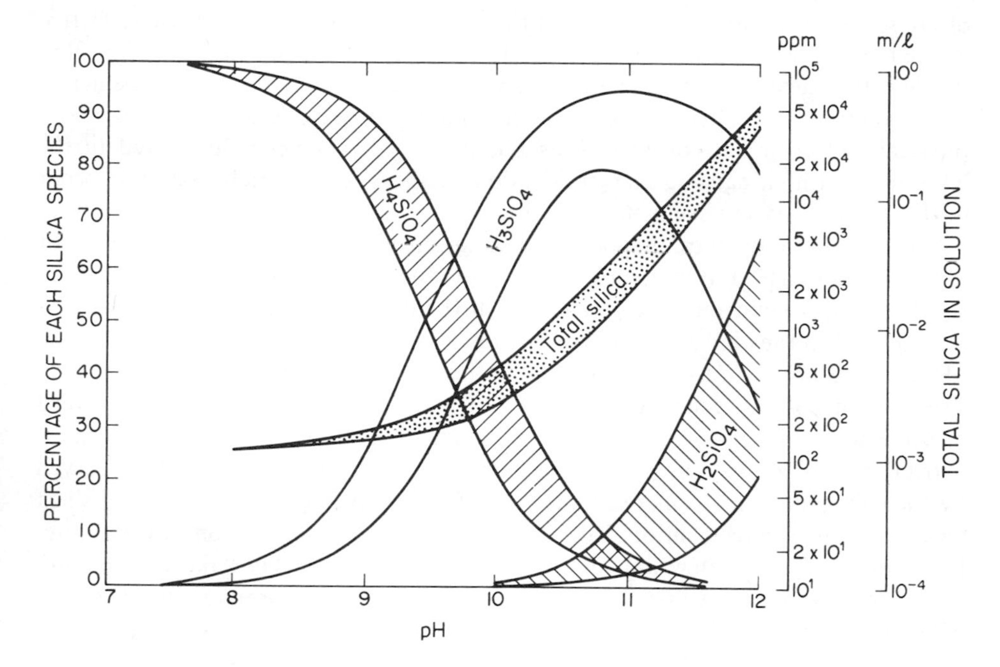

Fig. 10-10 Total silica and relative amounts of silica species in solution in water at 25°C as a function of pH.

One important source of silica is provided by dissolution of diatom and radiolarian tests at the increased subsurface temperatures accompanying burial of the sediment. Local concentrations of opaline tests may cause saturation of slowly moving ground water solutions, resulting in precipitation of crystalline silica, usually as quartz. In some Paleozoic sandstones siliceous sponge spicules are important as silica sources.

Inorganic sources of dissolved silica exist and are of varying importance, depending on the detrital mineral composition of the sediment. In highly quartzose sandstones there often is evidence of "pressure solution" (the Riecke Principle), the observation that adjacent quartz grains show sutured contacts, indicating that solution and resultant interpenetration have occurred. It has been inferred that this occurs under conditions of great stress such as occur at tangential grain contacts during compaction of unyielding materials. This idea is supported petrographically by the common occurrence of sutured contacts between quartz grains in metamorphic rocks and in intensely sheared sedimentary rocks, and theoretically by the fact that stressed solids are more soluble than unstressed ones. The mechanism of pressure solution has been discussed by Weyl (1959). He considers that a film of water a few molecules in thickness exists between quartz grain surfaces and that the attraction of the quartz grain surfaces for the film of water allows the water to have some rigidity and persist through the compaction process. This film serves as the migration path for diffusion of silica dissolved at points of stress concentration. The rate of diffusion varies with grain size, the effective normal stress between the grains, the diffusion constant in the diffusion film, the film thickness, clay concentration and distribution and the stress coefficient of solubility. The reality of the pressure solution process is supported by both physical and chemical considerations but it has not yet been duplicated in the laboratory without causing recrystallization of the grains, which does not occur in the sedimentary environment.

In many sandstones, a positive correlation exists between the presence of illitic clay laminae and pressure solution of quartz. Thomson (1959) has theorized that potassium ions are leached from the clay by underground waters and are replaced in the clay structure by hydrogen ions, resulting in an increase in the pH of the solution. In this way, a pH halo forms, with alkalinity decreasing away from the zone of clay. Silica will dissolve at the clay-quartz contact, diffuse outward into the solution, and precipitate as secondary quartz on the surfaces of nearby detrital quartz grains. Weyl, however, proposes that the main function served by the clay laminae is to increase the rate of diffusion of pressure-dissolved silica away from the site of solution. The clay film consists of a collection of clay platelets with associated water films, and the thickness of each clay flake with its film is on the order of 20 A. Thus, a clay film 10 μ thick between two quartz grains will contain about 5000 water films instead of the 1 or 2 water films between clean grains, resulting in a considerable increase in the rate of diffusion of silica and in the rate of pressure solution. Weyl suggests that the film between quartz grains does not need to be clay. The only requirement is that the film be composed of material of very small size compared with the size of the grains to be dissolved by pressure solution. Also, the film must be porous, saturated with water, and the material itself must not be subject to rapid pressure solution.

As noted earlier, Sippel, using luminescence petrography, has shown that some sutured grain contacts in orthoquartzites are related to growth of secondary quartz rather than to dissolution by pressure solution. Apparently, many quartz grains are scrubbed clean by formation waters prior to precipitation of secondary quartz on their surfaces and when the overgrowths abut in pore space, they do so in a digitate manner. The appearance of interdigitation between adjacent grains is, therefore, sometimes deceptive but the frequency of "suturing" produced in this way is not yet known.

Silica is also generated by diagenetic dissolution of silicate minerals, a phenomenon of considerable quantitative importance in some sandstones. Both feldspar crystals and heavy mineral grains can be dissolved by percolating groundwaters, with the amount of silica released depending on both the degree to which the minerals are attacked and the proportion of silicon in the minerals.

In sandstones containing abundant volcanic detritus, particularly glass, devitrification may occur, resulting in smectite clays with release of surplus silica to underground waters. Commonly, sufficient silica is generated in this way to saturate the water and, as devitrification usually occurs at shallow depths, opal nucleates rather than quartz. In ancient sandstones it is common to find that the opal has recrystallized to chert.

Calcium Carbonate Cements

Occurrence

Calcium carbonate is abundant in the sedimentary environment in both of its polymorphic forms but aragonite is almost unknown as a cementing agent in silicate sands. It is, however, a common cement in Quaternary carbonate sands, (Chap. 13).

In ancient sandstones, calcite probably is the most common chemical cement, appearing as a mosaic of anhedral subequant crystals 20 μ or more in diameter. In most sandstones the volume of calcite does not exceed about 30%, the original porosity of the sediment, but sometimes the percentage is much higher, greater than any possible original porosity. In such cases the carbonate clearly is replacing the detrital grains. The quartz particles are etched into irregularly shaped fragile grains that could not survive transport and thus could not be detrital. Sometimes the "force of crystallization" of the growing calcite crystals acts as a wedge splitting detrital silicate grains apart along previously existing planes of weakness such as partially healed fractures in quartz or cleavages in feldspars. In some sandstones the etching of the detrital grains has proceeded to the point that the rock is converted to a carbonate bed formed almost entirely of calcite. Such a bed may be difficult or impossible to distinguish from a bed of quartz-bearing limestone.

Chemistry

The chemistry of calcium carbonate precipitation is considerably more complex than that of the silica minerals but intensive field and laboratory studies during the past 15 years have succeeded in resolving some of the major uncertainties.

In order to saturate an aqueous solution with respect to calcite, it is necessary that the ion activity product $[Ca^{2+}]$ $[CO_3^{2-}]$ exceed the thermodynamic solubility product for this compound. Consequently, the activity of either calcium ion or carbonate ion may be increased with identical results but the amount of calcium in marine and subsurface waters is relatively stable. It can be increased independently of an increase in carbonate ion only by diagenetic "weathering" of calcic plagioclase, which normally is not sufficiently abundant in sandstones for this mechanism to be generally effective. In nearly all documented cases of calcium carbonate precipitation, saturation of the water has been caused by an increase in the activity of carbonate ion.

Carbon dioxide in solution under normal surface conditions is present almost entirely in the form of bicarbonate ion, HCO_3^-. Equilibrium constants in the two-component system H_2O-CO_2 have been determined as follows.

$$H_2O_{(liq)} + CO_{2(gas)} \rightleftharpoons H_2CO_{3(aq)}$$

$$K = \frac{[H_2CO_3]}{[H_2O][CO_2]} = 10^{-1.47}$$

$$H_2CO_{3(aq)} \rightleftharpoons H^+ + HCO_3^-$$

$$K_1 = \frac{[H^+][HCO_3^-]}{[H_2CO_3]} = 10^{-6.35}$$

$$HCO_3^- \rightleftharpoons H^+ + CO_3^{2-}$$

$$K_2 = \frac{[H^+][CO_3^{2-}]}{[HCO_3^-]} = 10^{-10.33}$$

As with silica in solution, these data can be used to determine both the "total CO_2" in solution at any pH and the proportion of the total CO_2 represented by each molecular or ionic species.

$$KK_1K_2 = \frac{[H_2CO_3][H^+][HCO_3^-][H^+][CO_3^{2-}]}{[H_2O]P_{CO_2}[H_2CO_3][HCO_3^-]}$$

Therefore,

$$[CO_3^{2-}] = \frac{KK_1K_2P_{CO_2}}{[H^+]^2} \qquad \left(\begin{array}{l}P_{CO_2} \text{ is the partial pressure of carbon dioxide in} \\ \text{water in equilibrium with the atmosphere, } 10^{-3.5}\end{array}\right)$$

Also,

$$KK_1 = \frac{[H_2CO_3][H^+][HCO_3^-]}{[H_2O]P_{CO_2}[H_2CO_3]}$$

Therefore,

$$[HCO_3^-] = \frac{KK_1P_{CO_2}}{[H^+]}$$

Also,

$$KK_2 = \frac{[H_2CO_3][H^+][CO_3^{2-}]}{[H_2O]P_{CO_2}[HCO_3^-]}$$

Therefore,

$$[H_2CO_3] = \frac{KK_2P_{CO_2}[HCO_3^-]}{[H^+][CO_3^{2-}]}$$

The abundance of the carbonate ion is inversely proportional to the *square* of the hydrogen ion concentration and the carbonate ion will increase rapidly in abundance as pH increases (Fig 10-11). Assuming a normal seawater pH of 8.2, the graph reveals that total CO_2 is present 97.9% as bicarbonate ion, 1.4% as undissociated carbonic acid, and only 0.7% as carbonate ion. Some modification of these percentages must be made for seawater, the original pore water of most sediments, however, because of ion-pair formation. Although carbonic acid as an aqueous species is not ion-paired, both bicarbonate ion and carbonate ion are ion-paired to a great extent as noted previously. Bicarbonate ion is 69% "free" and carbonate ion only 9% "free," changing the actual total available CO_2 from 7.1×10^{-4} to 4.9×10^{-4} moles/liter, the proportions of the three species to 97.9% bicarbonate ion, 2.0% undissociated carbonic acid, and 0.1% carbonate ion, and reducing carbonate ion from 5.1×10^{-6} to 4.6×10^{-7} moles/liter.

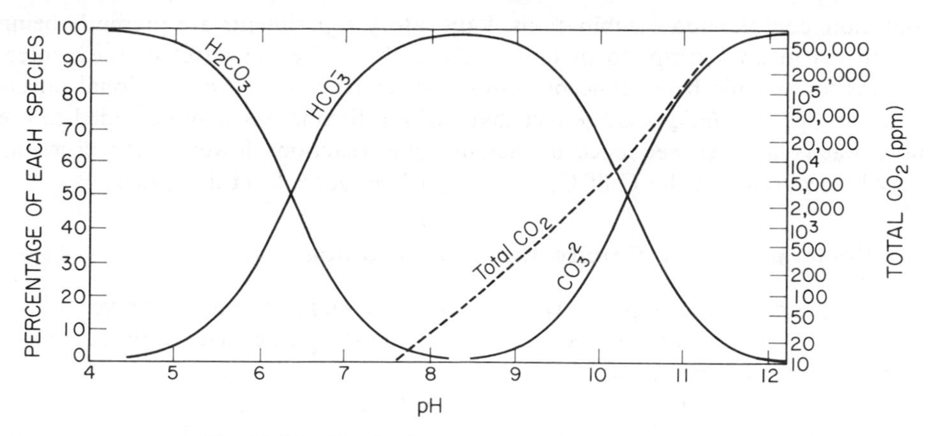

Fig. 10-11 Dissolved carbonate species in relation to pH.

To determine the approach to saturation of the seawater pore fluid, we now need to estimate the activity of calcium and the percent of this that is present as free ion. The activity coefficient of calcium in seawater has been most recently measured to be 0.24; the concentration of calcium in seawater is 0.01 moles/liter; and the percent free ion is 84%. Therefore, the actual availability of calcium in the pore water is (0.01)(0.24)(0.84) or 2.0×10^{-3} moles/liter. The effective ion activity product $[Ca^{2+}][CO_3^{2-}]$ is then $(4.6 \times 10^{-7})(2.0 \times 10^{-3})$ or 9.2×10^{-10} moles/liter, and this is comparable to the known thermodynamic solubility product $[Ca^{2+}][CO_3^{2-}]$ value of 4.0×10^{-9} moles/liter. With the errors inherent in measurements of activity coefficients and percentage of free ion in solutions as complex as seawater, the small difference between calculated ion activity product and the solubility product is not significant and we are only safe in concluding that original pore water in marine sediments is approximately saturated with respect to calcite. Aragonite is about 1.5 times more soluble than calcite.

Upon burial of the sediment, however, the concentration of the pore solution changes, the amount of ion-pair formation and percentage of free ions changes, temperature increases, and organic matter decomposes to release carbon dioxide, making it impossible to do more than generalize concerning the detailed chemistry of diagenetic precipitation of calcite. Unless diluted by downward moving meteoric and surface waters, filtration of the pore waters by clay membranes increases the molality of calcium ion, which may either lower or raise its activity coefficient. Anions are retained below the membrane as monovalent cations diffuse upward but the upward flow of metallic cations is compensated for by a downward diffusion of hydrogen ions. Because of this, it is not clear whether the amount of carbonate ion increases or only the percent of bicarbonate ion. The number and complexity of ion pairs increase as the molality of pore solutions increases and species composed of more than two ions may form, for example, $Na_2CO_3^0$, $K_2HCO_3^+$, or $Ca(SO_4)_2^{2-}$. On the other hand, temperature increase causes increased dissociation of carbonic acid and bicarbonate ion and dissociation of ion pairs and more complex ion combinations. Laboratory experiments are presently being conducted in an attempt to quantitatively evaluate the net effect of these interactions for specific brine concentrations and temperatures. An additional uncertainty in the calculations is the fact that carbon dioxide gas may be added to the pore waters as a consequence of metamorphic reactions lower in the crust, as may be occurring in the Gulf Coast Geosyncline (Jam, L. et al., 1969).

Replacement of Calcite Cement by Silica

Chemical relationships between calcite and quartz have for some years been the object of laboratory studies because of the common occurrence in sandstones of replacement of one mineral by the other. The results of these laboratory investigations suggest pH and temperature as the most common controlling parameters.

As we saw earlier, the solubility of silica is unaffected by pH at values below about 9 but, as is well-known, calcite is very soluble in acidic solutions. Therefore, changes in hydrogen ion concentration may cause solution of one phase and precipitation of the other (Fig. 10-12), a reaction made quite common by the fact

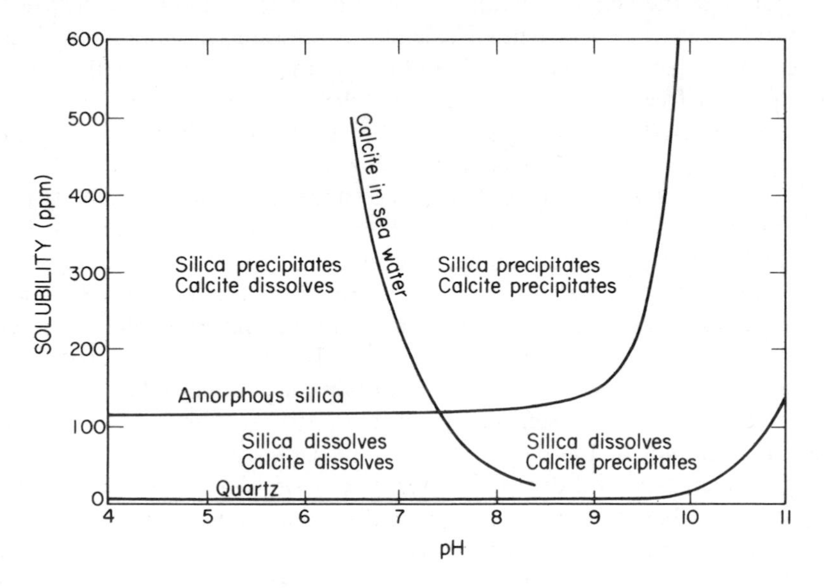

Fig. 10-12 Relationship between pH and the solubilities of calcite, quartz, and amorphous silica.

that the solubility curves as a function of pH intersect at a pH value quite commonly measured in groundwaters. It is worth noting, however, that we have not explained pseudomorphous replacement but only simultaneous solution of one mineral and precipitation of another. The mechanism by which a calcitic fossil shell is silicified (or vice versa) retaining the structure of the shell, is not understood although Harker (1971) has recently succeeded in pseudomorphing a diatom skeleton with calcite under hydrothermal conditions.

As is the case for most substances, the solubility of silica rises as temperature increases (Fig.10-8). But the heating of a solution of water and calcite generates a gas phase, carbon dioxide, and its escape causes the solubility of calcite to decrease with increasing temperature. That is, as temperature increases, carbonic acid increasingly dissociates into carbon dioxide and water (also into bicarbonate ions and hydrogen ions) and the carbon dioxide gas escapes, forcing continued formation of additional gas according to LeChatelier's principle. The net result of the interactions among the "CO_2 species" is a decreased solubility of calcium carbonate.

Siderite Cement

The formation of siderite cement is controlled by Eh, the quantitative measure of the oxidizing potential of the environment in which the reaction occurs. Under oxidizing conditions, iron ions are oxidized to the extent that the solubility product $[Fe^{2+}][CO_3^{2-}]$, which equals $10^{-10.5}$ at 25°C, cannot normally be exceeded and the common cementing agent in sandstones is hematite or its hydrated equivalents, the limonite group of substances, rather than siderite. Primary siderite can only form in association with reducing conditions such as exist at the surface in peat bogs or swamps, or it may exist in association with organic matter in the subsurface. The presence of siderite is commonly associated with other low-Eh minerals such as pyrite and marcasite.

When groundwater containing ferrous ions comes in contact with the atmosphere or is oxygenated by other means, the following reaction can occur.

$$2\,Fe^{2+} + 4\,HCO_3^- + H_2O + \tfrac{1}{2}O_2 \rightleftharpoons 2\,Fe(OH)_3 + 4\,CO_2$$

This may lower the pH somewhat but the solubility of ferric hydroxide is so low in the normal pH range that most of the iron will be precipitated, probably as an amorphous colloid. Subsequently, the hydroxide dehydrates to hematite.

As a cement, siderite sometimes forms as replacement of calcite and the equilibrium relationships for this reaction may be evaluated.

$$CaCO_3 + Fe^{2+} \rightleftharpoons FeCO_3 + Ca^{2+}$$

$$K = \frac{[Ca^{2+}]}{[Fe^{2+}]} = \frac{[Ca^{2+}][CO_3^{2-}]}{[Fe^{2+}][CO_3^{2-}]} = \frac{4.7 \times 10^{-9}}{3.1 \times 10^{-11}} = 150$$

Thus, at equilibrium, there is 150 times more calcium ion than ferrous ion in solution. Calcite will be dissolved and replaced by siderite if it is in contact with a solution containing ferrous iron ions in a concentration more than 1/150 that of calcium ions. Conversely, there will be desideritization by calcite if the pore solution has more than 150 times as much calcium ion as ferrous ion.

Hematite Cement

Red pigmenting matter in sandstones is present either as a primary pore filling precipitated from solution, as microcrystalline particulate iron oxide deposited with the clastic grains, or as iron ions adsorbed on clay mineral surfaces. The sources of the iron atoms in this pigment are the common iron-bearing accessory minerals of igneous and metamorphic rocks, hornblende, chlorite, biotite, ilmenite, and magnetite. Common hornblende in silicate rocks averages 15% iron oxide; chlorite, 21%; biotite, 22%; ilmenite, 46%; and magnetite, more than 95%; although most of this iron is present in these minerals in the ferrous form, oxidation during weathering converts the iron to the ferric state. There is some question, however, whether most ferrous iron is released during soil formation or during diagenesis as a result of intrastratal solution of iron-bearing heavy minerals. In moist tropical areas much ferric iron is present in the soils as tan to brown amorphous iron hydroxide and red oxides that approximate $Fe(OH)_3$ in composition. Traditionally, geologists have envisioned the erosion of this material in humid upland regions, transport to drier lowlands, burial, and "aging" of the amorphous material to crystalline ferric oxide, hematite (Van Houten, 1968). The fact that essentially impermeable shales are usually redder than associated sandstones has been cited as evidence that the iron oxide is depositional rather than diagenetic in origin. The existence of the "aging" process is, by its very nature, however, difficult to prove conclusively. The variables involved are uncertain and therefore thermodynamic and kinetic arguments concerning the chemical feasibility of the process are of little value. Small variations in the crystal size and state of the solid material; diagenetic temperatures; and the presence of other ions, organic matter, and perhaps other factors may be critical to whether the amorphous material ages to goethite (which is brown), ages to hematite with or without a goethite precursor, or in fact ages to hematite at all.

A common field relationship in ancient red beds is their association with extensive evaporite deposits, suggesting that lateritic terranes or humid uplands are not likely to occur in the source area highlands and, therefore, the pigment probably has some origin other than lateritic soils. In a series of papers, Walker (1967; et seq.) has investigated the origin of the red pigment in texturally and mineralogically immature arkosic sediments in arid Baja California, Mexico. Faunal and floral evidence and the presence of thick evaporite deposits indicate there has been no appreciable climatic change in the region since at least Early Pliocene time. Walker found that the sandstones in this Late Cenozoic sequence were reddened by the passage of time and microscopic and electron microprobe studies revealed that the increase in pigmentation resulted from oxidation of iron released by progressive diagenetic alteration of detrital hornblende, biotite, and other iron-bearing minerals derived from the granitic source rocks exposed nearby. In thin sections, the hornblende grains are visibly etched into cockscomb shapes and often are replaced by iron-rich montmorillonite (5 to 10% Fe_2O_3) formed as an alteration product of the hornblende. (Montmorillonite normally contains about 3% total iron). Walker maintains that the iron oxide pigment in red sandstones forms by redistribution of iron outward from the decomposing grains by oxygenated pore waters during diagenesis (Fig. 10-13) and that the climate during deposition of the sandstone may be irrelevant to the formation of hematite-cemented sandstones. He further finds that as little as 0.1% of hematite is sufficient to cause a bright red color in red beds and, inasmuch as both detrital and authi-

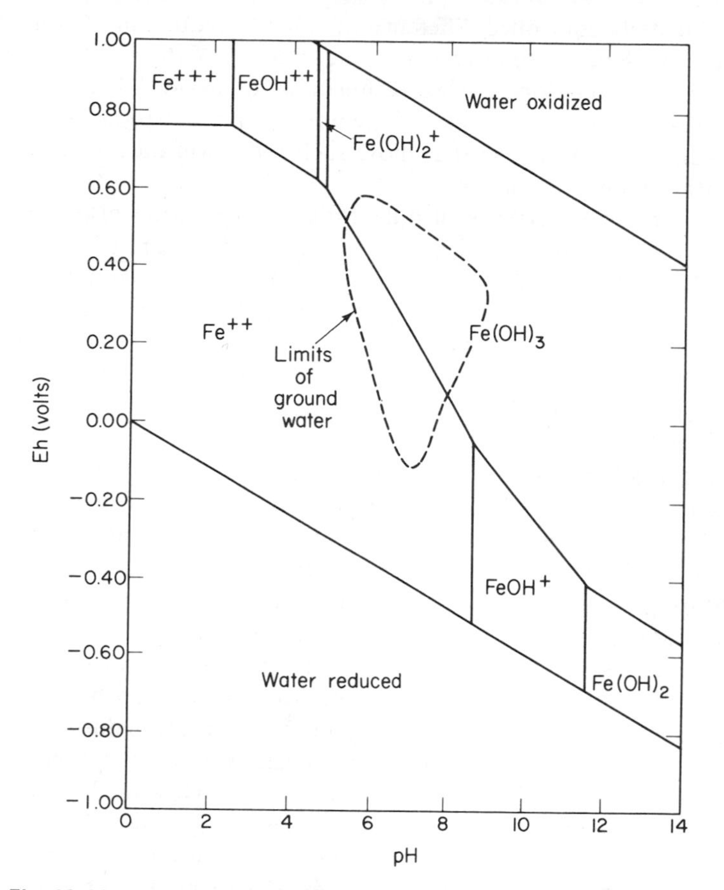

Fig. 10-13 Stability field diagram for aqueous ferric-ferrous system. (Limits for ground water are from data by Baas Becking, et al., 1960. Remainder of diagram from Hem and Cropper, 1959.)

genic clays normally contain 3 to 5% iron, the relative impermeability of shales is not a barrier to the formation of the normal even distribution of coloring matter in thick red shale sequences. The iron oxide does not need to migrate through pore waters in the shales but needs only to be transferred from within the clay mineral structure to its exterior. Therefore, Walker concludes that red shales, like red sandstones, carry no necessary inference concerning the climate during erosion and deposition of the rocks. He cites the red beds in the Pennsylvanian Fountain Formation as an ancient analogue of the red beds that are forming today in Baja California. Perhaps the red sediments in the Triassic fault basins of northeastern United States also became red during diagenesis.

As convincing as Walker's hypothesis seems for some hematitic rocks, however, it appears less satisfactory when applied to some other types of geologic occurrences. For example, it is quite common to find that continental rocks are red but their marine equivalents are drab gray or green. Yet there is no apparent difference in heavy mineral content between the two. It appears more reasonable to attribute this environmentally controlled difference in color to reduction of colloidal ferric iron by the abundant carbonaceous organic matter on the sea floor and redistribution of the ferrous ions during authigenesis of pyrite, glauconite, etc. The amount of organic matter in marine sediments decreases sharply a few inches below the water-sediment interface so that most such reduction must be a surface phenomenon rather than a diagenetic one.

Another type of occurrence of differential pigmentation also poses difficult problems to the universal application of diagenetic origin of red coloration. In western Oklahoma it is common to find red and green shales regularly interstratified at 3- to 10-ft intervals with fairly sharp boundaries between adjacent rock layers. The green units contain 4 to 5% total iron, nearly all of which is structurally bound in iron-rich chlorite or chamosite crystals, as revealed by leaching experiments and X-ray diffraction studies. The red shales consistently contain about 3% more iron than the green shales and leaching experiments indicate that this additional iron is located between the mineral grains in the shale. Three percent of interstitial ferric iron exceeds by an order of magnitude the few tenths of 1% reported by Walker in the Baja California muds and there is no mechanism known by which this much iron can be extracted from clay mineral structures during diagenesis. The detrital mineral compositions of the shale units are identical and the beds are distinguishable only by differences in their coloration. How could such regular interstratification develop during diagenesis of these rocks? As yet, there is no satisfactory answer to this question but, on the other hand, it also is not easy to envision the origin of the coloration during depositional processes. The nearby presence of evaporites suggests a fairly high regional water salinity and a corresponding paucity of organic matter. How might the alternating red and drab color sequence be achieved in such an environment?

In summary, there is no doubt that red coloration and hematite cement in sandstones may be produced during diagenesis. For rocks such as well-sorted and rounded hematite-cemented quartz sandstones (part of the Hickory Formation of central Texas and parts of the Potsdam Formation of northern New York), the diagenetic origin of the cement seems certain. For texturally immature sandstones and many shales the decision concerning the origin of the pigment is less certain and each unit must be examined with both possibilities in mind, origin by aging of amorphous iron compounds deposited with the sediment and introduction of the pigment during diagenesis.

10.5 CEMENTATION PATTERNS IN ANCIENT ROCKS

One aspect of cementation that has received very little attention to date is the pattern of cementation within lithologic units. Maps of this type are potentially very useful as indicators of ancient hydrodynamic patterns of subsurface waters, geologic structure at the time of cementation, and perhaps also the types of rocks exposed on the surface. For example, Warner (1965) studied the Duchesne River Formation (Eocene) of the Uinta Basin in northeastern Utah and found the percentage of authigenic cement in the sandstones to range from about 10 to 25%, the amount increasing in a regular manner from south to north (Fig. 10-14). The type of cement changed in an east-west pattern, however, being dominantly calcite to the east, iron oxide and silica in the central part of the outcrop, and a mixture of all types to the west. These and other petrographic observations led

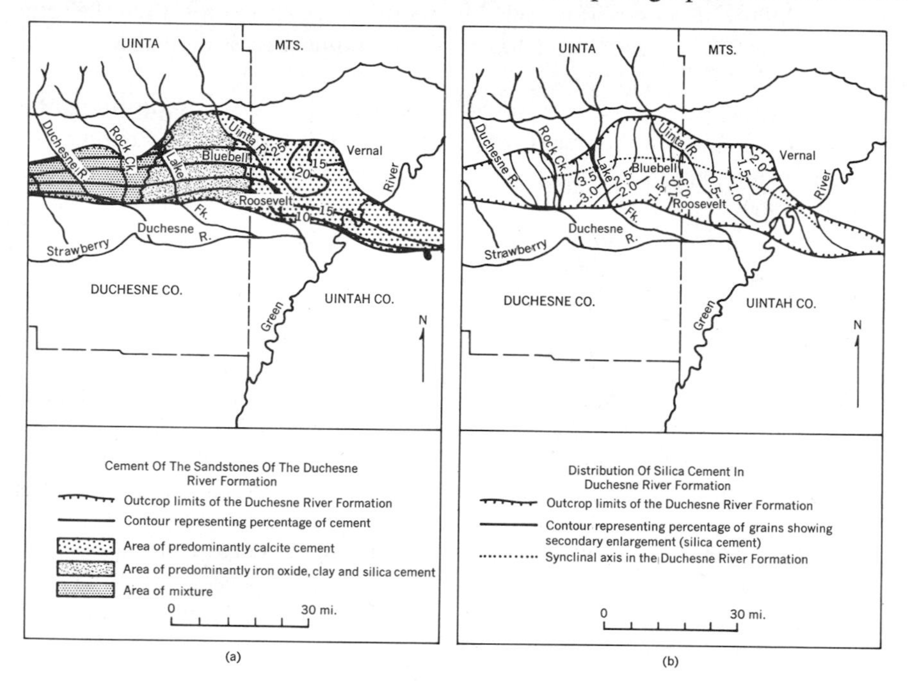

Fig. 10-14 Areal distribution of cement in sandstones of the Duchesne River Formation. (From Warner, 1965.)

him to conclude that the principal control on the cementation pattern was proximity to a source rock containing appropriate ions. The carbonate cement was derived from solutions made rich in calcium and carbonate ions from weathering of limestones immediately adjacent to the eastern edge of the Duchesne River Formation. Silica in the siliceous central part of the unit was obtained by weathering of the quartzitic core of the Uinta Mountains and desilication of volcanic ash beds on the north edge of the outcrop. In the far western part of the Duchesne River outcrop, the cement is a mixture and so is the source rock.

The distribution of the calcite cement also shows a close relationship to present-day structure, implying groundwater influence, because groundwater flow patterns are normally influenced strongly by structural patterns. The quantitative distribution pattern of silica cement, on the other hand, shows no relationship to the structural pattern of the basin, which implies a relationship to surface drainage patterns and distance of transport (short, to avoid dilution). The cement decreases in the general direction of drainage away from the source area.

In general, then, we may conclude that (a) if distribution patterns of cements are closely related to structural patterns, precipitation of the cement must have been influenced by groundwater after deformation and the cement distribution pattern may be used as an index to primary permeability. (b) If the distribution pattern of cements is more closely related to topography, as indicated by a correlation with flow directions as revealed by sedimentary structures, then the cement is probably predeformational and has not been influenced appreciably by subsequent groundwater.

Cementation patterns may also be related to the bulk composition of the host sandstone and environment of deposition (Carrigy and Mellon, 1964), petroleum accumulation (Todd, 1963), and the presence of microdisconformities (Freeman, 1969).

REFERENCES

ADDISON, W. E. and J. H. SHARP, 1962, "A Mechanism for the Oxidation of Ferrous Iron in Hydroxylated Silicates," *Clay Miner. Bull.*, **5**, pp. 73–79. (Consideration of the mechanics of ion transfer during the conversion of ferrous to ferric iron.)

ATWATER, G. I. and E. E. MILLER, 1965, "The Effect of Decrease in Porosity with Depth on Future Development of Oil and Gas Reserves in South Louisiana" (abst.), Amer. Assn. Pet. Geol. Prog. Ann. Mtg., New Orleans, p. 48.(Porosity decreases linearly with depth.)

BAAS BECKING, L. G. M., I. R. KAPLAN, and D. MOORE, 1960, "Limits of the Natural Environment in Terms of pH and Oxidation-Reduction Potentials," *Jour. Geol.*, **68**, pp. 243–284. (The definitive paper on this topic, based on a very large amount of data.)

BACK, WILLIAM, 1961, "Techniques for Mapping of Hydrochemical Facies," *U. S. Geol. Sur. Prof. Paper 450-B*, pp. 380–382. (Classification of hydrochemical facies.)

BACK, WILLIAM and B. B. HANSHAW, 1965, "Chemical Geohydrology," in *Advances in Hydroscience*, V. T. CHOW, ed., **2**, New York: Academic Press, pp. 49–109. (An excellent introduction to the principles and methodology of studies of subsurface brines.)

BILLINGS, G. B., B. HITCHON, and D. R. SHAW, 1969, "Geochemistry and Origin of Formation Waters in the Western Canada Sedimentary Basin, 2. Alkali Metals," *Chem. Geol.*, **4**, pp. 211–223. (Study of the origin of subsurface brines with particular attention to bromine and the alkali metals.)

BREDEHOEFT, J. D., C. R. BLYTH, W. A. WHITE, and G. B. MAXEY, 1963, "A Possible Mechanism for the Concentration of Brines in Subsurface Formations," *Am. Assoc. Pet. Geol. Bull.*, **47**, pp. 257–269. (Mathematical treatment of the hydrodynamics of salt sieving of subsurface waters.)

BUCKLEY, S. E., C R. HOCOTT, and M. S. TAGGART, JR., 1958, "Distribution of Dissolved Hydrocarbons in Subsurface Waters," in *Habitat of Oil*, Amer. Assn. Pet. Geol., Tulsa, Okla., L. G. Weeks, ed., pp. 850–882. (Extensive sampling program and analysis of hydrocarbons dissolved in brines from one of the largest petroliferous regions in the world.)

CARRIGY, M. A. and G. B. MELLON, 1964, "Authigenic Clay Mineral Cements in Cretaceous and Tertiary Sandstones of Alberta," *Jour. Sed. Petrology*, **34**, pp. 461–472. (Areal distribution of authigenic clay cements in relation to mineral composition and depositional environment of the host sandstones.)

CARROLL, DOROTHY, 1962, "Rainwater as a Chemical Agent of Geologic Processes—A Review," *U. S. Geol. Sur. Water-Supply Paper 1535-G*, 18 pp. (Survey of available data and analysis of the effects of rainwater on the composition of soil water, weathering, and ion exchange reactions in the soil.)

CHAVE, K. E., 1960, "Evidence on History of Sea Water from Chemistry of Deeper Subsurface Waters of Ancient Basins," *Amer. Assn. Pet. Geol. Bull.*, **44**, pp. 357–370. (Discussion of the quality of chemical analyses of subsurface waters and analysis of changes in ionic ratios of selected high quality water analyses.)

CLARKE, F. W., 1924, *Data of Geochemistry*, 5th ed. U.S. Geol. Sur. Bull. 770, 841 pp. (Now partially updated as Professional Paper 440 in many sections, Clarke's work still stands as a classic for basic chemical data concerning sedimentary materials.)

DAPPLES, E. C., 1967, "Silica as an Agent in Diagenesis," in *Diagenesis in Sediments*, G. Larsen and G. V. Chilingar, ed., New York: Elsevier Pub. Co., pp. 323–342. (Discussion of the precipitation of silica in sedimentary rocks, based on the results of laboratory investigations and field studies. See also pp. 91–126 for a discussion of sandstone diagenesis.)

DEGENS, E. T., J. M. HUNT, J. H. REUTER, and W. E. REED, 1964, "Data on the Distribution of Amino Acids and Oxygen Isotopes in Petroleum Brine Waters of Various Geologic Ages," *Sedimentology*, **3**, pp. 199–225. (Data illustrating that the δO^{18} values of formation waters are related to the salinity of the waters.)

FREEMAN, TOM, 1969, "Cement-Composition Discontinuity in the Cambrian of Texas and Its Stratigraphic Implications," *Geol. Soc. Amer. Bull.*, **80**, pp. 2095–2096. (Existence of a questioned disconformity proved by electron microprobe analysis of minor elements in calcite cements on the two sides of the unconformity.)

GARRELS, R. M. and F. T. MACKENZIE, 1971, *Evolution of Sedimentary Rocks*. N. Y., W. W. Norton and Co., Inc., 397 pp. (Historical geochemistry of sediments.)

GRAF, D. L., W. F. MEENTS, I. FRIEDMAN, and N. F. SHIMP, 1966, "The Origin of Saline Formation Waters, III. Calcium Chloride Waters," *Ill. Geol. Sur. Cir. 397*, 60 pp. (Analysis of the evolution of formation waters in the Illinois and Michigan Basins.)

HARKER, R. I., 1971, "Synthetic Calcareous Pseudomorphs Formed from Siliceous Microstructures," *Science*, **173**, pp. 235–237 (The first experimental success in duplicating a common natural occurrence.)

HEM, J. D. and W. H. CROPPER, 1959, "Survey of Ferrous-Ferric Iron Equilibria and redox potentials," *U.S. Geol. Sur. Water-Supply Paper 1459-A*, 31 pp. (Derivation of the Eh-pH stability fields for the various species of ionic iron found in natural waters.)

HEMBREE, C. H., B. R. COLBY, H. A. SWENSON, and J. R. DAVIS, 1952, "Sedimental and Chemical Quality of Water in the Powder River Drainage Basin, Wyoming and Montana," *U.S. Geol. Sur. Circ. 170*, 92 pp. (Detailed analysis and explanation of water chemistry in a large drainage basin.)

ILER, R. K., 1955, *The Colloid Chemistry of Silica and the Silicates*. N. Y., Cornell Univ. Press, 324 pp. (Classic text summarizing knowledge in this field.)

JAM L., PEDRO, P. A. DICKEY, and E. TRYGGVASON, 1969, "Subsurface Temperature in South Louisiana," *Amer. Assn. Pet. Geol. Bull.*, **53**, pp. 2141–2149. (Collation of subsurface temperature data for an extensively drilled modern geosyncline.)

KISCH, H. J., 1969, "Coal-Rank and Burial-Metamorphic Facies," in *Adv. in Org. Geochem.*, 1968, P. A. Schenck and I. Havenaar, ed.: N. Y., Pergamon Press, pp. 407–425. (Field and experimental data correlating coal rank with zeolite facies metamorphism.)

KRAUSKOPF, K. B. 1959, "The Geochemistry of Silica in Sedimentary Environments," in *Silica in Sediments*. Soc. Econ. Paleont. & Miner. Spec. Pub. No. 7, pp. 4–19. (Review of silica geochemistry.)

LAGERSTRÖM, GOSTA, 1959, "Equilibrium Studies of Polyanions III. Silicate Ions in $NaClO_4$ Medium," *Acta Chemica Scandinavica*, **13**, pp. 722–736. (Experimental data concerning the variety and abundance of silica polymers in solution.)

LIVINGSTONE, D. A., 1963, "Data of Geochemistry, Chapter G. Chemical composition of rivers and lakes," *U. S. Geol. Sur. Prof. Paper 440-G*, 64 pp. (Latest data and summary of information on this topic.)

MACKENZIE, F. T. and R. GEES, 1971, "Quartz: Synthesis at Earth-Surface Conditions," *Science*, **173**, pp. 533–535. (The first successful attempt.)

MASON, BRIAN, 1966, *Principles of Geochemistry*, 3rd. ed. New York: John Wiley & Sons, Inc. 329 pp. (The latest edition of the first English language textbook on this subject. Still a useful reference text.)

MAXWELL, J. C., 1964, "Influence of Depth, Temperature, and Geological Age on Porosity of Quartzose Sandstone," *Amer. Assn. Pet. Geol. Bull.*, *48*, pp. 697–709. (Summary of field and laboratory studies bearing on this topic.)

MCAULIFFE, CLAYTON, 1969, "Determination of Dissolved Hydrocarbons in Subsurface Brines," *Chem. Geol.*, **4**, pp. 225–233. (Analytical techniques.)

MURCHISON, DUNCAN, and T. S. WESTOLL, ed., 1968, *Coal and Coal-Bearing Strata*. New York: Elsevier Pub. Co., 418 pp. (A collection of articles on geologic aspects of coal formation, with valuable reviews in English of recent German studies.)

PITTMAN, E. D., 1972, "Diagenesis of Quartz in Sandstones as Revealed by Scanning Electron Microscopy," *Jour. Sed. Petrology*, **42**, (in press). (Superb scanning electron photomicrographs illustrating different stages of development of secondary growths of quartz on detrital quartz grains).

ROLLER, P. S. and G. ERVIN, JR., 1940, "The System Calcium Oxide-Silica-Water at 30°. The Association of Silicate Ion in Dilute Alkaline Solution," *Jour. Amer. Chem. Soc.*, **62**, pp. 461–471. (Early experimental determination of dissociation constants of orthosilicic acid.)

RUBEY, W. W., 1951, "Geologic History of Sea Water," *Geol. Soc. Amer. Bull.*, **62**, pp. 1111–1148. (Classic paper employing geology, geochemistry, and biology to suggest that the composition of seawater has remained essentially constant since fairly early in the earth's history.)

RYZHENKO, B. N., 1967, "Determination of Hydrolysis of Sodium Silicate and Calculation of Dissociation Constants of Orthosilicic Acid at Elevated Temperatures," *Geochem. Internat.*, **4**, pp. 99–107 (Experimental data.)

SIEVER, RAYMOND, 1962, "Silica Solubility, 0°–200°C, and the Diagenesis of Siliceous Sediments," *Jour. Geol.*, **70**, pp. 127–150. (Excellent summary of the topic.)

SIPPEL, R. F., 1968, "Sandstone Petrology, Evidence from Luminescence Petrography," *Jour. Sed. Petrology*, **38**, pp. 530–554. (A paper containing many photomicrographs illustrating the usefulness of this newly developed tool for diagenetic studies.)

SPRY, ALAN, 1969, *Metamorphic Textures*. New York: Pergamon Press, 350 pp. (Chapter 3 is concerned with the effects of grain boundary energies on rock coherence and is very applicable to sedimentary textures.)

STOPES, M. C., 1919, "On the Four Visible Ingredients in Banded Bituminous Coal. Studies in the Composition of Coal," *Proc. Roy. Soc.* (London), Ser. B, **90**, pp. 470–487. (Discussion and definition of macerals.)

THOMPSON, A. M., 1970, "Geochemistry of Color Genesis in Red-Bed Sequence, Juniata and Bald Eagle Formations, Pennsylvania," *Jour. Sed. Petrology*, **40**, pp. 599–615. (Detailed investigation of the origin of red and drab-colored zones in Appalachian Ordovician sandstones.)

THOMSON, ALAN, 1959, "Pressure Solution and Porosity," in *Silica in Sediments*. Soc. Econ. Paleon. & Miner. Spec. Pub. No. 7, pp. 92–110. (Interpretation of the phenomenon of chemical interaction between illite and detrital quartz grains during diagenesis.)

TODD, T. W., 1963, "Post-Depositional History of Tensleep Sandstone (Pennsylvanian), Big Horn Basin, Wyoming," *Amer. Assn. Pet. Geol. Bull.*, **47**, pp. 599–616. (The distribution of quartz, dolomite, and anhydrite cement is related to the hydrodynamic system generated by Laramide Orogeny.)

VAN HOUTEN, F. B., 1968, "Iron Oxides in Red Beds," *Geol. Soc. Amer. Bull.*, **79**, pp. 399–416. (Argument supporting the view that a large proportion of the hematite pigment in sediments, particularly in shales, results from the "aging" of amorphous iron compounds deposited with the sediment.)

VON ENGELHARDT, WOLF, 1967, "Interstitial Solutions and Diagenesis in Sediments," in *Diagenesis in Sediments*, G. Larsen and G. V. Chilingar, ed., New York: Elsevier Pub. Co., pp. 503–521. (Discussion of the interrelationships among filtration of underground waters by clay membranes, groundwater composition, and authigenesis of minerals during diagenesis.)

WALKER, T. R., 1967, "Formation of Red Beds in Ancient and Modern Deserts," *Geol. Soc. Amer. Bull.*, **78**, pp. 353–368. (The first of an important and well-documented series of papers demonstrating that red sediments may form by intrastratal solution of iron-bearing heavy minerals.)

WARNER, M. M., 1965, "Cementation as a Clue to Structure, Drainage Patterns, Permeability, and Other Factors," *Jour. Sed. Petrology*, **35**, pp. 797–804. (A brief but important paper using observed patterns of cementation in sandstone units to decipher time of cementation and source rocks of the elements forming the cements.)

WEYL, P. K., 1959, "Pressure Solution and the Force of Crystallization-A Phenomenological Theory," *Jour. Geophys. Res.*, **64**, pp. 2001–2025. (Clear discussion of the probable mechanism of pressure solution.)

WHITE, D. E., 1965, "Saline Waters of Sedimentary Rocks," in *Fluids in Subsurface Environments*, Amer. Assn. Pet. Geol. Mem. 4, Tulsa, Okla., pp. 342–366. (Excellent summary of present understanding of variations in subsurface waters and suggested explanations.)

WHITE, D. E., J. D. and HEM, G. A. WARING, 1963, "Data of Geochemistry, Chapter F, Chemical Composition of Subsurface Waters," *U.S. Geol. Sur. Prof. Paper 440-F*, 67 pp. (Compilation of chemical analyses in relation to sample source and variation in dissolved constituents.)

CHAPTER ELEVEN

MUDROCKS

11.1 INTRODUCTION

The terminology of the fine grained sedimentary rocks has been somewhat confused by the absence of any generally accepted terms for distinguishing clearly between texture and mineral composition. Most of the terms available (clay, mud, mudstone, etc.) are defined by texture but are commonly used only for sediments containing substantial amounts of clay minerals. Some authors use the qualifier "argillaceous" to mean "rich in clay minerals." *Clay* is defined as material finer than 1/256 mm; *silt*, as material in the range from 1/256 to 1/16 mm; and *mud*, as a general term including silt, clay, and mixtures of the two. The suffix *-stone* is used for lithified sediments. Some geologists restrict terms such as mudstone to rocks that show no fissility but others include both fissile and nonfissile rocks. The term *mudrock* is preferred as the general term (see Table 11-1). Among geologists working in Precambrian terrains, the term *argillite* is frequently used for a mudrock hardened by incipient metamorphism but showing no slaty cleavage. *Shale* is the term used for fissile mudrock and, more generally, for the entire class of fine grained sedimentary rocks that contain substantial quantities of clay minerals.

TABLE 11-1 CLASSIFICATION OF MUDROCKS

Ideal size definition	Field criteria	Fissile mudrock	Nonfissile mudrock
$> 2/3$ silt	Abundant silt visible with hand lens	Silt-shale	Siltstone
$> 1/3 < 2/3$ silt	Feels gritty when chewed	Mud-shale	Mudstone
$> 2/3$ clay	Feels smooth when chewed	Clay-shale	Claystone

Modern muds average approximately 15% sand, 45% silt, and 40% clay (Picard, 1971) but little more is known about their size distributions. Even less is known concerning the size distribution of mudrocks because it is difficult to disaggregate them and still more difficult to interpret the size distribution if the disaggregation is carried out. Even in the case of modern muds, the sediment is first dispersed by using a peptizing agent. This treatment breaks down any flocculated structure that may have been present when the clays were being deposited. The details of the resulting size distribution measured by sedimentation methods (see Chap. 3) probably reflect mainly the effectiveness of the dispersion rather than the nature of the original "sedimentation" size distribution, except for the coarse silt and sand fractions. In ancient mudrocks new clay minerals may be present as a result of diagenesis, further modifying the original size distribution.

11.2 MINERAL COMPOSITION

The most common techniques used to study muds or mudrocks are to determine the mineral composition of the fraction finer than 2 μ by X-ray diffraction and differential thermal analysis and to determine the bulk composition by chemical analysis. Modern X-ray techniques are capable of giving a quantitative modal analysis of the clays in a mudrock. Such an analysis can have considerable precision but the accuracy is probably low in most cases and there has generally been little attempt at cross-laboratory standardization so that it is difficult to compare results obtained in different studies. Owing to the variable composition of clay minerals, it is also difficult to calculate true modes from chemical analyses.

Few studies have been made of the petrology of the whole mudrock. An exception is an investigation of more than 400 mudrock samples, mainly from the Phanerozoic of the United States, by Shaw and Weaver (1965). There was a considerable variation in proportion of quartz to clay minerals and of quartz to feldspar in the samples studied. The proportion of quartz to clays decreases with distance from land off the Mississippi Delta. The quartz/feldspar ratio was related to that of the associated sandstones. The overall average mode was clay minerals, 61%; quartz and chert, 31%; feldspar, 4.5%; carbonates, 3.6%; organic matter, 1%; iron oxides, less than 0.5%.

Several compilations of data for the chemical composition of the "average shale" have been made (Table 11-2). However, there are substantial differences in the chemical and mineralogical composition of different mudrocks and there are some indications of change in composition with stratigraphic age. Until the population of mudrocks being studied is more precisely defined, it is difficult to determine the adequacy of the sample on which such "average shales" are based.

Clay Minerals

Modern Marine Sediments

In the discussion of weathering (Chap. 7), it was shown that metallic cations are progressively lost from clay structures as temperature and rainfall increase. Illite may be stripped of potassium ions and converted to montmorillonite; montmorillonite may be further degraded to kaolinite; and under extreme conditions kaolinite may be supplanted by gibbsite. Because the type of clay in soils reflects climatic control so strongly, it is reasonable to expect the distribution of clay mineral facies in modern depositional basins to be related to the climatic patterns on adjacent land masses. This expectation has been tested by many investigators and found to be generally valid. For example, Griffin (1962) studied clay minerals in soils and streams in the eastern part of the United States and in the northeastern Gulf of Mexico, the depocenter for most of this region. He found that the type of clay in the nearshore gulf sediments clearly reflects its source. The Apalachicola River, which drains the subtropical region of southeast United States, carries 60 to 80% kaolinite. In contrast, the Mississippi River, which receives most of its suspended sediment from the temperate northern half of the United States, contains only 10 to 20% kaolinite. Kaolinite is the dominant clay mineral in the surficial gulf sediments for 100 mi offshore of the mouth of the Apalachicola River.

The close control of clay facies in modern marine muds by climatic conditions on the adjacent landmasses also is evident on a larger scale. Maps of the facies patterns of clay mineral species in the world ocean have been prepared by Rateev et al. (1969) and by Biscaye (1965) and these reveal patterns analogous to those found by Griffin in the Gulf of Mexico (Figs. 11-1 to 11-4).

The generally low percentages of chlorite outside high latitude temperate and polar regions of the world ocean reflect the ease of oxidation of divalent iron in the brucite layer of chlorite during weathering. It can survive on the land surface only in areas where chemical weathering is inhibited, such as in glacial or arid regions. Chlorite ranges from 10 to 18% of the clay mineral assemblage in the various ocean basins and averages about 13%.

Kaolinite abundance ranges from 8 to 20% of the total clay suite in the different ocean basins and averages about 12%. Abundance maxima occur in areas where tropical rivers empty into the ocean; for example, off the coast of tropical west Africa, immediately seaward of Brazil in South America and into the equatorial Atlantic Ocean, and near Madagascar. The high percentages north of Australia reflect drainage and atmospheric transport from tropical soils of Miocene age. In brief, kaolinite is a low-latitude clay mineral.

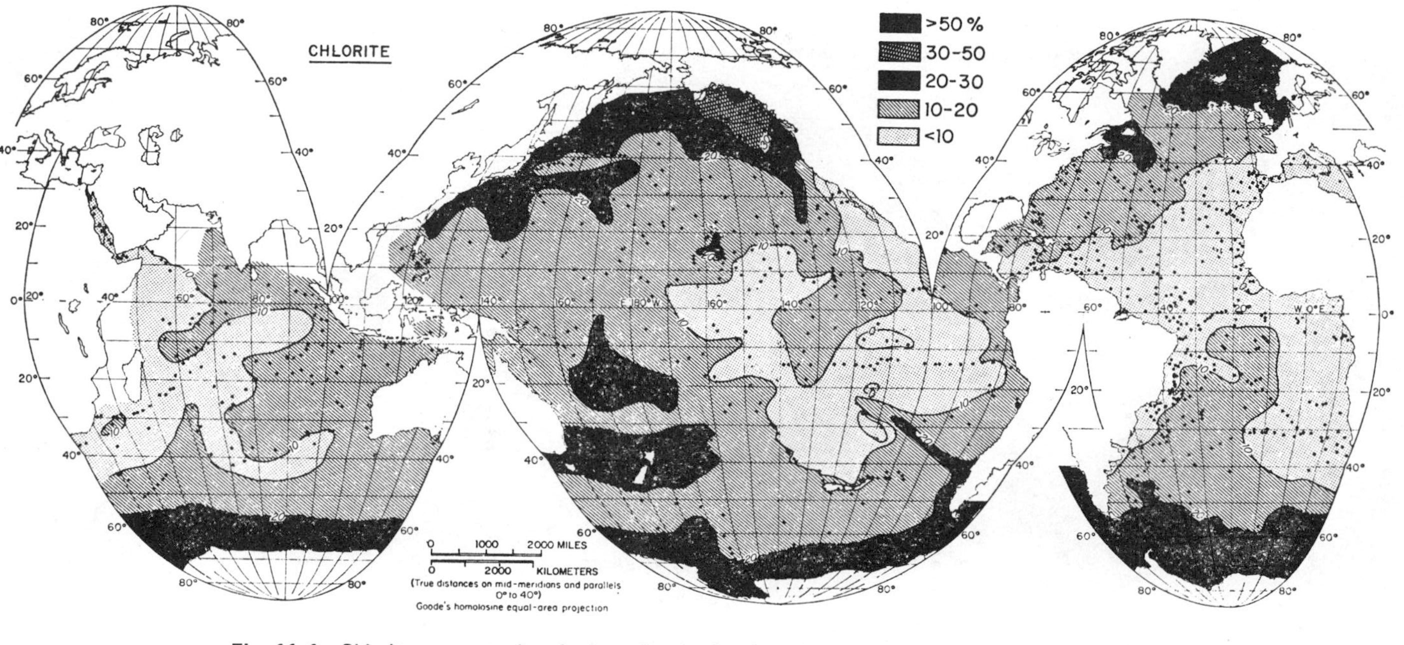

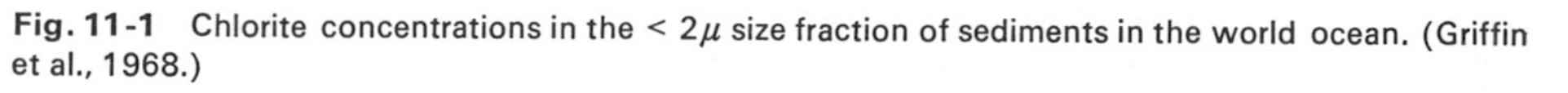
Fig. 11-1 Chlorite concentrations in the < 2μ size fraction of sediments in the world ocean. (Griffin et al., 1968.)

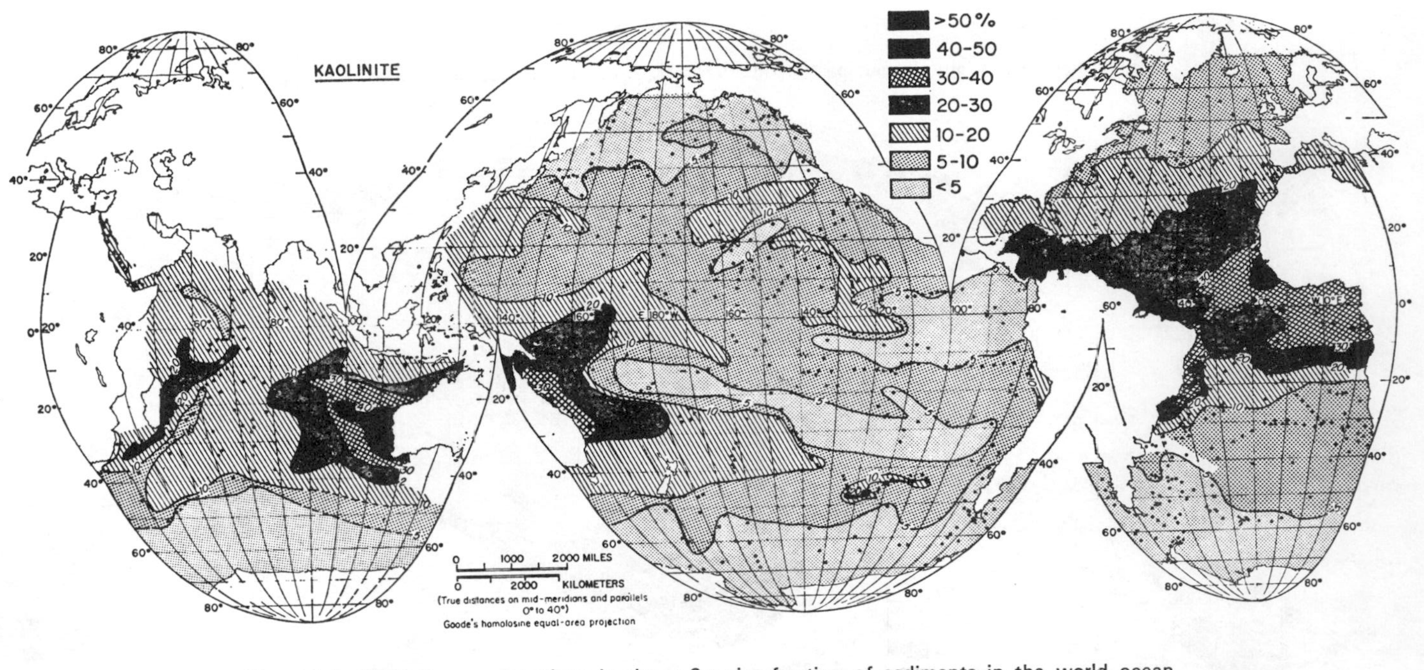

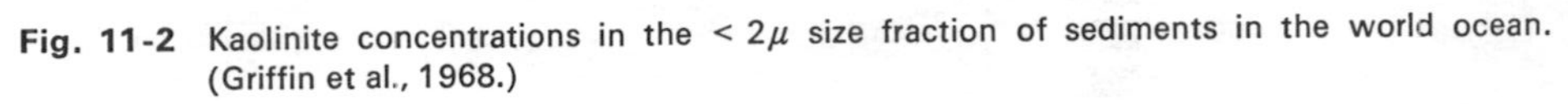
Fig. 11-2 Kaolinite concentrations in the $< 2\mu$ size fraction of sediments in the world ocean. (Griffin et al., 1968.)

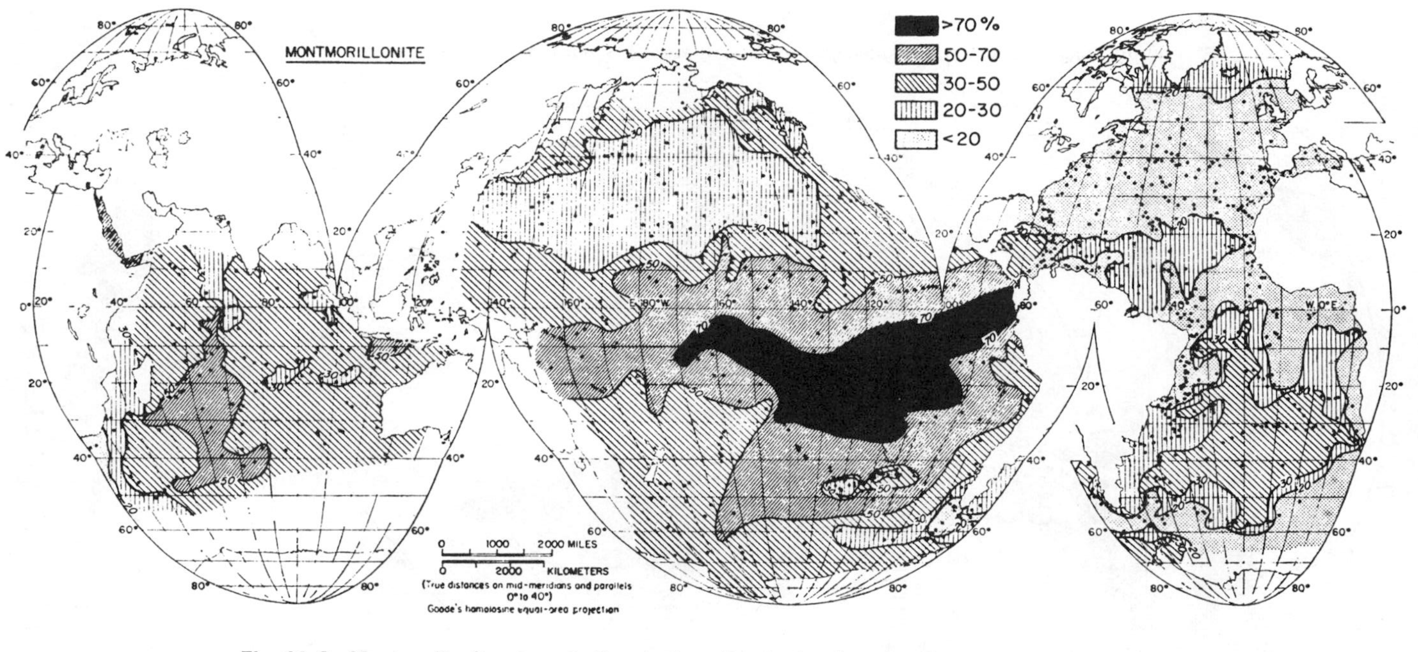

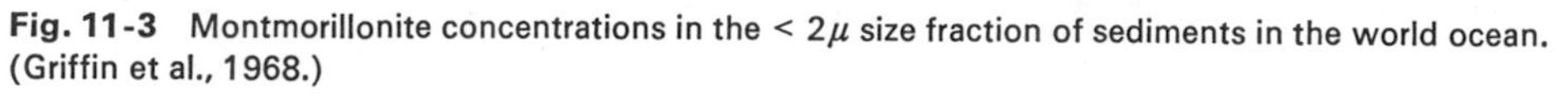

Fig. 11-3 Montmorillonite concentrations in the < 2μ size fraction of sediments in the world ocean. (Griffin et al., 1968.)

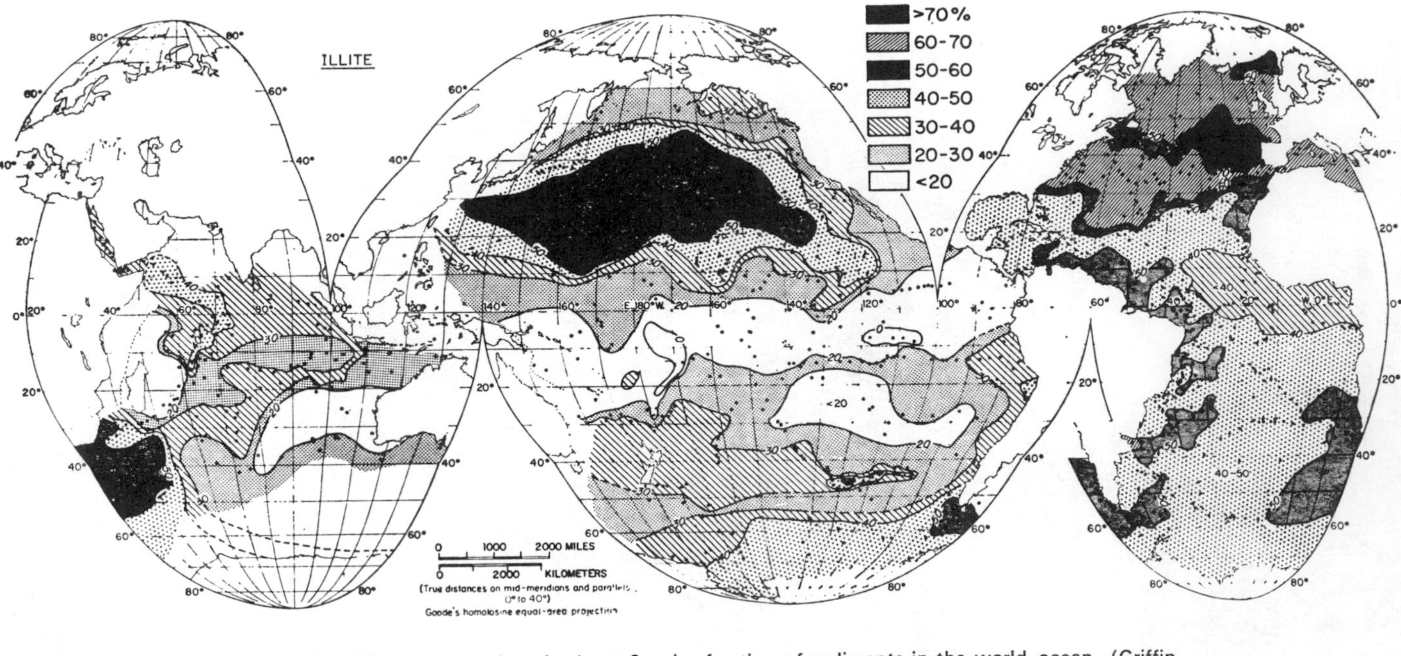

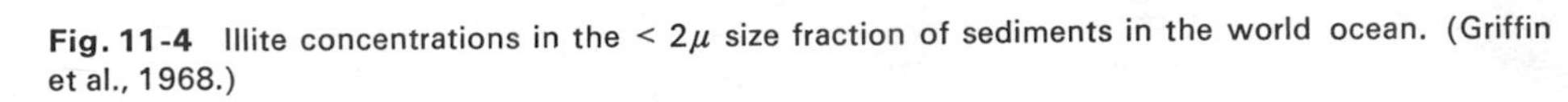
Fig. 11-4 Illite concentrations in the < 2μ size fraction of sediments in the world ocean. (Griffin et al., 1968.)

Most montmorillonite derives from volcanic sources and, therefore, it is most abundant in the South Pacific Ocean where it forms more than 50% of the oceanic clay mineral suite. The high in the central part of the South Atlantic Basin probably reflects alteration of Mid-Atlantic Ridge extrusive materials. The absence of a high montmorillonite area in the North Atlantic reflects dilution by illite from the western European and North American landmasses. The northern hemisphere is more continental than the southern hemisphere. The north-south trending high in the Indian Ocean is coincident with the location of the Mid-Indian Ocean Rise. Considering the world ocean as a whole, montmorillonite abundance averages 38%, ranging between 16 and 53% in the different basins.

Illite averages 37% of the oceanic clay mineral suite, ranging from 26 to 55% in the different ocean basins. Distinct highs occur in the North Atlantic because of stream runoff from adjacent landmasses and jet stream transport; in the North Pacific, because of jet stream transport from Asian areas; and off the south coast of Africa, because of the prevailing westerlies and some runoff from semiarid south Africa.

Besides the clearly demonstrable effect of source area, including both rock type and climate, there has been much discussion among clay mineralogists of other possible factors influencing the distribution of clays in the modern marine environment. Two factors might be of importance: (a) the grain size distribution of individual flakes of different clay minerals and the varying way in which the "effective" size of clay particles is increased by flocculation as slightly acid, low salinity river water mixes with basic highly saline seawater, and (b) the possible effect of transformation of one clay mineral to another, either while it is still suspended in seawater or after it settles to the sea bottom. Transformations that might theoretically be expected are the formation of illite, chlorite, or montmorillonite from kaolinite. It has also been suggested that illite and chlorite are formed from montmorillonite. Such transformations are generally called "diagenesis" by clay mineralogists, though it is supposed that they can take place while the clays are still suspended in seawater—this is a modification and extension of the term diagenesis as generally used by other sedimentologists.

Electron micrographs of clay mineral flakes in modern muds reveal that a size difference does indeed exist among the major clay mineral species. Normally, kaolinite is coarsest, with flake sizes commonly several microns in length. Illite flakes usually are 0.1 to 0.3 μ in length, and montmorillonite flakes are normally so small that transmission electron microscopy is inadequate to distinguish them within aggregates. Montmorillonite surfaces have a very high ion exchange capacity and are therefore very reactive; consequently, the mineral almost invariably occurs as floccules rather than as separate flakes. The size of these floccules may be greater or less than the size of an illite flake but normally does not exceed the size of a kaolinite flake. Of course, both kaolinite and illite can occur as floccules as well and it is only meaningful to consider the relative tendencies of the different clay species toward flocculation.

Many field studies of modern muds confirm the observations of electron microscopy. In nearly all studies of clay mineral distributions in shallow marine sediments, kaolinite is most abundant closest to shore. Montmorillonite commonly becomes more abundant farther offshore. In the Gulf of Paria, for example, montmorillonite tends to be more abundant in the open gulf than close to the Orinoco Delta. The same type of relationship has been observed offshore of the Niger Delta (Fig. 11-5).

The importance of "diagenesis" in the formation or modification of clay minerals during settling in the sea or at the sea floor (halmyrolysis) is more controversial. Most workers now agree that the effect is generally small compared with the effects of source and hydraulic sorting. Diagenesis might be expected to be more effective in areas of slow sedimentation, such as the deep ocean basin, but the clear effect of source area in these muds indicates that the effect of penecontemporaneous diagenesis must be relatively small.

This inference is further supported by the results of potassium-argon dating of illitic muds. For example, the age of a sample of illitic material of dominantly silt size from the Mississippi River Delta was found to average 280 million years with little vertical or horizontal variation. The clay size fraction of the same sample, however, averages only 166 million years, a result susceptible of three interpretations. The Mississippi River drains both an illitic source (the Ohio River Basin drains largely illitic Paleozoic sediments) and a source rich in finer-grained mixed-layer illite-montmorillonite (the Missouri River Basin contains Mesozoic and Cenozoic sediments rich in expandable clays). Hence, the young age of the finer grained part of the Mississippi Delta clays may simply be a mixing effect. A second possible interpretation is that finer grained and more reactive degraded or stripped illites have been rapidly regraded with potassium ions from the Gulf of Mexico waters, resulting in a younger age for the finer fraction. The third possibility is spontaneous authigenesis of illite flakes in the gulf waters, which would have an identical result. There is no way known to determine the correct alternative among the three possibilities. Authigenic illites are known to crystallize in a polytype (stacking sequence of structural layers) different from detrital illites, but no method has yet been developed to determine reliably the proportion of each polytype in a clay mixture. At present, we can conclude only that source area and climate are the dominant controls on clay mineral facies in modern seas but regrading and/or authigenesis of clays may occur in small amounts.

Mackenzie and Garrels (1966) conducted experiments in which they placed various clay minerals in seawater devoid of silica and monitored the pH and silica content of the water with time. They found that the clays released silica to seawater only up to the values commonly found in bottom waters, about 6 ppm. Therefore, the basic structural framework of clay minerals probably is stable in seawater. They further suggest that a "reverse-weathering reaction" of the type

$$\text{Amorphous aluminum-silicate} + HCO_3^- + H_4SiO_4 + \text{cations} \longrightarrow$$
$$\text{cation aluminum-silicate (mainly clay)} + CO_2 + H_2O$$

has played an important role through geologic time in removing silica and bicarbonate from seawater. Presumably the amorphous materials and poorly crystalline

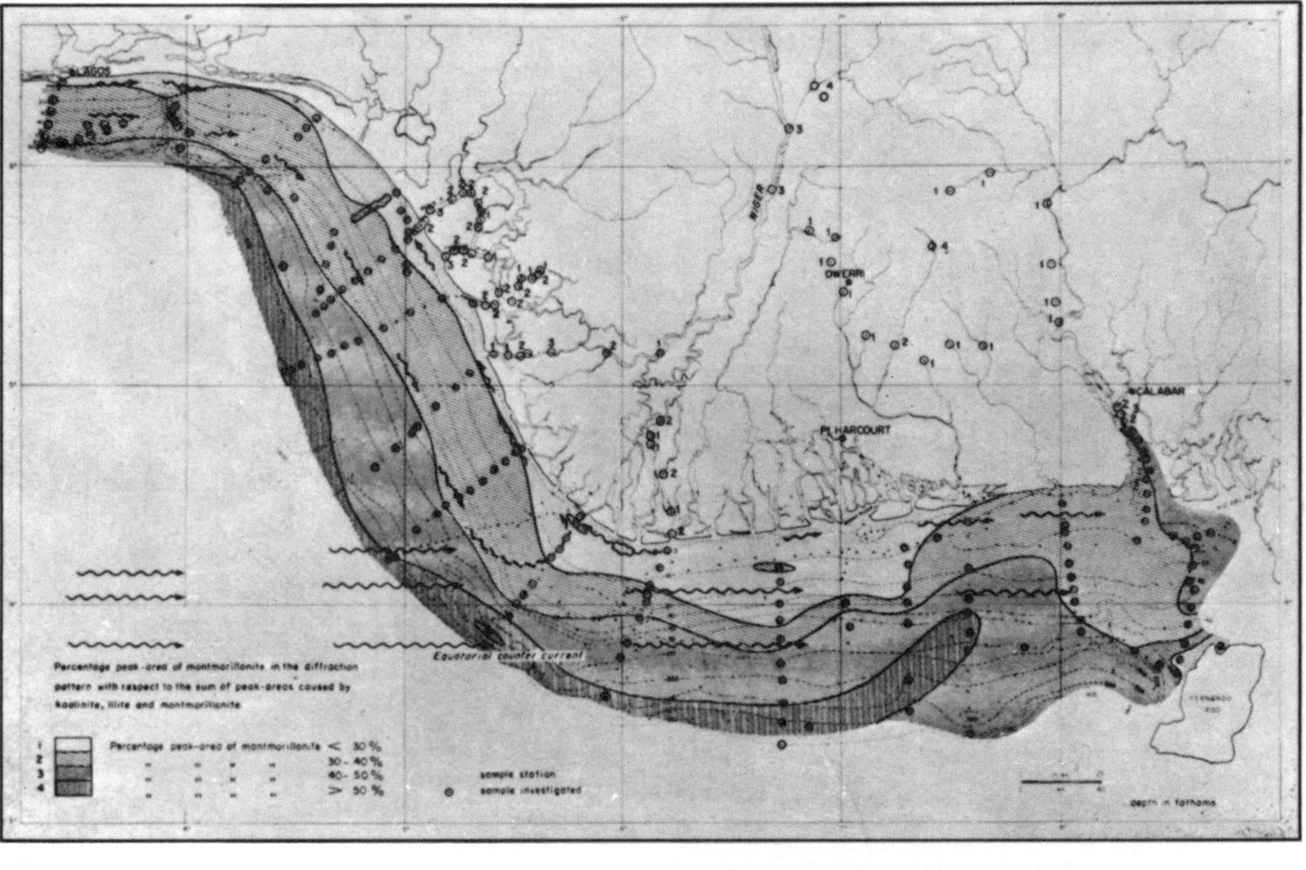

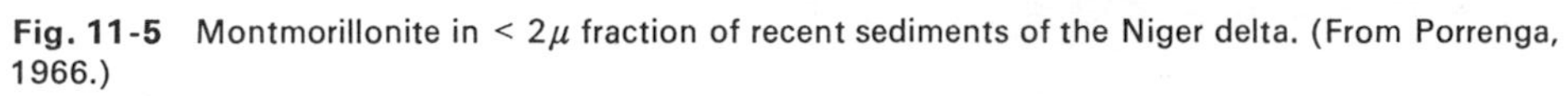

Fig. 11-5 Montmorillonite in $< 2\mu$ fraction of recent sediments of the Niger delta. (From Porrenga, 1966.)

substances are produced on the land surface during weathering. The quantitative importance of the Mackenzie-Garrels reaction is limited however, by the abundant data demonstrating parallelism between climate and modern clay mineral facies in the world ocean.

Modern Nonmarine Sediments

There are few data concerning clay minerals in fluvial and lacustrine environments, in contrast to the voluminous amount of data concerning modern marine clays. Therefore, few conclusions are possible concerning the distribution of clay species in freshwater deposits but, as with marine clays, the distribution is nearly always found to result from source area variations. This is to be expected because the salinity of river water averages only 0.3% that of normal seawater. With specific reference to the cations required for authigenic growth of clay minerals, average river water contains only 0.06% as much sodium as seawater, 0.32% as much magnesium, 0.61% as much potassium, and 3.75% as much calcium. The amount of ferrous iron is, of course, negligible in both environments. As authigenesis appears of only minor importance in seawater, there is no reason to anticipate it in river waters.

Clays in modern saline lakes also reflect their source areas. In the arid climates in which these lakes exist, weathering is slight and, assuming chlorite and illite are available in the surrounding highlands, they will be delivered intact to the lake basin. There is no evidence that kaolinite in these lakes is changed into smectites, illites, or chlorites, despite high contents of monovalent and divalent metallic cations in the water. Perhaps the reaction is too slow to be detected in these lakes, many of which may be only a few thousand years old.

Clay Minerals in Ancient Sediments

The major groups of clay minerals are known from sedimentary rocks of all ages and occur in almost all proportions, and these proportions may vary both laterally and vertically in a stratigraphic section.

The only major direct environmental control of clay mineral type that is well documented is the preferred occurrence of kaolinite in terrestrial rather than marine environments. As we noted in Chap. 7, however, kaolinite often is not present in abundance in terrestrial environments either, as its formation depends on the existence of suitable climatic conditions. It is possible that part of the kaolinite in terrestrial sediments is authigenic and that the formation of authigenic kaolinite in marine rocks is prevented by the presence of abundant potassium in connate waters. The idea is supported by one experimental study that indicated formation of a potassium phyllosilicate rather than kaolinite under conditions of high K^+/H^+ ratio. Unfortunately, this investigation was conducted at temperatures above 250°C and extrapolation of the phase boundaries to lower temperatures is speculative.

In many localities in Europe there exist massive clay beds known as tonsteins. These units are of terrestrial origin and very commonly occur in association with coal beds. The problem concerning their origin stems from the fact that they are often nearly pure kaolinite; are very thin, with thicknesses of only a few inches or less; and have extents of several thousand square miles. They are among the thinnest widespread stratigraphic units known and it is this characteristic that forms the heart of the problem concerning tonstein formation. What type of distributive mechanism is capable of spreading a thin layer of fine grained material over an extensive nonmarine area? Tonsteins are commonly associated with durain macerals and cannel coals, which indicates that they did not form in a flat peat bog but in an area of luxuriant tree growth. In such a lowland delta/swamp environment normal distributive processes result in considerable lateral variation in grain size, lensing of stratigraphic units, and a more complex mineralogy than found in tonsteins, which appears to rule out the possibility of a fluvial origin for these rocks.

The only mechanism capable of solving the distribution problem is a fall of volcanic ash. This mechanism receives support from the finding in some tonsteins of relict volcanic detrital materials such as felsitic fragments and a restricted heavy mineral suite consisting of idiomorphic zircon, apatite, and brown biotite. No vitric textures have been reported from tonsteins, presumably because of the high degree of alteration they have undergone since deposition. The content of soluble major and minor elements in most tonsteins is low and the only iron present occurs in cubes of pyrite, which is to be expected in the reducing environment represented by a coal swamp. Aluminum oxide minerals such as gibbsite, diaspore, and boehmite are absent but the presence in tonsteins of many authigenic worms and books of kaolinite permits the inference that aluminous minerals were originally present but have been silicified. Indeed, gibbsite is rare in ancient rocks and apparently is unstable in nearly all diagenetic environments (Curtis and Spears, 1971). In summary, it appears that most tonsteins are probably of ultimate volcanic origin but the interaction between organic matter, poor drainage, and the diagenesis of the volcanic materials is poorly understood.

Vertical changes in clay mineral facies may be considered in two groups: (a) Those taking place soon after burial. If major changes take place, they should be readily observable in long cores taken from recent sediments. (b) Those taking place after deep burial. In most investigations, these have been inferred from studying a succession of shales exposed in a single borehole. This is a dangerous procedure because changes could be due to differences in source or lateral changes. The same formation may be studied at different depths of burial but there is still the possibility of confusing lateral with diagenetic changes. However, the observation of similar vertical trends in studies from different formations and basins and correlation of the trends with changes known to be diagenetic gives some confidence that the observed changes are diagenetic.

The clay minerals in most cores of modern sediment show few vertical changes that cannot be attributed to changes in sediment source. The exceptions appear to be related to the presence of easily altered volcanic ash in the sediment.

Evidence for substantial changes in clay minerals with deep burial is more conclusive. Based on studies in several basins, Weaver (1959) concluded that montmorillonite altered below a depth of 10,000 ft to mixed layer montmorillonite-illite, with the proportion of illite increasing with further increase in depth of burial. Burst (1969) has documented the change from swelling to nonswelling clays for the Eocene of the Gulf Coast. The mineralogical change is accompanied by dehydration of the clays and takes place mainly between depths of 8000 to 13,000 ft, corresponding to temperatures of 95 to 140°C. In the Upper Cretaceous of the Douala Basin, Cameroun, it has been possible to trace a progressive increase in chlorite and illite and a decrease in montmorillonite, montmorillonite-illite mixed layer minerals, and kaolinite downward over a depth of more than 12,000 ft (Fig. 11-6). There is a corresponding decrease in water content (shown in the figure by an increase in density) and increase in the crystallinity of illite. In thc Upper Carboniferous in Germany, changes in the clay minerals have been compared with changes in the rank of coal that have been studied in the same borehole. Kaolinite is the common clay mineral in the upper 10,000 ft but is replaced lower in the section by illite, muscovite, and chlorite. The change from kaolinite to illite and chlorite corresponds to the change from semianthracite to anthracite coals and is accompanied by sericitization of feldspars and an increase in crystallinity of the clay minerals. From these examples, therefore, there is little doubt that substantial diagenetic changes in clay minerals accompany burial below 10,000 ft and/or increase in temperature to about 100°C. There also is evidence from experimental studies that indicates that montmorillonite loses its interlayer water at temperatures between 100 and 130°C and, in the presence of K^+/H^+ ratios approximating those of normal seawater, is converted to illite.

Diagenesis of clays may also result from their exposure to the action of circulating groundwater. This is shown by observed differences between the clay fraction in sandstones and in associated mudrocks. In the Illinois Basin, for example, the pre-Pennsylvanian shales contain mainly well-crystallized illite and chlorite but the associated sandstones contain kaolinite and much degraded illite and chlorite. As another example, Miocene mudrocks in the Los Angeles Basin contain much montmorillonite but interbedded sands in the same formation contain relatively little montmorillonite and a much larger fraction of chlorite. Kaolinite appears from field and petrographic studies to be the clay mineral species most frequently produced by spontaneous nucleation in ancient rocks but all the clay mineral groups have been reported. Presumably one of the critical factors for growth of these minerals from subsurface waters is the metallic cation/hydrogen ion ratio, with the type of metallic ion controlling the species synthesized. But because smectites, illites, and chlorites are three-layer structures, they cannot be synthesized as easily in the natural environment as can kaolinite, a two-layer structure. Chlorite, in particular, seems difficult to synthesize under most diagenetic conditions and reports of authigenic chlorite are largely confined to areas of incipient metamorphism of eugeosynclinal sediments.

Bentonites

The term bentonite is usually restricted to aggregates of clay formed by *in situ* alteration of volcanic ash and they may be either marine or nonmarine. Such

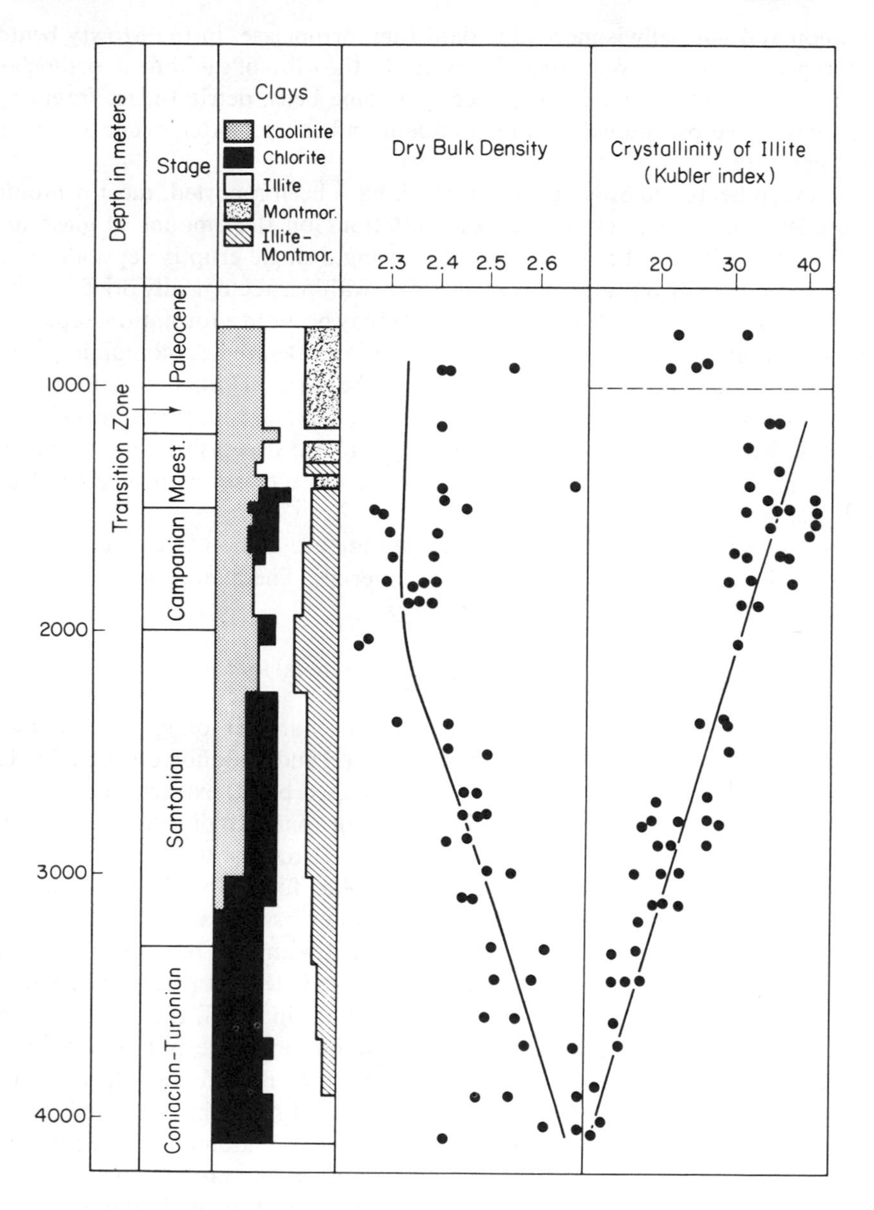

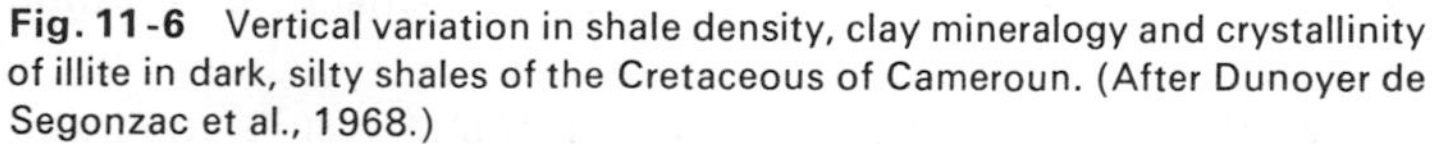
Fig. 11-6 Vertical variation in shale density, clay mineralogy and crystallinity of illite in dark, silty shales of the Cretaceous of Cameroun. (After Dunoyer de Segonzac et al., 1968.)

clays are further characterized by being composed largely or entirely of smectite clays that are highly colloidal and plastic. Uncontaminated bentonites contain, in addition to smectite, only igneous detritus such as euhedral biotite flakes, quartz fragments with beta crystal form, magnetite, euhedral zircon, and a variety of feldspars. If the original ash was sufficiently potassic in composition, sanidine is common and normally is more abundant than orthoclase. In the Mowry bentonites (Upper Cretaceous, Wyoming) for example, the ratio of sanidine to orthoclase is about 7: 1 (Slaughter and Earley, 1965). In some beds, detrital glass fragments or their devitrified pseudomorphs are abundant and zeolites often occur in cavities in the bentonite.

Although bentonite beds up to 50 ft thick have been reported, most bentonite beds are less than 1 ft thick because of limitations on the amount of glass and mineral matter that can be ejected upward during a single eruptive episode. Such episodes are normally repeated numerous times within a geologically brief interval, however, so that several dozen bentonitic units may occur in a formation, separated by either impure tuffs or other detrital materials. Bentonites containing relict sand-size ejecta are usually seen to be graded because of differences in settling velocity during their descent in the atmosphere. Detailed mapping reveals that these beds also are size graded laterally, with grain size decreasing logarithmically away from the volcanic source. Bed thickness also decreases with distance from the source.

The alteration of volcanic ash to smectite may occur soon after accumulation because of the marked instability of glassy materials. The alteration may be considered as a hydrolysis reaction summarized as

$$\text{Glass} + H_2O \longrightarrow \text{smectite} + \text{zeolite} + \text{silica} + \text{metal ions in solution.}$$

This reaction is promoted by the highly unstable character of glassy materials. The details of the transformation are not understood and undoubtedly are complex but may be considered to occur in steps characterized by (a) extraction of cations from the glass and their replacement by hydrogen ions; (b) disintegration of the disordered silica-alumina framework remaining; (c) reconstitution of the framework to clay minerals, largely smectites; (d) zeolite formation from cation-rich pore solutions; and (e) removal or precipitation of excess silica.

The composition of the smectite varies greatly among bentonites and the variation may be within the smectite structure itself or in the nature of the exchangeable cations. Depending on the composition of the original tuff, the smectites may be either calcic, sodic, or potassic but calcic montmorillonite is most common. Assuming approximately equal activities in solution, it is to be expected that calcium will replace sodium or potassium in smectites (see Chap. 7). Only a few bentonites are known in which sodic montmorillonite is predominant, the extensive Mowry Formation bentonites of Wyoming being the main example. In many cases where calcium is the dominant ion, magnesium is present in small amounts as an exchangeable ion.

The conversion of silicic glass to montmorillonite results in the production of excess silica, which may be either removed in pore solutions or crystallize interstitially as cryptocrystalline or microcrystalline quartz. Probably the pH of the pore solutions is the critical factor that controls removal of the dissolved silica (see Chap. 10); the higher the pH, the more likely is the silica to be removed. If silica is deposited as quartz, for every gram of montmorillonite formed, approximately one-sixth of a gram of silica would result from alteration of a latitic glass.

Proportionately more silica would be generated from a rhyolitic glass. Many bentonites in the Mowry Formation are gradationally succeeded stratigraphically by a few inches of siliceous smectite or porcellanite.

Clay Mineral Abundance and Time

Weaver (1967a) has constructed a graph illustrating the abundance of clay minerals in shales as a function of time, based on many thousands of analyses of Phanerozoic rocks in the United States (Fig. 11-7). More than half the clays

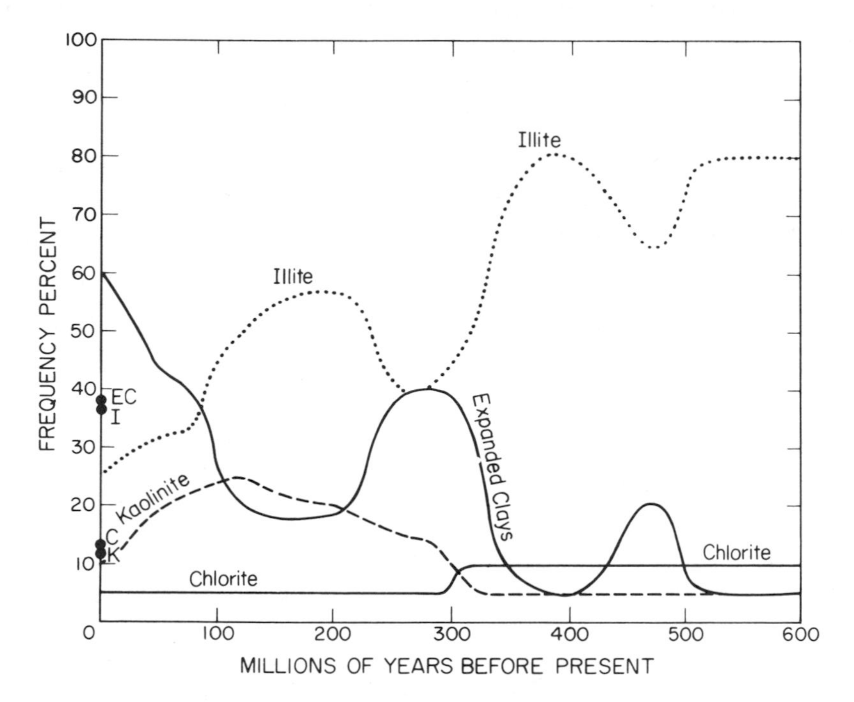

Fig. 11-7 Relative abundances of the major groups of clay minerals in Phanerozoic mudrocks (generalized from Weaver, 1967a. The dots indicating clay percentages in the modern ocean are estimates by Griffin et al., 1968.)

(57%) are illite. About one-quarter (24%) are expandable clays, either pure or mixed layer; 12% are kandites, and only 7% are chlorites. More significant than the averages, perhaps, are the trends in abundance of each clay mineral group with time. Kandites and expandable clays are less abundant in older rocks. The amount of illite varies inversely with the amount of expandable clays. And there seems to be a marked increase in illite in pre-Pennsylvanian rocks.

Possible explanations for these relationships include (a) changing climates; (b) changing paleogeographic patterns that control the supply of clays from the source area; (c) changes in the nature of the source rocks resulting from volcanic activity or uplift and erosion; (d) diagenetic changes affecting the older, more deeply buried sediments; and (e) changes in biologic controls of weathering.

Based on the abundant evidence demonstrating the formation and stability of illite during diagenesis of clay minerals, it seems best to attribute the predominance of illite in mudrocks to postburial processes. The inverse relationship in abundance of expandable clays and illite also is explained by diagenetic processes. Structural considerations suggest the expandable clays as the most likely precursor of illite crystals and, indeed, a large proportion of these swelling clays are mixed layer montmorillonite-illite crystals rather than pure montmorillonites.

The reason for the abrupt increase in illite in pre-Pennsylvanian rocks is less clear. Weaver (1967b) has suggested that a change in vegetation patterns is the most probable explanation, noting that the increase coincides with the widespread development of terrestrial vegetation. In the absence of well-developed vegetation it is likely that weathering would proceed at a reduced rate and that abundant stoichiometric illite might survive "soil" formation and be carried to the sea. The effect of the evolution of land plants would be a worldwide effect and might be expected to affect the chemical composition as well as the clay mineralogy of mudrocks. Weaver has made use of data on the composition of thousands of samples of mudrocks from the Russian platform to demonstrate that the date of the mineralogic change in America coincides with a change toward lower potassium and higher sodium in Russian mudrocks, which tends to confirm the hypothesis. Further confirmation awaits detailed studies from other areas and an attempt to control some of the other variables listed above.

Quartz

Quartz percentages in mudrocks are nearly always underestimated during hand specimen and thin section studies because of the shape and relatively high birefringence of most clay flakes as compared to quartz. The estimate of 31% quartz in the average mudrock was determined using a combination of chemical and X-ray techniques.

The size distribution of quartz in mudrocks is approximately 15% fine and very fine sand and 85% silt. Clay size quartz fragments rarely total more than a few tenths of 1% by weight. The mean grain size of quartz in mudrocks is about 5ϕ and nearly all quartz in mudrocks is monocrystalline because of the decrease in abundance of all polycrystalline fragments with decreasing grain size.

The average shape of quartz grains in mudrocks is speculative because quartz grain boundaries are commonly "fuzzy" in thin sections of mudrocks. This effect results not only from "smearing" of clay minerals during manufacture of the slide but also from the presence in most mudrocks of minute particles of iron oxide and organic matter. Frequently, however, the shape of quartz in mudrocks is reported to be elongate in thin section, an observation that, if accurate, has two possible interpretations. The first is that a large proportion of these grains has been derived from foliated metamorphic rocks such as slates, phyllites, and perhaps fine grained schists. The second possible interpretation is that these grains originate in the sedimentary environment by chipping of coarser quartz grains of any origin. Chipping of small fragments from subequant grains would yield either elongate or platy fragments, depending on the length of the chip, and either of these shapes appears elongate in thin sections cut normal to shale fissility, as most thin sections are. If this latter interpretation is correct, the three-dimensional shape of quartz in shales (and of silt size quartz generally) has no provenance implications. At present, there is no way to evaluate the relative importance of these two possible origins of quartz silt but the enormous amount of size reduction of detrital quartz by sedimentary processes (Blatt, 1970) suggests the second explanation as more generally applicable.

Despite the fuzziness of quartz grain boundaries in mudrocks, the degree of rounding of these grains can be safely inferred. Quartz fragments generally show a decrease in roundness with decreasing particle size because of the different modes of transport of gravel and sand as contrasted to silt. The first two size groups are transported either in traction or saltation; the latter, in suspension. Hence, rounded fine sand size quartz grains are uncommon and rounded silt size quartz is rare. Therefore, quartz in mudrocks is quite angular.

Feldspar

As noted in Chap. 8, very little is known concerning the relative abundances of microcline, orthoclase, and plagioclase in sandstones. Nothing is known concerning the presence of these species in mudrocks. Determinations would have to be made by a combination of optical petrography (to determine microcline percentage) and electron microprobe studies (to determine chemical compositions) because of the small grain size of the feldspar crystals. Small grain size, in combination with the known structural and chemical weakness of feldspar composition planes, suggests that Carlsbad and albite twins will be less common in feldspar grains in mudrocks than in sandstones.

Carbonate Minerals

There are few quantitative data concerning the relative abundances of the various carbonate minerals in mudrocks, although calcite, dolomite, ankerite, and siderite have been reported. Accurate determination of carbonate mineralogy in mudrocks has been essentially restricted to carbonate showing unusual morphology, such as concretions or cone-in-cone and, as a result, present evaluations of carbonate phases in fine grained silicate rocks must be based on relative abundances in sandstones and limestones. Under this assumption, calcite would predominate over the other carbonates. The origin of the calcite in mudrocks is speculative, as it may be either chemically precipitated or originate as particulate matter derived from the shells of organisms living in the mud. The boring and grinding activities of predators might easily produce the small amount of fine grained calcite present in the average mudrock.

One of the few detailed mineralogical and chemical studies of a calcareous shale is that made by Campbell and Oliver (1968) of the Devonian shales associated with the Leduc "reefs" of Alberta. Based on analysis of 800 samples, they found that the composition varied from 13 to 39% illite, 3 to 15% chlorite, 1 to 32% dolomite, 16 to 76% calcite, and 3 to 26% quartz. There was an almost complete range from pure limestone to calcareous shale. A peculiar feature of these rocks is the high content of dolomite. Shales deposited on the basin slope have a lower content of carbonate and clay (clay/quartz ratio of 2: 1) compared with shales from the deeper parts of the basin (clay/quartz ratio of 3: 1 to 4: 1). Most of the constituents were thought to be detrital, with clays and quartz (and some carbonate) derived from the basin edge and carbonate supplied from organic carbonate buildups within the basin. The origin of the dolomite is not known but may be diagenetic.

Pyrite

In dark colored organic-rich mudrocks it is common to find authigenic pyrite crystals ranging in size from microcrystalline and cryptocrystalline to several millimeters. This association between organic matter and pyrite is not surprising in view of the fact that both iron and sulfur are abundant in most depositional environments. Under oxidizing conditions the iron occurs in the ferric form, largely as hematite or as iron adsorbed on clay minerals or newly deposited organic matter. The sulfur exists mostly in the pore waters as sulfate ion but some is also present structurally bound as sulfur atoms in organic compounds. Under reducing conditions, however, ferrous iron, hydrogen sulfide, and native sulfur are the stable chemical species, and these species may react to produce iron sulfides. In modern black muds, the iron sulfides that are most common are noncrystalline FeS, mackinawite (a nonstoichiometric iron monosulfide), and greigite, Fe_3S_4. Unanalyzed aggregates of such materials in modern sediments are commonly referred to as hydrotroilite. In ancient shales these minerals are rarely found and pyrite, FeS_2, is almost the only species of iron sulfide present. (See further discussion in Chap. 19).

11.3 ORGANIC MATTER

The preservation of organic matter in a sediment requires a low-Eh environment to prevent complete oxidation of the organic compounds to carbon dioxide plus water. In the natural setting, the requirement of low Eh means that the rate of influx of organic matter to the site of deposition must exceed the quantity of oxygen available to oxidize it, a condition that can be met in a variety of quiet water environments. On land, organic matter commonly is preserved in abundance in floodplain muds, lake bottoms, and marshy or swampy areas; in the sea, there are silled basins formed by structural features (e.g., southern California Borderland area) or by geomorphic features (many Scandanavian fiords, Baltic Sea, Gulf Coastal bays, estuaries along eastern United States, and parts of Hudson Bay, Canada). All these areas have restricted water circulation. Although normal marine or fresh waters contain at least 5 to 8 ml/liter of dissolved oxygen and an Eh of + 0.3 volt, the water in areas of accumulating organic matter has dissolved oxygen contents near zero and Eh values in the area of − 0.3 volt. Restricted water circulation implies the absence of strong currents and, therefore, the absence of coarse clastic and extensive carbonate deposits (because nearly all carbonate deposits are organic in origin). Thus, organic matter accumulates with silt and clay size clastic detritus. The total amount of organic matter in sediments is estimated to be 3.8×10^{15} metric tons, of which approximately 95% is present in mudrocks (Degens, 1967). It is for these reasons that the "source rock" for petroleum is commonly a mudrock. Hydrocarbons produced by living organisms and the much larger amounts produced by diagenesis of organic matter in mudrocks are forced out of the mudrocks and into adjacent reservoir rocks largely by compaction.

The amounts of organic residue in mudrocks range from essentially zero to about 40%, with a geometric average of 1.1% (Fig. 11-8). Analyses of many modern

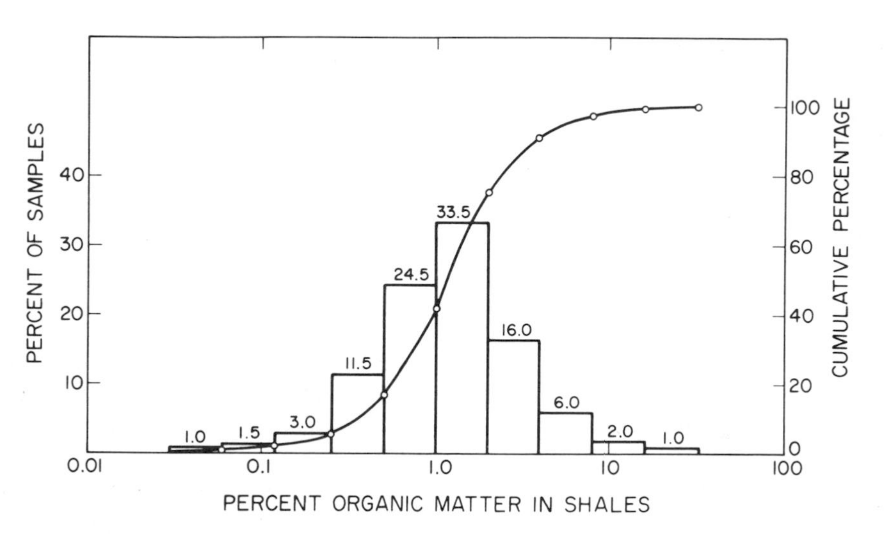

Fig. 11-8 Frequency distribution of organic matter in shales. (After Gehman, 1962.)

muds result in an identical average amount of organic matter, but chemical determinations of the types of organic *molecules* present in ancient mudrocks and modern muds reveal striking differences. The biochemical macromolecules abundant in recent muds are greatly reduced or absent from ancient rocks as a result of microbiological activity and chemical hydrolysis. Proteins, polysaccharides, fats, and nucleic acids are rapidly destroyed in the early stages of diagenesis. The organic molecules in pre-Quaternary muds and shales are those that normally result from decomposition of these complex units: amino acids from the proteins, simple sugars and phenols from the polysaccharides, fatty acids from the fats, and purine and pyrimidine bases from the nucleic acids. These breakdown products can interact to produce humic acids, fulvic acids, and other organic compounds commonly reported from ancient sediments. The reason these compounds accumulate is that they have no nutritive value for bacteria. Also, they interact with clay minerals in the sediment, are adsorbed, and immobilized. Reactive groups remaining in these molecules are often made inaccessible by the adsorption process.

As the quantity of available food materials decreases, bacterial activity diminishes correspondingly and further diagenesis of the organic matter occurs as a result of temperature increases associated with increased depth of burial. Functional groups and aliphatic side chains are lost from aromatic and aliphatic compounds, ultimately resulting in the formation of kerogens, hydrocarbons, and other economically valuable compounds. The changes that occur in organic matter between the time of death of an organism and its ultimate conversion to graphite and paraffinic petroleum are described in some detail in a summary paper by Degens (1967).

11.4 SEDIMENTARY STRUCTURES

Mudrocks may contain a great variety of sedimentary structures; some are visible in the field or hand specimen and some are visible only in thin section. Occurrences are known of graded bedding, small-scale cross-bedding, slumping, cut and fill structures, animal burrowing, and autobrecciation, as well as the more obvious features such as lamination and fissility.

Fissility

Fissility is a property of a shale that causes it to part easily or break parallel to stratification. It is readily observed either in outcrop or in thin section. Until recently there had been few studies of the origin of shale fissility and the variables involved are only now becoming apparent.

Of primary importance are the clay minerals in the rock—their abundance, degree of crystallinity, type, and orientation. As is normal in geologic problems, all these variables are interrelated to a greater or lesser degree but it appears that fissility is favored by high proportions of well-crystallized clays whose interlayer bonding sites are located on the clay surfaces rather than on their edges. The necessity for maximizing the proportion of sheet-structure minerals to achieve fissility is obvious, as all other structural types form less platy minerals, such as quartz, feldspar, or calcite. Similarly, poorly crystalline or amorphous solids hinder the development of fissility in a rock.

The importance of the clay species present as an influence on rock fissility results from structural considerations noted in the discussion of weathering earlier in this book. Unlike the other clay mineral groups, kandites lack an intrastructural charge deficiency and their unfilled electron orbitals are located largely on sheet edges. Therefore they normally will bond to other flakes in an edge-to-face arrangement and kaolinitic clays tend to be massive (claystones or mudstones) rather than fissile. Floccules of kaolinite can be seen in scanning electron micrographs of shale fabrics, appearing as poorly defined lumps in a matrix of parallel-oriented clay flakes. The areas showing parallel orientation of flakes may consist either of clays deposited in a dispersed state or of compacted floccules. Presumably these well-foliated regions are more illitic or montmorillonitic than the edge-to-face regions but present data are inadequate to test this hypothesis.

There is no consistent correlation between fissility and depth of burial, suggesting that sedimentary pressures alone are not sufficient to orient clay flakes in floccules with random internal clay mineral orientations. In 1943, Fairbairn pointed out the relatively poor fissility of kaolinitic slates in contrast to illitic slates, indicating that even low grade metamorphic stresses may be inadequate to effect reorientation.

Organic-rich black shales commonly are exceptionally fissile, often being characterized as "paper shales." One important cause of the well-developed fissility is believed to be an abundance among the organic molecules of straight-chain aliphatic hydrocarbons whose terminal hydrogen atoms have been replaced by reactive groups. These molecules are thus made polar and will orient themselves between clay flakes as shown in Fig. 11-9. The nonpolar ends of the molecules are only weakly attracted, resulting in a thinly fissile shale.

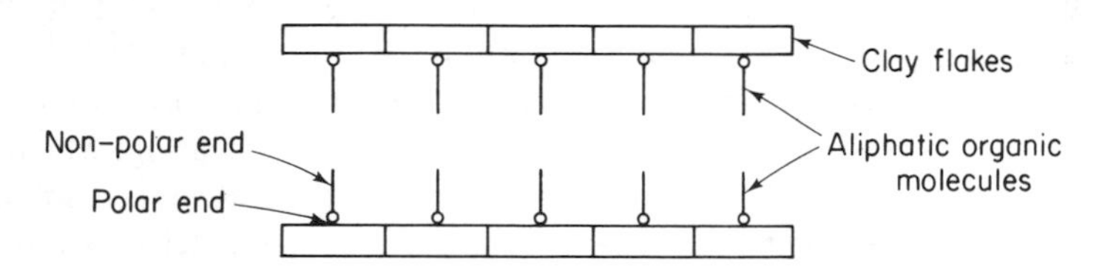

Fig. 11-9 Schematic diagram showing the relationship between polar aliphatic molecules, clay flakes, and fissility in paper shales. The forces of attraction between non-polar ends of the organic molecules are very weak.

Another significant factor favoring the development of fissility in shales is the absence of burrowing organisms (bioturbation), a condition often found in the stagnant areas in which black shales develop. In more aerated environments, however, burrowing activities frequently churn the mud, destroy laminations, and disorient the clays, resulting in the absence of fissility after lithification.

11.5 COMPACTION OF MUDROCKS AND POROSITY

Porosity in mudrocks is quite variable, as we would anticipate from our knowledge of the problems associated with the development of fissility. Indeed, the definition and measurement of porosity in a mudrock presents problems not encountered in sandstones. In a sandstone composed predominantly of quartz and similar minerals, the boundary between solid grain and pore space is reasonably well defined. If the pore space is filled with water, then this free or movable water represents the porosity. If the solid grains contain no water, then methods that measure the amount of total water measure porosity. In most sandstones, methods that measure total hydrogen concentration, such as neutron logging, also effectively measure porosity. Mudstones present a far more complex problem. Many of the clay minerals contain water as part of their structure and this water certainly should be considered part of the solid as opposed to being included as part of the pore space. In addition, water adsorbed on the surface of the clay flakes is not generally free to move and this water may form a large percentage of the total water between clay flakes in the mudrock. Within the space between the grains and their adsorbed water, there exists free water capable of moving or being easily removed by compaction. Thus, when we speak of mudrock porosity, we are usually referring to the percent of the total volume of the rock that contains free or easily moveable water. This is usually measured by mechanically compacting the rock and measuring the amount of fluid removed or the percent of volume reduction. These methods are at best an approximation of the true pore volume because of the possibility of altering the water content of the clay flakes or the amount of adsorbed water during the analysis.

Burst (1969) has suggested that the compaction of clays proceeds in three main stages. In the first, pore-water and water interlayers beyond two are removed by the action of overburden pressure. At the time of deposition, muds may have water contents of the order of 70 to 90%. After a few thousand feet of burial, the mud retains only about 30% water by volume, of which 20 to 25% is interlayer water and 5 to 10% is residual pore water. In the second stage, pressure is relatively ineffective as a dehydrating agent. Dehydration proceeds by heating, which removes another 10 to 15% of the water. The third stage of dehydration is also controlled by temperature but is apparently also very slow, requiring tens to hundreds of years to reach completion. Interlayer water is removed completely, leaving only a few percent of pore water in the mudrock. As discussed on p. 386, the second stage begins at temperatures close to 100°C and may be accompanied by diagenetic changes in clay mineralogy.

Factors affecting the rate of compaction in the first stage have been studied experimentally and reviewed by Meade (1966). In the early stages compaction may depend strongly on several factors besides depth of burial: grain size, rate of deposition, clay mineralogy, organic matter content, and geochemical factors. Fine clays have a higher initial porosity but compact more readily than silty clays, so that at depths of burial greater than about 1000 ft the grain size effect is slight. High rates of deposition result in excess pore pressures and underconsolidation of muds, as discussed in Chap. 5. These effects are eventually eliminated if sufficient time is allowed for movement of water out of the sediment. Montmorillonitic muds generally contain more water than illitic or kaolinitic muds. The concentration of interstitial waters in electrolytes has different effects depending on the mineralogy. High concentrations of sodium salts reduce the porosity of fine grained illites or montmorillonites but increase the porosity of coarse grained illites or kaolinites. Heling (1969) showed that mudstones from several different Tertiary formations in the Rhine graben showed different depth-porosity relationships. All formations had approximately the same clay mineral composition (illite-kaolinite-chlorite mixtures) but differed somewhat in grain size. The initial porosity was obtained by extrapolating the compaction curves back to 1-m depth of burial. After allowing for differences in grain size, it was observed that there was a direct relationship between initial porosities and the paleosalinities, which are known from paleontologic evidence. High initial porosities were correlated with high paleosalinities.

11.6 CHEMICAL COMPOSITION OF MUDROCKS

Fine grained rocks are difficult to study petrographically and, because of this fact, they frequently are analyzed chemically. Several data summaries have been published that show the proportions of the major oxides in mudrocks and slates; some of the more widely quoted ones are given in Table 11-2. As expected, silica and alumina are by far the most abundant oxides in mudrocks and slates because of the dominance of clays and quartz in their detrital fractions. The abundances of the other major oxides also seem appropriate to shale mineralogy: potassium from illite, ferric iron from hematite, calcium from calcite (high carbon dioxide percentage), ferrous iron and magnesium from chlorite, and sodium from plagioclase feldspar and montmorillonite. It is apparent, however, that there are substantial differences between the different estimates of "average shales." Part of the trouble is that there has been no serious effort by geochemists to define the mudrock population whose composition is being estimated. The collection of samples of analyses from the

TABLE 11-2 AVERAGE CHEMICAL COMPOSITION OF MUDROCKS

	A	B	C	D	E	F
SiO_2	58.1	58.5	55.4	60.2	56.2	53.4
Al_2O_3	15.4	17.3	13.8	16.4	15.1	16.4
Fe_2O_3	4.0	3.0	4.0	4.0	3.4	3.4
FeO	2.4	4.4	1.7	2.9	2.3	2.8
MgO	2.4	2.6	2.7	2.3	2.1	2.4
CaO	3.1	1.3	6.0	1.4	4.4	5.8
Na_2O	1.3	1.2	1.8	1.0	1.1	1.1
K_2O	3.2	3.7	2.7	3.6	2.6	2.7
TiO_2	0.6	0.8	0.5	0.8	0.8	0.7
P_2O_5	0.2	0.1	0.2	0.2	0.1	0.2
MnO	tr.	0.1	tr.	tr.	0.1	0.1
CO_2	2.6	1.2	4.6	1.5	3.3	4.3
SO_3	0.6	0.3	0.8	0.6	0.2	0.2
C	0.8	1.2	0.7	0.9	0.8	0.8
H_2O	5.0	3.9	5.6	4.7	5.0	4.5
Misc.	—	1.2	0.1	tr.	—	—
Total	99.7	100.8	100.6	100.5	100.5	100.8

A—Average shale (78 samples, average of C and D): Clarke, 1924, p. 24.
B—Average slate (69 samples): Pettijohn, 1957, p. 344.
C—Average of Mesozoic and Cenozoic shales (27 samples, 1 analysis): Clarke, 1924, p. 552.
D—Average of Paleozoic shales (51 samples, 1 analysis): Clarke, 1924, p. 552.
E—Mudrocks of Russian platform (4030 specimens, 290 analyses): Ronov et al., 1966, pp. 596–597.
F—Mudrocks of the Great Caucasus geosyncline (11,151 specimens, 455 analyses) Ronov et al., 1966, pp. 596–597.

literature has been haphazard rather than random. Clarke, for example, gives no details of the source of the samples in the two composites used to obtain his famous "average shale." The Russian work should be excepted from these criticisms. Russian geochemists have made a serious effort to produce meaningful averages for the composition of well-defined volumes of sedimentary rock, something not yet attempted elsewhere.

Comparison of columns E and F is particularly instructive because these two analyses represent the Russian averages for the mudrocks of the Mesozoic to Cenozoic of the Russian platform (a relatively stable shelf area) and the mudrocks of the same age in the Caucasian geosyncline. The geosynclinal mudrocks are slightly richer in alumina, ferrous iron, magnesium, calcium, and carbonates. Many of these differences (with the exception of calcium and carbonates) are also seen between Clarke's average "shale" and Pettijohn's average slate, suggesting that Clarke's average represents mainly United States "platform" mudrocks.

The Russian data show strong differences in composition related to stratigraphic position and also to geographic position relative to the source of the detrital material. For example, the alkalis generally tend to increase with distance from source and the K_2O/Na_2O ratio decreases. Not all formations show the same trends and the trends are generally clearer in the platform than in the geosynclinal deposits, as might perhaps be expected. Both platform and geosynclinal muds show a maximum in calcium content in the uppermost Cretaceous. The interpretation of these trends in terms of source, environment, tectonics, organic factors, and possible worldwide secular changes in the nature of the oceans and shelf seas is very difficult at the present stage of knowledge. The existence of such trends does, however, emphasize the futility of attaching much weight to the existing estimates of "average shale composition."

Examples of some extreme variations in mudrock composition are shown in Table 11-3. Siliceous mudrocks (not rich in detrital quartz) may contain more than 80% silica because of their content of glass shards or opaline diatom or radiolarian shells. Mudrocks containing authigenic feldspar may contain 10% potassium. Black shales commonly contain 5 to 10% organic carbon. Similarly, phosphatic mudrocks are exceptionally rich in calcium and phosphorus; many red mudrocks have high iron contents; and many mudrocks deposited in eugeosynclinal areas contain much sodium because of the presence of volcanic detritus.

Colors

Most pigmentation in mudrocks results from the presence of either iron-bearing compounds (green, red, brown, yellow) or free carbon (gray, black). Figure 11-10 illustrates the relationships between the pigmenting components in mudrocks and the colors they cause.

Iron Oxides and Silicates

The amount of iron oxide in mudrocks averages 6 to 7% and is nearly always less than 10% (except in iron-rich formations; see Chap. 19). The importance of

TABLE 11-3 ANALYSES OF MUDROCKS ILLUSTRATING UNCOMMONLY HIGH PERCENTAGES OF PARTICULAR OXIDES

	A	B	C	D	E	F
SiO_2	84.14	56.29	60.65	10.9	54.30	58.9
Al_2O_3	5.79	19.22	11.62	0.75	15.02	17.6
Fe_2O_3	1.21	4.39	0.36	1.2	9.48	1.8
FeO	—		—		—	5.7
MgO	0.41	1.65	1.90	0.04	2.86	4.5
CaO	0.31	0.09	1.44	46.5	4.08	1.6
Na_2O	0.99	0.19	0.60	0.20	0.43	4.3
K_2O	0.50	10.85	3.10	0.28	2.64	2.4
TiO_2	0.22	0.64	0.62	0.06	—	0.4
P_2O_5	—	—	0.18	35.0	—	tr.
MnO	—	—	0.04	—	—	0.1
CO_2	—	—	1.65	0.88	—	—
SO_3	—	0.72	3.20	—	—	—
C	—	—	9.20	0.3	—	—
F	—	—	—	3.50	—	—
H_2O	5.56	5.58	4.96	1.0	10.79	2.9
Total	100.03	99.62	99.52	100.61	99.60	100.2

A—Mowry Shale. Cretaceous, South Dakota: Pettijohn, 1957, p. 364.
B—Glenwood Shale, Ordovician, Minnesota: Pettijohn, 1957, p. 370.
C—Ohio Shale, Devonian, Ohio: Pettijohn, 1957, p. 362.
D—Retort Phosphatic Shale Mbr., Phosphoria Fm., Montana: Gulbrandsen, 1966, p. 772.
E—Pierre Shale, Cretaceous, Colorado: Hill et al., 1967, p. 16.
F—Argillite, Keweenawan, Batchawana Bay: Ontario, Canada: Macpherson, 1958, p. 76.

the state of oxidation of the iron in determining the color of red or purple as opposed to green, gray, or black shales has been known for many years. Tomlinson (1916) in a classic paper on the Paleozoic slates of New York and Vermont showed that for these rocks there was little difference in *total* iron between slates of different colors, but in red and purple slates ferric/ferrous ratios were greater than unity and in green and black slates the ratios were less than 0.5.

Many other studies, however, have shown that this is by no means always the case (see Van Houten, 1961). Many red and green mudrock sequences show

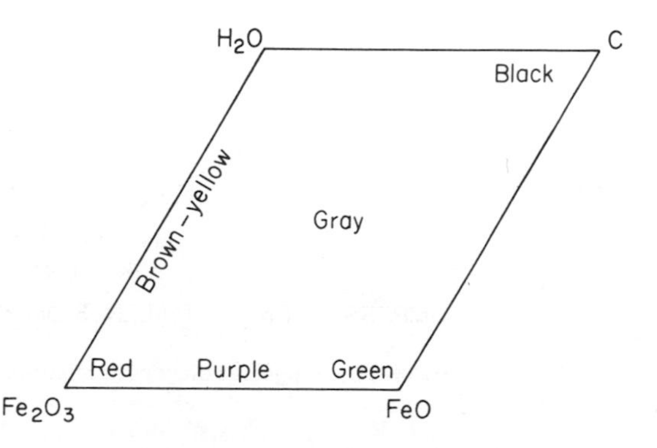

Fig. 11-10 Relationships between rock color and pigmenting agent in mudrocks.

clear evidence of diagenetic reduction of originally red mudrocks. For example, green ovoid spots are found in predominantly red mudrocks and green borders are developed along joints in red mudrocks. Comparing analyses of adjacent red and green mudrocks frequently shows that total iron is much lower in the green mudrocks, while content of ferrous iron does not differ greatly. In these cases it is clear that ferric iron has been reduced during diagenesis and the soluble ferrous iron has been lost to formation waters or moved elsewhere.

The iron-bearing minerals in mudrocks include oxides, sulfides, carbonates, and phyllosilicates. With the exceptions of sulfides and the carbonate mineral siderite, which are almost exclusively authigenic, the minerals may be either detrital or authigenic. Most iron is probably supplied to muds originally in the form of phyllosilicates such as illite, chlorite, and biotite; or in the form of hydrated ferric iron oxides ("limonite" or "goethite") that may form thin coatings on clay particles. Some iron is probably also supplied in organic colloidal complexes. In the marine environment, reducing conditions generally occur a short distance below the depositional interface and sulfur is available in solution and in organic matter. The result is reduction of iron, first to monosulfides and then to pyrite (see discussion in Chap. 19). Finely disseminated pyrite may give the mudrock a dark grey color.

The red color of mudrocks is caused by finely divided iron oxides, generally hematite. The problem of the origin of this red color is a part of the general problem of red beds, discussed in Chap. 10.

The greenish color of mudrocks results from their content of green phyllosilicates such as illite, chlorite, and, in some cases, glauconite. Green mudrocks

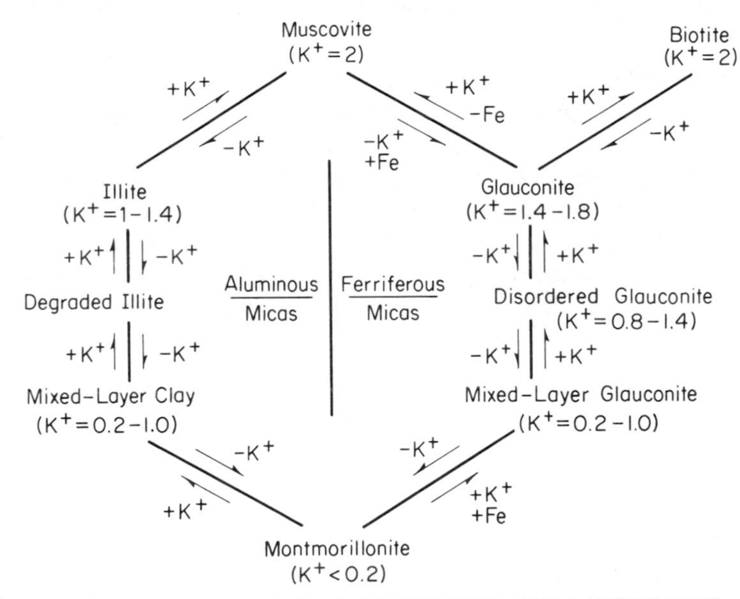

Fig. 11-11 Suggested diagenetic relationships in micas. (From Burst, 1958.)

generally also contain pyrite but in relatively small quantities and relatively large crystals. They frequently also contain some carbonate, though apparently not generally as siderite. Many examples are known where there is a lateral facies change from red fluvial or deltaic mudrocks into green marine mudrocks, and in most examples there is a corresponding reduction in total and ferric iron with little change in ferrous iron. The change in iron content may be explained either as due to flocculation and hydraulic sorting, with oxide particles deposited close to the shore, or as early diagenetic reduction and removal of iron in the marine environment.

The green color of a few mudrocks may be attributed to the presence of glauconitic mud, which forms largely by alteration of fecal pellets, by alteration of detrital phyllosilicate minerals such as illite and biotite, or by direct precipitation from seawater. Nonmarine authigenic glauconite has been reported but is extremely uncommon. Authigenesis of glauconite on the sea floor (halmyrolosis) requires only the presence of a phyllosilicate structure to modify, adequate supplies of both potassium and iron, and a favorable oxidation potential. As glauconite contains both ferric and ferrous iron, it is apparent that the immediate locality of glauconite formation ("the microenvironment") cannot be highly oxidizing. Burst (1958) has illustrated the chemical relationships between glauconite and phyllosilicates from which it is known to form (Fig. 11-11).

It is important to note that compounds containing ferric iron are not necessarily red, brown, or yellow. The oxide of ferric iron occurs in these colors but silicates with high ferric/ferrous ratios may be green, for example, illite and glauconite. The explanations for mineral colors are exceedingly complex and can be traced to factors including defects in the crystal structure of the mineral, the amount of the pigmenting atom, its position in the mineral structure, its bonding

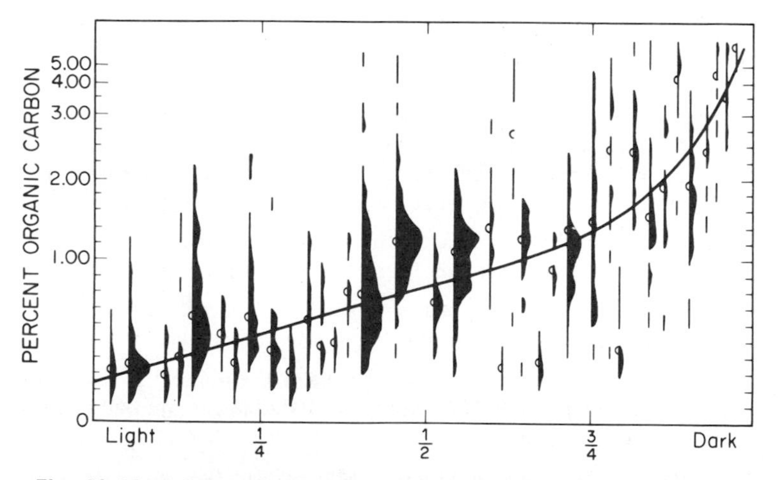

Fig. 11-12 Relation of color to content of organic carbon in sediments, based on 1000 analyses of organic matter in sediments of various lithologies. (From Trask and Patnode, 1937.)

characteristics in the structure, and the state of subdivision of the host mineral aggregate.

Carbon

Most darkening of rock colors through shades of gray to black results from the presence of organic matter but the darkening is not closely tied to the percent of organic carbon (Fig. 11-12). The scatter results from several causes. (a) The proportion of carbon in organic matter varies with the structure of the organic molecules present. For example, peat and oil shales are brown although both are very rich in organic matter. (b) Finely divided iron sulfides such as pyrite, marcasite, and melnikovite also impart a black color to sedimentary rocks and may be the cause of the dark, low carbon rocks present in Fig. 11-12. (c) Sediments may contain pigmenting agents that partially mask the black color of the organic carbon in the rocks.

REFERENCES

BERNER, R. A., 1970, "Sedimentary Pyrite Formation," *Amer. Jour. Sci.*, **268**, pp. 1–23. (Summary of investigations bearing on the origin of authigenic pyrite in sediments.)

BISCAYE, P. E., 1965, "Mineralogy and Sedimentation of Recent Deep-Sea Clay in the Atlantic Ocean and Adjacent Seas and Oceans," *Geol. Soc. Amer. Bull.*, **76**, pp. 803–832. (Examination of the types of clay minerals in Atlantic Basin bottom sediments illustrating control by weathering processes on surrounding continental masses.)

BURST, J. F., 1958, "Mineral Heterogeneity in Glauconite Pellets," *Amer. Miner.*, **43**, pp. 481–497. (Analysis of the possible variations in structure and chemical composition in green grains called glauconite in field and thin section studies.)

———, 1969, "Diagenesis of Gulf Coast Clayey Sediments and Its Possible Relation to Petroleum Migration," *Amer. Assn. Pet. Geol. Bull.*, **53**, pp. 73–93. (Evaluation of the relationship between percentage of expandable clays, temperature, and burial depth in Gulf Coast sediments, based on data from many thousands of wells. Hydrated structures decrease markedly above a temperature of about 95°C.)

CAMPBELL, F. A., and T. A. OLIVER, 1968, "Mineralogic and Chemical Composition of Ireton and Duvernay Formations, Central Alberta," *Bull. Canadian Pet. Geol.*, **16**, pp. 40–63. (Detailed study of shales associated with Leduc reefs.)

CLARKE, F. W., 1924, *Data of Geochemistry*, 5th ed. U.S. Geol. Sur. Bull. 770, 841 pp. (Now partially updated as Professional Paper 440 in many sections.)

CURTIS, C. D. and D. A. SPEARS, 1971, "Diagenetic Development of Kaolinite," *Clays and Clay Minerals*, **19**, pp. 219–227. (Solubility data indicate that gibbsite is unstable in most diagenetic environments.)

DEGENS, E. T., 1967, "Diagenesis of Organic Matter," in *Diagenesis in Sediments*, G. LARSEN and G. V. CHILINGAR, ed., New York: Elsevier Pub. Co., pp. 343–390. (Good survey of the topic.)

DUNOYER de SEGONZAC, G., J. FERRERO, and B. KUBLER, 1968, "Sur la Cristallinité de l'Illite dans la Diagénèse et l'Anchimetamorphose," *Sedimentology*, **10**, pp. 137–143. (Gives data on vertical changes in clay mineral composition and crystallinity from several different basins.)

GEHMAN, H. M., JR., 1962, "Organic Matter in Limestones," *Geochim. Cosmochim. Acta*, **26**, pp. 885–897. (Analyses of 1400 ancient rocks indicate that limestones contain only one-fifth the amount of organic matter as an equal amount of shale but the same amount of hydrocarbons.)

GIPSON, MACK, JR., 1966, "A Study of the Relations of Depth, Porosity and Clay Mineral Orientation in Pennsylvanian Shales," *Jour. Sed. Petrology*, **36**, pp. 888–903. (Statistical analysis of the interrelationships among these variables. The degree of association between any two of the variables is poor.)

GRIFFIN, G. M., 1962, "Regional Clay-Mineral Facies—Products of Weathering Intensity and Current Distribution in the Northeastern Gulf of Mexico," *Geol. Soc. Amer. Bull.*, **73**, pp. 737–768. (An excellent paper illustrating the ways by which clay mineral facies develop in a modern sedimentary basin.)

GRIFFIN, J. J., H. WINDOM, and E. D. GOLDBERG, 1968, "The Distribution of Clay Minerals in the World Ocean," *Deep-Sea Res.* **15**, pp. 433–459. (Summary of the current status of knowledge concerning the areal distributions of clay mineral species on the ocean bottom.)

GRIM, R. E., 1968, *Clay Mineralogy*, 2nd ed. New York, McGraw-Hill Book Co., 596 pp. (Chapters 13 and 14 summarize the literature on observed clay mineral variations in both recent and ancient sediments.)

GULBRANDSEN, R. A., 1966, "Chemical Composition of Phosphorites of the Phosphoria Formation," *Geochim. Cosmochim. Acta*, **30**, pp. 769–778. (Results of chemical analyses of sixty samples of the most extensive phosphate deposit in the United States.)

HELING, DIETRICH, 1969, "Relationships between Initial Porosity of Tertiary Argillaceous Sediments and Paleosalinity in the Rheintalgraben (sw-Germany)," *Jour. Sed. Petrology*, **39**, pp. 246–254. (Original porosity values of argillaceous sediments deposited in different environments can be used as indicators of paleosalinity.)

HILL, T. P., M. A. WERNER, and M. J. HORTON, 1967, "Chemical Composition of Sedimentary Rocks in Colorado, Kansas, Montana, Nebraska, North Dakota, South Dakota, and Wyoming," *U.S. Geol. Sur. Prof. Paper* **561**, 241 pp. (Compilation of analyses.)

MACKENZIE, F. T. and R. M. GARRELS, 1966, "Silica-Bicarbonate Balance in the Ocean and Early Diagenesis," *Jour. Sed. Petrology*, **36**, pp. 1075–1084. (Quantitative analysis of data leads the authors to conclude that orthosilicic acid reacts with bicarbonate ions and amorphous alumino-silicates in the sea to produce clay minerals.)

MACPHERSON, H. G., 1958, "A Chemical and Petrographic Study of Pre-Cambrian Sediments," *Geochim. Cosmochim. Acta*, **14**, pp. 73–92. (Chemical analyses of more than ninety Precambrian argillaceous rocks.)

MEADE, R. H., 1966, "Factors Influencing the Early Stages of the Compaction of Clays and Sands—Review," *Jour. Sed. Petrology*, **36**, pp. 1085–1101. (A review of both experiments and observations in boreholes.)

MILLOT, GEORGES, 1970, *Geology of Clays*, N. Y., Springer-Verlag, 429 pp. (Translation of the 1964 text on the geological factors influencing clays by a leading French investigator, with many examples not in the English literature.)

MÜLLER, GERMAN, 1967, "Diagenesis in Argillaceous Sediments," in *Diagenesis in Sediments*, G. Larsen and G. V. Chilingar, ed., New York: Elsevier Pub. Co., pp. 127–178. (An extensive summary and analysis of the topic by a leading European worker in the field.)

O'BRIEN, N. R., 1970, "The Fabric of Shale—An Electron Microscope Study" *Sedimentology*, **15**, pp. 229–246. (Transmission and scanning electron micrographs illustrating the relationship between fissility and clay mineral orientation.)

PARHAM, W. E., 1966, "Lateral Variations of Clay Mineral Assemblages in Modern and Ancient Sediments," *Proc. Internat. Clay Conf., Jerusalem*, **1**, pp. 135–145. (A useful compilation of known examples of lateral facies changes.)

PETTIJOHN, F. J., 1957, *Sedimentary Rocks*, 2nd ed. New York: Harper & Bros., 718 pp. (Chapter 8 discusses mudrocks.)

PICARD, M. D., 1971, "Classification of Fine-Grained Sedimentary Rocks," *Jour. Sed. Petrology*, **41**, pp. 179–195. (Compilation of data and suggested usage of terminology.)

PORRENGA, D. H., 1966, "Clay Minerals in Recent Sediments of the Niger Delta," *Clays and Clay Minerals*, Proc. 14th Natl. Conf., pp. 221–233. (Describes lateral facies variations off the Niger Delta.)

PRICE, N. B. and P. MCL. D. DUFF, 1969, "Mineralogy and Chemistry of Tonsteins from Carboniferous Sequences in Great Britain," *Sedimentology*, **13**, pp. 45–69. (Review of the hypotheses for the origin of tonsteins and their applicability to British examples.)

RATEEV, M. A., Z. N. GORBUNOVA, A. P. LISITZYN, and G. L. NOSOV, 1969, "The Distribution of Clay Minerals in the Oceans," *Sedimentology*, **13**, pp. 21–43. (Determination of clay mineral facies in the world ocean demonstrates a climatically controlled latitudinal pattern.)

RONOV, A. B., Y. P. GIRIN, G. A. KAZAKOV, and M. N. ILYUKHIN, 1966, "Sedimentary Differentiation in Platform and Geosynclinal Basins," *Geochem. Internat.*, **3**, pp. 595–608. (Gives data on chemical composition of mudrocks and sands in the Mesozoic and Cenozoic rocks of the Russian platform and Caucasian geosyncline and discusses chemical trends.)

SCHRAYER, G. J., and W. M. ZARRELLA, 1963, "Organic Geochemistry of Shales—I. Distribution of Organic Matter in the Siliceous Mowry Shale of Wyoming," *Geochim. Cosmochim. Acta*, **27**, pp. 1033–1046. (Percentages of organic matter in the Mowry Shale increase in the direction of occurrence of known petroleum deposits in sandstones of the same age, suggesting that the shales may have been source rocks. This result also suggests a new prospecting method.)

SHAW, D. B. and C. E. WEAVER, 1965, "The Mineralogical Composition of Shales," *Jour. Sed. Petrology*, **35**, pp. 213–222. (Landmark investigation of shale mineralogy using 400 shale samples of Paleozoic through Tertiary age from the United States. Samples from many different types of tectonic settings were included.)

SLAUGHTER, M., and J. W. EARLEY, 1965, "Mineralogy and Geological Significance of the Mowry Bentonites, Wyoming," *Geol. Soc. Amer. Spec. Paper 83*, 116 pp. (Field description and interpretation of the most extensive bentonite formation in North America.)

TOMLINSON, C. W., 1916, "The Origin of Red Beds," *Jour. Geol.*, **24**, pp. 153–179. (Classic early study of the origin of pigmentation in red and green sedimentary rocks.)

TRASK, P. D., and H. W. PATNODE, 1937, "Means of Recognizing Source Beds," *in Drilling and Production Practice*, **1936**, Amer. Petroleum Inst., pp. 368–384. (Classic paper clarifying the many variables involved in the origin and recognition of source beds of petroleum.)

VAN HOUTEN, F. B., 1961, "Climatic Significance of Red Beds," in *Descriptive Paleoclimatology*, A.E.M. Nairn, ed., New York: Interscience Pub., pp. 89–139. (Besides discussion of climatic factors, this fine paper contains a comprehensive compilation of chemical and mineralogical data for both sandstones and mudrocks.)

VENKATARATHNAM, K. and W. B. F. RYAN, 1971, "Dispersal Patterns of Clay Minerals in the Sediments of the Eastern Mediterranean Sea," *Marine Geol.*, **11**, pp. 261–282. (The distribution of clay mineral types reflects modern source areas and directions of current flow at various depths in the eastern Mediterranean Sea.)

WEAVER, C. E., 1959, "The Clay Petrology of Sediments," *Clays and Clay Minerals, Proc. 6th Natl. Conf.*, pp. 154–187. (Masterly discussion of the factors controlling the occurrence of clay minerals in sedimentary rocks. For an account more favorable to diagenesis, see Grim, 1968.)

———, 1967a, "The Significance of Clay Minerals in Sediments," in *Fundamental Aspects of Petroleum Geochemistry*, B. Nagy and U. Colombo, ed. New York, Elsevier Pub. Co., pp. 37–76. (An extensive discussion of clay minerals in sediments in relation to age, depositional environment, and diagenesis.)

———, 1967b, "Potassium, Illite, and the Ocean," *Geochim. Cosmochim. Acta*, **31**, pp. 2181–2196. (Interesting speculations concerning the effect caused by colonization of the land surface by plants.)

WHITEHOUSE, U. G., L. M. JEFFREY, and J. D. DEBRECHT, 1960, "Differential Settling Tendencies of Clay Minerals in Saline Waters," *Clays and Clay Minerals, Proc. 7th Natl. Conf.*, pp. 1–80. (Describes an extensive series of experiments on the effect of mineralogy and dissolved salts on the settling of clays.)

PART FOUR

CARBONATE ROCKS AND EVAPORITES

Sedimentation of these rocks requires an environment from which detrital silicate grains have generally been excluded. How has such an environment been produced in the past and the present? The sedimentary materials which make carbonate rocks are produced by organisms, and to a lesser extent, inorganic chemical processes. What are these organisms and what processes exist and have existed to produce carbonate sediment? Deposition of evaporites requires a limited but continuous supply of water and a climate suitable for intense concentration of that water by evaporation. How has nature met these conditions through time? The ancient carbonate and evaporite rocks we observe in outcrop and subsurface are the products of both sedimentation and diagenesis. What are the diagenetic processes that turn sediment into rock? How and where do these processes take place?

CHAPTER TWELVE

ORIGIN OF LIMESTONES

12.1 INTRODUCTION

Sedimentary carbonate rocks are composed predominantly of calcite, aragonite or dolomite. However, a rock consisting of 40% detrital quartz, 25% calcite fossil fragments and 35% calcite cement is properly considered a fossiliferous sandstone despite the fact that it is composed of 60% carbonate. Limestones form approximately 10% of the exposed sedimentary record and are known from rocks as old as 2.7 billion years.

Carbonate rocks have important economic significance because their pore space acts as a host for petroleum and natural gas. Approximately 20% of the hydrocarbons found in North America exist in carbonate rocks and the world volume is close to 50% because of the prolific Near Eastern carbonate fields. Carbonate rocks commonly act as reservoirs for groundwater. The porosity, permeability, and ease of reaction of the carbonate minerals allow carbonate rocks to serve as hosts for ore deposits. The lead-zinc minerals of the Mississippi Valley and of Pine Point, N.W.T., Canada, are examples of this type of deposit. Vast amounts of limestone and dolomite are used for agricultural lime, cement, building stone and concrete aggregate.

Our understanding of carbonate rocks has increased dramatically in the past 25 years. This change has resulted, as with so many advances in sedimentation, from studies of modern sediments and correlation of the results of these studies with observations made on ancient rocks. During the late 1940's and early 1950's there began a significant series of studies of modern carbonate sediments. These studies aimed at understanding the origin of carbonate particles, processes of deposition, and the mechanisms of early diagenesis. The impetus for these studies came in large part from the interest of the petroleum industry in finding and exploiting oil and gas in carbonate rocks.

Several generalizations stand out as important to a modern understanding of the carbonate rocks. (a) Thc bimodal grain size distribution of most carbonate sediments and rocks. Folk (1959) (see Ham, 1962) and the work that followed his pioneering classification gave prominence to the idea that most carbonates could be considered as mixtures of sand-size grains of calcium carbonate and carbonate mud. (b) The importance of organisms in producing carbonate sediment, both mud and coarser particles. The contribution of fossils and fossil fragments to limestone had long been recognized. However, the role invertebrate organisms play in generating fecal pellets, comminuting skeletal material, and producing sedimentary structures by burrowing activity has been brought sharply into focus. The role the calcareous algae and the filamentous blue-green algae play in producing carbonate mud, grains, and structures such as stromatolites has gained wide recognition. (c) The deposition of carbonate sediments in very shallow water. Coupled with the recognition of the important part organisms play in carbonate sedimentation, the studies of modern carbonate environments have yielded the generalization that the prolific production of organisms necessary to the development of carbonate rock requires extremely shallow water. Much of the deposition of modern carbonate sediment is in less than 40 ft of seawater (Fig. 12-1). It must be recognized, however, that some carbonate material is being deposited in deep seas. When we examine the ancient record, most workers are impressed with the high abundance of shallow water shelf carbonates. Indeed, the final site of deposition of much of the ancient carbonate rock material may very well have been above mean high tide on supratidal flats. This is not surprising considering a shallow water shelf or bank will soon fill up to sea level if sedimentation is rapid and subsidence is relatively slow. At this point sediment will accumulate on high tidal flats. Much of the sedimentary record of carbonate rocks may represent deposition above mean low tide. (d) Differential sedimentation of carbonate material to produce abrupt thickening and thinning of a carbonate rock unit. It has long been recognized that the living coral-algal reefs of the Carribean, Pacific atolls, and the Great Barrier Reef of Australia could produce a mound-like deposit of limestone whose lateral time-equivalent sedimentary section would be relatively thin. Carbonate mound-like features or bioherms are well-known in the ancient record. Many of them contain abundant organic remains. Closer inspection of many of these ancient carbonate "reefs" reveals that they are composed largely of carbonate mud with the larger skeletal particles "floating" within the mud matrix. Conclusive evidence for a rigid organic framework does not exist in most

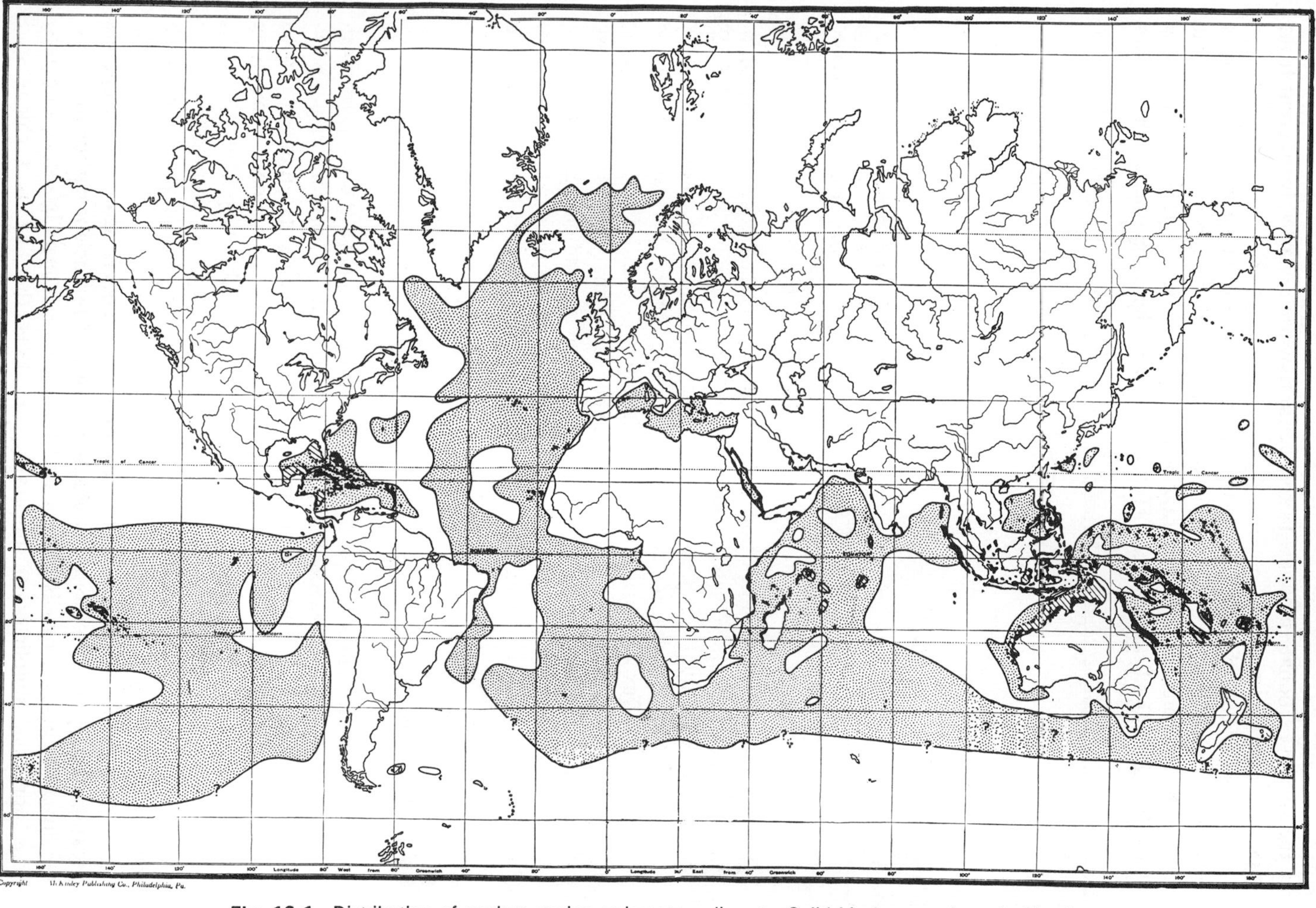

Fig. 12-1 Distribution of modern marine carbonate sediments. Solid black—organic reefs. Lined—larger areas of shallow-water carbonate sediments. Stippled—areas of other sediments, especially globigerina ooze, containing more than 30 percent $CaCO_3$. (From Rodgers, 1957.)

of the ancient carbonate mounds. In this sense they are remarkably different from modern coral-algal reefs.

It is significant that all the modern occurrences of carbonates are in areas essentially free of silicate detritus and this also seems to have been true of ancient carbonate environments. The average ancient limestone generally contains less than 5% detrital silicate particles (Fig. 12-2) and the antipathetic relationship

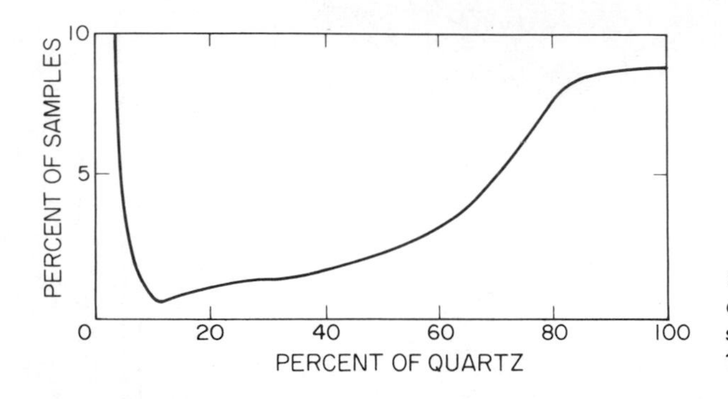

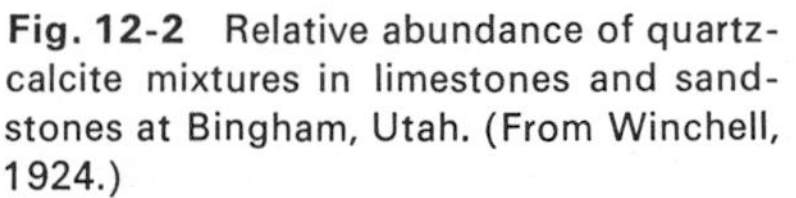
Fig. 12-2 Relative abundance of quartz-calcite mixtures in limestones and sandstones at Bingham, Utah. (From Winchell, 1924.)

between silicate detritus and carbonate deposition suggests a genetic relationship. Carbonate particles are very soft compared to quartz and are abraded to "dust" and redissolved if quartz is present in significant amounts. In addition, because most carbonate particles are either directly or indirectly the result of organic processes, the influx of detrital silicate sediments will drastically curtail organic production and thus reduce the rate of carbonate formation. For example, we can picture clams and oysters being "choked" by quartz and clay, a circumstance known to occur in the lagoons behind barrier bars along the Texas Gulf Coast. This chapter details these ideas and others that give us our present understanding of carbonate rocks.

12.2 MINERALOGY

Approximately sixty minerals occur in nature that have the CO_3 group in common. Of these, hexagonal $CaCO_3$—calcite—and its orthorhombic polymorph—aragonite—are the most common in modern sediments. In ancient rocks calcite and dolomite [$CaMg(CO_3)_2$] are by far the most common, comprising almost 100% of the carbonate minerals in carbonate rocks. Magnesite—$MgCO_3$, natron—$Na_2CO_3 \cdot 10\ H_2O$, and trona—$Na_3H(CO_3) \cdot 2H_2O$ are common in some evaporites and siderite—$FeCO_3$—occurs in association with peat and iron-rich rocks.

Although calcite and aragonite are common in modern sediments, aragonite because of its greater solubility is essentially absent in ancient carbonate rocks, having been removed by dissolution or through replacement by other minerals. This apparent simplicity of mineralogy neglects the fact that calcite and aragonite in nature are almost never compositionally ideal. Other divalent cations commonly substitute in varying amounts for the Ca^{2+} in the crystal structure, which produces a whole series of materials with different solubilities and properties. Those ions smaller than Ca^{2+} (0.99 A), such as Mg^{2+} (0.66 A), Fe^{2+} (0.74 A), and Mn^{2+} (0.80 A) are more readily accepted into the hexagonal calcite structure, while those ions whose radii are greater than Ca^{2+}, such as Ba^{2+} (1.32 A) and Sr^{2+} (1.12 A), are more readily accepted by the aragonite structure. In order to understand the minor element composition of natural calcite or aragonite, we must first ascertain whether the mineral was the product of biochemical processes within the tissue of organisms and was thus determined by the metabolic processes of the organism or whether the minor element composition was determined by physical-chemical processes taking place in a solution without direct biochemical influence. Any crystal that grows in contact with the surrounding water, rather than within plant tissues, should reflect the conditions existing in the water and not within the plant.

The amount of Mg^{2+} or Sr^{2+} substituted for Ca^{2+} in a carbonate skeleton is a function of phylogeny, shell mineralogy, water temperature, water composition, and the type of organic tissue. In fact, the nature of the organic matrix in which the crystal grows and the presence of the enzyme, carbonic anhydrase, probably determine whether calcite or aragonite will be secreted by the organism and, thus, may be a fundamental control on the abundance of magnesium and strontium in the shell. As far as known, magnesium and strontium serve no function in the skeletal material and probably constitute a structural defect. We would anticipate that these elements would be discriminated against during formation of a shell. This expectation is borne out by consideration of the composition of seawater and the average aragonitic and calcitic skeleton. Magnesium is particularly depleted in skeletal material relative to seawater. The ratio mol percent Mg/mol percent Ca is 5.2 in seawater but averages only about 0.05 in calcitic skeletal material. There also appears to have been a biochemical evolution in the ability of organisms to discriminate among crystallochemically similar ions during shell construction. More primitive organisms usually have less ability to discriminate, although there are exceptions (Table 12-1). Some modern calcareous algae (*Porolithon*, *Goniolithon*, and *Lithophyllum*) contain as much as 26 mol percent $MgCO_3$ in their hard parts, and a combination of X-ray and electron microprobe analyses of these materials suggests the presence of cryptocrystalline brucite, $Mg(OH)_2$, in the skeleton. As we would anticipate from consideration of ionic radii and crystal chemistry, there is some magnesium present in all calcitic skeletons and also some strontium present in all aragonite shells. In calcitic skeletal parts, two groups of organisms may be distinguished: (a) those containing a small amount of $MgCO_3$ (1 to 2%) such as articulate brachiopods, mollusks, Cirripedia (subclass of Arthropoda), and the perforate foraminifera; (b) those containing a large amount (10% or more) $MgCO_3$ such as Calcarea (class of Porifera), imperforate foraminifera, Octocorallia, Bryozoa, worm tubes, echinoderms, and Corallinaceae. The distinction between *low magnesium* and *high magnesium* calcite is commonly drawn at 4% $MgCO_3$. This grouping can be further subdivided into those containing a consistently high content of $MgCO_3$ and those containing a variable amount.

TABLE 12-1 AMOUNTS OF MAGNESIUM AND STRONTIUM IN MODERN CARBONATE SKELETONS†

Taxon	No. of specimens	Mineral composition	Mean % $MgCO_3$	High value	Low value
Foraminifera	29	Calcite	10.1	15.9	0.4
Homotrema rubrum (foram)	10	Calcite	13.1	13.7	11.7
Sponges	3	Calcite	8.5	14.1	5.5
Corals: subclass *Zooantharia Hexacorallia* order *Madreporaria*	30	Aragonite	0.5	1.1	0.1
Corals: subclass *Octocorallia Alcyonaria* (excluding *Heliopora*, which is always very low in Mg)	29	Calcite	12.3	15.7	0.4
Echinoidea (spines removed)	46	Calcite	10.6	15.9	4.5
Echinoid Spines	12	Calcite	6.6	10.2	2.0
Crinoidea	34	Calcite	11.1	15.9	7.3
Bryozoa (lowest percentages in compact coralline forms; highest in fernlike forms)	10	Calcite	3.3	6.9	0.2
Brachiopoda	10	Calcite	1.0	3.4	0.0
Pelecypoda	22	Calcite plus aragonite	0.6	2.8	0.0
Gastropoda	27	Calcite plus aragonite	0.4	2.4	0.0
Cephalopoda	10	Calcite plus aragonite	2.1	7.0	0.1
Red algae; family *Corallinaceae:*‡				$Mg(OH)_2$	
Lithothamnion-Lithophyllum-Goniolithon	30	Calcite	17.2	28.8	7.7
Green algae: family *Codiaceae: Halimeda*	8	Aragonite	1.3	5.5	0.0

†Data on magnesium from Chave, 1954, and Clarke and Wheeler, 1922; *Halimeda* data from Johnson, 1961; data on strontium from Graf, 1960, pp. 55–58.

‡Up to 10% cryptocrystalline brucite has been found in some *Goniolithon* specimens using the microprobe.

TABLE 12-1 (cont.)

Taxon	No. of specimens	Mineral composition	Mean % $SrCO_3$
Foraminifera	8	Calcite	0.2
Sponges	11	Calcite	0.2
Alcyonaria	8	Calcite	0.4
Madreporaria	23	Aragonite	1.1
Echinoidea	15	Calcite	0.3
Crinoidea	2	Calcite	0.4
Bryozoa	12	Calcite	0.3
Brachiopoda	7	Calcite	0.1
Pelecypoda	76	Calcite plus Aragonite	0.2
Gastropoda	68		0.2
Cephalopoda	6		0.5
Corallinaceae	9	Calcite	0.3
Halimeda	1	Aragonite	1.1

To the consistently high group belong the Corallinaceae, echinoderms, Octocorallia, and probably Calcarea. The amount of substitution of Mg for Ca in the calcite structure is usually determined by measuring the shift of the d(211) line for calcite on X-ray diffraction patterns. Use of an internal standard is usually required to achieve satisfactory data.

When calcite or aragonite grows by inorganic processes under equilibrium conditions, the amount of minor ion substitution is determined by the properties of the crystal surface in contact with the water, the composition of the water, pressure, and temperature. Therefore, in dealing with such minerals, it is often possible to determine much about the environment of formation from analysis of the minor element composition. The basic concept in understanding minor element coprecipitation is that of the *partition* or *distribution* coefficient.

When a substance is precipitated, foreign ions in the solution distribute themselves between the solid and liquid phases. Because of restrictions imposed by the crystal structure of the precipitated solid phase, ions other than the "correct" one are discriminated against during precipitation and the residual solution becomes enriched in these elements relative to the solid precipitate; however, discrimination is never complete.

Assuming precipitation is not too rapid, and equilibrium is maintained, the process of inclusion of trace elements in a crystal can be represented symbolically as

$$A_{\text{xl}} + B_{\text{aq.}} \rightleftharpoons A_{\text{aq.}} + B_{\text{xl}}$$

where A is the structurally preferred ion, B is the foreign ion, and the equation from left to right signifies relative enrichment of the crystal by the foreign ion.

From this it follows that

$$K = \frac{[A_{aq.}][B_{xl}]}{[B_{aq.}][A_{xl}]}$$

which by rearrangement becomes

$$\left(\frac{[B]}{[A]}\right)_{xl} = K\left(\frac{[B]}{[A]}\right)_{aq.}$$

K is referred to as the *partition coefficient* or *distribution coefficient* for the foreign ion and is simply another use of the equilibrium constant. As equilibrium precipitation was assumed, the trace element must be homogeneously distributed throughout the crystal. Just as plagioclase crystallizing from a magma may be zoned due to rapid cooling (i.e., nonequilibrium precipitation), however, so it also is possible to precipitate calcium carbonate rapidly from solution so that the concentration of foreign ions varies from the center to the periphery of the crystal. We would expect the concentration of foreign ions most commonly to be greater in the center of the crystal, trapped by an increment of enclosing solid before they can diffuse out. Solid diffusion is extremely slow at low temperatures.

Assuming nonequilibrium precipitation with respect to the entire crystal, the last equation must be modified to

$$\left(\frac{[B]}{[A]}\right)_{\substack{\text{surface}\\ \text{of crystal}}} = K\left(\frac{[B]}{[A]}\right)_{aq.}$$

The equation says that equilibrium is maintained between the crystal *surface* and the solution during precipitation but not necessarily with the interior of the crystal. If we assume dB and dA, the increments of foreign and preferred ions deposited on the surface of the growing crystal, to be proportional to their respective concentrations in the solution, then

$$\frac{dB}{dA} = K\frac{(B_0 - B)/\mathrm{V}}{(A_0 - A)/\mathrm{V}}$$

where B_0 and A_0 are the initial quantities in solution, B and A are the quantities deposited in the crystal, and V is the volume of the liquid phase. Integration of this equation yields

$$\log\frac{B_0}{B_f} = K\log\frac{A_0}{A_f}$$

where B_f and A_f are the final amounts of foreign ion and preferred ion in solution. Neither the *K* value for homogeneous crystallization nor the *K* value for nonhomogeneous crystallization should vary with the amount of solid precipitated if the equation in which it occurs describes the precipitation process. Both of these *K* values will vary with the temperature at which the precipitation occurs, however, as do all equilibrium constants. To determine in experimental studies whether the foreign ion is homogeneously distributed in the crystal or is zoned, we only need to calculate which equilibrium constant really is a constant regardless of the amount of precipitate.

The partition coefficients for strontium in aragonite and calcite have been determined and it has been shown that early-formed crystals do not react with the solution and that only surface equilibrium is maintained between the precipitates and the solution. That is, strontium in aragonite is zoned in a manner similar to the frequent zonation of calcium concentration in plagioclase feldspars. It is noteworthy that the value of the equilibrium constant for surface temperatures is about 1.1 indicating that the ratio mol Sr^{2+}/mol Ca^{2+} in aragonite precipitated from seawater should approximate the ratio in the water.

The determination of trace element content is made using normal analytical techniques such as emission spectrography, atomic absorption spectrography, or X-ray fluorescence spectrography. The determination of mineralogy is usually made by optical means, X-ray diffraction, or density. Calcite has a specific gravity of 2.72; dolomite, 2.85; and aragonite, 2.93. Various staining techniques selective for calcite or aragonite have been proposed, and some are widely used. Not uncommonly the results have been unsatisfactory because of fine crystal size. Many workers find that if a broken or polished surface of a carbonate rock or an uncovered thin section of a carbonate rock is viewed under a microscope, dolomite can be distinguished easily from calcium carbonate minerals by applying a small drop of dilute HCl. The acid is then quickly removed by blotting with paper tissue, leaving an etched surface. The location of the initial effervescence and the calcium carbonate minerals standing as depressions permit reasonably accurate determination of the amount of the minerals.

12.3 CARBONATE SAND

Most carbonate rocks can be thought of as consisting of calcium carbonate sand and mud. The sand, commonly 0.02 mm and larger, is formed by many processes, for the most part at or near the site of deposition. Our attention originally was focused on the significance and abundance of carbonate sand in modern sediments by Illing (1954) in his classic study of the Bahamas. In ancient rocks these grains usually can be easily recognized by microscopic examination either in reflected light or in thin section. Where diagenetic alteration has been severe, the internal structure of a grain may have become so severely modified that identification is impossible. The purpose of the following sections is to describe the various types of common particles and to discuss their origin and significance.

Skeletal Particles

The whole and broken skeletons of invertebrates and calcareous algae may produce carbonate sand. The type of grain and its mineralogy naturally depend on the organisms available (Fig. 12-3). Therefore, the age of the rock as well as the specific environment of formation determine the types of grains found. Flourishing crinoid thickets during the Late Paleozoic produced abundant grains for crinoidal limestones. Great patches of the green calcareous alga *Halimeda* on the shallow

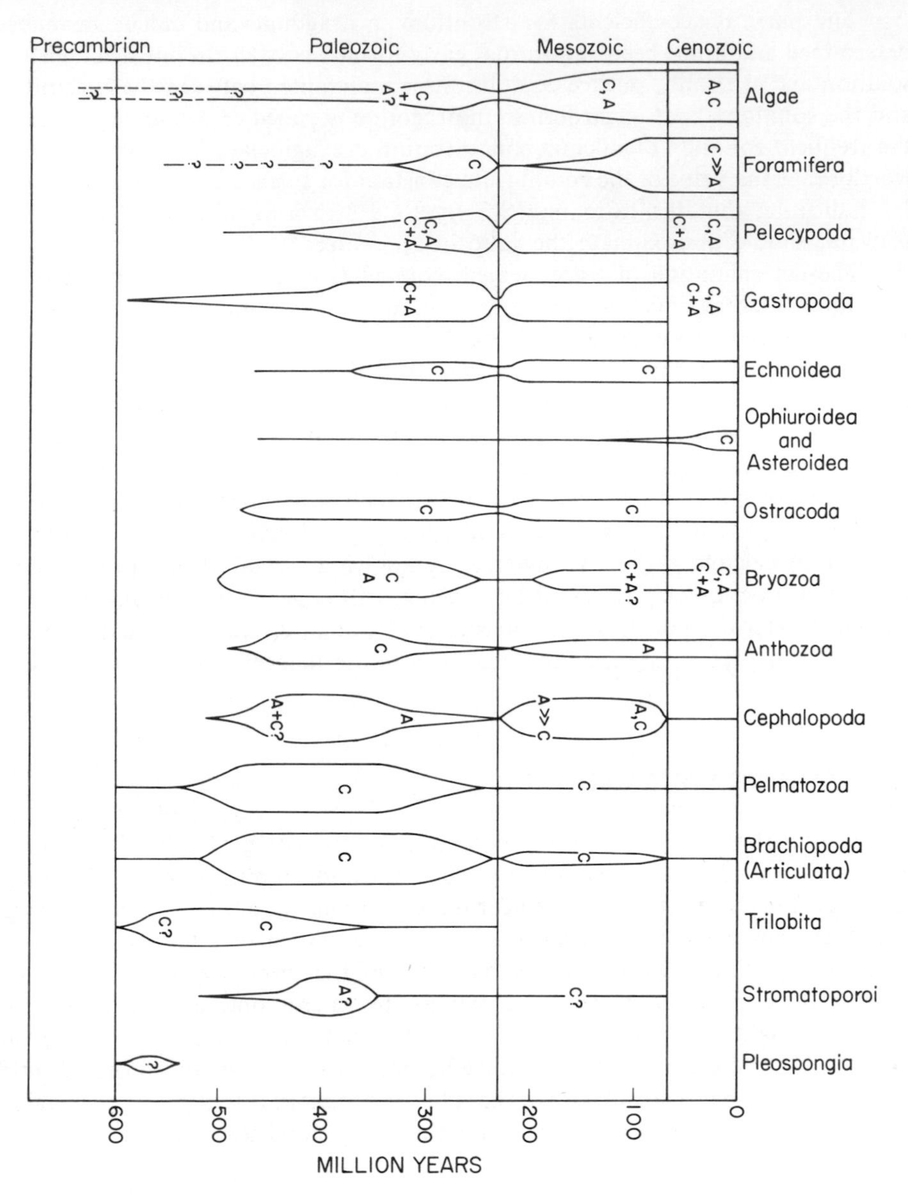

Fig. 12-3 Time-stratigraphic distribution of important carbonate-secreting organisms. Width of bars indicates relative importance as contributors to sediment. C = Calcite, A = Aragonite. (After Lowenstam, 1963.)

water banks and lagoons of modern seas produce grains that accumulate as carbonate sediment. Skeletal particles may be broken by several mechanisms. Transport by waves and currents may break skeletons into smaller grains. Invertebrates such as crabs may use a claw to crush and thus fragment shelled organisms.

Organisms that eat sediment for its organic content may crumble and break down skeletal material internally. Organisms such as the boring sponges and algae may perforate a larger particle, thus weakening it for easy crumbling by some other mechanism such as burrowing organisms. Decomposition of the muscle fibers and ligaments in bivalves results in the separation of the shell into two fragments. Thus pelecypods are rarely found in the articulated state.

A grain size difference of more than a thousandfold was found in the carbonate beach sediments on Isla Perez, a carbonate reef near the Yucatan Peninsula, as a result of the skeletal structure of the carbonate fragments. Elkhorn coral occurs dominantly as either long sticks the size of the joints in the parent colony or as 2ϕ grit, the size of the crystal packets composing the skeleton. *Halimeda*, a calcareous alga, disintegrates into 0ϕ flakes, the size of the skeletal elements, and into 10ϕ carbonate mud, the size of the aragonite crystals secreted by the alga (Folk and Robles, 1964).

Ooliths

The term oolith is applied to subspherical sand-size carbonate particles that have concentric rings of calcium carbonate surrounding a nucleus of another

Fig. 12-4 Modern ooliths from the Bahama Platform. Photo by R. J. Dunham.

particle (Fig. 12-4). The thickness of the concentric rings may vary from over 50% of the diameter of the oolith particle to a superficial concentric coating or rind on another grain. When originally formed in a marine environment, the oolitic coatings are composed of aragonite. The individual crystals of aragonite are oriented, at the time of formation, tangential to the surface of the grain or oolitic sheath. The oolitic coatings commonly show a series of concentric shells usually a few microns thick. These individual coatings may be interrupted by discontinuous patches, crescents, or continuous rings of cryptocrystalline aragonite. These breaks in the concentric oolitic rings or sheaths are normally produced by blue-green algal accretions or the diagenetic effect of small boring organisms. In a sense these interruptions represent micro-unconformities in the deposition of the oolitic coatings. In the ancient rocks the structure of the oolith may be changed by diagenesis to destroy the simple original concentric arrangement and substitute for it a radial or mosaic crystal fabric.

Today ooliths are found only where strong bottom currents exit. Commonly this represents areas of tidal bar accumulation or within tidal deltas. In the ancient record the oolite commonly shows abundant evidence for current transport such as large-scale cross-bedding (Fig. 12-5). There appears to be little doubt that the

Fig. 12-5 Cross bedded Pleistocene oolite near Miami, Florida,

marine oolitic coatings represent a direct inorganic precipitate of aragonite deposited with brief interruptions on moving particles. The independent evidence that the particles were moving during formation of the coating is that the oolitic rings are continuous around the grain. The reason for the formation of oolitic coatings has long been the subject of study. Earlier workers emphasized proximity of modern ooliths to upwelling currents, the role of organisms, and the accretion of sedimentary aragonite. Interesting data on the origin of ooliths comes from two sources; (a) the association of ooliths with high velocity current transport, and (b) observations by Weyl (1967) on the solution behavior of carbonate minerals in seawater. Weyl (1967) developed an instrument called a carbonate saturometer, a sensitive pH meter that records the change in pH when a carbonate solid is placed in the liquid to be tested. If dissolution of the carbonate mineral occurs, then the liquid is undersaturated with respect to the mineral and the pH will rise. If precipitation takes place, the liquid is supersaturated with respect to the solid and the pH of the liquid will decrease. In a solution such as seawater, when ooliths in suspension are tested with the saturometer, it has been found that precipitation is taking place at a very slow rate. The mineral appears to be almost in equilibrium with the water. As the concentration of the water does not change, then the phase precipitated must behave as a very soluble carbonate. When a carbonate grain is immersed in warm surface seawater, precipitation takes place. After a few minutes, the rate of precipitation slows down abruptly and the surface of the grain, i.e., the newly precipitated mineral, appears to be essentially in equilibrium with the seawater. When an oolith becomes buried under a few centimeters of other grains, the surface of the oolith is in contact with seawater in pore space. After only a few hours, a change occurs in the newly precipitated carbonate mineral and it becomes less soluble with respect to the seawater. If this grain is then exhumed and placed once again in suspension in seawater, precipitation takes place at a rapid rate for a few minutes. Once again the rate of precipitation becomes very slow and little more will be added until the grain has had a chance to become buried in the sediment. Again, the surface will become adjusted and the newly precipitated material will become less soluble. Observations in the field on Bahama ooliths indicate precipitation of only 2 ppm of calcium carbonate from the water in 3 min for ooliths moving along on top of a tidal bar. Ooliths that had been stationary for some time showed a precipitation of approximately 19 ppm calcium carbonate. For comparison, dry ooliths give readings that average about 23 ppm calcium carbonate.

From the previous discussion, the significance of the formation of ooliths in tidal bars and tidal deltas becomes obvious. The tidal setting is one in which a grain can be moved periodically and still stay within the same depositional setting for a sufficiently long period of time to develop a thick oolitic coating. The theory also confirms the generalization that an oolitic coating on a particle is evidence for current transport in an environment where the grain is periodically and repeatedly buried and exhumed from the sediment. The thickness of the coating is important only in indicating the length of time the grain has been subjected to the process. Oolitic coatings form on all kinds of grains, even quartz grains. They appear to be relatively uncommon, however, in rocks that have abundant quartz. It has been suggested that even if the coating does form, it can be removed easily by the abrasive effect of the quartz on the newly formed aragonite coatings.

Pisoliths

These particles have been considered in the past as large ooliths. Evidence is accumulating, however, that many if not all of the pisoliths are a diagenetic feature produced within the vadose zone. Some become separated from the sediment or rock in which they are formed and are redeposited in new sediment. See Chap. 13.

Pellets and Pelletoids

The most common carbonate particle in modern sediments and ancient carbonate rocks is round to oval, generally 30 to 100 μ in diameter, has a smooth outer surface, and is composed predominantly of microcrystalline carbonate. It is devoid of internal structure but may contain very small fragments of skeletal material. These grains resemble small rounded aggregates of carbonate mud that indeed many of them are. The term *pellet* will be reserved for fecal pellets produced by carbonate mud-ingesting organisms. In the modern sediments of the Bahama Platform polychaete worms, the gastropod *Batillaria minima* and some of the crustaceans contribute fantastic numbers of fecal pellets. Because organisms can produce fecal pellets in place from carbonate mud, the size of a pellet in a sediment is obviously not a measure of current energy. Only if the pellet is transported by a current does its size take on hydraulic significance. Many modern pellets are poorly consolidated friable aggregates that are bound only by organic matter and can be destroyed easily with the point of a pin under the microscope. In ancient carbonate rocks, examination of carbonate mudstones reveals that most of them show the vague outlines of diffuse pellets. This probably means that most carbonate mud has been worked many times over by organisms. However, the presence of well-defined pellets in modern and ancient rocks presents a problem. How do the pellets become lithified? No satisfactory general explanation has yet been advanced.

In both modern sediments and ancient rocks, similar appearing particles have been generated in an entirely different way. These particles, which we shall call *pelletoids*, form by the recrystallization of other particles such as skeletal grains or ooliths. In most cases some vague residual internal structure of the original grain can be seen to indicate its identity. It should be recognized that in the advanced stages of recrystallization however, the original particle may be converted into a uniform homogeneous mass of microcrystalline carbonate and be indistinguishable from a fecal pellet (Fig. 12-6).

Algal Accretionary Grains

Carbonate sand grains may act as the growth surface for primitive blue-green algae and related organisms. The algae then act as a sticky surface much like fly-paper for the entrapment of fine carbonate sediment. The algal structure is seldom if ever preserved and only irregular, crinkly, sometimes discontinuous layers of carbonate sediment surround the original particle. The layers commonly are discontinuous because sediment cannot be added on the bottom unless the grain is moved during formation of the layered coating. These grains are usually called *oncolites*. Oncolites of large size, up to 3 in. in diameter, have been found and are sometimes called *algal biscuits*. These grains are analogous in origin to algal stromatolites (see p. 436).

Intraclasts

These particles are fragments of partially lithified carbonate sediment that was eroded from the sea bottom or adjacent tidal flats. They may be of any size or shape and become incorporated within new sediment. The prefix "intra" indicates that they formed as particles within the general area of deposition of the host sediment. Several mechanisms serve to produce these grains.

1. Carbonate mud under an overburden of several feet of sediment may become sufficiently compacted that an eroded lump may be transported for short distances. This might occur along the bank of a tidal channel where earlier-deposited carbonate mud is being eroded as wafers and chunks. Such particles on redeposition commonly demonstrate their plastic nature with vague mashed boundaries and evidence of plastic deformation.

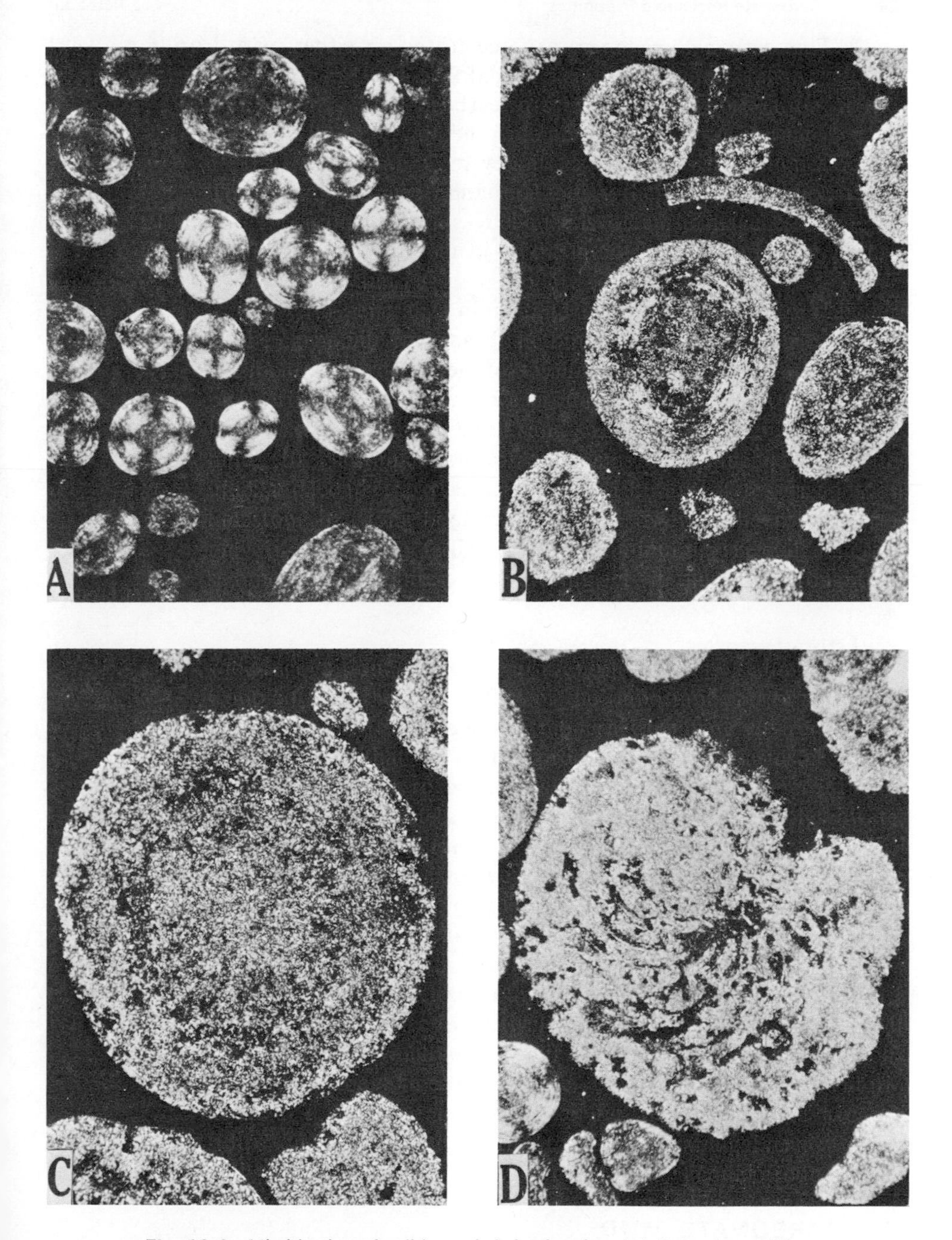

Fig. 12-6 Micritization of ooliths and skeletal grains. (A) Bahamian ooliths showing microcrystalline nuclei and oriented crystals in oolitic sheaths. Crossed nicols; 30×. (B) Partially micritized ooliths with some of the concentric structure preserved; 125×. (C) Micritized ooliths with only a vague outline of the original concentric structure; 225×. (D) Peneroplid that has been micritized. Vague outline of the original structure is still visible; 125×. (From E. G. Purdy, 1963.)

2. Fine carbonate sediment exposed to the air either by a drop in sea level or on high intertidal or supratidal flats will develop mud cracks and mud curls. These dried out wafers of carbonate sediment may receive some small amounts of carbonate cement in the subaerial environment, thus ensuring their existence as particles. They may range in size from a fraction of a millimeter to several feet across and commonly are covered by later sediment or washed into the nearby shallow marine environment. The flat pebble conglomerates common in the ancient record probably originated in this way.
3. Carbonate sand on beaches within the intertidal zone is easily cemented by fine fibrous aragonite into "beach rock." In addition, evidence is now accumulating that carbonate sediment may be cemented locally in a submarine environment. Eroded fragments of these materials will produce intraclasts. The origin of grapestone lumps that are intraclasts composed of loosely cemented carbonate sand grains is imperfectly known. These grains are common in most modern carbonate deposits and the cement is fine fibrous aragonite. Many of them may have originated as eroded beach rock. Some may have been cemented while still within the marine environment. See Chap. 13.

Lithoclasts

The implication of the term *intraclast* is that the lithification of the particle and its disruption from its original setting and redeposition took place essentially contemporaneously with the sedimentation of the stratigraphic unit in which it is found. *Lithoclast* is used to imply a rock fragment derived from outside the basin of carbonate deposition by erosion and transport. Where this distinction can be made, it is a useful concept. There is significant genetic difference in terms of hydraulics of transport and history between a mud crack flake or an eroded fragment of contemporary beach rock and an older limestone pebble. Fossiliferous sedimentary rock fragments that can be dated as significantly older than the host rock are lithoclasts. The distinction becomes more difficult, however, with limestone fragments lacking internal evidence for their age. In general, lithoclasts are very uncommon in most carbonate rocks. Where a carbonate platform terminates landward against a sea cliff of older rocks, then the eroded products of that cliff may contribute lithoclasts to the new sediment. Both intraclasts and lithoclasts have been termed *limeclasts* where the distinction between distant and local derivations cannot be made.

12.4 CARBONATE MUD

Micrite or microcrystalline carbonate is by far the most common constituent in carbonate rocks. The individual crystals in ancient rocks usually are less than 5 μ in diameter and are now commonly calcite. In modern carbonate sediments, most carbonate mud is composed of individual crystals averaging approximately 3 μ in length and less than 0.5 μ in width. These crystals are predominantly aragonite and needle-like in shape. It is presumed that the aragonite needles that form much of the carbonate mud today represent the direct analogue of the microcrystalline components of ancient carbonate rocks.

The origin of clay size carbonate mud has been a controversial topic for some years. Three general sources are possible.

1. *Mechanical or biologic abrasion* of larger carbonate particles to yield fine material. Clearly, some clay size carbonate mud in modern sediments and in ancient rocks was produced by this mechanism. Larger carbonate particles can be mechanically abraded by waves and currents to form carbonate mud. Biologic abrasion occurs when organisms ingest carbonate grains and abrade them internally. After the parrot fish eats stony coral for the contained organic material, he rejects the finely disintegrated carbonate from the skeleton of the coral. With the possible exception of the breakdown of the fibrous aragonite layers of some shells, mechanical or biologic abrasion will not produce needle-shaped crystals of aragonite.
2. *Direct inorganic precipitation* of aragonite from seawater. There is no question that such a precipitate, if it formed, could produce a sediment composed of aragonite needles. Moreover, it appears that the surface water of much of the oceans is supersaturated with respect to aragonite. Evaluation of the quantitative importance of this potential source of aragonite needles has long been clouded by the "whiting" phenomenon. The seas over most of the modern shallow carbonate platforms sometimes exhibit spectacular white masses of water carrying abundant carbonate material in suspension (Fig. 12-7). These patches of muddy water are commonly elongate in the direction of wind or tidal currents and are actually being moved by these currents. It is not uncommon to observe several of these features on any flight over the Bahama Platform. It has been presumed that these masses of seawater with suspended carbonate material represent areas of spontaneous precipitation of carbonate crystals. The suspension formed then is moved along from its point of origin by currents. Examination of the suspended material usually reveals that it has a composition similar to the present bottom sediment. Both the fine bottom sediment and the suspended material are commonly composed of aragonite needles. Thus it might be presumed that the source of the present bottom sediment was material from earlier whitings that had settled to the sea floor. The alternative possibility, however, that the whitings represent stirred up bottom material of any origin is equally attractive. It has been observed that fish tend to be found in greater numbers within these masses of muddy water. Do the fish simply seek out the whitings for food or to hide for protection? Or have the fish themselves stirred up the bottom sediment? Certainly many schools of fish as part of their normal living and eating process do stir up fine sediment on the bottom of the sea.

Fig. 12-7 Whiting viewed from the air over the Bahama Platform. The individual streaks of carbonate sediment laden water are over one mile long.

Several approaches have been used to determine the origin of the suspended material in the whitings. Where the calcium content of seawater filtered from a whiting has been analyzed, there is no evidence that it differs from normal seawater as one might expect if the volume of carbonate existing in suspension had been removed by precipitation. In addition, most whitings contain small amounts of other minerals such as high magnesium calcite and quartz that would not be precipitation products from normal surface seawater. During the time of the testing of atomic devices, the shallow seawater of the Bahama Platform contained excess carbon 14 derived from the blasts. Analyses of the carbon 14 in the filtered water from a whiting, from the suspended sediment and the existing bottom sediment, demonstrated that the carbon 14 content of the bottom sediment and the suspended material were similar, while that of the suspended material was inconsistent with the carbon 14 content of the associated seawater. There is no question in these cases that the suspended material in these whitings was simply stirred up bottom sediment. In fact, no evidence exists from normal or near-normal seawater that inorganic precipitation of aragonite needles is taking place today. If it is, it must be on a very small scale and has not been observed. Whitings produced by precipitation of aragonite from bodies of hypersaline brine such as the Dead Sea do occur. In the Dead Sea this phenomenon takes place about every 5 years when the lake is at its highest temperature.

3. *Production of aragonite needles within the tissues of calcareous algae*. Lowen-

stam and Epstein (1957) observed that the organic tissue of living calcareous algae, such as *Halimeda*, *Rhipocephalus*, and *Penicillus*, contains abundant needles of aragonite that are similar to those found in the bottom sediments of the Bahama Platform (Fig. 12-8). Indeed, they are also remarkably similar to needles produced by laboratory synthesis of aragonite by inorganic precipitation. These needles can easily be removed from the algae

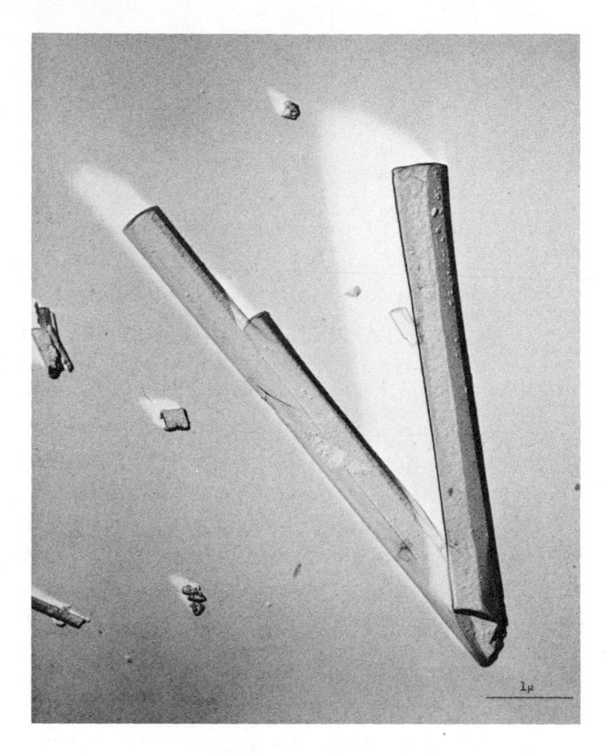

Fig. 12-8 Aragonite, from the stem of Penicillus. Shadowed. 30,000×.

by Clorox treatment. When the plant dies and the organic tissue decays, the needles are freed and are added to the bottom sediment. Lowenstam and Epstein recognized that the O^{18}/O^{16} in the carbonate would be indicative of the origin of the aragonite. Inorganic precipitation imparts a stable oxygen isotope ratio to the crystal that is a function of the temperature and the isotopic composition of the water. In addition, organisms such as calcareous algae generate crystals within their tissues that have O^{18}/O^{16} compositions that reflect in part the ability of the organism to fractionate the oxygen, and the resulting isotopic ratio existing in the crystal will depend in large part on the organism. Analyses of the needles removed from the algae, the sedimentary aragonite needles in the Bahaman sediment, and ooliths that do represent inorganic precipitates (Fig. 12-9) indicate that the carbonate mud from the Bahama Platform is consistent in oxygen isotopic composition with needles derived from algae. The oxygen isotope

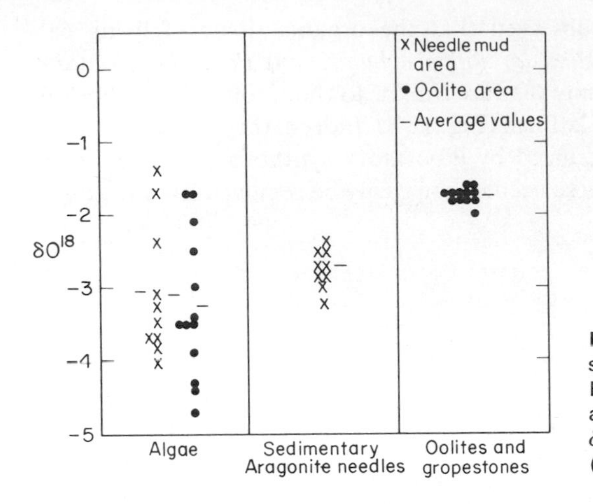

Fig. 12-9 The range of δO^{18} values for sedimentary materials from the Bahama Platform. Values are for different carbonates in different areas after corrections for δO^{18} of water of the areas was made. (After Lowenstam and Epstein, 1957.)

ratios of aragonite needles in the sediment, however, appear inconsistent with the ratios of the inorganically precipitated aragonite of the ooliths. The stable isotope data cannot give an unique answer because of some uncertainty about the temperature at which the growth of the crystals took place. Studies of the growth rates and abundance of the alga, *Penicillus*, an important aragonite needle producer in the Florida Reef Tract, indicate that *Penicillus* by itself is capable of producing the present mass of fine carbonate sediment. Congeneric species of the aragonite needle-producing algae of the Bahama Platform are widely distributed in the tropical seas of the world between the 15°C isotherms for the coldest month of the year.

12.5 CARBONATE FRAMEWORK

Although their volumetric contribution to modern carbonate sediments is small, the coral-algal reefs of the tropical seas represent a spectacular example of carbonate sedimentary processes (Fig. 12-10). Stony corals, hydrocorallines, massive alcyonarians, some coralline algae, foraminifera, and worms grow together in thick colonies and cement to each other, thus providing a rigid framework of carbonate material that may act as a wave resistant structure. Within the framework provided by these organisms, particulate carbonate sediment normally accumulates. In the final sedimentary product, the interstitial sediment may exceed the rigid frame in volume. This is certainly true in the Pleistocene reefs now exposed in the Florida Keys. The distinguishing characteristic, however, is the organic binding that produces the framework and rigid carbonate mass. These coral-algal reefs exist today on the edge of carbonate platforms in the tropical areas around the world generally between 30°N and 25°S latitude. Most of the organisms require sunlight either because they are plants as is the case with the algae or because of the zooxanthellae that live in and have a symbiotic relationship with the corals.

Fig. 12-10 A modern coral-algal reef community. Great Barrier Reef, Australia. (Photo by Jon Weber.)

After the initial colonization of a suitable shallow marine substrate, reefs grow toward increasing sunlight. Upward growth is limited by exposure to the air. Many coral-algal reefs, however, are exposed at low tide. The zone of maximum growth and production of calcium carbonate is a few feet below low tide because the ultraviolet light from the sun can penetrate thin layers of seawater and is lethal to many species. Prolific reef growth normally takes place on the windward side of carbonate platforms at depths of 60 ft or less. The majority of reef corals cannot survive at temperatures below 18.5°C and the optimum temperature lies between about 25 and 29°C. They tolerate salinities between 27 and 50 ppt. The influx of fresh water or hypersaline brine for short periods, however, is often fatal. The rates of growth for the modern and Pleistocene corals depend largely on local environmental factors, such as availability of food and clarity of the water. The spheroidal massive coral *Montastrea annularis* can grow upward 5 to 7 mm/yr and the elkhorn coral *Acropora palmata* can grow 2 to 3 mm/yr.

In general, the moosehorn coral *Acropora palmata* predominates in very shallow water, sometimes being exposed to the air at low tide. In slightly deeper water, generally 5 to 15 ft below the surface, vast thickets of the elkhorn coral *Acropora cervicornis* are common. Below, in slightly deeper water, the head and brain corals become the principal frame builders. These include *Montastrea*, *Siderastrea*, and *Diploria*. Many other calcareous organisms live in and contribute to the reef. Of special note are the encrusting forms of coralline algae such as *Lithothamnion*. The modern coral-algal reefs of the Caribbean differ from the Pacific and Indian Ocean reefs by being slightly less well-developed and by having a less significant contribution from the coralline algae. Barbados, standing on the Atlantic edge of the Caribbean, has had a modern and Pleistocene history of uplift. Thus coral-algal reefs formed during the relatively low stages of sea level during the Late Pleistocene are exposed above sea level on the island. These well developed reefs show the same zonation of organisms seen in living reefs.

Although prolific continuous reef development takes place today along the edge of carbonate platforms, small patches of reef growth are common in the shallow water behind the reef front. The concentration of coral-algal reefs on the windward margins of modern platforms reflects the need for nutrients by sedentary organisms.

Most of the carbonate buildups in the ancient record are carbonate mud mounds. When examined in detail, they are found to be composed of carbonate mud with carbonate sand-size and larger particles floating within the carbonate mud. In addition, they not uncommonly show stages of subaerial exposure followed by repeated growth by sedimentation at the same location. In a general sense they are miniature Bahama Platforms within a large carbonate platform. Examples of these features include the Mississippian buildups in the Sacramento Mountains of New Mexico and Swimming Woman Canyon, Montana (Fig. 12-11); Devonian "reefs" in western Alberta; and Pennsylvanian platey algal mounds in Utah, New Mexico, and the midcontinent area. Some, such as the Mississippian buildups in the Sacramento Mountains of New Mexico, have a core of carbonate mud surrounded by carbonate sediment rich in crinoid particles. The mud mounds of Florida Bay and the Florida reef tract provide a possible modern example of these structures. The mud mounds in this area reach a thick-

Fig. 12-11 Mississippian carbonate mounds near Swimming Woman Canyon, Montana.

ness of at least 12 ft compared with a few inches of sediment over much of the rest of the sea bottom. Some have built up to sea level and have acquired their own small tidal flats. They are partially stabilized by the turtle grass *Thallasia* and mangrove. However, accumulation has resulted from local production of carbonate sediment with some contribution of carbonate mud transported and trapped by turtle grass and mangrove. Many mounds are rich in skeletal material. For example, the coral *Porites* and the coralline alga *Goniolithon* are quite common. These forms grow in mud and do not form a rigid framework. Indeed, they can be pulled up from the mud as individual specimens while living.

There is much to be learned about ancient and modern carbonate buildups of the mud mound type. Little if anything is known about the selection of a given site for their growth, the role of *in situ* production versus current transport and accumulation, and the influence of periodic subaerial exposure in determining their stability and shape.

12.6 ORGANIC MATTER IN LIMESTONES

The amount of organic matter in modern carbonate sediments is quite variable, ranging from 0.1 to about 10% with a geometric mean of close to 1%. The source of the organic substances is the organisms in the environment and, therefore, it is reasonable to suppose that bioclastic materials will contain more organic matter than nonskeletal particles. In a study of organic matter in modern carbonate sediments and in ancient limestones, Gehman (1962) compared the relative amounts in lime mud, sand-size skeletal grains, and nonskeletal grains. His results clearly illustrate the similarity of organic contents in sand-size shell fragments and in lime mud, a result to be expected if the mud results from disaggregation of algal skeletons. However, the addition of nonskeletal particles such as ooliths reduces the organic content of the sediment greatly.

Ancient limestones average 0.2% organic matter and nearly all limestones contain less than 1.0%, and Gehman found no significant difference in amount between microcrystalline limestones, skeletal limestones, and nonskeletal limestones. Examination of the lithified Pleistocene limestones that underlie the modern Gulf of Batabano (Cuba) sediments shows their organic content to range from 0.06 to 0.31%, essentially identical to the values in the Cambrian-Tertiary limestones. This suggests that the loss of organic matter is closely tied to the processes that lithify carbonate sediments. See Chap. 13.

12.7 SEDIMENTARY PROCESSES

Current-Transported Deposits

Like their detrital silicate counterparts, carbonate grains respond to available currents and can be moved from their site of formation and redeposited. The relationship between current energy and size of grain will naturally differ from those derived for quartz. Quartz has a specific gravity of 2.65, whereas the specific gravity of calcite is 2.72 and aragonite is 2.95. Thus the hydraulic equivalent grain size for the carbonate minerals is smaller for these grains than quartz. A current of a given velocity will transport a smaller carbonate grain than a quartz grain. This generalization is misleading, however, because most of the carbonate grains at the time of production contain organic matter and/or pore space. The carbonate minerals exist only as a porous frame (Fig. 12-12). This reduces the effective

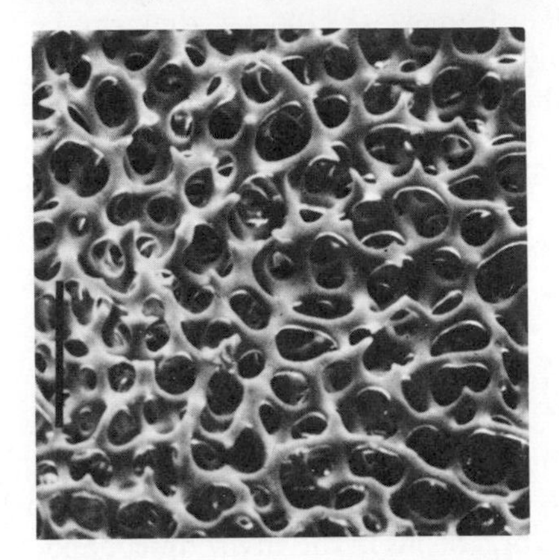

Fig. 12-12 Skeletal ossicle of the asteroid *Acanthaster planci.* Note the large amount of original void space. Scale = 70 microns. (Photo by Jon Weber.)

density of the grain. Indeed, some of the carbonate particles may have densities much less than that of quartz. In cases where decomposition of the organic matter has generated gas and the gas is trapped within the grain or where the organic matter makes up a large part of the grain, the grain may actually have a bulk density less than water and float. This appears to happen in some foraminifera. In a study of one modern crinoid species, it was found that the specific gravity of skeletal fragments decreased to 1.05 during organic decomposition and later increased to 1.30 (Cain, 1968).

The criteria for current transport are oolitic coatings and demonstrated abrasion accompanied by size sorting of abraded grains from different sources. Sedimentary structures, such as cross-bedding, are commonly present in these current-transported deposits. The presence of detrital grains of quartz and other silicate minerals is highly suggestive of current transport of the associated carbonate grains. However, these grains may be introduced into the carbonate depositional environment by wind or by floating.

The size and shape of a carbonate rock body deposited by currents normally reflect the type of current and period of time involved in deposition. Because many carbonate rocks are not deposited by currents, these current-deposited rocks commonly appear as individual isolated microfacies. For example, a beach or tidal bar will appear as a distinct carbonate sand body with a shape that reflects the distribution of the current at the time of deposition. As with all sedimentary rock accumulations, the effect of major storms is significant in producing current-transported accumulations. From observations of the effect of hurricanes on modern carbonate sediments in south Florida, it appears that carbonate sands and framework material such as modern reefs undergo significant disruption during these storms. However, carbonate mud and carbonate sediment with abundant carbonate mud are only slightly affected. Apparently, the fine carbonate mud with its contained organic matter is not easily disrupted by wave action in shallow water even during major storms.

Current transport normally produces a preferred orientation of the long axes of individual grains. Some particles such as crinoid stems commonly exhibit remarkable examples of these directional features. Preferential orientation of the long axes of grains may be used to determine the direction of current transport in the same way that similar grain orientation is used in the detrital silicate rocks.

12.8 *IN SITU* ACCUMULATION OF GRAINS

Fundamental to the understanding of carbonate rocks is the recognition that sedimentary materials, particles, mud, and frame are formed within the basin of deposition. In strong contrast to the detrital silicate rocks where we are required to think in terms of source and long distance transport, carbonate materials are produced at or near the site of final deposition. For example, the framework reefs are generated in place and the interstitially deposited particles may have been moved only very short distances or represent a lag deposit of broken reef material.

Many of the particulate sediments may also have been formed essentially in place (Fig. 12-13). With the exception of the ooliths and lithoclasts, we have seen that it is possible to form all the other common particles essentially in place and that carbonate mud is formed by organisms locally on the shallow sea bottom (Fig. 12-14). Thus with no current transport of material, it is possible to produce *in situ* a particulate carbonate sediment with any ratio of carbonate mud to carbonate sand. Under these conditions the fabric of the sediment or rock results from a competition between the rate of production of sand and the rate of

Fig. 12-13 Shallow water bottom sediments in an area of predominantly carbonate mud accumulation. Algae, isolated corals and turtle grass predominate. (Photo by Gray Multer.)

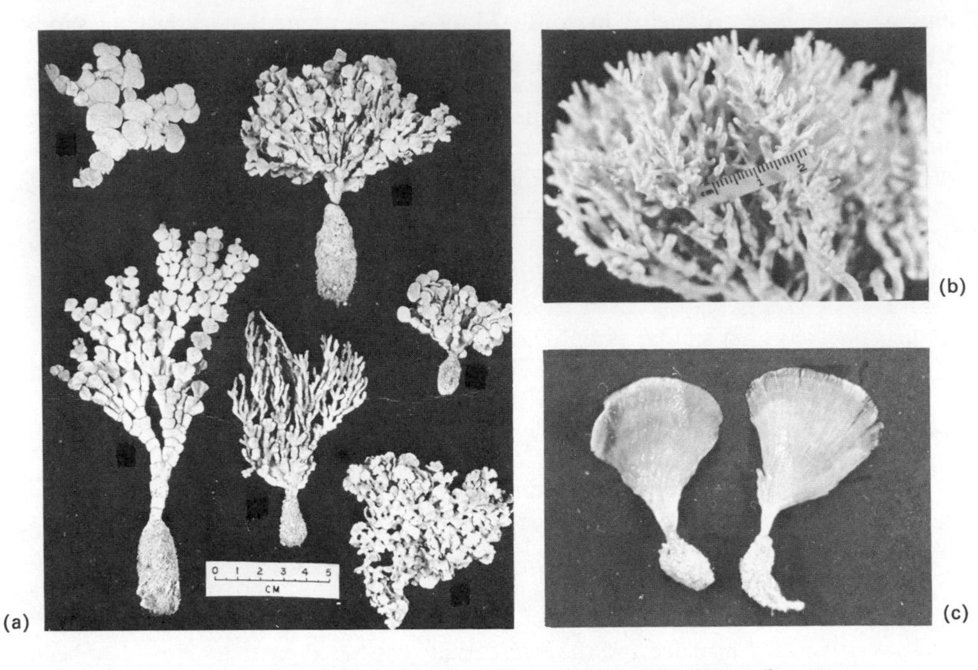

Fig. 12-14 Common calcareous algae in modern carbonate sedimentary areas. (a) *Halimeda,* (b) *Goniolithon,* (c) *Udotea.*

deposition of carbonate mud. The rate of accumulation of carbonate mud is determined by the rate of production of that mud at a given place reduced by the amount of mud that is carried off by current activity; or expanded by the amount of mud that settles from suspension. If the production of sand is rapid with respect to the deposition of carbonate mud, a rock rich in coarse particles will result. In the extreme case a sediment locally free of mud could be formed. The essentially *in situ* accumulation of sand-size and larger particles with little or no interstitial carbonate mud appears to be well illustrated by the deeper lagoon sediments of Kapingamarangi Atoll in the Pacific Ocean (McKee, et al., 1959). Here accumulations of grains of different types appear to be determined by depth of water. If much mixing of grains by currents had taken place, the boundaries between the sediment of different grain types would not exist. Some skeletal particles typified by the crinoids and the platey calcareous algae such as the Pennsylvanian form *Ivanovia* appear to be ideally suited for this type of *in situ* production and accumulation. It is not uncommon to find in the ancient record accumulations of these grains with little carbonate mud. Associated articulated fossils and other evidence indicate the grains themselves have not been moved from the site of production. Many accumulations of pelleted carbonate rocks probably had a similar depositional history.

In the detrital silicate rocks, properties such as size of the grains and sorting of the grains have immense significance because they reflect the hydraulic history of the material. This significance is completely lost in those carbonate rocks in which production and accumulation of grains has been essentially *in situ*. The size of a skeletal grain reflects only the size of the organism that lived at that place. The size of fecal pellets tells us little about the velocity of currents within the depositional environment when these pellets have accumulated *in situ*. Their size, like the size of the skeletal particle, reflects the organisms that generated the particles.

Rounding of grains may take place by current transport or by organic abrasion. However, many of the carbonate particles such as pellets, intraclasts, algal accretionary grains, and some skeletal particles have a generally rounded shape. This roundness in itself demonstrates nothing about the history of the grain but may represent an inherent characteristic of the grain. Crinoid particles are a good example. If these single crystal particles are incorrectly identified, they might be assumed to have become round by mechanical abrasion.

The prime criteria for *in situ* accumulations of carbonate grains result from lack of evidence for current transport. These include the absence of oolitic coatings and absence of demonstrated rounding by abrasion accompanied by size sorting of abraded grains from different sources. In addition, *in situ* accumulations of carbonate grains lack sedimentary structures such as cross-bedding or preferential grain orientation, which result from current transport. The presence of organisms in growth position or delicate organisms such as crinoids with long segments still intact argues for lack of current transport of the grains.

The size and shape of a sedimentary body made up of carbonate grains accumulated *in situ* depends entirely on the area of deposition and the length of time such deposition persisted. When mapped, these *in situ* accumulated carbonate rock bodies commonly have irregular shapes unless the shape is determined by an adjacent current-transported deposit. For example, this material may accumulate between carbonate tidal bar deposits. The linear tidal bars may restrict the distribution of the *in situ* accumulation, imparting a linear trend to the latter.

12.9 SEDIMENTARY STRUCTURES

Laminations

The presence of laminations implies that the nature or size of the carbonate material deposited varied through time. A change such as this may be achieved by current deposition or by variations in the type of organisms that lived and accumulated at a given place. Because of the many variables in nature, most carbonate sediments should show laminations. They do not, however, and laminated carbonate rocks, although occurring in some ancient rocks, are generally uncommon. The fundamental control on the preservation of laminations is the activity of burrowing organisms. During sedimentation, there is competition between the rate of deposition of sedimentary material and the rate at which organisms burrow and disrupt the sediment. This is the most important determinant in the preservation or destruction of laminations. The laminations produced in rapid traction current deposition are not usually completely destroyed. There is seldom a large enough population of burrowing organisms to keep pace with sedimentation. Under these conditions for example, the cross laminations in oolitic carbonate sands commonly are preserved. Four other settings are conducive to the preservation of laminations in carbonate rocks. They are (a) An environment that lacks water circulation, and thus oxygen, will have few burrowing organisms. Carbonate sediment, because it is largely the product of organisms, is seldom produced in such a setting. However, carbonate sediment may be current or gravity transported into such an environment. (b) Ephemeral lakes, such as playa lakes, where the lake dries up and removes the possibility of burrowing organisms. Many playa lake sediments are strikingly laminated. (c) The supratidal environment where burrowing organisms are uncommon because water covers most of these areas only after a major storm and sediment is deposited at this time. Individual storm-deposited layers commonly are preserved. Individual sediment layers representing the deposits of individual hurricanes are recognized in the supratidal sediments of south Florida. (d) Algae bind the sediment into a tough leathery algal mat and commonly preserve laminations. Algal mats are not easily penetrated by burrowing organisms and relatively few such organisms are found in them. The latter two environments are by far the most common ways in which laminations have been preserved in carbonate rocks.

Stromatolites and Algal Mats

The term *stromatolite* has been applied to laminated structures that occur as bulbous heads or stacks and commonly are found in carbonate rocks. Algal stromatolites so common in the ancient record have received considerable attention (Fig. 12-15). Studies of well-developed stromatolites in the Belt series (Precambrian of Montana), the Wilberns (Upper Cambrian) carbonate mounds on the Llano River in Texas, and others combined with a search for modern analogues have been extremely productive. Previously these forms had been considered organic forms despite the fact that they contain no evidence of a preserved organic skeleton. Only laminated carbonate sediment can be seen when they are examined. The laminations commonly follow the outline of the structure or are terminated at the edge of individual heads or stacks. Individual laminae appear thicker on the top of the structure and drape over the edge of the individual head or stack. It is not uncommon to observe that a given lamina in draping over the edge of a head appears to have been deposited at an angle that would be impossible considering the expected angle of repose of the sediment. In this lies the clue to their origin. The sediment was deposited onto a sticky organic surface that trapped and held the grains from sliding in much the same way flies adhere to flypaper. Following deposition, the organic mat decomposed and only the laminated sediment remains. The organic film active in the formation of most algal stromatolites has probably been a complex of filamentous and unicellular green (Chlorophyta) and blue-green (Cyanophyta) algae.

Fig. 12-15 Stromatolites exposed on a horizontal surface. Pethei Group, Precambrian on the east arm of Great Slave Lake, Canada. (Photo by R. G. Walker.)

The blue-green algal mats found commonly on tidal flats in modern carbonate settings serve the same function today. These noncalcareous algal mats contain many species. Twenty-eight actual species have been recorded in the algal mats from Florida. The organic filaments trap and bind carbonate sediment, then proceed to grow up around the sediment grains, and produce a new surface for trapping and binding the next sediment layer (Fig. 12-16).

Logan, Rezak, and Ginsburg (1964) recognized that application of biologic names to structures that were in fact masses of particulate carbonate sediment was meaningless and that the forms observed resulted in large part from local

Fig. 12-16 Curled dessicated algal mats on the supratidal flats of Crane Key, Florida Bay. (Photo by Gray Multer.)

environmental effects rather than from the growth form of an organism. They proposed a classification of algal stromatolites based on their geometric form and related these forms to sites of growth with respect to sea level and erosional effectiveness of the water. The classification uses the arrangement of the geometric shapes, specifically hemispheroids and spheroids, to define the different types (Fig. 12-17). Three different general geometric structures commonly appear in modern algal stromatolites: (a) laterally linked hemispheroids; (b) discrete, vertically stacked hemispheroids; and (c) discrete spheroids. They use the abbreviations LLH, SH, and SS to designate these three different structures. In nature, combinations of the types are possible. Environmental interpretations have been suggested assuming that similar structures were formed in the past under similar conditions because of the strong physical rather than organic influence in determining the shape of these structures. For example, protected intertidal mud flats, where wave action is slight, may be inferred by the presence of laterally linked hemispheroids. These have been termed the *Collenia* types in the past. Exposed, intertidal mud flats, where scouring action of waves and other interacting factors prevent growth of algal mats between stromatolites, are inferred by the presence of the vertically stacked hemispheroids. These have been termed *Cryptozoon* in the past. Low intertidal areas that are exposed to waves and agitated shallow water below low tide mark are inferred by the presence of the concentrically arranged spheroids. These concentrically arranged spheroids are oncolites. Because they move during deposition and behave as coarse detrital particles, they have been discussed (under *algal accretionary grains* in this chapter) as contributing sand grains to particulate carbonate sediment.

Current Structures

Current-transported carbonate rocks composed of sand show most of the sedimentary structures that result from current movement of sand that is found in detrital silicate rocks. These range from graded bedding where carbonate material falls off the edge of a carbonate platform and moves as a turbidity current into deeper water to cross-bedding produced by current transport of grains on a carbonate platform. A common feature of the shallow water carbonate platform is the tidal bar and tidal channel. By the very nature of tidal currents, it is possible to produce in these deposits cross-bedding dip directions that are oriented in clusters that are 180 degrees opposed to each other. These two directions represent the ripples produced by both the flood and ebbing tide. In addition, the low angle cross-bedding of carbonate beach deposits occurs in the ancient record as does the high angle cross-bedding of subaerial dunes composed of carbonate particles. The latter with their soil zones and nonmarine fossils are well exhibited on the islands of the Bahamas. These islands represent deposits of ooliths produced in the marine environment at a Pleistocene low water stage and current-transported onto the land. Reworking of the ooliths by wind and formation of aeolian dunes created relatively high areas that now stand above the present sea level as islands.

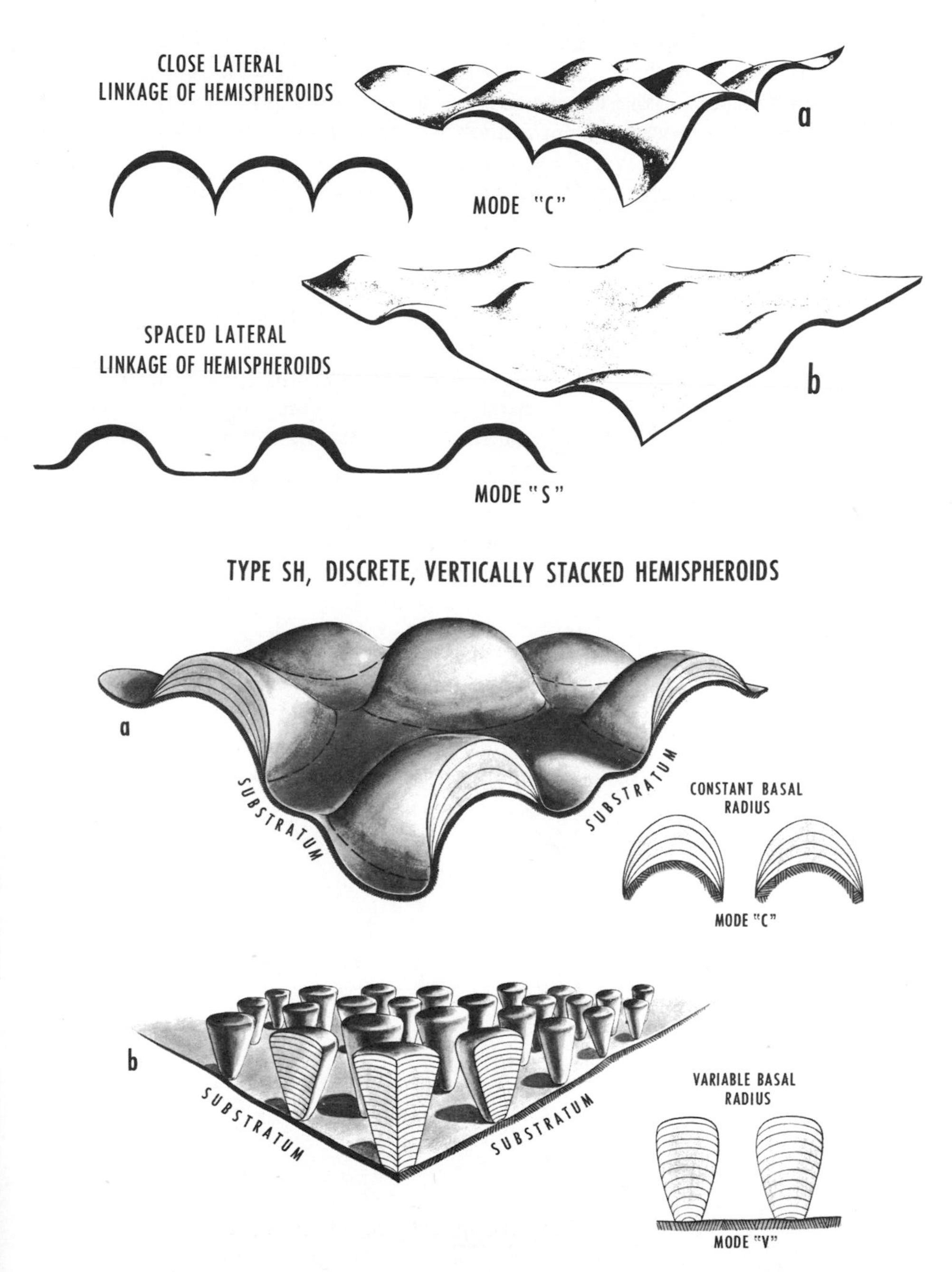

Fig. 12-17 Stromatolite types of Logan, Rezak and Ginsburg (1964).

12.10 THE CARBONATE PLATFORM

Deposition of shallow water carbonate sediments today is taking place on platforms of two types. The first is attached to landmasses such as south Florida and the south coast of the Persian Gulf. The second is isolated shallow water banks that rise from oceanic depths such as the Bahama Platform or the coral atolls of the Pacific. Both types have existed in the past. For example, during part of the Permian, an attached platform existed in west Texas and southeastern New Mexico that developed into the Permian "reef" complex and associated carbonate sediments. To the south the Central Basin Platform stood as an isolated carbonate platform during part of its history (Fig. 12-18). These two carbonate depositional areas of the Permian are analogous in that sense to the modern carbonate areas of south Florida and the Bahamas.

Fundamental to the formation of a carbonate platform is an area where carbonate-producing organisms can grow prolifically. This implies very shallow water, certainly shallow enough to accommodate the light-requiring organisms.

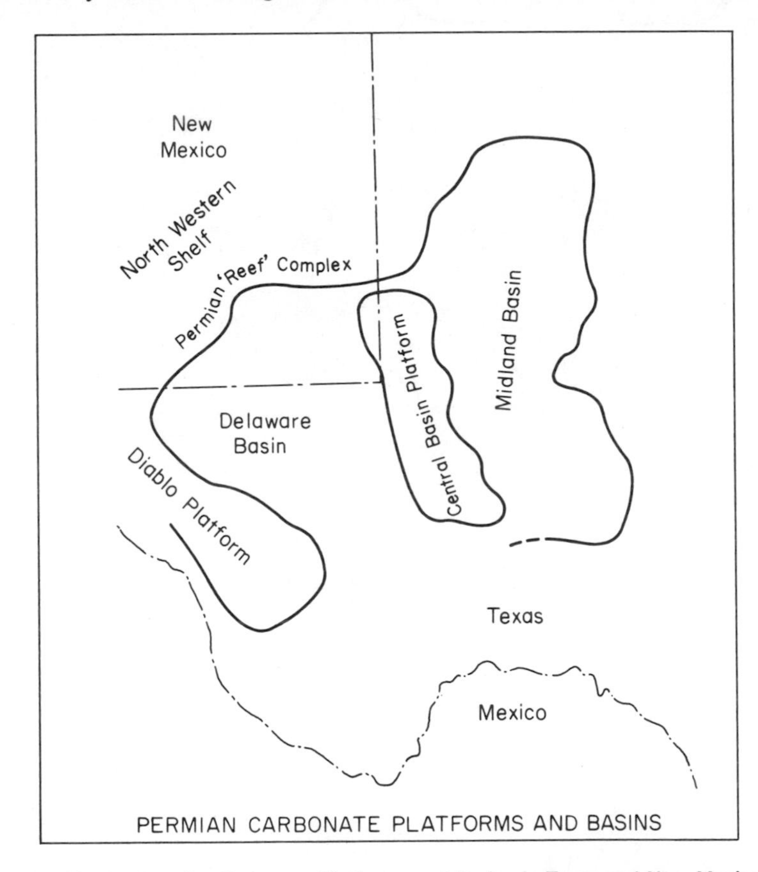

Fig. 12-18 Permian Carbonate Platforms and Basins in Texas and New Mexico.

In addition, there should be little influx of terrigenous detritus that would dilute the carbonate contribution and produce impure carbonate rocks. Far more important, the introduction of detrital silicate material has the effect of muddying the water, reducing the depth of light penetration, and making the environment inhospitable for organic life. Thus, prolific introduction of silicate detritus is generally inconsistent with the production and deposition of carbonate sediments and rocks. The elimination of the detrital material is easily achieved on an isolated platform rising from oceanic depths. Bottom transport of detrital material across oceanic depths and up the side of the platform is generally impossible. Isolated carbonate islands such as Pacific atolls generally lack silicate detritus except that which is introduced by volcanic activity or wind. Where the carbonate platform is attached directly to a landmass, then the absence of large rivers in the area and the interruption of longshore transport of detrital silicate material by sediment traps such as bays and lagoons is essential. Mixing of silicate and carbonate material will naturally occur at the boundary. This is taking place along the west coast of Florida.

Once a carbonate platform begins, the shallow water must be maintained by subsidence of the platform if thick sections of carbonate rock are to accumulate. Unless subsidence occurs, the shallow water will become filled rapidly with the carbonate sediment, while slightly deeper water will receive relatively little sediment. With subsidence, the slightly deeper water will become deeper and the new carbonate sediment will accumulate on previously deposited carbonate sediment in shallow water. This has a net effect of generating platforms with relatively steep sides and abrupt margins. The effect may be exaggerated by relative drops in sea level accompanied by subaerial exposure of the earlier-deposited carbonate sediment. At such times marine erosion will cut sea cliffs in this sediment, thus forming even steeper sides. With renewal of subsidence, the boundary between the shallow water and the deeper water becomes even more abrupt and the boundary between rapid and extremely slow carbonate sedimentation becomes abrupt. Thus the ancient rock record exhibits many examples of carbonate rock sections that change laterally and abruptly to other rocks of equivalent or younger age. This occurs when the carbonate platform becomes incorporated in the geologic record by sedimentation in the surrounding areas.

The Bahama Platform represents an excellent and well-studied example of a modern isolated carbonate depositional area. This area contains twenty major islands and thousands of small islands and rocks that rise from water generally less than 20 to 30 ft deep. The flat shelf on which these islands are located has a surface area of approximately 60,000 sq mi and rises from ocean depths in excess of 1500 ft on the west and 12,000 ft on the east, with local slope angles in excess of 45 degrees. Deep borings that penetrate to depths of approximately 15,000 ft encountered shallow water carbonate and associated rocks (Fig. 12-19). These indicate that the general area of the Bahama Platform has undergone subsidence and carbonate sedimentation since at least early Cretaceous time. The origin of the deep water areas such as the Tongue of the Ocean and Exuma Sound within the platform is the subject of some controversy. They appear to have

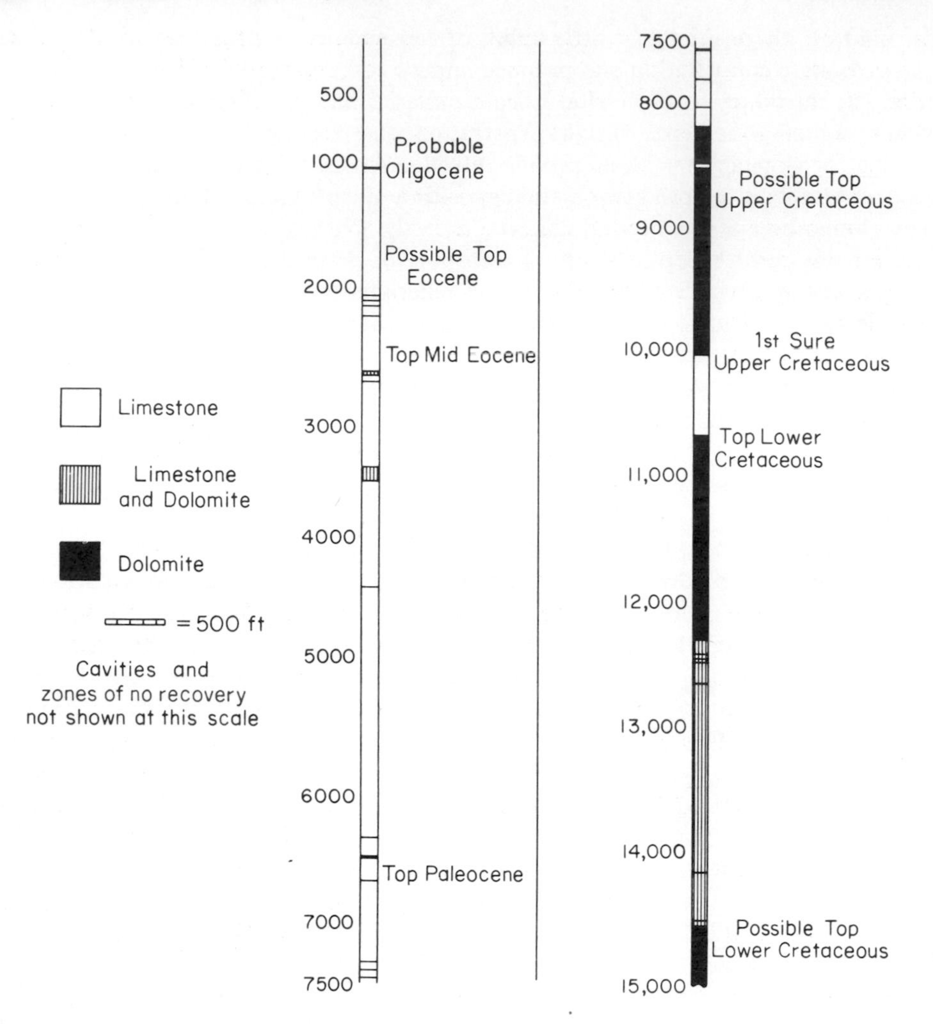

Fig. 12-19 Geologic Log, Bahamas Oil Co. Ltd. Andros No. 1. Andros Island, Bahamas.

become slightly deeper water, however, because of either structural or erosional reasons sometime during the Mesozoic and since have received relatively little sediment. Deep sea drilling and analysis of recovered sediments from the Tongue of the Ocean indicate this body of water has received deep water sediments from at least the Cretaceous. These deep water areas failed to stay close to sea level and were left behind by the continuing process of subsidence and rapid carbonate sedimentation on the platforms.

The air temperature on the platform during the winter months averages about 70°F. while the summer temperature ranges between 80 and 90°F. Water temperature ranges between 68 and 86°F with an average temperature of approximately 77°F. Much of the platform has a relatively humid climate with an annual rainfall of between 40 and 60 in/yr. This rainfall is seasonal and most of it falls during the summer months. The southern part of the platform is more arid and gypsum is forming on islands such as Inagua. Much of the Bahama area lies within the hurricane belt and these storms have a dramatic effect on the sediments. The prevailing wind is from the east and northeast and ranges from southeast to northeast. Wind direction has an effect on the carbonate sediment distribution. Specifically, most of the well-developed coral-algal reefs are on the windward edges of the platform. Mean tidal range is about 2 to 3 ft but tides of 4 ft are not uncommon. Despite these small tidal ranges, tides play an important role in redistribution of sediments.

Certain generalizations may be made about the carbonate sediment distribution on the Bahama Platform (Fig. 12-20). The coral-algal reefs occur on the windward edge of the platform. These reefs exist as a thin veneer on eroded Pleistocene limestones that were exposed subaerially prior to the rise in sea level that covered the platform approximately 4,500 years ago. Spectacular examples of these frame-

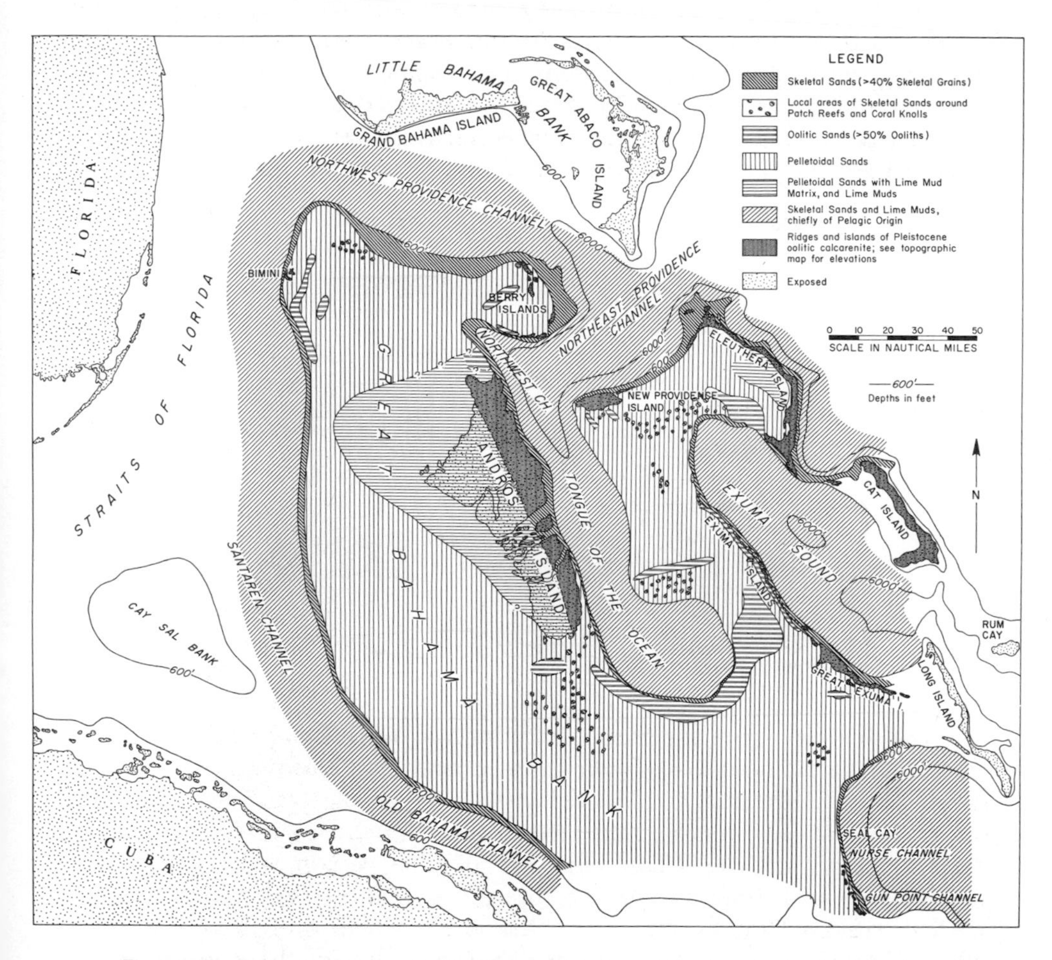

Fig. 12-20 Recent sediments of the Bahama Platform. (Courtesy R. N. Ginsburg.)

work deposits can be seen along the west edge of the Tongue of the Ocean east of Andros Island. Ooliths require strong current action, preferably tidal action, and are found in great areas of tidal bars at the south end of the Tongue of the Ocean where tidal currents become focused because of the long narrow indentation of this feature into the platform. A similar phenomenon accompanied by the formation of ooliths occurs at the north end of Exuma Sound. Strong tidal currents have produced spectacular tidal bars composed of ooliths along the edge of the platforms near Bimini on the Straits of Florida. Carbonate sands composed predominantly of skeletal particles occur in a band several miles wide along the edge of the platform. It appears that nutrient-rich waters rising from the oceanic depths have a pronounced effect on the productivity of calcareous skeletal producers. Because of proximity to the edge of the platform, many of these carbonate sands have undergone some current transport. The vast interior of the platform is composed of carbonate sand and carbonate mud. The carbonate sand-size particles are predominantly pellets and pelletoids with relatively minor amounts of skeletal grains. On the west side of Andros Island, protected from the prevailing wind, the modern marine sediment is carbonate mud or pelleted carbonate mud, much of which is probably produced in place. In this area protected from the wind-driven currents, however, the mud is not easily removed and mud carried in suspension has the opportunity to settle and accumulate (Fig. 12-21).

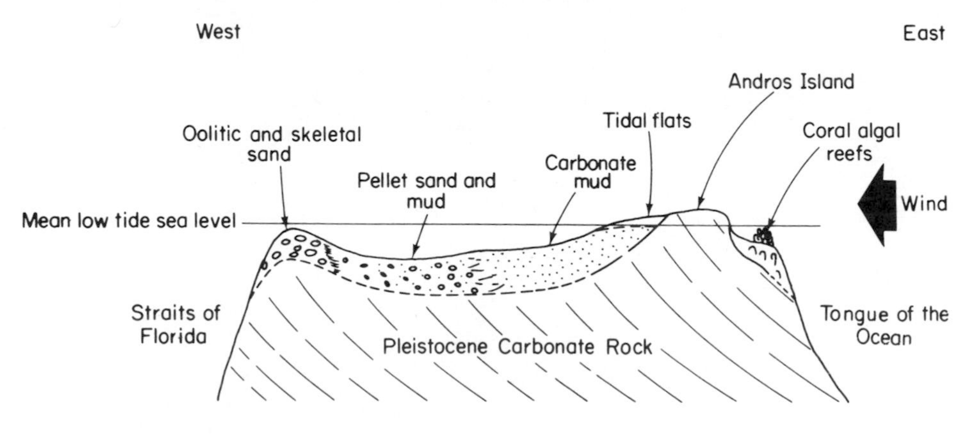

Fig. 12-21 Diagram illustrating the distribution of recent carbonate sediments on the Bahama Platform.

The south coast of the Persian Gulf from Iraq to the Oman represents a modern carbonate platform that is attached to a land mass, specifically the Arabian subcontinent. This area has received carbonate sediments in much the same setting since at least Jurassic time. The present Persian Gulf is today relatively shallow, being nowhere deeper than 300 ft. The deepest water occurs close to the northern or Iranian shore. The prevailing wind is northwesterly and thus onshore. Periodically, storms off the deserts of Arabia may bring wind-transported detrital silicate material into the shallow waters of the southern coast of the gulf. Unlike the Bahama Platform, this area is relatively arid with much of it having an annual precipitation of less than 2 in. The modern marine sediment distribution is in large part determined by two factors: (a) bottom topography and (b) the presence of barrier islands that separate the shelf carbonate sediments from lagoon sediments. The outer shelf sediments are predominantly skeletal sands and carbonate mud. The slightly higher elevations on the sea bottom appear to be relatively enriched in carbonate sand-size and larger particles, while the submarine depressions are enriched in carbonate mud. In the lagoon areas behind the barrier islands, carbonate mud and pelleted carbonate mud predominate. The barrier islands themselves commonly have coral-algal reefs developed along their outer, windward-facing margins. In the tidal passes that separate the barrier islands, tidal deltas are developed. The common grain type in these tidal deltas is the oolith.

12.11 TIDAL FLAT SEDIMENTATION

In the preceding discussion we have considered primarily subtidal marine sediments, i.e., sediment deposited below mean low tide. In the area west of Andros Island (Fig. 12-22) in the Bahamas and onshore along the south coast of the Persian Gulf, a tremendous amount of sediment is accumulating in that part of the coast

Fig. 12-22 Aerial view of the sedimentary environments on the west shore of Andros Island, Bahamas. (Photo by E. A. Shinn.)

between mean low tide and mean high tide, the intertidal zone. In addition, much carbonate sediment is accumulating above mean high tide on that part of the coast that is covered by seawater only during the largest storms, the supratidal zone. These "onshore" areas have long been neglected as sedimentary environments. As study of these depositional sites has proceeded and criteria from modern sediment studies for these environments have developed, however, a significant change has taken place in our understanding of the ancient carbonate rock record. Many carbonate rocks that were once thought to have been formed in shallow marine water are now recognized to be the products of the "onshore environments", both intertidal and supratidal (see Laporte, 1967, and Roel, 1967). Not only have these environments been recognized as important contributors to the ancient carbonate record but because they represent sediment deposited in an environment that experiences periodic subaerial exposure, the carbonate material is subject to drying and change in water chemistry that produces a wide variety of early diagenetic changes within the sediment.

At a given stand of sea level the production of carbonate sediment in shallow water will quickly lead to even shallower water. Rates of *in situ* carbonate accumulation today commonly exceed 1 ft/1000 yr. Current-transported sediment locally can accumulate much faster. As the water becomes shallow, some areas will be exposed to the air at low tide. This may take place by accretion to shorelines or by the development of offshore shoals that expand and coalesce to produce broad tidal flats. Such intertidal flats soon build up and out and create areas that are exposed to the air at normal high tides. Wave action accompanying storms carries carbonate sediment onto these flats, thus building them higher above the normal tidal range. At this point, modern carbonate sediment deposited above mean high tide exists exposed to the air until the next big storm. Thus the normal history of a carbonate platform is to make the water shallow by sedimentation and to build out the shoreline with the development of sediments deposited within the intertidal and supratidal zones. For example, in the past several thousand years the carbonate tidal flats in places along the south shore of the Persian Gulf have prograded seaward by sedimentation at a rate of 1 to 2 m/yr. This type of regressive accumulation by sedimentary processes will proceed until almost the entire platform becomes land. When only a small area of shallow marine water is left, there will develop a competition between the rates of sediment production in the marine environment and the rate of destruction of the tidal flat sediments by erosion. Thus some marine sediment should always exist on the shelf edge. At a constant position of sea level, a vertical sequence at any place on the supratidal flats will show sediments deposited in the supratidal zone on top that cover material deposited in the intertidal area. An infratidal zone is sometimes recognized but is difficult to identify in rock. This environment represents the area below mean low tide and the lowest waterline achieved with a strong offshore wind. It is an area of periodic subaerial exposure. The thickness of the intertidal sediments, ignoring compaction and assuming a constant position of sea level, will approximate the tidal range at the site of deposition. These intertidal sediments in turn will be underlain by sediments deposited in the marine environment. Such a sequence will produce layered rocks in which the layers represent sediments deposited in the three environments. The layers by the very nature of the sedimentation will be time transgressive. Just as sediments west of Andros Island are being deposited in the three settings today and thus are time equivalent and laterally continuous, a time surface in an ancient carbonate rock cuts across the deposits of the three environments. Naturally, these simple relationships will be altered by absolute changes in sea level or tectonic movement. It is interesting to speculate that, assuming no change in the position of sea level on the Bahama Platform, much of this great shallow water marine shelf will be land within several tens of thousands of years by development of supratidal flats. At that time little new carbonate sediment will accumulate and further addition to the sedimentary record must wait until the sea level rises. Thus the ancient sedimentary record always must be incomplete in this type of sedimentation. Because we live in a time of relatively recent rise in sea level, it is difficult for us to visualize the vast tidal flats that existed in the past. Once a supratidal flat has become so wide that even the largest storms can no longer carry water and sediment across them, however, the earlier-deposited sediments assume a character more like continental deserts or coastal plains. They cease to be directly influenced by marine water. In the Persian Gulf, modern onshore sedimentation has produced supratidal flats (sabkhas) 20 mi wide since the last rise in sea level, and the process is just beginning. The frequency of flooding by marine water on these flats today gives us an appreciation of the extent of marine influence (Fig. 12-23). Thus we must recognize that when we observe sediments of essentially the same age deposited in the supratidal zone occurring over areas of tens of thousands of square miles that the rocks are time transgressive and that these flats did not necessarily exist as supratidal flats at the same time. They were

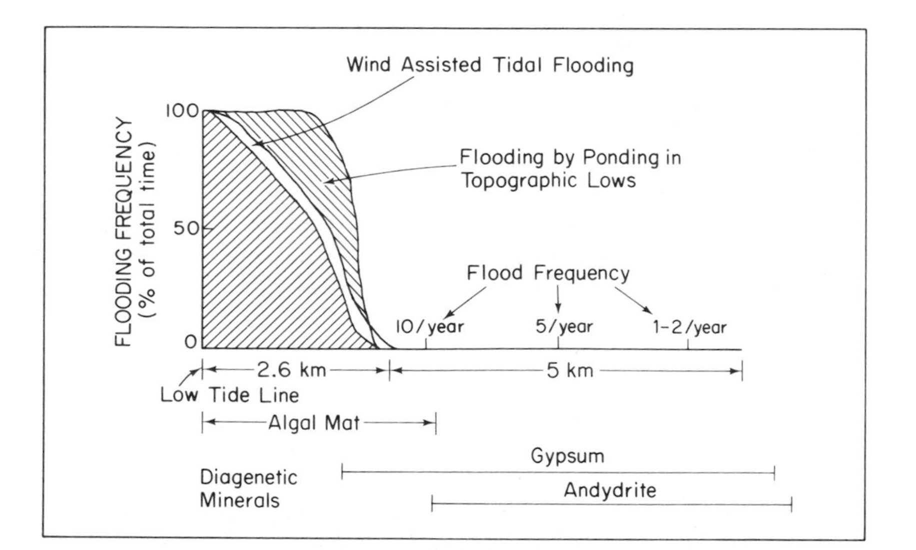

Fig. 12-23 Frequency of tidal flooding of the intertidal and supratidal flats, Trucial Coast, Persian Gulf. (Modified from data prepared by D. J. J. Kinsman and R. K. Park.)

being added to at the frontal edge and those areas far removed from the sea had long since ceased to receive new marine material.

Recognition of the tidal flat sedimentary complex comes in large part from recognizing the vertical sedimentary sequence. Only generalizations can be made here and the reader is advised to consult the rapidly expanding literature for specific criteria that are being used to recognize sediments deposited in each environment. In general, the sediments deposited in the marine environment, unless current transported, will be mixtures of carbonate sand, mostly pellets and skeletal particles, and carbonate mud. The marine sediments are commonly homogenized by burrowing organisms. The intertidal zone appears to be characterized by carbonate muds sometimes showing fine rippling or remnants of fine rippling. In the higher parts of the intertidal zone, stromatolites predominate today and are presumed to have existed here in the past. Commonly, tidal channels cut the intertidal zone producing linear bodies of carbonate sands that sometimes show reversible current direction and lag carbonate gravels near the base. These carbonate tidal channel deposits vertically interrupt the normal intertidal sedimentary record. The supratidal sediments usually are laminated and exhibit mud cracks and other evidence of subaerial exposure. Flat pebble conglomerates composed of intraclasts are produced in the high intertidal zone and predominate in the lower supratidal zone. Evidence of repeated early diagenetic phenomena such as vadose pisoliths and nodular anhydrite or gypsum are common. Horizons of whole fossils lying on a single bedding surface are common on intertidal and low supratidal flat sediments. This occurs when these organisms are thrown up onto the flats during a large storm.

12.12 DEEP-SEA CARBONATE SEDIMENTS

Our discussion of carbonate rocks and modern carbonate sediments has explored the shallow water carbonate shelf material. However, more than a third of the deep oceanic sea floor today is covered by oozes containing more than 30% calcium carbonate.

By far, the most common deep sea carbonate sediment is an ooze composed of the remains of planktonic foraminifera and calcareous nannofossils (dissociated parts of planktonic algae). Such ooze has been referred to as globigerina ooze because a major constituent is the tests of the foraminifera *Globigerina*. Actually this is not always true because other foraminifera may be more abundant in some cases. Furthermore, calcareous nannofossils may predominate over foraminifera and, especially in older rocks, carbonate sediment may be composed entirely of calcareous nannofossils. In places, the remains of pteropods and heteropods may be major constituents. Foraminifera generally occur as sand-size particles, whereas nannofossils are less than 50 μ in size. Foraminifera may also appear in smaller sizes as broken particles. Calcareous nannofossils and foraminifera are composed of calcite; pteropods, being molluscs, produce shells of aragonite.

Deep-sea deposits that are relatively rich in calcium carbonate grade laterally and commonly abruptly into red clays or siliceous oozes. The change from carbo-

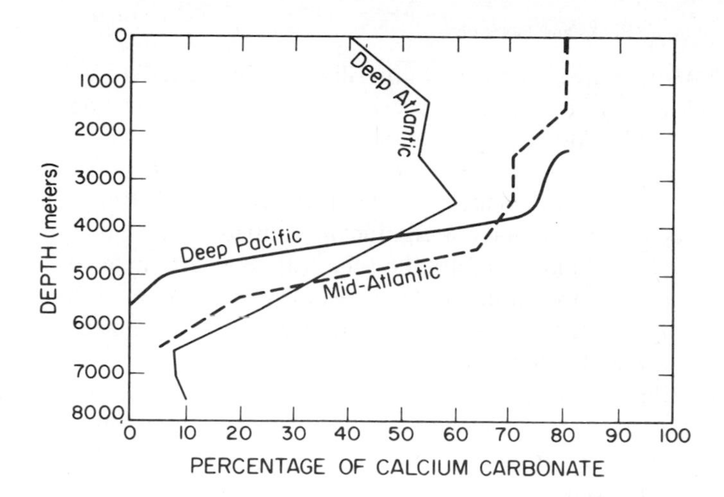

Fig. 12-24 The amount of calcium carbonate in deep oceanic sediments as a function of depth of water. (After Turekian, 1968.)

nate-rich sediment to sediment with little or no carbonate is commonly determined by depth of water. This depth has been termed the *compensation level* and occurs between 4000 and 7000 m below the surface (Fig. 12-24). However, laboratory and field data suggest that the oceans are undersaturated with respect to calcium carbonate below a depth of approximately 500 m. This change from supersaturation to undersaturation with depth results from CO_2 produced by respiration and decay of organisms in the deep oceans. In addition, there is a small contributing effect from increased hydrostatic pressure. We can ask, What determines the depth of water in which calcium carbonate particles will accumulate? A large particle may fall to the bottom, be buried, and removed from contact with moving seawater and be preserved. For example, if a carbonate particle 12 μ in diameter falls through seawater undersaturated with respect to calcium carbonate, it will be dissolved before it has fallen 1000 m. Larger particles will survive the fall through deep water and continue to be dissolved on the bottom. Organic coatings on the shells certainly inhibit the rate of carbonate dissolution by not allowing the seawater to come in contact with the carbonate minerals. This may help to explain the accumulation of calcareous nannofossils, many of which are smaller than 12 μ. In the foraminifera, there is a selective process of solution whereby the smaller forms with thinner test walls, or even larger ones with delicately structured tests, are dissolved first, leaving behind the more resistant tests of other species. The accumulation of calcium carbonate in the deeper oceanic waters requires relatively large or more resistant particles and sufficiently rapid sedimentation so that the particle is buried and removed from contact with the deep ocean water. The position of the compensation level in the oceans is a response to all these factors.

Other factors also influence the calcium carbonate content of the deep ocean sediments. *Globigerina* thrives only in the relatively warm waters of the tropics, especially in the equatorial zones of nutrient upwelling, where there is very high productivity of calcareous plankton. This has led, for instance, to the accumulation of a thick blanket of calcareous ooze below the Pacific tropical zone. Productivity is also high in other areas of nutrient upwellings in the higher latitudes but accumulation of calcareous ooze may be prevented because the sea floor is below the compensation level. Along the edges of the ocean basins where detrital material is being added by rivers or where there is a significant contribution from oceanic volcanic material, the carbonate particles produced may become diluted to the point that they no longer make a significant contribution to the sediment. This dilution factor is well illustrated by the relatively low carbonate concentrations in modern sediments in deep water off the coast of Argentina. In addition, calcareous sediment may be diluted where there is a large contribution made by the remains of siliceous planktonic organisms.

The contribution of calcareous plankton to deep marine sediments has existed since the Jurassic Period. Calcareous nannofossils were the major contributor to Early Cretaceous and Jurassic sediments. Foraminifera became major contributors by Late Cretaceous time. With some spectacular exceptions in rocks of Mesozoic age found in California and the Alps, deep-sea carbonate sedimentary rocks are very uncommon in the preserved record on the continents.

Along the edge of carbonate platforms, carbonate sediment produced in shallow water may be carried off the platform into deeper water and accumulate in the nearby deeper ocean. Indeed, turbidity currents may develop and carbonate rocks exhibiting graded bedding are found interbedded with deep marine sediments. Such material in ancient rocks has been reported from the Marathon geosyncline in west Texas.

12.13 CALCIUM CARBONATE DEPOSITION ON THE LAND

Travertine and Tufa

These crust-like deposits of calcium carbonate form by both inorganic and plant-influenced precipitation from fresh water in rivers, lakes, and springs (Fig. 12-25). Thermal springs commonly produce these materials and local but spectacular deposits result. When the material is porous and spongy, the term *tufa* is normally applied. When massive, relatively dense, and sometimes banded, the term *travertine* is used. However, travertine has sometimes been used to describe the deposits of calcium carbonate formed in limestone caves. Speleothems, the calcium carbonate deposits in limestone caves, are common in modern caves and have been recognized in ancient rocks. The deposition of this material as stalactites, stalagmites, and other cave ornamentation can take place by at least two mechanisms: (a) Evaporation of rainwater that has percolated through and dissolved carbonate rock. Evaporation of this water causes the precipitation of the calcium carbonate on the roof and walls of the cave. (b) Change in the partial pressure of carbon dioxide in the water during its movement from the surface to inside the cave. Rainwater comes to equilibrium with the high carbon dioxide partial pressure existing within the vadose zone of the soil. When this water reaches the cave, which has an atmosphere similar to that of the outside air, it will reequi-

Fig. 12-25 Massive travertine deposited from hot springs along the shore of Jemez Creek, New Mexico.

librate with this atmosphere. This results in a loss of carbon dioxide from the water and a consequent precipitation of calcium carbonate as cave ornamentation. The change in partial pressure of carbon dioxide from soil atmosphere to cave atmosphere appears to be the more common mechanism. This has been demonstrated in caves by analyzing waters collected from the surface of growing stalactites. Ions in the water that do not contribute to the growing calcium carbonate have the same concentration as exists in the groundwater. This would not be true if significant evaporation had taken place. Evaporation of water within caves is generally insignificant because the relative humidity is usually high.

12.14 FRESHWATER LAKE DEPOSITS

The nonmarine environment produces many of the particles and features that are found in the marine setting. Ooliths, skeletal particles of freshwater organisms, pellets, intraclasts, and freshwater algal structures are common. The diversity of organisms is lacking and recognition of the nonmarine setting commonly comes from recognition that the organisms that contributed to the sediment were nonmarine. In other examples the regional setting of interfingering continental deposits makes a given limestone an unlikely candidate for a marine or marine-associated deposit. The terms freshwater lake and nonmarine deposit are somewhat misleading. Many lakes today contain water with extremely high dissolved salt concentrations: sodium chloride, sodium sulfate, or sodium carbonate. It is appropriate to limit the term marine to those bodies of water that have the composition of the sea at the time of deposition or bodies of water derived by evaporation or slight dilution of seawater. We have seen in the discussion of tidal flat sediments and shall see again in our discussion of evaporites that large volumes of sediment are deposited on the land under marine influence. Indeed, as a tidal flat becomes wider by accretion, the sea no longer directly influences the earlier-deposited material because the big storms can no longer carry seawater all the way across the flats. At this time and in this setting the definition of marine versus nonmarine again loses its significance.

Calcium carbonate is being deposited today in lakes with salinities that range from those of the Dead Sea and Great Salt Lake to the marl-producing fresh water lakes of Minnesota and Wisconsin. The Dead Sea is one of the most hypersaline bodies of water in the world. Its salinity ranges from 285 to 330 ppt. The dissolved material is in large part derived by dissolution of older rocks over which the Jordan River flows on its way to the lake. Gypsum is being precipitated today but unlike the Pleistocene when gypsum accumulated, it is now being immediately replaced by calcite. This calcite is settling to the bottom together with aragonite formed by direct precipitation. The sediment found on the bottom today appears rhythmically banded with black and white laminae. The white bands contain abundant aragonite, while the black layers lack abundant aragonite.

Great Salt Lake, Utah, occupies a tectonic depression and is representative of many of the salt lakes in western United States that have been reduced to their present size with a corresponding increase in concentration of dissolved salts by the reduction in rainfall that has accompanied the climatic change since the last glaciation. The present salinity of the water is approximately 230 ppt, mostly sodium chloride and sodium sulfate. Dissolved salts are being added to the lake, which has a volume of 15.85 cu km, at a rate of 1.1×10^6 tons/yr. The lake receives detrital silicates from the surrounding land producing an impure carbonate sediment in the deeper parts of the lake and along part of the shore area. Locally, oolitic carbonate sands and pellet carbonate sands occur along the shore and out to depths of water of about 10 ft. The pellets are produced for the most part by the brine shrimp *Atrema*. Algal mats produced by blue-green colonial algae carpet the shore and shallow water of much of the lake.

Marl, a loose term for carbonate mud containing some silicate mud, is currently being deposited in many lakes in north-central United States. Certain plants, for example, *Chara*, are able to produce calcium carbonate in their stems, leaves, and fruits. This source of freshwater calcium carbonate has been generally available since the Mesozoic and examples have been reported from the Lower Devonian. Other plants and microorganisms may also contribute calcium carbonate to these sediments.

In the ancient record there are many examples of carbonate rocks that were deposited under nonmarine conditions in lakes. They include the early Cenozoic Green River Formation, famous for the preservation of fossil fish and other organisms and organic structures. The Early Cenozoic Horse Springs Formation found in southern and central Nevada represents the deposits of an early intermountain lake. The Triassic carbonate lake deposits of eastern United States were formed in a setting similar to that found in the Basin and Range valleys of western United States.

REFERENCES

The reader is encouraged to examine "A Selective, Annotated Bibliography on Carbonate Rocks" by Paul E. Potter, published in the *Bulletin of Canadian Petroleum Geology*, **16**, no. 1, March, 1968. (Outstanding statement of the carbonate literature up to 1967.)

BALL, M. M., 1967, "Carbonate Sand Bodies of Florida and the Bahamas," *Jour. Sed. Pet.* **37**, pp. 556–591. (Origin and significance of carbonate sand accumulations in two platform areas.)

BATHURST, R. G. C., 1972, *Carbonate Sediments and Their Diagenesis*. New York: American Elsevier Co. Inc., 700 pp. (Excellent discussion of sedimentation, diagenesis and particle microstructure.)

CAIN, J. D. B., 1968, "Aspects of the Depositional Environment and Paleoecology of Crinoidal Limestones," *Scottish Jour. Geol.* 4, pp. 191–208. (Experimental and petrographic study of crinoid rocks.)

CAYEUX, LUCIEN, 1970, *Carbonate Rocks*, translated and updated by Albert V. Carozzi, Hafner Publishing Co., 506 pp. (A classic work now available in English.)

CHAVE, K. E., 1954, "Aspects of the Biogeochemistry of Magnesium: Calcareous Marine Organisms," *Jour. Geol.* **62**, pp. 266–283. (Magnesium substitution in the calcite of marine organisms. See also discussion of sediments and rocks on pp. 587–599 of the same volume.)

———, 1962, "Mineralogy of Carbonate Sediments," *Limnol. Oceanogr.*, **7**, pp. 218–223. (Fraction analysis of modern carbonates shows sediment mineralogy to be controlled by skeletal mineralogy and organisms.)

CLARKE, F. W. and W. C. WHEELER, 1922, "The Inorganic Constituents of Marine Invertebrates, *U. S. G. S. Prof. Paper* **124**, 61 pp. (A classic paper on the chemistry of marine organisms.)

DAPPLES, E. C., 1938, "The Sedimentational Effects of the Work of Marine Scavengers," *Amer. Jour. Sci.*, **36**, pp. 54–65. (The earliest and still a valuable survey of the effects of predators on fragmentation of organic marine skeletons.)

DEGENS, E. T., 1965, *Geochemistry of Sediments*, Englewood Cliffs, N. J.: Prentice-Hall, Inc., 342 pp. (A good general reference with emphasis on problems.)

EARDLEY, A. J., 1938, "Sediments of Great Salt Lake, Utah," *Am. Assoc. Petroleum Geol. Bull.* **22**, pp. 1359–1387. (Discussion of these nonmarine carbonate sediments.)

FISCHER, A. G., S. HONJO, and R. E. GARRISON, 1967, *Electron Micrographs of Limestones and Their Nannofossils*. Princeton, N. J.: Princeton Univ. Press, 141 pp. (A collection of superb electron micrographs illustrating floral diversity on a submicroscopic scale in sensibly homogeneous micrite.)

FOLK, R. L., and R. ROBLES, 1964, "Carbonate Sands of Isla Perez, Alacran Reef Complex, Yucatan," *Jour. Geol.*, **72**, pp. 255–292. (Detailed investigation of the genesis, size, shape, and sorting of carbonate sands on a reef in a well-studied carbonate platform.)

GEHMAN, H. M., JR., 1962, "Organic Matter in Limestones," *Geochim. Cosmochim. Acta*, **26**, pp. 885–897. (Analyses of 1400 ancient rocks indicates that limestone contains only one-fifth the amount of organic matter as an equal amount of shale, but the same amount of hydrocarbons.)

GRAF, D. L., 1960 "Geochemistry of Carbonate Sediments and Sedimentary Carbonate Rocks," *Ill. Geol. Survey Circulars* 297, 298, 301, 308, and 309. (An excellent source of data on carbonate geochemistry.)

HAM, W. E. (ED.), 1962, "Classification of Carbonate Rocks, a Symposium," *Tulsa, Amer, Assn. Petroleum Geologists*, Mem. 1, 279 pp. (Papers by Dunham and Folk and practically all pre-1962 papers on classification are included.)

HOROWITZ, A. S. and P. E. POTTER, 1971, *Introductory Petrography of Fossils*. New York: Springer-Verlag, 202 pp. (An outstanding discussion and collection of photographs on identifying fossils in thin section.)

ILLING, L. V., 1954, "Bahaman Calcareous Sands," *Amer. Assn. Pet. Geol. Bull.*, **38**, pp. 1–95. (The first modern description of the carbonate sands of the Bahamas.)

ISENBERG, H. D., S. D. DOUGLAS, E. S. LAVINE, and H. WEISSFELLNER, 1967, "Laboratory Studies with Coccolithophorid Calcification", in *Stud. in Trop. Oceanog.*, no. 5, Univ. Miami, Florida, pp. 155–177. (Laboratory experiments exploring the physiology and biochemistry of skeletal formation in one species of coccolith.)

KINSMAN, D. J. J., and H. D. HOLLAND, 1969, "The Co-Precipitation of Cations with $CaCO_3$-IV. The Co-Precipitation of Sr^{2+} with Aragonite between 16° and 96°C," *Geochim. Cosmochim. Acta*, **33**, pp. 1–17. (Experimental determination of the variation of the equilibrium constant in this system at low temperatures and atmospheric pressure.)

LAPORTE, L. F., 1967, "Carbonate Deposition Near Mean Sea Level and Resultant Facies Mosaic: Manlius Formation (Lower Devonian) of New York State," *Amer. Assn. Pet. Geol. Bull.* **51**, pp. 73–101. (One of the first modern petrologic and environmental studies of the classic Devonian section of western New York.)

LEBLANC, R. J., and J. G. BREEDING (EDS.),1957, "Regional Aspects of Carbonate Deposition," *Tulsa, Soc. Econ. Paleont. Mineral.*, Spec. Pub. 5, 178 pp. (A symposium on carbonates. Papers by Rogers, Newell and Rigby, Ginsburg, Moore, and Fairbridge.)

LOGAN, B. W., R. REZAK, and R. N. GINSBURG, 1964, "Classification and Environmental Significance of Algal Stromatolites," *Jour. Geol.*, **72**, pp. 68–83. (Discussion of the origin and environmental significance of these puzzling carbonate structures.)

LOWENSTAM, H. A., 1950, "Niagaran Reefs of the Great Lakes Area," *Jour. Geol.*, **58**, pp. 430–486. (Definition and discussion of what constitutes a "true reef.")

LOWENSTAM, H. A., and S. EPSTEIN, 1957, "On the Origin of Sedimentary Aragonite Needles of the Great Bahama Bank," *Jour. Geol.*, **65**, pp. 364–375. (An early use of oxygen isotope ratios demonstrating the probability that disintegration of calcareous algae is the source of the abundant aragonite needles sedimented in this area.)

MCKEE, E. D., J. CHRONIC, and E. B. LEOPOLD, 1959, "Sedimentary Belts in Lagoon of Kapingamarangi Atoll," *Amer. Assn. Pet. Geol. Bull.*, **43**, pp. 501–562. (Origin of sedimentary facies among carbonate sediments on a Pacific Atoll.)

MILLIMAN, J. D., M. GASTNER, and J. MULLER, 1971, "Utilization of Magnesium in Coralline Algae," *Bull. Geol. Soc. of America*, **82**, pp. 573–579. (Demonstration of localized excess magnesium in coralline algae over that indicated by the shift in X-ray diffraction peaks.)

PRAY, L. C., 1958, "Fenestrate Bryozoan Core Facies. Mississippian Bioherms, Southwestern United States," *Jour. Sed. Petrology*, **28**, pp. 261–273. (Discussion of carbonate mud mound buildups.)

PURDY, E. G., 1963, "Recent Calcium Carbonate Facies of the Great Bahama Bank 1. Petrography and Reaction Groups. 2. Sedimentary Facies," *Jour. Geol.*, **71**, pp. 334–355 and 472–497. (Discussion of modern carbonate particles and their origin on this extensive carbonate platform.)

ROEHL, P. O., 1967, "Stony Mountain (Ordovician) and Interlake (Silurian) Facies Analogs of Recent Low-Energy Marine and Subaerial Carbonates, Bahamas," *Amer. Assn. Pet. Geol. Bull.*, **51**, pp. 1979–2031. (Recognition of subareal, supratidal, intertidal, and subtidal environments in lower Paleozoic rocks.)

SCHOLLE, PETER, 1971, "Diagenesis of Deep-Water Carbonate turbidites, Upper Cretaceous Monte Antola Flysch, Northern Apennines, Italy," *Jour. Sed. Pet.*, **41**, pp. 233–250. (Discussion of carbonate Turbidite with emphasis on diagenesis.)

STOCKMAN, K. W., R. N. GINSBURG, and E. A. SHINN, 1967, "The Production of Lime Mud by Algae in South Florida," *Jour. Sed. Petrology*, **37**, pp. 633–648. (An excellent integrated study using data from several allied sciences to demonstrate the sufficiency of calcareous algae as the source of the lime mud in the south Florida reef tract.)

STORR, J. F., 1964, "Ecology and Oceanography of the Coral-Reef Tract, Abaco Island, Bahamas," *Geol. Soc. Amer. Spec. Paper* **79**, 98 pp. (Excellent study of a modern coral reef environment by a marine biologist.)

TUREKIAN, K. K., 1968, *Oceans*. Englewood Cliffs, N. J.: Prentice-Hall, Inc., 120 pp. (An elementary introduction to oceanography.)

WELLS, J. W., 1957, "Coral Reefs," in *Treatise on Marine Ecology and Paleoecology*, **1**, *Ecology*, *Geol. Soc. Amer. Mem.* **57**, pp. 609–631. (Discussion of the occurrence and variety of coral reefs in modern seas.)

———, 1957, "Corals," in *Treatise on Marine Ecology and Paleoecology*, **1**. *Ecology*, *Geol. Soc. Amer. Mem.* **57**, pp. 1087–1089. (Brief review of coral ecology.)

WEYL, P. K., 1961, "The Carbonate Saturometer," *Jour. Geol.*, **69**, pp. 32–44. (Discussion of a useful instrument utilizing a pH meter that measures whether or not a carbonate mineral is dissolved or a carbonate mineral is precipitated when carbonate material is placed in a given water.)

———, 1967, "The Solution Behavior of Carbonate Materials in Sea Water," *Studies in Tropical Oceanography*, no. 5, University of Miami, Inst. of Marine Sciences, pp. 178–228. (The solution chemistry involved in the origin of ooliths and other carbonate mineral behavior.)

WINCHELL, A. N., 1924, "Petrographic Studies of Limestone Alterations at Bingham," *Amer. Inst. Mining and Metal. Eng. Trans.*, **LXX**, pp. 884–903. (Basically a study of contact metamorphism but containing an analysis of the relative abundances of quartz-calcite mixtures based on petrographic study of more than 1400 thin sections.)

CHAPTER THIRTEEN

DIAGENESIS AND CLASSIFICATION OF LIMESTONES

13.1 INTRODUCTION

In its broadest sense, diagenesis of carbonate sediments and rocks includes all those processes that act on these materials after their initial deposition but before elevated temperatures and pressures create minerals and structures normally considered within the realm of metamorphism. It is generally recognized that physical, chemical, and organic processes begin acting on carbonate sediments immediately following deposition and have tremendous influence in changing the mineralogy, generating sedimentary structures and textures, and, most significantly, making rocks from sediments.

It is important to recognize that the interpretation of the diagenetic history of a carbonate rock may make a significant contribution to our understanding of the environment of deposition of the sediment and the various environments it has been subjected to following deposition. In this sense it is possible to generalize that with few exceptions the nature of the pore water and the rate of pore water movement are the prime determinants in diagenesis. Thus the interpretation of the diagenetic history of a carbonate rock is an interpretation of the pore water history. For example, the large scale dissolution of calcium carbonate material to create vugs cannot take place in shallow seawater that is saturated with respect to the common carbonate minerals. The presence of vugs in a carbonate rock implies that water of markedly different composition, commonly meteoric water, has moved through the rock in large volumes. This in turn implies subaerial exposure in order to provide both access to the fresh water and the hydraulic head needed to move the required volume of this water through the rock. The geologist in deciphering the diagenetic history attempts to reconstruct the time of this subaerial exposure within the context of the total history of the rock.

Our knowledge of carbonate rock diagenesis has been derived from studies of ancient rocks and modern carbonate sediments. In the ancient rocks we see the final product and attempt to learn the diagenetic history, commonly with the aid of petrographic studies and chemical measurements. In modern sediments we see the original mineralogy and structures and use similar techniques. Many significant advances in our understanding of diagenetic processes have resulted from chemical studies of the sedimentary material and the reactions that take place between these materials and natural waters. The complexity in a natural system that occurs when the moving pore water changes its composition by reactions with a wide variety of solids within the rock makes duplication in the laboratory difficult.

13.2 DIAGENETIC PROCESSES

Most of the processes that act on carbonate rock during diagenesis accomplish one or more of the three following operations: *dissolution*, *cementation*, and *replacement*. *Dissolution* produces void space within the rock by removal in solution of previously existing carbonate minerals. *Cementation* is generally considered to be the growth of new crystals into void space, whatever the origin of the void space. *Replacement* involves the essentially simultaneous dissolution of existing minerals and the precipitation of a new mineral.

Exceptions to the generalizations above are found in processes such as the collapse of carbonate material into dissolved voids that is commonly accompanied by local fracturing, internal sedimentation within voids, dessication, and organic disruption. There is some evidence that the composition of carbonate crystals, especially the minor element content, might be changed by solid state leaching. Such a process might cause the increase or decrease in the content of any given element, without a solution-precipitation reaction.

In one study, Pilkey and Goodell (1964) petrographically examined shells of five species of mollusks of Miocene, Pliocene, Pleistocene, and Holocene age from a variety of environments and found no visible recrystallization of the shell structures. Sharp contacts between aragonitic and calcitic layers were preserved and there was no clear evidence of diagenesis visible. They then determined for each shell the contents of magnesium, manganese, strontium, barium, and iron and found significant variations with time for each element in the shell. The abundance of some elements increased; others decreased. In aragonitic shells, magnesium and manganese decreased with time; strontium, barium, and iron increased. In calcitic shells, strontium and magnesium decreased but the amounts of the other elements remained unchanged. It is possible that diffusion of ions had occurred along intercrystalline films of water in response to concentration gradients in the pore solutions but there is also the possibility that delicate replacement had occurred that was not visible using the petrographic microscope. A reexamination of these fossils using the scanning electron microscope might resolve this uncertainty.

Dissolution

In most natural waters high magnesium calcite and aragonite will be more soluble than low magnesium calcite. Shallow seawater generally is saturated with respect to both aragonite and calcite. Given sufficient time, all the calcium carbonate minerals found in modern sediments, with the exception of low magnesium calcite, will be dissolved and/or replaced by low magnesium calcite.

Ancient carbonate rocks commonly contain abundant voids. Many of these may be produced by simple dissolution of limestone that makes void spaces of all sizes. Some may be as small as an individual ring of an oolith; others may be the size of the largest caves (Fig. 13-1). The dissolution may be selective and only the aragonite part of a mollusk may be removed, for example; or nonselective where water moves along joints and enlarges the joints and locally produces enlarged cavities. In the latter case, there appears to be little discrimination by the solution for one part of the rock fabric as opposed to another. In all cases

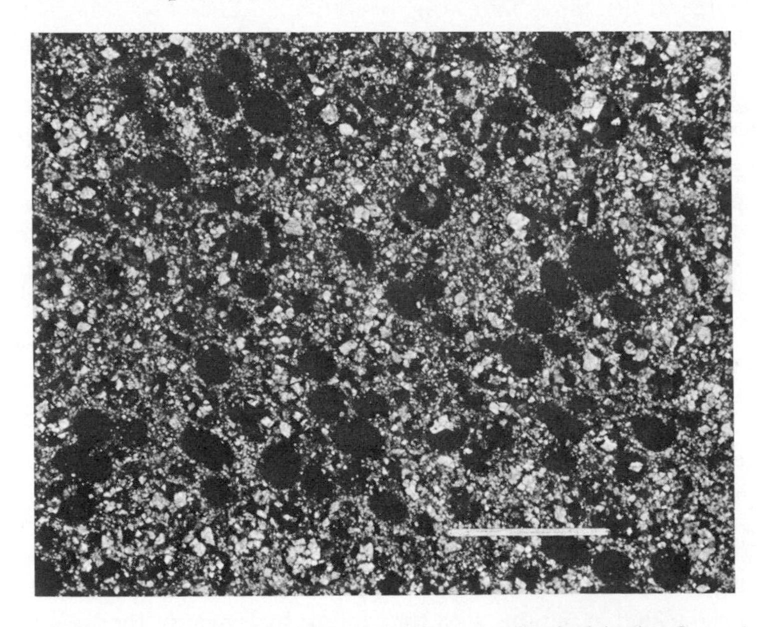

Fig. 13-1 Vugs produced by the dissolution of ooliths in dolomite. Crossed nicols. Scale on photo = 1 mm.

of dissolution, it must be presumed that the pore water was not shallow seawater and was, at the time of the dissolution, undersaturated with respect to the minerals removed. In addition it can also be presumed, because of the relatively low solubility of calcium carbonate in most natural waters, that a very large number of pore volumes of the fluid passed through the rock. The term *pore volume* is used to express the volume of fluid that exists within the rock at any one time. It is equivalent to the term porosity. Because the dissolution can only be accomplished by a large volume of moving water, the required volume is expressed conveniently as the number of times the pore space within the rock is filled with new water. For example, a surface water with a CO_2 content of 0.50% capable of dissolving 1.25 g of $CaCO_3$/1000 g of water would remove from the walls of the pore space in a carbonate rock with 50% porosity sufficient material to increase the porosity to only 50.02%. Thus a volume of water, equivalent to many times the volume of water existing within the rock at any one time, is required in order to accomplish significant dissolution.

The dissolution of the interior of ooliths, both nucleus and many of the precipitated coating rings, is a common phenomenon. Where only the outer rings and cement remain to outline the original spherical particle, it is presumed that the material removed was either more soluble, still aragonite at the time of dissolution, or more porous. The same is true for skeletal particles in limestone removed to create molds. Additional factors may affect selectivity during dissolution of skeletal materials in limestones. (a) What is the character of the enclosing matrix? Is the interskeletal material homogeneously distributed; is it microcrystalline or coarsely crystalline; is it porous, permeable, dolomitic, siliceous, etc.? These differences can be decisive in determining the ease of penetration of groundwaters into the rock. (b) How much and what kind of organic matter is present in the enclosing rock and in the skeletal fragments themselves? The decay of organic matter may locally generate acidic conditions and cause dissolution of carbonate minerals. Three localities have yielded unaltered aragonitic fossils of Paleozoic age and each host rock contains abundant organic matter. One is an organic-rich shale of Mississippian age; the second, a black shale of Pennsylvanian age; the third, an asphaltic limestone of Pennsylvanian age. The ability of organic matter to preserve original shell material results not only from the reduction in permeability it causes but also from the ability of amino acids derived from breakdown of proteins in the shell to keep water molecules from contact with the calcareous material. (c) Is the skeletal material calcitic or aragonitic; and to what degree is each crystal structure distorted by the presence of intracrystalline minor elements? Low magnesium calcite is more stable than aragonite which, in turn, is more stable than high magnesium calcite. (d) What are the size, shape, and orientation of the original crystals forming the skeleton? Small, elongate crystals have large surface/volume ratios and, therefore, are more susceptible to solution. (e) What is the "ultrastructure" of the crystals within the shell? Some are porous; some are not.

Of particular interest to the diagenetic interpretation of carbonate rocks is the feature called *vadose silt* (Fig. 13-2). In many ancient rocks, especially those that were built up into carbonate mounds, the pore space within the rock, both

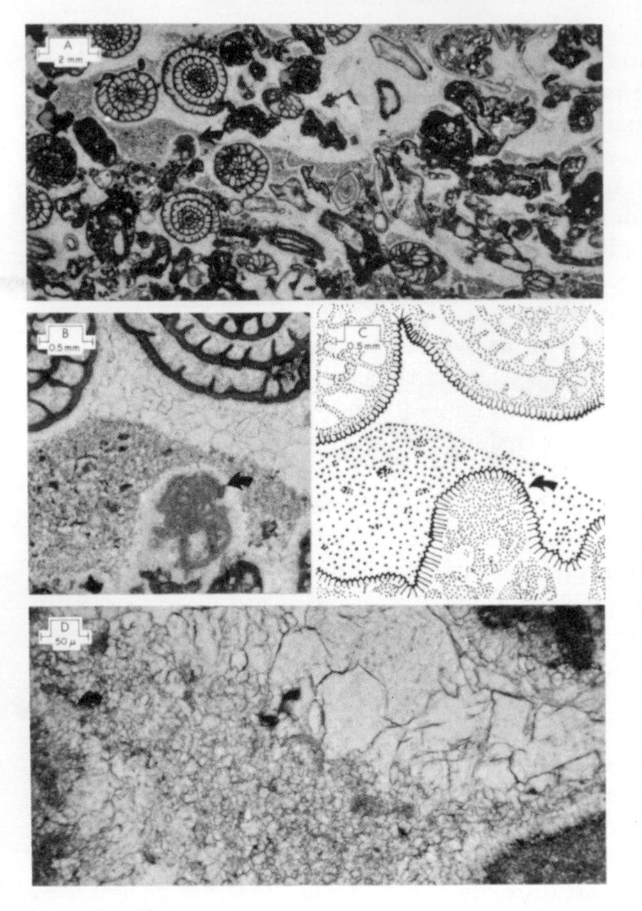

Fig. 13-2 Crystal silt in cement-rimmed interstices. Crystal silt containing bits of unidentifiable skeletal debris lies in the bottom of interstices between fusulinids and other grains. The resulting sedimentary floors are roughly parallel. Beneath the crystal silt is drusy cement, which completely rims the interstices; above is blocky cement. Shell Hilburn 1, 10, 504A, plane light, ×7, ×25, ×25, ×100. This perm plug is unoriented perpendicular to bedding, but adjoining samples confirm the position shown. Symbolism: fine stippling, original sediment; parallel lines, drusy cement; coarse stippling, crystal silt (or pellet silt); white, blocky cement. (From R. J. Dunham, 1969a.)

original voids and those created by dissolution, is partially filled with an internal sediment. This sediment generally consists of silt-size calcite crystals. Larger particles are usually missing and well-preserved microfossils are also absent. Most significant is the observation that the silt was deposited on top of an earlier drusy calcite cement (see Sec. 13.4) and thus postdates this cement. In addition, the presence of the sedimented silt within vugs demands that the deposition of the silt postdate the dissolution of the voids. Because the silt was deposited within the void space of the rock and commonly shows micro cross lamination, it is presumed that it was deposited from a current having a moderately high velocity, probably in excess of 40 cm/sec. Such current velocities are uncommon within rocks that have their pore space completely filled with water. Dunham (1969a), in his study of the Townsend carbonate mound in New Mexico, noted that relatively high velocities are generated where water moves down through the vadose zone on its path from an exposed surface down to the groundwater table. Thus it is presumed that this silt is internal sediment washed down through the vadose zone and accumulated within the pore space of previously existing rock. This conclusion is significant to the interpretation of the diagenetic history of a carbonate rock, especially the recognition of periods of subaerial exposure and the diagenetic events that take place when a carbonate rock is exposed to subaerial conditions.

The carbonate literature contains many descriptions of vugs with flat bottoms and digitate tops, commonly arranged in bands and usually found within mound-like carbonate deposits. They are well developed in the Silurian mounds of the upper midwest in the United States and the Carboniferous mounds found in the Republic of Ireland and also in England. These vugs often exhibit internal sediment of silt size deposited by currents of moderately high velocity and the internal sediment commonly was deposited on top of an earlier drusy calcite cement. Such features have been given the name *stromatactis* (Fig. 13-3) because

Fig. 13-3 Stromatactis from a Carboniferous mound in Ireland. The rock was deposited as a carbonate mud with minor amounts of sand size carbonate particles. Note the flat bottoms and irregular tops.

of the presumption that they were of organic origin or that the void space was created by the dissolution of an organism. In many examples it is impossible to determine the nature of the material removed to produce the void space in which the internal sediment and cement were deposited. In some cases it may have been produced by the removal of skeletal or organic material. In others the vugs were certainly generated by removal of carbonate rock by dissolution.

Cementation

Cementation in carbonate rocks normally refers to the growth of crystals into pore space from carbonate rock surfaces. The new crystal occupies space that formerly was occupied by fluid, commonly water or air plus water where the cementation takes place within the vadose zone. This pore space may be *primary*, i.e., cavities that are a result of the depositional process. They may occur between carbonate grains, within fossil shells, or within the cavities created by organisms such as corals when they build a partially open framework. In addition, much cementation takes place within *secondary* pore space that is produced by dissolution or fracturing of a carbonate rock.

The mineralogy of the carbonate cements naturally depends on the composition of the fluid from which the crystals grow. In general, when the cementing crystals grow from seawater or a similar solution, the mineral aragonite and sometimes high magnesium calcite form. When the cement grows from a freshwater solution, the mineral calcite forms. This is well illustrated by studies of cementation of beach rock and growth of cements in carbonate rocks exposed to rainwater.

Beach rock is carbonate sand found along some beaches within the intertidal zone and sometimes below present low tide. It has been cemented by needle-like crystals of aragonite that have grown from the grain surfaces into pore space between the grains (Fig. 13-4). These crystals interlock locally to bind the grains

Fig. 13-4 Beach rock, Bonaire, N. A. Strombus shells have been cemented into the beach sand. These shells have been broken by man with knives. Thus the cementation is very recent.

together. The discontinuous layers of cemented carbonate beach sand are generally no more than a few feet thick. The aragonite grew from seawater and the many pore volumes of seawater needed for cementation are brought to the beach sand with each rise of the tide and with each splash of the waves. Evaporation of the seawater while in contact with air within the pore space appears to contribute to the precipitation of the cement.

The young age of much of the present-day beach rock is documented on Bonaire in the Netherlands Antilles where beach rock includes shells of the conch *Strombus* firmly cemented into the beach sediment. Many of these shells have small holes made by fishermen in breaking the shell in order to cut the muscle of the conch. The introduction of the emptied shells into the beach sand must postdate the use of steel knives by fishermen and probably took place during this century. Beach rock containing glass bottles has been reported. Analysis of bottle types indicates that cementation can be no older than 10 or 20 years. Carbonate rocks, deposited during relatively high stands of the sea during the Pleistocene and now well above sea level, exhibit the development of calcite cement at the expense of dissolved aragonite. This diagenesis is taking place in fresh rainwater within and immediately below the vadose zone. The Miami Oolite in south Florida and the uplifted coral reefs of Barbados clearly demonstrate this type of calcite growth in fresh water.

Although aragonite has been observed under conditions where it has been shielded from contact with water by an organic coating, low magnesium calcite is the almost universal carbonate mineral in the ancient record. Three types of calcite cement are commonly observed in ancient carbonate rocks. They are *drusy*, *blocky*, and *rim cement* (Fig. 13-5). The term *drusy* or *drusy mosaic* refers to elongate or acicular crystals rimming or filling voids. The crystals have grown with their long axes essentially perpendicular to the void wall. Under a polarizing microscope these crystals show undulatory extinction either because they are distorted or, more commonly, because they have grown perpendicular to a curved surface. *Blocky* cement refers to a mosaic of crystals in which the individuals have roughly the same diameter in all directions. They tend to be granular and to fill void space. Some of them may have resulted from the growth of several drusy crystals at the expense of others and filled the void space with relatively large crystals. *Rim cements* or *syntaxial overgrowths* are large crystals of calcite that have grown in optical continuity with original single crystal grains such as crinoid particles. In the past many authors have observed crinoidal limestones with rim cement and concluded from the large crystal size that the rock had been "recrystallized," forgetting that the bulk of the rock was composed of grains that were themselves single crystals. Stages in the growth of rim cements can be dramatically illustrated by using iron-sensitive staining techniques as reported by Evamy and Shearman (1965). All three types of cement grow, meet other crystals, and form crystal boundaries that do not conform to any particular crystallographic plane and are termed *compromise boundaries.*

Pisoliths are concentrically laminated particles of calcium carbonate ranging from 1 to 10 mm in diameter (Fig 13-6). They generally have been considered to be sedimentary particles with an origin similar to that of the algal oncolites (see Chap. 12). For example, in the Permian Reef Complex of Texas and New Mexico (Dunham, 1969b), the strata containing pisoliths generally have been considered to be depositional facies. However, many of the strata of pisolitic rock exhibit inverse graded bedding but fail to show evidence of current transport such as cross-bedding and other carbonate grains are usually absent. Unlike ooliths, many pisoliths have no foreign nucleus. Evidence has been accumulated that shows that pisoliths have formed within carbonate rocks as a diagenetic structure. Most significant is the fact that pisoliths commonly fit together with polygonal boundaries that would not be possible if the grains were brought together by sedimentation. Moreover, sedimentary particle inclusions are found within the pisoliths on the upper side of each individual pisolith. This would be impossible if the grains were transported sedimentary particles. Thus, pisoliths are not detrital particles but are structures formed in

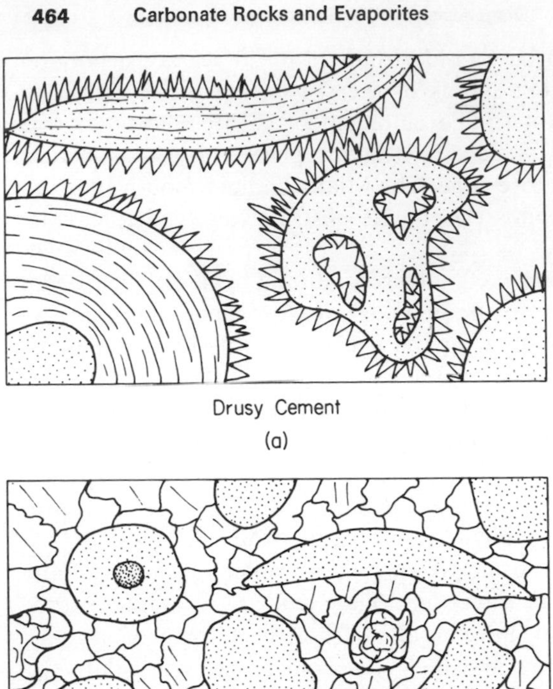

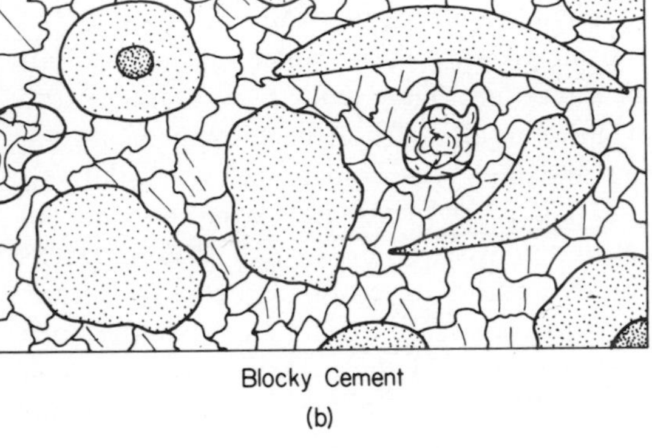

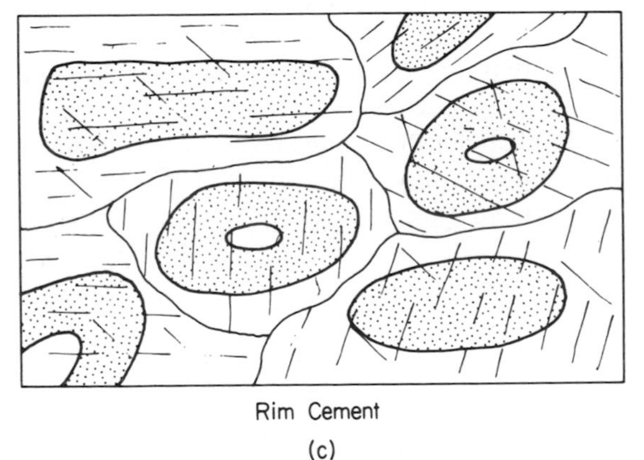

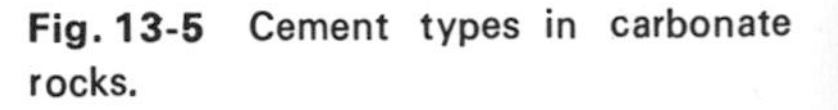
Fig. 13-5 Cement types in carbonate rocks.

place by a diagenetic process. They appear to be concretionary in origin and are similar to structures being formed in present-day caliche (Chap. 7). Pisoliths have been reported from the supratidal carbonate sediment and rock on Bonaire in the Netherlands Antilles. There is some evidence that these pisoliths have been formed, eroded, and redeposited in the supratidal sediment. When pisoliths are viewed as

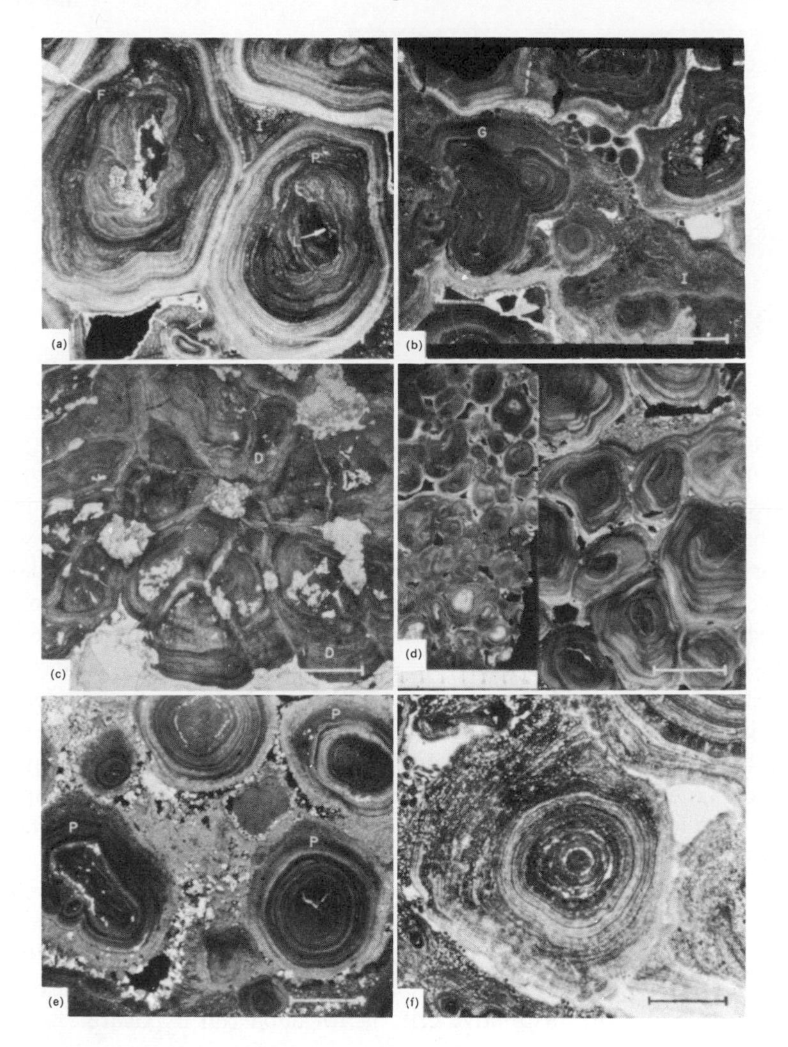

Fig. 13-6 (a) Pisolith coated with laminated carbonate, then leached (L), then recoated. Note fracturing (F), internal sediment (I), and perched inclusions (P). Scale is 5 mm. (b) Laminae of internal sediment (I) and cement between pisoliths. Note gradation (G) of these laminae into those composing pisoliths. Scale is 5 mm. (c) Fitted polygonal structure and downward elongation (D). Scale is 5 mm. (d) Fitted polygonal structure and downward elongation in slab and thin section. Scale on this section is 5 mm. (e) Perched inclusions (P) inside a group of pisoliths. Note that the laminae containing inclusions in their upper parts are free of inclusions in their lower parts. Scale is 5 mm. (f) Closer view of perched or lopsided inclusions inside pisolith. Scale is 2 mm. (From R. J. Dunham, 1969b.)

vadose concretionary features rather than as depositional particles, our understanding of the history of the strata that contain them is changed.

Subaerial-laminated crusts represent a distinctive diagenetic structure common in modern and ancient rocks (Fig. 13-7). These crusts reach thicknesses of 1 to

Fig. 13-7 Laminated crust developed on exposed coral reef rock, Bonaire, N. A. The crust has a maximum thickness of 2 inches.

10 cm in subaerially exposed Pleistocene carbonate rocks, such as the Miami Oolite and Key Largo Formation of south Florida, and coral reefs exposed on the five-meter terrace on Bonaire in the Netherlands Antilles. They conform to the eroded carbonate rock surface, even following that surface down into sink holes. The upper exposed surface of the crust is commonly smooth, while the lower surface, against the original limestone, is bulbous or mamillary. The millimeter-scale laminations of the crusts are produced by layering of stained crystals with diffuse boundaries. Examination of the crusts reveals that they contain not only material from the underlying rock but foreign grains similar to those found in the overlying soil. Thus, much of the crust material was added to the underlying carbonate rock surface by cementation of the lower part of the overlying soil. For this reason, they have sometimes been termed *soilstone*. It is not uncommon to find the vague outlines of original grains of the underlying carbonate apparently in place in the lower part of the crust, which suggests that part of the crust represents solution alteration and cementation accompanied by laminar-like staining of the top of the underlying limestone. Because of the appearance of these features developed on subaerially exposed carbonate rocks today and the relation to the soil-forming process, it has been suggested (Multer and Hoffmeister, 1968) that they are indicative of subaerial vadose zone diagenesis. These crusts are common in ancient rocks and are distinctive criteria for subaerial exposure.

Replacement

Carbonate petrologists have long recognized that primary carbonate rock particles such as ooliths, structures such as stromatolites, and unstable minerals such as aragonite that were originally in a carbonate sediment are represented in an ancient rock by a different mineralogy or different crystal size. This follows from the recognition of ghosts of the primary material, represented by inclusions, pseudomorphs, or reproduction of an original outline. It is obvious, in most of these, that the original material could not have simply been dissolved away and the new phase precipitated as cement. In order to preserve the evidence of the original material, the growth of the new mineral must have taken place essentially simultaneously with the step-by-step dissolution of the earlier material. This implies that the water must be in a delicate balance so that it will dissolve one phase and precipitate another at the same time. The excess dissolution at any one time cannot produce a void sufficiently large that the relics of the original material are obliterated. The replacing phase may be found in space occupied by original carbonate minerals and the pore space between those original minerals. For further discussion see Chap. 14.

Modern carbonate muds are deposited almost entirely as aragonite or high magnesium calcite. The relative amounts depend, for the most part, on the types of organisms contributing their hard parts to the sediment. Calcareous algae are the dominant organic contributors of aragonite needles. Presumably ancient carbonate muds were formed similarly but the matrix carbonate in ancient limestones is not aragonite needles 1 to 4 μ in length. It consists of subequant blocks of low magnesium calcite 1 to 3 μ in diameter. Electron micrographs indicate that the average aragonite needle is about $0.3 \times 0.3 \times 3.0\ \mu$, a volume of 0.3 cu μ and a surface area of 3.8 sq μ. The calcite blocks, on the other hand, have a volume of about 8 cu μ and a surface area of 24 sq μ. Therefore, the ratio of surface to volume has decreased from nearly 13 in the original aragonite crystals to 3 in the recrystallized calcitic product. Between twenty-five and thirty needles have been dissolved to produce a single calcite crystal, which reduces the free energy of the aggregate because metastable aragonite has been replaced by stable low magnesium calcite; and the part of the free energy due to surface energy is reduced by decreasing the surface area of the aggregate by more than three-fourths. Replacement involves the two processes (a) dissolution of original minerals and (b) growth of the replacing mineral in the small dissolved space plus continued growth of crystals as cement in preexisting void space.

13.3 MICROSPAR

The crystals of calcite cement that occupy the pores of ancient limestones are nearly always more than 20 to 30 μ in size. Thus, they further emphasize the bimodal distribution of crystal and grain sizes noted earlier. In some rocks, however, carbonate crystals and/or clastic grains of 5 to 20 μ size are abundant. For example, parts of the famous Solenhofen lithographic limestone of Bavaria are composed entirely of such crystals, usually termed *microspar*. Considerable disagreement exists concerning the origin of microspar and for specific rocks there may be no way to be certain whether the microspar crystals are detrital carbonate silt or calcite crystals that have grown by a combination of pore-filling cement and replacement of adjacent crystals. Limestones composed largely of microspar are often called calcisiltites, a term that may or may not imply detrital origin, depending on the personal definition of the user.

Folk (1959, 1965) considers that most microspar is replaced micrite and offers the following lines of evidence to support his contention: (a) In many microsparite rocks, the microspar occurs as irregular patches grading by continual decrease of grain size into areas of normal micrite. In some cases the microspar occurs as a halo around fossil fragments or other coarse clastic carbonate particles. Occasionally the microspar crystals seem to be oriented with long dimensions normal to the surfaces of the coarse clastic grains. (b) Limestones formed of microspar generally lack lamination or cross-bedding or graded bedding, structures common in transported carbonate rocks. Further, the extremely uniform crystal size in such rocks is beyond the sorting capabilities of all transporting agents. (c) In thin sections microspar is often seen to transect depositional structures and grains. For example, it cuts across fecal pellets, retaining the organic matter of the pellets although the crystal size is greatly coarsened. (d) The insoluble residue of microspar limestones usually consists largely of very fine grained silicate material hydraulically equivalent to aragonite mud.

Some limestones are constructed of a few sand-size or larger detrital grains "floating" in a coarse sparry calcite mosaic. The very large amount of sparry calcite clearly precludes its origin as pore-filling cement and an origin involving replacement of micrite seems to be required. The key concept here is grain support. If it is lacking, and the remainder of the rock is formed of crystals coarser than 3 to 4 μ, recrystallization of original clay-sized carbonate usually can be inferred. It must always be kept in mind, however, that detrital calcisiltites do exist and in some areas are abundant. Every effort must be made to determine the origin of "intergranular" material before an origin by chemical processes is concluded.

13.4 SILICIFICATION

The chemical conditions favorable for the simultaneous solution of calcite and precipitation of silica as chalcedony, chert, or macrocrystalline quartz, are discussed in Chaps. 10 and 16. The process is favored by relatively low pH, low temperature, and, of course, saturation of the pore solution with silica. The silicification reaction may be represented symbolically as

$$CaCO_3 + H_2O + CO_2 + H_4SiO_4 \rightleftharpoons SiO_2 + Ca^{2+} + 2\,HCO_3^- + 2\,H_2O$$

although it probably is more complex than this. The filling of voids produced by solution of carbonate crystals is not difficult to visualize and the abundance of structureless chert nodules in bedded limestones suggests that it is a common process. More difficult to understand is the preservation of delicate organic laminations in carbonate skeletons and general pseudomorphing of shells by microcrystalline quartz crystals. Such chertification is common in siliceous limestones, despite the fact that very little similarity exists between the two minerals. Chertified carbonate shells are not deformed in any way evident to petrographic observations. Perhaps useful insights might be gained by electron microscopic study of these fossils.

13.5 DIAGENETIC ENVIRONMENTS

The three general environments are (a) the subsea on or just below the sea floor, (b) subaerial exposure in the vadose or shallow groundwater zone, and (c) the deep subsurface (Fig. 13-8).

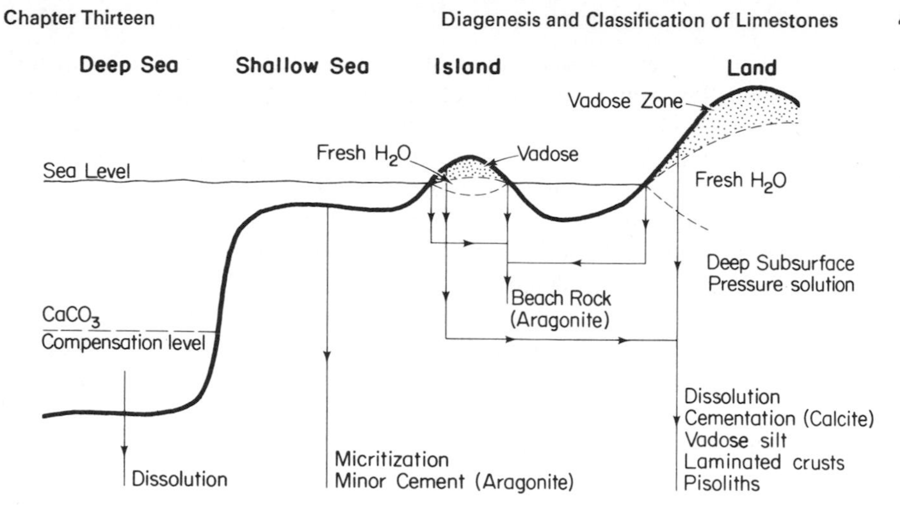

Fig. 13-8 Environments of calcium carbonate diagenesis.

Carbonate diagenesis in the subsea appears limited to relatively few processes. This apparent limitation may reflect the relatively little attention these processes have received. Most workers, however, are impressed with the small amount of diagenesis that has taken place in carbonate sediments that have remained under the sea. Because diagenesis is a response of carbonate minerals to the water within the pore space, it is logical that minerals formed within a given water will undergo relatively little change until such time as the composition of the water changes. This logic has certain limitations. For example, many of the carbonate minerals produced by organisms are aragonite or high magnesium calcite. These relatively unstable minerals should be replaced by low magnesium calcite without changing the water or introducing many pore volumes of the same water.

In general, modern marine carbonate sediments, most of which are less than 5000 years old, have not been lithified in a marine environment. The thickness of modern marine carbonate sediment deposited since the recent rise in sea level can be determined in most places with a pointed steel rod. The rod easily penetrates the uncemented sediment and stops at the cemented Pleistocene rock that was exposed to subaerial conditions during the low stands of the sea during Wisconsin time. The boundary between modern marine carbonate sediments and the cemented Pleistocene rock can easily be mapped using sparker or shallow seismic techniques. Therefore any mineral transformations that take place in the marine environment must be very slow.

The formation of microcrystalline fabric in skeletal grains and ooliths discussed under "Pellets and Pelletoids" on p. 422 certainly occurs. There appears, although the data are limited, to be little change in mineralogy, however, when this *micritization* process takes place.

Dissolution of carbonate material below the compensation level in the deep sea is a common phenomenon (see Chap. 12). The growth of minor amounts of drusy aragonite and high magnesium calcite cement in the shallow subsea is most interesting. This phenomenon is generally restricted to areas below tidal flats, coral reefs, and very shallow water, although cementation in the deep sea has been reported (Fischer and Garrison, 1967). Shallow water submarine cementation has been documented along the south shore of the Persian Gulf (Shinn, 1969) and in the Caribbean. Because the sediments on the bottom of the sea are filled with water of the same density as seawater, there is no common mechanism for moving large volumes of fluid down through the sediments. Study of cemented carbonate rock in the present shallow seas is complicated by the fact that we often see on the sea floor relict rock patches that were exposed during low stands of the sea during Wisconsin time. In many cases these have not been covered by modern sediments and often the age of the rock is not easily determined. Because the cementing crystals are small, it is almost impossible to remove them and obtain C^{14} age dates for the time of cement growth.

During subaerial exposure, there is ample opportunity for change of fluid and the movement of fluid. Examples range from beach rock formed by seawater constantly moving in and out of the periodically exposed beach to dissolution by rainwater occurring in an ancient carbonate rock exposed on a mountainside. Most interesting are those processes such as pisolith formation, dissolution, growth of calcite cement, introduction of vadose silt, and the formation of subaerial laminated crusts that commonly take place just after the rock has been exposed to the air. Considerable diagenesis takes place on carbonate islands with their vadose zone, freshwater lens, and proximity to the sea.

In the deep subsurface environment, the main diagenetic process affecting carbonate rocks appears to be pressure solution, which causes dissolution and cementation of grains as well as formation of stylolites. The lack of other large-scale diagenetic changes in carbonate rocks in this environment is related again to the difficulty of changing the composition of the water and the lack of processes for causing large-scale movement of fluid. Some calcite is precipitated as a result of changes in the solubility of this mineral with changes in pressure and temperature. Water is moved through carbonate rocks, after being expelled by compaction. The changes and volumes involved, however, are relatively minor compared with the diagenetic potential of the subaerial environment.

An ancient rock that began as a sediment with an average of 50% or more porosity may have less than 10% porosity today. This required either the introduction of calcium carbonate approximately equal to the original solid volume of the rock or a loss of one-half of the original volume of the rock by compaction. There is abundant evidence that most carbonate rocks have undergone relatively little compaction. Thus, the introduction of calcium carbonate from an outside source is required. The source of this calcium carbonate and its means of transportation and deposition within the rock is one of the main problems of limestone diagenesis. It appears that only subaerial exposure offers the potential for dissolving, transporting, and precipitating this calcium carbonate on a large scale. Like many ideas, however, this one may undergo significant change as we see more study of the diagenetic processes, especially the interpretation of carbonate cements.

13.6 CHEMICAL COMPOSITION OF ANCIENT LIMESTONES

Many analyses have been made of carbonate rocks to determine the types and amounts of minor and trace elements they contain but these data are difficult to evaluate in terms of the diagenesis of the rock. Nearly all the analyses are of bulk samples and, therefore, there is no certain way to determine the proportion of the minor element content that results from inclusions in the carbonate rock rather than from inclusion in the calcite crystal structure. For example, chlorine in ancient carbonates averages about 200 ppm, undoubtedly located in water bubble inclusions in the rocks. Boron averages 100 ppm, probably as ions adsorbed on clay minerals in limestones; manganese approximates 500 ppm, partly adsorbed on organic matter and partly as ions substituting for magnesium or calcium. And, of course, most of the magnesium in ancient limestones probably exists in dolomite rather than in metastable high magnesium calcites. For example, the magnesium content of an ancient limestone can vary by an order of magnitude because of dolomite amounts that are not easily detectable by normal X-ray diffraction analysis. Strontium averages about 500 ppm and is probably present within the crystal structure of calcium carbonate minerals because celestite and strontianite are rare in limestones. In summary, the amounts of most of the minor elements in limestones are extremely variable among samples and the differences are due largely to liquid or solid inclusions among the calcite or aragonite crystals rather than to structural or interstitial substitution of the minor element for calcium ions.

A decrease in strontium content of limestones with increasing age has been noted by numerous workers and it is generally agreed that the decrease results from conversion of aragonite to calcite during diagenesis. The crystal structure of aragonite is much more receptive to large cations such as strontium than is calcite. The conversion from aragonite to calcite is generally a solution-precipitation reaction that is greatly accelerated by movement of relatively fresh waters through intergranular voids in the carbonate sediment. Therefore, it is to be anticipated that most changes will occur very early in the history of the sediment, before deep burial. But, in either event, the content of strontium in the diagenetic calcite must be low because the partition coefficient for coprecipitation of strontium with calcite at 25°C is only 0.14 and decreases to 0.08 at 100°C. For aragonite the analogous values are 1.1 and 0.8. The capacity of calcite to accommodate strontium in its crystal structure is thus about 10% that of aragonite so that calcitic limestone should contain no more than about 1000 ppm strontium, compared to nearly 10,000 ppm in aragonitic limestone. The rapidity of the change was well illustrated by Stehli and Hower (1961), who showed that modern carbonate sediments in south Florida contain an average of 71% aragonite, 19% high magnesium calcite, and 11% low magnesium calcite, based on forty samples ranging in origin from coarse reef debris to lime muds. The Pleistocene analogues of these modern sediments average 21% aragonite, 2% high magnesium calcite, and 77% low magnesium calcite, based on nineteen samples.

13.7 CLASSIFICATION OF THE CARBONATE ROCKS

Prior to the publication by Folk in 1959 of his carbonate rock classification, most classifications and terms in common use relied on words such as "recrystallized," or specified the size of crystals or noted the presence of easily seen fossils as "coquina." At best, the terms Grabau published in 1904 and 1913, "calcilutite;" "calcarenite," and "calcirudite" for carbonate rocks in the clay, sand, and gravel size ranges of carbonate particles were the common descriptive words applied by most workers. Folk's classification initiated a revolution in our way of looking at and thinking about these rocks and, together with the other classifications that followed his, has set the stage for numerous studies that recognize that the carbonate rocks are generally bimodal in size distribution. That is, they are composed of grains of calcium carbonate, usually of coarse silt or sand size and carbonate mud, generally of clay size.

Folk recognized that most carbonate rocks are made up of three components: (a) discrete carbonate aggregates or *allochems*. These include particles such as intraclasts, ooliths, skeletal grains, and pellets. (b) Microcrystalline calcite ooze, *micrite*, or carbonate mud, and (c) *sparry calcite* that is normally a chemically precipitated, pore-filling cement. The first part of the rock name is contributed by the allochem and the second word by the micrite or sparry calcite. For example, a rock is termed *oosparite* for a sparry calcite-cemented oolite and *pelmicrite* for a rock composed of pellets and lime mud. Figure 13-9 illustrates some of the variations used in this classification. The simple rock name may be modified by using appropriate descriptive adjectives such as dolomitic crinoidal biomicrite. Framework reefs are termed *biolithite*.

	OVER 2/3 LIME MUD MATRIX				SUBEQUAL	OVER 2/3 SPAR CEMENT		
Percent Allochems	0–1%	1–10%	10–50%	OVER 50%	SPAR AND LIME MUD	SORTING POOR	SORTING GOOD	ROUNDED AND ABRADED
Representative Rock Terms	Micrite and Dismicrite	Fossili ferous Micrite	Sparse Biomicrite	Packed Biomicrite	Poorly Washed Biosparite	Unsorted Biosparite	Sorted Biosparite	Rounded Biosparite
1959 Terminology	Micrite and Dismicrite	Fossiliferous Micrite	Biomicrite			Biosparite		
Terrigenous Analogues	Claystone		Sandy Claystone	Clayey or Immature Sandstone		Submature Sandstone	Mature Sandstone	Supermature Sandstone

Lime mud matrix   Sparry calcite cement

CARBONATE TEXTURAL SPECTRUM

Fig. 13-9 Folk's classification of carbonate rocks.

TABLE 13-1 DUNHAM'S CLASSIFICATION ACCORDING TO DEPOSITIONAL TEXTURE†

<table>
<tr><th colspan="5">Depositional texture recognizable</th><th>Depositional texture not recognizable</th></tr>
<tr><td colspan="4">Original components not bound together during deposition</td><td rowspan="4">Original components were bound together during deposition. . . as shown by intergrown skeletal matter, lamination contrary to gravity, or sediment-floored cavities that are roofed over by organic or questionably organic matter and are too large to be interstices.</td><td rowspan="4">Subdivide according to classifications designed to bear on physical texture or diagenesis.</td></tr>
<tr><td colspan="3">Contains mud (Particles of clay and fine silt size)</td><td>Lacks mud</td></tr>
<tr><td colspan="2">Mud supported</td><td colspan="2">Grain supported</td></tr>
<tr><td>Less than 10% grains</td><td>More than 10% grains</td><td></td><td></td></tr>
<tr><td>Mudstone</td><td>Wackestone</td><td>Packstone</td><td>Grainstone</td><td>Boundstone</td><td>Crystalline carbonate</td></tr>
</table>

†After Dunham, 1962.

The reader should consult *Classification of Carbonate Rocks*, published in 1962 by the American Association of Petroleum Geologists and edited by W. E. Ham. The introduction written by Ham and L. C. Pray discusses not only modern classifications but offers many significant ideas on the reasons for classification and the types of classifications.

Another classification with a most useful concept was published by Dunham in the Ham volume (Table 13-1). Dunham aimed his classification at determining the original depositional texture and makes a significant distinction between those carbonate rocks in which the particles or grains are touching and thus support the rock and those in which the grains are floating within a matrix of carbonate mud. This boundary cannot be simply a function of the ratio of grains to mud because of the many and varied shapes found among carbonate grains. For example, in a rock composed of ooliths, which are generally spherical, grain support may be achieved with about 62% grains. Whereas in a carbonate rock containing abundant platey algae or thin-shelled brachiopods, grain support may be obtained with only 20 to 25% grains. Grains are defined as particles greater than 20 μ. Mud is composed of crystals less than 20 μ and usually much smaller.

Most of the effect of these modern classifications has been positive and has served the dual purpose of organizing collections of carbonate rocks into meaningful packages and causing workers to examine their rocks with care in order to recognize the criteria for classification. As with all descriptive classifications, however, there has been abuse. There are too many papers that simply conclude that the carbonate rocks of a given study can be arranged according to the classification of one or more authors. All too often these workers have failed to recognize that a classification is simply a tool for organizing information, not the source of a conclusion. In attempting to determine the origin of a carbonate rock, a ready-made classification may distract the person studying the rocks from his goal by giving him a false sense of accomplishment. It is for this reason that many workers prefer to go directly to genetic classifications of their own making. They seek out the best questions they can phrase about the origin of the rock and apply the best available criteria or devise new criteria from their own studies. In this way progress is made in our overall understanding of carbonate rocks and the role they play in contributing to our knowledge of earth history. For even if the genetic conclusions ultimately prove to be incorrect, they serve to stimulate other workers to seek better answers.

REFERENCES

ALLEN, R. C., E. GAVISH, G. M. FRIEDMAN, and J. E. SANDERS, 1969, "Aragonite-Cemented Sandstones from Outer Continental Shelf off Delaware Bay: Submarine Lithification Mechanism Yields Product Resembling Beachrock," *Jour. Sed. Petrology*, **39**, pp. 136–149. (Detailed study of rock sample dredged from continental shelf to determine whether cementation was subaerial or submarine.)

BANNER, F. T., and G. V. WOOD, 1964, "Recrystallization in Microfossiliferous Limestones," *Geol Jour.*, **4**, pp. 21–34. (Detailed petrographic study demonstrating selective recrystallization of skeletal carbonate fragments.)

BATHURST, R. G. C., 1958, "Diagenetic Fabrics in Some British Dinantian Limestones," *The Liverpool and Manchester Geol. Jour.*, **2**, pp. 11–36. (A discussion with criteria for the types of carbonate cement.)

BEALES, F. W., 1965, "Diagenesis in Pelleted Limestones," in *Dolomitization and Limestone Diagenesis*, Soc. Econ. Paleon. Miner. Spec. Pub. No. 13, pp. 49–70. (Excellent discussion of the origin and lithification of these interesting rocks.)

CHAVE, K. E., K. S. DEFFEYES, R. M. GARRELS, M. E. THOMPSON, and P. K. WEYL, 1962, "Observations on the Solubility of Skeletal Carbonates in Aqueous Solutions," *Science*, **137**, pp. 33–34. (Measurements of the solubility of carbonate minerals using the saturometer.)

DUNHAM, R. J., 1962, "Classification of Carbonate Rocks According to Depositional Texture," in *Amer. Assn. Pet. Geol. Mem. No. 1*, pp. 108–121. (An original and widely used classification scheme.)

———, 1969a, "Early Vadose Silt in Townsend Mound (Reef), New Mexico," in *Depositional Environments in Carbonate Rocks*, G. M. FRIEDMAN, ed., Soc. Econ. Paleon. & Miner. Spec. Pub. No. 14, pp. 139–181. (Detailed petrographic studies indicate abundant internal detrital sedimentation in the vadose groundwater zone during the formation of this Permian carbonate island.)

———, 1969b, "Vadose Pisolite in the Capitan Reef (Permian), New Mexico and Texas," in *Depositional Environments in Carbonate Rocks*, G. M. FRIEDMAN, ed., Soc. Econ. Paleon. & Miner. Spec. Pub. No. 14, pp. 182–191. (Discussion of the vadose origin of Pisolite.)

EVAMY, B. D., and D. J. SHEARMAN, 1965, "The Development of Overgrowths from Echinoderm Fragments," *Sedimentology*, **5**, pp. 211–233. (The combination of staining techniques and thin section petrography may reveal significant details of the process of rim cementation. Excellent photomicrographs.)

FISCHER, A. G., and R. E. GARRISON, 1967, "Carbonate Lithification on the Sea Floor," *Jour. Geol.*, **75**, pp. 488–496. (Survey of reported occurrences and description of new examples.)

FOLK, R. L., 1959, "Practical Petrographic Classification of Limestones," *Amer. Assn. Pet. Geol. Bull.*, **43**, pp. 1–38. (Classic paper on the origin of grains, matrix, and cement in limestones and their use in limestone classification.)

———, 1962, "Spectral Subdivision of Limestone Types," in *Amer. Assn. Pet. Geol. Mem. No. 1*, pp. 62–84. (Additional thoughts and an expansion by Folk of his 1959 carbonate classification.)

———, 1965, "Some Aspects of Recrystallization in Ancient Limestones," in *Dolomitization and Limestone Diagenesis*. Soc. Econ. Paleon. & Miner. Spec. Pub. No. 13, pp. 14–48. (Extensive discussion of the problems associated with the recrystallization of limestones.)

FRIEDMAN, G. M., 1964, "Early Diagenesis and Lithification in Carbonate Sediments," *Jour. Sed. Petrology*, **34**, pp. 777–813. (Mineralogic and textural changes in carbonate materials during diagenesis.)

GROSS, M. G., 1964, "Variations in the O^{18}/O^{16} and C^{13}/C^{12} Ratios of Diagenetically Altered Limestones in the Bermuda Islands," *Jour. Geol.*, **72**, pp. 170–194. (Documentation of diagenesis and cementation of limestones by calcite precipitated from fresh waters. Classic example of the usefulness of stable isotope ratios in the solution of diagenetic enigmas.)

LAND, L. S., 1967, "Diagenesis of Skeletal Carbonates," *Jour. Sed. Petrology*, **37**, pp. 914–930. (Discussion of the changes in mineralogy as carbonate rocks undergo diagenesis.)

MACKENZIE, F. T., L. S. LAND, R. N. GINSBURG, and O. P. BRICKER, (eds.), 1969, *Carbonate Cements*, St. George's West, Bermuda, Bermuda Biol. Sta. for Res., Spec. Pub. No. 3, 325 p. (Collection of 62 abstracts or papers with abundant photomicrographs of both modern and ancient cementation phenomena in carbonate rocks.)

MULTER, H. G., and J. E. HOFFMEISTER, 1968, "Subareal Laminated Crusts of the Florida Keys," *Bull. Geol. Soc. Amer.*, **79**, pp. 183–192. (An examination of the soil-related processes that are responsible for these features.)

PILKEY, O. H., and H. G. GOODELL, 1964, "Comparison of the Composition of Fossil and Recent Mollusk Shells," *Geol. Soc. Amer. Bull.*, **75**, pp. 217–228. (Examination of unrecrystallized mollusk shells of Late Tertiary to Recent age reveals significant differences in content of metallic cations with age. The differences are attributed to effects of weathering and diagenesis.)

PRAY, L. C. and P. W. CHOQUETTE, 1970, "Geologic Nomenclature and Classification of Porosity in Sedimentary Carbonates," *Bull. Amer. Assoc. Petroleum Geol.* **54**, pp. 207–250. (An outstanding summary of porosity in carbonate rocks.)

SHINN, E. A., 1969, "Submarine Lithification of Holocene Carbonate Sediments in the Persian Gulf," *Sedimentology*, **12**, pp. 109–144. (A well-documented example of submarine cementation in carbonate sediment.)

SIPPEL, R. F. and E. D. GLOVER, 1965, "Structures in Carbonate Rocks Made Visible by Luminescence Petrography," *Science*, **150**, pp. 1283–1287. (Textural features not visible in carbonate rocks using normal thin section techniques can be studied using this technique.)

STEHLI, F. G., and J. HOWER, 1961, "Mineralogy and Early Diagenesis of Carbonate Sediments," *Jour. Sed. Petrology*, **31**, pp. 358–371. (The relative stability of carbonate materials in near-surface conditions is low magnesium calcite, aragonite, and high magnesium calcite.)

WEYL, P. K., 1967, "The Solution Behavior of Carbonate Materials in Sea Water," in *Studies in Tropical Oceanography*, no. 5, Univ. Miami, Florida, pp. 178–228. (Series of experiments demonstrating that carbonate minerals in seawater do not behave as homogeneous thermodynamic phases.)

CHAPTER FOURTEEN

DOLOMITE

14.1 INTRODUCTION

The term dolomite refers both to the mineral species dolomite, $CaMg(CO_3)_2$, and to the carbonate rock that contains more than 50% of the mineral. The term dolostone has sometimes been applied to the rock composed of the mineral dolomite. The use of the same word, however, for the mineral and the rock has been used widely for many years without causing much confusion, and the term dolomite will probably continue to be used for both.

The magnesium, calcium, and carbonate ions within the ideal dolomite crystal structure exhibit an ordered three-layered arrangement with planes populated entirely by magnesium ions alternating with planes of carbonate ions and planes of calcium ions in the sequence $Mg\text{-}CO_3\text{-}Ca\text{-}CO_3$. The resulting oxide analysis is 21.9% MgO, 30.4% CaO, and 47.7% CO_2 by weight.

Natural crystals, however, often depart significantly from this ideal composition. In many modern dolomites, and in some ancient ones ranging in age to at least the Devonian Period, the dolomite is not stoichiometric but ranges in composition from approximately 56 mole % calcium, and 44 mole % magnesium to the ideal 50 % value. The larger calcium ions cause a widening of the lattice spacings of certain planes in the dolomite crystal structure and thus a shift in the position of certain X-ray diffraction peaks. The shift in the position of the 2.88 A X-ray peak for dolomite results from the substitution of some calcium ions in layers of magnesium ions and is measurable with a precision of 0.02 mole % magnesium when sodium chloride is used as an internal standard (Fig. 14-1). Because this method is independent of the amount of calcite in the rock, it can be made on bulk samples of carbonate rock containing as little as 5 % dolomite.

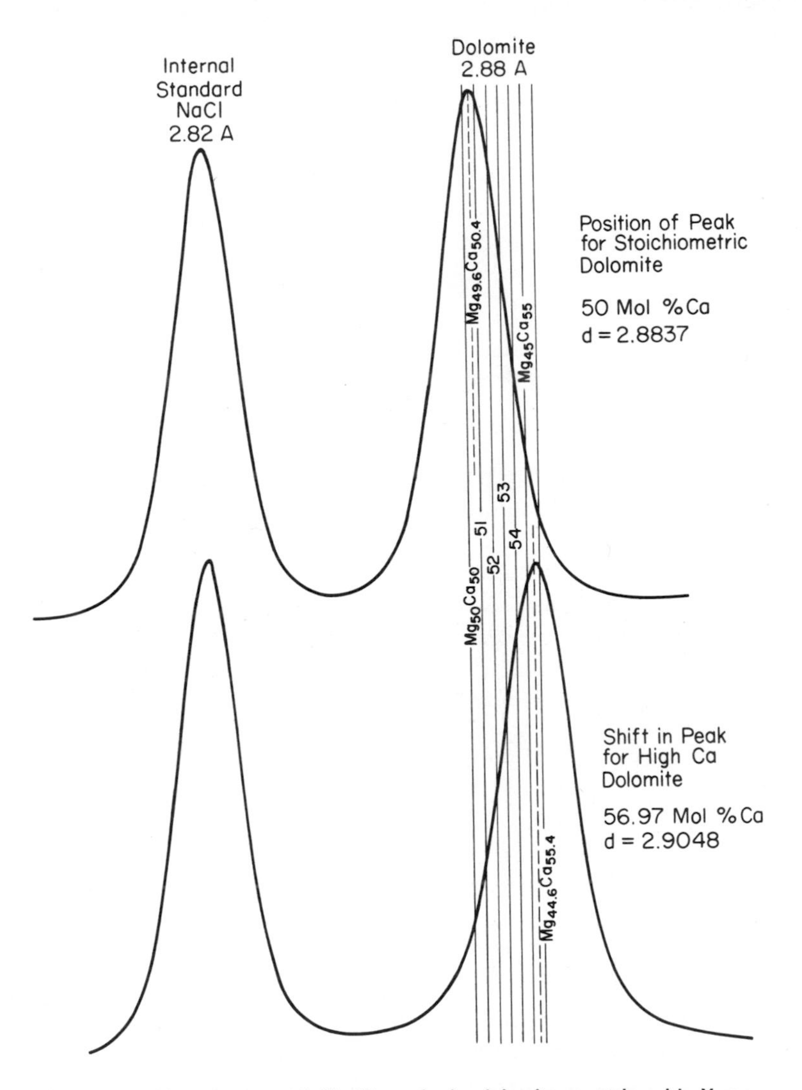

Fig. 14-1 Measurement of Mg/Ca ratio in dolomite crystals with X-ray diffraction.

A second cause of departure from the ideal composition in dolomite crystals is isomorphous substitution of other divalent ions for calcium or magnesium in the structure. The most common and abundant substitute is ferrous iron for magnesium, and a dolomite containing subequal amounts of magnesium and ferrous iron is called ankerite—$Ca(Mg, Fe)(CO_3)_2$. The compound $CaFe(CO_3)_2$ is not known to occur naturally.

Dolomite occurs in rocks ranging in age from Late Precambrian to Holocene. Compilation of nearly 9000 analyses of carbonate rocks from the Russian Platform and North America indicates that the calcite/dolomite ratio decreases from nearly infinity in Quaternary rocks to 80:1 in Cretaceous rocks to 3:1 in Early Paleozoic rocks to 1:3 in Precambrian rocks. This may in part reflect the increased opportunity with time for a carbonate rock to become dolomitized. It could, however, reflect a greater prevalence of the depositional and diagenetic environments conducive to the formation of dolomite in the Paleozoic and Late Precambrian as contrasted to later times. Available data are inadequate to distinguish between these two alternatives.

14.2 PRIMARY VERSUS SECONDARY DOLOMITE

There has long existed a controversy that is centered around the question of whether any of the dolomite that we see in the ancient record was formed by precipitation of the mineral from bodies of water, followed by settling of the grains and sedimentation, or in some other way. An accumulation of such a sediment would be termed a *primary dolomite*. The problem can be viewed in two ways: (a) Can such a process take place? (b) How would we recognize such a rock? Several lines of evidence are used commonly to demonstrate that dolomite has grown by replacement of preexisting limestone. Examination of ancient dolomite in thin section often reveals that individual dolomite rhombs or clusters of rhombs penetrate original calcium carbonate particles such as skeletal grains. In addition, the distribution of inclusions within individual dolomite crystals or clusters of crystals suggests original calcium carbonate particles. These inclusion patterns are usually seen as ghosts within the dolomite when the crystals are turned to extinction under a petrographic microscope (Fig. 14-2).

In some rocks, dolomite crystals have selectively replaced earlier organic structures such as worm tubes or skeletal fragments and, in many such cases, the dolomite exists as patches or mottles that clearly show the outline of the earlier material. Of particular interest are the transitions from limestone to dolomite that exist in many carbonate rock sections. In these cases, layered carbonate rocks change abruptly from dolomite to limestone. The boundary between the two rock types cuts across bedding (Fig. 14-3). Thus the dolomitization has been superimposed on an earlier calcium carbonate sediment or rock. Where these and related kinds of evidence exist, the secondary or replacement nature of the dolomite is demonstrable. The crystal size of the dolomite appears to have no necessary relation to whether the dolomite is secondary or not, as these relationships can be

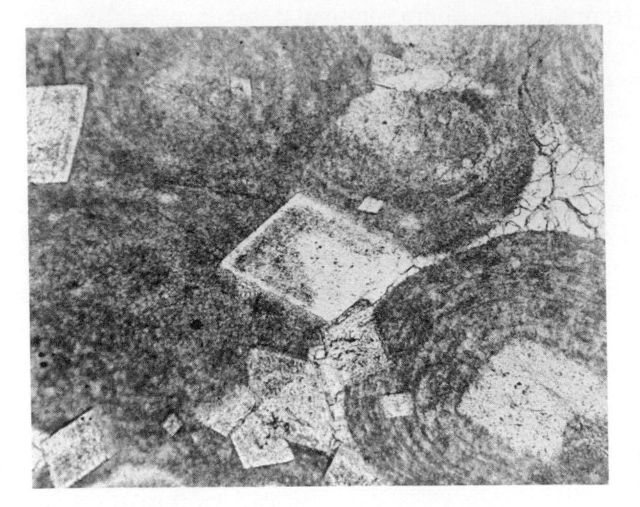

Fig. 14-2 Dolomite rhombs replacing oolitic rock. The large rhomb in the center is 100 microns in diameter.

demonstrated in dolomite having crystal sizes ranging from less than a micron to several hundred microns.

On the other hand, there exists dolomite in which textural evidence of a replacement origin appears to be absent. The reason for the absence may be insufficient study, lack of preservation of critical evidence, an original sediment such as a carbonate mud that does not lend itself to producing many of the petrographic criteria, or, alternatively, that the rock was deposited as a primary dolomite. With this in mind, we need to consider the conditions necessary for the formation of primary dolomite.

No conclusive evidence exists that *de novo* growth of dolomite crystals from a body of water is taking place today. Only a few years ago, however, prior to the discovery of modern dolomite of replacement origin, the apparent absence of dolomite in Holocene sediments was used as an argument for dolomitization being a phenomenon associated with deep burial. In considering the type of evidence that would be acceptable, the "whiting" phenomenon discussed in Chap. 12 immediately comes to mind. If a cloud of dolomite crystals could be observed forming in a body of water, and it could be demonstrated that the dolomite was not resuspended bottom sediment, this would be quite convincing. Whitings indicate rapid nucleation and initial growth and, therefore, the rate of growth of dolomite crystallites is of interest.

One locality in which the rate of growth of such crystallites has been studied is Deep Spring Lake, located on the California-Nevada state line (Peterson et al., 1966). No whitings have been seen but dolomite crystals, mostly less than 1 μ in size, are abundant on the lake bottom. Detailed C^{14} measurements have been made of this dolomite. The age of various size fractions from the sediment was determined and, in every case, the coarser sizes had a greater average age. The rate of growth of the crystals ranged from 0.05 to 0.09 μ/1000 y, which is extremely slow compared to growth rates of most substances from saturated solutions.

Fig. 14-3. Limestone/dolomite transition near Madison, Wisconsin.

X-Ray diffraction studies of different layers within the dolomite crystals revealed that the surficial layer(s) on the growing crystal is (are) calcium-rich but that 1:1 stoichiometry of magnesium and calcium ions is achieved by short-range solid state diffusion before the inner part of the crystal is buried too deeply inside the crystal and effectively isolated from the growth medium.

Deep Spring Lake is perhaps the closest approach to documented primary dolomite yet reported but if primary dolomite is defined as spontaneous nucleation from a body of natural water, not even the Deep Spring occurrence qualifies. The dolomite crystals in the lake appear to result from metasomatism of calcium carbonate at the sediment-water interface and the process here seems to require time periods of tens or hundreds of years to occur. Clearly, a time interval of this duration is irresolvable in ancient rocks. For this reason, some geologists would refer to the Deep Springs dolomite as primary and the correctness of this usage is largely a semantic question.

In summary, there exists some dolomite in Holocene sediments and much dolomite in ancient carbonate rocks that lacks visible evidence of a secondary or replacement origin. If primary dolomite is defined as spontaneous nucleation and growth from water of dolomite crystals followed by their sedimentation, then no primary dolomite is known to exist. If we are willing to include as primary dolomite the crystals formed by metasomatism while the calcium carbonate crystals were still influenced by the composition of surface waters in their environment of origin, then several primary dolomites are known from modern sediments. The Deep Spring occurrence is one; a locality in south Australia known as the Coorong is a second. In neither case has a whiting of dolomite been observed. But in both cases extremely fine grained dolomite crystals are forming by metasomatism. In addition, we are faced with the problem that no known criterion exists for recognizing a primary dolomite in ancient rocks.

14.3 REQUIREMENTS FOR DOLOMITE FORMATION

Consideration of the requirements for the formation of dolomite in sedimentary rocks may be approached from two points of view. (a) Why is it so difficult to form dolomite crystals *de novo* from natural waters? (b) What are the conditions under which dolomitization of calcium carbonate occurs?

Spontaneous Nucleation

In view of the apparent lack of dolomite whitings in natural solutions, we may ask whether such waters are saturated with respect to dolomite. The geologic evidence on this point is equivocal. We have seen that no *de novo* dolomite is known, suggesting undersaturation. On the other hand, detrital dolomite in the Gulf of Mexico and off the coast of south Florida exhibits perfect rhombohedra despite long periods of time in seawater. The dolomite is known to be detrital only as a result of C^{14} dating. So apparently dolomite neither precipitates nor dissolves in seawater, suggesting that seawater is approximately at saturation so that dissolution or precipitation reactions are imperceptibly slow.

We next turn to laboratory experimentation and the attempts to precipitate dolomite from solutions like seawater at low temperatures. The reaction of interest is

$$CaMg(CO_3)_2 \rightleftharpoons Ca^{2+} + Mg^{2+} + 2\,CO_3^{2-}$$

and the equilibrium constant is calculated from the relationship

$$K = \frac{[Ca^{2+}][Mg^{2+}][CO_3^{2-}]^2}{[CaMg(CO_3)_2]}$$

Unfortunately, no one has yet succeeded in synthesizing dolomite from seawater solution at low temperatures. Therefore, the value of the equilibrium constant is not known accurately. Several lines of evidence, however, suggest that the value is close to 10^{-17}. For example, chemical analyses of slow moving groundwaters in contact with dolomite result in an equilibrium constant of about 2×10^{-17} or $10^{-16.7}$; extrapolation of data from concentrated solutions from which dolomite has been precipitated at temperatures of several hundred degrees suggests an equilibrium constant on the order of 10^{-17}. Solution of dolomite crystals in aqueous solutions results in an apparent cessation of reaction when the product $[Ca^{2+}][Mg^{2+}][CO_3^{2-}]^2$ is on the order of 10^{-17}. Therefore, although the accuracy with which the equilibrium constant is known is not good, we may tentatively accept 10^{-17} as an approximation.

To determine whether seawater is saturated with respect to dolomite, we need to know for each of the three ionic species the molar concentration in the sea, the activity coefficient, and the percent of each species that is present as free ion. That is, we need to know the percentage of each species that is "free-floating" in the solution and not ion-paired. Although the molar concentration of all the major ions in seawater is accurately known, this is not true for the activity coefficients and percentages of free ion. For example, within the last few years the activity coefficient of calcium ion in seawater has been given as both 0.28 and 0.24; the percent free ion for calcium as both 91 and 84; the percent free ion for magnesium as both 87 and 90. Thus, the calculation to determine whether seawater is saturated with respect to dolomite suffers from the same inadequacies of data as the estimate of the equilibrium constant for the mineral. It is useful, however, to calculate the state of saturation of seawater for dolomite to see whether it comes close. Our available data appear sufficiently good for this purpose. We shall use the most recently published figures in the calculation.

Ion	Molality in seawater (c)	Activity coefficient (γ)	Activity (γc)	Percent free ion	Product
Ca^{2+}	9.98×10^{-3}	0.24	2.40×10^{-3}	84	2.02×10^{-3}
Mg^{2+}	5.23×10^{-2}	0.36	1.88×10^{-2}	90	1.69×10^{-2}
CO_3^{2-}			5.13×10^{-6}	9	4.62×10^{-7}
$(CO_3^{2-})^2$					2.13×10^{-13}

$$K = (2.02 \times 10^{-3})(1.69 \times 10^{-2})(2.13 \times 10^{-13}) = 7.26 \times 10^{-18} \text{ or } 10^{-17.1}$$

Therefore, the chemical data indicate that seawater is approximately saturated with respect to dolomite, as suggested earlier by geologic observations.

We must turn to kinetics for an explanation of the apparent lack of dolomite growth in normal seawater. The hypothesis most generally accepted is that nucleation of highly ordered structures such as dolomite is a slow process at low temperatures and when in competition for calcium and carbonate ions in a solution saturated with respect to calcite, dolomite precipitation is prevented from occurring. Calcium ions and magnesium ions have sufficiently similar behavior in aqueous solution so that it is very difficult for them to completely separate into different layers in a crystal structure, as is required in dolomite. Hence, nucleation of dolomite is slow. At higher temperatures, however, the thermal energy of the ions is sufficiently large so that they can migrate rapidly to ordered positions and dolomite is easily precipitated in laboratory experiments at temperatures of several hundred degrees centigrade.

Dolomitization of Calcium Carbonate

Plots of frequency distributions have been made of calcite-dolomite mixtures occurring in carbonate rocks and these plots reveal a markedly bimodal distribution (Fig. 14-4). If most dolomite is indeed secondary, this distribution suggests that

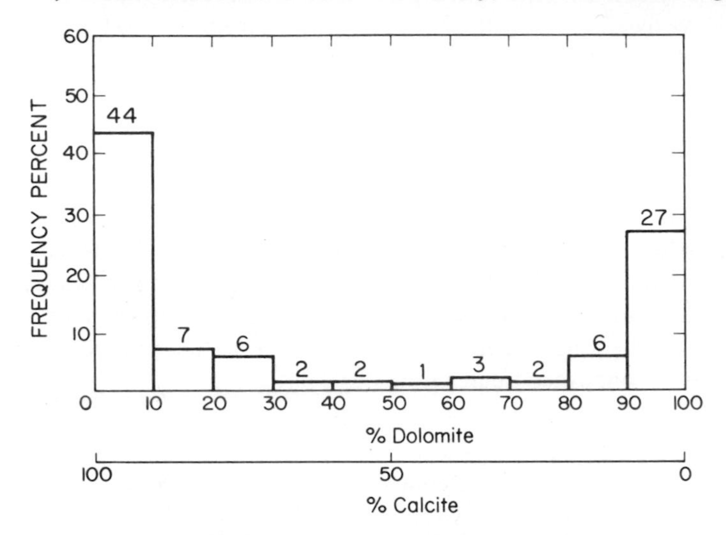

Fig. 14-4 Computed percentages of calcite and dolomite for 1148 analyses of North American carbonate rocks. (Steidtmann, 1917.)

a threshold of some sort must be exceeded before dolomitization can proceed but that once the threshold has been passed, dolomitization commonly goes to completion.

At low temperatures, diffusion of ions in crystals is extremely slow, even by geologic standards and, therefore, dolomitization will not be a solid state reaction. It will occur by the simultaneous dissolution of calcium carbonate and precipitation of dolomite from an aqueous solution passing through the rock. In general, two requirements must be met in order to cause dolomitization. (a) The Mg/Ca concentration of the water must be sufficiently high to permit the reaction to take place. (b) There must be a mechanism capable of flushing a sufficient volume of the dolomitizing fluid through the rock so that the reaction can be completed and a dolomite rock produced.

The reaction of interest is

$$2CaCO_3 + Mg^{2+} \rightleftharpoons CaMg(CO_3)_2 + Ca^{2+}$$

$$F^\circ = 2(-269.9) \quad -108.9 \qquad -518.7 \quad -132.2$$

$$\triangle F^\circ = \text{products} - \text{reactants} = -650.9 + 648.7 = -2.2 \text{ kcal}$$

Adjusting this standard free energy value to the activity ratio Ca^{2+}/Mg^{2+} in seawater, the free energy change is raised to -3.6 kcal because the activity of magnesium ion is significantly larger than that of calcium ion. So dolomitization of calcite appears to be favored thermodynamically in normal seawater.

The question of the Mg^{2+}/Ca^{2+} ratio required for dolomitization to proceed can be approached by another path. We can express the equilibrium constant (K) for calcite and dolomite as

$$(a\mathrm{Ca}^{2+})(a\mathrm{Mg}^{2+})(a\mathrm{CO}_3^{2-})^2 = K_{\mathrm{dol}}$$
$$(a\mathrm{Ca}^{2+})(a\mathrm{CO}_3^{2-}) = K_{\mathrm{cal}}$$

In a solution in which both calcite and dolomite are in equilibrium, there is a single carbonate activity. Solving these simultaneous equations to eliminate carbonate as a variable, we obtain

$$\frac{(a\mathrm{Mg}^{2+})}{(a\mathrm{Ca}^{2+})} = \frac{K_{\mathrm{dol}}}{(K_{\mathrm{cal}})^2} = K_{dz}$$

where K_{dz} represents the ratio of magnesium to calcium above which the reaction will proceed in the direction of dolomite formation at the expense of preexisting calcite. From data previously given,

$$K_{dz} = \frac{10^{-17.1}}{(10^{-8.4})^2} = 10^{-0.3}$$

or a magnesium/calcium activity ratio of 0.5. Obviously, this value is subject to the same uncertainty as is present in the estimate of the equilibrium constant for dolomite. If, for example, K_{dol} were $10^{-16.9}$ rather than $10^{-17.1}$, as it well may be, K_{dz} would equal $10^{-0.1}$ or a magnesium/calcium ratio of 0.8.

Another way to evaluate the critical ratio is to analyze groundwaters in limestones, dolomitic limestones, and dolomite rocks. Due to its slow flow rate, a groundwater might maintain a prolonged contact with its host rock under similar temperature and pressure conditions and it is not unreasonable to hypothesize that equilibrium between water and host may be achieved. What is the range in magnesium/calcium activity ratios in pore waters in these three types of rocks? Based on a fairly large number of analyses (Hsu, 1967), the critical K_{dz} value seems to lie between 0.7 and 0.9, a range quite compatible with a K_{dol} estimate of about 10^{-17}. Because we have calculated the free energy for dolomitization of calcite in seawater to be negative (-3.6 kcal/mole), the activity ratio of magnesium to calcium in seawater should be significantly above the critical value of slightly less than unity. And indeed it is. The ratio in sea water is

$$\frac{1.69 \times 10^{-2}}{2.02 \times 10^{-3}} \quad \text{or} \quad 8.37$$

This value is so much greater than the critical ratio as estimated from either laboratory determinations of dolomite solubility or from groundwater equilibrium data that we can be reasonably certain seawater is a dolomitizing solution. Despite this, no known dolomitization is taking place today in normal seawater. This is well illustrated on Pacific atolls where seawater is being continually pumped through porous modern reefs by the tides and yet no dolomite has been formed.

The second requirement, that of a sufficiently large volume of water with an activity Mg to activity Ca ratio above the critical value, is equally obscure. Without knowing the original aMg/aCa, the efficiency of the reaction, and the rate of water flow, it is impossible to estimate for a given rock with any accuracy the number of volumes of fluid equal to the volume of the pore space that must pass through the rock before it is completely replaced by dolomite. Certainly, the value must be of the order of thousands or perhaps millions. Thus the requirements for dolomitization are twofold. Water must have a sufficiently high aMg/aCa. In a marine-like water the activity ratio must be above 8.4 for efficient dolomitization. In addition, a driving force must exist that will move thousands of pore volumes of this fluid through the calcium carbonate sediment or rock.

14.4 MECHANISMS OF DOLOMITIZATION

Evaporative Reflux

In 1960 Adams and Rhodes suggested a mechanism based on observations they had made on the Permian Reef Complex in west Texas and New Mexico. The idea met the two requirements for dolomitization by suggesting that the evaporation of seawater to the point of precipitation of gypsum ($CaSO_4 \cdot 2H_2O$) at a concentration slightly in excess of three times seawater (see Chap. 15) would remove calcium from the water, thus automatically raising the Mg/Ca activity ratio above the 8.4 found in seawater. Evaporation also has the effect of creating a denser water that would tend to sink down through earlier deposited sediment and rock whose pore space was once filled with the lighter seawater. The velocity of this water flow will depend naturally on the density contrast and the permeability of the underlying material. In this way a water having a high aMg/aCa could be flushed in large volumes through earlier deposited limestone. Adams and Rhodes (1960) termed the process *seepage refluction.* The concept was in large part arrived at intuitively by observing that dolomite in the Permian Reef Complex is generally in front of and under the evaporites.

Recognition of dolomite being formed in modern sediments had taken place as early as 1953 in Lake Balkhash. However, Deffeyes, Lucia, and Weyl (1965) reported in their studies of the south end of the island of Bonaire in the Netherlands Antilles (Fig. 14-5) an example of modern dolomitization in association with gypsum precipitation. The south end of Bonaire is underlaid by Pleistocene limestone composed for the most part of coral-algal reef. A living coral-algal reef thrives offshore. Storm-driven waves have produced a beach ridge of coral rubble along the south and southwest shore of the island. This ridge separates the south end of the island, part of which is below sea level, from the sea. In the low areas behind the ridge, shallow hypersaline lakes exist. Landward beyond the lakes are exposed areas

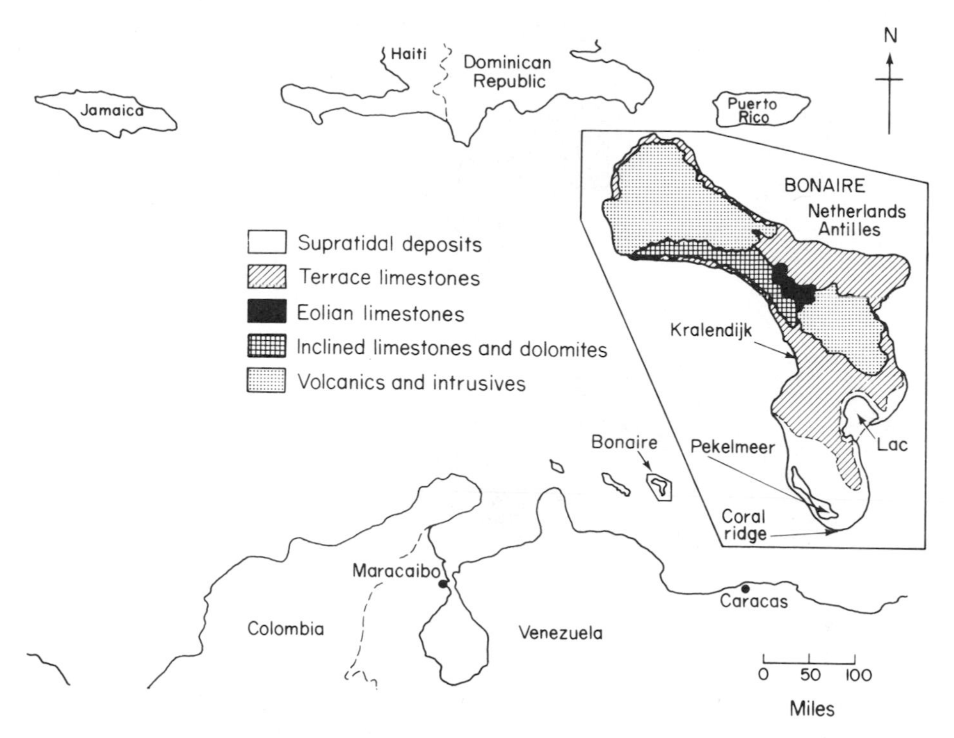

Fig. 14-5 Bonaire, Netherlands Antilles.

of carbonate sediment that periodically become inundated by the lake water. The modern sediment found on the flats and under the lake consists of carbonate sediment and rock, some of which has been dolomitized.

The main lake, called the Pekelmeer, deposited normal marine sediment prior to the full development of the barrier ridge. For the last several hundred years, it has accumulated gypsum and is precipitating gypsum today. Of particular interest is the fact that seawater enters the lake via springs throughout most of the year. This is possible because the lake is below sea level. Despite the fact that there is constant addition of new seawater and the lake has no surface outlet, the salinity of the water in the natural lake never reaches the high concentrations necessary for the formation of halite. This is remarkable considering that evaporation is constantly taking place from the surface of the lake and from the pore space in the surrounding flats. This can be explained only by some method of removing brine from the lake and the only possible exit routes are down through the underlying sediment and Pleistocene rock. Thus south Bonaire provides an example of modern dolomite in association with a brine concentrated by evaporation beyond the point of precipitation of gypsum. The precipitation of gypsum has caused an increase in the Mg/Ca molar ratio of from 5.2 in seawater to values in excess of 20 in the brine that is being flushed down through the underlying sediment because

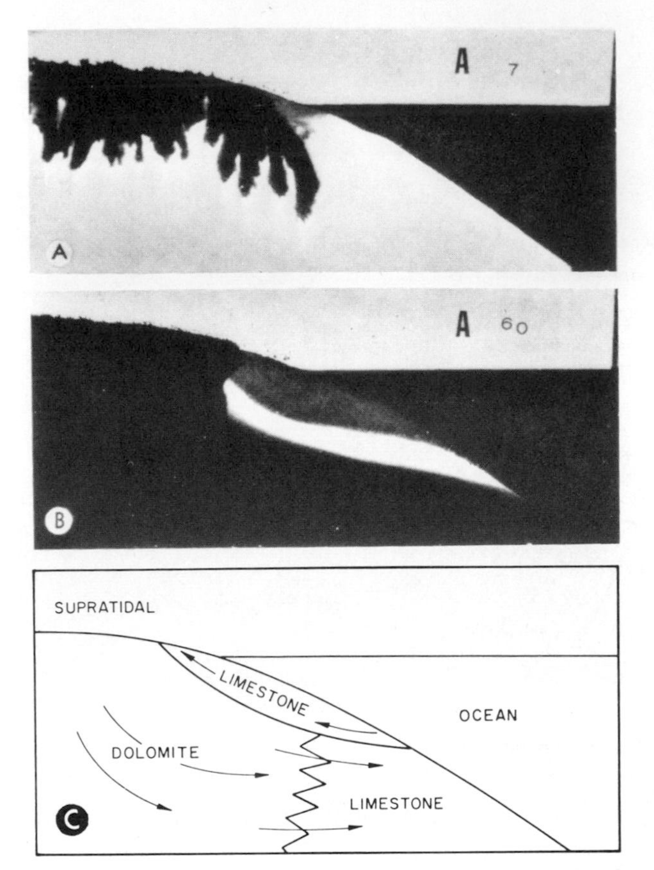

Fig. 14-6 Model reflux experiments. (A) Initial downward displacement of dense water after 7 minutes. (B) Steady-state flow after 60 minutes. (C) Hypothetical relationship between supratidal area, refluxing hypersaline water, influxing ocean water, and dolomitization, based on model experiments and observations on Bonaire. (From Deffeyes et al., 1965.)

of its density (Fig. 14-6). This mechanism that provides a means for meeting the conditions for dolomitization has been termed *evaporative reflux*. Radiocarbon dates of the dolomite indicate that it is modern and probably formed within the last 1000 years.

Other examples of modern dolomite exist in similar settings—tidal flats and desert playas where the evaporative-reflux process can operate, for example, within the tidal flat sediments on the south shore of the Persian Gulf and within the Andros Island supratidal flats in the Bahamas. In the latter case no gypsum has been found with the dolomite. Perhaps gypsum has been removed by rainwater dissolution in this relatively humid setting or perhaps another explanation may prove ultimately to be the case. Radiocarbon dating of the dolomite, however, on the Andros Island supratidal flats and dolomite found under the present marine sediment offshore has demonstrated that the dolomite layer is transgressive and formed during a rising sea level. Dolomite formed under supratidal conditions and now found below sea level was formed over 2000 years ago (Fig. 14-7). The association of modern dolomite with supratidal flats and desert playas is a reality.

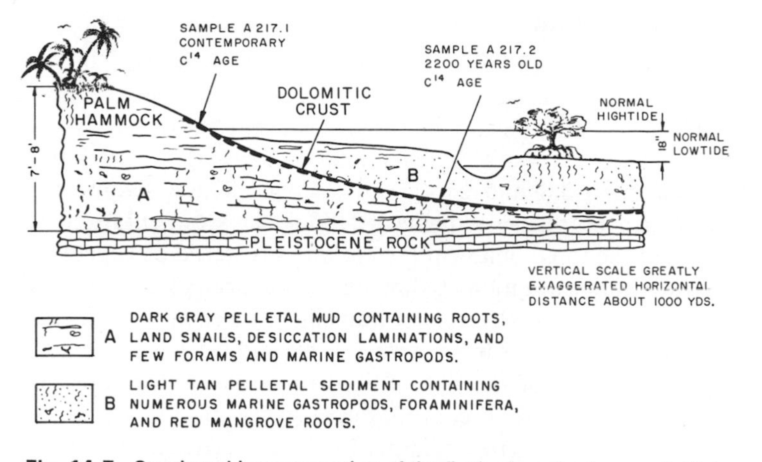

Fig. 14-7 Stratigraphic cross section of the flank of a palm hammock. Note that the crust layer is continuous but is contemporary in age where exposed above normal high tide level, and 2200 years old where buried and approximately 4 feet below normal high tide level. Sediments below the crust have a different character than those above the crust. (From Shinn et al., 1965.)

Other Mechanisms of Dolomitization

Dolomitization by evaporative reflux is currently believed to be the major mechanism by which dolomite is formed in carbonate rocks. This mechanism is, however, not the only one operative in the natural environment. For example, dolomite occurs in clay-rich Pleistocene and Holocene sediments on the continental shelf off the Louisiana coastline of the United States and in these sediments the most abundant clay mineral is calcium smectite (montmorillonite). However, in adjacent calcite-rich sediments the abundant clay mineral is sodium smectite. We may infer that dolomitization occurs in this area as a consequence of calcification of sodium smectite, which locally raises the Mg/Ca ratio of the water. The presence of dolomite in deep oceanic sediments may result from a similar process.

It was shown on p. 485 that normal sea water is a dolomitizing solution and that the reason dolomite is less common than we would expect in modern sediments is kinetic. It follows, therefore, that an increase in temperature associated with burial of calcium carbonate and associated pore waters to a depth of several thousand feet would stimulate dolomitization without the increase in Mg/Ca ratio required to achieve the same result at surface temperatures. Examples of dolomitization attributable to the geothermal gradient have not yet been documented, although dolomite rhombs in some eugeosynclinal sediments may result from this mechanism.

Other suggested mechanisms of dolomitization are very speculative. Dissolution of accessory minerals during the diagenesis of sandstones may locally increase the Mg/Ca ratio of pore waters. Subsurface leaching of chlorite might achieve the same result.

Examples from Ancient Rocks

The upper part of the Madison Formation (Mississippian) in the Williston Basin and adjacent areas contains vertically repeated carbonate-evaporite cycles. Deposition of these cycles began in the marine environment. With continued deposition, carbonate marine sediments were followed by carbonate sediments deposited in the intertidal and supratidal settings. These sediments were thus formed as part of a regressive sequence of tidal flat sedimentation. Gypsum or anhydrite occurs at the top of many of the cycles. This would be expected as gypsum normally forms within supratidal sediments in an arid climate (see Chap. 15). Dolomitized limestone generally occurs immediately beneath the evaporites. The thickness of the dolomite varies from cycle to cycle and from place to place within a given cycle. The position of the dolomitized limestone immediately beneath the evaporites is consistent with the evaporative-reflux model for dolomitization.

In the Devonian carbonate buildups in the subsurface of western Alberta, the Redwater Reef contains almost no dolomite. In addition, anhydrite or evidence for dissolved anhydrite is absent. The Sturgeon Lake Reef, on the other hand, is highly dolomitized and contains abundant anhydrite and evidence for dissolved evaporites. The association of dolomite with evaporites and the absence of dolomite where evaporites apparently were not formed suggests a model similar to that proposed from Bonaire. This is especially true because the evaporites within these pinnacle-like buildups cannot be marine and must have been deposited at times of subaerial exposure.

At the beginning of this chapter, it was noted, on the basis of 9000 analyses of carbonate rocks from the Russian Platform and North America, that dolomite tends to be more common in the older rocks of the Paleozoic than those deposited in more recent times. We can now examine the possible significance of that observation. It is possible that the composition of seawater was different during the Paleozoic than at present. There are no hard data to support the idea, however. Alternatively, within the context of the idea of evaporative reflux, it is possible for any carbonate rock to be uplifted, exposed by erosion, and then covered by evaporative tidal flat sediments. Under these conditions, it is possible that the older rock would be subjected to refluxing brines produced in the tidal flat sediments. The older a rock, the greater the number of times it could have been placed in this setting. Thus older rocks should have had many more opportunities to be dolomitized.

There is a third possibility that has not been adequately explored. If dolomitization is related to evaporative tidal flat and desert playa conditions, then carbonate rocks deposited in this environment should become dolomite. Wholly marine carbonate rocks or those deposited on tidal flats under humid climate conditions may be more common in the Mesozoic of North America and the Russian Platform than in the Paleozoic. This possibility should be examined together with the distribution of dolomite and the environmental and climatic depositional setting of carbonate rocks with respect to time for the rest of the world. It may well be that in a given area the distribution of dolomite with time reflects the distribution of evaporative tidal flat sedimentation with time. Data on Paleozoic carbonate rocks from North American oil fields indicate that dolomite is almost nonexistent in the Pennsylvanian compared to other periods. Dolomite is present in association with evaporites in the Pennsylvanian of the Paradox Basin in Utah. It appears to be generally absent in the marine carbonate rocks without evaporites of the midcontinent and American southwest.

Dolomitization by evaporative reflux may operate without leaving evaporite beds. And it certainly is true that the geologic record contains many extensive dolomite units from which no evaporites have been reported, for example, the Ellenberger Formation and its equivalents in west Texas and Oklahoma. These rocks however, contain abundant evidence of tidal flat sedimentation. In general, it is very difficult to completely eliminate evaporative reflux as the mechanism of dolomitization. This is true because nearly all carbonate sediments in the geologic record were deposited in shallow water, which makes intertidal sedimentation a real possibility in most cases of dolomitization of limestone. In addition, evaporites are easily removed by dissolution and often an obvious record of their past presence is missing. There is much need for additional study of the relationship between dolomite in the ancient record and evaporites or evidence of past evaporites such as molds of gypsum crystals; pseudomorphs of other minerals, such as calcite after gypsum; or evaporite solution breccias.

The generation of a water capable of dolomitizing a limestone or lime sediment is not enough to produce dolomite rock. The influence of *dolomite selectivity* is seen in many ancient carbonate rocks. A given limestone may not be dolomitized even though it lies within the path of a dolomitizing fluid, for example, beneath an evaporitic tidal flat. Several factors related to the rock itself become important. (a) The restriction of dolomite to rock surrounding fractures and faults, as in the Arbuckle Limestone of Oklahoma, strongly suggests that the permeability created by the fractures determined the distribution of the dolomite. (b) Mississippian carbonate rocks in western Alberta and Devonian crinoidal rocks in the subsurface of west Texas show a strong preference for dolomitization of those rocks that initially had abundant carbonate mud: wackestones and mudstones. The selective growth of dolomite in carbonate mud has been noted by many workers. In most cases, it is impossible to determine whether this reflects the greater surface area available for reaction, the availability of more soluble carbonate minerals at the time of dolomitization, or even the higher permeability for dolomitizing fluid transport at the time of dolomitization. The latter would only be true if the carbonate mud-free rocks had been cemented by earlier calcite prior to dolomitization.

Dolomite in Sandstone

Many quartz sandstones such as the Saint Peter are both underlaid and overlaid by dolomite. In general, dolomite is absent in these detrital silicate rocks. Modern dolomite is also scarce or absent within tidal flats composed of silicate sediments even when gypsum is being formed. Despite these generalizations, some sandstones do have dolomite within the pore space that acts as cement. Examination of this dolomite often reveals ghosts of earlier calcite or aragonite cement. Thus, the sandstone was originally cemented by calcium carbonate cement and that cement was later dolomitized. In other cases, evidence for replacement of an earlier carbonate appears to be lacking. Dolomite in these sandstones poses a problem. If the sandstone is a deep water deposit, then the question can be asked Is the dolomitization possible by evaporative reflux? If the sandstone was later covered by evaporative tidal flat sediments or by deep water evaporites, then the evaporative reflux mechanism remains possible. If these conditions have never been met, then we must seek another explanation of the requirements for growing the dolomite.

14.5 THE EVOLUTION OF DOLOMITE FABRICS AND POROSITY

The mineral dolomite has a strong tendency to develop and grow in the unit rhombohedra. Many dolomite rocks exhibit a sucrosic texture that resembles brown sugar, produced by the relatively loose fitting together of dolomite rhombohedra (Fig. 14-8). Ancient dolomite in the subsurface is much more commonly a porous rock than limestone. Indeed, approximately 80% of the oil and gas that will be recovered from carbonate rocks in North America will come from dolomite. This indicates that the dolomitization process tends to produce and preserve pore space better than limestone. Much of this pore space exists between the rhombohedral dolomite crystals.

A dolomite rhomb that grows by replacement of a calcium carbonate sediment or rock commonly occupies space that previously was taken up by finer carbonate mud or parts of polycrystalline grains. Porosity may occur between the crystals of the mud and in some of the polycrystalline grains. The growing rhomb develops in space previously used by both calcium carbonate minerals and pore space.

If the volume of calcium carbonate rock is more than 57% $CaCO_3$, then that volume will be able to provide sufficient calcium to produce a solid dolomite crys-

Fig. 14-8 Sucrose dolomite. The individual rhombs are approximately 75 microns in diameter. The porosity of the rock is approximately 18 percent.

tal (Murray, 1960, and Weyl, 1960). Magnesium is derived directly from the water. In a sense, the calcium is also derived from the water because dolomitization is a solution-precipitation reaction. However, this calcium is replaced by dissolution of calcium carbonate rock. Natural waters contain small amounts of total carbon dioxide, compared to calcium and magnesium, therefore carbonate must be added in order to generate a solid dolomite crystal from porous calcium carbonate (Weyl, 1960). The excess carbonate is neccessary because the solid dolomite rhomb has a molar volume 12 to 13% smaller than that of calcite. Thus dissolution of a solid block of calcite would produce a block of dolomite of 12 to 13 percent less volume. If the dolomite rhomb grows to fill the total volume of dissolved calcium carbonate, excess carbonate must be added to the water because no calcium carbonate existed in the pore space that is taken up by the dolomite rhomb. For this reason, the needed carbonate must be derived from the rock by excess dissolution of calcium carbonate. This means that a much larger volume of calcium carbonate must be dissolved from around the growing crystal than just the volume of the dolomite itself. The amount of excess dissolution is equal to the volume of pore space being occupied by the growing rhomb plus the molar volume difference of 12 to 13%. This use of local carbonate derived by dissolution of adjacent calcium carbonate has been termed *local source dolomitization.*

The final product of such a process is a rock composed of dolomite rhombs largely separated by pore space that was created by the dissolution of limestone from beyond the limits of each dolomite rhomb. This is a sucrosic dolomite texture. Particles within the original limestone that resisted the growth of dolomite in the early stages of dolomitization, such as dense crinoid particles, commonly are dissolved to provide the carbonate to grow the adjacent dolomite crystals. These particles then appear as molds or vugs in the rock. Because the necessary excess carbonate is locally derived from the immediately adjacent material, the absence of calcium carbonate appears to inhibit the growth of dolomite. Vugs or large holes in the rock that existed at the time of dolomitization are commonly lined with dolomite crystals. These crystals began growth within the carbonate rock matrix. Growth of a second layer of pore-filling dolomite crystals is uncommon however, and vugs originally present in limestone are commonly preserved in dolomite rock. For a similar reason, growth of dolomite in silicate sandstone is uncommon. Most of the dolomite cement in these rocks appears to have formed by replacement of an earlier calcium carbonate cement or fossils. With continued flow of water, dolomite may be added as cement. This dolomite may be added as new crystals or as overgrowths on earlier-formed crystals.

14.6 RECOGNITION OF ORIGINAL LIMESTONE FABRICS IN DOLOMITE

One major purpose for studying ancient dolomite is to reconstruct the depositional texture and fabric of the calcium carbonate sediment. The dolomitization process tends to obliterate original textures and fabrics on which interpretation of the ancient sedimentary environments must be based. Like most generalizations, however, this one has many exceptions (Murray, 1964). Dolomite rocks commonly have preserved with remarkable detail the character of the original material. Because of this preservation, excellent work has been done in studies of the depositional environment and paleoecology of ancient dolomite. Preservation of evidence of the original, predolomitization sediment or rock takes place in a number of ways and results from a number of processes. The following sections illustrate some of these mechanisms.

Selective Replacement of Skeletal Material with Preservation of Organic Voids

Colonial organisms such as tabulate corals and bryozoans that build skeletons of calcium carbonate are commonly dolomitized with a distinct preference for replacement of the original carbonate. Dolomite grows by replacement of existing calcite or aragonite and avoids growth as new crystals in existing void space. This type of preservation is especially noticeable when the cavity is considerably larger than the dolomite rhomb. In the tabulate corals, the cavity within the corallite is commonly preserved with only terminations of crystals penetrating into it. Such a preference for replacement produces a fairly accurate gross reproduction of the organism. The preference of dolomite to grow by replacement of existing carbonate and to avoid the growth of new crystals in preexisting pore space suggests that the dolomite grew by utilizing a local source of carbonate ion. Where carbonate minerals were absent in the void space, dolomite did not grow.

Paleozoic examples of this type of preservation in corals are shown in Figs. 14-9 and 14-10. Spectacular examples of preferential replacement in the dolomitized corals have been observed in dolomite recovered from the drilling on Funafuti Atoll in the Pacific.

Fig. 14-9 *Syringopora*. This completely dolomitized colony exhibits growth of dolomite within the sediment between the corallites. The body chambers and corallites are very well preserved as voids. (Published with the permission of John Wiley Co.)

Fig. 14-10 *Favosites*. The skeleton has been completely dolomitized, with preservation of the original voids. These voids occupy more than half the volume of the specimen. (Published with the permission of John Wiley Co.)

Selective Removal of an Organism That Is Resistant to Dolomization

Dissolution of skeletal material to produce molds may accompany dolomitization or may follow partial dolomitization. The latter case would require selective replacement of the rock by dolomite that avoided growth within a particular organism. The organism, still calcite or aragonite, would be available for later selective dissolution, which would leave a mold in the dolomitized rock. The selective dissolution might be accomplished during exposure to meteoric waters. This post-dolomitization leaching may be poorly selective and produce poor preservation. Such imperfect preservation may result from dissolution of matrix dolomite.

Perhaps the most common example of production of molds of fossils that selectively resist dolomitization is the preservation of crinoid columnals and calyxes in completely dolomitized rock. In such rocks, other skeletal debris, such as bryozoan material that may be less resistant to dolomitization, is commonly replaced. In the later stages of dolomitization, the calcite in some of the crinoid fragments is dissolved to provide carbonate for production of the final dolomite. The result is the preservation of these organisms as molds. These molds sometimes are marred only by internal surfaces of impinged rhombs. Internal casts of axial canals are often preserved through dolomite replacement of carbonate mud that was deposited in these openings prior to dolomitization (Fig. 14-11). The susceptibility

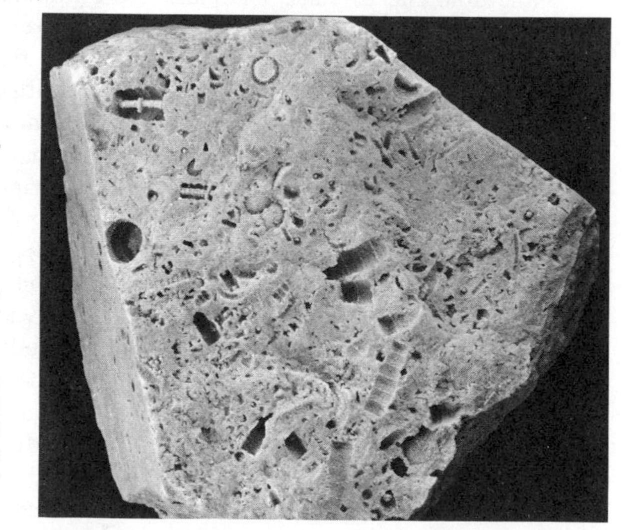

Fig. 14-11 Crinoid molds. These molds are found only in completely dolomitized crinoidal mudrock. The external surfaces of the columnals and internal casts of the axial canals are well preserved. (Published with the permission of John Wiley Co.)

of some organisms to dolomitization and the resistance of others certainly depends on their porosity, crystal size, and crystal surface area. In addition, the mineralogy of the skeletal particle at the time of dolomitization is very important as calcite, aragonite, and high magnesium calcite have different solubilities. The difficulty of evaluating predolomitization diagenetic changes that have altered the original mineralogy has hampered studies of susceptibility of organisms to dolomitization.

Delicate Replacement of Skeletal Carbonate

Dolomite sometimes reproduces details of shell structure by delicate and subtle differences in crystal size, clarity, and crystallographic orientation (Fig. 14-12). This appears to be especially true when the rock is replaced by relatively

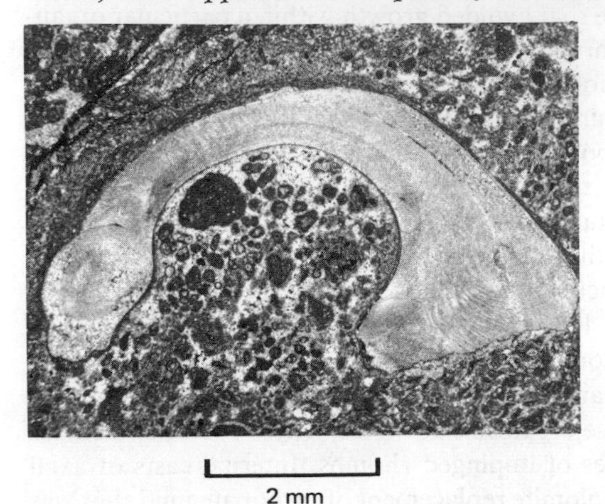

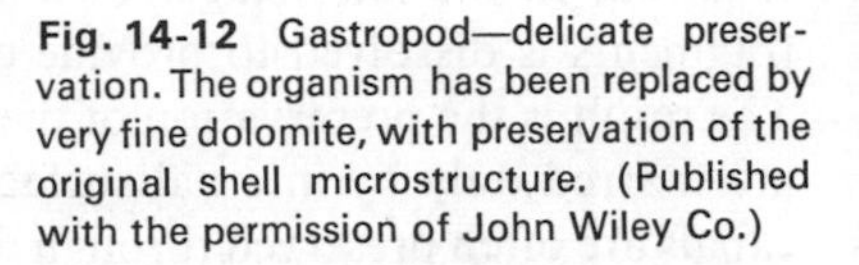

Fig. 14-12 Gastropod—delicate preservation. The organism has been replaced by very fine dolomite, with preservation of the original shell microstructure. (Published with the permission of John Wiley Co.)

fine dolomite crystals. Specifically, the important relationship is the size of the rhomb with respect to the size of the structure to be preserved.

Dolomite rhombs tend to grow with their crystallographic axes oriented, within the limit of observation, coincident with those of the replaced calcium carbonate. This concurrence is observed in dolomite that has replaced drusy calcite or aragonite cement or single-crystal crinoid particles. If the orientation of rhombs in a dolomite rock was controlled by an original calcium carbonate crystal in a carbonate mud, the orientation of the mass of rhombs may appear random. This would be true because the orientation of the original calcium carbonate crystals in which the rhombs began to grow would have been random.

Many ancient limestones composed of carbonate mud are nearly barren of visible organic remains. Such carbonate muds with relatively few organic sand-size or larger particles are commonly dolomitized. Therefore, it cannot be simply assumed that in a dolomite lacking obvious evidence of skeletal particles this evidence was destroyed by the dolomitization. Conversely, it is dangerous to interpret any dolomite that is apparently devoid of organic remains as having been deposited as an unfossiliferous carbonate mud.

Preservation of a Carbonate Sand Fabric

Analogous to the preservation of the tabulate coral skeletal pattern is the preservation of a carbonate sand fabric that has been only slightly cemented with calcite or aragonite prior to dolomitization (Fig. 14-13). In such cases, it is not

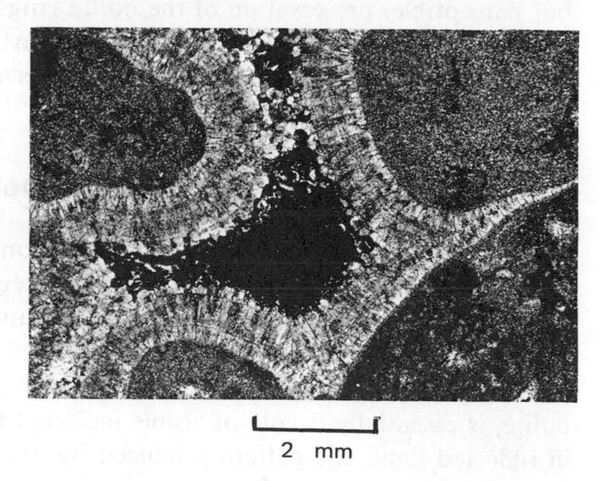

Fig. 14-13 Dolomitized carbonate pellet sand. These particles and the incomplete drusy cement have been replaced by fine crystals of dolomite. The acicular pattern, so common in early $CaCO_3$ drusy cements, is preserved because of a tendency of the rhombs to grow with their crystallographic axes coincident with those of the replaced oriented cement. There are associated particulate sediments that have not been dolomitized and that still retain an acicular calcite drusy cement. (Published with the permission of John Wiley Co.)

uncommon to find the particles and early drusy cement replaced by dolomite and the interparticle void space preserved. Such a preference for replacement results in good preservation of the fabric of the rock. For example, an originally cross-bedded carbonate sand may appear as a cross-bedded dolomite. Without careful study, such a rock might be misinterpreted as having been deposited as a sand composed of dolomite grains. The grains are commonly replaced by several dolomite crystals, however, that individually exhibit evidence of replacement of discrete original calcite or aragonite particles.

Preservation of Carbonate Sand Particles as Inclusions in Dolomite Crystals

Outlines of particles and the internal structure of particles such as ooliths are sometimes preserved as inclusions within single dolomite crystals. Figure 14-14 illustrates an oolite partially replaced by large dolomite rhombs with imperfect,

Fig. 14-14 Inclusions within a dolomite rhomb. This dolomite rhomb has grown partially within two ooliths and partially within the interoolith area. Inclusions within the interoolith part of the rhomb suggest that some clear calcite existed there prior to dolomitization and has been replaced. The rings of the oolith are reproduced crudely as inclusions within the ooliths. Note the clear rim on the rhomb within the area of the two ooliths. (Published with the permission of John Wiley Co.)

but perceptible, preservation of the oolite rings as inclusions within the replacing rhomb. Recognition of the identity of the particles and the original fabric of the rock where preserved as inclusions may permit interpretations of the original sedimentary environment.

Clear-Rimmed, Cloudy-Centered Dolomite Crystals

Figure 14-14 also illustrates a phenomenon commonly observed in dolomite—clear rims and cloudy centers. The cloudy center results from inclusions of the original minerals trapped within the dolomite. Because of the preservation of the pattern of the oolite, these inclusions are interpreted as nonreplaced patches of oolite. The clear rim, despite the fact that it exists within ooliths and represents replaced oolite, is essentially devoid of visible included material. When dolomite is studied in reflected light, the pattern produced by abundant dolomite crystals with clear rims and cloudy centers is sometimes mistaken for that produced by a carbonate pellet sand deposited free of interstitial lime mud. The cloudy centers appear to be particles when the rhombic outline is obscured at low magnification.

14.7 DETRITAL DOLOMITE

It has been observed by several workers (Amsbury, 1962; Sabins, 1962) that dolomite crystals, sand-size aggregates of dolomite crystals, and pebbles of dolomite rock exist in sediments and sedimentary rocks that are younger than the time of formation of the dolomite. Dolomite normally is dissolved easily during weathering. Sometimes it is preserved, however, and following transportation becomes part of a new sediment. This may take place just shortly after deposition when wind or water removes recently formed dolomite and redeposits the grains. Alternatively, the dolomite may be derived from the erosion of older rock. Dolomite crystals found in Florida Bay, when dated by radiocarbon, have an age much older than the modern carbonate sediment that has accumulated in the bay. It is presumed that these crystals, many of which appear as rhombohedra, like the quartz grains that are also found in this sediment, were derived from older rocks on the west coast of Florida and carried to the western edge of Florida Bay by ocean currents. Grains of dolomite within Cretaceous sandstones from western United States sometimes exist in positions that would normally be occupied by a quartz grain and are of the same size as known detrital grains. It is presumed these are of detrital origin.

14.8 DEDOLOMITIZATION

Some calcitic beds and calcite-cemented sandstones consist of large rhombs composed of polycrystalline aggregates of calcite crystals that appear to have formed by replacement of dolomite by calcite (Scholle, 1971). Even where the rhombs exist as a single crystal of calcite, the rhomb shape suggests a dolomite precursor despite the fact that rhomb-shaped calcite crystals are known to exist. *Dedolomite*, i.e., a dolomite that has been replaced by calcite, commonly occurs in association with gypsum or oxidized pyrite. It appears to have formed at very near-surface conditions. Meteoric water falling on outcropping gypsum generates a solution with a very high Ca/Mg ratio. The exact ratio necessary for the replacement of dolomite by calcite is not known. The higher the ratio, however, the faster the rate at which the reaction should proceed. It is possible, but by no means certain, that dedolomite is indicative of subaerial exposure. Because the existence of dedolomite has generally not been recognized in the past, it is important that this possibility be considered in future studies of many limestones. This is especially true where lateral transitions from limestone to dolomite are studied in a single bed. The question should be asked, Is the boundary produced because this was the terminus of dolomitization or, alternatively, was the limestone once dolomite and has been replaced by calcite?

REFERENCES

ADAMS, J. E., and M. L. RHODES, 1960, "Dolomitization by Seepage Refluction," *Amer. Assn. Pet. Geol. Bull.*, **44**, pp. 1912–1920. (The first statement of dolomitization by evaporative reflux of concentrated seawater.)

AMSBURY, D. L., 1962, "Detrital Dolomite in Central Texas," *Jour. Sed. Pet.*, **32**, pp. 3–14. (Discussion of origin, transport and destruction of detrital dolomite.)

CULLIS, C. G., 1904, "The Mineralogical Changes Observed in the Cores of the Funafuti Borings, in the Atoll of Funafuti," *Royal Soc. London*, sec. xiv, pp. 392–420. (Classic early study of dolomitization.)

DEFFEYES, K. S., F. J. LUCIA, and P. K. WEYL, 1965, "Dolomitization of Recent and Plio-Pleistocene Sediments by Marine Evaporite Waters on Bonaire, Netherlands Antillies," in *Dolomitization and Limestone Diagenesis*. Soc. Econ. Paleon. & Miner. Spec. Pub. No. 13, pp. 71–88. (Fine study of the hydrology and geology of dolomitization in a modern environment.)

EPSTEIN, SAMUEL, D. L. GRAF, and E. T. DEGENS, 1964, "Oxygen Isotope Studies on the Origin of Dolomites," in *Isotopic and Cosmic Chemistry*, H. Craig, S. L. Miller, and G. J. WASSERBURG, ed. Amsterdam, North-Holland Pub. Co., pp. 169–180. (Isotopic studies indicate that most dolomite is of replacement origin.)

HSU, K. JINGHWA, 1967, "Chemistry of Dolomite Formation," in *Carbonate Rocks Developments in Sedimentology* **9** *B*, Amsterdam,: Elsevier Pub. Co., pp. 170–191. (A good summary of the geochemistry of dolomite.)

ILLING, L. V., A. J. WELLS, and J. C. M. TAYLOR, 1965, "Penecontemporary Dolomite in the Persian Gulf," *Soc. Econ. Paleon. & Miner. Spec. Pub. No.* **13**, pp. 89–111. (Modern dolomite reported from the supratidal flats of the south coast of the Persian Gulf.)

MURRAY, R. C., 1960, "Origin of Porosity in Carbonate Rocks," *Jour. Sed. Petrology*, **30**, pp. 59–84. (Origin of the sucrose dolomite fabric.)

MURRAY, R. C., 1964, "Preservation of Primary Structures and Fabrics in Dolomite," in *Approaches to Paleoecology*, New York: John Wiley & Sons, Inc., pp. 388–403. (Statement of criteria for recognizing original particles and structures in dolomite.)

PETERSON, M. N. A., C. C. VON DER BORCH, and G. S. BIEN, 1966, "Growth of Dolomite Crystals," *Amer. Jour. Sci.*, **264**, pp. 257–272. (Detailed study of the growth mechanism and growth rate of primary dolomite crystals in Deep Spring Lake, California.)

SABINS, F. F. Jr., 1962, "Grains of Detrital, Secondary and Primary Dolomite from Cretaceous Strata of the Western Interior," *Geol. Soc. America Bull.*, **73**, pp. 1183–1196. (Well developed arguments for recognizing detrital dolomite.)

SCHOLLE, P. A., 1971, "Diagenesis of Deep-Water Carbonate Turbidites, Upper Cretaceous Monte Antola Flysch, Northern Apennines, Italy," *Jour. Sed. Pet.*, **41**, pp. 233–250. (Discussion of detrital dolomite and dedolomitization.)

SHINN, E. A., R. N. GINSBURG, and R. M. LLOYD, 1965, "Recent Supratidal Dolomite from Andros Island, Bahamas," in *Dolomitization and Limestone Diagenesis*. Soc. Econ. Paleon. & Miner. Spec. Pub. No. 13, pp. 112–123. (Penecontemporaneous replacement of calcium carbonate deposits by dolomicrite.)

STEIDTMANN, E., 1917, "Origin of Dolomite as Disclosed by Stains and Other Methods," *Geol. Soc. Amer. Bull.*, **28**, pp. 431–450. (Field and laboratory studies of many dolomites reveals most to be of replacement origin.)

WEYL, P. K., 1960, "Porosity through Dolomitization: Conservation-of-Mass Requirements," *Jour. Sed. Petrology*, **30**, pp. 85–90. (Chemical requirements for sucrose dolomite.)

CHAPTER FIFTEEN

EVAPORITES AND NATIVE SULFUR

15.1 THE CALCIUM SULFATE MINERALS AND ROCKS

Introduction

Calcium sulfate minerals may appear in nature as single, isolated crystals or clusters of crystals in a host carbonate or detrital rock. In addition, they often form the major part of layered rocks with only minor amounts of contained carbonate, silicate, or carbonaceous material. Two minerals, gypsum ($CaSO_4 \cdot 2\,H_2O$) and anhydrite ($CaSO_4$), commonly appear in nature. Hydrated forms containing 0.5 molecule H_2O or less have been reported from almost dry sediment exposed to the hottest earth surface temperatures. These minerals, bassanite, or calcium sulfate hemihydrate are rapidly converted to gypsum in the presence of water.

The original minerals generally form by precipitation from concentrated brines. The concentration necessary for precipitation usually is attained by evaporation at the air-water interface that is generally the surface of a standing body of water such as an isolated sea, lagoon, or saline lake. The air-water interface may also occur within the vadose zone where the pore space in the rock or sediment is partially saturated with brine and partially saturated with air. In either case the brine becomes increasingly concentrated when net evaporation exceeds addition of a less dilute solution such as fresh seawater or rainwater. The original brine is usually generated in one of two ways, seawater and reconstructed brine. Reconstructed brine is produced as rainwater or groundwater passes through rocks that contain soluble calcium sulfate minerals. For example, the waters that form the gypsum at White Sands, New Mexico, are produced when rainwater falls on the nearby San Andres Mountains and flows through gypsum deposits of Permian age. These waters then flow down and join the groundwater system and emerge under the nearby valley flats. Evaporation of this brine in the valley floor sediment or in playa lakes increases the concentration of the brine to the point of precipitating gypsum. When normal seawater is concentrated by evaporation to approximately 3.35 times the original salinity (at 30°C), gypsum begins to precipitate. The kinetics of this reaction are moderately rapid and some gypsum will form within a few minutes when the concentration is increased above saturation.

We normally think of a hot climate as the only setting for evaporite deposition. In general this is true. However, the only requirement is that evaporation be sufficient to cause a significant concentration of the brine. The evaporation rate must be sufficient to counteract the inflow of new water and loss of brine from the system before it has had an opportunity to be concentrated to the point of calcium sulfate precipitation. High evaporation rates are commonly found in areas of high temperatures and relatively low rainfall. These conditions are also found in arctic and antarctic regions and modern gypsum has been reported from these areas. Locally, the freezing of seawater leaves a brine that may be concentrated to the point of gypsum formation in cold climates.

15.2 MECHANISMS OF BRINE CONCENTRATION

Usiglio in 1849 evaporated a 5-liter sample of Mediterranean seawater and determined the composition and amount of the salts precipitated. These data are reproduced in Table 15-1. Clarke (1924) observed that the values for bromine are too high and potassium low. Moreover the precipitation of calcium carbonate during the early stages of evaporation in nature is questionable. Otherwise Usiglio's original data are in general agreement with more recent measurements. Much of the carbonate we see associated with evaporite sequences was probably produced by calcareous organisms (see Chap. 12). Of particular interest is the coprecipitation of gypsum and halite during the intermediate stages of evaporation because in ancient rocks the occurrence of calcium sulfate minerals in halite beds is common.

The concept of *reflux*, which was discussed in Chap. 14, is important to the understanding of the formation of calcium sulfate minerals, especially beds of layered rocks that are predominantly gypsum or anhydrite. It has long been recognized that evaporation to dryness of a column of normal seawater 1000 ft thick will produce only 0.75 ft of solid gypsum and approximately 13.7 ft of halite. Gypsum

TABLE 15-1 SALTS LAID DOWN DURING CONCENTRATION OF SEAWATER*

Density†	Volume	Fe_2O_3	$CaCO_3$	$CaSO_4 \cdot 2H_2O$	NaCl	$MgSO_4$	$MgCl_2$	NaBr	KCl
1.0258	1.000	0.0030‡							
1.0500	0.533		0.0642						
1.0836	0.316		Trace						
1.1037	0.245		Trace						
1.1264	0.190		0.530	0.5600					
1.1604	0.1445			0.5620					
1.1732	0.131			0.1840					
1.2015	0.112			0.1600					
1.2138	0.095			0.0508	3.2614	0.0040	0.0078		
1.2212	0.064			0.1476	9.6500	0.0130	0.0356		
1.2363	0.039			0.0700	7.8960	0.0262	0.0434	0.0728	
1.2570	0.0302			0.0144	2.6240	0.0174	0.0150	0.0358	
1.2778	0.023				2.2720	0.0254	0.0240	0.0518	
1.3069	0.0162				1.4040	0.5382	0.0274	0.0620	
Total deposit		0.0030	0.1172	1.7488	27.1074	0.6242	0.1532	0.2224	
Salts in last bittern					2.5885	1.8545	3.1640	0.3300	0.5339
Sum			0.0030	0.1172	1.7488	29.6959	3.3172	0.5524	0.5339

* After Stewart, 1963.
† Given by Usiglio in Baumé degrees. Restated here in specific gravities.
‡ Values given in grams for one liter of sea water.

begins to precipitate when the seawater has been concentrated to slightly more than one-third of its original volume. Halite begins to precipitate when the seawater has been concentrated to approximately one-tenth of its original volume. The deposition of relatively thick sections of gypsum or anhydrite with relatively little halite implies that there was a source of new seawater or other brine being made available for evaporation and that the brine was seldom allowed to concentrate to the point of halite precipitation. If a container of seawater is allowed to evaporate and new seawater is constantly added to maintain the original volume, the brine will soon become concentrated to the point of halite precipitation. If, however, the concentrated brine is removed from the container at the same time the new seawater is added, then it is possible to maintain the concentration at an equilibrium value and precipitate only gypsum.

King (1947) recognized, after study of the Permian Castile evaporites in west Texas and New Mexico, that these thick deposits of gypsum and anhydrite must have accumulated in a basin where new seawater was constantly added and concentrated brine was constantly being removed. He postulated a narrow seaway connecting the open sea and the evaporite basin. Lighter normal seawater moved into the basin from the sea as an upper layer and replenished water lost by evaporation. Denser concentrated brine produced within the basin moved back to the sea, underneath the lighter seawater through the narrow seaway. This mechanism demands a surface seaway or channel with flow in two directions. This type of flow exists in many channels or straits today and undoubtedly existed in the past. For example, this type of two-way flow exists today between the Atlantic Ocean and the Mediterranean Sea at the Strait of Gibraltar despite the fact that the salinity is not sufficiently high to cause precipitation of gypsum.

An alternative mechanism for producing the same result is subsurface reflux. Here the new seawater is added either by surface flow through a narrow channel or seaway or alternatively by seawater springs. Return flow of the concentrated brine takes place down through the sediments and rocks that underlie and surround the evaporite basin. The driving force that causes the concentrated brine to flow is the density contrast between the heavy concentrated brine produced within the evaporite basin and lighter water within the pore space in underlying sediment and rock. For example, seawater has a density of approximately 1.03, whereas brine precipitating gypsum commonly exceeds 1.13 gr/cc.

This mechanism is also applicable to concentration of brine within the vadose zone. New seawater or brine may be added on the surface. For example, a large storm may drive seawater onto normally exposed high tidal flats. In addition, new brine may be added by capillary rise within the vadose zone to replenish water lost by evaporation. Once the brine becomes concentrated within the vadose zone pore space, it will tend to sink because of its relatively high density at those times when the saturation of brine within the vadose zone becomes high. Thus, within the pore space of the high phreatic zone, a countercurrent flow is possible. Heavy brine sinks in tongues and new lighter water rises to replenish the loss due to evaporation and reflux. This appears to be a common mechanism on modern tidal flats where gypsum is being formed with little or no halite. The permeability of the underlying sediment and rock is critical to this mechanism. If the bulk permeability of the underlying material is high, then only a slight concentration will cause the brine to seep down rapidly and be removed. Under these conditions, the brine may never become concentrated to the point of gypsum precipitation. If the permeability is low, then reflux of brine will be slow, a highly concentrated brine will develop, and the more soluble salts will be deposited. The equilibrium concentration will also depend on the rate of evaporation and the rate of new brine addition.

15.3 THE ORIGINAL MINERAL—GYPSUM OR ANHYDRITE

When we examine an ancient sedimentary rock containing gypsum or anhydrite, we must ask the question: What was the mineralogy at the time of deposition? There is no simple answer and the problem must be examined in reference to all the evidence available on these deposits and any specific criteria that come from the deposit itself. The question is significant not only for understanding the origin and history of the rock but also for appreciating several aspects of the total geology associated with these deposits. If the deposit is now anhydrite but began as gypsum, then there should have been a 38% reduction in the solid volume at the time of the replacement of the gypsum by anhydrite. Such a significant loss of solid volume has several consequences. An abnormally high fluid pressure results in the pore space because the volume of water plus anhydrite is greater than the volume of gypsum and because the 38% loss of solid volume should produce a nongrain-supported material. This fluid pressure may approach the lithostatic pressure if the permeability is insufficient to permit the brine to dissipate. Such high fluid pressure may permit the sediment to flow and thus create deformation structures within individual beds. In addition, a sedimentary horizon with abnormal fluid pressure may act as a locus for the generation of a thrust fault. After the pore fluid has been removed by compaction, the final average sediment thickness will naturally be reduced by 38% of the original gypsum within the rock. At the time of replacement of gypsum by anhydrite, the flow of brine that was induced by the abnormal fluid pressure may cause brine to move out into surrounding sediment or rock and precipitate anhydrite in associated rocks that originally were devoid of evaporite minerals and were not deposited under evaporite conditions. If the original mineral was anhydrite, these consequences would not be anticipated.

Physical Evidence

Two lines of evidence argue that the original mineral is gypsum and that gypsum is the mineral to be expected at the earth's surface under conditions as they exist today. (a) In almost all cases of modern calcium sulfate deposition, gypsum is the mineral that is being produced. (b) In general, anhydrite in ancient rocks is found in the deep subsurface and gypsum is found in outcrop and the shallow subsurface.

When we examine the modern deposits where calcium sulfate is being produced, gypsum is found in almost all cases. These examples occur throughout the world and represent areas of extremely hot and extremely cold mean annual temperatures. Several examples of modern anhydrite have been reported. Perhaps the best known of these is within the supratidal flat sediments of the Persian Gulf coast of the Trucial States. Here, clusters of fine crystals in the form of nodules exist within the supratidal carbonate sediments. Some controversy exists as to whether this anhydrite grew as anhydrite or represents replaced gypsum crystals similar to the gypsum crystals that are found associated with the anhydrite and in nearby areas. In the lower part of the tidal flats, anhydrite replacing gypsum has been found. In the extreme upper part of the flats, gypsum is replacing earlier anhydrite. In the middle part of the flats, however, where little or no gypsum exists in the upper few feet of sediment, strong evidence is lacking to show whether the anhydrite is a secondary replacement of gypsum or grew as anhydrite.

Gypsum has been reported to occur as deep as 3900 ft in the subsurface. Anhydrite alone appears below this depth (Murray, 1964). In outcrop, gypsum is the common, almost universal, calcium sulfate mineral. The transition from gypsum at the outcrop to anhydrite in the subsurface generally takes place over the first 100 to 200 ft. This is true because we usually are observing rocks that have been buried to greater depths, have been uplifted, and currently are being exposed by erosion. When gypsum and anhydrite are observed in deep quarries and shallow borings, it can be shown that the gypsum is replacing an earlier anhydrite. The depth of the transition in these cases normally depends on the permeability of the rock and the rainfall. For gypsum to replace anhydrite, water must be added to the system and the local availability of water commonly determines the rate at which the transition will take place. Replacement of anhydrite by gypsum usually begins along bedding planes and fractures and finally alters the whole rock to gypsum. As the gypsum quarries become deeper, it is not uncommon to find more and more anhydrite that has escaped replacement by gypsum.

There does not appear to be an example on the earth today of deposition of calcium sulfate minerals accompanied by continual subsidence. If such a place were found, it would be possible to observe the change from gypsum to anyhdrite with depth of burial. In the Trucial states along the Persian Gulf, deep wells encounter gypsum to depths of 200 to 800 ft below the surface (Fig. 15-1).

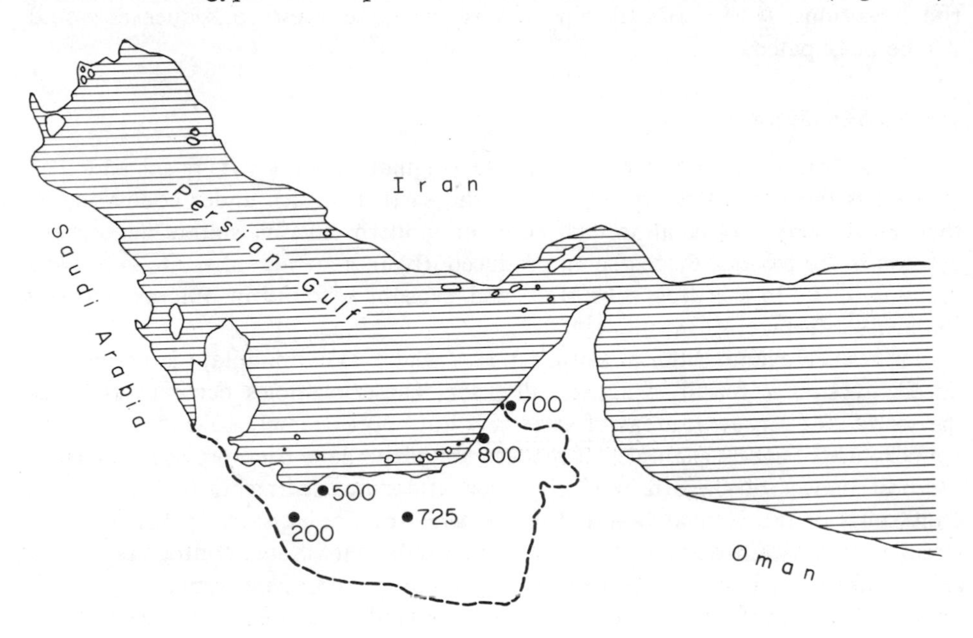

Fig. 15-1 Depth below the surface of the gypsum/anhydrite boundary, lower Fars Fm (Miocene).

The change to anhydrite below this depth takes place over a few tens of feet. In this area, however, there appears to have been relatively recent local uplift and the depth of the transition may be higher than the depth at which it originally took place. Indeed, it is not possible to determine accurately whether the subsurface boundary represents a transition from gypsum to anhydrite or slightly earlier-formed anhydrite to gypsum.

Chemical Evidence

Laboratory investigation of the precipitation of gypsum and anhydrite has been complicated by two problems. The first has been the inability to produce an equilibrium diagram for the $CaSO_4$-H_2O system, for which both dissolution and precipitation of the solid phases has been achieved in brines. The second problem is that formation of gypsum often takes place under nonequilibrium conditions and kinetic factors commonly determine whether gypsum will be formed under any set of conditions.

Three chemical factors should determine the form of the equilibrium diagram for the $CaSO_4$-H_2O system. They are (a) temperature; (b) activity of water, a factor related to the concentration of the brine; and (c) pressure. Anhydrite should be favored at higher temperatures and concentrated brine, which means lower values of activity of water. Gypsum should be favored by an increase in hydrostatic pressure. This is true because a given volume of anhydrite plus water is larger than the volume of an equivalent amount of gypsum. However, pressure is a relatively unimportant variable in the natural system. The increase in the value of the equilibrium temperature from thermodynamic calculations at 40°C, for example, to a value of 41°C would require a pressure:

$$\frac{dP}{dT}_{\text{at } 40^\circ\text{C}} = 85.4 \text{ bars/degree.}$$

Because it has thus far been impossible to synthesize anhydrite under equilibrium conditions out of solutions similar to seawater, Posnjak (1938 and 1940) attempted to approximate values for the transition temperature and the effect of concentration of the solution by plotting the solubility of the two solids, gypsum and anhydrite, as a function of temperature. The least soluble phase was considered the stable phase. This method, which suffers from an inability to actually precipitate the two phases under equilibrium conditions, has led to considerable controversy and misunderstanding.

On Posnjak's plot of solubility of gypsum and anhydrite as a function of temperature the lines intersect at 42°C. Below 42°C, gypsum is less soluble; above 42°C, anhydrite is less soluble. Thus it has been concluded that gypsum would precipitate from concentrated calcium sulfate solutions at temperatures below 42°C and that anhydrite would form at higher temperatures. Using the same method, Posnjak determined that if seawater were evaporated at 30°C, gypsum would form when a concentration of 3.35 times the original salinity had been reached and continue to form until a concentration of 4.8 times normal salinity had been reached. Above this concentration, anhydrite should form. These values have been quoted widely despite their limited value.

Hardie (1967), attempting to escape the problem of the inability to precipitate anhydrite under equilibrium conditions from seawater-like solutions, recognized that the two most significant variables were temperature and activity of water. Gypsum and anhydrite can be grown under equilibrium conditions in solutions such as concentrated Na_2SO_4 and H_2SO_4. He produced equilibrium data

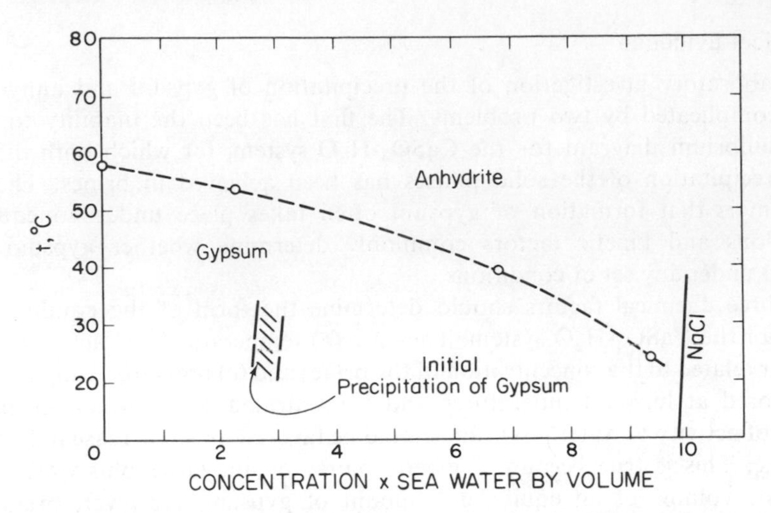

Fig. 15-2 Gypsum-Anhydrite equilibrium diagram. (Calculated from data of Hardie, 1967.)

for gypsum and anhydrite using these solutions and plotted his data as a function of temperature and activity of water. It is possible to determine the activity of water as function of concentration of seawater and thus convert the data to a diagram that expresses gypsum and anhydrite stability fields as a function of temperature and concentration of seawater (Fig. 15-2). This is the best approximation available for the gypsum-anhydrite system. The conditions needed to produce anhydrite exist several places on the earth today, where very high temperatures are found and where the hydrologic conditions for high brine concentration prevail. However, gypsum is found commonly in modern sediments even under these extreme conditions because of metastable precipitations.

The difficulty with the approach above is that anhydrite does not usually form under equilibrium conditions. Metastable gypsum is common. Indeed, if seawater is evaporated at any temperature below that required for the formation of the calcium sulfate hemihydrate, and at any rate that has ever been attempted in the laboratory, gypsum and only gypsum is formed. Once gypsum precipitates, it can be slowly replaced by anhydrite. The reverse reaction of replacement of anhydrite by gypsum is relatively rapid in the presence of water and this commonly takes place in modern sediments and ancient rocks.

A conclusion may be drawn that under most earth surface conditions gypsum should be the original mineral. Under extreme earth surface conditions of temperature and brine concentration, anhydrite should form but gypsum is commonly produced metastably.

15.4 DEPOSITIONAL ENVIRONMENTS

Layered sedimentary rocks composed of calcium sulfate minerals appear to have been deposited in two general types of environments. These settings are large standing bodies of water that have been evaporated to the point of gypsum precipitation and tidal flats or desert playas where brine within the vadose zone has become concentrated by evaporation to the point of precipitation of these minerals.

There does not exist today a large standing body of water that has been evaporated to the point of gypsum precipitation and where gypsum is accumulating. Indeed, to sufficiently isolate a body of seawater from the vast reservoir of the sea in order to create the conditions for evaporite deposition may have been a relatively uncommon occurrence in the past. If we examine the degree of restriction necessary today, then it becomes obvious why such situations may be relatively uncommon. A part of the sea must be separated, either by tectonic movement or the building of a sedimentary or volcanic barrier, from the remainder of the sea. If a major evaporite deposit is to be produced, however, then there must be a way for new seawater to enter the isolated basin continually. This entry may be either a very small surface connection or the evaporite basin may be completely isolated at the surface and be fed only by seawater springs. If the basin is fed by a surface channel, then the cross-sectional area of the channel would have to be something less than one-millionth of the surface area of the evaporite basin. This number is based on an examination of the sizes of entryways to present seas, their surface area, and salinity. A look at the hydrographic chart for the Red Sea and its associated gulfs gives us an idea of the degree of restriction necessary. Near-normal seawater that supports a flourishing coral reef exists at the head of the Gulf of Akaba despite the extreme restriction of this body of water and the aridity of the surrounding land climate. No hypersaline brine generated by evaporation exists here. The size of the entryway to the Red Sea is simply too large to permit a high degree of brine concentration.

In his summary of the deep, standing body of water evaporite basins, Schmalz (1969) discusses the Piano del Sale (Danakil depression) in eastern Ethiopia. This evaporite basin was formerly connected to the Red Sea by a highly restricted opening. It has been isolated by Tertiary and Holocene lava flows and has ceased to exist as an active evaporite basin. Originally the body of water was deeper than 200 m and now contains anhydrite, halite, and some potash minerals.

Another factor is the permeability of the underlying sediments. If they are highly permeable, then subsurface reflux of brine will take place at low concentrations and the body of water never will become sufficiently concentrated to produce gypsum. It is possible that complete surface isolation was achieved in some ancient evaporite basins. Marine water seeps through the sediments of narrow land barriers and emerges as springs within the evaporite basin, thus providing the source of new seawater. The supply of new seawater by springs has an advantage over surface channelways in that springs provide a greater degree of stability and allow conditions to remain constant for longer periods of time. This would account for uninterrupted, relatively thick sedimentary sections of evaporites. A surface entryway is always at the mercy of sedimentation, erosion, tectonic activity, and volcanism that could easily modify the configuration of the channel and thus change the conditions in the basin.

Two additional models have been proposed for a standing body of water evaporite basin. They are (a) a very large, very shallow sea that, because of its size and the depth of water, is in poor communication with the rest of the sea; and (b) the assumption that the surface water is less saline than the underlying water. In the first model it would appear very difficult to accumulate thick sections of evaporites that show continuous sedimentation of evaporite and associated sediments. This is especially true because gypsum and salt depositional rates in excess of 0.5 mm/yr would be expected and any shallow body of water would soon become filled and the sediment exposed. It is unlikely that subsidence under most conditions could maintain the body of water. In the second model we must remember that evaporation can only take place at the air-water interface. Seawater entering an evaporite basin will become concentrated in the surface layer and form a brine that immediately tends to sink because of its density. The heavy brine will accumulate at the bottom of the basin but it should not be more concentrated

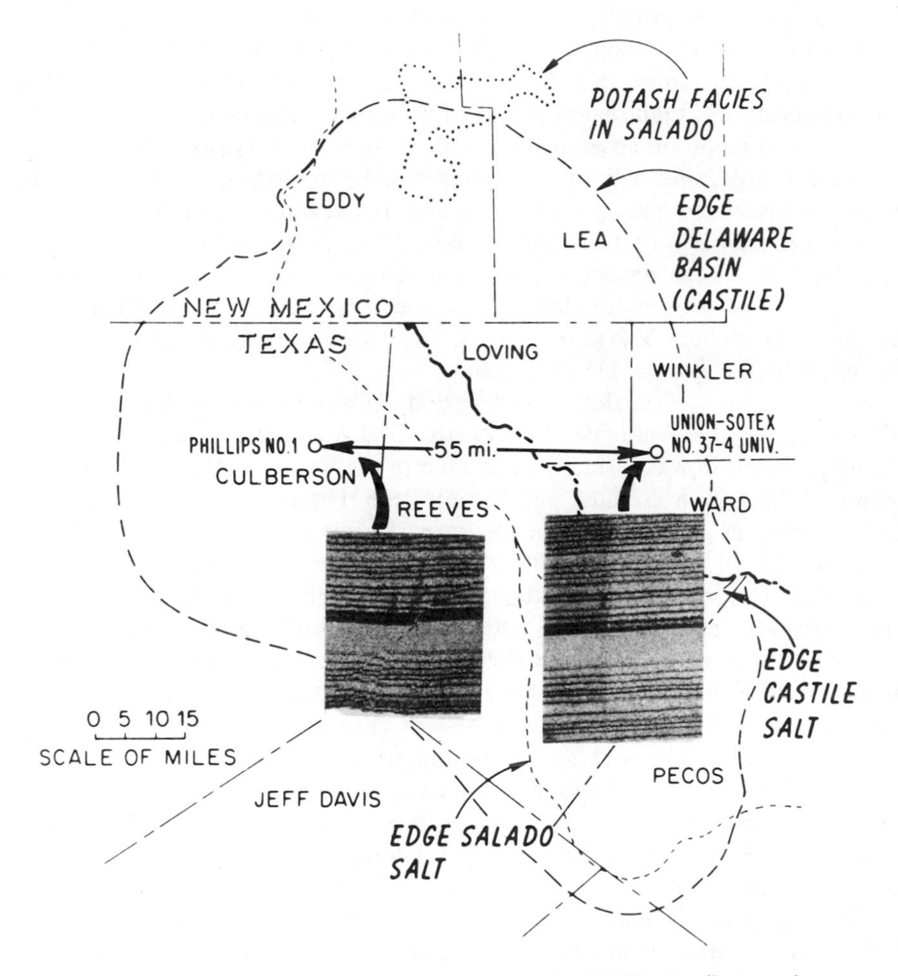

Fig. 15-3 Correlation of varves near base of anhydrite II in Castile evaporite. (Prepared by W. T. Holser and Roger Anderson.)

than the uppermost surface layer. If crystals form at the surface, they will sink. During their descent they will encounter undersaturated brine and will tend to dissolve. Only when the total brine column becomes saturated with respect to the calcium sulfate mineral will significant sedimentation take place. Models that assume a flow of seawater along the surface of the evaporite basin together with continuous concentration neglect sinking of surface-produced brine down through the body of water as a continual process. For these reasons, standing body of water evaporite basins probably were filled with concentrated brine. Thus the possibility of organic production on their shores, or the development of a living carbonate reef as a barrier to circulation, is highly unlikely.

Despite the absence of a modern example of a deep standing body of water evaporite, we can go to the Quaternary record to examine the characteristics of this type of deposit. The Pleistocene Lisan Formation was deposited in the Dead Sea Basin. This basin is not representative of an isolated basin fed by seawater but it does represent an example of gypsum deposition in a standing body of water. The sediment is laminated with individual layers that range in thickness from a few millimeters to over a centimeter and that alternate between gypsum and clay. The material is devoid of a recognizable fauna and the laminae are persistent laterally over wide areas. In addition, soft sediment deformation structures are common with small faults and deformed bedding occurring within beds and overlaid by undeformed sediment. These same characteristics are found in most standing body of water evaporite deposits (Fig. 15-3). In other ancient examples

Fig. 15-4 Gypsum crystals growing by displacement of sand and gravel near White Sands, New Mexico.

the sedimentary material between the gypsum or anhydrite may be shale, limestone, or dolomite.

The second common depositional setting for the calcium sulfate minerals is within the vadose zone and upper phreatic zone on supratidal flats and desert playas. Most modern gypsum is found within this setting. The gypsum occurs as crystals that have grown within the sediment. If the sediment is moderately soft and easily pushed aside, then the crystals grow by displacement (Fig. 15-4). That is, their force of crystallization will push aside the surrounding host sediment and the crystals form relatively free of inclusions of the host sediment (Fig. 15-5).

Fig. 15-5 Individual displacement gypsum crystals. The center crystal is 5 inches long.

In more rigid host sediment the gypsum will grow in pore space and produce sand crystals, which are single crystals that poikilitically include the host sediment. This occurs because the force of gypsum crystallization is insufficient to physically displace the host sediment grains. Crystals of this type are common at depths of 10 ft or more in the sand under the Laguna Madre of south Texas. The host sediment may be either silicate material or carbonate and commonly will contain a fauna indicative of its source and depositional setting.

Evaporation can take place within the vadose zone where air and water exist together. The brine within the pore space will become concentrated and commonly supersaturated with respect to gypsum. Some gypsum will be precipitated. Because a dense brine is produced by evaporation, however, it tends to sink or reflux down through the underlying sediment beneath the groundwater table. Gypsum will continue to precipitate as long as the brine remains supersaturated with respect to that mineral. Thus some crystals will grow well below a saline groundwater table. Today crystals are forming at depths as great as 10 ft below the groundwater table. In a regressive tidal flat environment, where the intertidal zone is moving seaward because of the addition of sediment, a vertical section of sediment from marine to supratidal may be observed under the supratidal flats. The earlier-deposited marine sediment will be beneath the groundwater level if sea level remained constant. Gypsum crystals that grow from refluxing brine are found within sediments that were originally deposited within the intertidal and marine settings and in sediment that originally contained a pore fluid of normal or near-normal seawater (Fig. 15-6). These sediments have a fauna indicative of their environment, which appears to be completely inconsistent with the presence of abundant evaporite minerals. When these gypsum crystals are buried and reach a subsurface temperature, activity of pore water and pressure environment favorable for anhydrite, they are replaced by anhydrite. This results in the production of small anhydrite

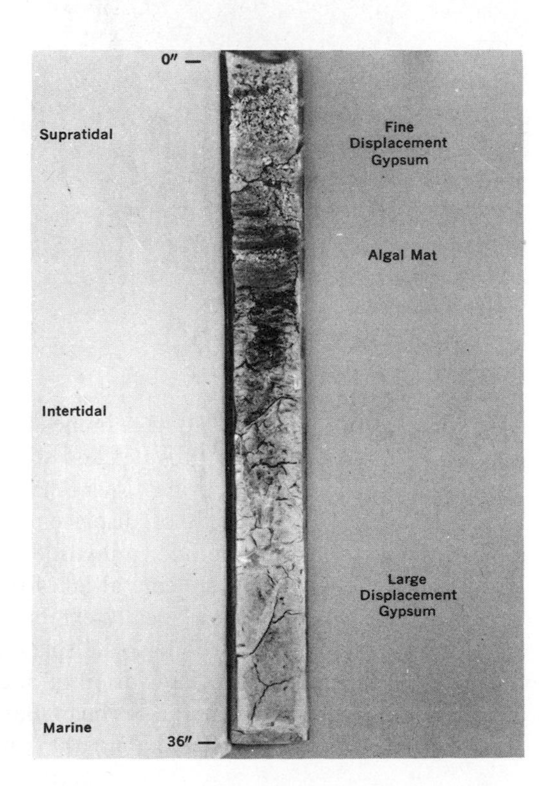

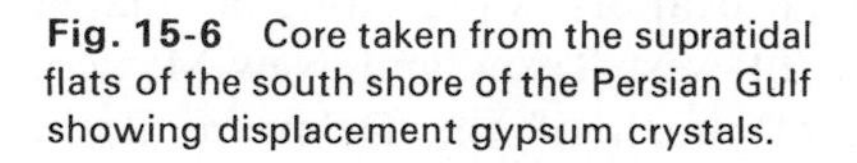
Fig. 15-6 Core taken from the supratidal flats of the south shore of the Persian Gulf showing displacement gypsum crystals.

crystals from a single large gypsum crystal and a 38% loss of solid volume. The result is a mush of fine anhydrite crystals that compact into semispherical nodules. Thus nodules of anhydrite are formed from earlier displacement gypsum crystals. If the original gypsum crystals were relatively free of included material, the nodule will be relatively pure anhydrite. If the gypsum crystals were small, then small nodules will result. If they were large, then the rock will contain large nodules. If the matrix was sufficiently rigid that it could maintain the crystal shape, then pseudomorphs of anhydrite after gypsum will be produced. This requires addition of calcium sulfate to the anhydrite in order to make up the volume lost during the replacement of gypsum.

Nodular anhydrite is extremely common in the ancient sedimentary record. It may occur as individual isolated nodules or the nodules may be so closely packed that only thin sheaths of the host sediment vaguely outline the individual nodules. This structure is commonly called chicken wire and is easily recognized (Fig. 15-7).

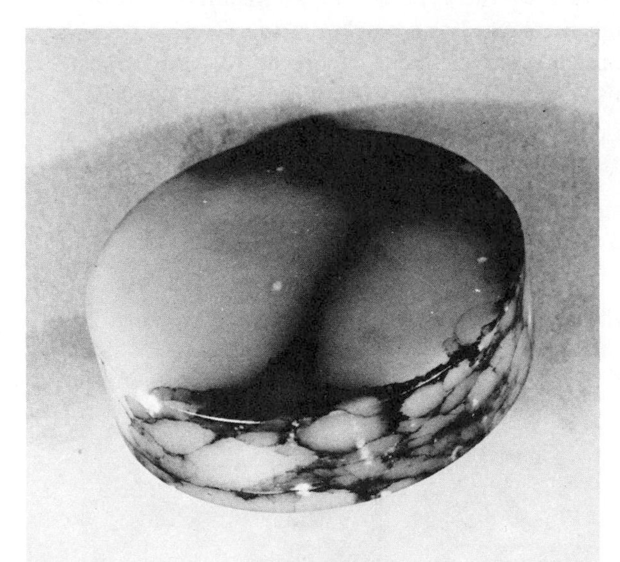

Fig. 15-7 Nodular anhydrite, Mississippian of Montana.

It is not uncommon to find nodules immediately underneath deposits of evaporites formed in a standing body of water and sometimes within these laminated rocks. Reflux of supersaturated brine down through earlier-deposited sediments and accompanied by the growth of displacement crystals probably produced these nodules. Within beds of nodular anhydrite, there commonly exist thin local beds of anhydrite deposited in ephemeral lakes that existed on the supratidal flats or desert playas. Some of these locally exhibit laminations.

Several ancient evaporite basins display a relationship between the laminated and nodular types. Such an association of nodular evaporites preceding laminated evaporites is clearly seen in the Permian Basin of west Texas and New Mexico. The carbonate rocks that make up the rim of much of the Delaware Basin contain abundant nodular anhydrite. They are found in the Permian rocks of the Central Basin Platform and the northern shelf. Following deposition of these carbonate and evaporite rocks, the Delaware Basin, which was starved for sediment during the deposition of the surrounding platforms, was filled with laminated gypsum and anhydrite (the Castile Formation) and finally halite and the more soluble salts. The implication of this relationship appears to be that the deep water basin that received the laminated evaporites was formed in part by subsidence, during which the margins remained at or near sea level and received sediments in shallow water and on tidal flats. The center of the basin received relatively little sediment during deposition at the margins, became deeper, and later was filled with laminated evaporites. The western edge of the Central Basin Platform, illustrated in Fig. 15-8, represents the boundary between the older nodular anhydrite of the platform and the later laminated anhydrite of the Castile Formation.

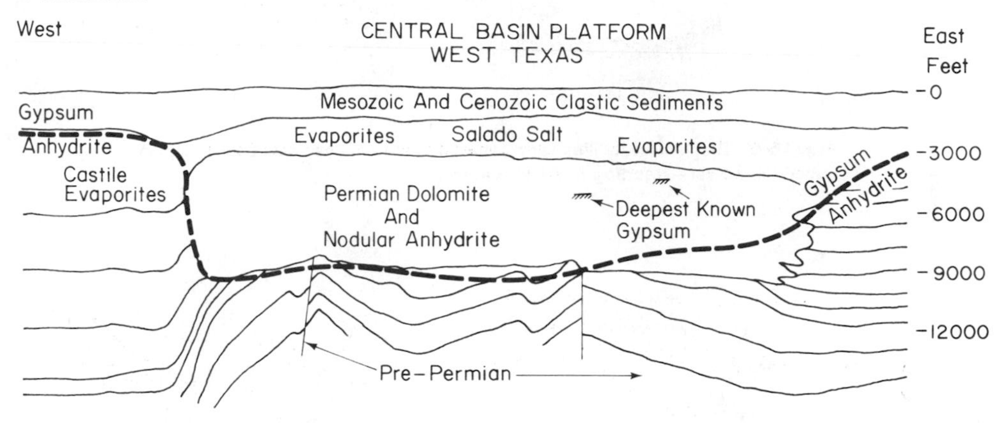

Fig. 15-8 Calculated depth to the present-day gypsum/anhydrite boundary under equilibrium conditions.

15.5 THE GYPSUM-ANHYDRITE CYCLE

A cycle exists in nature for the calcium sulfate minerals. This cycle results from the stability relationship found in the calcium sulfate-water system and the normal sequence of deposition-burial-uplift-erosion. The original mineral at the time of initial formation appears to be gypsum. There may be exceptions to this rule despite the arguments in its favor. The gypsum may be deposited either in large standing bodies of water evaporated to precipitation of gypsum or within and below the vadose zone of tidal flats and desert playas (Fig. 15-9). Gypsum formed within the vadose or phreatic zone grows predominantly as displacement crystals within earlier deposited detrital or carbonate sediment. Locally this gypsum may be replaced by anhydrite under near-surface conditions. This anhydrite should persist with burial if the material is maintained within the anhydrite stability field. This field is determined by the temperature, activity of water of the associated

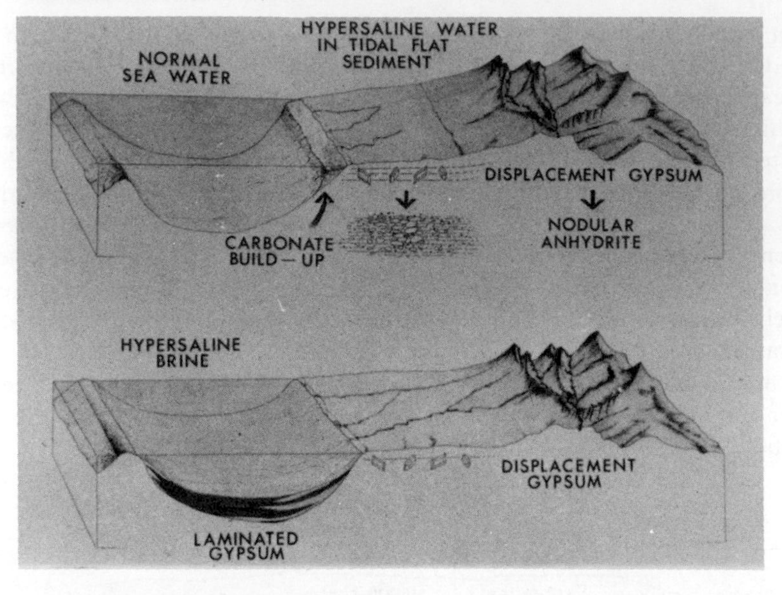

Fig. 15-9 Block diagram illustrating the two types of evaporite basins. Upper subaerial. Lower—standing body of water.

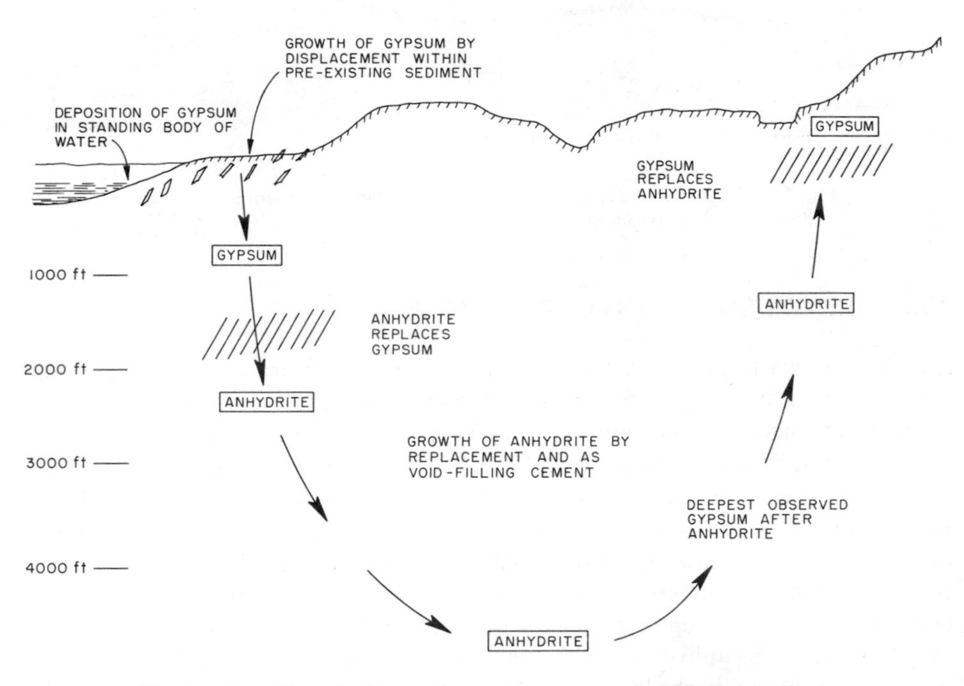

Fig. 15-10 Schematic diagram illustrating the gypsum-anhydrite diagenetic cycle.

solution, and the pressure. The back replacement of the anhydrite to gypsum within the first few feet of burial is more common. Original gypsum or near-surface gypsum after anhydrite will be carried into the subsurface until conditions for replacement by anhydrite are met. Depending on the salinity of the enclosing brine and the geothermal gradient, this will occur at a depth of 1000 to 2000 ft below the surface. The anhydrite will be brought back to the surface with uplift and erosion. During this phase, the composition of the groundwater tends to be less saline than during deposition, and the depth of the anhydrite-gypsum transition will be lessened. For example, if we calculate the expected depth for the transition in the Central Basin Platform in west Texas, we find that it occurs at depths ranging from over 9000 ft to as shallow as 2000 ft (Fig. 15-8). The difference depends largely on the salinity of the groundwater, which is greater on the west side of the platform because of proximity to evaporitic sedimentary rocks of the Delaware Basin. The actual presence of gypsum after anhydrite does not appear until much higher in the section because the transition from anhydrite to gypsum requires the addition of water and water is not easily available to the dense anhydrite of the deeper subsurface. Indeed, some anhydrite actually reaches the surface without being replaced by gypsum and appears in outcrop (Fig. 15-10).

15.6 COMPACTION

Anhydrite in the deep subsurface is universally dense and contains no measurable porosity. In this respect, it differs from most other sedimentary rocks. The lack of porosity is a measure of the volume reduction in the anhydrite. Once removed from the surface environment, there is little opportunity to add calcium sulfate as cement on a large scale. This does not imply that each unit volume of anhydrite will remain constant. There may indeed be transport of calcium sulfate locally within a gross evaporite unit.

If laminated gypsum is carried into the subsurface, it should compact physically and lose pore water until a depth is reached where the composition of the water, the temperature, and the pressure allow the replacement of the gypsum by anhydrite. The replacement will cause a loss of 38% of the solid volume. This loss will take place over a narrow range of overburden thickness and represents a delayed but rapid period of compaction. The anhydrite will then continue to compact physically. Dissolution at grain contacts will further decrease the gross volume and the section will thin until no porosity remains. If, for example, the original gypsum sediment had a porosity of 60% and the gypsum comprised 80% of the sediment, the final rock would represent only 40% of the original, assuming no compaction of the nonanhydrite layers. When we remember that the replacement of gypsum by anhydrite generates abnormal fluid pressure and outflow of water with loss of calcium sulfate and the ever present possibility of later dissolution of the anhydrite by introduced fresh water, then the thicknesses of laminated anhydrite that are observed in the subsurface probably represent one-third or less of the original depositional thickness.

In the case of a nodular anhydrite that develops from a solid gypsum crystal in the subsurface, the only compaction to be expected is that which results from the replacement of gypsum by anhydrite, which is 38% of the solid volume. Thus a sediment containing 80% displacement gypsum crystals would be reduced by 30%, assuming no compaction of the nonanhydrite material.

Bulk dissolution of evaporites by groundwater is a common phenomenon. There are many examples where collapse of overlying beds has occurred, sometimes with the production of collapse breccias and sometimes with a simple letting down of the overlying section as the evaporite is removed. In many cases no evaporite rock remains. In some examples, however, the breccias found in outcropping rocks can be correlated with evaporite beds that still exist in the subsurface.

15.7 ROCK SALT AND THE MORE SOLUBLE MARINE EVAPORITES

Sedimentary rock sections up to 3000 ft thick, composed predominantly of the mineral halite (NaCl), occur in the geologic record (Table 15-2). These beds may be associated with thin beds of gypsum or anhydrite or, alternatively, these minerals may be dispersed within the bedded halite. When thin beds of gypsum or anhydrite occur, they are commonly dark gray and define the sedimentary bedding. This association of gypsum and anhydrite with halite is to be expected because they coprecipitate over a wide range of seawater concentrations. Structural studies of deformed salt in salt diapirs have relied on these anhydrite or gypsum bands as mapping horizons. Rock salt deposits are commonly found associated in the sedimentary record with red shale but also occur with sandstone and carbonates.

The formation of halite requires essentially the same conditions as those of the calcium sulfate minerals. However, the degree of total restriction of the basin from the sea must be even greater. In addition, the formation of thick sedimentary sequences of rock salt requires the continuous addition of new seawater, accompanied by limited reflux of the heavy brine out of the basin. Seawater must be concentrated to approximately one-tenth of its original volume for the precipitation of halite. Thus the gross permeability of the floor of the basin to reflux must generally be less than in a basin where gypsum is precipitated. A large standing body of water evaporated to the point of halite precipitation does not exist today, although they have existed in the past as witnessed by the thick uninterrupted sections of halite.

Halite is being formed under natural conditions in some desert playas, such as in parts of the Rub al Khale Desert in southern Saudi Arabia and the Oman. Here brine produced by the dissolution of earlier evaporites is being concentrated by evaporation within the vadose zone and in shallow ephemeral lakes. The salt produced is relatively impure because of associated deposition of detrital material. In addition, salt occurs in beds ranging in thickness from a few inches to several tens of feet in the upper part of carbonate-evaporite cycles. In the upper part of the Madison Formation in the Williston Basin, for example, halite occurs above nodular anhydrite. There is independent evidence in these evaporites for repeated subaerial exposure, such as mud cracks, and it is presumed that the salt was deposited under high supratidal flat or desert playa conditions.

One form of halite associated with the vadose zone is particularly interesting. In silicate or carbonate sand, the evaporation of seawater within the vadose zone to the point of halite precipitation causes the growth of crystals in the small interstices between individual clastic grains. Seawater in the vadose zone exists in the small pore space surrounding the points of grain contact. With evaporation and further concentration, the halite crystals grow adjacent to the points where the grains meet. The force of crystallization of the halite has the effect of pushing each grain away from its neighbor and creating a surface sediment with a puffy appearance that looks very much like freshly plowed ground. The original grains are now no longer in grain contact. If the salt is dissolved, the grains will fall back into grain contact. The new grain positions will commonly have a different packing, however, and keystone vugs may be formed. That is, the grains will not efficiently fill the space and one grain may act as a keystone holding open a larger void space within the sandstone or grainstone. If the rock receives an early permanent cement at this stage, this structure may be preserved. A similar type of structure may be produced locally on a beach where the sand is disrupted by the escape of air after a large wave has smashed upon the beach and trapped air within the sediment.

When salt is produced in artificial solar salt operations, some crystals form at the surface of the water. Dellwig (1955) has shown that a similar phenomenon existed during halite precipitation in ancient standing bodies of water. Many crystals of halite found in ancient deposits are cloudy because of the presence of abundant fluid inclusions and the crystals have a characteristic hopper shape. Evaporation at the surface produces initial halite crystals. Surface tension holds the crystal temporarily floating on top of the brine. With continued growth of the crystal, it tends to sink although its position at the surface is maintained by surface tension. Under these conditions the upper ends of the four vertical faces of the cube are in contact with the continually evaporating brine and growth of the crystal continues on the edges of these faces. As the process proceeds, the crystal continues to sink and new halite is added on the outer and upper edge. This results in the development of an inverted hollow pyramid of halite. These crystals are fragile and may be broken when disturbed by wave action. When broken or disturbed, they sink and accumulate on the bottom as sediment. Several crystals may grow together to form a mat on the water surface that may sink as a unit. These crystals later develop overgrowths of halite and assume the common cubic form. This later material is generally free of inclusions and the crystals are clear. The clear halite was probably added during the descent of the crystal through the water or immediately after deposition. Deposition of abundant surface-formed crystals followed by growth of crystals within the body of water may impart a layered cyclic banding to the salt deposit of alternating clear and cloudy salt beds. Clear and cloudy salt banding appears to be present in the central part of the Michigan Basin and almost absent on the margins. As such, it can be used as a facies indicator and its presence suggests relatively deep water.

TABLE 15-2 MARINE EVAPORITE DEPOSITS OF THE CONTERMINOUS UNITED STATES

Thicknesses are approximate and include interbedded nonevaporite sediments. Evaporite types are calcium sulfates, including gypsum and anhydrite, S; halite, H; and bedded bittern salts of potassium and magnesium, B. (After Stewart, 1963)

Age	Area	Formation or interval	Thickness of evaporite section (ft)	Evaporite type
Ordovician	Williston Basin	Whitewood-Bighorn	25	S
	Illinois Basin	Joachim	50	S
Silurian or Devonian	Williston Basin	Niagaran	25	S
	Michigan Basin	Bass Island, Salina	3,000	S, H
	New York	Camillus, Syracuse	500	S, H
	West Virginia	Salina	800	S
	Iowa-Missouri	Niagaran-Cayugan?	100	S
Devonian	Williston Basin	Potlatch, Jefferson	400	S
		Prairie	600	S, H, B
	Iowa	Cedar Valley-Wapsipinicon	50	S
	Michigan Basin	Detroit River	1,200	H
	Southwestern Montana	Three Forks	50	S
Mississippian	Williston Basin	Otter, Charles	1,000	S
	Southwestern Montana	Kibbey	100	S
	Iowa	Keokuk	30	S
	Illinois Basin	St. Louis	200	S
	Michigan Basin	Michigan	350	H
	Western Virginia	Maccrady	1,000?	S, H
Pennsylvanian	Paradox Basin, Utah	Paradox	4,000	S, H, B
	Gypsum Basin, Colorado	Maroon	500+	S
	Black Hills region	Minnelusa	50	S
or Permian	Michigan Basin	Virgilian?	50+	S

	Black Hills	Minnekahta, Opeche	100	S
	Gypsum Basin, Colorado	Maroon	500	S
	Grand Canyon area	Kaibab, Toroweap	900	S, H
	Northern New Mexico	San Andres	100	S
	South central New Mexico	Yeso, Abo	2,000	S, H
		Chalk Bluff, Whitehorse	1,000	S, H
	Southeastern New Mexico	Rustler, Salado, Castile	4,500	S, H, B
	Central Texas	San Angelo, Clear Fork, Wichita	1,500	S
Permian	Texas Panhandle	Pease River, Clear Fork, Wichita	2,000	S
	Oklahoma Panhandle	Dog Creek, Blaine, Cimarron	1,000	S
	Western Oklahoma	Blaine, Clear Fork, Wichita	1,500	S, H
	West central Kansas	Harper, Wellington, Marion	800	S, H
	Iowa	Leonardian?	50	S
or Jurassic	Gulf Coast	Louann, Werner	1,500±	S, H
Triassic	Wyoming, Nebraska, North and South Dakota.	Chugwater	several hundred	S
	Central Wyoming	Gypsum Springs	200	S
	Central Montana	Gypsum Springs	100	S
	Southeastern Idaho	Preuss	450	S, H
	Central Utah	Arapien	1,000?	H
	South central Utah	Carmel	400	S
Jurassic	North central New Mexico	Todilto	100	S
	East central Colorado	Summerville equivalent	200	S
	Southern Arkansas	Buckner	350	S
	Northeastern Texas	Buckner	700±	S, H
	Southwestern Alabama	Buckner	950	S, H
	Northwestern Louisiana	Ferry Lake	300±	S
Cretaceous	Southwestern Texas	Fredericksburg	150	S, H
	South central Florida	Comanchean	6,000	S
	Southeastern Florida	Comanchean	4,000+	S, H
Tertiary	Florida	?	several hundred	S

Evaporation of seawater to approximately one-twentieth of its original volume causes the precipitation of polyhalite, $Ca_2K_2Mg(SO_4) \cdot 2\ H_2O$. Polyhalite and halite should precipitate together until the solution becomes saturated with respect to potassium- and calcium-free magnesium sulfates. The last stages of evaporation are very complex; the minerals produced depend on whether the earlier-formed crystals are removed from contact with the solution or whether the residual brine is allowed to react with the early minerals. For a discussion of the later stage evaporite minerals, the reader should begin with Stewart (1963).

Although none of the large deposits of the potassium and magnesium salts of the world, such as the Zechstein, Salado, or Prairie evaporites, exhibit a simple predictable succession of salts, most of the differences are explainable in terms of dilution of the water body during evaporation, diagenetic reactions, and dissolution. In general, the more soluble salts appear in the upper part of the evaporite sequence. Figure 15-11 illustrates a comparison of a theoretical succession from euxinic bottom conditions developed on initiation of brine concentration, through the filling of a basin, with the deposits found in the Elk Point evaporites of Alberta and the basin center evaporites of the Zechstein.

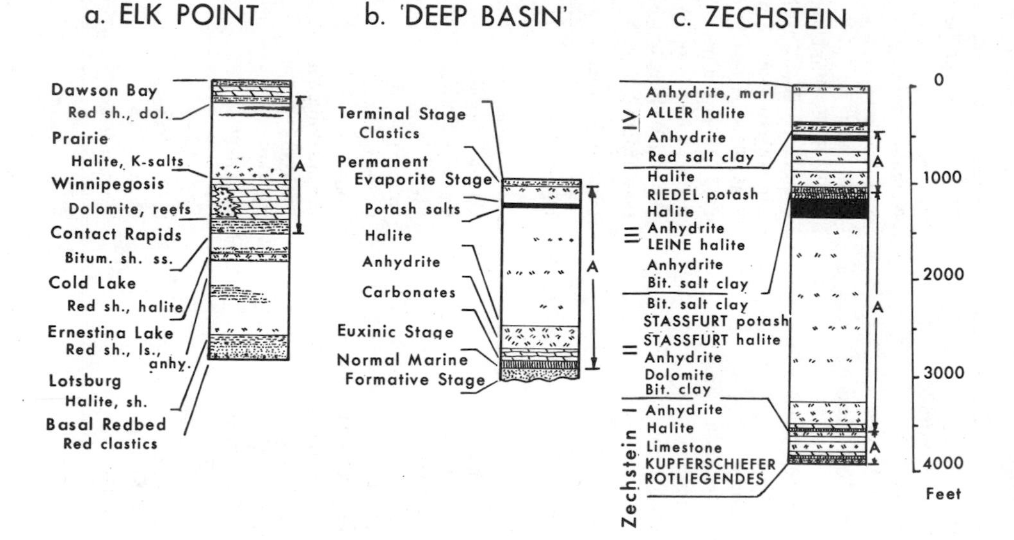

Fig. 15-11 Stratigraphic sections of two ancient evaporite deposits compared with hypothetical stratigraphy of evaporite sediments in deep basin. Typical deep-basin sequence shown in (b) (marked A) is observed in upper Elk Point subgroup (a), and is repeated three times in Zechstein deposit (c). (After Schmalz, 1969.)

15.8 NATIVE SULFUR

Thus far in our consideration of diagenetic phenomena among the major evaporite minerals, we have considered only hydration-dehydration and recrystallization reactions. An additional reaction path of importance in the sedimentary environment is chemical reduction of sulfate minerals to hydrogen sulfide and free sulfur. This process is of interest from both a scientific and an economic point of view. The economic importance is documented by the fact that natural reduction of sulfate minerals provides about 90% of the world's sulfur reserves and 95% of present production. The scientific importance of the reduction process is that it seems to be almost entirely the result of bacterial activities. Because of this, it affords an unusually clear insight into one aspect of the relationship between organic activity and lithogenesis.

Occurrence

Sedimentary sulfur deposits are always found in association with evaporite deposits containing gypsum or anhydrite because of the need for a more oxidized form of sulfur to serve as the raw material. For this reason, sulfur deposits are more abundant in Permian rocks than in rocks of any other period. In the United States, extensive deposits are found in Permian carbonate rocks in west Texas and southeastern New Mexico. The sulfur commonly occurs in limestone units (Fig. 15-12) and there is evidence to indicate that the largest deposits occur in zones of porosity associated with old oil traps. Dead oil and oil staining are common in all deposits. It is possible that hydrogen sulfide released from the petroleum

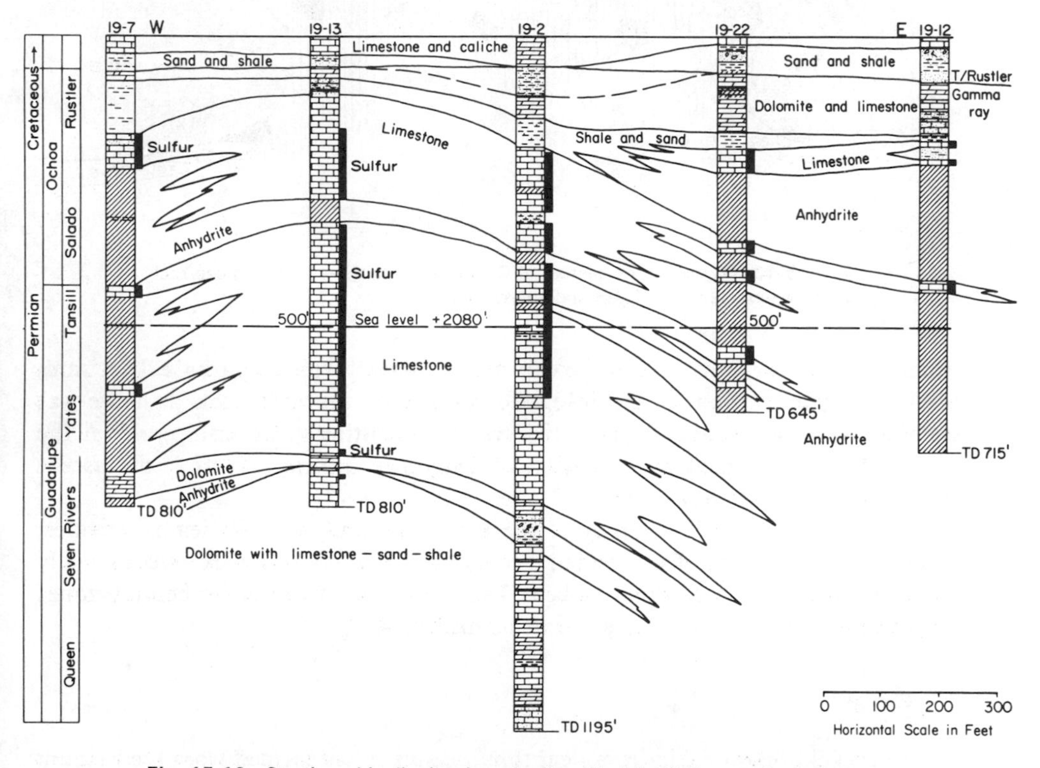

Fig. 15-12 Stratigraphic distribution of sulfur, Sinclair Oil Corporation's Fort Stockton area, Pecos County, Texas. (From Zimmerman and Thomas, 1969.)

was a partial source material for sulfur formation as petroleum is produced in the same areas and from the same formations that contain the sulfur. In some localities in west Texas, secondary sulfur occurs in *gypsite* deposits (disintegrated and weathered gypsum) and in alluvial conglomerates, clays, sands, and caliche mantling the surface.

The most extensive and commercially important sulfur deposits in the United States are those associated with salt domes in the Gulf of Mexico coastal area. The deposits occur in the cap rocks of intrusive plugs of halite at depths ranging from 300 to 3200 ft. Most typically, the sulfur is concentrated along the contact between gypsum and limestone beds (Fig. 15-13). Stratigraphic and petrographic

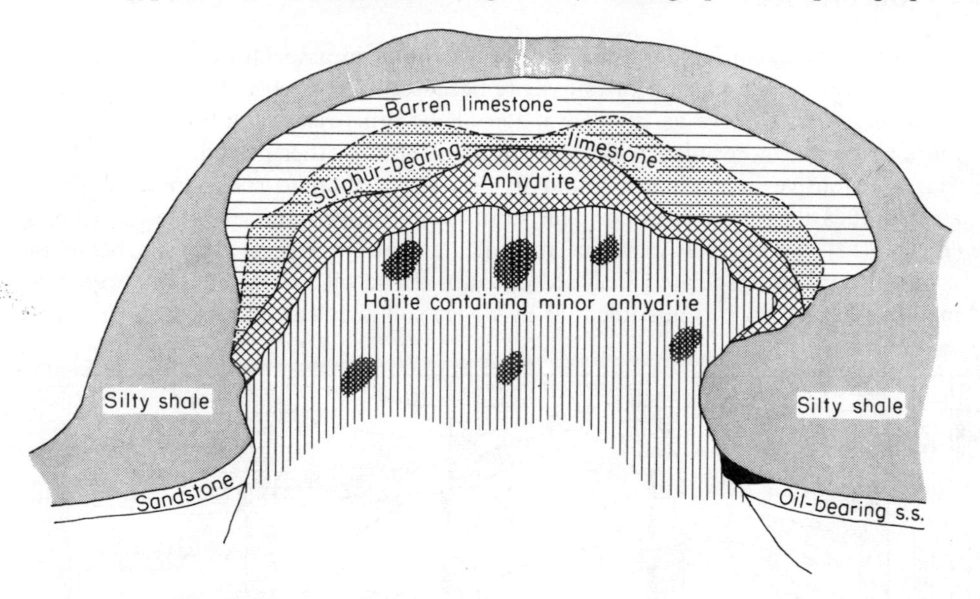

Fig. 15-13 Typical cross-section of Gulf coast salt dome showing cap rock with development of sulphur and limestone.

evidence reveal the anhydrite in the cap rock to have formed by residual accumulation and cementation of the insoluble residue of the salt stock as halite was leached away by circulating groundwaters. Crosscutting relationships reveal the gypsum to have formed by hydration of anhydrite during and after the ascent of the salt plug to shallower depths.

The sulfur in the lower part of the limestone unit may fill seams, cavities, or fissures or may be disseminated throughout the porous rock. Occasionally the sulfur occurs in a continuous bed. The thickness of the sulfur-bearing zones ranges up to 300 ft with sulfur contents as high as 40%.

Origin

From field relationships it is clear that the sulfur has formed from the calcium sulfate minerals. The most extensive sulfur accumulations occur in porous limestone immediately overlying the gypsum or anhydrite but the upper parts of these limestones are usually barren. It also is significant that there is little sulfur present within the body of the gypsum or anhydrite units because this suggests that the presence of the limestone is related to the origin of the sulfur. Also, the sulfur is observed to transect Pleistocene units. Therefore it did not originate penecontemporaneously with the salt bed, which is of Mesozoic age.

As a working hypothesis, we may suppose that the sulfur is formed rather rapidly by reactions that can be generalized as

$$6\,CaSO_4 + 4\,H_2O + 6\,CO_2 \longrightarrow 6\,CaCO_3 + 4\,H_2S + 2\,S + 11\,O_2$$

If the calcite formed during diagenesis, which would involve circulating meteoric waters, we would expect it to be reflected in the stable isotope ratios of carbon and oxygen in the calcite. Carbonates formed in association with evaporite deposits have δC^{13} values that range between -3 and $+5$ per mil and average approximately $+2$. The carbonate in freshwater limestones averages about -10 per mil. The calcite in salt dome caps, however, ranges in $\delta\ C^{13}$ from -15 to -60 per mil and averages about -45. Carbon with δ values below -20 can only be derived from organic sources, presumably from the petroleum associated with the salt domes.

The other question concerning the origin of the sulfur is the nature of the reducing agent that transformed the sulfate. Two possibilities seem to be present. Thermodynamic calculations reveal that petroleum can serve as the reducing agent but experiments indicate that the temperatures required for significant amounts of sulfur to form are at least several hundred degrees centigrade. The shallow depth of formation of some of the deposits precludes the possibility of such high temperatures during sulfur formation. The key to the origin of the sulfur is the presence in the cap rock of sulfate-reducing bacteria, most commonly *Desulphovibrio desulphuricans*. In the laboratory, these organisms can produce H_2S at a rate of 1000 ppm/day and, at this rate, even the thickest sulfate cap known (about 1000 ft thick) could be reduced in 100 years. This rate assumes an adequate supply of petroleum and optimum conditions over the entire cap so the natural rate will be less than 1000 ppm/day. But the extraordinary potential of *Desulphovibrio* is clear.

The clinching argument for the origin of sulfur deposits by bacterial reduction of gypsum or anhydrite lies in the S^{32}/S^{34} ratios of the sulfates and the coexisting native sulfur. Comparison of $\delta\ S^{34}$ values of modern seawater with modern evaporite deposits and the sulfate in salt dome cap rock reveals very similar values, on the order of $+20$ per mil. Reduction of sulfate to sulfide causes a large isotope fractionation with the sulfide becoming enriched in the lighter isotope. The sulfur in cap rock averages $\delta\ S^{34}$ of -5 per mil, compared to an average of $+20$ for the coexisting sulfate minerals. The magnitude of this fractionation is similar to that obtained in laboratory experiments using sulfate-reducing bacteria. Of course, the amount of isotopic enrichment depends on the magnitude and efficiency of the bacteria and residual sulfate in coastal Gulf of Mexico salt domes has been reported with δ values as high as $+66$ per mil, an enrichment of about 45 per mil over the ocean water in which they formed.

In summary, study of the genesis of sedimentary sulfur provides an excellent example of the fruits of a multidisciplinary approach to the solution of geologic problems. Field relationships tell us that the sulfur forms from sedimentary sulfate minerals. Stable isotope geochemistry reveals that the limestone cap rock is not of evaporitic origin. Organic geochemistry reveals that although the petroleum associated with the sulfur supplies the carbon dioxide in the reducing process, the petroleum itself is not the reducing agent. And, finally, geological microbiology establishes that sulfate-reducing bacteria are the reducing agents.

REFERENCES

ADAMS, J. E., and M. L. RHODES, 1960, "Dolomitization Refluxion," *Am. Assoc. Petroleum Geol. Bull.* **44**, pp. 1912–1920. (The first statement of the concept of subsurface reflux.)

CLARKE, F. W., 1924, "The Data of Geochemistry," 5th ed., *U.S.G.S. Bull.*, **770**, 841 p. (Classic source of geochemical data.)

DEFFEYES, K. S., F. J. LUCIA, and P. K. WEYL, 1964, "Dolomitization: Observations on the Island of Bonaire, Netherlands Antilles," *Science*, **143**, pp. 678–679. (Subsurface reflux and its role in producing gypsum in a modern example.)

DEGENS, E. T., 1965, *Geochemistry of Sediments.* Englewood Cliffs, N. J.: Prentice-Hall, Inc., pp. 170–183. (A discussion of the geochemistry of evaporites and a statement of the problems.)

DELLWIG, L. F., 1955, "Origin of the Salina Salt of Michigan," *Jour. Sed. Pet.*, **25**, pp. 83–110. (A discussion of halite sedimentation with emphasis on the significance of surface growth of the crystals.)

FEELY, H. W., and J. L. KULP, 1957, "Origin of Gulf Coast Salt-Dome Sulphur Deposits," *Am. Assoc. Petroleum Geol. Bull.*, **41**, pp. 1802–1853. (The definitive study of this topic, containing both experimental and field data.)

HARDIE, L. A., 1967, "The Gypsum-Anhydrite Equilibrium at One Atmosphere Pressure," *Amer. Mineralogist*, **52**, pp. 171–200. (The best data on the gypsum-anhydrite system with emphasis on the significance of the role of activity of water.)

IVANOV, M. V., 1964, "Microbiological Processes in the Formation of Sulfur Deposits," *Israel Prog. Sci. Trans.* (*1968*), Jerusalem, 298 pp. (Summary of extensive Soviet work on the importance of bacteria in sulfur formation.)

KERR, S. D., and A. THOMPSON, 1963, "Origin of Nodular and Bedded Anhydrite in Permian Shelf Sediments, Texas and New Mexico," *Am. Assoc. Petroleum Geol. Bull.* **47**, no. 9, pp. 1726–1732. (Origin of nodular anhydrite.)

KING, R. H., 1947, "Sedimentation in Permian Castile Sea," *Am. Assoc. Petroleum Geol. Bull.* **31**, pp. 470–477. (The first statement of the significance of surface reflux to the deposition of evaporites.)

KINSMAN, D. J. J., 1969, "Modes of Formation, Sedimentary Associations, and Diagnostic Features of Shallow-Water and Supratidal Evaporites," *Am. Assoc. Petroleum Geol. Bull.*, **53**, pp. 830–840. (Discussion of evaporites being deposited on the supratidal flats of the Persian Gulf.)

MacDonald, G. J. F., 1953, "Anhydrite-Gypsum Equilibrium Relations," *Amer. Jour. Sci.* **251**, pp. 884–898. (The role of pressure in the gypsum-anhydrite system.)

Maiklem, W. R., D. G. Bebout and R. P. Glaister, 1969, "Classification of Anhydrite—A Practical Approach," *Bull. Canadian Petroleum Geology*, 17, pp. 194–233. (A well illustrated classification of anhydrite types.)

Murray, R. C., 1964, "Origin and Diagenesis of Gypsum and Anhydrite," *Jour. Sed. Pet.*, **34,** pp. 512–523. (A summary of the deposition of gypsum and anhydrite with emphasis on deposits that form in the vadose zone and those that form in standing bodies of water.)

Posnjak, E., 1938, "The System, $CaSO_4$–H_2O," *Amer. Jour. Sci.* **235–A,** pp. 247–272. (The significance of temperature in the gypsum-anhydrite system.)

———, 1940, "Deposition of Calcium Sulfate from Sea Water," *Amer. Jour. Sci.*, **238,** pp. 559–568. (The significance of the composition of the solution to the gypsum-anhydrite system.)

Schmalz, R. F., 1969, "Deep-Water Evaporite Deposition: A Genetic Model," *Amer. Assn. Pet. Geol. Bull.* **53**, pp. 798–823. (A clear statement of deposition of evaporites in deep water.)

Stewart, F. H., 1963, "Marine Evaporites," in *Data of Geochemistry. U.S. Geol. Surv. Prof. Paper* **440** *Y*, pp. Y1–Y52. (An excellent summary of the formation and geochemistry of the marine evaporites.)

Zimmerman, J. B., and E. Thomas, 1969, "Sulfur in West Texas: Its Geology and Economics," *Univ. Texas Bur. Econ. Geol., Geol. Circ.* 69–2, 35 pp. (Summary of occurrence, potential, and economics of production.)

PART FIVE

OTHER SEDIMENTARY ROCKS

Many sedimentary sequences contain nearly pure deposits of chert, apatite, zeolite minerals, or hematite. How do such rocks form? Are cherts organic or inorganic in origin? How might we distinguish between these alternatives? The textures and structures of phosphatic rocks display a remarkable resemblance to those seen in carbonate rocks. How is this similarity to be interpreted? Zeolites form and are abundant in many localities on the deep ocean floor and in subaerially altered rocks. What are the environmental and chemical controls on the formation of zeolite minerals? The thickest sequences of sedimentary iron ores are of Precambrian age and are commonly very siliceous. Does this mean that the chemical composition of ocean water has changed drastically since the Precambrian?

CHAPTER SIXTEEN

CHERT

16.1 INTRODUCTION

Chert is a rock composed largely or entirely of microcrystalline or cryptocrystalline quartz. Most cherts are nearly pure silica, crystalline impurities usually totaling less than 10% and consisting largely of clay minerals, calcite, and hematite. Extracrystalline water is present in chert in amounts averaging less than 1%. Chert often contains chalcedony, which also is microcrystalline quartz but has a distinctive radiating texture visible in thin section. Quartz crystal size in cherts is quite variable, ranging from a few tenths of a micron (determined by electron microscopy) to a few tens of microns, and often it is not possible to distinguish between coarsely crystalline chert fragments and metaquartzite fragments in a clastic rock. Categorization based on crystal size is entirely arbitrary.

Cherts have been given many varietal names based on impurities, the more common names being *jasper* (red because of included hematite), *flint* (gray to black because of included organic matter), *novaculite* (very pure and white because of relatively high content of extracrystalline water), and *porcellanite* (porcellanous texture resulting from argillaceous and calcareous impurities). *Opal* often is present in cherts and some Tertiary cherts show all gradations between pure crystalline silica and pure opal or amorphous silica. The opaline cherts usually contain microscopically visible siliceous shells of diatoms, radiolaria, or sponge spicules, indicating the chert has formed from crystallization of the chemically unstable amorphous silica.

16.2 TEXTURE OF CHERT

Crystal Size

Electron microscopy of fractured chert surfaces reveals the crystals of quartz to be polyhedral and subequant to elongate in shape. Euhedral quartz crystals are absent, presumably because there was mutual interference among the crystals during growth. The size of the crystals is variable within a single chert bed on the scales of observation of both the thin section and the electron microscope but usually is larger than 1 μ. The smallest crystal size reported is 0.2 μ and there is a thermodynamic reason for believing this to be near the minimum possible crystal size of microquartz. Although solubility data can for most purposes be considered as independent of grain or crystal size, the two are inversely related at very small

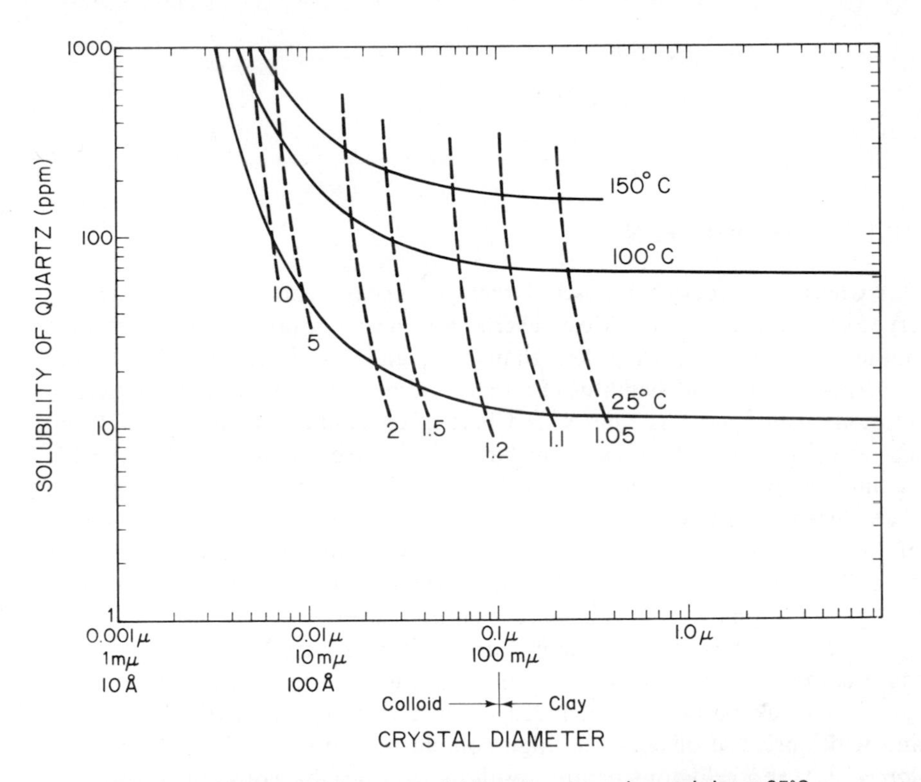

Fig. 16-1 Relationship between solubility of quartz and crystal size at 25°C, 100°C, and 150°C. Dashed lines are contours indicating solubility increase as a function of crystal size and temperature.

sizes. The explanation for the relationship is that solubility depends on the free energy of the crystal and at very small crystal sizes, such as are present in chert, the surface free energy of the quartz crystals rises to the point where it becomes quantitatively significant, thus raising the total free energy of the aggregate. Hence, the solubility of an aggregate of very small crystals may be many times greater than of a large crystal of the same substance. The crystal size at which the surface free energy becomes important depends in part on temperature. The formula used to determine solubility of tiny crystals is known as the Ostwald-Freundlich equation.

$$\ln \frac{S_r}{S} = \frac{2EV}{rRT}$$

where

S_r = solubility of a small particle of radius r in moles/liter

S = solubility of a very large particle in moles/liter

E = surface free energy of the large particle (interfacial surface energy between solid and solvent) in ergs/square centimeter

V = molar volume of the solid in cubic centimeters

R = gas constant (8.31×10^7 ergs/mole/degree)

T = absolute temperature

r = particle diameter in centimeters

For quartz

$$S_{298°\mathrm{K}} = 10.7 \text{ ppm} \quad \text{or} \quad 1.8 \times 10^{-4} \text{ moles/liter}$$

$$E = 416 \text{ ergs/sq cm}$$

$$V = \frac{60.1}{2.65} = 22.7 \text{ cc}$$

Figure 16-1 illustrates the effect of decreasing crystal size on quartz solubility for temperatures of 25°C, 100°C, and 150°C and indicates that crystals of quartz in chert should have a minimum stable size of a few tenths of a micron. The larger solubility of smaller crystals probably would keep them from persisting in rocks over geologically significant lengths of time. In view of the facts that the equation assumes a pure chemical system rather than the contaminated systems of nature and that the equation assumes spherical crystals rather than the polygonal elongate crystals present in chert, correspondence between theory and observation is quite satisfactory. As a single SiO_2 unit has a diameter of approximately 0.5 mμ, it appears that several hundred unit cells must be polymerized before the unit will have long-term stability.

Aggregate Texture

The textures of cherts are complex (Fig. 16-2) and include "metamorphic" textures such as granoblastic and porphyroblastic; "igneous" textures such as

Fig. 16-2 Transmission electron micrographs of chert cryptotextures. (A) Hypidiomorphic granular or granoblastic; (B) Seriate; (C) Spongy; (D) Porphyritic or porphyroblastic. (From Folk and Weaver, 1952.)

seriate and merocrystalline; and sedimentary-hydrothermal textures and structures such as spherulitic, botryoidal, and color banded. In addition, a unique spongy textured surface commonly is seen in cherts using an electron microscope. It is clear from the great variety of textures present that these rocks have a complex crystallization history; studies aimed at discovering the meaning of these variations are presently in their infancy. Important problems to be solved include the following. (a) Does the bimodal grain size of quartz crystals in some cherts result from partial recrystallization of a solid; varying degrees of saturation of an aqueous solution; or different temperatures of crystallization? (b) What causes the growth of quartz crystals in the pattern of elongate fibers, such as in chalcedony? (c) What effect does the presence of mineral inclusions and trace element impurities have on the crystallization pattern of chert? The increasing use now being made of the scanning electron microscope promises at least tentative replies to these and other similar questions in the near future.

16.3 IMPURITIES IN CHERT

Mineral impurities in cherts are largely confined to clay minerals, calcite, and hematite because cherts either form far from sources of terrigenous detritus or else form by replacement of limestones, which generally contain little detrital quartz, feldspar, or lithic fragments. Black cherts contain organic matter formed from the remains of the soft parts of organisms. As most cherts are nearly all quartz, petrographic and X-ray techniques are unsatisfactory as methods for determining the amounts of impurity materials; chemical analyses are a necessity. Because of the different chemical compositions of the few common impurities, mineralogic interpretation of these analyses is greatly simplified. All alumina usually can be assigned to clay minerals; all carbon dioxide, to calcite or dolomite; and at least most of the iron, to hematite. Potassium percentage can be used to evaluate the illite content of the clay assemblage. Nevertheless, it must be remembered that these assignments are not accurate because iron may be in chlorite as well as hematite; illites vary in potassium content depending on their degree of degradation; and the alumina percentage in clay minerals is variable both within and between clay mineral groups. Also, some cherts contain feldspar grains, volcanic rock fragments, or detrital heavy minerals such as augite or magnetite. Petrographic work must accompany mineralogic interpretations of chemical analyses.

In general, diatomaceous and radiolarian cherts are relatively enriched in clay minerals, while nodular and spiculiferous cherts are higher than average in calcite, a clear reflection of their most common environments of formation. The former types are commonly associated with pyroclastic rocks; the latter two occur most often in carbonate sequences. Nodular cherts nearly always contain less than 1% total iron because of their origin as replacement for limestone; bedded cherts generally contain more than 1% iron and values as high as 9% have been reported. Table 16-1 contains chemical analyses of natural cherts and illustrates the rela-

TABLE 16-1 CHEMICAL COMPOSITION OF CHERTS

	1	2	3	4	5	6	7
SiO_2	99.82	83.67	89.15	86.9	82.94	70.78	95.11
Al_2O_3	0.11	3.30	3.45	4.6	0.1	0.45	0.14
Fe_2O_3	0.00	7.74	0.58	2.7	3.4	0.02	0.40
FeO	0.07	—	2.19	—	—	0.30	0.44
MgO	0.00	0.50	1.05	0.8	0.19	1.88	0.71
CaO	0.00	0.40	0.24	0.1	1.60	12.90	1.11
Na_2O	0.00	1.00	0.22	0.4	0.65	0.05	<0.01
K_2O	0.03	1.42	0.53	1.0	1.40	0.06	0.01
P_2O_5	—	0.10	0.04	0.04	0.8	0.16	0.01
TiO_2	—	0.40	0.22	0.14	0.27	0.03	0.03
MnO	—	0.30	0.03	0.03	—	0.02	0.01
CO_2	0.00	—	—	—	0.40	12.04	1.38
H_2O^+	0.03	0.85	1.17	2.3	0.33 (combined H_2O^+ and H_2O^-)	0.48	0.25
H_2O^-	0.00	0.10	—	0.3		0.32	0.43
C	—	—	0.11	0.65	—	0.33	0.05
Totals	100.06	99.78	99.71	99.96	92.38	100.09	100.29

All analyses from Cressman, 1962, p. 6, 8, 10.

1. White novaculite from Arkansas.
2. Red radiolarian chert, Jurassic, Val d'Err, Switzerland.
3. Red and green radiolarian chert, Pumpernickel Fm., Lower Permian, Antler Peak Quad., Nevada.
4. Black radiolarian chert, Lower Carboniferous (Culm), Wallace, Hesse, Germany.
5. Chert in Phosphoria Fm., Permian, Utah.
6. Chert nodule, Delaware Limestone, Devonian, Ohio.
7. Chert layer in dolomite, Bisher Fm. of Foerste, Silurian, Ohio.

tionship between the content of impurity elements and the more readily observed characters of the rocks.

16.4 FIELD RELATIONSHIPS

Areally significant accumulations of chert in stratigraphic sections are of two general types: (a) nodules in carbonate sequences and (b) bedded chert associated with shales or iron formations. The relative volumetric importance of these types in the earth's crust is unknown.

Chert nodules in limestones are usually shaped like a lumpy ellipsoid several inches in size and flattened in the plane of the bedding. The purer the limestone, the more irregular the shape of the nodules. Commonly nodules are concentrated along particular bedding planes and are nearly absent along adjacent ones; spacing along bedding surfaces appears to be random. In some areas chert nodules are so abundant they coalesce in the plane of the bedding, forming either discrete beds with uneven surfaces or networks several inches thick and perhaps tens of feet in length. Cherts in limestones are usually structureless and translucent and contain many microscopic specks of carbonate inclusions that are interpreted as evidence of incomplete replacement of the host rock by the silica. Some nodules are black (organic matter) inside but white outside, a result of surface weathering. One of the most spectacular examples of replacement chert in nodules and layers of nodules is found in the Cretaceous chalks of western Europe and England. The nodules, most of which are solid (although some are hollow or contain nonreplaced chalk centers), commonly have strange and exotic shapes. They generally are concentrated along bedding planes; however, occasionally veins or stringers have developed vertically along joints.

The growth of opal, chalcedony, or jasper in wood commonly produces excellent preservation of the original wood tissue structures. Especially in opal the reproduction of the original tissue may be so perfect that even the cell structure is preserved. Locally the growth of larger crystals may cause the obliteration of the original structure or alternatively those parts of the wood may have been removed or obliterated prior to the growth of the minerals.

When associated with carbonate bank deposits, chert nodules or replacement chert areas occur largely in shelf and basinal areas rather than in the more pure carbonate shelf edge or flank talus deposits. The Permian reef complex in west Texas and southeastern New Mexico provides an excellent illustration of these relationships. Fossils in the basinal limestones have been selectively replaced by silica and chert nodules are abundant. The same is true of some of the shelf deposits. The silica content of the carbonate buildup itself is low, however, averaging approximately 2%. The silica concentration appears to be unrelated to the original fabric of the carbonate rocks.

Bedded cherts are common in eugeosynclinal sequences. These cherts are even-bedded, thinly laminated to massive, often black or green in color, and usually are several feet thick. Thicknesses of hundreds of feet are known. Interbedded rocks are confined to thin partings of dark shale, are commonly siliceous and sometimes contain pyrite. When examined petrographically, most of these chert beds are found to contain scattered remains of sponge spicules, radiolarian tests, or diatom frustules but most thin sections show no discernible organic remains. Bedded cherts are also found associated with Precambrian iron formations (see Chap. 19).

Deep-sea drilling has encountered Cretaceous and Tertiary bedded cherts in both the Atlantic and Pacific Ocean Basins. These chert beds are associated with turbidity current deposits and siliceous oozes containing diatoms, radiolaria, and other siliceous organisms. The inability to find these chert layers at all locations and the variation in age of the first chert layer encountered in drilling suggests that individual chert beds underneath the present deep sea are not continuous on the scale of hundreds of miles.

The colors of bedded cherts, the occurrence of pyrite in associated dark shales, the often high content of organic matter, and the even bedding and lamination all suggest origin in quiet waters of low Eh far from any source of coarse clastic detritus. In the context of eugeosynclinal tectonics this is interpreted to mean "deep" water. The evidence for water depth is largely circumstantial, however, as all fossils in the rocks are planktonic. Depths of 400 to 600 ft are as consistent with the available data as are several thousand feet. Recently it has been suggested that these cherts may be tidal in origin.

16.5 ORIGIN OF CHERT

Organic

The key to the origin of most Phanerozoic cherts is the chemical activity of organisms, the most important of which are diatoms (Triassic to present; mostly marine), radiolaria (Cambrian to present; marine only), and siliceous sponges (Cambrian to present; both marine and nonmarine). Despite the fact that both river waters and seawaters are undersaturated by one or two orders of magnitude with respect to amorphous silica, these organisms are able to remove additional silica from the waters. River waters contain approximately 10 times as much silica as surface ocean waters and for this reason the concentration of diatoms commonly is greater immediately seaward of river mouths than in the open ocean. In the open sea where diatoms and radiolaria are abundant in the surface waters, the silica content fluctuates seasonally between 0.5 and 2.0 ppm as the population of siliceous plankton waxes and wanes. Under optimum natural conditions, diatoms will reproduce by cell division once each day. Jørgensen (1953) demonstrated experimentally that two species of diatoms in an aquarium were able to reduce silica concentrations from initial values of 0.65 to 1.25 ppm to values of 0.065 to 0.085 ppm. The silica molecules are adsorbed onto an organic substrate provided by the diatom and then polymerize. In this way the silica-secreting plankton build shells of opal and, while alive, have the ability to maintain and enlarge their shells in contact with a medium that should dissolve them. Some diatoms are able to obtain silica by attacking and decomposing kaolinite but the relative importance of this mechanism is unknown. Presumably its importance increases as the silica content of the water decreases.

Shell maintenance in an undersaturated medium such as seawater or fresh water requires an energy expenditure by the organism and hence is possible only during the life of the creature. Upon death, the organic tissues are oxidized and the opaline silica is free to dissolve. This dissolution, however, is extremely slow at 25°C and even slower at the lower temperatures of bottom waters. As a result, diatom and radiolarian oozes exist in the present ocean depths and no doubt similar accumulations existed in ancient seas. Deposits consisting almost entirely of siliceous skeletons require for their formation only the near-absence of detrital material and calcareous plankton. Siliceous oozes have no necessary relationship to water depth and hence chert beds are dangerous criteria to use as evidence of "deep water."

If siliceous remains form only a relatively minor part of a carbonate sediment, silica is available for the eventual formation of chert nodules. The only significant difference between the formation of chert nodules in limestone deposits and chert beds many feet in thickness often is simply the volume of silica available. As we shall see, siliceous remains are not the only source of silica for the formation of chert in Phanerozoic rocks but probably are volumetrically the most important source.

Upon burial, circulating pore waters react with the opaline tests causing crystallization of the opal to microcrystalline quartz. Opal of siliceous organisms contains 2 to 13% of water and even in the absence of pore water this gel water is sufficient to initiate the solution-precipitation process that characterizes the opal-quartz transformation. As quartz is anhydrous, it is apparent that water present in the opaline shell is expelled during crystallization. The activation energy of the crystallization reaction has been determined to range between 14 and 23 kcal/mole depending on the characteristics of the opal. The higher value is less than half that of the aragonite-calcite transformation (57.4 kcal/mole). Ernst and Calvert (1969) determined the rate constant for the opal-quartz transformation in neutral solutions with no circulation of water. At 200°C, complete conversion of opal to quartz takes 47 years; at 100°C, 36,000 years; at 50°C, 4.3 million years; and at 20°C, 180 million years. Because subsurface temperatures always exceed surface temperatures, the figure of 180 million years for complete conversion should be considered a maximum. Nearly all presently existing opaline rocks are Cenozoic in age and the oldest known is Albian, approximately 100 million years old.

Diatoms and radiolaria may be either disk-shaped, elongate, or spherical and the volume occupied by a living diatom ranges from 10^{-11} to 10^{-5} cc. This volume is occupied almost entirely by cytoplasm, however; perhaps 10^{-9} cc (10^{-13} cu ft) is a reasonable estimate of the solid volume that could be formed by the amorphous silica in a diatom skeleton. Therefore, to form a chert bed 1 ft thick and 1 sq mi in extent would require approximately 10^{17} diatoms. As chert beds commonly are tens and sometimes hundreds of feet thick and extend over many hundreds of square miles, it is apparent that astronomical numbers of siliceous plankton have given their all to the geologic record.

Inorganic

Field observations suggest that some chert deposits have resulted from direct precipitation of amorphous silica from marine or lake waters. The evidence consists of the occurrence of slump structures and intraformational breccias within the chert unit, indicating nontectonic deformation essentially contemporaneous with sedimentation. As deposits of siliceous organisms cannot be lithified so rapidly, the presence of these slump structures enables geologists to infer the presence of a silica gel. Although these observations and interpretations are not new, it is only recently that a modern locality has been found (Peterson and von der Borch, 1965) where nonbiologic precipitation of opal is occurring in an otherwise fairly common geologic and geographic setting. The precipitation occurs in some Australian ephemeral lakes that seasonally develop pH values greater than 10 because of active photosynthesis by indigenous algae. At these high pH values, detrital quartz grains and possibly clay minerals are corroded, saturating the lake waters with respect to amorphous silica. The algal activity fluctuates seasonally and, in addition, the lake basins are dry during part of the year. The decrease in pH and decrease in volume of solvent (water) that result from these changes cause the silica in solution to precipitate, forming an "amorphous" gel containing cristobalite crystallites. Carbonate minerals are also forming in the lake basins and it has been suggested that much of the nodular chert in limestone beds originated in this type of depositional environment. The common presence of carbonate inclusions in nodular chert is usually taken to indicate *replacement* of carbonate by silica during diagenesis.

Silica necessary to form siliceous rocks may also be derived, either directly or indirectly, through volcanic activity. The direct mechanism involves devitrification of volcanic glass fragments (shards). This process may result either in the formation of a felsitic fragment with no release of silica or in the formation of montmorillonitic clay (bentonite) with release of surplus silica to surrounding waters. If the amount of silica released is sufficient to saturate the water with respect to amorphous silica, opal will precipitate. The high silica content of most siliceous shales may result from this process, as does the opaline cement in many lithified volcanic sandstones.

The indirect mechanism has been discovered by oceanographic investigations in areas where volcanic activity is frequent, such as in the area of the Aleutian Islands. The waters in these areas contain uncommonly large numbers of siliceous organisms and these numbers are swelled still further during periods of active volcanism. Apparently, alteration of the volcanic glass associated with submarine volcanic activity increases the silica content of the overlying waters, enabling expansion of the population of siliceous plankton (diatom blooms). It has been demonstrated experimentally that if adequate amounts of organic nutrients are available, the yield of diatoms is proportional to the amount of silica added to the water. Many ancient chert beds are stratigraphically associated with debris of volcanic origin and, in such cases, it may not be possible to determine whether the chief silica source was alteration of volcanic glass or diagenetic recrystallization of organic skeletons. This distinction is particularly difficult to make in pre-Tertiary rocks because shell structure may be completely obliterated by crystallization to microcrystalline quartz. Often, however, "ghosts" of the shells are visible in thin section.

The origin of the volumetrically abundant Precambrian cherts poses a problem that is unresolved. Siliceous sponge spicules and radiolarian tests are known from the Cambrian, but reports of Precambrian siliceous organisms are rare and doubted by many paleontologists. This may be due in part to the small amount of petrographic study the rocks have received. We can ask whether this means that these organisms were absent or numerically scarce during this period or does it simply reflect the more complete destruction of the delicate shells as a function of time, circulating meteoric waters, and recrystallization. If siliceous organisms were not present to continually remove silica from surface waters, we probably would have a much greater amount of direct precipitation of amorphous silica than now seems to occur and the chemistry of such a process has been described from modern occurrences in Oregon and in east-central Africa (Eugster, 1967). As in south

Australia, the geomorphic setting is a series of closed lake basins but in Oregon and Kenya the geology of the surrounding area is volcanic. The lakes are essentially sodium carbonate brines. Evaporation raises the pH of the lake waters above 10, which simultaneously causes corrosion of the volcanic rocks and clastic fragments in the lake and allows the water to retain the released silica in solution. Concentrations as high as 2700 ppm of silica have been reported. Flood runoff from surrounding streams periodically lowers the pH of the lake waters causing supersaturation with respect to amorphous silica, and chert is formed by diagenesis of a hydrated sodium silicate precursor. Extensive volcanic terranes possibly were much more abundant during Precambrian times than subsequently and, therefore, many lacustrine cherts may have formed in this manner. The chert-ironstones of the Mesabi Range in the Lake Superior region may be an example (See Chap. 19).

A lacustrine environment probably was not the setting for all Precambrian chert deposits and many cherts of probable shallow marine origin are known. At present there is no known way to determine whether their origin was organic or inorganic.

When field relationships are not definitive, the question of whether a chert bed is of marine or nonmarine origin may be answered by oxygen isotopic analysis. Relative to PDB, modern freshwater deposits of silica have δO^{18} values that average approximately 19 and range from 14 to 23; marine deposits average 31. Unfortunately, the δ values for marine cherts show a linear decrease with time and average only 27 in Carboniferous rocks and 23.5 in Cambrian rocks. Most Precambrian cherts have δ values within the range of freshwater deposits. The decrease in δ values with time has been interpreted as indicating equilibration with isotopically lighter meteoric waters. It appears that distinction between marine and nonmarine origins for Precambrian chert beds cannot be made by means of stable isotope geochemistry.

The origin of the bedded cherts in general has long been the subject of controversy. The present-day weight of evidence favors the conclusion of an organic origin. That is, they have resulted from the recrystallization of or the mobilization of silica from siliceous oozes. Where bedded cherts are intimately associated in the geologic record with volcanic detritus or pillow lava, however, the possibility of an inorganic source for the silica must be seriously considered.

REFERENCES

Bramlette, M. N., 1946, "The Monterey Formation of California and the Origin of Its Siliceous Rocks," *U.S. Geol. Sur. Prof. Paper 212*, 55 pp. (Classic study of the origin and diagenesis of an extensive series of opaline cherts of Tertiary age.)

Calvert, S. E., 1966, "The Accumulation of Diatomaceous Silica in the Sediments of the Gulf of California," *Geol. Soc. Amer. Bull.*, **77**, pp. 569–596. (Quantitative evaluation of the input of biogenic silica using modern oceanographic concepts.)

Cavaroc, V. V., Jr., and J. C. Ferm, 1968, "Siliceous Spiculites as Shoreline Indicators in Deltaic Sequences," *Geol. Soc. Amer. Bull.*, **79**, pp. 263–272. (A modern study of the paleogeographic and ecologic implications of the areal distribution of spiculite beds in some Pennsylvanian rocks in the Appalachian Plateau.)

CRESSMAN, E. R., 1962, "Data of Geochemistry, Chap. T, Nondetrital Siliceous Sediments." *U. S. Geol. Sur. Prof. Paper 440–T*, 23 pp. (Summary of data.)

ERNST, W. G., and S. E. Calvert, 1969, "An Experimental Study of the Recrystallization of Porcellanite and Its Bearing on the Origin of Some Bedded Cherts," *Amer. Jour. Sci.*, **267-A**, pp. 114–133. (Experimental study of the origin of chert.)

EUGSTER, H. P., 1967, "Hydrous Sodium Silicates from Lake Magadi, Kenya: Precursors of Bedded Chert;" *Science*, **157**, pp. 1177–1180. (Modern chert formation.)

FANNING, K. A., and D. R. SCHINK, 1969, "Interaction of Marine Sediments with Dissolved Silica," *Limnol. and Oceanog.*, **14**, pp. 59–68. (Quantitative analysis of the interaction between detrital clay and dissolved silica in seawater reveals that absorption of silica is of minor importance.)

FOLK, R. L. and C. E. WEAVER, 1952, "A Study of the Texture and Composition of Chert," *Amer. Jour. Sci.*, **250,** pp. 498–510. (Electron photomicrographs of chert.)

JØRGENSEN, E. G., 1953, "Silicate Assimilation by Diatoms," *Physiologia Plantarum*, **6,** pp. 301–315. (Laboratory study in an aquarium.)

KRAUSKOPF, K. B., 1959, "The Geochemistry of Silica in Sedimentary Environments," in *Silica in Sediments*, H. A. Ireland, ed., Soc. Econ. Paleon. & Miner. Spec. Pub. 7, pp. 4–19. (Summary of the state of knowledge concerning the opal and chert in sediments.)

LABERGE, G. L., 1967, "Microfossils and Precambrian Iron-Formations," *Geol. Soc. Amer. Bull.*, **78,** pp. 331–342. (The author cites evidence indicating that remains of siliceous organisms are not as uncommon in Precambrian bedded cherts as is generally believed and many such cherts may be demonstrably of biologic origin.)

LEWIN, J. C., 1961, "The Dissolution of Silica from Diatom Walls," *Geochim. Cosmochim. Acta*, **21,** pp. 182–198. (Experiments using freshly killed diatoms indicate that adsorbed inorganic ions retard the rate of dissolution of skeletons.)

———, 1962, "Silicification," in *Physiology and Biochemistry of Algae*, R. A. Lewin, ed. New York: Academic Press, pp. 445–455. (Summary of the biochemistry of growth of diatom tests from natural waters.)

MCBRIDE, E. F., and A. THOMSON, 1970, "The Caballos Novaculite, Marathon Region, Texas," *Geol. Soc. Amer. Spec. Paper 122*, 129 pp. (Detailed study of one of the thickest and most extensive sequences of massive chert beds in the western hemisphere.)

OLDERSHAW, A. E., 1968, "Electron Microscopic Examination of Namurian Bedded Cherts, North Wales (Great Britain)," *Sedimentology*, **10**, pp. 255–272. (Electron microscopy reveals differences in texture that may be related to content of impurities and rate of precipitation of silica.)

PETERSON, M. N. A., and C. C. VON DER BORCH, 1965, "Chert: Modern Inorganic Deposition in a Carbonate-Precipitating Locality," *Science*, **149**, pp. 1501–1503. (The only documentation of inorganic chert precipitation in a common geographic and geologic setting.)

SIEVER, RAYMOND, 1962, "Silica Solubility, 0°–200°C., and the Diagenesis of Siliceous Sediments," *Jour. Geol.*, **70**, pp. 127–150. (New solubility data and a comprehensive review of the subject.)

SIEVER, RAYMOND, and R. A. SCOTT, 1963, "Organic Geochemistry of Silica," in *Organic Geochemistry*, I. A. Breger, ed, New York: Pergamon Press, pp. 579–595. (Examination of the interaction between organic substances and silica, with a consideration of the origin of petrified wood.)

CHAPTER SEVENTEEN

PHOSPHATES

17.1 INTRODUCTION

Sedimentary phosphate occurs in many marine and some nonmarine rocks and, in many ways, its range and types of occurrences are comparable to those of sedimentary calcite. Phosphatic minerals may occur as microcrystalline mud and are known from continental, shallow marine, and pelagic deposits. Phosphate cement (nearly always microcrystalline) occurs in either highly phosphatic rocks or rocks composed of other kinds of grains; sand-size phosphate particles are known as oolites, pellets (non-fecal), fecal pellets, bioclastic materials, and phosphoclasts. Sedimentary phosphate does not form reefs but may occur in amorphous form, a phenomenon unknown for substances composed of calcium carbonate. Primary phosphate deposits may be secondarily enriched by leaching of interstitial materials to produce a surficial residuum comparable to caliche in many respects.

The amount of sedimentary phosphate in the crust is not known accurately because phosphate rock has not been intensively searched for until very recently. The world production of phosphate rock rose from only 34 million metric tons in 1953 to 42 in 1960, 63 in 1964, and 93 in 1968, largely in response to increasing agricultural needs. The most important use of phosphate is in agriculture, which accounts for about two-thirds of world consumption. Other uses include chemicals for detergents and other cleaning compounds, leavening agents, food preservatives, insecticides, and many other products. In some areas of the world, waste phosphate in effluent from industrial and household uses has caused serious pollution problems.

17.2 MINERALOGY

About 200 minerals are known that contain 1% or more P_2O_5 but most of the phosphorus in the earth's crust and in sedimentary rocks occurs in varieties of the mineral apatite. The most common varieties occur in isomorphous series, with end-member compositions:

$Ca_5(PO_4)_3F$ fluorapatite

$Ca_5(PO_4)_3Cl$ chlorapatite

$Ca_5(PO_4)_3OH$ hydroxyapatite

Fluorapatites containing hydroxyl ions and several percent of carbonate ions are commonly termed francolite, and apatites rich in carbonate ion are often called dahllite but these two terms are poorly defined and seem unnecessary.

Fluorapatite is the common apatite of most igneous rocks and also is the stable phase under most sedimentary conditions. However, apatite rich in hydroxyl ions is characteristic of modern bones and teeth. Authigenic marine phosphorites and phosphatic rocks are composed of carbonate fluorapatite in which CO_3^{2-} substitutes for PO_4^{3-} in amounts seldom exceeding a few percent. Charge balance in the crystal structure is maintained by substitution of monovalent for divalent cations.

Manganese, magnesium, ferrous iron, and strontium may partially replace calcium in apatite as they do in calcite and aragonite but the total of these substituents normally does not exceed 1 to 2% of an oxide analysis in sedimentary apatites. The more important elemental substitutions in sedimentary apatite are shown in Table 17-1.

In Chap. 12 we considered the partition coefficient for the coprecipitation of strontium with aragonite and calcite in modern seawater, illustrating that the strontium content can be used to infer the Sr/Ca ratio in the growth environment. The same principle may be used for sedimentary apatites. Russian geologists have been particularly interested in this aspect of phosphate rocks and have made several

TABLE 17-1 IONIC SUBSTITUTIONS IN SEDIMENTARY FLUORAPATITE

Constituent ion	Substituent ion
Ca^{2+}	Na^+, Sr^{2+}, Mn^{2+}, K^+, U^{4+}, Mg^{2+}, rare earths (Y, Yt, Ce)$^{2+,3+}$
P^{5+}	C^{4+}, S^{6+}, Si^{4+}, As^{5+}, V^{5+}, Cr^{6+}
F^-	OH^-, Cl^-
O^{2-}	F^-, OH^-

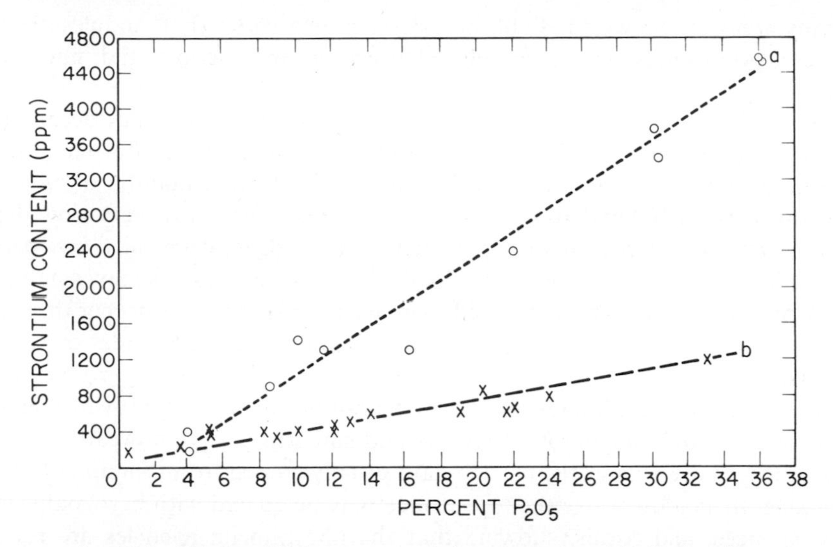

Fig. 17-1 Relationship between strontium content and P_2O_5 in primary authigenic phosphorites. (a) Maardu and Kingisapp deposits, northwestern USSR, Mesozoic; (b) phosphorites of Gornaya Shoriya, USSR, Precambrian. (Redrawn from Bliskovskiy et al., 1967.)

hundred chemical analyses of strontium content in sedimentary phosphates. Figure 17-1 illustrates the results of analyses of two typical groups of phosphorites from the Soviet Union. To avoid the difficulty of partitioning calcium in the sediment between calcite and apatite, P_2O_5 (phosphorus pentoxide) is plotted on the abscissa rather than calcium. In the Maardu and Kingisepp deposits, the phosphatic beds consist of detrital quartz and phosphatic shells of brachiopods and contain almost no syngenetic calcium carbonate. It was therefore inferred by Bliskovskiy et al. (1967) that the calcium content of the water in equilibrium with this sediment was relatively low and the Sr/Ca ratio in the phosphate is high. In contrast, the phosphorites of Gornaya Shoriya form part of a carbonate formation and contain 65 to 70% calcite and dolomite. The Sr/Ca ratio in the environment of authigenesis was therefore inferred to be rather low and this is reflected in the trace element content of the phosphate precipitated.

In some phosphorites the phosphate material is reported to occur largely as a cryptocrystalline or X-ray amorphous substance called collophane, a sack term analogous to "limonite" as used for amorphous ferric iron compounds in sedimentary rocks. The usage of the term collophane should be restricted to occurrences in which the apatite-like phase cannot be identified.

17.3 MODERN MARINE DEPOSITS

Phosphatic nodules were first found on the modern abyssal ocean floor by scientists of the Challenger Expedition in the 1870's and extensive deposits of this type are now known to occur on the sea bottom at both deep and shallow depths in many areas of the world. Published descriptions show that nodules from all of these occurrences are essentially identical in megascopic and microscopic character.

Probably the best known accumulations of phosphate nodules occur off the coast of southern California, U.S.A., and adjacent Baja California, Mexico, between latitudes 34°N and 25°N. Off southern California, nodules occur largely at depths between 100 and 1000 ft and in areas characterized by very slow deposition of clastic sediment, for example, bank tops, ridges, deep hills, and parts of the mainland shelf that are shallower than their surroundings. The few small pieces of phosphorite that have been dredged from basin or slope muds deeper than about 3000 ft appear to have been transported from shallower areas by slides or turbidity currents.

The external morphology of the phosphorite consists of brownish nodules and slabs commonly having flat bottoms and smooth, glazed, unweathered, hummocky tops. In deep water the particles are usually coated by a thin film of manganese oxide. In shallow water the top surface may be coated with bryozoans, worm tubes, sponges, and corals, showing that the phosphorite particles are rarely, if ever, rolled about on the sea floor.

The purity of the phosphorite is variable but most are composed largely of carbonate fluorapatite. Some nodules contain more than 95% phosphate but others contain abundant locally available mineral grains, partly phosphatized calcareous shells, planktonic siliceous shells, glauconite, and organic matter.

The internal structure of the nodules is complex and varied. Although some are composed of pure homogeneous apatite, some are conglomeratic or oolitic and many nodules are layered in an irregular and nonconcentric manner. These phosphatic layers vary in thickness from a few millimeters to a few centimeters and commonly show color grading. The lower portion of each layer is light brown near its base and grades upward into darker phosphorite that ends in a glazed surface often capped by a film of manganese oxide. These films represent times during which growth of the nodule was temporarily arrested and their very common presence on the outer surface of nodules dredged from the modern sea floor suggests that formation of phosphorite in this environment is negligible at present. This inference has recently been supported by radioactive dating of the outermost portion of forty phosphorite samples from the sea floor in many areas of the world, including southern California. The U^{234}/U^{238} disequilibrium method was used and all the U^{234}/U^{238} values were unity or less, indicating that the surfaces of the nodules are older than the dating limit of the method, about 800,000 years.

17.4 ANCIENT MARINE DEPOSITS

Ancient phosphorites are known from all geologic periods including the Precambrian and most deposits are demonstrably of shallow marine origin. Evidence includes the presence of shelf-dwelling calcareous organisms, cross-bedding, abrasion features on phosphoclasts, remnants of reef-building algae, and lateral interdigitation of phosphatic beds with quartz sandstones having similar shallow water characteristics.

Phosphoria Formation

Because of its economic importance as a source of phosphate for agricultural uses, the Phosphoria Formation (Permian) of northwestern United States is probably the best studied and understood shallow marine phosphorite unit. It is quite common in the literature to find the origin of other marine phosphorite units discussed in terms of the structural setting and stratigraphic variations of the Phosphoria and, for this reason, it is useful to consider the geology of the formation in some detail.

The Phosphoria phosphorites were deposited over an area of about 135,000 sq mi in both the platform and miogeosynclinal portions of the Mesozoic Cordilleran structural belt (Fig. 17-2). The regional stratigraphy of the Phosphoria reflects its structural setting not only in the great increase in thickness toward the miogeosyncline but also by lithologic changes. Chert and cherty mudstone dominate in the deeper parts of the depositional basin and are successively succeeded eastward by limestones and calcareous sandstones and ultimately by continental red shales and sandstones. Highly phosphatic beds occur in both the geosynclinal and platform facies and appear to reach a maximum thickness near the "hinge line" separating the two facies.

Apatite in the Phosphoria Formation is concentrated in two units, the Meade Peak Member and the thinner overlying Retort Member of similar lithology and petrography. In its type area, the Meade Peak is composed mainly of dark carbonaceous, phosphatic, and argillaceous rocks; some dark, dolomite and limestone also are present. Carbonate fluorapatite occurs in a variety of rock types and forms, including finely pelletal phosphorite, oolitic phosphorite, pisolitic phosphorite, nodular phosphorite, and bioclastic phosphorite generally composed of fish scales or brachiopod shells. The dolomites and limestones also are phosphatic. Individual beds vary in thickness from millimeters up to several feet. Eastward, toward shallower depositional areas, the phosphorite beds become coarser grained, less argillaceous, and more calcareous. Westward, phosphatic mudstones and cherts become increasingly abundant at the horizon of the Meade Peak Member.

The variety of rock types in the Meade Peak and their vertical and lateral sequences reflect differing environments of deposition. Relatively nearshore deposition in well-aerated waters is suggested by the noncarbonaceous, cross-bedded, well-sorted oolitic and pisolitic phosphorites that contain abraded fossils and lineated elliptical pellets. Antipathetic to this lithology are the carbonaceous, pyritic phosphatic mudstones, whose few fossils show no evidence of current transport. It is apparent that the facies variation of the Phosphoria phosphorites is similar to that encountered in many limestone formations, where the units grade from black micrite to bioclastic and oolitic rocks from the basin toward the shoreline.

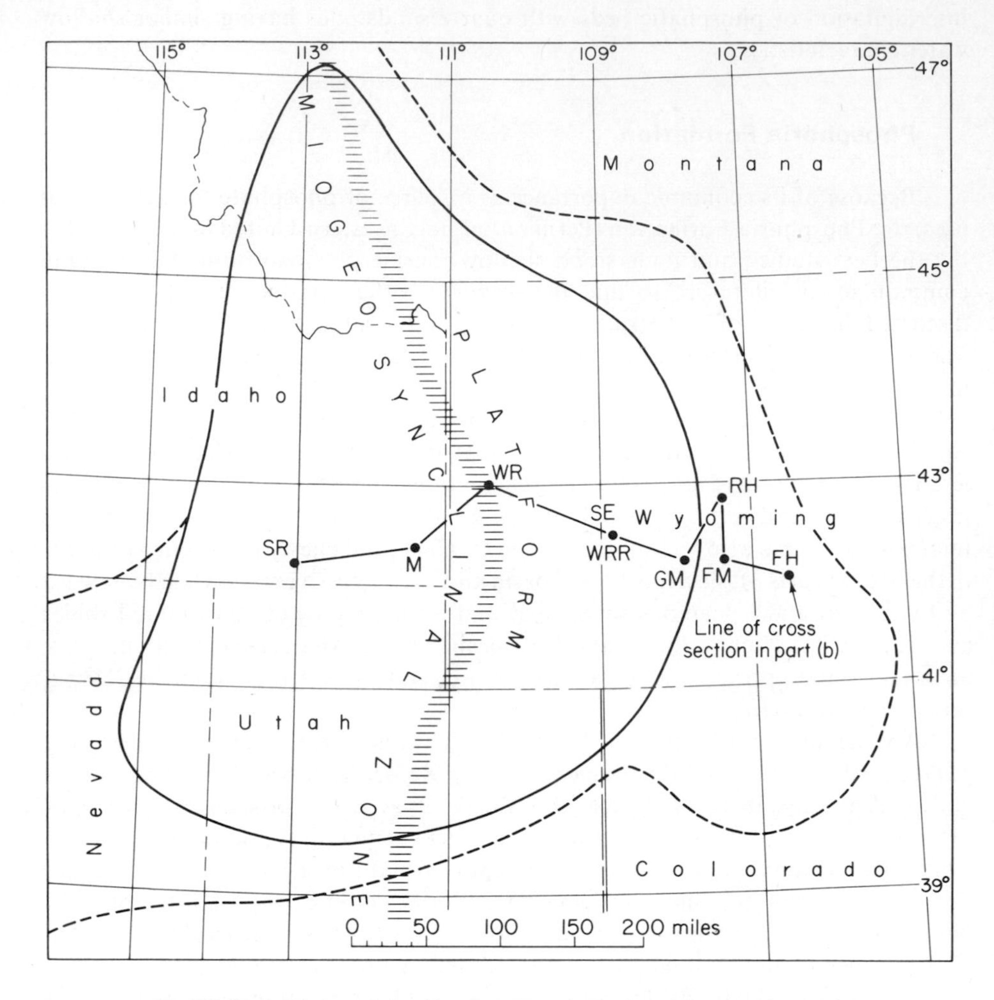

Fig. 17-2(a) Limits of the Permian Phosphoria Formation and its partial stratigraphic equivalents (dashed line) and their phosphorite deposits (solid line). Cordilleran miogeosynclinal zone and platform are separated by cross-hatched band.

The average Phosphoria phosphorite is composed of approximately 80% apatite, 10% quartz, 5% illite, 2% organic matter, and 1% each of dolomite-calcite, iron oxide, and other constituents. The average of chemical analyses of the phosphorite reflects this mineral composition: CaO, 44.0%; P_2O_5, 30.5%; SiO_2, 11.9%; F, 3.1%; CO_2, 2.2%; organic matter, 2.1%; SO_3, 1.8%; Al_2O_3, 1.7%; Fe_2O_3, 1.1%; Na_2O, 0.6%; K_2O, 0.5%; MgO, 0.3%; oil, 0.2%; TiO_2, 0.1%. Some of the oxide abundances are closely correlated. For example, a plot of weight percent Na_2O versus SO_3 for individual samples results in a straight line with a 1:1 atom ratio of Na/S. This is interpreted to indicate a coupled substitution of Na^+ and

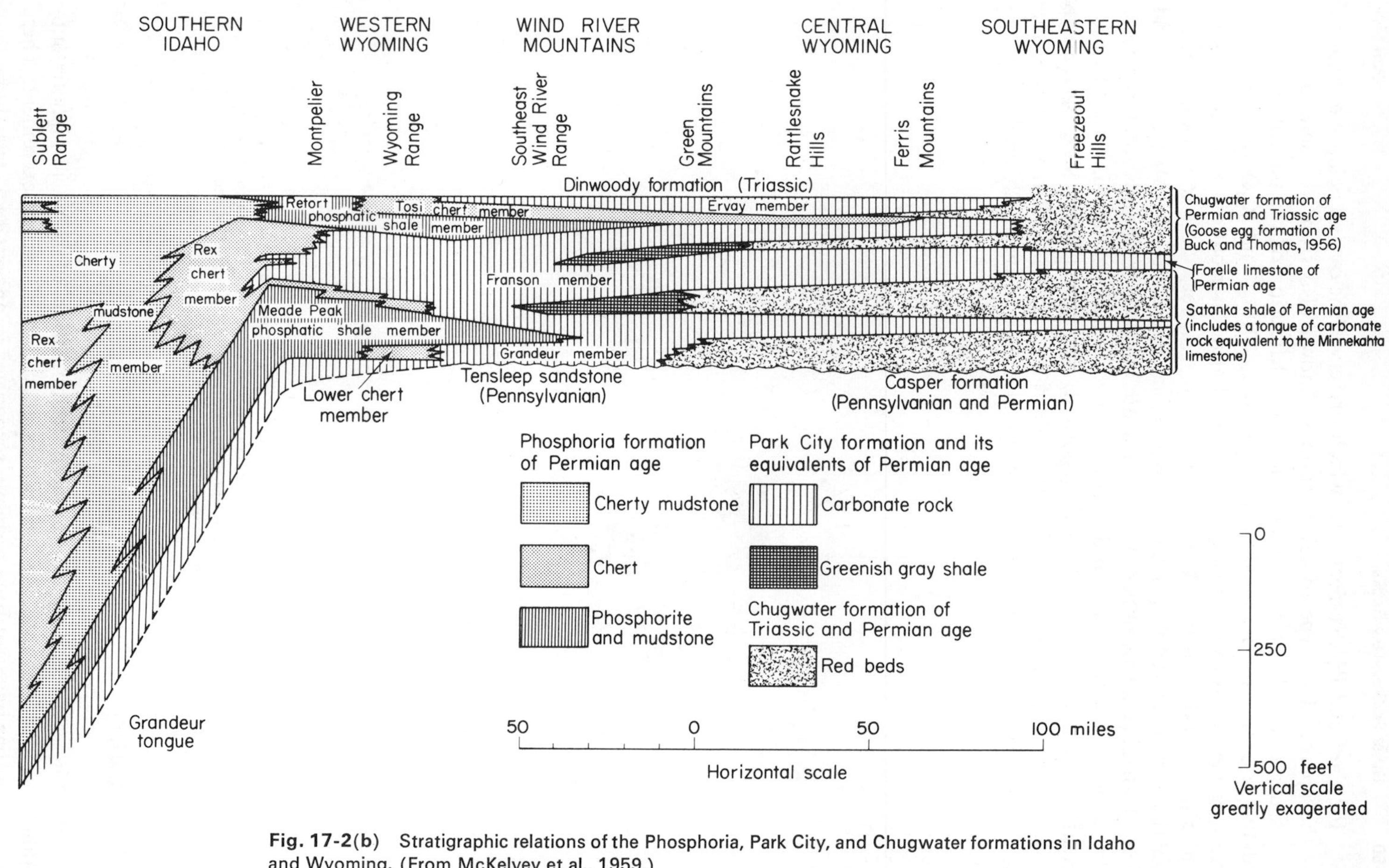

Fig. 17-2(b) Stratigraphic relations of the Phosphoria, Park City, and Chugwater formations in Idaho and Wyoming. (From McKelvey et al., 1959.)

S^{6+} for Ca^{2+} and P^{5+}, respectively, which preserves the overall balance of charges in the apatite. This type of substitution is known also from synthetic apatites.

Quartz and illite form most of the mineral fraction of the phosphorites, excluding apatite, and a triangular plot of SiO_2, Al_2O_3, and K_2O yields interesting and meaningful results (Fig. 17-3). The linear distribution of sample points indicates a nearly constant Al_2O_3/K_2O ratio of 3.7, very close to the value of 3.9 obtained in analyses of quartzose clastic silicate rocks of the Meade Peak and Retort Members. This implies that the silicate minerals in the two rock types are the same and function solely as dilutents, unmodified or unaffected by the environment in which the phosphate was deposited. The variation in SiO_2 percentage reflects both variation in the amount of detrital quartz in different samples and addition of silica as chert formed by recrystallization of originally opaline sponge spicules.

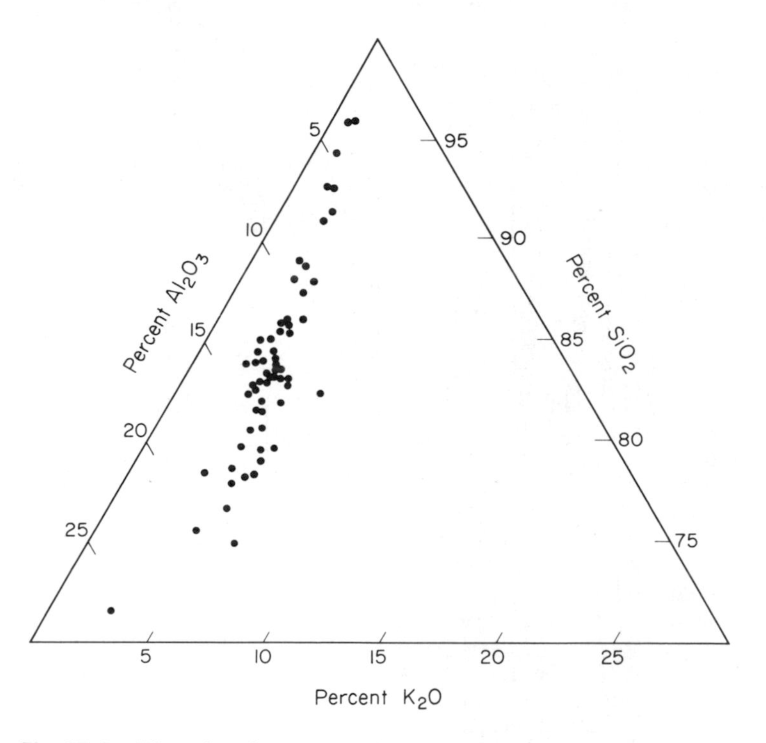

Fig. 17-3. Triangular diagram showing the relations of Al_2O_3-K_2O-SiO_2 in phosphorites of the Phosphoria Formation. Constituents recalculated to 100 weight percent. Diagram represents only the SiO_2 corner of the full 3-component system. (After Gulbrandsen, 1966.)

17.5 GUANO

Guano is a relatively rare modern accumulation of richly phosphatic substances formed by penecontemporaneous alteration of animal excrement. These accumulations have not been certainly recognized in ancient rocks, partly because of their very restricted distribution and partly because alteration quickly destroys the evidence of their true character. Most of the large guano caches are formed at the surface by seafowl but smaller quantities are formed by bats and to a lesser extent by cave-dwelling mammals and birds. The largest deposits are found along coastlines in the vicinity of upwelling ocean currents, which increase nutrient supply at the sea surface, thus stimulating the plankton blooms on which the seafowl graze.

Fresh seafowl droppings contain only 4% P_2O_5 but this is quickly beneficiated to 10 to 12% by small amounts of moisture that decompose organic matter in the droppings. Subsequent leaching increases the P_2O_5 content to 20 to 32% and stimulates crystallization of the calcium phosphate minerals monetite ($HCaPO_4$) and whitlockite [$Ca_3(PO_4)_2$] but presumably these minerals are metastable in the sedimentary environment with respect to apatite. Solutions rich in guano extracts may permeate underlying rocks, resulting in metasomatism. Where the substrate is limestone, as on many coral atolls, the mineral formed is apatite. On islands of volcanic origin, aluminum or aluminum-iron phosphates such as wavellite are produced.

17.6 ORIGIN OF MARINE PHOSPHORITES

Inorganic Precipitation

The areal distribution of Late Tertiary and Quaternary phosphorites is not random. They occur only in certain geographic locations, specifically in warm climates between the fortieth parallels (Fig. 17-4a), mainly on the west coasts of continents but also in small part on other coasts. A key common characteristic of these relatively recent phosphate accumulations is their location on one side of a basin where deep phosphate-rich waters are upwelling adjacent to a shallow shelf. In about three-quarters of the locations, the upwelling is caused by divergence, which is generally found in the trade wind belt where surface waters are blown offshore by the trade winds augmented by Coriolis force. In the other locations, upwelling results from deep-flowing currents being forced up over shallow submarine areas. The paleogeography and paleobathymetry of Late Tertiary to Holocene times is sufficiently well-known to establish that upwelling accompanied phosphorite deposition in cases such as the Monterey Formation (Miocene) of southern California, the Miocene phosphorite in southeastern United States, and the Miocene phosphorite of Peru. As noted on p. 546, uranium dating techniques have established that no detectable amount of phosphate is forming in these areas today so it is apparent that upwelling, although necessary, is not sufficient for precipitation of phosphorites.

Ancient phosphorite has a latitudinal distribution much different from that of young phosphorite, ranging from 6 to 70 degrees, with the bulk of the forty-nine known localities in the 30 to 45 degree range (Fig. 17-4(b)). When the present latitudes of these phosphorites are replaced by their paleolatitudes as determined using paleomagnetic data (Fig. 17-4(c)), however, the latitudes of formation of

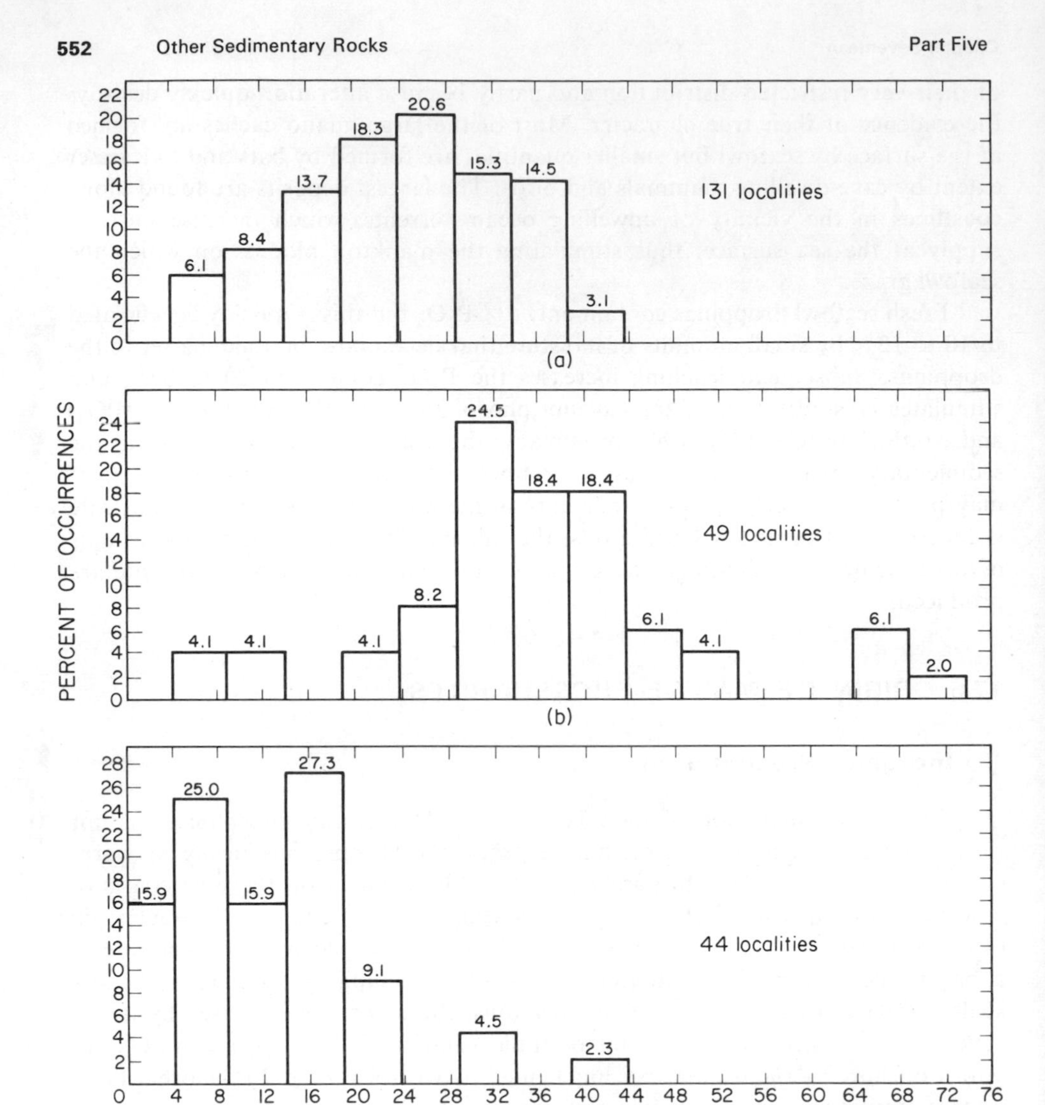

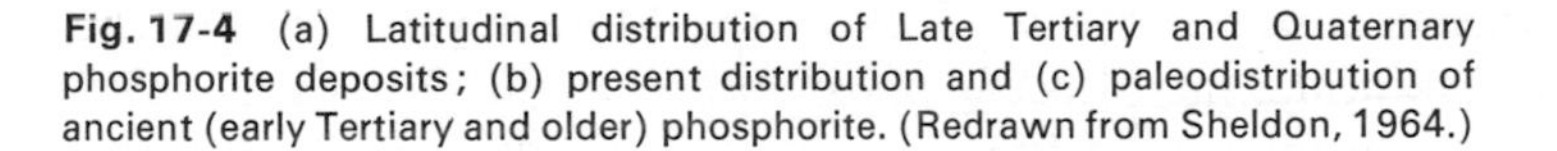

Fig. 17-4 (a) Latitudinal distribution of Late Tertiary and Quaternary phosphorite deposits; (b) present distribution and (c) paleodistribution of ancient (early Tertiary and older) phosphorite. (Redrawn from Sheldon, 1964.)

ancient phosphorites have essentially the same range as modern phosphorites. In view of the inexactness of paleomagnetic data and the small sample size, the correspondence between Figs. 4(a) and 4(c) seems close enough to permit the conclusion that latitudinal control of phosphate formation through upwelling ocean waters is as valid for ancient deposits as it is for more recent ones (Sheldon, 1964).

The importance of upwelling waters to phosphate precipitation was first recognized by A. V. Kazakov in 1937 (see review by McKelvey et al., 1953, pp.

55-56) and although some additions and modifications have since been made in Kazakov's concept of the genesis of marine phosphates, it remains today as a classic example of the integration of geologic, oceanographic, and chemical data to solve a significant sedimentologic problem. Using oceanographic data collected by other workers, Kazakov observed that the P_2O_5 content of marine waters is at a maximum at depths between about 100 and 1500 ft. At shallower depths, PO_4^{3-} is rapidly consumed by phytoplankton during photosynthesis; at greater depths, supersaturation of the water with respect to apatite is prevented by increased partial pressure of CO_2. Supplementing these data with results of his experiments in the system CaO-P_2O_5-HF-CO_2-H_2O at concentrations similar to those of seawater, he concluded that solid phosphate can be chemically precipitated on shelving bottoms at depths between 150 and 600 ft. In this depth zone the pH of ascending cold waters rises as their temperature increases and the partial pressure of CO_2 decreases. The water first becomes saturated with respect to calcium carbonate, which Kazakov said precipitated first, and then with calcium phosphate, which precipitated as a variety of apatite. Subsequent geochemical studies (Roberson, 1966) have confirmed that seawater is approximately saturated with respect to fluorapatite, just as it is with respect to calcium carbonate.

The apparent lack of precipitation of apatite in present-day areas of upwelling, however, suggests that additional factors are involved. In particular, a mechanism is required to depress the formation of calcium carbonate, whose precipitation is favored by approximately the same chemical conditions that favor apatite precipitation but which occurs at a much faster rate. Organic production of calcite and aragonite is so rapid compared with plausible rates of formation of apatite by inorganic means that any apatite precipitation would be completely masked. Gulbrandsen (1969) has proposed that depression of calcium carbonate formation is accomplished through increase in phosphate ion activity as a consequence of the decomposition of organic matter. This source of phosphate is known to be adequate in areas of nutrient-rich waters with high organic matter productivity. The organisms, mainly phytoplankton, incorporate the phosphate into their tissues during growth in the upper few tens of feet of the sea, sink to the shallow sea bottom on death, and their tissues are partly or completely oxidized by oxygenated waters. To prevent complete stagnation, water must circulate from the surface to the phosphate-rich bottom waters (Fig. 17-5) but the rate of oxygen renewal normally is inadequate to oxidize all the organic matter. Hence, nearly all ancient marine phosphate deposits, such as the Phosphoria Formation, contain some organic matter and a few contain some pyrite as well. Perhaps the reason apatite is not forming in modern areas of upwelling is that the water is not now shallow enough to ensure that sinking organic remains reach the bottom. Or perhaps the rate of influx of clastic detritus is at present excessive. More probably, the true reason is as yet unknown.

Metasomatism of Calcium Carbonate

The absence of apatite formation by inorganic means in modern seas has led to the search for an alternative mechanism. As noted earlier, the oceanographic and geochemical conditions that are conducive to phosphate formation also favor

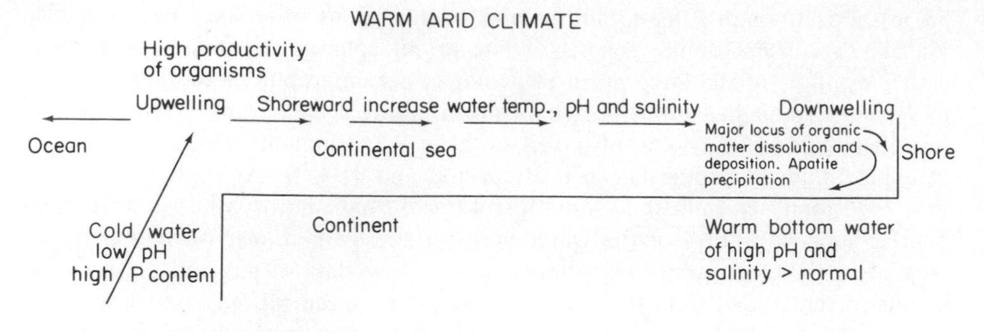

Fig. 17-5. Diagrammatic representation of optimum conditions for the formation of marine apatite and phosphorite. (After R. A. Gulbrandsen, 1969.)

supersaturation with respect to calcium carbonate, and it has been proposed that most if not all marine phosphorites form by replacement of a calcite or aragonite precursor. In this regard, it has been observed in petrographic studies by Soviet geologists that aragonite is more easily replaced by apatite than is calcite. As we observed in Chap. 12, modern carbonate sediments are commonly composed largely of aragonite.

Apatitic mud, either amorphous or microcrystalline, would presumably be formed from micrite ooze; structureless phosphate pellets might form from either aragonitic pellets or grapestone. Phosphate ooliths also are known and it is perhaps significant that they occur in the shallowest areas of phosphate formation, as do aragonitic ooliths in carbonate environments.

Some phosphate deposits contain calcareous fossils in all stages of replacement by apatite and in these cases there can be no doubt that phosphate metasomatism has occurred. However, evidence for its operation on a large scale is lacking in most phosphorite deposits. And in addition, it is difficult to account for deep-sea phosphate nodules by this mechanism as they occur as obviously untransported particles in areas lacking calcareous ooze. On the other hand, the mechanism of inorganic precipitation of phosphate suggested by Kazakov is also inadequate to explain the deep-sea phosphorites.

The mechanism of phosphatization of calcite was studied experimentally by Ames (1959) using radioactive tracers in the system Na_3PO_4-$CaCO_3$-H_2O at low temperature. The experiment consisted essentially of passing a sodium phosphate solution through a tube containing calcite fragments and then determining the minerals that grew in the tube during the experiment. These determinations revealed the simultaneous disappearance of calcite and appearance of apatite according to the idealized reaction

$$NaOH + 3\,Na_3PO_4 + 5\,CaCO_3 \longrightarrow Ca_5(PO_4)_3OH + 5\,Na_2CO_3$$

Calcite fragments in the tube were not only replaced but perfectly pseudomorphed by the apatite formed, which was a carbonate apatite containing several percent carbonate ion in place of phosphate ion in the apatite structure. Most natural sedimentary apatites also contain carbonate ion in solid solution. Ames' experiments, then, duplicated most of the important factors in field occurrences of shallow marine phosphorites: nondepositional environment, calcium carbonate available for replacement in a solution saturated with respect to calcite, and alkaline pH. Ames also established that phosphatization could occur when PO_4^{3-} concentrations were only 0.1 ppm, a value normally exceeded below the photic zone in the modern world ocean. In areas of oceanic upwelling, the phosphate content of seawater is usually 0.3 to 0.8 ppm. Some marine environments, such as estuaries and silled basins, function as phosphate (and other nutrient) traps and commonly contain 0.1 to 1.0 ppm dissolved inorganic phosphorus. It has been suggested that the Miocene phosphorites of the Atlantic coastal plain are estuarine deposits formed by replacement of a calcium carbonate precursor.

As noted earlier, phosphate concentrations are very low within the photic zone of the sea because of consumption by living organisms. And it is precisely in this zone that nearly all modern carbonate environments are located. In areas such as south Florida and the Bahamas, phosphate content of the waters is only 0.01 to 0.03 ppm and penecontemporaneous phosphatization is unknown.

The range of phosphate concentrations in subsurface brines is unknown.

17.7 ALTERATION OF PHOSPHATIC ROCKS

Although most phosphatic rocks are composed essentially of carbonate fluorapatite, other phosphate minerals may occur. All that is required is an adequate supply (activity) of phosphate ions and appropriate cations. For example, the calcium-uranium-phosphate mineral autunite occurs as a secondary mineral in uraniferous rocks and the mineral vivianite, a ferrous iron phosphate, forms as a weathering product in sedimentary iron ores and in areas where pyrite and pyrrhotite are abundant. Perhaps most widely disseminated, although rarely abundant, are the aluminous phosphates, particularly wavellite.

The importance of secondary phosphate minerals rests on the facts that many of them are known and that phosphate minerals, like zeolites (see Chap. 18), are hardly ever detrital. Therefore, the different mineral species may be used as sensitive indicators of chemical environment (Altschuler et al., 1956).

17.8 PROBLEMS OF PHOSPHATE GENESIS

In addition to the preeminent problem concerning the primary or replacement origin of major phosphate units such as the Phosphoria Formation, many other unanswered questions exist concerning phosphatic rocks. As with all questions of geologic interest, these have been generated by field observations. For example, Braithwaite (1968) has described Quaternary phosphatization of Pleistocene calcareous sediment caused by leaching of guano. Excellent pseudomorphing of shell fragments by apatite was reported, similar to the experimental results reported by Ames. Petrographic study of the phosphatic rocks reveals the presence of whitlockite [$Ca_3(PO_4)_2$] and suggests the former presence of brushite ($CaHPO_4 \cdot 2\,H_2O$). Both of these minerals are unstable relative to apatite but the conditions under which they may persist are unknown.

Phosphatic chalks in England are associated with extraordinary abundances of recognizable fragments of fish skeletons. In what way are they related to the occurrence of the phosphate nodules?

The average shale contains 1600 ppm of phosphorus according to recent estimates of element abundances in different parts of the earth's crust. Despite this large content of phosphate, which is presumably present as adsorbed ions on clay flakes, most shales lack authigenic apatite. Under what conditions can the phosphate in shales be mobilized to centers of nucleation?

These are only a few of the unsolved significant problems in the geology of phosphate. Clearly, these problems will be solved only through a combination of stratigraphic, petrographic, and geochemical data, and the increasing interest in phosphate deposits for agricultural uses throughout the world guarantees that the study of phosphate rocks will be of much interest for the foreseeable future.

REFERENCES

ALTSCHULER, Z. S., E. B. JAFFE, and F. CUTTITTA, 1956, "The Aluminum Phosphate Zone of the Bone Valley Formation, Florida, and Its Uranium Deposits," *U.S. Geol. Sur. Prof. Paper 300*, pp. 495–504. (Description and analysis of the occurrence of unusual sedimentary phosphate minerals in a well-known phosphorite deposit.)

AMES, L. L., JR., 1959, "The Genesis of Carbonate Apatites," *Econ. Geol.*, **54**, pp. 829–841. (Often cited experimental replacement of calcite by apatite.)

BLISKOVSKIY, V. Z., V. A. YEFIMOVA, and L. V. ROMANOVA, 1967, "The Strontium Contents of Phosphorites," *Geochem. Internatl.*, **4**, pp. 1186–1190. (Summary of extensive Russian work on this topic.)

BRAITHWAITE, C. J. R., 1968, "Diagenesis of Phosphatic Carbonate Rocks on Remire, Amirantes, Indian Ocean," *Jour. Sed. Petrology*, **38**, pp. 1194–1212. (Richly phosphatic solutions derived from leaching of guano have caused precipitation of apatite in Pleistocene calcareous sediment. Pseudomorphing of both internal and external structure of bioclasts occurs and primary phosphate occurs as a cementing agent.)

EMERY, K. O., 1960, *The Sea off Southern California*. New York: John Wiley & Sons, Inc., 366 pp. (Comprehensive survey of a continental borderland area, its marine geology, oceanography, and geochemistry.)

GULBRANDSEN, R. A., 1966, "Chemical Composition of Phosphorites of the Phosphoria Formation," *Geochim. Cosmochim. Acta*, **30**, pp. 769–778. (Results of analyses of sixty samples of phosphorite, forty from the Meade Peak Member and twenty from the Retort Member.)

———, 1969, "Physical and Chemical Factors in the Formation of Marine Apatite," *Econ. Geol.*, **69**, pp. 365–382. (Analysis of the chemical factors important in the inorganic precipitation of apatite from seawater, stressing the role of organic matter as a source of phosphorus.)

McKelvey, V. E., 1967, "Phosphate Deposits," *U.S. Geol. Sur. Bull.* **1252**-*D*, 21 pp. (Brief summary of ideas concerning the origin and distribution of phosphate beds, with particular emphasis on commercial aspects.)

McKelvey, V. E., et al., 1959, "The Phosphoria, Park City and Shedhorn Formations in the Western Phosphate Field," *U.S. Geol. Sur. Prof. Paper 313-A*, 47 pp. (Description of the stratigraphy, facies, and petrology of the Phosphoria Formation.)

McKelvey, V. E., R. W. Swanson, and R. P. Sheldon, 1953, "The Permian Phosphorite Deposits of Western United States," *19th Internatl. Geol. Cong., Algiers*, Sec. 11, pp. 45–64. (Summary of the stratigraphy, petrology, and origin of one of the world's most extensive deposits.)

Pettijohn, F. J., 1957, *Sedimentary Rocks*, 2nd ed. New York: Harper & Bros., 718 pp. (Pages 470–478 consider the origin and distribution of phosphatic sediments.)

Roberson, C. E., 1966, "Solubility Implications of Apatite in Sea Water," *U.S. Geol. Sur. Prof. Paper 500-D*, pp. 178–185. (Laboratory studies of the solubility of fluorapatite and marine phosphorite in artificial seawater demonstrate that normal seawater is approximately saturated with respect to fluorapatite.)

Sheldon, R. P., 1964, "Paleolatitudinal and Paleogeographic Distribution of Phosphorite," *U.S. Geol. Sur. Prof. Paper 501-C*, pp. 106–113. (The latitudinal range of ancient and modern phosphorites is identical when allowance is made for continental drift.)

CHAPTER EIGHTEEN

ZEOLITES

18.1 INTRODUCTION

The zeolites are a group of hydrated aluminosilicate minerals of the alkali and alkaline earth elements, similar in composition to feldspars. They are among the most common authigenic silicate minerals in sedimentary rocks and, as a group, rival authigenic quartz, calcite, and dolomite in abundance in many areas. In sandstones they nearly always occur in association with detrital volcanic constituents such as glass shards and felsitic fragments. In fine grained rocks they are characteristic of tuffs and shales formed in saline alkaline lake deposits. They also occur in limestones.

As noted by Hay (1966), zeolite minerals are particularly important in considering the diagenetic history of sedimentary rocks because they may be abundant in the rocks, but detrital zeolites are rare. Further, there are many mineral species in the group; therefore, zeolite assemblages are useful to differentiate diagenetic or low grade metamorphic facies. Unfortunately, however, many zeolite minerals form metastably because of kinetic factors resulting from the large variety of structures that are similar and have similar free energies. This makes accurate demarcation of facies boundaries particularly difficult and it is a handicap in the interpretation of laboratory studies.

18.2 MINERALOGY

Zeolite minerals are tectosilicates but their structures are more open than those of their close relatives, the feldspars and feldspathoids. The feldspars have the most compact structure among these three groups (sp. gr. 2.55 to 2.76) and the metallic cations are confined in relatively small cavities surrounded by framework oxygens. The cations cannot easily move without disrupting the framework and the replacement of sodium or potassium by calcium necessarily involves a change in the Si/Al ratio. The feldspathoids have somewhat more open structures (sp. gr. 2.15 to 2.55) in which the cations occupy but do not always fill larger cavities that are intercommunicating. Thus, metallic cations may pass through some of these minerals without disrupting the framework.

The zeolite frameworks (sp. gr. 1.92 to 2.37) are still more open than those of the feldspathoids, containing larger cavities and larger channels. In addition, the zeolites are hydrous minerals, with the water molecules located within the structural channels. The water is generally loosely bound to the framework and cations and, like the cations, the water can be removed and replaced without disrupting framework bonds. Because the zeolite framework is structurally almost independent of the metallic cations and since the cations do not fill all the cavities, replacements of the type Ca $\rightleftharpoons$ 2 (Na, K) can occur, a phenomenon not possible in the feldspars or feldspathoids.

About fifteen zeolite minerals have been reported from sedimentary rocks but the exact number is partly a matter of species definition. For example, there is some question whether heulandite and clinoptilolite are different minerals or simply low and high silica end-member compositions of a single mineral. Another moot point is the structural position of analcite, which may be grouped with either the feldspathoids or zeolites. As it is a common authigenic mineral in zeolite assemblages, we shall consider it as a zeolite in our discussion. In view of the ease of substitution of metallic cations, particularly calcium, sodium, and potassium, into zeolite structures (Table 18-1), it is to be expected that few of the zeolites can be adequately characterized chemically in terms of whole numbers of cations.

18.3 OCCURRENCE

Zeolites occur in sedimentary rocks deposited in nearly all environments and, although most common in tuffaceous sandstones and shales, they also are known from arkoses, quartzose sandstones, graywackes, carbonate rocks, coal, bauxite, and ironstone. In age, zeolites range from at least early Paleozoic to Holocene. Despite the wide range of rock types in which they are found, zeolitic environments can be meaningfully grouped into two broad categories: (a) occurrences in saline alkaline lakes and soils and (b) occurrences in normal freshwater and marine deposits.

TABLE 18-1 FORMULAE AND AVERAGE CHEMICAL COMPOSITIONS OF ZEOLITES MOST ABUNDANT IN SEDIMENTARY ROCKS (data from many sources). The compositions of all zeolites are extremely variable and, therefore, the percentages in the table can only be taken as approximations.

Mineral	SiO_2	Al_2O_3	CaO	Na_2O	K_2O	H_2O	Others	
**Laumontite* Ca $Al_2Si_4O_{12} \cdot 4H_2O$	50.7	21.9	12.1	0.5	0.2	13.7	1.0	Calcium-rich
Wairakite Ca $Al_2Si_4O_{12} \cdot H_2O$	55.9	23.0	11.7	1.1	0.2	8.5	0.1	Calcium-rich
**Stilbite* (Ca, Na_2, K_2) $Al_2Si_7O_{18} \cdot 7H_2O$	56.0	16.7	7.7	0.9	0.7	17.9	0.3	Calcium-rich
Chabazite Ca $Al_2Si_4O_{12} \cdot 6H_2O$	60.3	12.7	3.2	2.9	0.9	18.1	0.9	Sodium and Calcium subequal
Mordenite (Ca, Na_2, K_2) $Al_2Si_{10}O_{24} \cdot 7H_2O$	66.3	11.5	2.6	3.1	1.0	14.0	1.5	Sodium and Calcium subequal
**Natrolite* $Na_2Al_2Si_3O_{10} \cdot 2H_2O$	47.2	27.1	0.2	15.4	0.3	9.6	0.4	Sodium-rich
Analcite $NaAlSi_2O_6 \cdot H_2O$	64.0	16.9	0.1	9.5	0.2	8.7	0.5	Sodium-rich
Heulandite and Clinoptilolite (Ca, Na_2) $Al_2Si_7O_{18} \cdot 6H_2O$	64.7	12.4	1.5	3.8	3.1	13.2	1.4	Potassium and sodium subequal
Erionite (Ca, K_2, Na_2, Mg)$_{4.5}Al_9Si_{27}O_{72} \cdot 27H_2O$	56.8	13.8	0.2	6.2	4.1	15.1	3.8	Potassium and sodium subequal
Phillipsite *deep sea* ($\frac{1}{2}$ Na, Ca, K)$_3$ $Al_3Si_5O_{16} \cdot 6H_2O$	54.1	17.7	1.3	3.8	6.8	15.6	0.2	Potassium and sodium subequal
subaerially altered tuff	56.1	15.1	0.2	6.3	4.3	15.6	2.2	Potassium and sodium subequal

* = analyses from igneous rocks; no analyses available from sedimentary occurrences.

Saline Alkaline Lakes and Soils

Zeolites are very common and of wide extent in these nonmarine environments and are known in rocks ranging in age from Mississippian to Holocene. In some lacustrine sections zeolites occur as thin, relatively pure beds; in other sections they are disseminated throughout sequences nearly half a mile thick extending over more than 2000 sq mi. The most striking richly zeolitic beds are the "analcimolites," analcime-rich beds more than 100 ft thick and covering areas thousands of square miles in extent. For example, the Popo Agie Member of the Chugwater Formation (Triassic) in west-central Wyoming has a volume of 150 cu mi of which at least half is analcime-rich rock containing 40 to 60% analcime (High and Picard, 1965). Similarly extensive analcimolites are known from the Upper Jurassic and Lower Cretaceous of the Congo Basin and from the Lower Cretaceous of the central Sahara.

Zeolite-Volcaniclastic Association

The earliest report of areally extensive zeolites in ancient rocks was of analcite-rich units, in 1928 by Bradley, in a note concerning the lacustrine Green River Formation (Eocene) in western United States (Bradley, 1929). He described beds composed of 70% analcite; 14% chalcedony, quartz, and opal; 6% potassic feldspar; 8% limonite; and small amounts of pyrite and calcite. Most analcimic beds in the Green River Formation contain siliceous tuffaceous material. Bradley suggested, therefore, that the analcite formed by chemical reaction between dissolution products of the volcanic material that fell into the Green River lakes and dissolved species normally in the lake water. Subsequent studies of the formation support Bradley's conclusion and also have documented the presence of numerous saline minerals within the Green River sediments.

Several younger analogues of the Green River lakes have been studied (Hay, 1966). These include China and Searles Lakes in California; the Quaternary sequence of Olduvai Gorge, Tanzania; and the Miocene beds of Kramer, California. In each of these lacustrine environments, tuffaceous detritus has reacted with aqueous solutions at low temperatures to produce both zeolites and saline minerals. The Olduvai Gorge deposit has been particularly well studied with respect to the stratigraphy and geochemistry of the zeolite minerals present (Hay, 1966). Six mappable zeolitic facies have been recognized (Fig. 18-1), each of which was formed at near-surface conditions in contact with aqueous solutions rich in sodium, potassium, and carbonate ions at pH values between approximately 9.5 and 10.5. The climate was generally hot and arid and the Olduvai lake lacked a permanent outlet; hence the lake level fluctuated with variations in rainfall in a manner similar to both modern saline lakes and to the Green River lakes of Eocene time. At times of high rainfall the Olduvai lake increased its diameter by 6 mi, resulting in dilution of the lake water to the point where fish could survive. Zeolites are not present and even glassy tuffs are unaltered in beds that contain fossil fish.

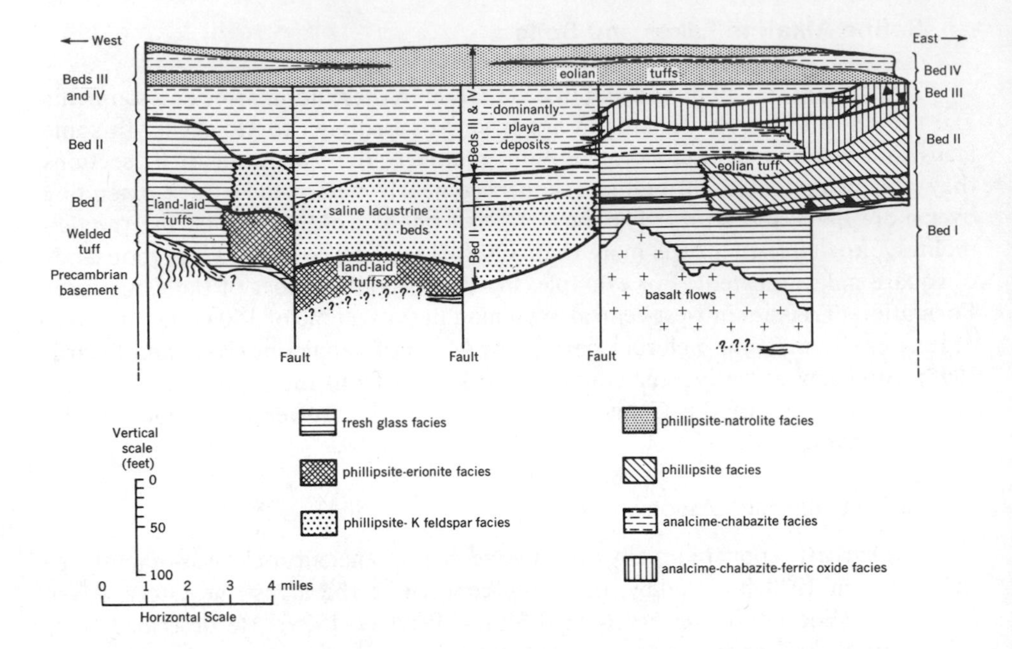

Fig. 18-1 Stratigraphic diagram showing the distribution of zeolitic mineral assemblages in Pleistocene beds of Olduvai Gorge, Tanzania. Mineral distribution is based upon field mapping and laboratory study. Diagram is an east-west section along the length of the gorge. The horizontal datum used in reconstructing the stratigraphic sequence is the base of the eolian tuffs of Bed IV. The indicated fault displacements took place during the accumulation of the Olduvai sequence, prior to the deposition of the eolian tuffs of Bed IV. (After Hay, 1966.)

The raw materials from which the zeolites were formed in the Pleistocene sediments in this region were vitric and lithic trachyte and nephelinite (nepheline plus sodic pyroxene) tuffs erupted from nearby volcanos. Sodium carbonate lavas are still being discharged from these volcanos. The lake waters were rich in the dissolution products of such rocks, sodium, potassium, and carbonate, and it was the interaction of these high pH waters with the chemically unstable volcanic materials that produced the Pleistocene zeolites. Nephelinite and trachyte glasses have been altered principally to phillipsite. Nepheline has been largely replaced by natrolite and the evaporite mineral dawsonite, by the reaction

$$\underset{\text{nepheline}}{3\,NaAlSiO_4} + 3\,H_2O + CO_2 \longrightarrow \underset{\text{dawsonite}}{NaAlCO_3(OH)_2} + \underset{\text{natrolite}}{Na_2Al_2Si_3O_{10}\cdot 2\,H_2O}$$

As shown in Fig. 18-1, erionite, analcite, and chabazite also are present in the Olduvai sediments, presumably in response to fluctuations in water chemistry during the life of the lake. However, the geochemical controls on the occurrence of these various minerals are poorly known. Hence, we are not yet able to describe in detail the variation in hydrochemical facies during the tens of thousands of years during which the lake existed.

Zeolite Formation in the Absence of Volcaniclastic Detritus

Zeolites may also form in saline alkaline lakes that lack volcanic detritus if the requisite chemical requirements are satisfied in some other manner. That is, the lake must have an adequate source of calcium, sodium, potassium, aluminum, and silica and these must be present in appropriate activity ratios.

Perhaps the best documented example of such an occurrence is Lake Natron, Tanzania (Hay, 1966). The lake bottom sediments consist of alternating layers of organic-rich clay and trona [$HNa_3(CO_3)_2 \cdot 2\,H_2O$], sometimes accompanied by other sodic saline minerals such as gaylussite and pirssonite. Coarser sediments around the lake margin are richly quartzo-feldspathic and are derived from highlands of Precambrian crystalline rocks 15 to 40 mi west of the lake. Lake Natron is generally only 5 to 10 ft deep and, as is to be expected in the arid climate of the region, is a highly concentrated brine. One analysis from the lake contained about 15% dissolved solids, of which 40% was soda, 37% was carbonate ions, 12% was chloride, and 8% was bicarbonate ions. Dissolved silica also was high at 176 ppm, which is to be expected in view of the measured pH value of 9.7.

There are several apparent sources for the ions in the lake waters. Alkaline springs are present near the margin of the lake and contribute dissolved salts, some of which may be recycled from the salt deposits and brine beneath the surface. Other sources are normal surficial weathering in the drainage basin and the natrocarbonatite lavas at the south end of Lake Natron. Fragments of volcanic glass or nepheline have not been found in the lake sediments despite intensive examination.

The only zeolite mineral present in the lake clays is analcite, which forms up to 25% of the sediment and averages about 10%. Nearly all the crystals are less than 2 μ in diameter and some are only 0.1 μ. Because of the dominance of sodium among cations in the lake waters, it is not surprising that analcite is present. The absence of natrolite can probably be attributed to the high activity of silica. It seems reasonable to suppose that a high silica/soda ratio favors formation of analcite, the more silica-rich mineral.

The lacustrine Lockatong Formation (Triassic) of New Jersey and Pennsylvania is the best known ancient analogue of modern Lake Natron. As described by Van Houten (1962), the formation is composed of a series of lacustrine deposits up to 3750 ft thick, whose depositional extent is more than 2000 sq mi. The lacustrine setting was, therefore, much larger than modern Lake Natron (400 sq mi) but the two are similar in that neither lake contains evidence of volcaniclastic detritus. The Lockatong sediments consist of cyclic deposits of "detrital" and "chemical" sequences. The analcite is clearly not an alteration product of volcanic glass because no pyroclastic detritus of any type has been found in the sediments and the analcite consistently recurs at the same stratigraphic position in each chemical sequence. Analcite is present in calcareous, highly indurated mudstones (argillites) and may form 40% of an argillite unit. These analcite-rich units contain 7% soda and as little as 47% silica. Van Houten (1962) suggested that the chemical sequences were deposited during times of limited rainfall and/or high evaporation. During these times, the amount of soda and the Na/Ca and Na/Mg ratios would be increased as the formation of calcium carbonate and dolomite continually removed calcium and magnesium from the lake waters. During the final stage of minimum rainfall, saline compounds were precipitated on mud flats but most of the evaporite minerals were subsequently removed as rainfall increased and the chemical cycle was succeeded by a detrital sequence.

Freshwater and Marine Deposits

The more common sedimentary environments differ in many ways from conditions in saline alkaline lakes, particularly in regard to environmental chemistry. Most significant from the viewpoint of zeolite formation are the lower salinities of the aqueous medium, presence of circulating water, generally lower pH, commonly higher temperature and pressure, and lower partial pressure of water. However, not all of these differences are present in each case of zeolite authigenesis in freshwater and marine deposits and, in fact, certain combinations are antipathetic. For example, the higher diagenetic temperatures that accompany deep burial drive off H_2O molecules, causing a drop in the partial pressure of water and affecting the amount and velocity of water movement in the pores of the sediment. On the other hand, certain combinations are preferred in the natural environment. For example, circulating waters usually imply inflow of low salinity meteoric and fresh waters into more saline formation waters, lowering not only salinity but pH as well.

Numerous examples are known of natural occurrences of zeolites in freshwater and marine environments that illustrate the effect of variations in chemical parameters on zeolite authigenesis. We consider briefly two of them.

1. The Oak Spring Formation (Tertiary) in southern Nevada is up to 2200 ft thick and unconformably overlies relatively impermeable rocks of Paleozoic age. The Oak Spring consists of bedded and massive silicic tuff and some flows of impermeable rhyolite and basalt. Authigenesis of zeolites occurs preferentially in nonwelded permeable tuff units that overlie either the Paleozoic "basement" or impermeable Oak Spring flow rocks, suggesting that the zeolitization was caused by solutions migrating along the contacts of the less permeable rocks. Supporting this inference is the notable rarity of alteration in firmly welded tuffs, which are nearly all fresh even where the surrounding nonwelded tuff is completely zeolitized.
2. Deep ocean basin sediments contain extensive accumulations of zeolites, which were first noted in the reports of the Challenger Expedition during the latter part of the nineteenth century. Phillipsite and, to a lesser degree,

its barium analogue harmotome cover extensive areas in the Pacific Basin, often with concentrations greater than 50% of the total sediment (Fig. 18-2). Zeolites apparently are forming in the Pacific Basin at a pH of about 8, at a temperature only slightly above 2°C, and at high hydrostatic pressure. The distribution of these minerals with depth in the sediment is not known. Petrographic and X-ray diffraction analyses of the zeolitic pelagic sediment reveal the phillipsite to be forming nearly always from palagonite, a mineraloid formed by alteration of glass in basaltic flows and pyroclastic deposits. Presumably, phillipsite, rather than other zeolites, forms from altered submarine basalt because phillipsite is low in silica and is highly hydrated. Phillipsite is a potassic zeolite and there is scant potassium in basaltic rocks (less than 1%) but apparently the amount in seawater (380 ppm) is adequate to support the growth of phillipsite.

Zeolites are reported less commonly from the other ocean basins, presumably because of more rapid rates of clastic sedimentation in these areas. For example,

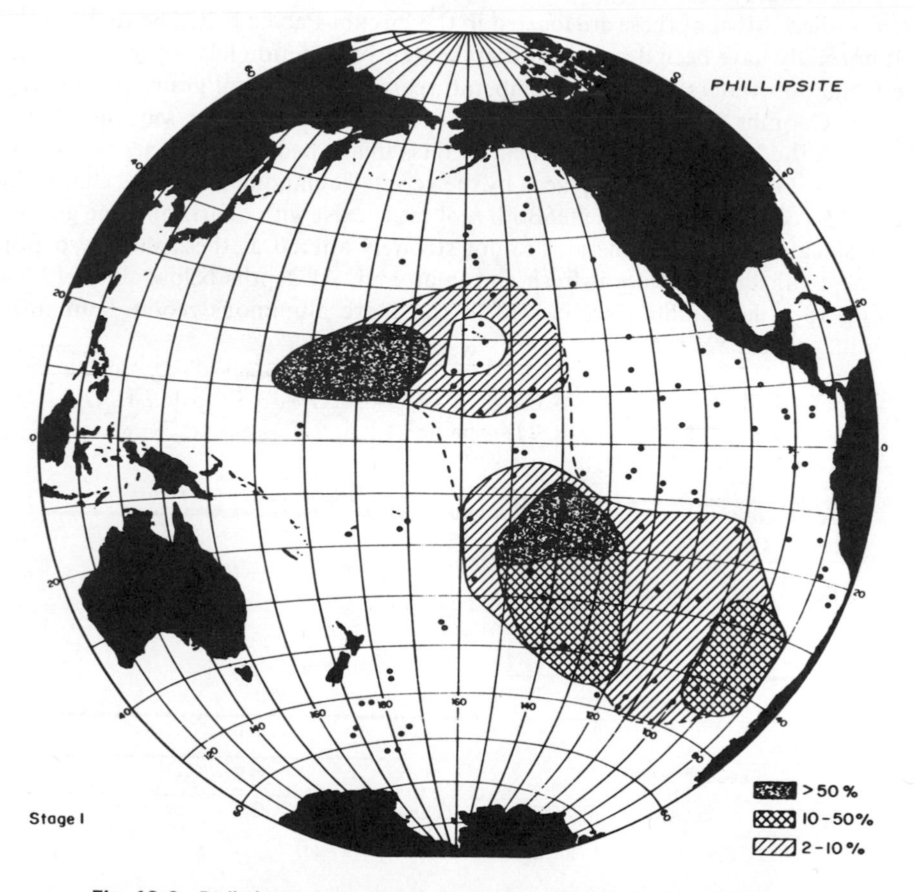

Fig. 18-2 Preliminary data on the distribution of phillipsite in the pelagic deposits of the Pacific (on a carbonate-free basis). (After Bonatti, 1963.)

the rapid influx of detritus into the Atlantic Basin during the Quaternary Period completely overshadows the slow rate of authigenic formation of zeolites in this environment.

18.4 ZEOLITES AND LOW GRADE METAMORPHISM

In our discussion of zeolite authigenesis in sedimentary rocks the most prominent relationship among variables clearly is that between zeolite formation and volcanic detritus. Tuffaceous beds are much more likely to contain zeolites than nontuffaceous beds. Volcanic materials, especially glass, are very unstable at the earth's surface and react rapidly. As ready sources of soda, potassia, alumina, and silica, they have no peer among naturally occurring particles.

Because of this relationship, zeolites and associated authigenic minerals such as prehnite, pumpellyite, potassic feldspars, and albite are major rock-forming minerals in many of the eugeosynclinal accumulations of volcanic debris that have been studied. Most of these are located in the circum-Pacific belt. The rocks in this belt normally have been deeply buried and extensive reaction has occurred between the formation waters at elevated temperatures and the detrital grains of the graywackes. Coombs (1954), for example, has described a graywacke sequence 30,000 ft thick in the New Zealand geosyncline. Glass in ash beds in the upper part of the section has been completely replaced by the zeolite heulandite (including clinoptilolite) or less commonly by analcite. Both zeolites coexist with quartz and fine grained phyllosilicates. Detrital minerals are not strongly altered at these shallow depths and most calcic plagioclase is fresh and unaltered. At depths below about 17,000 ft (Fig. 18-3) heulandite is converted to the more aluminous zeolite laumontite according to the reaction

$$\underset{\text{heulandite}}{CaAl_2Si_7O_{18} \cdot 6\,H_2O} \longrightarrow \underset{\text{laumontite}}{CaAl_2Si_4O_{12} \cdot 4\,H_2O} + \underset{\text{quartz}}{3\,SiO_2} + \underset{\text{water}}{2\,H_2O}$$

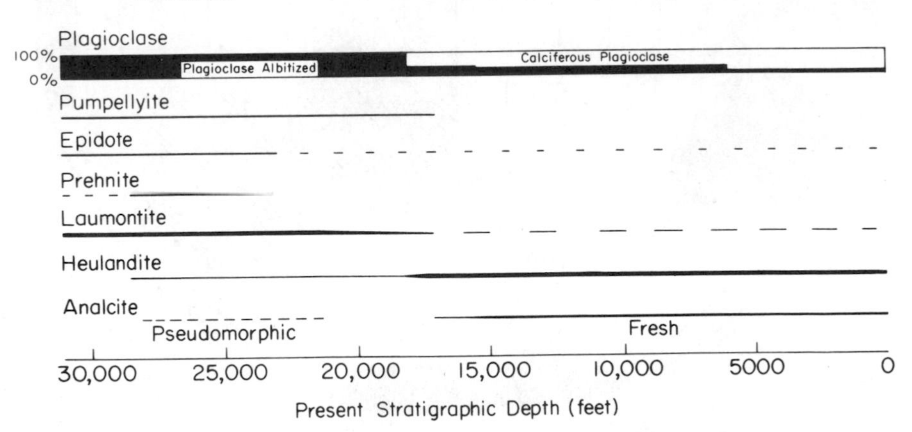

Fig. 18-3 Mineralogic features in relation to stratigraphic depth below the highest beds at present exposed, Taringatura, Southland, New Zealand. (From Coombs et al., 1959.)

Also, analcite is replaced by laumontite and the soda released albitizes existing calcic feldspars. Subsequently, laumontite reacts with celadonite in the rocks to produce pumpellyite, other aluminosilicates, quartz, and water. Alternatively, laumontite can react with prehnite and chlorite to produce pumpellyite, quartz, and water. Such reactions usually occur at temperatures between 150 and 250°C. Further reaction and dehydration at higher temperatures results in epidote and other common metamorphic minerals. The laumontite facies is generally regarded as the highest sedimentary condition and the pumpellyite facies as the lowest truly metamorphic facies, but the demarcation boundary is rather arbitrary. As indicated in Fig. 18-3, there is substantial overlap in the range of occurrence of most of the zeolite and "metamorphic" minerals. In many eugeosynclinal areas the overlap appears to be nearly complete because of spatial variations in the importance of parameters other than temperature.

18.5 SUMMARY OF ZEOLITE GEOCHEMISTRY

Zeolite minerals in sedimentary rocks are almost exclusively authigenic constituents. Therefore, their presence can be fully explained in terms of the interaction between aqueous solutions and other types of detrital grains. Field and laboratory studies of many zeolite localities indicate that the interactions are particularly sensitive to variation in temperature, the activity ratios of dissolved species, and the ratio aH_2O/aCO_2. Unfortunately, the stability field of each zeolite has not yet been defined within the multivariate space of these parameters. We are, therefore, only able to make general statements concerning, for example, the temperature range or range in aCa^{2+}/aNa^{+} in which wairakite will form in preference to analcite, its sodium analogue.

Zeolites are hydrous minerals and are, therefore, sensitive to temperature variations. Among the calcium-rich zeolites, for example, we would expect wairakite and laumontite to occur at greater depths than stilbite because the former contain fewer bound water molecules (Table 18-1). Zeolite sequences generally show consistent trends in hydrothermally altered rocks as well as in those modified diagenetically. For example, at Wairakei, New Zealand, mordenite (7 H_2O) and heulandite (6 H_2O) occur in boreholes in the temperature range 150 to 230°C; laumontite (4 H_2O), between 195 and 220°C; and wairakite (1 H_2O), between 200 and 250°C (Coombs et al., 1959, p. 77). Anhydrous feldspar appears at temperatures between 230 and 250°C.

The activities of calcium, sodium, and potassium ions and of silica in the pore fluid are particularly important determinants of which zeolite species will precipitate. The hydrogen ion concentration (pH) is also important because the hydrogen ion competes with the base cations for positions in an alumino-silicate framework. If hydrogen ions are too abundant, the law of mass action suggests that a phyllosilicate will form rather than a zeolite. High pH clearly favors zeolite formation. The importance of the relative activities of the different base cations and silica is equally clear. A high value for the ratio aCa^{2+}/aNa^{+} favors wairakite over analcite; high $aSiO_2$ favors mordenite over heulandite and heulandite over chabazite (Table

18-1); high aK^+/aCa^{2+} favors phillipsite over chabazite. The composition of chemically unstable tuffaceous detritus in a sediment is very important in this regard. For example, in the Teels Marsh playa in western Nevada, rhyolitic ash layers are present and the alteration product of these layers is largely phillipsite. In Lake Natron, Tanzania, tuffaceous detritus is absent but the brine contains 40% soda by weight because of alkaline spring discharge into the lake; the zeolite forming is very largely analcite.

The partial pressure of carbon dioxide is an important control on the stability of zeolite minerals. If it is too high, the activity of H_2O is lowered, which hinders zeolite formation; in addition, sufficient carbonate ion may be produced to cause calcite to form as the stable authigenic calcium-rich mineral rather than a zeolite. The thermodynamics of this relationship have been considered by Zen (1961). He demonstrated that common zeolite assemblages such as laumontite-quartz or heulandite-quartz should not coexist stably with an assemblage such as calcite-kaolinite-quartz. Zen's analysis explains chemically a long-recognized field observation, namely, that deeply buried miogeosynclinal beds do not contain zeolites in contrast to eugeosynclinal beds. Apparently the abundance of carbonate beds in miogeosynclinal sections buffers P_{CO_2} at a value sufficiently high to prevent zeolite formation. In eugeosynclinal sections limestone units are relatively rare and, in addition, glassy and felsitic basaltic volcanic detritus is normally available to serve as a ready source of calcium and, to a lesser extent, sodium ions. Sufficient sodium to form sodic zeolites such as natrolite can be obtained from normal marine pore solutions.

In summary, the chemical conditions favoring the occurrence of zeolites in general and the occurrence of specific zeolite species are complex. Further, the variables are interrelated to a significant extent. For example, the activity of silica in solution is related to the amount and type of solid silica present and to hydrogen ion concentration. The ready availability of calcium depends on the abundance and the state of the volcanic fragments being altered, in particular whether they are glassy or coarsely crystalline. The availability of sodium varies with the salinity of subsurface waters, which varies in response to factors such as permeability, filtration through clay membranes, and other factors discussed Chap. 10. It is apparent, therefore, that although clear trends in zeolite mineralogy may be present in some areas in response to burial depth or sediment permeability, such trends will commonly be obscured. Each zeolite occurrence poses distinctive questions concerning its origin and a judicious combination of field and laboratory studies is required to obtain satisfactory answers.

REFERENCES

BONATTI, ENRICO, 1963, "Zeolites in Pacific Pelagic Sediments," *N.Y. Acad. Sci. Trans.*, **25**, pp. 938–948. (Survey of the occurrence and origin of phillipsite in the Pacific Ocean Basin.)

BRADLEY, W. H., 1929, "The Occurrence and Origin of Analcite and Meerschaum Beds in the Green River Formation of Utah, Colorado, and Wyoming," *U.S. Geol. Sur. Prof. Paper* **158**, pp. 1–7. (Earliest detailed description of a zeolite-rich sedimentary rock, and how it probably formed.)

COOMBS, D. S., 1954, "The Nature and Alteration of Some Triassic Sediments from Southland, New Zealand," *Trans. Roy. Soc. New Zealand*, **82**, pp. 65–109. (An early and still important study of the development of zeolitic and low grade metamorphic facies in a eugeosynclinal volcanic pile.)

COOMBS, D. S., A. J. ELLIS, W. S. FYFE, and A. M. TAYLOR, 1959, "The Zeolite Facies, with Comments on the Interpretation of Hydrothermal Syntheses," *Geochim. Cosmochim. Acta*, **17**, pp. 53–107. (Classic study using field, petrographic, and experimental evidence to establish the validity of the zeolite facies.)

HAY, R. L., 1966, "Zeolites and Zeolitic Reactions in Sedimentary Rocks," *Geol. Soc. Amer. Spec. Paper* **85**, 130 pp. (Excellent review paper.)

———, 1970, "Silicate Reactions in Three Lithofacies of a Semi-Arid Basin, Olduvai Gorge, Tanzania," *Mineral. Soc. Amer. Spec. Paper No.* 3, 237–255. (Detailed analysis of zeolite formation in a well-studied lacustrine environment.)

HIGH, L. R., JR., and M. D. PICARD, 1965, "Sedimentary Petrology and Origin of Analcime-Rich Popo Agie Member, Chugwater (Triassic) Formation, West Central Wyoming," *Jour. Sed. Petrology*, **35**, pp. 49–70. (Detailed description and analysis of the distribution and origin of this richly zeolitic unit.)

IIJIMA, AZUMA, and MINORU, UTADA, 1966, "Zeolites in Sedimentary Rocks, with Reference to the Depositional Environments and Zonal Distribution," *Sedimentology*, **7**, pp. 327–357. (Summary of the occurrence of zeolites in sedimentary rocks, with particular reference to their occurrence in the Mesozoic and Cenozoic pyroclastic rocks of Japan.)

SHEPPARD, R. A. and A. J. GUDE, 3rd, 1969, "Diagenesis of Tuffs in the Barstow Formation, Mud Hills, San Bernardino County, California," *U.S. Geol. Sur. Prof. Paper 634*, 35 pp. (Comprehensive field and laboratory study of the geology and geochemistry of zeolite development in a Miocene saline alkaline lake.)

VAN HOUTEN, F. B., 1962, "Cyclic Sedimentation and the Origin of Analcime-Rich Upper Triassic Lockatong Formation, West-Central New Jersey and Adjacent Pennsylvania," *Amer. Jour. Sci.*, **260**, pp. 561–576. (Stratigraphic and paleogeographic study of a part of the classic Triassic section of northeastern United States.)

ZEN, E-AN, 1961, "The Zeolite Facies: An Interpretation," *Amer. Jour. Sci.*, **259**, pp. 401–409. (Thermodynamic explanation for the absence of zeolite minerals under conditions of high partial pressure of carbon dioxide.)

CHAPTER NINETEEN

IRON-RICH ROCKS AND SEDIMENTARY MANGANESE DEPOSITS

19.1 INTRODUCTION

Iron is the fourth most abundant element in the earth's crust and most terrigenous sedimentary rocks contain substantial quantities of iron oxides or silicates. A few sedimentary rocks, however, contain more than 15% iron (Fe) (James, 1966) and have been termed iron-rich.

The iron-rich sedimentary rocks include the world's largest reserves of iron ore, most of which is contained in the Precambrian banded cherty iron formations often known as "taconite," which average 25 to 40% iron. In addition to their economic importance, the occurrence of *all* types of iron-rich sedimentary rocks raises a number of fascinating, unsolved, scientific problems. Modern examples of iron-rich deposits exist but they are clearly not fully comparable with the extensive and varied types of deposits observed in most of the geological record. Absence of modern analogues has hindered the search for ideas that adequately explain the ancient rocks.

Most workers have recognized two general groups of ancient iron-rich rocks: the Precambrian iron formations and the Phanerozoic ironstones. The characteristics of these two groups are listed in Table 19-1.

TABLE 19-1 DIFFERENCES BETWEEN IRONSTONE AND IRON FORMATION (after James, 1966)

	Ironstone	Iron formation
Age	Pliocene to middle Precambrian. Principal beds are from the lower Paleozoic and the Jurassic.	Cambrian to early Precambrian; the principal formations are approximately 2000 million years old; the youngest age is Cambrian
Thickness	Major units are a few meters to a few tens of meters	Major units are 50 to 600 m
Original areal extent	Individual depositional basins are rarely more than 100 mi in maximum dimension	Difficult to determine; some deposits have continuity over many hundreds of miles
Physical character	Massive to poorly banded; silicate and oxide facies oolitic	Thinly bedded; has layers of dominantly hematite, magnetite, siderite, or silicate alternating with chert, which makes up approximately half the rock; oolites are rare
Mineralogy	Dominant oxide is goethite; hematite is very common; magnetite is relatively rare; chamosite is the primary silicate; calcite and dolomite are common constituents	No goethite; magnetite and hematite are about equally abundant; primary silicate is greenalite; chert is a major constituent; dolomite present in some units but calcite rare or absent
Chemistry	Except for high iron content, no distinctive aspects	Remarkably low content of Na, K, and Al
Associated rocks	Both are typically interbedded with shale, sandstone, or graywacke; yet the iron formation has little or no clastics as compared to the ironstone	
Relative abundance of facies	No gross differences apparent; probable order of abundance for ironstone: oxide, silicate, siderite, sulfide; for iron formation: the order is similar but siderite facies may be more abundant than silicate facies	

19.2 SOURCE OF IRON

The source of iron brought to a depositional basin has long remained a problem. Modern weathering is generally dominated by oxidizing conditions and iron in the oxidized state is highly insoluble. Seawater contains less than 0.01 ppm dissolved or colloidal iron and average river water contains 0.67 ppm iron (see Chap. 10). The ideal conditions for supply of iron to surface waters would appear to be found in tropical, humid areas of low relief and intense chemical weathering. In such regions complexing of iron with organic acids may increase the delivery of dissolved or colloidal iron to the basin of deposition. For an iron-rich rock to develop, however, the amount of detrital silicate deposition must be kept to a minimum so that the iron-bearing minerals will not be diluted by other sediments. The Amazon River drains a region of intense tropical weathering and contains 2 to 3 ppm iron but the sediments deposited at the mouth of this river contain less than 3% iron. Because of these problems, alternative sources for the iron within the depositional basin such as dissolution of iron-bearing minerals on the bottom of the sea or volcanic exhalations have commonly been proposed.

19.3 MINERALOGY AND MINERALOGIC FACIES

Four main mineralogic facies of iron-rich sedimentary rocks have been recognized: (a) oxides, (b) carbonates, (c) silicates, and (d) sulfides. Several of these facies may be found within a single deposit and in some cases the different facies may be rhythmically repeated in the stratigraphic succession. Lateral facies changes are also well documented and textural evidence indicates that each of the facies may be "primary" in the sense that it is not necessarily formed by diagenetic alteration of one of the other facies.

Oxides

The main oxides are goethite, hematite, and magnetite. Lepidocrocite and maghemite are rare.

Goethite (FeOOH) is the main oxide in the Jurassic ores of Europe. It is commonly found in ooliths, where it may form alternating shells with the silicate, chamosite. The ooliths are commonly embedded in a matrix that may be chamosite, clay, siderite, or calcite. Goethite does not occur in Precambrian iron formations. *Hematite* (Fe_2O_3) forms oolites in many Paleozoic and Precambrian deposits as a primary constituent, which suggests that the hematite was originally goethite that has been transformed diagenetically. Magnetite is more common in the Precambrian than in Phanerozoic rocks but it occurs, for example, in the Silurian Clinton iron ores and in the European Jurassic oolites both as oolitic coatings and as isolated crystals.

Carbonates

The carbonate mineral *siderite* ($FeCO_3$) forms one of the main components of both ironstones and iron formations. It forms "primary" deposits of fine grained mud and nodules and cement of diagenetic origin. It also occurs as oolitic coatings between layers of silicate minerals but does not form entire primary ooliths.

Silicates

Chamosite is the main iron silicate in Phanerozoic ironstones. The term refers to a green phyllosilicate of disputed structure with a 7 A basal spacing. Apparently, it readily recrystallizes to a chlorite with 14 A spacing. In the Precambrian iron formations, *greenalite* is the common silicate. It is closely related to chamosite, generally lacks the concentric oolitic structure of chamosite, and is associated typically with magnetite. True *glauconite* is a mica-type mineral with reflections at 3.3 and 10 A but the term has been loosely applied to other greenish mineral aggregates. Iron-rich chlorite (thuringite), iron talc (minnesotaite), and the brittle mica (stilpnomelane) are other iron silicates that may be locally important.

Sulfides

Pyrite is very common and forms a major constituent of the pyritic black shale facies of Precambrian iron formations. *Marcasite* is found associated with nodular siderite in coal measures. The "hydrotroilite" of older authors is now believed to be noncrystalline FeS, mackinawite (tetragonal $Fe_{1.05}S$), greigite (cubic Fe_3S_4), or mixtures of these. All are fine grained, black, and soluble in hot concentrated HCl, in contrast to pyrite, which is not soluble in acid and has a gray rather than black color when fine grained. The "monosulfides" are rarely found except in modern sediments.

In theory, the most important environmental factors controlling the occurrence of the different mineral facies are *pH*, *Eh*, and activity. Diagrams that show these relations for the iron-bearing minerals are valuable in that they indicate the general nature of the stability relationships between the different iron mineral species. Their application to natural environments is, however, limited by the complex composition of many natural mineral phases, which depart widely from the simple compositions assumed for the construction of the diagrams, and by the slow rates of reaction and common formation of metastable phases in low temperature environments. The reader should examine the diagrams in James (1966) and Curtis and Spears (1968).

19.4 MODERN ENVIRONMENTS OF IRON DEPOSITION

Although there are no examples of modern sedimentary iron deposits fully comparable with ancient deposits, there are a number of examples of modern environments in which iron minerals are known to be forming. Study of these examples helps us to understand the general principles governing the deposition of iron minerals and, therefore, contributes to some extent toward the understanding of ancient iron-rich rocks.

Iron minerals are forming today in several environments: (a) tidal flats and bog lakes, (b) oceanic shelves in tropical climates, and (c) deep basins with restricted circulation, such as fjords and the Black Sea.

The formation of *pyrite* in muds has recently been the subject of several theoretical, experimental, and field studies. Sulfur isotope analysis has helped to throw light on the origin of the sulfur. These studies have been summarized by Berner (1970) and his conclusions are given below. He deduced that the sulfur in the pyrite of modern muds comes from two sources: (1) organic matter; (2) sulfate dissolved in seawater. Although organic matter forms up to 10% of some modern marine muds this carbonaceous material contains only 1% sulfur. Yet the muds frequently contain more than 1% pyrite so that it is clear that an additional source of sulfur is required. The other source is the bacterial reduction of dissolved sulfate, which has been shown by sulfur isotope studies to be supplied by diffusion into the sediment from the overlying seawater. Where dissolved H_2S exists, pyrite is the thermodynamically stable form of iron; however, pyrite is not formed directly. The first iron compounds to appear are black, finely disseminated iron monosulfides (mackinawite, greigite, and noncrystalline forms). They contribute much of the black color commonly seen just beneath the surface of tidal flats and organic-rich lakes and stagnant basins. These compounds are thermodynamically metastable and they are observed to change to pyrite a few inches below the sediment-water interface. Berner has presented evidence that indicates that the formation of pyrite takes place mainly by the reaction of monosulfides with elemental sulfur produced from the oxidation of H_2S either inorganically or through the action of sulfur-oxidizing bacteria. If the bottom waters contain dissolved oxygen, much elemental sulfur is produced by the reaction of FeS and H_2S with dissolved oxygen stirred into the sediment by storms, currents, and burrowing organisms.

The fact that free H_2S is present in a sediment, or even in the deeper parts of basins with restricted circulation, does not necessarily mean that iron sulfides will be the only iron minerals present in the sediment. In the Black Sea, deeper waters have a slightly higher salinity (24 ppt) than the surface waters (18 ppt). The result of the density stratification has been to prevent mixing between the surface, oxygenated waters, and the waters below a depth of 150 m with the result that the waters from 150 m to depths in excess of 2000 m contain dissolved H_2S. Muds in the deeper parts of the Black Sea contain 4 to 5% iron, of which 47 to 80% is in the form of carbonates or chlorite and, at the most, only 44% is present as iron sulfide. Not all the iron is present as pyrite because over 95% of the iron supplied to the Black Sea is in the form of very fine clastic particles, which are not converted to sulfides as they settle through the deeper waters, despite the high content of H_2S in these waters. Once buried below the sediment interface, the conversion of the iron compounds to sulfides is limited by the availability of organic matter, sulfur, and possibly also by slow reaction rates. This is the case in spite of the high content of organic matter, which reaches values in excess of 2% organic carbon.

Chamosite, the most abundant primary iron silicate in Phanerozoic ironstones, has been recognized and locally makes up 60% of the sediment in the shallow waters of the Orinoco shelf, the shelf off Sarawak, and the Niger Delta (Porrenga, 1966). This chamosite develops in fecal pellets and within the remains of organisms and is generally found in waters less than 60 m in depth. Glauconite is found in deeper and thus colder waters in these same areas.

The iron-rich mineral *glauconite* occurs in water depths from 30 to 2000 m but is a minor constituent in modern marine sediments. Bright green pellets, 0.01 to 0.5 mm in diameter, are common. Glauconite also occurs as irregular fillings, crusts, replacements, and canal fillings in carbonate shells. Burst (1958) has shown that the term "glauconite" is commonly used rather loosely to include at least four mineral classes, of which only the first corresponds closely to the true mineral species: (a) well-ordered, high potassium, mica-type structures rich in ferric iron, (b) disordered, nonswelling low potassium mica-type structures, (c) extremely disordered montmorillonite-type structures, and (d) clay mineral mixtures.

In many cases, organic matter appears to play an important role in glauconite genesis, presumably by locally generating a reducing microenvironment while the overall marine environment in which glauconite forms is oxidizing. It appears that the immediate starting material for glauconite formation can be any degraded (expandable) layer lattice mineral (see discussion in Chap. 11). Slow sedimentation is necessary to permit the alteration of organic matter, mica, or clays to glauconite on the sea floor.

A common type of ancient iron-rich deposit, though now of minor economic importance, is the "*blackband*" and "*clayband*" ironstones, consisting of fine grained *siderite* nodules or thin nodular beds in shales and commonly associated with coal measures. The siderite nodules are found in both marine and nonmarine shales and may be associated with pyrite. Ho and Coleman (1969) have described incipient diagenetic pyrite replacements and siderite nodules that appear to be modern analogues in sediments below the swamps of the Atchafalaya Basin, Mississippi Delta region. Pyrite occurs replacing rootlets or other organic matter or as in well-crystallized cubes. The siderite is present in very diffuse, poorly defined nodules, seen as light areas a few centimeters in diameter on X-radiographs of lacustrine or swamp clays. In this modern example, and probably in most ancient examples, the nodules are formed by diagenetic mobilization of iron under reducing, organic-rich diagenetic conditions. pH values obtained in the cores are generally in the range of 7 to 8, which is the range favoring the precipitation of siderite or pyrite. The mobilization of iron under these conditions is attributed by the authors to locally lower pH and soluble organic complexes.

The "*bog iron ores*" of modern swamps and lakes represent one of the few examples of modern iron-rich sediments. They are found in poorly drained, recently glaciated regions or areas of older iron-rich rocks. Two general groups have been recognized: lake ores and bog ores. Lake ores consist of oolitic or pisolitic grains, cemented together into disks up to a foot or more across. They are commonly found in water only a few feet deep, at the border of the lake. Bog ores form thin layers of earthy to pisolitic iron oxides at the surface or below several feet of peat. Both types of ores appear to be formed by the migration of acidic organic-rich groundwater. Precipitation of ferric iron takes place where the migrating soil or swamp-derived waters enter a more oxidizing, less acidic environment.

Iron minerals may be produced by certain organisms. For example, marine chitons (*Polyplacophora*) form teeth of magnetite and marine gastropods have been shown to produce teeth composed of goethite (*Patella vulgata*, *Nomaeopelta dalliana*, *Lottia gigantea*, and *Acmaea mitra*).

Concentration of heavy minerals such as magnetite in sedimentary placer deposits is occurring today as in the past and represents a secondary development of iron-rich rocks that locally are important.

19.5 PHANEROZOIC IRONSTONES

The ironstones have been described by Gross (1965) as belonging to two main types: the Minette type (named for the Minette beds of Jurassic age in eastern France) and the Clinton type (named for the Clinton iron ores of Silurian age in the region from New York south to Alabama).

The *Clinton* ores are typically deep red to purple, massive hematite-chamosite-siderite beds with oolitic textures. They are rich ores, containing 50% or more iron. There is generally a fairly high content of clastic particles, including quartz. It has been shown that the oolitic hematite-chamosite ironstones of the Clinton grade east into semicontinental hematitic sandstones and west into glauconite-bearing marine beds. In New York the ironstones form thin lenticles in limestone-shale formations. They are demonstrably shallow marine because many consist of fragments of marine fossils replaced by or coated with hematite. Ooliths with alternating coatings of hematite and chamosite around quartz or calcareous nuclei are also present.

Minette-type ores occur in several Jurassic formations in Europe, including the Lorraine ores of France, the Middle Jurassic Northampton sand, and the Lower Jurassic Cleveland and Frodingham ironstones of England. These are also oolitic ores but oxides are less important than in the Clinton type and the ores are predominantly composed of chamosite and siderite.

The Northampton sand ironstones were the subject of a classic petrographic study by Taylor (1949). They display three facies: (a) a carbonate "phase," dominantly nonoolitic and composed of sideritic mudstones, siltstones, sandstones and limestones; (b) a mixed "phase" composed of chamosite-oxide ooliths in a matrix of siderite; (c) an aluminosilicate "phase" characterized by an absence of primary carbonate and a dominance of chamositic and kaolinitic oolites and sandstones. The three phases follow each other in stratigraphic sequence, with a minor erosional break between phases (b) and (c). Another erosional break is followed by repetition of the rhythm. These ironstones are also shallow marine and have structures such as cross-bedding, minor erosional breaks, and intraformational conglomerates. The aluminosilicate phase, however, does not contain fossil remains. The formation becomes sandy to the east, suggesting the proximity of a shoreline. The commonest ooliths have coatings composed of tangentially arrayed, very fine chamosite flakes around a core of structureless chamosite; but types with cores of broken ooliths, fragments of chamosite mud, and limonite are common and the chamosite coatings may alternate with kaolinite, siderite, and, rarely, magnetite. Many ooliths are flattened or distorted and must have been soft prior to diagenetic cementation by chamosite, siderite, calcite, or, more rarely, a carbonate apatite.

19.6 PRECAMBRIAN IRON FORMATIONS

Gross (1965) distinguished two major types of Precambrian iron formations. The *Algoma type*, named for the Algoma district of Ontario, was deposited with volcanic and sedimentary rocks of the type commonly found in eugeosynclines. Typical beds are composed of gray or red jasper cherts interbedded with magnetite- and hematite-rich layers but massive siderite, pyrite-pyrrhotite, and other beds also occur. The deposits vary in thickness but are commonly less than a few miles in length. Carbon-bearing schists with iron sulfides also are common with these rocks. The association with volcanic rocks has led several workers to suppose that the iron had a volcanic source. The *Superior type*, named for the Lake Superior region of the United States and Canada and also found in the Labrador trough, is typically composed of thinly banded cherty rocks with a granular or oolitic texture and contains very little clay or quartzose clastic material. Cherty magnetite and hematite rocks and cherty iron silicate and carbonate rocks form facies, generally separated into distinct stratigraphic zones. Associated beds are quartzite, dolomite, black ferruginous slates, argillite, and volcanic rocks. The iron formations are hundreds of feet thick and extend as much as 1000 mi continuously along the strike.

Trendall (1968) has summarized the main characteristics of three major banded iron formations: the Animikie Basin of Lake Superior, the Hamersley Basin of Western Australia, and the Transvaal System Basin of South Africa. He finds that these three formations resemble each other in basin size (each is about 300 mi long), age, chemical composition and mineralogy, deposition of chert preceding as well as contemporaneous with the iron formations, restriction of the iron formations to the central part of the stratigraphic succession in the basin, and association with volcanic rocks. The basins do, however, show some differences, particularly between the Superior and the Australian and South African basins. The Superior iron formations vary greatly in thickness over short distances in some parts of the basins. Both the Australian and South African formations show better developed fine banding than the Superior formation. In Australia, the iron formation consists of alternating bands (called mesobands) 6.4 to 50 mm. thick, within which there is a regular microbanding on a submillimeter scale. The silica-rich mesobands consist of a quartz mosaic (chert) with grains 10 to 30 μ in diameter and they are sharply divided from the intervening and intergradational magnetite and quartz-iron oxide mesobands. Mesobands with siderite, stilpnomelane, and riebeckite are less common. It is claimed that individual mesobands, and even microbands, can be correlated over almost the entire basin, for distances up to 185 mi. Such extreme stratigraphic continuity has not been demonstrated for other basins.

The origin of Precambrian iron formations remains obscure, despite years of study and abundant exposures provided by mining operations. The source of the iron and silica in these formations has been attributed to volcanism where these rocks are associated with volcanic rocks or to intense weathering of the landmass where low relief of the landmass has been deduced from absence of detrital sediments. Chert associated with iron formation may have had an organic source as does most of the Phanerozic chert (see Chap. 16). This possibility has been considered following the recognition of many Precambrian organisms. However, the existence of abundant Precambrian siliceous organisms remains uncertain. One problem of greatest importance is the identification of primary mineral facies as opposed to rocks composed of minerals produced by diagenesis. Much remains to be learned and it is possible that some minerals now considered to be "primary" may ultimately prove to have been formed after deposition by diagenesis.

Varve-like banding has suggested the need for a homogeneous environment existing over large areas with a mechanism for repeatedly and abruptly changing from predominantly iron to predominantly silica deposition. Mechanisms such as seasonal overturn of lakes, seasonal changes in temperature, or supply of nutrients to siliceous organisms have been suggested but an acceptable unified theory of origin remains to be presented.

19.7 SEDIMENTARY MANGANESE NODULES

The occurrence of concretions or nodules of manganese oxides has been known for approximately 100 years. First discovered by the Challenger Expedition in 1873 and reported in 1891 by Murray and Renard, these unusual chemical sedimentary deposits aroused considerable scientific interest. The early finds were predominantly from deep water in the range of 4 to 5 km. However, they were also discovered in relatively shallow water in the Firth of Clyde on the west coast of Scotland. By 1940 they had been recognized on Pacific seamounts and are now known from many inland lakes. Because of contained copper, nickel, and cobalt in addition to the manganese, these features have long been considered a future potential source of these metals. Manganite ($MnO \cdot OH$) is the most common manganese oxide mineral found in these concretions. Many of the manganese oxide minerals have disordered layers that permit the inclusion of Fe, Na, Ca, Sr, Cu, Cd, Co, Ni, and Mo.

Manganese nodules occur in a wide variety of sizes. Micronodules range in size up to 1 mm and are generally found disseminated in sediments. These may be accreted together to form crusts on the sea bottom. Nodules range up to over 20 cm in diameter and sometimes are almost spherical. Flatter nodules or slabs have been found with dimensions in excess of 1 m. All these concretions have developed by the growth of manganese minerals in concentric bands around detrital particles or organic tests such as coccoliths. The common core material is pebbles or boulders of volcanic rock. Encrustations as thick as 11 cm have been reported and they often include foreign material such as clay minerals, skeletal calcium carbonate, and volcanic rock fragments. Now recognized from most of the ocean bottoms of the world as a result of dredging and sea bottom photography, these manganese concretions make up approximately one-third of the rocks, as opposed to sediment, exposed on the ocean floors. Because they represent concretionary growth on the sea bottom, they are naturally most commonly exposed in areas of relatively slow sedimentation. Even in areas of more rapid sedimentation, abundant data from cores indicate that they sometimes exist below the surface within the sediment.

The rate of concretionary growth of manganese oxide nodules varies considerably. Naval gun shell fragments found off the coast of southern California have developed, in this century, layers several inches thick. This material may be unusual because of the unusually high iron content. Upper Cretaceous fossils from mid-Pacific mountains show 5.5 cm of encrustation. Rates of 1 mm/1 million yr or less are probably common in the deep sea.

The manganese found in these nodules was derived from at least two sources: (a) weathering of continental rocks and (b) reaction of submarine volcanic material with seawater. It is impossible to determine the relative contribution of these two sources. Many of the nodules, however, are associated with submarine volcanic debris. The mechanism of precipitation involves the oxidation of manganese to Mn^{4+} from the Mn^{2+} state. This can take place by reaction with oxygen-rich seawater. The possibility that iron oxide surfaces produce a catalytic effect that leads to the oxidation of the manganese is generally suspected.

Manganese nodules are very common on the bottom of lakes and seas in both deep and shallow water. They are uncommon but have been reported from beneath the sediment-water interface in shallow cores. However, ancient sedi-

Fig. 19-1 Jurassic manganese nodules from Sicily. (Photo by H. C. Jenkyns.)

mentary rocks containing manganese nodules are very rare. It appears that manganese nodules may be dissolved once they become buried. Jenkyns (1967) reported manganese nodules in the Jurassic of Sicily very similar to those found in modern sediments (Fig. 19-1). The section containing these nodules is much condensed compared with sedimentary sections of similar age and appears to have formed in relatively shallow water on seamounts.

REFERENCES

BERNER, R. A., 1970, "Sedimentary Pyrite Formation," *Amer. Jour. Sci.*, **268**, pp. 1–23. (Pyrite forms by diagenetic reaction of monosulfides with sulfur.)

BORCHERT, H., 1960, "Genesis of Marine Sedimentary Iron Ore," *Inst. Mining and Metallurgy Trans.*, Bull. 640, **70**, pp. 261–279. (Literature review and statement of theory of derivation of iron from sea bottom reactions.)

BURST, J. F., 1958, "Glauconite Pellets: Their Mineral Nature and Application to Stratigraphic Interpretation," *Amer. Assn. Pet. Geol. Bull.*, **42**, pp. 310–327. (The origin and interpretation of glauconite.)

CURTIS, C. D., and D. A. SPEARS, 1968, "The Formation of Sedimentary Iron Minerals," *Econ. Geol.*, **63**, pp. 258–270. (Eh-concentration diagrams, and their application to British Carboniferous and Jurassic iron-bearing rocks.)

DEGENS, E. T., and D. A. ROSS, (ed.), 1969, *Hot Brines and Recent Heavy Metal Deposits in the Red Sea*. New York: Springer-Verlag, 600 pp. (Symposium on the extraordinary Red Sea discovery.)

DIMROTH, ERICH, 1968, "Sedimentary Textures, Diagenesis, and Sedimentary Environments of Certain Precambrian Ironstones," *Neues Jahrb. Geol. Paleont. Abh.*, **130**, pp. 247–274. (Petrographic description and classification of Labrador iron formation.)

DUNHAM, K. C., 1960, "Syngenetic and Diagenetic Mineralization," *Proc. Yorkshire Geol. Soc.*, **32**, pp. 229–284. (Discussion and good description of ironstones.)

GOLDBERG, E. D., 1961, "Chemistry in the Oceans," in *Oceanography*, M. Sears, ed. Washington, D.C.: Amer. Assn. Advancement Sci. Pub., no. 67, pp. 583–597. (A good summary of the chemistry of marine sedimentary manganese with the argument for the role of iron in inducing the oxidation of the manganese.)

GROSS, G. A., 1965, "Geology of Iron Deposits in Canada, I, General Geology and Evaluation of Iron Deposits," *Geol. Sur. Canada Econ. Geol. Rept.* 22. (Detailed study with excellent synthesis.)

HALLAM, ANTHONY., 1966, "Depositional Environment of British Liassic Ironstones Considered in the Context of their Facies Relationships," *Nature*, **209**, pp. 1306–1309. (Good discussion of the evidence for the origin of ironstones.)

HALLIMOND, A. F., et al., 1951, "The Constitution and Origin of Sedimentary Iron Ores: A Symposium," *Yorkshire Geol. Soc. Proc.*, **28**, pp. 61–101. (Discussion of iron-rich rocks.)

HO, CLARA, and J. M. COLEMAN, 1969, "Consolidation and Cementation of Recent Sediments in the Atchafalaya Basin," *Geol. Soc. Amer. Bull.*, **80**, pp. 183–192. (Pyrite and siderite nodule formation.)

JAMES, H. L., 1966, "Data of Geochemistry," 6th ed., Chapter W, *Chemistry of the Iron-rich Sedimentary Rocks.* U. S. Geol. Sur. Prof. Paper 440-W, 61 pp. (A comprehensive review.)

JENKYNS, H. C., 1967, "Fossil Manganese Nodules from Sicily," *Nature*, **216,** pp. 673–74. (Reported occurrence of ancient sedimentary manganese nodules.)

MELLON, G. B., 1962, "Petrology of Upper Cretaceous Oolitic Iron-Rich Rocks from Northern Alberta," *Econ. Geol.*, **57**, pp. 921–940. (A careful petrographic study and good discussion of genesis of ooliths.)

MERO, J. L., 1962 "Ocean Floor Manganese Nodules," *Econ. Geol.*, **57**, pp. 747–767. (Occurrence and analytical data on manganese nodules with emphasis on the economic potential of these materials.)

PORRENGA, D. H., 1966, "Glauconite and Chamosite as Depth Indicators in the Marine Environment," *Marine Geol.*, **5**, pp. 495–501. (Report of recent chamosite and occurrence of recent glauconite and chamosite.)

STRAKHOV, N. M., 1958, "Sur les formes du fer dans les sediments de la Mer Noire," *Eclogae Geol. Helvetiae*, **51**, pp. 753–761. (Source of the data on the Black Sea discussed in this chapter.)

TAYLOR, J. H., 1949, "Petrology of the Northampton Sand Ironstone Formation," *Geol. Sur. Great Britain Mem.*, 111 pp. (A classic study. For a brief summary, see the reference by Hallimond et al. listed above.)

TRENDALL, A. F., 1968, "Three Great Basins of Precambrian Banded Iron Formation Deposition: A Systematic Comparison," *Geol. Soc. Amer. Bull.*, **79**, pp. 1527–1544. (Good descriptive summary, with new information on Australian iron formation.)

PART SIX

CONCLUSION

Most of the work done on sedimentary rocks, and most of this book, has been concerned with detailed observation and interpretation. It remains to consider some of the broader questions of sedimentary rock genesis, to which only speculative answers can be given at this time. Is there a larger scale pattern to the development of sedimentary rocks through geological time? What is the importance of controls such as changes in sea level, climate and tectonic activity? Were processes of sedimentation different in Precambrian times from those observed today?

CHAPTER TWENTY

MAJOR EXTERNAL CONTROLS OF SEDIMENTATION

20.1 INTRODUCTION

In the preceding chapters of this book, the origin of many features of sedimentary rocks has been discussed. The question still remains whether or not there is any larger pattern to the development of sedimentary rocks on the continents through geological time. Can any major controls of sedimentation be identified, which would provide a rational explanation of the major associations of sedimentary rocks produced by the operation of the geological cycle over the last 3.5 billion years?

There appear to be three such major controls: (a) changes in sea level, (b) changes in climate, and (c) the effect of large-scale tectonic movements of the earth's crust. The first of these, changes in sea level, may well result from one of the other two. Sea level may rise, as it did some 18,000 to 7000 years ago, as a result of a major climatic change, associated with the melting of large ice sheets. Alternatively, it is possible that large areas of a continent may rise or sink or that there may be worldwide fluctuations in sea level resulting from major tectonic events, particularly those affecting large parts of the ocean floors. At the present state of knowledge, however, it is generally very difficult to determine whether or not ancient sea level changes were worldwide (eustatic) and whether or not they originated as a result of climatic or of tectonic factors (see discussion in Duff, Hallam, and Walton, 1967). Thus it appears necessary at present to consider sea level variation as an independent major control of sedimentation.

20.2 SEA LEVEL CHANGE

The importance of sea level as a major control of erosion and sedimentation has been recognized for over 100 years. Changes in sea level greatly affect sedimentation in alluvial plains. For example, during the Pleistocene low stand of the sea, the lower Mississippi River had a relatively steep gradient and was a braided stream transporting coarse sand and gravel to the Gulf of Mexico. It has been estimated that more than 1000 cu mi of sediment were eroded from the entrenched valley of the Mississippi at this time, and probably most of this sediment was deposited in the large subaqueous cone extending from the base of the continental shelf onto the abyssal plain of the Gulf of Mexico. The subsequent rise in sea level resulted in rapid alluvial filling of the entrenched lower valley of the Mississippi with deposits becoming finer upward as the river gradient became less steep and the channel pattern changed from braided to meandering.

The effects of a rise in base level on a major river system are not restricted to the vicinity of the shoreline but extend for many miles upstream. For some miles upstream of base level the flow is not uniform but gradually varied in depth with the water surface adjusted to a level called by hydraulic engineers the "backwater profile." The backwater profile is well-known to engineers and is described in most books on the hydraulics of open channels and its practical effects are known from observing the changes in rivers upstream from major dams. These effects have been observed to extend to distances tens of miles above the dam site and to elevations far above the water level in the reservoir. The backwater effect results in deceleration of the flow and increase in depth upstream. In a river moving sediment on its bed, this results in aggradation of the bed and adjustment of slope, which in turn extends the effect of the base level change even further upstream.

The marine environment also responds to change in sea level but in ways that are not easily predicted. Much of the older geomorphology of coastal regions was based upon deductive models of the response of coasts to changes in sea level. Modern geomorphological studies have invalidated many of the criteria for shorelines of "emergence" or "submergence" that were deduced from those older models. Studies of sediments on the continental shelf, particularly on the east coast of North America, have shown that much of the shelf is underlaid by Pleistocene relict sediments. These sediments and many of the topographic features on the shelf were formed during lower stands of the sea and drowned by the rapid Holocene rise of sea level. Thus, for many shelves, an equilibrium between sediment supply and dynamic processes operating on the shelves has not yet been established. There is no general agreement among sedimentologists studying modern shelves on the origin of present shoreline features, such as barrier islands, or on the nature of the ultimate equilibrium, assuming that sea level remains stable long enough for it to be reached. Curray (1965) has argued that continuing slow subsidence of modern shelves will ultimately lead to a "profile of deposition," with sandy sediments at the shoreline grading out to finer sediments at the shelf edge. Ultimately sediments carried out and deposited along this "profile of deposition" will bury a thin discontinuous unit of sandy sediment reworked from Pleistocene sediments drowned during the rise in sea level. On broad shelves without a major fluvial source of sediment, this burial of relict sediments has only just begun so that fine grained modern sediments are confined to a belt immediately seaward of the shore and relict, sandy sediments are typical of the shelf edge. Curray believes that only fine sediments can be carried out onto the deeper parts of the shelf, though he has stressed the importance of stirring of the shelf bottom by infrequent severe storms. Swift (1970) has suggested that the "relict" sands on the outer shelf are being slowly reworked by storm waves, tides, and ocean currents. There is evidence from mineral composition, for example, that at least part of the littoral sands of the east coast of the United States are derived by reworking of deeper water shelf sands.

In all such discussions, it is important to remember the importance of the rate and nature of sediment supply from the land, as well as the postulated movements of sea level. Barrier beaches formed since sea level became almost stable, about 7000 years B.P., show sequences of transgression (Sapelo Island, Georgia), vertical buildup (Padre Island, Texas), or regression (Galveston Island, Texas, see Fig. 20-1). The main factor controlling transgression or regression in these cases is not rise or fall of sea level but the balance between sediment movement and supply.

The role played by rising sea level in the formation of barrier beaches themselves is still disputed. At least three origins for modern barriers have been suggested: (a) They were formed offshore by wave buildup of longshore bars. (b) They were formed by growth out from headlands of spits, resulting from longshore drift of sand, and subsequent breaching of the spits during storms formed barriers. (c) They were formed from beaches by a combination of slow sea level rise and a supply of sand in sufficient quantities to allow buildup of a barrier, which became progressively separated from the land by a wider and wider lagoon as sea level continued to rise. The last explanation is favored by Hoyt (1967) for Sapelo Island and many other barriers because there is no evidence in the Holocene stratigraphic record that the area landward of barriers, now occupied by the lagoon, was previously a marine environment with littoral and shallow neritic sediments and faunas, as would be required by either of the other two postulated origins.

In the stratigraphic record many transgressive and regressive sands form more or less continuous "blankets" of wide areal extent. It is frequently supposed that the transgressive blanket sands are diachronous units formed by the migration and reworking of a beach–dune, or lagoon–barrier in response to slowly rising sea level. There appear to be few if any modern analogues of these transgressive units and one possible explanation for their absence is that sea level has not been sufficiently stable to permit their formation.

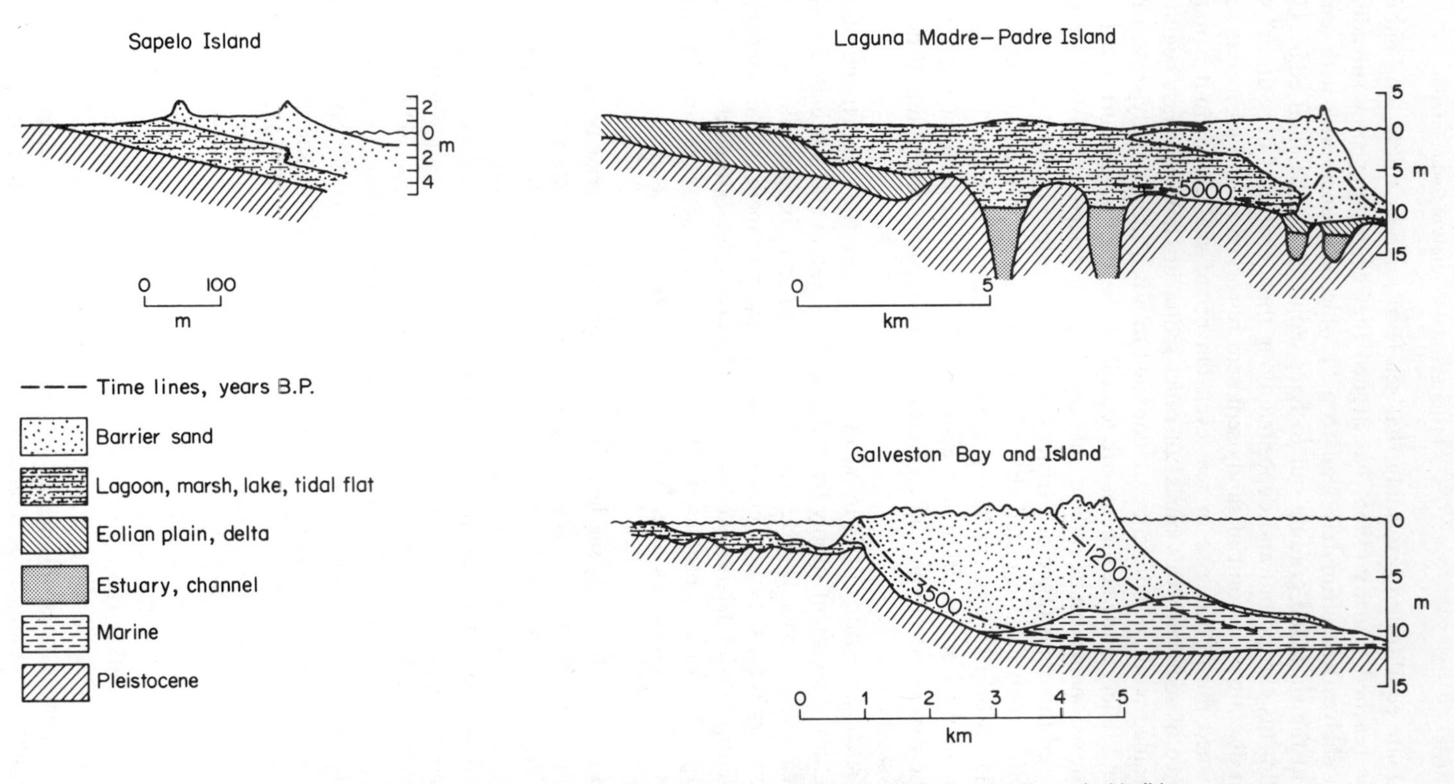

Fig. 20-1 Cross-sections of modern barriers, showing transgression (Sapelo Island), vertical buildup (Padre Island) and regression (Galveston). (After Curray *in* Stanley, 1969, based on studies by Hoyt, Fisk and Bernard and others.)

Sea level changes affect not only shallow water but also deep water sedimentation. In southern California, some submarine canyons active during lower stands of sea level are no longer active channels for supplying sediment to deep sea fans because the head of the canyon does not extend sufficiently far into shallow water to trap any of the sandy sediment moved by longshore drift. It seems likely, as originally supposed by Daly, that the periods of lower sea level during the Pleistocene may have been especially favorable to the supply of coarse sediment into submarine canyons and consequently to the formation of coarse turbidites in the deep sea.

20.3 CLIMATIC CHANGE

The evidence for the Pleistocene ice ages is striking testimony to the fact of major climatic changes in geological history. Some of the major effects that climate may have on the type and rate of sedimentation have been remarked many times in earlier pages of this book. They include the following: (a) Ice ages produce a lowering of sea level because of withdrawal of ocean water into large ice caps. In the Pleistocene it is estimated that the maximum lowering was of the order of 145 m. If all the existing ice sheets were to melt, sea level would rise by an additional 70 m. Not only do ice ages change the level of the sea, they also produce major changes in the distribution of climatic zones in equatorial regions, reducing the size and degree of aridity of the subtropical arid zones and displacing them somewhat toward the equator (Fig. 20-2). Thus the arid regions of the southwest United

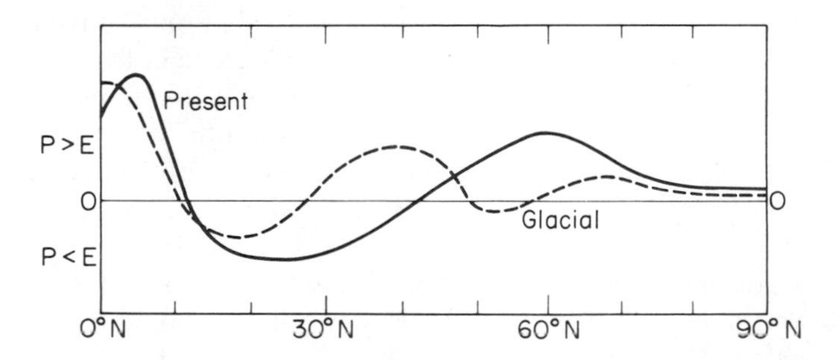

Fig. 20-2 Location of humid and arid zones in glacial and interglacial (present) times. (After Flohn, 1953, Erdkunde, v. 7, p. 270.)

States were the sites of lakes during the ice ages, testifying to "pluvial" climates with about 10 in. more rainfall than at present. (b) Climate is a major factor controlling weathering, soils, and land vegetation. It is also a major factor in the development of different landforms, although the precise nature of the differences in landforms between different climatic zones is still controversial. As discussed in Chap. 2, climate is a major control of the rate of erosion, which appears to be greatest in semiarid regions. The major agents of sediment movement on land are strongly affected by climate. Mass movement is effective under cold and tropical wet climates; wind action is most effective under cold or hot arid conditions; and rivers are most effective under temperate to tropical conditions with moderate rainfalls. (c) Climate is obviously of great importance in the formation of organic or chemical sediments such as limestones and evaporites. Formation of carbonate requires a warm climate that increases organic productivity and saturation of calcium carbonate in the surface waters of the sea. Evaporites require arid climates. (d) Climate indirectly affects clastic as well as organic and chemical sedimentation in shallow seas, mainly by controlling the supply of sediment from glaciers or rivers. Hayes (1967) found that, on the modern continental shelves, gravel is most abundant in cold regions and mud is most abundant in the wet tropics. Climatic variation can produce marked changes in littoral sediments by controlling the supply of sand from rivers in semiarid regions. In southern California, for example, the United States Army Engineers found that there was an empirical relationship between the rate of longshore movement of sand and the average rainfall during a 2-year period, $3\frac{1}{2}$ years earlier (Norris, 1964). A change in runoff may therefore induce beach growth (regression) or erosion (transgression). Climatic changes may also affect littoral sedimentation by altering the intensity or dominant direction of wave action and by altering the paths of major storms.

A prominent aspect of climatic change is its cyclic character (see Duff, Hallam, and Walton, 1967). This appears to be well demonstrated on both short and long time scales. Besides seasonal variations, there are demonstrated fluctuations on several larger scales, detectable both from climatic records and from study of varved clays and evaporite deposits. Relatively short-term cycles, such as the sunspot cycle (11 years) or cycles of various periods up to 3000 years, are generally poorly developed. A cycle of about 21,000 years has been reported from the varved lacustrine clays of the Eocene Green River Formation and from several other formations. This cycle is the right length to correspond to the precession of the equinoxes, which may be expected to produce a substantial effect on the climate.

Of more geological importance are glacial-interglacial cycles with a periodicity of about 40,000 to 50,000 years. At least eight major temperature minima have been recognized during the last 400,000 years (Emiliani, 1970). For the Permo-Carboniferous glaciation, it has been claimed that there were at least fifty periods of glacial advance, as evidenced by fifty distinct tillites exposed in Victoria, Australia (Wanless and Cannon, 1966).

On an even longer time scale, there is evidence from oxygen isotopic studies of a fall in average temperature at middle latitudes of 10 to 12°C since the beginning of the Tertiary. Major ice ages were apparently restricted to Pleistocene, Permo-Carboniferous, Siluro-Ordovician, and at least two periods in the Precambrian. It is doubtful that there existed major ice sheets during other periods, though it is possible that evidence of more continuous glaciation will be discovered, particularly for the Paleozoic (see the review by Crowell and Frakes, 1970). At periods when there were no major ice sheets, it is possible that the climate was more uniform over the whole earth than at present.

The cyclic nature of climatic change has suggested to many geologists that cycles of sedimentation should be explained by climatic controls. Examples include the suggested control of Pennsylvanian cyclothems in terms of direct climatic control of sediment supply or of indirect control by eustatic rise and fall of sea level; the climatic control of cyclic lacustrine "detrital" and "chemical" cycles in the Triassic Lockatong Formation of New Jersey (see Chap. 18); and climatic control of the larger 5- to 10-ft cycles in the Green River Formation.

The major difficulty in proving climatic control for sedimentary cycles is to develop paleoclimatic criteria of sufficient sensitivity to measure relatively small fluctuations in climate and to distinguish the effects of climatic variation from the effects of other controls. Oxygen isotope paleotemperatures may be distorted by diagenetic isotope exchange and paleontological methods are rarely reliable for periods earlier than the Late Tertiary. Climatic changes in an area of erosion might have important effects in the basin of deposition hundreds of miles away. Thus, even though climate is known to be an important control of sedimentation, it will probably continue to be difficult to prove its importance in any particular stratigraphic sequence.

20.4 TECTONIC CONTROL

The major control of all sedimentation is tectonics. If there were no uplift, there would soon be no land to erode and hence no sediments to deposit. The question, therefore, is not whether or not tectonics controls sedimentation but how the control operates.

The major theories of tectonics and sedimentation are very closely linked so that a discussion of tectonic control must be as much a discussion of major theories of structural geology as of the major observations of stratigraphy and sedimentology. At the present time, the theories are undergoing very rapid changes consequent upon the development of the hypothesis of plate tectonics. Some of the older "well-established" theories are being rejected or strongly modified. In the discussion that follows, therefore, it must be borne in mind that there is a need for rethinking about all traditional concepts in the light of the new theories.

Tectonics controls sedimentation at every stage of the geological cycle: (a) by controlling the rate of uplift and erosion, (b) by controlling the gradient of the major topographic surfaces across which sediment is transported, (c) by controlling the rate of subsidence in the basin of deposition, and (d) by influencing the changing pressures and temperatures produced in the sediment by burial, folding, and faulting.

Early stratigraphers classified sedimentary rocks into major units (systems) separated by major unconformities. Each unit was regarded as recording a major cycle of transgression and regression of the sea over a relatively stable land mass (the "shield" or "craton" of later writers). At first it was supposed by many geologists including Edward Suess, who popularized the idea in his influential work *The Face of the Earth* (first published in German in 1883 to 1908), that the cause of the major onlaps and offlaps of sediment was worldwide eustatic change in sea level.

Other geologists doubted that the major transgressions were synchronous. It was gradually accepted that the basis for definition of major stratigraphic units, such as systems, should be by type section and faunal succession rather than by tracing major unconformities. Geologists laid more stress on diastrophic rather than eustatic controls of vertical movements of land masses and many doubted that major rock units bounded by extensive unconformities had any real existence, even on a continental scale. In 1963, Sloss revived the concept of large-scale unconformity-bounded units and called them *sequences*. Sloss distinguished six Phanerozoic sequences on the North American craton (Fig. 20-3). Besides the major sequences, sedimentation on the cratons is modified locally by subsidence of interior basins, such as the Michigan and Williston Basins, and linked subsidence and uplift in fault-controlled "yoked basins" such as the Colorado Pennsylvanian Basin and Front Range uplift. Sedimentation on the craton is also affected by events taking place in adjoining mobile belts. During or following orogeny, floods of clastic sediments, forming "clastic wedges," spread out from the mobile belt onto the borders of the craton (Fig. 20-3). Sloss failed to discover major cyclicity in the six sequences, which are composed of varying lithological associations.

In contrast to the behavior of the stable cratons, the mobile or orogenic belts are characterized by greater thicknesses of sediment, more rapid deposition, folding,

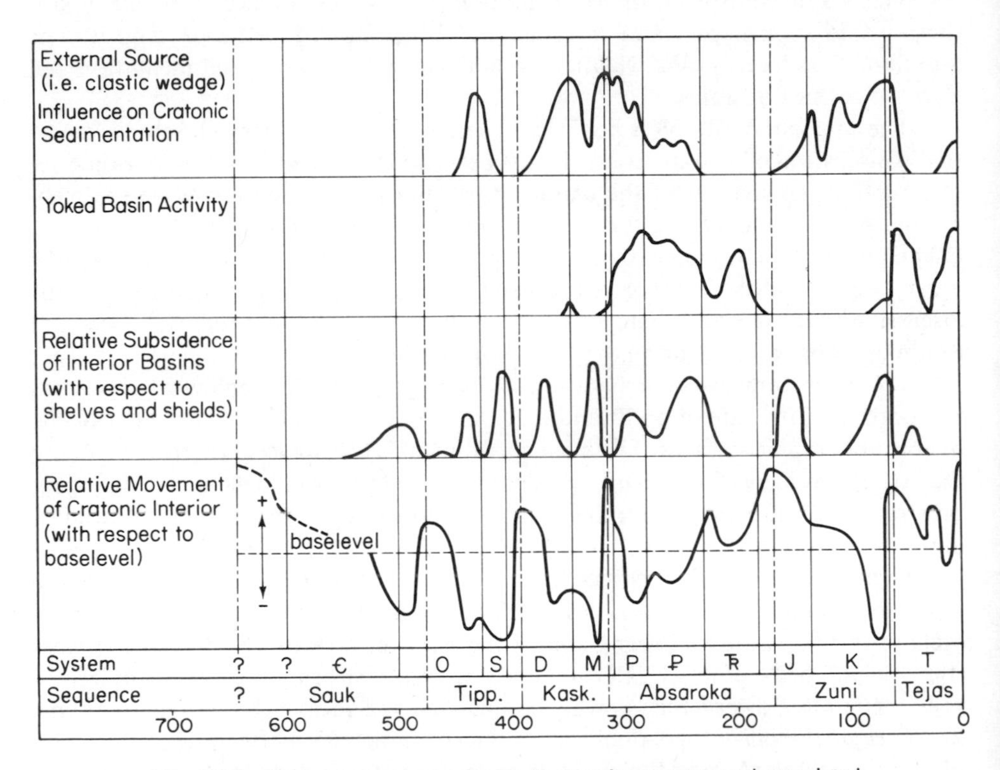

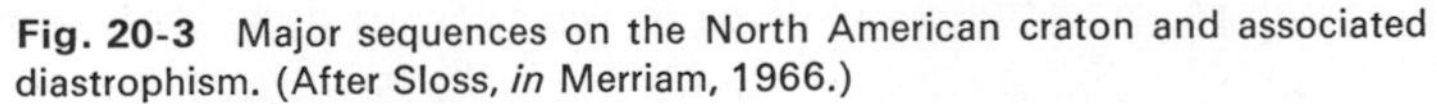
Fig. 20-3 Major sequences on the North American craton and associated diastrophism. (After Sloss, *in* Merriam, 1966.)

thrusting, metamorphism, and igneous activity. The major concept relating to sedimentation in orogenic regions has been the concept of geosynclines, introduced by Hall in 1859. (The term, originally "geosynclinal," was introduced by Dana in 1873.) Hall recognized that some parts of the crust were characterized by very great thicknesses of sediment accumulated over long periods of time. He proposed what was later called the "geosynclinal theory," "that mountain ranges were coincident with lines of great sedimentary accumulation." The geosynclinal theory was then taken up and elaborated mainly by European geologists, notably Marcel Bertrand and Edward Haug. These and other geologists introduced the concept of a distinct geostrophic cycle, characteristic (with some variation) of all geosynclines and mountain belts. Not only subsidence and orogeny but certain types of sedimentation, volcanism, and plutonism (intrusion and metamorphism) were supposedly characteristic of each stage in the developing cycle. A further systematization was introduced by Hans Stille, who distinguished two parts of a typical geosyncline. The eugeosyncline was that part of an "orthogeosyncline" characterized by strong volcanism during sedimentary accumulation and by synorogenic plutonism during deformation. Further subdivision of geosynclinal types was attempted by Marshall Kay.

If a normative geostrophic cycle exists, it must have important consequences for the larger-scale aspects of sequences of sedimentary rocks, particularly those in or adjacent to geosynclinal belts. The intimate connection of certain sequences of sedimentary rocks with orogeny was first proposed for rocks in the Alps and terms that were originally used in the Alps for large stratigraphic units (groups) have subsequently been used elsewhere, and for rocks of different ages, to designate broad facies types thought to be related to orogeny.

In the Alps, Triassic to Jurassic evaporites, limestones, and shales pass up into Jurassic-Lower Cretaceous volcanic rocks (ophiolites), shales, and thin limestones (Bundnerschiefer or Schistes lustrés) thought to have been deposited in relatively deep water. In some parts of the Alps there are bedded radiolarian cherts of Jurassic age, thought by some geologists to represent sediments deposited very slowly in deep water. The term "*leptogeosynclinal*" was proposed by Trümpy (1955, see Trümpy, 1960, p. 865) for "thin deep-sea deposits representing a long time span," to emphasize the idea that not all geosynclinal deposits are thick or were rapidly deposited in shallow water.

Above these predominantly shaly deposits come the thick group of interbedded sandstones and shales called *Flysch*. Much, though not all, of the Alpine Flysch is Eocene in age. The characteristics of the rocks suggest to most sedimentologists that they were deposited in deep water with the sands introduced by turbidity currents. Others dispute the turbidite origin for the sandstones, preferring to invoke deep-sea currents and yet others believe the Flysch was deposited in shallow water. Different geologists have laid stress on different aspects of the Alpine Flysch in applying the term flysch more generally to groups of rocks elsewhere (see Lajoie, 1970). For some, flysch is defined by an assemblage of sedimentary (and biogenic) structures and rock types; for others, the term implies a particular tectonic setting: a thick sequence of preparoxysmal marine geosynclinal sediments, consisting predominantly of interbedded sandstones and shales. In the Alps, the main period of folding (the orogenic "paroxysm") varied from Late Cretaceous to Oligocene and it has been shown that the flysch facies in the Alps also varies in age from Upper Cretaceous to Lower Oligocene, becoming younger toward the north (Trümpy, 1960). It seems reasonable, therefore, to link the deposition of the flysch with the erosion of geanticlines beginning to rise in what had previously been the deeper part of the geosyncline. Studies of the petrography of flysch sandstones and breccias confirm that this is a reasonable hypothesis. In some areas the spectacular "Wildflysch," composed of immense blocks of sedimentary, igneous, and metamorphic rock types in a contorted shale matrix, is present and has been interpreted as material slumped or slid from submarine ridges or from the front of nappes as they advanced across a basin in which mud was deposited. Thus it is that flysch has come to be regarded as a general term designating a facies typical of early stages of orogenic activity, when sedimentary basins are broken by rising, steep, discontinuous cordilleras. Many geologists also believe that deposition of flysch is commonly followed by a major period of orogeny. It need hardly be added that the term has frequently been applied, outside the Alps, to groups that share few of these characteristics of the Alpine Flysch.

Following the deposition of Flysch in the Alps, another thick series of sandstones, conglomerates, and shales was deposited, known as *Molasse* (see Bersier, 1959; Füchtbauer, 1967). The Molasse is Middle Oligocene to late Miocene in age and reaches a thickness of more than 4000 m. Rarely, there is a continuous transition from Flysch to Molasse but generally the Molasse was deposited further to the north, outside the central parts of the geosyncline. Molasse sediments may be marine, brackish, or fresh water. Many are deltaic or fluviatile and show characteristic fining-upward sequences. The Molasse was deposited after the main orogenic movements in the Alps and the sediments were eroded from the high Alpine mountains of Oligocene-Miocene age and spread out to the north, encroaching on areas that were relatively stable. Much of the Molasse, therefore, has never been strongly folded, in contrast to the Flysch, most of which was deformed during the main or late stages of the Alpine folding. The term molasse has also been used outside the Alps for rocks of different age that form "clastic wedges" with a similar tectonic setting. For example, in the Appalachians, the Ordovician Martinsburg has been called flysch and the succeeding Juniata and Tuscarora of Ordovician-Silurian age have been called molasse, related to the Taconic orogeny and subsequent uplift. Also in the Appalachians, marine sandstones and shales in the Middle and Upper Devonian have been called flysch and the succeeding deltaic Catskill facies of the Upper-Devonian has been called molasse. Some writers have also called the Triassic beds of the Appalachians molasse but this seems inappropriate as these beds show neither the characteristic sedimentary features nor the same tectonic setting as the Alpine Molasse.

The concept of control of sediment type and sequence by the diastrophic cycle was further developed by Paul D. Krynine (1942 and later papers). Krynine suggested three major controls for sedimentation: latitude, diastrophism, and "local depositional conditions." Diastrophism controlled not only the major form and evolution of the basin, as discussed above, but also the mineralogy of the sediments. Climate, which also affected mineralogy, was only partly controlled by latitude and was partly a function of the distribution of land masses and mountain chains. Thus climate was also partly controlled by diastrophism. Mineralogy of sediments was a function of the type of rocks in the source area, the rate of uplift, type of weathering, relations between the source area and the basin of deposition, and the rate of subsidence in the basin, all of which were strongly affected by diastrophism. Krynine pointed out that "the control of the source area by regional diastrophism is twofold: (a) passive, by bringing into the zone of erosion through orogenesis and epeirogenesis a series of potential source rocks and (b) active, by determining what type of rock will be found in the source area."

The first stage of the cycle, according to Krynine, was a stage of "peneplanation" and tectonic quiescence, characterized by carbonates and quartzose sands (the orthoquartzite-carbonate series). Following this the geosynclinal stage developed, with moderate diastrophism supplying detritus into the basin of deposition. Most of the detritus supplied was eroded from geanticlines mantled by older sedimentary or low grade metamorphic rocks. Crystalline basement was ordinarily not exposed. Consequently the sandstones contained many micaceous or shaly rock fragments and little feldspar (Krynine's low rank graywackes). After the paroxysmal stage of orogeny, erosion stripped bare the deeper, high-grade metamorphic and granitic rocks, and the derived sandstones became richer in feldspar. Finally, postorogenic block faulting resulted in a combination of rapid uplift of crystalline source rocks, rapid erosion, short transport, and rapid burial in a deeply subsiding basin so that little chemical alteration of the detritus could take place. The typical sediment formed in such basins was a feldspar-rich sandstone or arkose.

This scheme was elaborated by Krynine in a number of papers and was modified by many other sedimentologists. It was the basis of Krynine's pioneer classification of sandstones and other sedimentary rocks. Krynine wrote about broad generalities but his own studies were mainly confined to the Appalachians. Some important details in his system have not been supported by later studies and in his early formulations Krynine tended to neglect "high rank graywackes" and volcanic source rocks. Subsequent theoretical and mineralogical studies have tended to raise problems and doubts rather than to advance the idea of tectonic control of sedimentary mineralogy.

One problem has been to devise real tests for hypotheses of tectonic control. Obviously, broad generalizations such as those described above cannot be invalidated by a few exceptions, even though the exceptions may be on a substantial scale. For example, Atkinson (1962) showed that the Tertiary sandstones deposited in a fault trough (taphrogeosyncline) on the west coast of Spitzbergen were predominantly low rank graywackes, not arkoses as predicted by Krynine's theory. Klein (1962) showed that even in the Triassic basins of the Maritime Provinces of Canada the sandstones were by no means invariably arkoses. In both of these examples it is clear that the composition of the sandstones is determined to a large extent by the local source rocks, which are not rich in feldspar. Such examples, however, can scarcely be held to invalidate Krynine's theories, which applied only to a higher level of generalization. The question is not "Do *all* postorogenic fault troughs contain arkoses?" but "Are postorogènic fault troughs a preferred site for the deposition of arkoses?". Answering this and similar questions raised by Krynine requires the detailed petrographic study of many different basins of deposition.

One of the consequences of Krynine's theory of diastrophic control of source rocks is that within a single basin of deposition there should be a progressive change in mineralogy of the sediments deposited in the basin as the degree of diastrophism changes and as deeper levels of the crust are exposed by erosion in the source area. This hypothesis was tested by Schwab (1969), who studied a single composite geosynclinal sequence, extending from the Cambrian to the Tertiary in western Wyoming. He concluded that the sandstones do exhibit systematic vertical changes in mineralogy, showing an upward decrease in mineralogical and textural maturity, from quartzites to low rank graywackes with 30% lithic fragments (see also the work of Dickinson, 1969, in California).

In the past, two different tectonic explanations have been suggested to explain the repeated tongues of coarse sediment composing a single major clastic wedge. The first and simplest of these is that each tongue represents a renewed period of diastrophic activity. The second explanation is that the tongues result from periods of uplift that are not necessarily produced by diastrophic activity. It has been suggested that, after orogeny, repeated periods of uplift result from isostatic readjustment. Following the initial uplift, erosion removes crustal load to the point where buoyant forces are sufficiently strong to overcome crustal strength and a renewed period of relatively rapid uplift takes place. This is followed by further erosion, triggering further uplift, and so on. A major objection to this theory, however, is that recent studies show that even very small crustal loads (such as the filling of the Lake Mead reservoir in Nevada and Arizona) result in isostatic readjustment, indicating a very low strength for the crust-mantle system. One possible alternative to these two explanations is that uplift in the central mountain region is more or less continuous but that gravity movement on thrusts resulting from sliding tectonics on the flanks of this uplift is discontinuous.

It must be recognized, however, that in many cases there is little evidence that a particular influx of coarse sediment into the basin is a direct result of renewed uplift in the source area. Alternative explanations are controls internal to the basin, such as delta shifting, and other major external controls, such as a lowering of sea level or a change in climate. The difficulty of distinguishing between these different possibilities is well documented by the continuing controversy over the origin of coal measure cyclothems.

Finally, brief consideration must be given to the implications of the "new global tectonics" for theories of tectonic control of sedimentation. The concept of plate tectonics implies that geologists have not been entirely correct in envisaging orthogeosynclines as long, narrow regions of sediment accumulation. It suggests instead that, far from being narrow, the eugeosyncline may have to be expanded to include the entire ocean basin from median ridge to continental margin (a concept first suggested long ago by Haug, see Chorley, 1963). Geologists have always been at a loss to point to modern geosynclines that convincingly fitted the models deduced from ancient mountain chains. Viewed from the perspective of the sea floor spreading hypothesis, the present seems at last to be more in tune with the past, but it also seems that prespreading models may have to be considerably modified and diversified. For example, Mitchell and Reading (1969) suggest the existence of four main types of geosynclines, based on the relation between ocean floor spreading direction and continental margins: (a) Atlantic-type geosynclines, developed on continental margins lacking submarine trenches and volcanism; (b) Andean-type geosynclines, developed at margins bordered by a trench; (c) island arc-type geosynclines, in some cases remote from continents; and (d) Japan Sea-type geosynclines, which are small ocean basins lying between continents and island arcs. Each type of geosyncline is characterized by a different type of differential movement between major crustal units (or absence of differential movement in the case of the Atlantic type). Orogeny may result from such movement or from continental collision. Depending on the sequence of spreading events, one type of geosyncline may change into another type. The authors do not deny the reality of the main geosynclinal sequences suggested by older workers but suggest that some concepts, such as mio- and eugeosyncline are oversimplified. In Atlantic geosynclines, preflysch deposits form on the mid-oceanic ridge and abyssal plains and hills. Flysch consists of compositionally mature turbidites deposited on the abyssal plains and is preorogenic. Ultimately there may be an orogenic phase, as a result of a change in spreading regime, but the flysch is not related to this orogeny in any way. In Andean geosynclines, there is no clear distinction between the eu- and miogeosyncline. Wildflysch should be well developed and molasse-type sediments will be contemporaneous with flysch. Island-arc geosynclines will show preflysch followed by flysch with much volcanic detritus. There is no miogeosyncline and no molasse.

These speculations are outlined not because there is any special reason to believe in their ultimate validity but because they show the way in which it may be necessary to rethink traditional ideas about tectonic control of sedimentation as the sea floor spreading theory is developed and expanded (see also Dewey and Bird, 1970).

20.5 MAJOR SECULAR VARIATIONS

Throughout most of this book, a strictly uniformitarian or actualistic approach to the origin of sedimentary rocks has been adopted. It has generally been assumed not only that the processes acting in the past were the same ones that can be observed today but that the relative magnitudes and rates of recent and ancient processes were comparable. Even in the discussion of the diastrophic cycle given above, it was assumed implicitly that not every part of the world is presently involved in the same part of the cycle. It is generally supposed that almost any stage in the diastrophic cycle can be examined in some part of the present earth. A number of exceptions to this general assumption have been noted. It appears that not all periods of geological time were characterized by the type of climate that now exists or by the presence of major ice sheets. The recent history of the shallow marine shelves is unusual in that there has been a major, rapid eustatic rise in sea level during the last 18,000 years. Doubts have arisen that there are any modern analogues, at least those comparable in scale, to the conditions that led to large deposits of ironstones, evaporites, and shallow carbonates in the geological past.

Many of the ideas suggested by geologists regarding long-term changes in the nature of the earth have been frankly speculative and supported by very little geological evidence. Nevertheless, they should be given serious consideration by stratigraphers and sedimentologists who study those aspects of the geological record in which the history of the earth is most clearly preserved. The earth has a geological history extending back 4.5 billion years, of which over 3.5 billion years is recorded in sedimentary rocks. Most sedimentologists study Phanerozoic rocks, representing only the last 0.6 billion years of this record and for this part of the record it is perhaps reasonable to assume that conditions did not differ much from those existing on the earth today. But the same assumption would seem to be unreasonable or at least to bear the onus of proof, for earliest Precambrian times. The major changes that might affect the type of sediments deposited through time include (a) changes in the size and composition of the continents; (b) changes in the composition of the oceans and atmospheres; (c) changes in the nature of life; and (d) astronomic changes, such as a closer position of the moon in earlier geological periods.

Changes in Size and Composition of the Continents

There is no general agreement among geologists and geophysicists about secular changes in the size of continents. Studies of Sr/Rb isotopes indicate accelerating growth in size of continents to a present rate of 1 cu km/my, a figure that agrees roughly with the estimated present rate of extrusion of volcanic rocks. Other studies, including those of Pb isotopes, indicate that almost all the crust was formed during the first half million years of earth history, with minor additions since then. The difference between these points of view may possibly reflect mainly a difference in the meaning of the term "continent" or "crust."

The concept of continental accretion has been widely accepted in North America and it suggests that not only have the sizes of continents increased but also the lengths of mobile belts (Clifford, *in* Scotford, 1969). Goodwin (1968) has suggested a trend from Archean thin-crustal unstable protocontinents to later thick-crustal, relatively stable continents. Archean sediments and volcanics accumulated in randomly distributed or sublinear basins, in contrast with the later relatively long, continuous troughs and stable platforms of Proterozoic times. A corresponding change in sediment type from a dominance of immature flysch-like sediments in the Archean to more mature, shelf-type sediments in the Proterozoic resulted.

Several studies of the relative abundance of different rock types and of the chemical composition of different age provinces in North America have suggested progressive changes through geological time. Engel (1963) claimed there was a progressive decrease in mafic volcanics and granitic rocks and increase in sedimentary rocks with decreasing age. Fahrig and Eade (1968), in a study based on composite analyses of 14,000 specimens, found small but apparently significant differences in chemical composition between Archean and Proterozoic rocks in the Canadian shield (Table 20-1).

TABLE 20-1 AVERAGE CHEMICAL COMPOSITION OF ARCHEAN AND PROTEROZOIC ROCKS OF THE CANADIAN SHIELD (from Fahrig and Eade, 1968)

	Proterozoic	Archean
SiO_2	65.8	65.1
Al_2O_3	15.5	16.0
Fe_2O_3	1.3	1.5
FeO	3.5	3.0
MgO	2.0	2.3
CaO	3.2	3.4
Na_2O	3.4	4.1
K_2O	3.4	2.7

Engel also estimated the average chemical composition of sedimentary rocks at three different periods. He claimed that his data show a distinct trend toward more calcareous sediments and a higher K_2O/Na_2O ratio. Estimated clastic/carbonate ratio decreased from 1000:1 to 15:1; dolomite/limestone ratio, from 3:1 to 1:4 over the same time period.

Many of these trends are not very marked (cf. Table 20-1) and the extent to which they are real and universal needs further study. Armstrong (1960) pointed out that the supposed absence of limestone from Archean rocks was not supported by data for the Canadian shield. To what extent are the trends merely a function of diagenetic or metamorphic change or the preservation potential of the sedimentary rock units? Or do they, in fact, represent real progressive changes in the composition of the continents and of the sediments derived from them? (See Garrels and Mackenzie, 1971.)

Changes in the Composition of the Oceans and Atmosphere

Rubey (1951, 1955) has demonstrated that the volatiles now present in the atmosphere, hydrosphere, and in rocks such as carbonates and coal cannot all have been present in the early atmosphere but must have been accumulated slowly over geological time by leakage from the interior of the earth ("outgassing"). It follows that there must have been some stage in early geological history when the composition of the atmosphere and oceans was substantially different from the present composition.

The probable composition of gases supplied from the earth's interior can be inferred by studying present volcanic gases and rocks. Comparison of modern and ancient volcanic rocks suggests that the gases probably had a similar composition in the past and that the most abundant components included H_2O, CO_2, CO, and N_2 but not free oxygen. Thus it seems probable that the early atmosphere contained no oxygen. It probably also contained little methane because it has been shown that in the absence of oxygen the dissociation of methane should yield abundant carbon, which is not seen in known early Precambrian sediments. Oxygen can be supplied to the atmosphere from two main sources: (a) photo-dissociation of water vapor. It has been shown (Brancazio and Cameron, 1964) that this is capable of providing oxygen sufficient to raise the partial pressure to only about one thousandth of its present value because the presence of oxygen and ozone effectively screens out the radiation capable of producing further dissociation. (b) Photosynthesis by plants.

At what stage in geological history did the atmosphere and oceans begin to approximate to their present composition? For information relevant to this problem, an examination must be made of the oldest known sedimentary rocks, which date back over 3.5 billion years (b.y.). Some of the earliest sedimentary rocks do contain indications of an atmosphere that was not strongly oxidizing. Conglomerates of the Huronian and Witwatersrand (with ages over 2 b.y.) contain grains of pyrite and uraninite interpreted as detrital grains. Such minerals cannot survive as detrital grains in the present atmosphere. The earliest true red beds do not have ages in excess of about 2 b.y. It has been pointed out that only very low partial pressures of oxygen are necessary to produce ferric oxides but if red beds are produced by diagenesis in the vadose zone (as suggested by Walker—see p. 366), it may be that relatively high partial pressures of oxygen in the atmosphere would be required to supply sufficient dissolved oxygen to vadose waters to form red beds.

Banded iron formations, containing abundant iron oxides, are found in strata with ages of 3 to 2 b.y., which seems to indicate a strongly oxidizing atmosphere in contrast to the absence of oxygen indicated by detrital pyrite and uraninite. Cloud (1968) has suggested that possibly early aquatic organisms served as a source of oxygen and that the rhythmic banding resulted from a fluctuating balance between oxygen-producing biota and supply of ferrous iron in solution from chemical weathering. We return to some of these problems in the next section.

Other evidence supplied by ancient sedimentary rocks includes the following: (a) There are essentially no Precambrian evaporites, though salt casts are known from very old sediments such as the Belt series (about 1 b.y.). Does this indicate that seawater salinity was low or that tectonic and climatic conditions were not appropriate for evaporite formation or simply that Precambrian evaporites have been lost by solution or metamorphism? Boron in shales has been used as a paleosalinity index and Reynolds (1965) found no evidence for a change in boron content of marine shales from the Precambrian to the present.

Most authors agree that the composition and possibly also the volume of seawater have changed little since the late Precambrian but there are few data bearing on the early Precambrian as yet. Most of the discussion by geochemists has been based on theoretical models (see references under Brancazio and Cameron, 1964 and Holland, 1965).

Changes in the Nature of Life

The major changes in the nature of life that might be expected to have an important effect on sedimentation include (a) development of plants capable of photosynthesis, (b) development of land plants, (c) development of organisms secreting carbonate skeletons, and (d) development of abundant calcareous plankton.

The age and nature of the earliest life is still a matter for speculation but in recent years there have been a large number of well-documented discoveries of primitive life-forms in the Precambrian. Algae of blue-green affinities have been reported from rocks in South Africa dated at 2.7 b.y. and abundant algae capable of constructing widespread stromatolites are reported by Hoffman (1967) from rocks dated at about 2 b.y. in the Great Slave Lake region of Canada. Well-preserved microflora are preserved in the Gunflint cherts of about the same age, on the north shore of Lake Superior (see Barghoorn *in* Scotford, 1969).

It is generally supposed that the development of the first life-forms required an atmosphere free from oxygen. As noted above, there is some evidence that such an atmosphere existed prior to about 2 b.y. ago. The first life-forms would have been dependent on external food sources (heterotrophs) and an important stage in development would be marked by the first forms capable of manufacturing their own organic substances (autotrophs). Present microorganisms include those that use chemical and those that use light energy—the latter, organisms capable of photosynthesis, were probably the first autotrophs to develop, according to Cloud (1968). Not all types of photosynthesis release free oxygen, however, so it does not follow that the atmosphere would at once start to change over to one with free oxygen. In any event, probably much of the oxygen first produced would be used in weathering and not permitted to accumulate in the atmosphere. Living and fossil plants have been shown to selectively absorb the lighter isotope of carbon, C^{12}. It has been found that some of the oldest known organic carbon, in the Fig Tree Chert of the Swaziland System of South Africa, shows a similar carbon isotope fractionation. This has led Barghoorn (*in* Scotford, 1969) to conclude that photosynthesis had already started 3 b.y. ago. Cloud (see Fig. 20-4) believes that oxygen accumulated only slowly at first and Berkner and Marshall (*in* Brancazio and Cameron, 1964) have suggested that the development of abundant oxygen coincided with, and was responsible for, the abundant development of metazoan life at the beginning of the Cambrian. One condition for the development of abundant animal life was the development of cells with nuclei (eucaryotic cells) as opposed to those lacking nuclei (procaryota), which include the blue-green algae. According to Cloud, the oldest eucaryotes are probably older than 0.7 b.y.

Organisms secreting a carbonate skeleton appeared in the Cambrian, following shortly on the sudden explosive development of forms that took place at the beginning of that period. The cause of the appearance of Metazoa at the beginning of

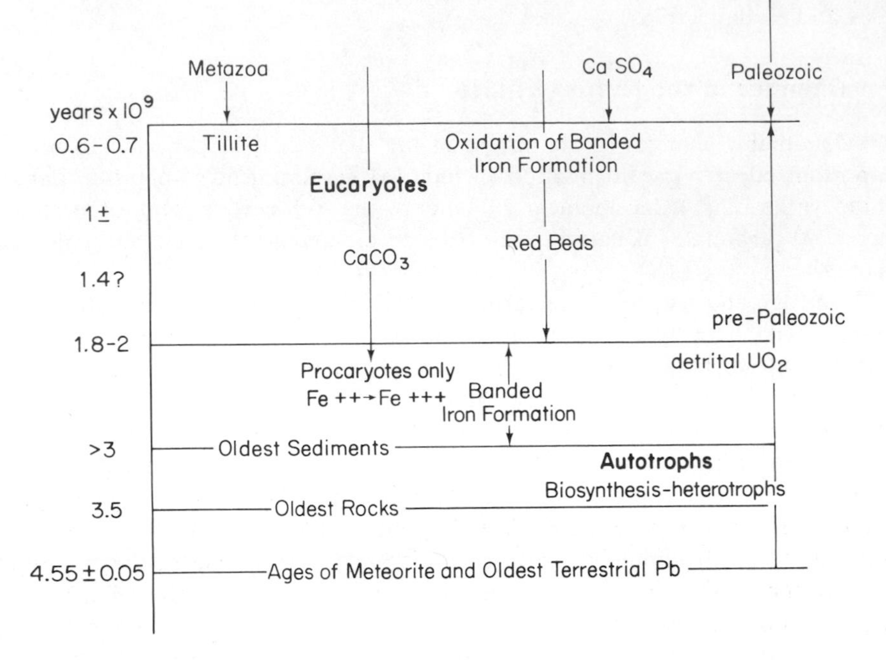

Fig. 20-4 Main features of biospheric, atmospheric and lithospheric evolution in the Precambrian. (After Cloud, 1968.)

the Cambrian is still not known. Among suggested causes are the evolution of sexual modes of reproduction; increase in oxygen content of the atmosphere to the point that it was adequate for animal respiration, particularly in the well-mixed shallow waters of the continental shelves; and changeover from acid to basic seawater. The problem remains unsolved. The development of skeletal carbonates undoubtedly resulted in a marked change in the pattern of sedimentation in shallow seas and the pattern has continued to change in lesser ways as groups of organisms evolved, flourished, and declined, to be replaced by others occupying a similar ecological niche and playing a similar sedimentological role.

Astronomic Changes

The most important of these changes is the gradual recession of the moon from the earth, resulting from the action of tidal friction, but possibly some other factors might have importance. These include capture of the moon; a secular decrease in the value of G, the gravitational constant; and the effect of a period of weak magnetic field and consequent high exposure of the earth's surface to electromagnetic radiation during reversals of the earth's magnetic field.

Calculations based on computed tidal friction and modern astronomical observations indicate that the day is lengthening at a rate about 20 sec every million

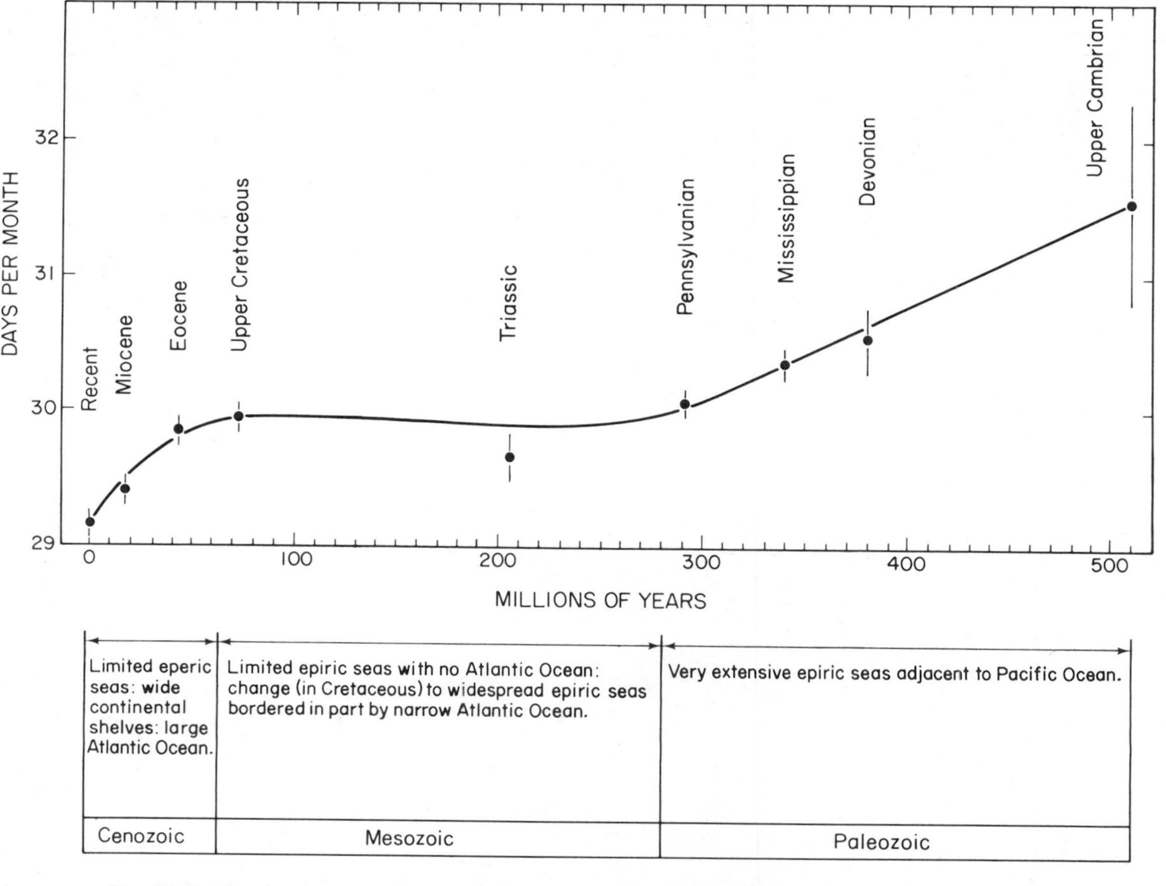

Fig. 20-5 Changes in the number of days per synodic month through the Phanerozoic. Error bars indicate one standard deviation. (From Pannella and others, 1968 with permission of the American Association for the Advancment of Science.)

years. Observations on supposed diurnal, monthly, and annual growth rings on corals, mollusks, and stromatolites permit measurement of the number of days in the year or in the synodic month throughout Phanerozoic time. Data assembled by Pannella et al. (1968) are shown in Fig. 20-5. The data show two main periods of slowing down of the earth's rotation with a period of little change from Pennsylvanian to Upper Cretaceous. These changes are interpreted in terms of variation in the rate of dissipation of tidal friction, with this rate being highest during periods with wide continental shelves or broad epeiric seas.

Tidal friction not only results in lengthening of the day but also in a gradual increase of the distance of the moon from the earth. Consequently there is a gradual decrease in the lunar component of the tide-generating force and the tides must have been stronger, on the average, in the geologic past than they are at present. At some time in the geologic past the moon must have been much closer to the earth. Varying estimates have been made of the period when the moon must have been so close to the earth that catastrophic effects would have resulted. George Darwin, in the late nineteenth century, set this time at about 60 million years ago and this calculation was used by Lord Kelvin as evidence, together with calculations on the cooling of the earth, that the earth could be no older than 100 million years. It is now known that the age of the earth is at least 4.5 billion years but astronomic calculations still suggest that the moon was very close to the earth only about 1.0 ($\pm$0.5) b.y. ago. For this reason, some believe that the moon must have been captured by the earth relatively late in their history. The event of capture should have left marked traces on both the earth and moon.

Speculations such as these, though they may have little permanent value, do suggest some of the major earth and planetary problems, for whose solution sedimentologists could provide some badly needed data.

REFERENCES

ARMSTRONG, H. S., 1960, "Marbles in the "Archean" of the Southern Canadian Shield," *Report 21st Sess. Internatl. Geol. Cong.*, Norden (1960), pt. IX, pp. 7–20. (Documents occurrences of limestones in older Precambrian sediments.)

ATKINSON, D. J., 1962, "Tectonic Control of Sedimentation and the Interpretation of Sediment Alternations in the Tertiary of Prince Charles Foreland, Spitzbergen," *Geol. Soc. Amer. Bull.*, **73**, pp. 343–364. (Good discussion of the role of tectonics and provenance in controlling sandstone composition and sequence in a fault trough.)

AUBOUIN, JEAN, 1965, *Geosynclines*. Amsterdam: Elsevier Pub. Co., 335 pp. (Discusses the historical development of the concept and describes the Alpine geosyncline in the eastern Mediterranean.)

BERSIER, A., 1959, "Sequences Detritiques et Divagations Fluviatiles," *Eclogae Geol. Helv.*, **51**, pp. 854–893. (A classic description and interpretation of the Alpine Molasse.)

BRANCAZIO, P. J., and A. G. W. CAMERON (ed.), 1964, *The Origin and Evolution of Atmospheres and Oceans*. New York: John Wiley & Sons, Inc., 314 pp. (A collection of papers, including an important paper by Berkner and Marshall on the history of oxygen in the atmosphere.)

CHORLEY, R. J., 1963, "Diastrophic Background to Twentieth-Century Geomorphological Thought," *Geol. Soc. Amer. Bull.*, **74**, pp. 953–970. (Review of the rise and decline of theories of eustatic control.)

CLOUD, P. E., JR., 1968, "Atmospheric and Hydrospheric Evolution on the Primitive Earth," *Science*, **160**, pp. 729–736. (Speculative synthesis of the evidence. See also Cloud's paper in the book edited by Scotford.)

CROWELL, J. C., and L. A. FRAKES, 1970, "Phanerozoic Glaciation and the Causes of Ice Ages," *Amer. Jour. Sci.*, **268**, pp. 193–224. (Arrangement of continents near the poles is important as one factor in causing ice ages.)

CURRAY, J. R., 1964, "Transgressions and Regressions", *in* R. L. MILLER (ed.), *Papers in Marine Geology* (Shepard Commemorative Vol.), New York: Macmillan Co., pp. 175–203

———, 1965, "Late Quaternary History, Continental Shelves of the United States," *in* H. E. Wright, Jr., and D. G. Frey (ed.), *The Quaternary of the United States*. Princeton, N. J.: Princeton University Press, pp. 723–735. (In these two papers Curray reviews the Quaternary transgression and derives general models for transgression and regression. See also his lectures in the volume edited by Stanley.)

DEWEY, J. F., and J. M. BIRD, 1970, "Mountain Belts and the New Global Tectonics," *Jour. Geophys. Res.*, **75**, pp. 2625–2647. (An example of the new approach to theories of tectonic control of sedimentation.)

DICKINSON, W. R., 1969, "Evolution of Calc-Alkaline Rocks in the Geosynclinal System of California and Oregon," *Oregon Dept. Geol. Min. Ind. Bull.*, **65**, pp. 151–156. (Gives data on variation in sandstone composition throughout Upper Jurassic to Cretaceous time in the Great Valley of California.)

DUFF, P. M. D., A. HALLAM, and E. K. WALTON, 1967, *Cyclic Sedimentation*. Amsterdam: Elsevier Pub. Co., 280 pp. (Discusses criteria for distinguishing the major causes of cycles and gives many examples and references.)

EMILIANI, C., 1970, "Pleistocene Paleotemperatures," *Science*, **168**, pp. 822–825. (The latest in a series of papers giving results from paleontological and oxygen isotope studies of deep-sea cores.)

ENGEL, A. E. J., 1963, "Geological Evolution of North America," *Science*, **140**, pp. 143–152. (Summarizes evidence for accretion and changing composition of the continent.)

FAHRIG, W. F., and K. E. EADE, 1968, "The Chemical Evolution of the Canadian Shield," *Canadian Jour. Earth Sci.*, **5**, pp. 1247–1252. (Data based on analysis of thousands of samples from areas of the Shield.)

FÜCHTBAUER, HANS, 1967, "Die Sandsteine in der Molasse nordlich der Alpen," *Geol. Rundschau*, **56**, pp. 266–300. (Summary of much modern petrographic work on the Alpine Molasse.)

GARRELS, R. M., and F. T. MACKENZIE, 1971. *Evolution of Sedimentary Rocks*, New York: W. W. Norton Co., 397 pp. (An introductory textbook that treats the earth as a single huge geochemical factory in discussing sedimentary rocks.)

GLAESSNER, M. F. and C. TEICHERT, 1947, "Geosynclines: A Fundamental Concept in Geology," *Amer. Jour. Sci.*, **245**, pp. 465–482, 571–591. (A thorough historical review.)

GOODWIN, A. M., 1968, "Archaean Protocontinental Growth and Early Crustal History of the Canadian Shield," *Rept. 23rd Sess. Internatl. Geol. Cong.*, Czechoslovakia (1968), Proc. Sec. 1, pp. 69–89. (One of many papers by the author discussing changing composition and structural behavior of the primitive Canadian Shield.)

HAYES, M. O., 1967, "Relationship between Coastal Climate and Bottom Sediment Type on the Inner Continental Shelf," *Marine Geol.* **5**, pp. 111–132. (Based on a study of sediment distribution charts for the entire world.)

HOFFMAN, PAUL, 1967, "Algal Stromatolites: Use in Stratigraphic Correlation and Paleocurrent Determination," *Science*, **157**, pp. 1043–1045. (Describes stromatolites from rocks about 2 b.y. old exposed near Great Slave Lake.)

HOLLAND, H. D., 1965, "The History of Ocean Water and Its Effect on the Chemistry of the Atmosphere," *Proc. Natl. Acad. Sci.*, **53**, pp. 1173–1183. (CO_2 buffered by carbonates and silicates so that large deviations from present pCO_2 in atmospheres of the past are unlikely; mineral reactions also control the major cation ratios and pH of ocean water.)

HOYT, J. H., 1967, "Barrier Island Formation," *Geol. Soc. Amer. Bull.*, **78**, pp. 1125–1136. (Barriers do not form from offshore bars, but from progressive submergence of a coastal dune or of a beach ridge.)

KLEIN, G. DE V., 1962, "Triassic Sedimentation, Maritime Provinces, Canada," *Geol. Soc. Amer. Bull.*, **73**, pp. 1127–1146. (Sandstone types are directly related to source area rock types, not to tectonics.)

KNOPF, ADOLF, 1948, "The Geosynclinal Theory," *Geol. Soc. Amer. Bull.*, **59**, pp. 649–670.

———, 1960, "Analysis of Some Recent Geosynclinal Theory," *Amer. Jour. Sci.*, **258-A**, pp. 126–136. (These two papers give an exceptionally clear historical review of European concepts of geosynclines and orogeny.)

KRYNINE, P. D., 1942, "Differential Sedimentation and Its Products during One Complete Geosynclinal Cycle," *Proc. 1st Pan Amer. Cong. Mining Eng. Geol.*, pt. 1, **2**, pp. 537–560. (The first extended statement of Krynine's views on diastrophic control of sedimentation.)

LAJOIE, JEAN (ed.), 1970, "Flysch Sediments of North America," *Geol. Assn. Canada Spec. Paper 7, 272 pp.* (Symposium with reviews of the definition, characteristics, and interpretation of flysch and North American examples.)

MARSDEN, B. G. and A. G. W. CAMERON, (ed.), 1966, *The Earth-Moon System.* New York: Plenum Press, 288 p. (A collection of articles on cosmological, geophysical, and paleontological aspects of the earth-moon system and changes in the length of the month and year.)

MERRIAM, D. F. (ed.) 1966, "Symposium on Cyclic Sedimentation," *State Geol. Sur. Kansas. Bull.*, 169, **1** and **2**, 636 p. (Many papers on types and causes of sedimentary cycles.)

MITCHELL, A. H. and H. C. READING, 1969, "Continental Margins, Geosynclines, and Ocean Floor Spreading," *Jour. Geol.*, **77**, pp. 629–646. (Reinterpretation of the geosynclinal concept and of flysch and molasse facies in the light of the new global tectonics.)

NAIRN, A. E. M. (ed.), 1961, *Descriptive Palaeoclimatology*, New York: Interscience Pub., 380 p.

———, 1964, *Problems in Palaeoclimatology.* New York: Interscience Pub., 705 pp. (Two symposia with many papers reviewing criteria for establishing ancient climates.)

Norris, R. M., 1964, "Dams and Beach-Sand Supply in Southern California," *in* R. L. Miller, (ed.), *Papers in Marine Geology*. (Shepard Commemorative Vol.), pp. 154–171. (Discusses effect on beaches of reducing sediment supply from rivers.)

Pannella, G., C. MacClintock, and M. N. Thompson, 1968, "Paleontological Evidence of Variations in Length of Synodic Month Since Late Cambrian," *Science*, **162**, no. 3855, pp. 792–796. (Data from growth rings on corals, molluscs, and stromatolites.)

Reynolds, R. L. Jr., 1965, "The Concentration of Boron in Precambrian Seas," *Geochim. Cosmochim. Acta*, **29**, pp. 1–16. (Precambrian illites in unmetamorphosed carbonate rocks have similar boron content to that in modern marine illite.)

Rubey, W. W., 1951, "Geological History of Sea Water: An Attempt to State the Problem," *Geol. Soc. Amer. Bull.*, **62**, pp. 1111–1147.

———, 1955, "Development of the Hydrosphere and Atmosphere with Special Reference to Probable Composition of the Early Atmosphere," in *Geol. Soc. Amer. Spec. Paper* 62, pp. 631–650. (Two classic papers demonstrating that the atmosphere and hydrosphere are not primitive but must have developed through geological time.)

Schwab, F. L., 1969, "Cyclic Geosynclinal Sedimentation: A Petrographic Evaluation," *Jour. Sedimentary Petrology*, **39**, pp. 1325–1343. (Change in sandstone composition indicates tectonic control.)

Scotford, D. M. (ed.), 1969, "The Primitive Earth, A Symposium." Oxford, Ohio, Miami University, Dept. of Geol. (Uncensored transcript of talks given by a lively panel of speakers.)

Sloss, L. L., 1963, "Sequences in the Cratonic Interior of North America," *Geol. Soc. Amer. Bull.*, **74**, pp. 93–114. (Recognizes six sequences separated by widespread unconformities. See also paper in Merriam, 1966.)

Stanley, D. J. (ed.), 1969, *The New Concepts of Continental Margin Sedimentation.* Washington, D. C.: Amer. Geol. Inst. Short Course Lecture Notes. (Reviews of shelf sedimentation by Curray and Swift.)

Swift, D. J. P., 1970, "Quaternary Shelves and the Return to Grade," *Marine Geol.*, **8**, pp. 5–30. (Develops model of a "graded shelf." See also lectures in Stanley, 1969.)

Trümpy, R., 1960, "Paleotectonic Evolution of the Central and Western Alps," *Geol. Soc. Amer. Bull.*, **71**, pp. 843–908. (Synthesis of structure and stratigraphy with emphasis on tectonic control of sedimentation.)

Wanless, H. R. and Julie R. Cannon, 1966, "Late Paleozoic Glaciation," *Earth Sci. Rev.*, **1**, pp. 247–286. (Review paper. Cites evidence for multiple glaciation.)

INDEX